PRINCIPLES OF POWER SYSTEM

AF819785

PRINCIPLES OF
POWER SYSTEM

[INCLUDING GENERATION, TRANSMISSION, DISTRIBUTION, SWITCHGEAR AND PROTECTION]

[For B.E./B.Tech., AMIE and other Engineering Examinations]

V.K. MEHTA
ROHIT MEHTA

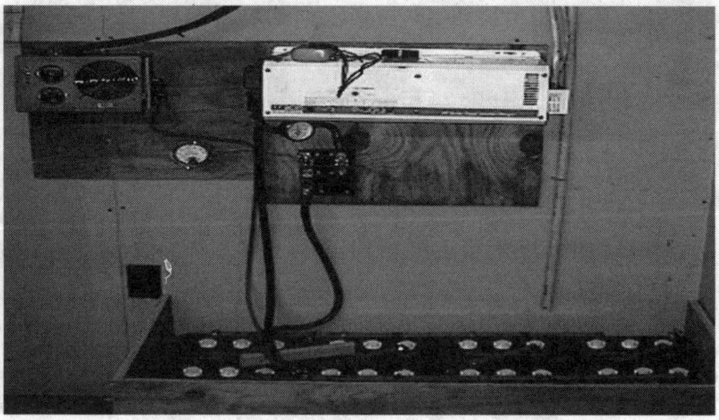

S. CHAND
PUBLISHING

S Chand And Company Limited
(ISO 9001 Certified Company)

S Chand And Company Limited
(ISO 9001 Certified Company)
Head Office: D-92, Sector–2, Noida – 201301, U.P. (India), Ph. 91-120-4682700
Registered Office: A-27, 2nd Floor, Mohan Co-operative Industrial Estate, New Delhi – 110 044
Phone: 011-49731800
www.**schandpublishing.com**; e-mail: **info@schandpublishing.com**

Marketing Offices:

Chennai	:	Ph: 23632120; chennai@schandpublishing.com
Guwahati	:	Ph: 2738811, 2735640; guwahati@schandpublishing.com
Hyderabad	:	Ph: 40186018; hyderabad@schandpublishing.com
Jalandhar	:	Ph: 4645630; jalandhar@schandpublishing.com
Kolkata	:	Ph: 23357458, 23353914; kolkata@schandpublishing.com
Lucknow	:	Ph: 4003633; lucknow@schandpublishing.com
Mumbai	:	Ph: 25000297; mumbai@schandpublishing.com
Patna	:	Ph: 4011400; patna@schandpublishing.com

© S Chand And Company Limited, 1982

All rights reserved. No part of this publication may be reproduced or copied in any material form (including photocopying or storing it in any medium in form of graphics, electronic or mechanical means and whether or not transient or incidental to some other use of this publication) without written permission of the copyright owner. Any breach of this will entail legal action and prosecution without further notice.

***Jurisdiction**: All disputes with respect to this publication shall be subject to the jurisdiction of the Courts, Tribunals and Forums of New Delhi, India only.*

First Edition 1982
Subsequent Editions and Reprints 1985, 87, 88, 90, 91, 92, 93, 94, 97, 98, 99, 2000, 2001, 2002, 2003, 2004
First Multicolour Edition 2005, Reprints 2006, 2007, 2008 (Twice), 2012, 2013 (Twice), 2014, 2015 (Twice), 2016 (Twice), 2017, 2018 (Twice), 2019 (Twice); Reprint 2020

Low Priced Students' Paperback Edition (LPSPE), 2022; Reprints 2022, 2023, 2024 (Thrice), 2025

Reprint 2026 (Twice)

ISBN: 978-93-5501-077-3 **Product Code:** H3PPS52POWS10ENAA21L

PRINTED IN INDIA

By Shri Shyamlal Printing Press Private Limited, Plot 20/4, Site-IV, Industrial Area Sahibabad, Ghaziabad – 201 010 and Published by S Chand And Company Limited, A-27, 2nd Floor, Mohan Co-operative Industrial Estate, New Delhi – 110 044.

PREFACE TO THE FOURTH EDITION

The general response to the third edition of the book was very encouraging. We, the authors feel that our work has been amply rewarded and wish to express our deep sense of gratitude, in common to the large number of readers who have used it, and in particular to those of them who have sent helpful suggestions from time to time for the improvement of the book.

Keeping in view the desire of many of the book users to make the book attractive and colourful, this Multicolour Illustrative Edition of the book has been published. We hope you will like it in its new format.

In the present edition, we have made sincere efforts to make the book up-to-date. A notable feature is the inclusion of Chapter on ``Neutral Grounding''. A large number of practical problems are given to make the book more useful to the students. It is our hope that you will enjoy reading this text and benefit from its content. After studying this book, it is hoped that you will have a good understanding of the ideas of Power System and its application in many real world situations.

We, as authors have taken considerable care to ensure that this text is as error-free and effective as possible. We welcome comments, suggestions and corrections from both students and teachers.

Welcome to the exciting world of Power System.

V.K. MEHTA
ROHIT MEHTA

PREFACE

The subject of power system has assumed considerable importance in the recent years. There has been a growing demand in the past for a book which deals with this vast but vital subject in a compact and lucid manner. The present text ``Principles of Power System'' has been written to achieve this goal. This text is intended primarily for the students preparing for diploma, A.M.I.E., degree and other engineering examinations. It also meets the needs of those readers who want to gain a sound understanding of the principles of power system.

Three outstanding features are claimed for this book. First is the styling—the author has used the same lecture style that prove successful in his previous texts. Second, the book is easy to read and stimulating in its direct personal approach. Third, the emphasis is on concepts as well as mathematical derivations. It is hoped that these features will help the readers to understand the theoretical and practical aspects of power system.

Author lays no claim to the original research in preparing the book. Liberal use of the materials available in the works of eminent authors has been made. What he may claim, in all modesty, is that he has tried to fashion the vast amount of material available from primary and secondary sources into coherent body of description and analysis.

The author wishes to thank his colleagues and friends who have contributed many valuable suggestions regarding the scope and content sequence of the book. Author is also indebted to M/s S. Chand and Company Ltd., for bringing out the book in a short time and pricing it moderately inspite of heavy cost of paper and printing.

Errors might have crept in despite utmost care to avoid them and author shall be grateful if these are pointed out along with other suggestions for improvement of the book.

V.K. MEHTA

CONTENTS

COSφ
0,8
0,5
CAP.
0,8
0,5
IND.

12. Distribution Systems—General 300—309

Distribution System—Classification of Distribution Systems—A.C. Distribution—D.C. Distribution—Methods of obtaining 3-wire D.C. System—Overhead *versus* Underground System—Connection Schemes of Distribution System—Requirements of a Distribution System—Design Considerations in Distribution System.

13. D.C. Distribution 310—355

Types of D.C. Distributors—D.C. Distribution Calculations—D.C. distributor fed at one end (concentrated loading)—Uniformly loaded distributor fed at one end—Distributor fed at both ends (concentrated loading)—Uniformly loaded distributor fed at both ends—Distributor with both concentrated and uniform loading—Ring Distributor—Ring main distributors with Interconnector—3-wire D.C. system—Current distribution in 3-wire D.C. System—Balancers in 3-wire D.C. system—Boosters—Comparison of 3-wire and 2-wire d.c. distribution—Ground detectors.

14. A.C. Distribution 356—373

A.C. Distribution Calculations—Methods of solving A.C. Distribution Problems—3-phase unbalanced loads—4-wire, star-connected unbalanced loads—Ground detectors.

Place magnet on
to test or reset
-FIND-

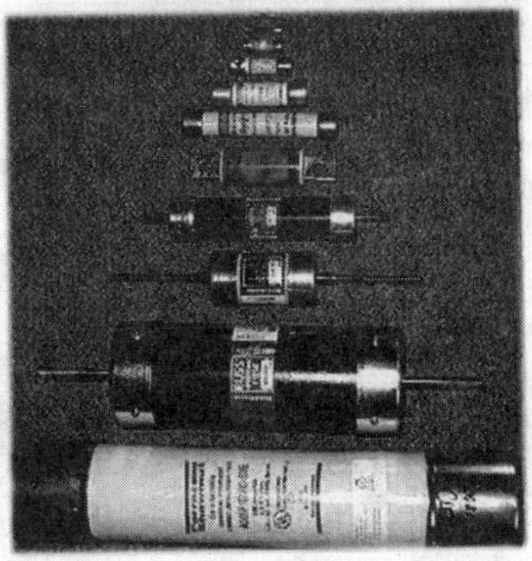

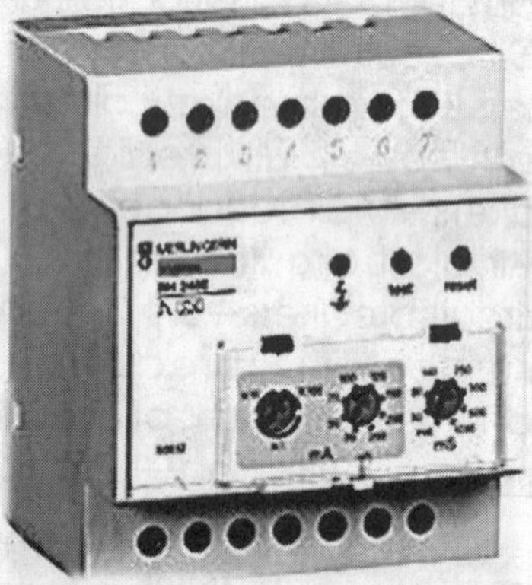

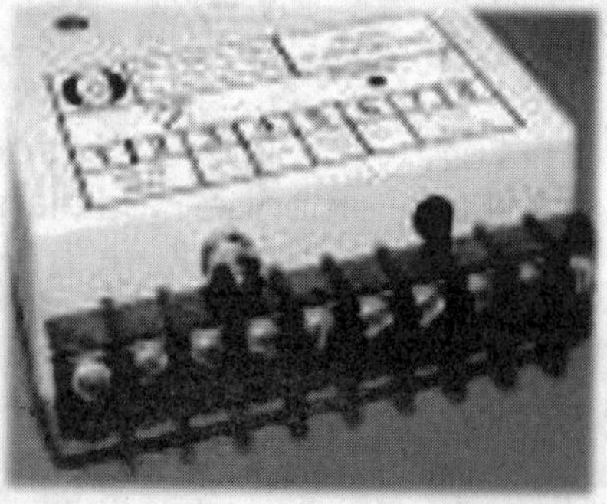

CHAPTER

1

Introduction

General

Energy is the basic necessity for the economic development of a country. Many functions necessary to present-day living grind to halt when the supply of energy stops. It is practically impossible to estimate the actual magnitude of the part that energy has played in the building up of present-day civilisation. The availability of huge amount of energy in the modern times has resulted in a shorter working day, higher agricultural and industrial production, a healthier and more balanced diet and better transportation facilities. As a matter of fact, there is a close relationship between the energy used per person and his standard of living. The greater the per capita consumption of energy in a country, the higher is the standard of living of its people.

Energy exists in different forms in nature but the most important form is the *electrical energy*. The modern society is so much dependent upon the use of electrical energy that it has become a part and parcel of our life. In this chapter, we shall focus our attention on the general aspects of electrical energy.

1.1 Importance of Electrical Energy

Energy may be needed as heat, as light, as motive power etc. The present-day advancement in science and technology has made it possible to convert electrical energy into any desired form. This has given electrical energy a place of pride in the modern world. The survival of industrial undertakings and our social structures depends primarily upon low cost and uninterrupted supply of electrical energy. In fact, the advancement of a country is measured in terms of per capita consumption of electrical energy.

Electrical energy is superior to all other forms of energy due to the following reasons :

(*i*) **Convenient form.** Electrical energy is a very convenient form of energy. It can be easily converted into other forms of energy. For example, if we want to convert electrical energy into heat, the only thing to be done is to pass electrical current through a wire of high resistance *e.g.*, a heater. Similarly, electrical energy can be converted into light (*e.g.* electric bulb), mechanical energy (*e.g.* electric motors) etc.

(*ii*) **Easy control.** The electrically operated machines have simple and convenient starting, control and operation. For instance, an electric motor can be started or stopped by turning on or off a switch. Similarly, with simple arrangements, the speed of electric motors can be easily varied over the desired range.

(*iii*) **Greater flexibility.** One important reason for preferring electrical energy is the flexibility that it offers. It can be easily transported from one place to another with the help of conductors.

(*iv*) **Cheapness.** Electrical energy is much cheaper than other forms of energy. Thus it is overall economical to use this form of energy for domestic, commercial and industrial purposes.

(*v*) **Cleanliness.** Electrical energy is not associated with smoke, fumes or poisonous gases. Therefore, its use ensures cleanliness and healthy conditions.

(*vi*) **High transmission efficiency.** The consumers of electrical energy are generally situated quite away from the centres of its production. The electrical energy can be transmitted conveniently and efficiently from the centres of generation to the consumers with the help of overhead conductors known as transmission lines.

1.2 Generation of Electrical Energy

The conversion of energy available in different forms in nature into electrical energy is known as **generation of electrical energy**.

Electrical energy is a manufactured commodity like clothing, furniture or tools. Just as the manufacture of a commodity involves the conversion of raw materials available in nature into the desired form, similarly electrical energy is produced from the forms of energy available in nature. However, electrical energy differs in one important respect. Whereas other commodities may be produced at will and consumed as needed, the electrical energy must be produced and transmitted to the point of use at the instant it is needed. The entire process takes only a fraction of a second. This instantaneous production of electrical energy introduces technical and economical considerations unique to the electrical power industry.

Energy is available in various forms from different natural sources such as pressure head of water, chemical energy of fuels, nuclear energy of radioactive substances etc. All these forms of energy can be converted into electrical energy by the use of suitable arrangements. The arrangement essentially employs (see Fig. 1.1) an alternator coupled to a prime mover. The prime mover is driven by the energy obtaimed from various sources

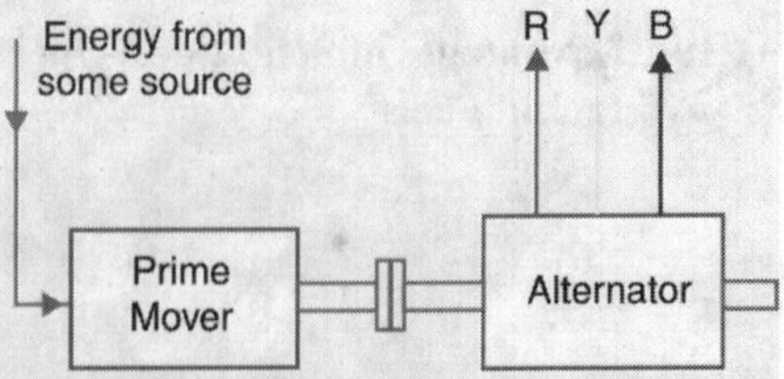

Fig. 1.1

such as burning of fuel, pressure of water, force of wind etc. For example, chemical energy of a fuel (*e.g.*, coal) can be used to produce steam at high temperature and pressure. The steam is fed to a prime mover which may be a steam engine or a steam turbine. The turbine converts heat energy of steam into mechanical energy which is further converted into electrical energy by the alternator. Similarly, other forms of energy can be converted into electrical energy by employing suitable machinery and equipment.

1.3. Sources of Energy

Since electrical energy is produced from energy available in various forms in nature, it is desirable to look into the various sources of energy. These sources of energy are :

(*i*) The Sun (*ii*) The Wind (*iii*) Water (*iv*) Fuels (*v*) Nuclear energy.

Out of these sources, the energy due to Sun and wind has not been utilised on large scale due to a number of limitations. At present, the other three sources *viz.*, water, fuels and nuclear energy are primarily used for the generation of electrical energy.

(*i*) **The Sun.** The Sun is the primary source of energy. The heat energy radiated by the Sun can be focussed over a small area by means of reflectors. This heat can be used to raise steam and electrical energy can be produced with the help of turbine-alternator combination. However, this method has limited application because :

(*a*) it requires a large area for the generation of even a small amount of electric power

(*b*) it cannot be used in cloudy days or at night

(*c*) it is an uneconomical method.

Nevertheless, there are some locations in the world where strong solar radiation is received very regularly and the sources of mineral fuel are scanty or lacking. Such locations offer more interest to the solar plant builders.

(*ii*) **The Wind.** This method can be used where wind flows for a considerable length of time. The wind energy is used to run the wind mill which drives a small generator. In order to obtain the electrical energy from a wind mill continuously, the generator is arranged to charge the batteries. These batteries supply the energy when the wind stops. This method has the advantages that maintenance and generation costs are negligible. However, the drawbacks of this method are (*a*) variable output, (*b*) unreliable because of uncertainty about wind pressure and (*c*) power generated is quite small.

(*iii*) **Water.** When water is stored at a suitable place, it possesses potential energy because of the head created. This water energy can be converted into mechanical energy with the help of water turbines. The water turbine drives the alternator which converts mechanical energy into electrical energy. This method of generation of electrical energy has become very popular because it has low production and maintenance costs.

(*iv*) **Fuels.** The main sources of energy are fuels *viz*., solid fuel as coal, liquid fuel as oil and gas fuel as natural gas. The heat energy of these fuels is converted into mechanical energy by suitable prime movers such as steam engines, steam turbines, internal combustion engines etc. The prime mover drives the alternator which converts mechanical energy into electrical energy. Although fuels continue to enjoy the place of chief source for the generation of electrical energy, yet their reserves are diminishing day by day. Therefore, the present trend is to harness water power which is more or less a permanent source of power.

(*v*) **Nuclear energy.** Towards the end of Second World War, it was discovered that large amount of heat energy is liberated by the *fission* of uranium and other fissionable materials. It is estimated that heat produced by 1 kg of nuclear fuel is equal to that produced by 4500 tonnes of coal. The heat produced due to nuclear fission can be utilised to raise steam with suitable arrangements. The steam can run the steam turbine which in turn can drive the alternator to produce electrical energy. However, there are some difficulties in the use of nuclear energy. The principal ones are (*a*) high cost of nuclear plant (*b*) problem of disposal of radioactive waste and dearth of trained personnel to handle the plant.

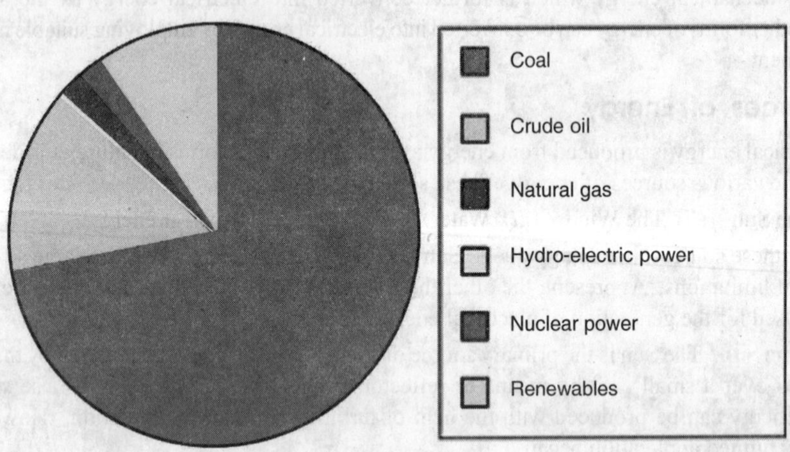

Energy Utilisation

1.4 Comparison of Energy Sources

The chief sources of energy used for the generation of electrical energy are water, fuels and nuclear energy. Below is given their comparison in a tabular form :

S.No.	Particular	Water-power	Fuels	Nuclear energy
1.	*Initial cost*	High	Low	Highest
2.	*Running cost*	Less	High	Least
3.	*Reserves*	Permanent	Exhaustable	Inexhaustible
4.	*Cleanliness*	Cleanest	Dirtiest	Clean
5.	*Simplicity*	Simplest	Complex	Most complex
6.	*Reliability*	Most reliable	Less reliable	More reliable

1.5 Units of Energy

The capacity of an agent to do work is known as its energy. The most important forms of energy are mechanical energy, electrical energy and thermal energy. Different units have been assigned to various forms of energy. However, it must be realised that since mechanical, electrical and thermal energies are interchangeable, it is possible to assign the same unit to them. This point is clarified in Art 1.6.

(*i*) Mechanical energy. The unit of mechanical energy is *newton-metre* or *joule* on the M.K.S. or SI system.

The work done on a body is one newton-metre (or joule) if a force of one newton moves it through a distance of one metre *i.e.*,

Mechanical energy in joules = Force in newton × distance in metres

(*ii*) Electrical energy. The unit of electrical energy is *watt-sec* or *joule* and is defined as follows:

One watt-second (or joule) energy is transferred between two points if a p.d. of 1 volt exists between them and 1 ampere current passes between them for 1 second *i.e.*,

Electrical energy in watt-sec (or joules)

= voltage in volts × current in amperes × time in seconds

Joule or watt-sec is a very small unit of electrical energy for practical purposes. In practice, for the measurement of electrical energy, bigger units *viz.*, watt-hour and kilowatt hour are used.

1 watt-hour = 1 watt × 1 hr

= 1 watt × 3600 sec = 3600 watt-sec

1 kilowatt hour (kWh) = 1 kW × 1 hr = 1000 watt × 3600 sec = 36×10^5 watt-sec.

***(iii)* Heat.** Heat is a form of energy which produces the sensation of warmth. The unit* of heat is calorie, British thermal unit (B.Th.U.) and centigrade heat units (C.H.U.) on the various systems.

Calorie. It is the amount of heat required to raise the temperature of 1 gm of water through 1°C *i.e.*,

1 calorie = 1 gm of water × 1°C

Sometimes a bigger unit namely *kilocalorie* is used. A kilocalorie is the amount of heat required to raise the temperature of 1 kg of water through 1°C *i.e.*,

1 kilocalorie = 1 kg × 1°C = 1000 gm × 1°C = 1000 calories

B.Th.U. It is the amount of heat required to raise the temperature of 1 lb of water through 1°F *i.e.*,

1 B.Th.U. = 1 lb × 1°F

C.H.U. It is the amount of heat required to raise the temperature of 1 lb of water through 1°C *i.e.*,

1 C.H.U. = 1 lb × 1°C

1.6 Relationship Among Energy Units

The energy whether possessed by an electrical system or mechanical system or thermal system has the same thing in common *i.e.*, it can do some work. Therefore, mechanical, electrical and thermal energies must have the same unit. This is amply established by the fact that there exists a definite relationship among the units assigned to these energies. It will be seen that these units are related to each other by some constant.

***(i)* Electrical and Mechanical**

1 kWh = 1 kW × 1 hr

= 1000 watts × 3600 seconds = 36×10^5 watt-sec. or Joules

∴ 1 kWh = 36×10^5 Joules

It is clear that electrical energy can be expressed in Joules instead of kWh.

***(ii)* Heat and Mechanical**

(a) 1 calorie = 4·18 Joules (By experiment)

(b) 1 C.H.U. = 1 lb × 1°C = 453·6 gm × 1°C

= 453·6 calories = 453·6 × 4·18 Joules = 1896 Joules

∴ 1C.H.U. = 1896 Joules

(c) 1 B.Th.U. = 1 lb × 1°F = 453·6 gm × 5/9 °C

= 252 calories = 252 × 4·18 Joules = 1053 Joules

∴ 1 B.Th.U. = 1053 Joules

It may be seen that heat energy can be expressed in Joules instead of thermal units *viz.* calorie, B.Th.U. and C.H.U.

---

* The SI or MKS unit of thermal energy being used these days is the *joule*—exactly as for mechanical and electrical energies. The thermal units *viz.* calorie, B.Th.U. and C.H.U. are obsolete.

(*iii*) **Electrical and Heat**

(*a*) $1 \text{ kWh} = 1000 \text{ watts} \times 3600 \text{ seconds} = 36 \times 10^5 \text{ Joules}$

$= \dfrac{36 \times 10^5}{4 \cdot 18} \text{ calories} = 860 \times 10^3 \text{ calories}$

$\therefore \quad 1 \text{ kWh} = 860 \times 10^3 \text{ calories} \quad \text{or} \quad 860 \text{ kcal}$

(*b*) $1 \text{ kWh} = 36 \times 10^5 \text{ Joules} = 36 \times 10^5/1896 \text{ C.H.U.} = 1898 \text{ C.H.U.}$

$[\because 1 \text{ C.H.U.} = 1896 \text{ Joules}]$

$\therefore \quad 1 \text{ kWh} = 1898 \text{ C.H.U.}$

(*c*) $1 \text{ kWh} = 36 \times 10^5 \text{ Joules} = \dfrac{36 \times 10^5}{1053} \text{ B.Th.U.} = 3418 \text{ B.Th.U.}$

$[\because 1 \text{ B.Th.U.} = 1053 \text{ Joules}]$

$\therefore \quad 1 \text{ kWh} = 3418 \text{ B.Th.U.}$

The reader may note that units of electrical energy can be converted into heat and *vice-versa*. This is expected since electrical and thermal energies are interchangeable.

1.7 Efficiency

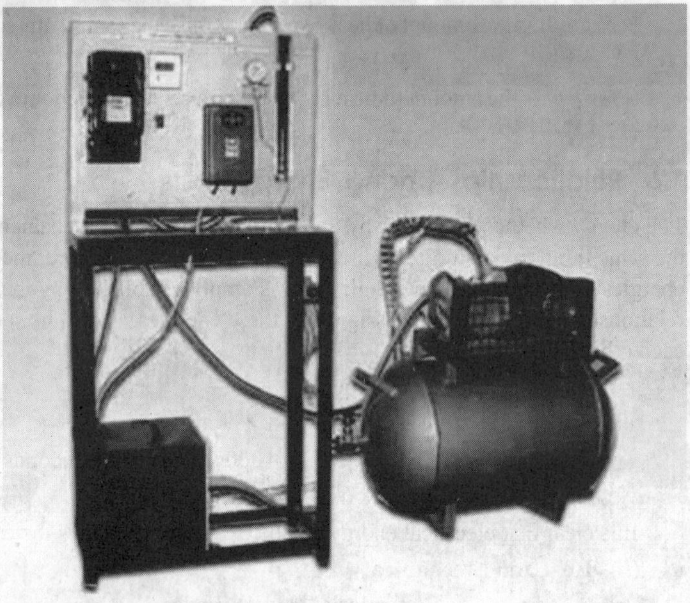

Measuring efficiency of compressor.

Energy is available in various forms from different natural sources such as pressure head of water, chemical energy of fuels, nuclear energy of radioactive substances etc. All these forms of energy can be converted into electrical energy by the use of suitable arrangement. In this process of conversion, some energy is *lost* in the sense that it is converted to a form different from electrical energy. Therefore, the output energy is less than the input energy. *The output energy divided by the input energy is called* **energy efficiency** *or simply* **efficiency** *of the system.*

$$\text{Efficiency, } \eta = \frac{\text{Output energy}}{\text{Input energy}}$$

As power is the rate of energy flow, therefore, efficiency may be expressed equally well as output power divided by input power *i.e.*,

$$\text{Efficiency, } \eta = \frac{\text{Output power}}{\text{Input power}}$$

Example 1.1. *Mechanical energy is supplied to a d.c. generator at the rate of 4200 J/s. The generator delivers 32·2 A at 120 V.*

(*i*) *What is the percentage efficiency of the generator ?*

(*ii*) *How much energy is lost per minute of operation ?*

Solution.

(*i*) Input power, $P_i = 4200 \text{ J/s} = 4200 \text{ W}$

Output power, $P_o = EI = 120 \times 32{\cdot}2 = 3864 \text{ W}$

∴ Efficiency, $\eta = \dfrac{P_o}{P_i} \times 100 = \dfrac{3864}{4200} \times 100 = \mathbf{92\ \%}$

(*ii*) Power lost, $P_L = P_i - P_o = 4200 - 3864 = 336 \text{ W}$

∴ Energy lost per minute (= 60 s) of operation

$= P_L \times t = 336 \times 60 = \mathbf{20160\ J}$

Note that efficiency *is always less than 1* (*or 100 %*). In other words, every system is less than 100 % efficient.

1.8 Calorific Value of Fuels

The amount of heat produced by the complete combustion of a unit weight of fuel is known as its **calorific value.**

Calorific value indicates the amount of heat available from a fuel. The greater the calorific value of fuel, the larger is its ability to produce heat. In case of solid and liquid fuels, the calorific value is expressed in *cal/gm* or *kcal/kg*. However, in case of gaseous fuels, it is generally stated in *cal/litre* or *kcal/litre*. Below is given a table of various types of fuels and their calorific values along with composition.

S.No.	Particular	Calorific value	Composition
1.	**Solid fuels**		
	(*i*) Lignite	5,000 kcal/kg	C = 67%, H = 5%, O = 20%, ash = 8%
	(*ii*) Bituminous coal	7,600 kcal/kg	C = 83%, H = 5·5%, O = 5%, ash = 6·5%
	(*iii*) Anthracite coal	8,500 kcal/kg	C = 90%, H = 3%, O = 2%, ash = 5%
2.	**Liquid fuels**		
	(*i*) Heavy oil	11,000 kcal/kg	C = 86%, H = 12%, S = 2%
	(*ii*) Diesel oil	11,000 kcal/kg	C = 86·3%, H = 12·8%, S = 0·9%
	(*iii*) Petrol	11,110 kcal/kg	C = 86%, H = 14%
3.	**Gaseous fuels**		
	(*i*) Natural gas	520 kcal/m^3	CH_4 = 84%, C_2H_6 = 10% Other hydrocarbons = 5%
	(*ii*) Coal gas	7,600 kcal/m^3	CH_4 = 35%, H = 45%, CO= 8%, N = 6% CO_2 = 2%, Other hydrocarbons = 4%

1.9 Advantages of Liquid Fuels over Solid Fuels

The following are the advantages of liquid fuels over the solid fuels :

(*i*) The handling of liquid fuels is easier and they require less storage space.

(*ii*) The combustion of liquid fuels is uniform.

(*iii*) The solid fuels have higher percentage of moisture and consequently they burn with great difficulty. However, liquid fuels can be burnt with a fair degree of ease and attain high temperature very quickly compared to solid fuels.

(*iv*) The waste product of solid fuels is a large quantity of ash and its disposal becomes a problem. However, liquid fuels leave no or very little ash after burning.

(*v*) The firing of liquid fuels can be easily controlled. This permits to meet the variation in load demand easily.

1.10 Advantages of Solid Fuels over Liquid Fuels

The following are the advantages of solid fuels over the liquid fuels :

(*i*) In case of liquid fuels, there is a danger of explosion.
(*ii*) Liquids fuels are costlier as compared to solid fuels.
(*iii*) Sometimes liquid fuels give unpleasant odours during burning.
(*iv*) Liquid fuels require special types of burners for burning.
(*v*) Liquid fuels pose problems in cold climates since the oil stored in the tanks is to be heated in order to avoid the stoppage of oil flow.

SELF-TEST

1. **Fill in the blanks by inserting appropriate words/figures.**
   (*i*) The primary source of energy is the ....
   (*ii*) The most important form of energy is the ....
   (*iii*) 1 kWh = ... kcal
   (*iv*) The calorific value of a solid fuel is expressed in ....
   (*v*) The three principal sources of energy used for the generation of electrical energy are .... and ....
2. **Pick up the correct words/figures from the brackets and fill in the blanks.**
   (*i*) Electrical energy is .... than other forms of energy. (*cheaper, costlier*)
   (*ii*) The electrical, heat and mechanical energies .... be expressed in the same units. (*can, cannot*)
   (*iii*) .... continue to enjoy the chief source for the generation of electrical energy. (*fuels, radioactive substances, water*)
   (*iv*) The basic unit of energy is .... (*Joule, watt*)
   (*v*) An alternator is a machine which converts .... into .... (*mechanical energy, electrical energy*)

ANSWERS TO SELF-TEST

1. (*i*) Sun, (*ii*) electrical energy, (*iii*) 860, (*iv*) cal/gm or kcal/kg, (*v*) water, fuels and radioactive substances.
2. (*i*) Cheaper, (*ii*) can, (*iii*) fuels, (*iv*) Joule, (*v*) mechanical energy, electrical energy.

CHAPTER REVIEW TOPICS

1. Why is electrical energy preferred over other forms of energy ?
2. Write a short note on the generation of electrical energy.
3. Discuss the different sources of energy available in nature.
4. Compare the chief sources of energy used for the generation of electrical energy.
5. Establish the following relations :
   (*i*) 1 kWh = 36×10^5 Joules (*ii*) 1 kWh = 860 kcal
   (*iii*) 1 B.Th.U. = 1053 Joules (*iv*) 1 C.H.U. = 1896 Joules
6. What do you mean by efficiency of a system ?
7. What are the advantages of liquid fuels over the solid fuels ?
8. What are the advantages of solid fuels over the liquid fuels ?

DISCUSSION QUESTIONS

1. Why do we endeavour to use water power for the generation of electrical energy ?
2. What is the importance of electrical energy ?
3. What are the problems in the use of nuclear energy ?
4. Give one practical example where wind-mill is used.
5. What is the principal source of generation of electrical energy ?

CHAPTER

2

Generating Stations

Introduction

In this modern world, the dependence on electricity is so much that it has become a part and parcel of our life. The ever increasing use of electric power for domestic, commercial and industrial purposes necessitates to provide bulk electric power economically. This is achieved with the help of suitable power producing units, known as *Power plants or Electric power generating stations.* The design of a power plant should incorporate two important aspects. Firstly, the selection and placing of necessary power-generating equipment should be such so that a maximum of return will result from a minimum of expenditure over the working life of the plant. Secondly, the operation of the plant should be such so as to provide cheap, reliable and continuous service. In this chapter, we shall focus our attention on various types of generating stations with special reference to their advantages and disadvantages.

2.1 Generating Stations

Bulk electric power is produced by special plants known as **generating stations** *or* **power plants.**

A generating station essentially employs a

prime mover coupled to an alternator for the production of electric power. The prime mover (*e.g.*, steam turbine, water turbine etc.) converts energy from some other form into mechanical energy. The alternator converts mechanical energy of the prime mover into electrical energy. The electrical energy produced by the generating station is transmitted and distributed with the help of conductors to various consumers. It may be emphasised here that apart from prime mover-alternator combination, a modern generating station employs several auxiliary equipment and instruments to ensure cheap, reliable and continuous service.

Depending upon the form of energy converted into electrical energy, the generating stations are classified as under :

(*i*) Steam power stations (*ii*) Hydroelectric power stations
(*iii*) Diesel power stations (*iv*) Nuclear power stations

2.2 Steam Power Station (Thermal Station)

A generating station which converts heat energy of coal combustion into electrical energy is known as a **steam power station.**

A steam power station basically works on the Rankine cycle. Steam is produced in the boiler by utilising the heat of coal combustion. The steam is then expanded in the prime mover (*i.e.*, steam turbine) and is condensed in a condenser to be fed into the boiler again. The steam turbine drives the alternator which converts mechanical energy of the turbine into electrical energy. This type of power station is suitable where coal and water are available in abundance and a large amount of electric power is to be generated.

Advantages

(*i*) The fuel (*i.e.,* coal) used is quite cheap.
(*ii*) Less initial cost as compared to other generating stations.
(*iii*) It can be installed at any place irrespective of the existence of coal. The coal can be transported to the site of the plant by rail or road.
(*iv*) It requires less space as compared to the hydroelectric power station.
(*v*) The cost of generation is lesser than that of the diesel power station.

Disadvantages

(*i*) It pollutes the atmosphere due to the production of large amount of smoke and fumes.
(*ii*) It is costlier in running cost as compared to hydroelectric plant.

2.3 Schematic Arrangement of Steam Power Station

Although steam power station simply involves the conversion of heat of coal combustion into electrical energy, yet it embraces many arrangements for proper working and efficiency. The schematic arrangement of a modern steam power station is shown in Fig. 2.1. The whole arrangement can be divided into the following stages for the sake of simplicity :

1. Coal and ash handling arrangement **2.** Steam generating plant
3. Steam turbine **4.** Alternator
5. Feed water **6.** Cooling arrangement

1. Coal and ash handling plant. The coal is transported to the power station by road or rail and is stored in the coal storage plant. Storage of coal is primarily a matter of protection against coal strikes, failure of transportation system and general coal shortages. From the coal storage plant, coal is delivered to the coal handling plant where it is pulverised (*i.e.*, crushed into small pieces) in order to increase its surface exposure, thus promoting rapid combustion without using large quantity of

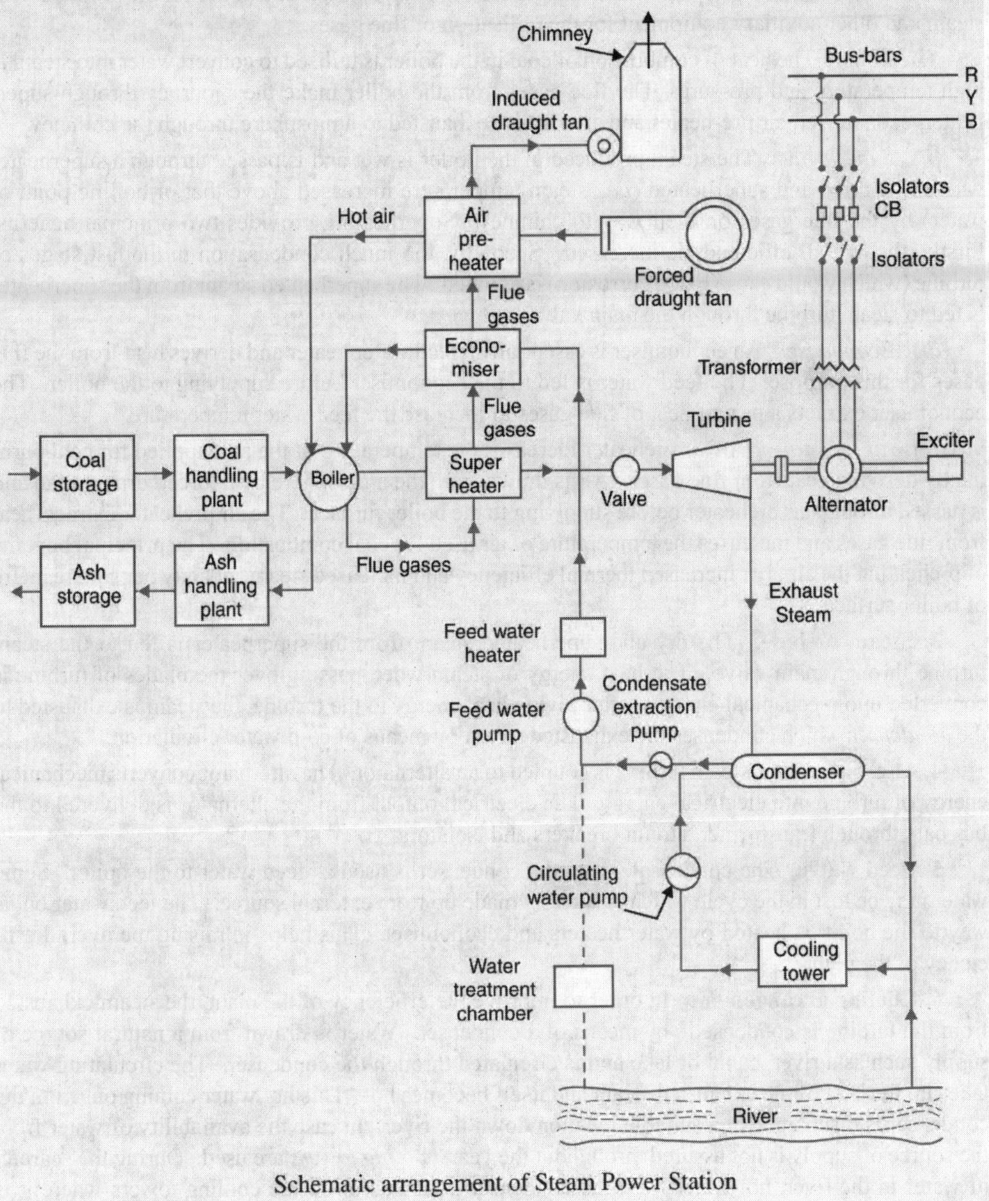

Schematic arrangement of Steam Power Station

Fig. 2.1

excess air. The pulverised coal is fed to the boiler by belt conveyors. The coal is burnt in the boiler and the ash produced after the complete combustion of coal is removed to the ash handling plant and then delivered to the ash storage plant for disposal. The removal of the ash from the boiler furnace is necessary for proper burning of coal.

It is worthwhile to give a passing reference to the amount of coal burnt and ash produced in a modern thermal power station. A 100 MW station operating at 50% load factor may burn about 20,000 tons of coal per month and ash produced may be to the tune of 10% to 15% of coal fired *i.e.*, 2,000 to 3,000 tons. In fact, in a thermal station, about 50% to 60% of the total operating cost consists of fuel purchasing and its handling.

2. Steam generating plant. The steam generating plant consists of a boiler for the production of steam and other auxiliary equipment for the utilisation of flue gases.

(*i*) *Boiler.* The heat of combustion of coal in the boiler is utilised to convert water into steam at high temperature and pressure. The flue gases from the boiler make their journey through super-heater, economiser, air pre-heater and are finally exhausted to atmosphere through the chimney.

(*ii*) *Superheater.* The steam produced in the boiler is wet and is passed through a superheater where it is dried and superheated (*i.e.*, steam temperature increased above that of boiling point of water) by the flue gases on their way to chimney. Superheating provides two principal benefits. Firstly, the overall efficiency is increased. Secondly, too much condensation in the last stages of turbine (which would cause blade corrosion) is avoided. The superheated steam from the superheater is fed to steam turbine through the main valve.

(*iii*) *Economiser.* An economiser is essentially a feed water heater and derives heat from the flue gases for this purpose. The feed water is fed to the economiser before supplying to the boiler. The economiser extracts a part of heat of flue gases to increase the feed water temperature.

(*iv*) *Air preheater.* An air preheater increases the temperature of the air supplied for coal burning by deriving heat from flue gases. Air is drawn from the atmosphere by a forced draught fan and is passed through air preheater before supplying to the boiler furnace. The air preheater extracts heat from flue gases and increases the temperature of air used for coal combustion. The principal benefits of preheating the air are : increased thermal efficiency and increased steam capacity per square metre of boiler surface.

3. Steam turbine. The dry and superheated steam from the superheater is fed to the steam turbine through main valve. The heat energy of steam when passing over the blades of turbine is converted into mechanical energy. After giving heat energy to the turbine, the steam is exhausted to the *condenser* which condenses the exhausted steam by means of cold water circulation.

4. Alternator. The steam turbine is coupled to an alternator. The alternator converts mechanical energy of turbine into electrical energy. The electrical output from the alternator is delivered to the bus bars through transformer, circuit breakers and isolators.

5. Feed water. The condensate from the condenser is used as feed water to the boiler. Some water may be lost in the cycle which is suitably made up from external source. The feed water on its way to the boiler is heated by water heaters and economiser. This helps in raising the overall efficiency of the plant.

6. Cooling arrangement. In order to improve the efficiency of the plant, the steam exhausted from the turbine is condensed* by means of a condenser. Water is drawn from a natural source of supply such as a river, canal or lake and is circulated through the condenser. The circulating water takes up the heat of the exhausted steam and itself becomes hot. This hot water coming out from the condenser is discharged at a suitable location down the river. In case the availability of water from the source of supply is not assured throughout the year, *cooling towers* are used. During the scarcity of water in the river, hot water from the condenser is passed on to the cooling towers where it is cooled. The cold water from the cooling tower is reused in the condenser.

2.4 Choice of Site for Steam Power Stations

In order to achieve overall economy, the following points should be considered while selecting a site for a steam power station :

(*i*) *Supply of fuel.* The steam power station should be located near the coal mines so that transportation cost of fuel is minimum. However, if such a plant is to be installed at a place

* Efficiency of the plant is increased by reducing turbine exhaust pressure. Low pressure at the exhaust can be achieved by condensing the steam at the turbine exhaust.

where coal is not available, then care should be taken that adequate facilities exist for the transportation of coal.

(*ii*) *Availability of water.* As huge amount of water is required for the condenser, therefore, such a plant should be located at the bank of a river or near a canal to ensure the continuous supply of water.

(*iii*) *Transportation facilities.* A modern steam power station often requires the transportation of material and machinery. Therefore, adequate transportation facilities must exist *i.e.*, the plant should be well connected to other parts of the country by rail, road. etc.

(*iv*) *Cost and type of land.* The steam power station should be located at a place where land is cheap and further extension, if necessary, is possible. Moreover, the bearing capacity of the ground should be adequate so that heavy equipment could be installed.

(*v*) *Nearness to load centres.* In order to reduce the transmission cost, the plant should be located near the centre of the load. This is particularly important if *d.c.* supply system is adopted. However, if *a.c.* supply system is adopted, this factor becomes relatively less important. It is because *a.c.* power can be transmitted at high voltages with consequent reduced transmission cost. Therefore, it is possible to install the plant away from the load centres, provided other conditions are favourable.

(*vi*) *Distance from populated area.* As huge amount of coal is burnt in a steam power station, therefore, smoke and fumes pollute the surrounding area. This necessitates that the plant should be located at a considerable distance from the populated areas.

Conclusion. It is clear that all the above factors cannot be favourable at one place. However, keeping in view the fact that now-a-days the supply system is *a.c.* and more importance is being given to generation than transmission, a site away from the towns may be selected. In particular, a site by river side where sufficient water is available, no pollution of atmosphere occurs and fuel can be transported economically, may perhaps be an ideal choice.

2.5 Efficiency of Steam Power Station

The overall efficiency of a steam power station is quite low (about 29%) due mainly to two reasons. Firstly, a huge amount of heat is lost in the condenser and secondly heat losses occur at various stages of the plant. The heat lost in the condenser cannot be avoided. It is because heat energy cannot be converted into mechanical energy without temperature difference. The greater the temperature difference, the greater is the heat energy converted* into mechanical energy. This necessitates to keep the steam in the condenser at the lowest temperature. But we know that greater the temperature difference, greater is the amount of heat lost. This explains for the low efficiency of such plants.

(*i*) **Thermal efficiency.** *The ratio of heat equivalent of mechanical energy transmitted to the turbine shaft to the heat of combustion of coal is known as* **thermal efficiency** *of steam power station.*

$$\text{Thermal efficiency,} \quad \eta_{thermal} = \frac{\text{Heat equivalent of mech. energy transmitted to turbine shaft}}{\text{Heat of coal combustion}}$$

The thermal efficiency of a modern steam power station is about 30%. It means that if 100 calories of heat is supplied by coal combustion, then mechanical energy equivalent of 30 calories will be available at the turbine shaft and rest is lost. It may be important to note that more than 50% of total heat of combustion is lost in the condenser. The other heat losses occur in flue gases, radiation, ash etc.

(*ii*) **Overall efficiency.** *The ratio of heat equivalent of electrical output to the heat of combustion of coal is known as* **overall efficiency** *of steam power station i.e.*

---

* Thermodynamic laws.

Overall efficiency, $\eta_{overall} = \dfrac{\text{Heat equivalent of electrical ouput}}{\text{Heat of combustion of coal}}$

The overall efficiency of a steam power station is about 29%. It may be seen that overall efficiency is less than the thermal efficiency. This is expected since some losses (about 1%) occur in the alternator. The following relation exists among the various efficiencies.

Overall efficiency = Thermal efficiency × Electrical efficiency

2.6 Equipment of Steam Power Station

A modern steam power station is highly complex and has numerous equipment and auxiliaries. However, the most important constituents of a steam power station are :

1. Steam generating equipment 2. Condenser 3. Prime mover
4. Water treatment plant 5. Electrical equipment.

1. Steam generating equipment. This is an important part of steam power station. It is concerned with the generation of superheated steam and includes such items as boiler, boiler furnace, superheater, economiser, air pre-heater and other heat reclaiming devices.

(*i*) *Boiler.* A boiler is closed vessel in which water is converted into steam by utilising the heat of coal combustion. Steam boilers are broadly classified into the following two types :

(*a*) Water tube boilers **(*b*)** Fire tube boilers

In a water tube boiler, water flows through the tubes and the hot gases of combustion flow over these tubes. On the other hand, in a fire tube boiler, the hot products of combustion pass through the tubes surrounded by water. Water tube boilers have a number of advantages over fire tube boilers *viz.*, require less space, smaller size of tubes and drum, high working pressure due to small drum, less liable to explosion etc. Therefore, the use of water tube boilers has become universal in large capacity steam power stations.

(*ii*) *Boiler furnace.* A boiler furnace is a chamber in which fuel is burnt to liberate the heat energy. In addition, it provides support and enclosure for the combustion equipment *i.e.*, burners. The boiler furnace walls are made of refractory materials such as fire clay, silica, kaolin etc. These materials have the property to resist change of shape, weight or physical properties at high temperatures. There are following three types of construction of furnace walls :

(*a*) Plain refractory walls
(*b*) Hollow refractory walls with an arrangement for air cooling
(*c*) Water walls.

The plain refractory walls are suitable for small plants where the furnace temperature may not be high. However, in large plants, the furnace temperature is quite high* and consequently, the refractory material may get damaged. In such cases, refractory walls are made hollow and air is circulated through hollow space to keep the temperature of the furnace walls low. The recent development is to use water walls. These consist of plain tubes arranged side by side and on the inner face of the refractory walls. The tubes are connected to the upper and lower headers of the boiler. The boiler water is made to circulate through these tubes. The water walls absorb the radiant heat in the furnace which would otherwise heat up the furnace walls.

(*iii*) *Superheater.* A superheater is a device which superheats the steam *i.e.*, it raises the temperature of steam above boiling point of water. This increases the overall efficiency of the plant. A superheater consists of a group of tubes made of special alloy steels such as chromium-molybdenum. These tubes are heated by the heat of flue gases during their journey from the furnace to the chimney.

* The size of furnace has to be limited due to space, cost and other considerations. This means that furnace of a large plant should develop more kilocalories per square metre of furnace which implies high furnace temperature.

The steam produced in the boiler is led through the superheater where it is superheated by the heat of flue gases. Superheaters are mainly classified into two types according to the system of heat transfer from flue gases to steam *viz.*

(*a*) Radiant superheater (*b*) Convection superheater

The radiant superheater is placed in the furnace between the water walls and receives heat from the burning fuel through radiation process. It has two main disadvantages. Firstly, due to high furnace temperature, it may get overheated and, therefore, requires a careful design. Secondly, the temperature of superheater falls with increase in steam output. Due to these limitations, radiant superheater is not finding favour these days. On the other hand, a convection superheater is placed in the boiler tube bank and receives heat from flue gases entirely through the convection process. It has the advantage that temperature of superheater increases with the increase in steam output. For this reason, this type of superheater is commonly used these days.

(*iv*) *Economiser.* It is a device which heats the feed water on its way to boiler by deriving heat from the flue gases. This results in raising boiler efficiency, saving in fuel and reduced stresses in the boiler due to higher temperature of feed water. An economiser consists of a large number of closely spaced parallel steel tubes connected by headers of drums. The feed water flows through these tubes and the flue gases flow outside. A part of the heat of flue gases is transferred to feed water, thus raising the temperature of the latter.

(*v*) *Air Pre-heater.* Superheaters and economisers generally cannot fully extract the heat from flue gases. Therefore, pre-heaters are employed which recover some of the heat in the escaping gases. The function of an air pre-heater is to extract heat from the flue gases and give it to the air being supplied to furnace for coal combustion. This raises the furnace temperature and increases the thermal efficiency of the plant. Depending upon the method of transfer of heat from flue gases to air, air pre-heaters are divided into the following two classes :

(*a*) Recuperative type (*b*) Regenerative type

The recuperative type air-heater consists of a group of steel tubes. The flue gases are passed through the tubes while the air flows externally to the tubes. Thus heat of flue gases is transferred to air. The regenerative type air pre-heater consists of slowly moving drum made of corrugated metal plates. The flue gases flow continuously on one side of the drum and air on the other side. This action permits the transference of heat of flue gases to the air being supplied to the furnace for coal combustion.

2. Condensers. A condenser is a device which condenses the steam at the exhaust of turbine. It serves two important functions. Firstly, it creates a very low *pressure at the exhaust of turbine, thus permitting expansion of the steam in the prime mover to a very low pressure. This helps in converting heat energy of steam into mechanical energy in the prime mover. Secondly, the condensed steam can be used as feed water to the boiler. There are two types of condensers, namely :

(*i*) Jet condenser (*ii*) Surface condenser

In a jet condenser, cooling water and exhausted steam are mixed together. Therefore, the temperature of cooling water and condensate is the same when leaving the condenser. Advantages of this type of condenser are : low initial cost, less floor area required, less cooling water required and low maintenance charges. However, its disadvantages are : condensate is wasted and high power is required for pumping water.

In a surface condenser, there is no direct contact between cooling water and exhausted steam. It consists of a bank of horizontal tubes enclosed in a cast iron shell. The cooling water flows through the tubes and exhausted steam over the surface of the tubes. The steam gives up its heat to water and is itself condensed. Advantages of this type of condenser are : condensate can be used as feed water, less pumping power required and creation of better vacuum at the turbine exhaust. However, disad-

---

* By liquidating steam at the exhaust of turbine, a region of emptiness is created. This results in a very low pressure at the exhaust of turbine.

vantages of this type of condenser are : high initial cost, requires large floor area and high maintenance charges.

3. Prime movers. The prime mover converts steam energy into mechanical energy. There are two types of steam prime movers *viz*., steam engines and steam turbines. A steam turbine has several advantages over a steam engine as a prime mover *viz*., high efficiency, simple construction, higher speed, less floor area requirement and low maintenance cost. Therefore, all modern steam power stations employ steam turbines as prime movers.

Steam turbines are generally classified into two types according to the action of steam on moving blades *viz*.

(*i*) Impulse turbines (*ii*) Reactions turbines

In an impulse turbine, the steam expands completely in the stationary nozzles (or fixed blades), the pressure over the moving blades remaining constant. In doing so, the steam attains a high velocity and impinges against the moving blades. This results in the impulsive force on the moving blades which sets the rotor rotating. In a reaction turbine, the steam is partially expanded in the stationary nozzles, the remaining expansion takes place during its flow over the moving blades. The result is that the momentum of the steam causes a reaction force on the moving blades which sets the rotor in motion.

4. Water treatment plant. Boilers require clean and soft water for longer life and better efficiency. However, the source of boiler feed water is generally a river or lake which may contain suspended and dissolved impurities, dissolved gases etc. Therefore, it is very important that water is first purified and softened by chemical treatment and then delivered to the boiler.

The water from the source of supply is stored in storage tanks. The suspended impurities are removed through sedimentation, coagulation and filtration. Dissolved gases are removed by aeration and degasification. The water is then *'softened'* by removing temporary and permanent hardness through different chemical processes. The pure and soft water thus available is fed to the boiler for steam generation.

5. Electrical equipment. A modern power station contains numerous electrical equipment. However, the most important items are :

(*i*) *Alternators.* Each alternator is coupled to a steam turbine and converts mechanical energy of the turbine into electrical energy. The alternator may be hydrogen or air cooled. The necessary excitation is provided by means of main and pilot exciters directly coupled to the alternator shaft.

(*ii*) *Transformers.* A generating station has different types of transformers, *viz*.,

- (*a*) main step-up transformers which step-up the generation voltage for transmission of power.
- (*b*) station transformers which are used for general service (*e.g.*, lighting) in the power station.
- (*c*) auxiliary transformers which supply to individual unit-auxiliaries.

(*iii*) *Switchgear.* It houses such equipment which locates the fault on the system and isolate the faulty part from the healthy section. It contains circuit breakers, relays, switches and other control devices.

Example 2.1. *A steam power station has an overall efficiency of 20% and 0·6 kg of coal is burnt per kWh of electrical energy generated. Calculate the calorific value of fuel.*

Solution.

Let x kcal/kg be the calorific value of fuel.

Heat produced by 0·6 kg of coal = 0·6 x kcal

Heat equivalent of 1 kWh = 860 kcal

$$\text{Now,}\quad \eta_{overall} = \frac{\text{Electrical output in heat units}}{\text{Heat of combustion}}$$

$$\text{or}\quad 0{\cdot}2 = \frac{860}{0{\cdot}6x}$$

$$\therefore\quad x = \frac{860}{0{\cdot}6 \times 0{\cdot}2} = \mathbf{7166{\cdot}67\ kcal/kg}$$

Example 2.2. *A thermal station has the following data :*

Max. demand	=	*20,000 kW*	;	*Load factor*	=	*40%*
Boiler efficiency	=	*85%*	;	*Turbine efficiency*	=	*90%*
Coal consumption	=	*0·9 kg/kWh*	;	*Cost of 1 ton of coal*	=	*Rs. 300*

Determine (i) thermal efficiency and (ii) coal bill per annum.

Solution.

(*i*) Thermal efficiency = $\eta_{boiler} \times \eta_{turbine} = 0{\cdot}85 \times {\cdot}9 = \mathbf{0{\cdot}765}$ **or 76·5 %**

(*ii*) Units generated/annum = Max. demand × L.F. × Hours in a year

= 20,000 × 0·4 × 8760 = 7008 × 10^4 kWh

$$\text{Coal consumption/annum} = \frac{(0\cdot 9)\left(7008\times 10^4\right)}{1000} = 63{,}072 \text{ tons}$$

∴ Annual coal bill = Rs 300 × 63072 = **Rs 1,89,21,600**

Example 2.3. *A steam power station spends Rs. 30 lakhs per annum for coal used in the station. The coal has a calorific value of 5000 kcal/kg and costs Rs. 300 per ton. If the station has thermal efficiency of 33% and electrical efficiency of 90%, find the average load on the station.*

Solution.

Overall efficiency, $\eta_{overall} = 0{\cdot}33 \times 0{\cdot}9 = 0{\cdot}297$

Coal used/annum = $30 \times 10^5/300 = 10^4$ tons = 10^7 kg

Heat of combustion = Coal used/annum × Calorific value

= $10^7 \times 5000 = 5 \times 10^{10}$ kcal

Heat output = $\eta_{overall}$ × Heat of combustion

= $(0{\cdot}297) \times (5 \times 10^{10}) = 1485 \times 10^7$ kcal

Units generated/annum = $1485 \times 10^7/860$ kWh

$$\therefore\quad \text{Average load on station} = \frac{\text{Units generated / annum}}{\text{Hours in a year}} = \frac{1485\times 10^7}{860\times 8760} = \mathbf{1971\ kW}$$

Example 2.4. *The relation between water evaporated (W kg), coal consumption (C kg) and kWh generated per 8-hour shift for a steam generating station is as follows :*

$$W = 13500 + 7{\cdot}5\ kWh \quad \text{....(i)}$$

$$C = 5000 + 2{\cdot}9\ kWh \quad \text{....(ii)}$$

(*i*) *To what limiting value does the water evaporating per kg of coal consumed approach as the station output increases ?* (*ii*) *How much coal per hour would be required to keep the station running on no load ?*

Solution.

(*i*) For an 8-hour shift, weight of water evaporated per kg of coal consumed is

$$\frac{W}{C} = \frac{13500 + 7 \cdot 5 \text{ kWh}}{5000 + 2 \cdot 9 \text{ kWh}}$$

As the station output (*i.e.*, kWh) increases towards infinity, the limiting value of *W*/*C* approaches 7·5/2·9 = 2·6. Therefore, the weight of water evaporated per kg of coal consumed approaches a limiting value of **2·6 kg** as the kWh output increases.

(*ii*) At no load, the station output is zero *i.e.,* kWh = 0. Therefore, from expression (*ii*), we get, coal consumption at no load

$$= 5000 + 2{\cdot}9 \times 0 = 5000 \text{ kg}$$

$\therefore$ Coal consumption/hour $= 5000/8 = \mathbf{625\ kg}$

Example 2.5. *A 100 MW steam station uses coal of calorific value 6400 kcal/kg. Thermal efficiency of the station is 30% and electrical efficiency is 92%. Calculate the coal consumption per hour when the station is delivering its full rated output.*

Solution.

Overall efficiency of the power station is

$$\eta_{overall} = \eta_{thermal} \times \eta_{elect} = 0{\cdot}30 \times 0{\cdot}92 = 0{\cdot}276$$

$$\text{Units generated/hour} = (100 \times 10^3) \times 1 = 10^5 \text{ kWh}$$

$$\text{Heat produced/hour, } H = \frac{\text{Electrical output in heat units}}{\eta_{overall}}$$

$$= \frac{10^5 \times 860}{0 \cdot 276} = 311 \cdot 6 \times 10^6 \text{ kcal} \qquad (\because \text{ 1 kWh} = 860 \text{ kcal})$$

$$\therefore \quad \text{Coal consumption/hour} = \frac{H}{\text{Calorific value}} = \frac{311 \cdot 6 \times 10^6}{6400} = \mathbf{48687\ kg}$$

TUTORIAL PROBLEMS

1. A generating station has an overall efficiency of 15% and 0·75 kg of coal is burnt per kWh by the station. Determine the calorific value of coal in kilocalories per kilogram. **[7644 kcal/kg]**
2. A 75 MW steam power station uses coal of calorific value of 6400 kcal/kg. Thermal efficiency of the station is 30% while electrical efficiency is 80%. Calculate the coal consumption per hour when the station is delivering its full output. **[42 tons]**
3. A 65,000 kW steam power station uses coal of calorific value 15,000 kcal per kg. If the coal consumption per kWh is 0·5 kg and the load factor of the station is 40%, calculate (*i*) the overall efficiency (*ii*) coal consumption per day. **[(*i*) 28·7% (*ii*) 312 tons]**
4. A 60 MW steam power station has a thermal efficiency of 30%. If the coal burnt has a calorific value of 6950 kcal/kg, calculate :

   (*i*) the coal consumption per kWh,

   (*ii*)the coal consumption per day. **[(*i*) 0·413 kg (*ii*) 238 tons]**
5. A 25 MVA turbo-alternator is working on full load at a power factor of 0·8 and efficiency of 97%. Find the quantity of cooling air required per minute at full load, assuming that 90% of the total losses are dissipated by the internally circulating air. The inlet air temperature is 20º C and the temperature rise is 30º C. Given that specific heat of air is 0·24 and that 1 kg of air occupies 0·8 m^3. **[890 m^3/minute]**
6. A thermal station has an efficiency of 15% and 1·0 kg of coal burnt for every kWh generated. Determine the calorific value of coal. **[5733 kcal/kg]**

2.7 Hydro-electric Power Station

A generating station which utilises the potential energy of water at a high level for the generation of electrical energy is known as a **hydro-electric power station.**

Hydro-electric power stations are generally located in hilly areas where dams can be built conveniently and large water reservoirs can be obtained. In a hydro-electric power station, water head is created by constructing a dam across a river or lake. From the dam, water is led to a water turbine. The water turbine captures the energy in the falling water and changes the hydraulic energy (*i.e.*, product of head and flow of water) into mechanical energy at the turbine shaft. The turbine drives the alternator which converts mechanical energy into electrical energy. Hydro-electric power stations are becoming very popular because the reserves of fuels (*i.e.*, coal and oil) are depleting day by day. They have the added importance for flood control, storage of water for irrigation and water for drinking purposes.

Advantages

(*i*) It requires no fuel as water is used for the generation of electrical energy.

(*ii*) It is quite neat and clean as no smoke or ash is produced.

(*iii*) It requires very small running charges because water is the source of energy which is available free of cost.

(*iv*) It is comparatively simple in construction and requires less maintenance.

(*v*) It does not require a long starting time like a steam power station. In fact, such plants can be put into service instantly.

(*vi*) It is robust and has a longer life.

(*vii*) Such plants serve many purposes. In addition to the generation of electrical energy, they also help in irrigation and controlling floods.

(*viii*) Although such plants require the attention of highly skilled persons at the time of construction, yet for operation, a few experienced persons may do the job well.

Disadvantages

(*i*) It involves high capital cost due to construction of dam.

(*ii*) There is uncertainty about the availability of huge amount of water due to dependence on weather conditions.

(*iii*) Skilled and experienced hands are required to build the plant.

(*iv*) It requires high cost of transmission lines as the plant is located in hilly areas which are quite away from the consumers.

2.8 Schematic Arrangement of Hydro-electric Power Station

Although a hydro-electric power station simply involves the conversion of hydraulic energy into electrical energy, yet it embraces many arrangements for proper working and efficiency. The schematic arrangement of a modern hydro-electric plant is shown in Fig. 2.2.

The dam is constructed across a river or lake and water from the catchment area collects at the back of the dam to form a reservoir. A pressure tunnel is taken off from the reservoir and water brought to the valve house at the start of the penstock. The valve house contains main sluice valves and automatic isolating valves. The former controls the water flow to the power house and the latter cuts off supply of water when the penstock bursts. From the valve house, water is taken to water turbine through a huge steel pipe known as *penstock*. The water turbine converts hydraulic energy into mechanical energy. The turbine drives the alternator which converts mechanical energy into electrical energy.

A surge tank (open from top) is built just before the valve house and protects the penstock from bursting in case the turbine gates suddenly close* due to electrical load being thrown off. When the

* The governor opens or closes the turbine gates in accordance with the changes in electrical load. If the electrical load increases, the governor opens the turbine gates to allow more water and *vice-versa*.

gates close, there is a sudden stopping of water at the lower end of the penstock and consequently the penstock can burst like a paper log. The surge tank absorbs this pressure swing by increase in its level of water.

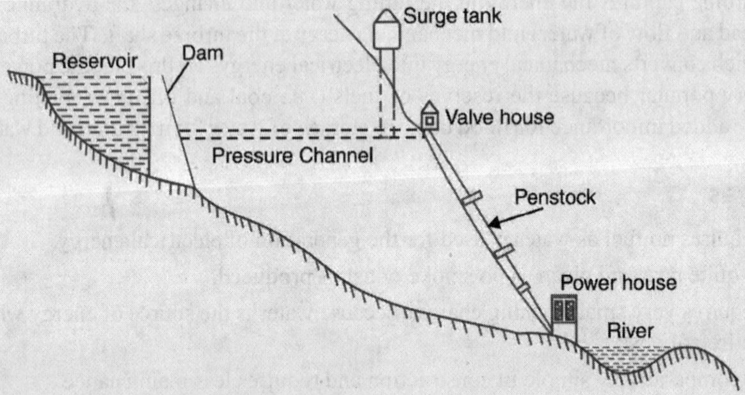

Schematic arrangement of a Hydro-electric plant

Fig. 2.2

2.9 Choice of Site for Hydro-electric Power Stations

The following points should be taken into account while selecting the site for a hydro-electric power station :

(*i*) *Availability of water*. Since the primary requirement of a hydro-electric power station is the availability of huge quantity of water, such plants should be built at a place (*e.g.*, river, canal) where adequate water is available at a good head.

(*ii*) *Storage of water.* There are wide variations in water supply from a river or canal during the year. This makes it necessary to store water by constructing a dam in order to ensure the generation of power throughout the year. The storage helps in equalising the flow of water so that any excess quantity of water at a certain period of the year can be made available during times of very low flow in the river. This leads to the conclusion that site selected for a hydro-electric plant should provide adequate facilities for erecting a dam and storage of water.

(*iii*) *Cost and type of land.* The land for the construction of the plant should be available at a reasonable price. Further, the bearing capacity of the ground should be adequate to withstand the weight of heavy equipment to be installed.

(*iv*) *Transportation facilities.* The site selected for a hydro-electric plant should be accessible by rail and road so that necessary equipment and machinery could be easily transported.

It is clear from the above mentioned factors that ideal choice of site for such a plant is near a river in hilly areas where dam can be conveniently built and large reservoirs can be obtained.

2.10 Constituents of Hydro-electric Plant

The constituents of a hydro-electric plant are (1) hydraulic structures (2) water turbines and (3) electrical equipment. We shall discuss these items in turn.

1. Hydraulic structures. Hydraulic structures in a hydro-electric power station include dam, spillways, headworks, surge tank, penstock and accessory works.

(*i*) *Dam*. A dam is a barrier which stores water and creates water head. Dams are built of concrete or stone masonary, earth or rock fill. The type and arrangement depends upon the topography of the site. A masonary dam may be built in a narrow canyon. An earth dam may be best suited for a wide valley. The type of dam also depends upon the foundation conditions, local materials and transportation available, occurrence of earthquakes and other hazards. At most of sites, more than one type of dam may be suitable and the one which is most economical is chosen.

(*ii*) *Spillways*. There are times when the river flow exceeds the storage capacity of the reservoir. Such a situation arises during heavy rainfall in the catchment area. In order to discharge the surplus water from the storage reservoir into the river on the down-stream side of the dam, spillways are used. Spillways are constructed of concrete piers on the top of the dam. Gates are provided between these piers and surplus water is discharged over the crest of the dam by opening these gates.

(*iii*) *Headworks*. The headworks consists of the diversion structures at the head of an intake. They generally include booms and racks for diverting floating debris, sluices for by-passing debris and sediments and valves for controlling the flow of water to the turbine. The flow of water into and through headworks should be as smooth as possible to avoid head loss and cavitation. For this purpose, it is necessary to avoid sharp corners and abrupt contractions or enlargements.

(*iv*) *Surge tank*. Open conduits leading water to the turbine require no* protection. However, when closed conduits are used, protection becomes necessary to limit the abnormal pressure in the conduit. For this reason, closed conduits are always provided with a surge tank. A surge tank is a small reservoir or tank (open at the top) in which water level rises or falls to reduce the pressure swings in the conduit.

A surge tank is located near the beginning of the conduit.

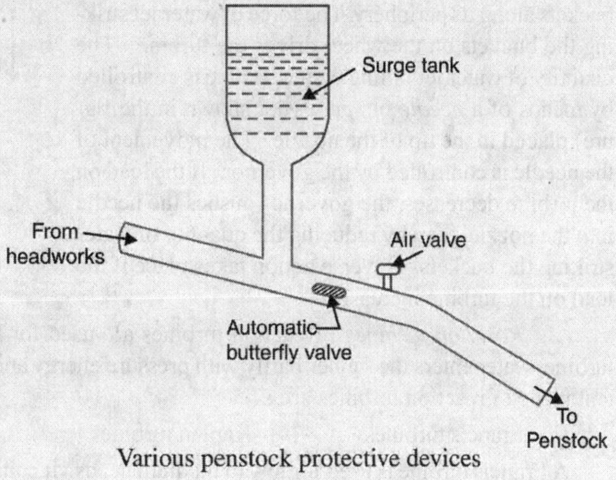

Various penstock protective devices

Fig. 2.3

When the turbine is running at a steady load, there are no surges in the flow of water through the conduit *i.e.*, the quantity of water flowing in the conduit is just sufficient to meet the turbine requirements. However, when the load on the turbine decreases, the governor closes the gates of turbine, reducing water supply to the turbine. The excess water at the lower end of the conduit rushes back to the surge tank and increases its water level. Thus the conduit is prevented from bursting. On the other hand, when load on the turbine increases, additional water is drawn from the surge tank to meet the increased load requirement. Hence, a surge tank overcomes the abnormal pressure in the conduit when load on the turbine falls and acts as a reservoir during increase of load on the turbine.

(*v*) *Penstocks*. Penstocks are open or closed conduits which carry water to the turbines. They are generally made of reinforced concrete or steel. Concrete penstocks are suitable for low

* Because in case of open conduits, regulating gates control the inflow at the headworks and the spillway discharges the surplus water.

heads (< 30 m) as greater pressure causes rapid deterioration of concrete. The steel penstocks can be designed for any head; the thickness of the penstock increases with the head or working pressure.

Various devices such as automatic butterfly valve, air valve and surge tank (See Fig. 2.3) are provided for the protection of penstocks. Automatic butterfly valve shuts off water flow through the penstock promptly if it ruptures. Air valve maintains the air pressure inside the penstock equal to outside atmospheric pressure. When water runs out of a penstock faster than it enters, a vacuum is created which may cause the penstock to collapse. Under such situations, air valve opens and admits air in the penstock to maintain inside air pressure equal to the outside air pressure.

2. Water turbines. Water turbines are used to convert the energy of falling water into mechanical energy. The principal types of water turbines are :

(*i*) Impulse turbines (*ii*) Reaction turbines

(*i*) *Impulse turbines.* Such turbines are used for high heads. In an impulse turbine, the entire pressure of water is converted into kinetic energy in a nozzle and the velocity of the jet drives the wheel. The example of this type of turbine is the Pelton wheel (See Fig. 2.4). It consists of a wheel fitted with elliptical buckets along its periphery. The force of water jet striking the buckets on the wheel drives the turbine. The quantity of water jet falling on the turbine is controlled by means of a *needle* or *spear* (not shown in the figure) placed in the tip of the nozzle. The movement of the needle is controlled by the governor. If the load on the turbine decreases, the governor pushes the needle into the nozzle, thereby reducing the quantity of water striking the buckets. Reverse action takes place if the load on the turbine increases.

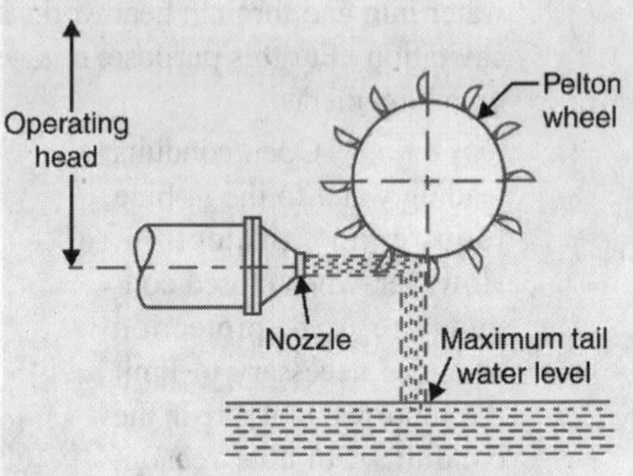

Fig. 2.4 Pelton Wheel

(*ii*) *Reaction turbines.* Reaction turbines are used for low and medium heads. In a reaction turbine, water enters the runner partly with pressure energy and partly with velocity head. The important types of reaction turbines are :

(*a*) Francis turbines (*b*) Kaplan turbines

A Francis turbine is used for low to medium heads. It consists of an outer ring of stationary guide blades fixed to the turbine casing and an inner ring of rotating blades forming the runner. The guide blades control the flow of water to the turbine. Water flows radially inwards and changes to a downward direction while passing through the runner. As the water passes over the "rotating blades" of the runner, both pressure and velocity of water are reduced. This causes a reaction force which drives the turbine.

A Kaplan turbine is used for low heads and large quantities of water. It is similar to Francis turbine except that the runner of Kaplan turbine receives water axially. Water flows radially inwards

Bhakra Dam

through regulating gates all around the sides, changing direction in the runner to axial flow. This causes a reaction force which drives the turbine.

3. Electrical equipment. The electrical equipment of a hydro-electric power station includes alternators, transformers, circuit breakers and other switching and protective devices.

Example 2.6. *A hydro-electric generating station is supplied from a reservoir of capacity 5×10^6 cubic metres at a head of 200 metres. Find the total energy available in kWh if the overall efficiency is 75%.*

Solution.

Weight of water available is

$$W = \text{Volume of water} \times \text{density}$$
$$= (5 \times 10^6) \times (1000) \quad (\because \text{ mass of } 1\text{m}^3 \text{ of water is 1000 kg})$$
$$= 5 \times 10^9 \text{ kg} = 5 \times 10^9 \times 9{\cdot}81 \text{ N}$$

$$\text{Electrical energy available} = W \times H \times \eta_{overall} = (5 \times 10^9 \times 9{\cdot}81) \times (200) \times (0{\cdot}75) \text{ watt sec}$$
$$= \frac{(5 \times 10^9 \times 9{\cdot}81) \times (200) \times (0{\cdot}75)}{3600 \times 1000} \text{ kWh} = \mathbf{2.044 \times 106 \text{ kWh}}$$

Example 2.7. *It has been estimated that a minimum run off of approximately 94 m^3/sec will be available at a hydraulic project with a head of 39 m. Determine (i) firm capacity (ii) yearly gross output. Assume the efficiency of the plant to be 80%.*

Solution.

Weight of water available, $W = 94 \times 1000 = 94000$ kg/sec

Water head, $H = 39$ m

Work done/sec $= W \times H = 94000 \times 9{\cdot}81 \times 39$ watts
$= 35,963 \times 10^3$ W $= 35,963$ kW

This is gross plant capacity.

(*i*) Firm capacity = Plant efficiency × Gross plant capacity
$= 0{\cdot}80 \times 35{,}963 =$ **28,770 kW**

(*ii*) Yearly gross output = Firm capacity × Hours in a year
$= 28{,}770 \times 8760 =$ **252×10^6 kWh**

Example 2.8. *Water for a hydro-electric station is obtained from a reservoir with a head of 100 metres. Calculate the electrical energy generated per hour per cubic metre of water if the hydraulic efficiency be 0·86 and electrical efficiency 0·92.*

Solution.

Water head, $H = 100$ m ; discharge, $Q = 1$ m^3/sec ; $\eta_{overall} = 0{\cdot}86 \times 0{\cdot}92 = 0{\cdot}79$

Wt. of water available/sec, $W = Q \times 1000 \times 9{\cdot}81 = 9810\ N$

Power produced $= W \times H \times \eta_{overall} = 9810 \times 100 \times 0{\cdot}79$ watts
$= 775 \times 10^3$ watts $= 775$ kW

$\therefore$ Energy generated/hour $= 775 \times 1 =$ **775 kWh**

Example 2.9. *Calculate the average power in kW that can be generated in a hydro-electric project from the following data*

Catchment area = 5×10^9 m^2 ; Mean head, H = 30 m

Annual rainfall, F = 1·25 m ; Yield factor, K = 80 %

Overall efficiency, η_{oveall} = 70 %

If the load factor is 40% , what is the rating of generators installed ?

Solution.

Volume of water which can be utilised per annum

$= \text{Catchment area} \times \text{Annual rainfall} \times {}^*\text{yield factor}$

$= (5 \times 10^9) \times (1{\cdot}25) \times (0{\cdot}8) = 5 \times 10^9 \text{ m}^3$

Weight of water available per annum is

$W = 5 \times 10^9 \times 9{\cdot}81 \times 1000 = 49{\cdot}05 \times 10^{12} \text{ N}$

Electrical energy available per annum

$= W \times H \times \eta_{overall} = (49{\cdot}05 \times 10^{12}) \times (30) \times (0{\cdot}7) \text{ watt-sec}$

$$= \frac{(49\cdot05 \times 10^{12}) \times (30) \times (0\cdot7)}{1000 \times 3600} \text{ kWh} = 2\cdot86 \times 10^8 \text{ kWh}$$

$\therefore$ Average power $= 2{\cdot}86 \times 10^8/8760 = \mathbf{32648\ kW}$

$$\text{Max. demand} = \frac{\text{Average demand}}{\text{Load factor}} = \frac{32648}{0\cdot4} = \mathbf{81620\ kW}$$

Therefore, the maximum capacity of the generators should be 81620 kW.

Example 2.10. *A hydro-electric power station has a reservoir of area 2·4 square kilometres and capacity 5×10^6 m³. The effective head of water is 100 metres. The penstock, turbine and generation efficiencies are respectively 95%,90% and 85%.*

(*i*) *Calculate the total electrical energy that can be generated from the power station.*

(*ii*) *If a load of 15,000 kW has been supplied for 3 hours, find the fall in reservoir level.*

Solution.

(*i*) Wt. of water available, W = Volume of reservoir × wt. of 1m³ of water

$= (5 \times 10^6) \times (1000) \text{ kg} = 5 \times 10^9 \times 9{\cdot}81 \text{ N}$

Overall efficiency, $\eta_{overall} = 0{\cdot}95 \times 0{\cdot}9 \times 0{\cdot}85 = 0{\cdot}726$

Electrical energy that can be generated

$= W \times H \times \eta_{overall} = (5 \times 10^9 \times 9{\cdot}81) \times (100) \times (0{\cdot}726) \text{ watt-sec.}$

$$= \frac{(5 \times 10^9 \times 9\cdot81) \times (100) \times (0\cdot726)}{1000 \times 3600} \text{ kWh} = \mathbf{9{,}89{,}175\ kWh}$$

(*ii*) Let x metres be the fall in reservoir level in 3 hours.

$$\text{Average discharge/sec} = \frac{\text{Area of reservoir} \times x}{3 \times 3600} = \frac{2\cdot4 \times 10^6 \times x}{3 \times 3600} = 222\cdot2x \text{ m}^3$$

Wt. of water available/sec, $W = 222{\cdot}2x \times 1000 \times 9{\cdot}81 = 21{\cdot}8x \times 10^5 \text{ N}$

Average power produced $= W \times H \times \eta_{overall}$

$= (21{\cdot}8x \times 10^5) \times (100) \times (0{\cdot}726) \text{ watts}$

$= 15{\cdot}84x \times 10^7 \text{ watts} = 15{\cdot}84x \times 10^4 \text{ kW}$

But kW produced = 15,000 (*given*)

$\therefore \quad 15{\cdot}84x \times 10^4 = 15{,}000$

or
$$x = \frac{15{,}000}{15\cdot84 \times 10^4} = 0\cdot0947 \text{ m} = \mathbf{9{\cdot}47\ cm}$$

Therefore, the level of reservoir will fall by 9·47 cm.

---

* The total rainfall cannot be utilised as a part of it is lost by evaporation or absorption by ground. Yield factor indicates the percentage of rainfall available for utilisation. Thus 80% yield factor means that only 80% of total rainfall can be utilised.

Alternative method

$$\text{Level of reservoir} = \frac{\text{Vol. of reservoir}}{\text{Area of reservoir}} = \frac{5 \times 10^6}{2 \cdot 4 \times 10^6} = 2 \cdot 083 \text{ m}$$

$$\text{kWh generated in 3 hrs} = 15000 \times 3 = 45{,}000 \text{ kWh}$$

If kWh generated are 9,89,175 kWh, fall in reservoir level = 2·083 m

If kWh generated are 45,000 kWh, fall in reservoir level

$$= \frac{2 \cdot 083}{9{,}89{,}175} \times 45{,}000 = 0 \cdot 0947 \text{ m} = \mathbf{9.47 \text{ cm}}$$

Example 2.11. *A factory is located near a water fall where the usable head for power generation is 25 m. The factory requires continuous power of 400 kW throughout the year. The river flow in a year is (a) 10 m³/sec for 4 months, (b) 6 m³/sec for 2 months and (c) 1·5 m³/sec for 6 months.*

(*i*) *If the site is developed as a run-of-river type of plant, without storage, determine the standby capacity to be provided. Assume that overall efficiency of the plant is 80%.*

(*ii*) *If a reservoir is arranged upstream, will any standby unit be necessary ? What will be the excess power available ?*

Solution.

(*i*) **Run of river Plant.** In this type of plant, the whole water of stream is allowed to pass through the turbine for power generation. The plant utilises the water as and when available. Consequently, more power can be generated in a rainy season than in dry season.

(*a*) *When discharge* $= 10 \text{ m}^3/\text{sec}$

Wt. of water available/sec, $w = 10 \times 1000 \text{ kg} = 10^4 \times 9{\cdot}81 \text{ N}$

Power developed $= w \times H \times \eta_{overall} = (10^4 \times 9{\cdot}81) \times (25) \times (0{\cdot}8)$ watts

$= 1962 \times 10^3$ watts $= 1962$ kW

(*b*) *When discharge* $= 6 \text{ m}^3/\text{sec}$

Power developed $= 1962 \times 6^*/10 = 1177{\cdot}2$ kW

(*c*) *When discharge* $= 1{\cdot}5 \text{ m}^3/\text{sec}$

Power developed $= 1962 \times 1{\cdot}5/10 = 294$ kW

It is clear that when discharge is 10 m³/sec or 6 m³/sec, power developed by the plant is more than 400 kW required by the factory. However, when the discharge is 1·5 m³/sec, power developed falls short and consequently standby unit is required during this period.

∴ Capacity of standby unit = 400 – 294 = **106 kW**

(*ii*) **With reservoir.** When reservoir is arranged upstream, we can store water. This permits regulated supply of water to the turbine so that power output is constant throughout the year.

$$\text{Average discharge} = \frac{(10 \times 4) + (2 \times 6) + (1 \cdot 5 \times 6)}{12} = 5 \cdot 08 \text{ m}^3/\text{sec.}$$

∴ Power developed = 1962 × 5·08/10 = 996·7 kW

Since power developed is more than required by the factory, no standby unit is needed.

∴ Excess power available = 996·7 – 400 = **596·7 kW**

Example 2.12. *A run-of-river hydro-electric plant with pondage has the following data :*
Installed capacity = 10 MW ; Water head, H = 20 m
Overall efficiency, $\eta_{overall}$ *= 80% ; Load factor = 40%*

---

* If discharge is 10 m³/sec, power devloped = 1962 kW
If discharge is 1 m³/sec, power devloped = 1962/10
If discharge is 6 m³/sec, power devloped = 1962 × 6/10

(i) *Determine the river discharge in m^3/sec required for the plant.*

(ii) *If on a particular day, the river flow is 20 m^3/sec, what load factor can the plant supply ?*

Solution.

(i) Consider the duration to be of one week.

Units generated/week = Max. demand × L.F. × Hours in a week

$= (10 \times 10^3) \times (0{\cdot}4) \times (24 \times 7)$ kWh

$= 67{\cdot}2 \times 10^4$ kWh ... (i)

Let Q m^3/sec be the river discharge required.

Wt. of water available/sec, $w = Q \times 9{\cdot}81 \times 1000 = 9810\,Q$ newton

Average power produced $= w \times H \times \eta_{overall} = (9810\,Q) \times (20) \times (0{\cdot}8)$ W

$= 156960\,Q$ watt $= 156{\cdot}96\,Q$ kW

Units generated/week $= (156{\cdot}96\,Q) \times 168$ kWh $= 26{,}369$ Q kWh ... (ii)

Equating exps. (i) and (ii), we get,

$$26{,}369\,Q = 67{\cdot}2 \times 10^4$$

$$\therefore \quad Q = \frac{67{\cdot}2 \times 10^4}{26{,}369} = \mathbf{25.48 m^3/sec}$$

(ii) If the river discharge on a certain day is 20 m^3/sec, then,

Power developed $= 156{\cdot}96 \times 20 = 3139{\cdot}2$ kW

Units generated on that day $= 3139{\cdot}2 \times 24 = 75{,}341$ kWh

Load factor $= \dfrac{75{,}341}{10^4 \times 24} \times 100 = \mathbf{31.4\%}$

Example 2.13. *The weekly discharge of a typical hydroelectric plant is as under :*

Day	*Sun*	*Mon*	*Tues*	*Wed*	*Thurs*	*Fri*	*Sat*
Discharge(m^3/sec)	*500*	*520*	*850*	*800*	*875*	*900*	*546*

The plant has an effective head of 15 m and an overall efficiency of 85%. If the plant operates on 40% load factor, estimate (i) the average daily discharge (ii) pondage required and (iii) installed capacity of proposed plant.

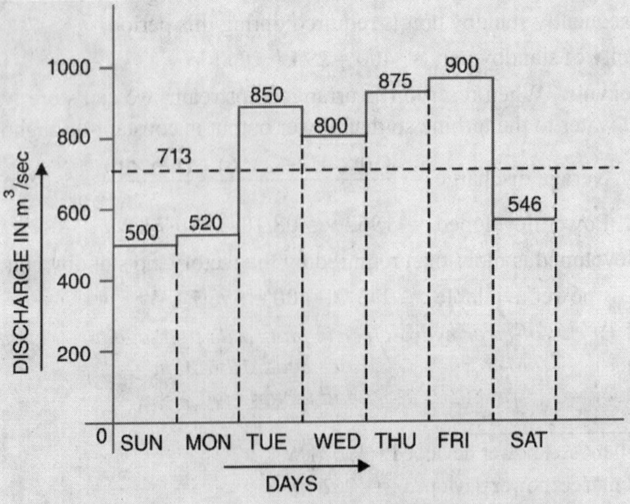

Fig. 2.5

Solution.

Fig. 2.5 shows the plot of weekly discharge. In this graph, discharge is taken along *Y*-axis and days along *X*-axis.

(*i*) Average daily discharge $= \dfrac{500 + 520 + 850 + 800 + 875 + 900 + 546}{7}$

$= \dfrac{4991}{7} = \mathbf{713\ m^3/sec}$

(*ii*) It is clear from graph that on three dyas (*viz*., Sun, Mon. and Sat.), the discharge is less than the average discharge.

Volume of water actually available on these three days

$= (500 + 520 + 546) \times 24 \times 3600\ m^3 = 1566 \times 24 \times 3600\ m^3$

Volume of water required on these three days

$= 3 \times 713 \times 24 \times 3600\ m^3 = 2139 \times 24 \times 3600\ m^3$

Pondage required $= (2139 - 1566) \times 24 \times 3600\ m^3 = \mathbf{495 \times 10^5\ m^3}$

(*iii*) Wt. of water available/sec, $w = 713 \times 1000 \times 9{\cdot}81$ N

Average power produced $= w \times H \times \eta_{overall} = (713 \times 1000 \times 9{\cdot}81) \times (15) \times (0{\cdot}85)$ watts

$= 89180 \times 10^3$ watts $= 89180$ kW

Installed capacity of the plant

$= \dfrac{\text{Output power}}{\text{Load factor}} = \dfrac{89180}{0{\times}4} = 223 \times 10^3\ \text{kW} = \mathbf{223\ MW}$

TUTORIAL PROBLEMS

1. A hydro-electric station has an average available head of 100 metres and reservoir capacity of 50 million cubic metres. Calculate the total energy in kWh that can be generated, assuming hydraulic efficiency of 85% and electrical efficiency of 90%. **[$10{\cdot}423 \times 10^6$ kWh]**
2. Calculate the continuous power that will be available from hydroelectric plant having an available head of 300 meters, catchment area of 150 sq. km, annual rainfall 1·25 m and yield factor of 50%. Assume penstock, turbine and generator efficiencies to be 96%, 86% and 97% respectively. If the load factor is 40% what should be the rating of the generators installed ? **[7065 kW, 17662 kW]**
3. A hydroelectric plant has a reservoir of area 2 sq. kilometres and of capacity 5 million cubic meters. The net head of water at the turbine is 50 m. If the efficiencies of turbine and generator are 85% and 95% respectively, calculate the total energy in kWh that can be generated from this station. If a load of 15000 kW has been supplied for 4 hours, find the fall in reservoir. **[$5{\cdot}5 \times 10^5$ kWh ; 27·8 cm]**
4. It has been estimated that a minimum run-off of approximately 94 m^3/sec will be available at a hydraulic project with a head of 39 m. Determine the firm capacity and yearly gross output.

   [3600 kW, $315{\cdot}36 \times 10^6$ kWh]

   Hint. Wt. of water flowing/sec $= \dfrac{94 \times (100)^3}{1000}$ kg
5. A hydroelectric power station is supplied from a reservoir having an area of 50 km^2 and a head of 50 m. If overall efficiency of the plant is 60%, find the rate at which the water level will fall when the station is generating 30,000 kW. **[7·337 mm/hour]**
6. A hydro-electric plant has a catchment are of 120 square km. The available run off is 50% with annual rainfall of 100 cm. A head of 250 m is available on the average. Efficiency of the power plant is 70%. Find (*i*) average power produced (*ii*) capacity of the power plant. Assume a load factor of 0.6.

   [(*i*) 3266 kW (*ii*) 5443 kW]

2.11 Diesel Power Station

A generating station in which diesel engine is used as the prime mover for the generation of electrical energy is known as **diesel power station.**

In a diesel power station, diesel engine is used as the prime mover. The diesel burns inside the engine and the products of this combustion act as the "working fluid" to produce mechanical energy. The diesel engine drives the alternator which converts mechanical energy into electrical energy. As the generation cost is considerable due to high price of diesel, therefore, such power stations are only used to produce small power.

Although steam power stations and hydro-electric plants are invariably used to generate bulk power at cheaper cost, yet diesel power stations are finding favour at places where demand of power is less, sufficient quantity of coal and water is not available and the transportation facilities are inadequate. These plants are also used as standby sets for continuity of supply to important points such as hospitals, radio stations, cinema houses and telephone exchanges.

Advantages

(*i*) The design and layout of the plant are quite simple.

(*ii*) It occupies less space as the number and size of the auxiliaries is small.

(*iii*) It can be located at any place.

(*iv*) It can be started quickly and can pick up load in a short time.

(*v*) There are no standby losses.

(*vi*) It requires less quantity of water for cooling.

(*vii*) The overall cost is much less than that of steam power station of the same capacity.

(*viii*) The thermal efficiency of the plant is higher than that of a steam power station.

(*ix*) It requires less operating staff.

Disadvantages

(*i*) The plant has high running charges as the fuel (*i.e.*, diesel) used is costly.

(*ii*) The plant does not work satisfactorily under overload conditions for a longer period.

(*iii*) The plant can only generate small power.

(*iv*) The cost of lubrication is generally high.

(*v*) The maintenance charges are generally high.

2.12 Schematic Arrangement of Diesel Power Station

Fig. 2.6 shows the schematic arrangement of a typical diesel power station. Apart from the diesel-generator set, the plant has the following auxiliaries :

(*i*) *Fuel supply system.* It consists of storage tank, strainers, fuel transfer pump and all day fuel tank. The fuel oil is supplied at the plant site by rail or road. This oil is stored in the storage tank. From the storage tank, oil is pumped to smaller all day tank at daily or short intervals. From this tank, fuel oil is passed through strainers to remove suspended impurities. The clean oil is injected into the engine by fuel injection pump.

(*ii*) *Air intake system.* This system supplies necessary air to the engine for fuel combustion. It consists of pipes for the supply of fresh air to the engine manifold. Filters are provided to remove dust particles from air which may act as abrasive in the engine cylinder.

(*iii*) *Exhaust system.* This system leads the engine exhaust gas outside the building and discharges it into atmosphere. A silencer is usually incorporated in the system to reduce the noise level.

(*iv*) *Cooling system.* The heat released by the burning of fuel in the engine cylinder is partially converted into work. The remainder part of the heat passes through the cylinder walls, piston, rings etc. and may cause damage to the system. In order to keep the temperature of the engine parts within the safe operating limits, cooling is provided. The cooling system consists of a water source, pump and cooling towers. The pump circulates water through cylinder and head jacket. The water takes away heat form the engine and itself becomes hot. The hot water is cooled by cooling towers and is recirculated for cooling.

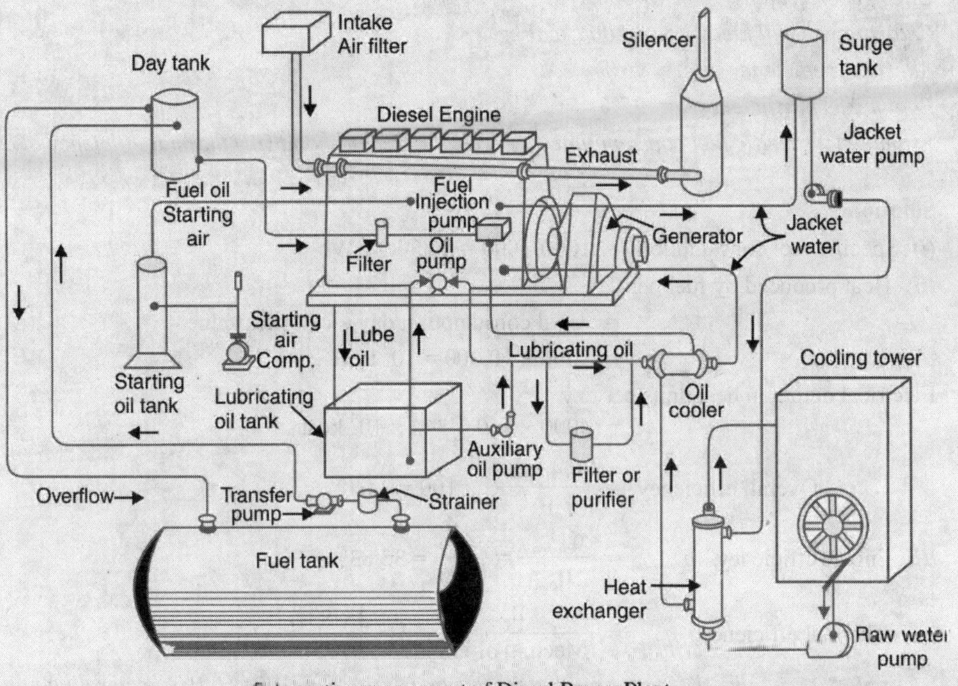

Schematic arrangement of Diesel Power Plant

Fig. 2.6

(*v*) *Lubricating system.* This system minimises the wear of rubbing surfaces of the engine. It comprises of lubricating oil tank, pump, filter and oil cooler. The lubricating oil is drawn from the lubricating oil tank by the pump and is passed through filters to remove impurities. The clean lubricating oil is delivered to the points which require lubrication. The oil coolers incorporated in the system keep the temperature of the oil low.

(*vi*) *Engine starting system.* This is an arrangement to rotate the engine initially, while starting, until firing starts and the unit runs with its own power. Small sets are started manually by handles but for larger units, compressed air is used for starting. In the latter case, air at high pressure is admitted to a few of the cylinders, making them to act as reciprocating air motors to turn over the engine shaft. The fuel is admitted to the remaining cylinders which makes the engine to start under its own power.

Example 2.14. *A diesel power station has fuel consumption of 0·28 kg per kWh, the calorific value of fuel being 10,000 kcal/kg. Determine* (*i*) *the overall efficiency, and* (*ii*) *efficiency of the engine if alternator efficiency is 95%.*

Solution.

Heat produced by 0·28 kg of oil = 10,000 × 0·28 = 2800 kcal

Heat equivalent of 1 kWh = 860 kcal

(*i*) Overall efficiency $= \dfrac{\text{Electrical output in heat units}}{\text{Heat of combustion}} = 860/2800 = 0{\cdot}307 = \mathbf{30{\cdot}7\%}$

(*ii*) Engine efficiency $= \dfrac{\text{Overall efficiency}}{\text{Alternator efficiency}} = \dfrac{30{\cdot}7}{0{\cdot}95} = \mathbf{32{\cdot}3\%}$

Example 2.15. *A diesel power station has the following data :*

Fuel consumption/day = 1000 kg

Units generated/day = 4000 kWh

Calorific value of fuel = 10,000 kcal/kg

Alternator efficiency = 96%

Engine mech. efficiency = 95%

Estimate (i) specific fuel consumption, (ii) overall efficiency, and (iii) thermal efficiency of engine.

Solution.

(*i*) Specific fuel consumption = 1000/4000 = **0·25 kg/kWh**

(*ii*) Heat produced by fuel per day

= Coal consumption/day × calorific value

$= 1000 \times 10{,}000 = 10^7$ kcal

Electrical output in heat units per day

$= 4000 \times 860 = 344 \times 10^4$ kcal

Overall efficiency $= \dfrac{344 \times 10^4}{10^7} \times 100 = \mathbf{34{\cdot}4\%}$

(*iii*) Engine efficiency, $\eta_{engine} = \dfrac{\eta_{overall}}{\eta_{alt.}} = \dfrac{34{\cdot}4}{0{\cdot}96} = 35{\cdot}83\%$

Thermal efficiency, $\eta_{ther} = \dfrac{\eta_{engine}}{\text{Mech. } \eta \text{ of engine}} = \dfrac{35{\cdot}83}{0{\cdot}95} = \mathbf{37{\cdot}71\%}$

Example 2.16. *A diesel engine power plant has one 700 kW and two 500 kW generating units. The fuel consumption is 0·28 kg per kWh and the calorific value of fuel oil is 10200 kcal/kg. Estimate (i) the fuel oil required for a month of 30 days and (ii) overall efficiency. Plant capacity factor = 40%.*

Solution.

(*i*) Maximum energy that can be produced in a month

= Plant capacity × Hours in a month

$= (700 + 2 \times 500) \times (30 \times 24) = 1700 \times 720$ kWh

Plant capacity factor $= \dfrac{\text{Actual energy produced}}{\text{Max. energy that could have been produced}}$

or $0{\cdot}4 = \dfrac{\text{Actual energy produced}}{1700 \times 720}$

∴ Actual energy produced in a month

$= 0{\cdot}4 \times 1700 \times 720 = 489600$ kWh

Fuel oil consumption in a month

$= 489600 \times 0{\cdot}28 =$ **137088 kg**

(*ii*) Output $= 489600$ kWh $= 489600 \times 860$ kcal

Input $= 137088 \times 10200$ kcal

$\therefore$ Overall efficiency $= \dfrac{\text{Output}}{\text{Input}} = \dfrac{489600 \times 860}{137088 \times 10200} = \mathbf{0{\cdot}3 \text{ or } 30\%}$

2.13 Nuclear Power Station

A generating station in which nuclear energy is converted into electrical energy is known as a **nuclear power station.**

In nuclear power station, heavy elements such as Uranium (U^{235}) or Thorium (Th^{232}) are subjected to nuclear fission* in a special apparatus known as a *reactor*. The heat energy thus released is utilised in raising steam at high temperature and pressure. The steam runs the steam turbine which converts steam energy into mechanical energy. The turbine drives the alternator which converts mechanical energy into electrical energy.

The most important feature of a nuclear power station is that huge amount of electrical energy can be produced from a relatively small amount of nuclear fuel as compared to other conventional types of power stations. It has been found that complete fission of 1 kg of Uranium (U^{235}) can produce as much energy as can be produced by the burning of 4,500 tons of high grade coal. Although the recovery of principal nuclear fuels (*i.e.*, Uranium and Thorium) is difficult and expensive, yet the total energy content of the estimated world reserves of these fuels are considerably higher than those of conventional fuels, *viz.*, coal, oil and gas. At present, energy crisis is gripping us and, therefore, nuclear energy can be successfully employed for producing low cost electrical energy on a large scale to meet the growing commercial and industrial demands.

Advantages

(*i*) The amount of fuel required is quite small. Therefore, there is a considerable saving in the cost of fuel transportation.

(*ii*) A nuclear power plant requires less space as compared to any other type of the same size.

(*iii*) It has low running charges as a small amount of fuel is used for producing bulk electrical energy.

(*iv*) This type of plant is very economical for producing bulk electric power.

(*v*) It can be located near the load centres because it does not require large quantities of water and need not be near coal mines. Therefore, the cost of primary distribution is reduced.

(*vi*) There are large deposits of nuclear fuels available all over the world. Therefore, such plants can ensure continued supply of electrical energy for thousands of years.

(*vii*) It ensures reliability of operation.

Disadvantages

(*i*) The fuel used is expensive and is difficult to recover.

(*ii*) The capital cost on a nuclear plant is very high as compared to other types of plants.

(*iii*) The erection and commissioning of the plant requires greater technical know-how.

(*iv*) The fission by-products are generally radioactive and may cause a dangerous amount of radioactive pollution.

---

* **Fission.** The breaking up of nuclei of heavy atoms into two nearly equal parts with release of huge amount of energy is known as nuclear fission. The release of huge amount of energy during fission is due to *mass defect i.e.* the mass of the final product comes out to be less than the initial product. This mass defect is converted into heat energy according to Einstein's relation, $E = mc^2$.

(*v*) Maintenance charges are high due to lack of standardisation. Moreover, high salaries of specially trained personnel employed to handle the plant further raise the cost.

(*vi*) Nuclear power plants are not well suited for varying loads as the reactor does not respond to the load fluctuations efficiently.

(*vii*) The disposal of the by-products, which are radioactive, is a big problem. They have either to be disposed off in a deep trench or in a sea away from sea-shore.

2.14 Schematic Arrangement of Nuclear Power Station

The schematic arrangement of a nuclear power station is shown in Fig. 2.7. The whole arrangement can be divided into the following main stages :

(*i*) Nuclear reactor (*ii*) Heat exchanger (*iii*) Steam turbine (*iv*) Alternator.

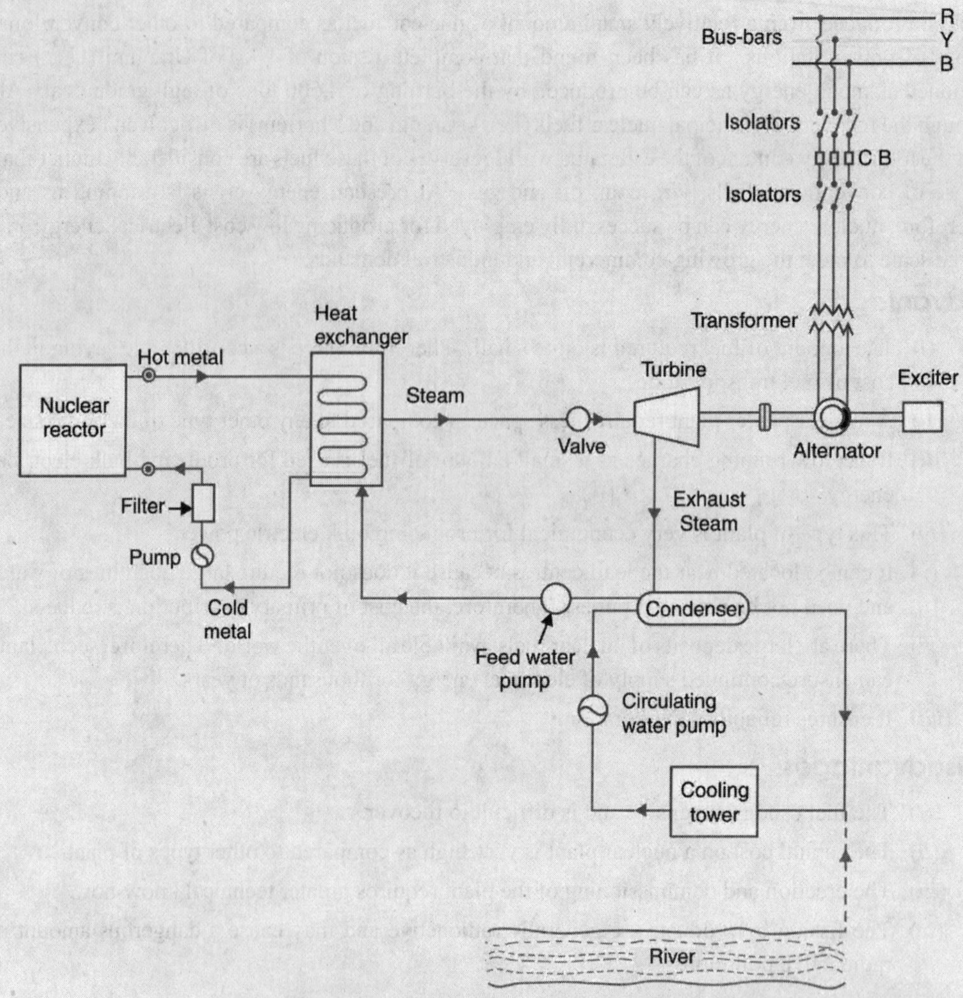

Schematic arrangement of Nuclear Power Station

Fig. 2.7

(*i*) **Nuclear reactor.** It is an apparatus in which nuclear fuel (U^{235}) is subjected to nuclear fission. It controls the *chain reaction** that starts once the fission is done. If the chain reaction is not controlled, the result will be an explosion due to the fast increase in the energy released.

A nuclear reactor is a cylindrical stout pressure vessel and houses fuel rods of Uranium, moderator and control rods (See Fig. 2.8). The fuel rods constitute the fission material and release huge amount of energy when bombarded with slow moving neutrons. The moderator consists of graphite rods which enclose the fuel rods. The moderator slows down the neutrons before they bombard the fuel rods. The control rods are of cadmium and are inserted into the reactor. Cadmium is strong neutron absorber and thus regulates the supply of neutrons for fission. When the control rods are pushed in deep enough, they absorb most of fission neutrons and hence few are available for chain reaction which, therefore, stops. However, as they are being withdrawn, more and more of these fission neutrons cause fission and hence the *intensity* of chain reaction (or heat produced) is increased. Therefore, by pulling out the control rods, power of the nuclear reactor is increased, whereas by pushing them in, it is reduced. In actual practice, the lowering or raising of control rods is accomplished automatically according to the requirement of load. The heat produced in the reactor is removed by the coolant, generally a sodium metal. The coolant carries the heat to the heat exchanger.

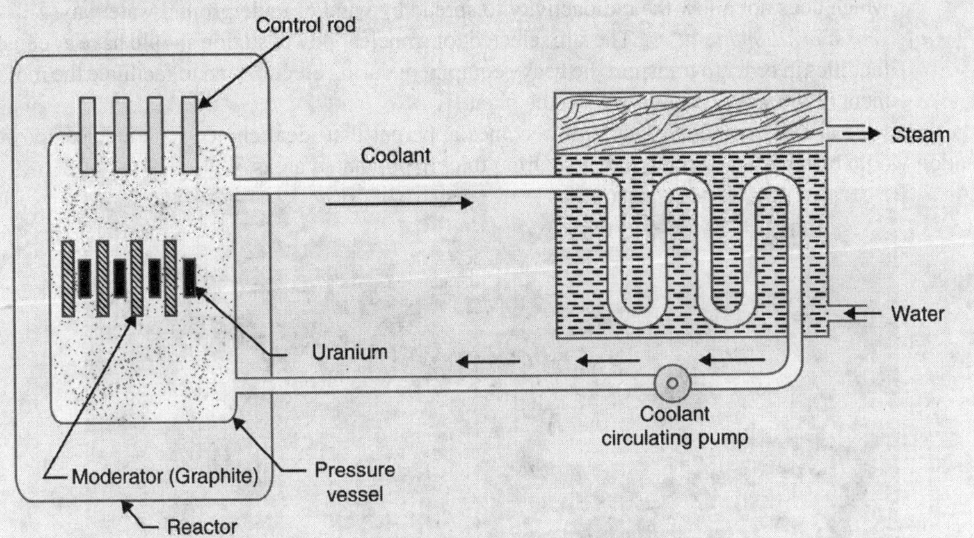

Nuclear Reactor
Fig. 2.8

(*ii*) **Heat exchanger.** The coolant gives up heat to the heat exchanger which is utilised in raising the steam. After giving up heat, the coolant is again fed to the reactor.

---

* **Chain reaction.** Nuclear fission is done by bombarding Uranium nuclei with slow moving neutrons. This splits the Uranium nuclei with the release of huge amount of energy and emission of neutrons (called fission neutrons). These fission neutrons cause further fission. If this process continues, then in a very short time huge amount of energy will be released which may cause explosion. This is known as *explosive chain reaction*. But in a reactor, controlled chain reaction is allowed. This is done by systematically removing the fission neutrons from the reactor. The greater the number of fission neutrons removed, the lesser is the intensity (*i.e.*, fission rate) of energy released.

(*iii*) **Steam turbine.** The steam produced in the heat exchanger is led to the steam turbine through a valve. After doing a useful work in the turbine, the steam is exhausted to condenser. The condenser condenses the steam which is fed to the heat exchanger through feed water pump.

(*iv*) **Alternator.** The steam turbine drives the alternator which converts mechanical energy into electrical energy. The output from the alternator is delivered to the bus-bars through transformer, circuit breakers and isolators.

2.15 Selection of Site for Nuclear Power Station

The following points should be kept in view while selecting the site for a nuclear power station :

(*i*) *Availability of water.* As sufficient water is required for cooling purposes, therefore, the plant site should be located where ample quantity of water is available, *e.g.*, across a river or by sea-side.

(*ii*) *Disposal of waste.* The waste produced by fission in a nuclear power station is generally radioactive which must be disposed off properly to avoid health hazards. The waste should either be buried in a deep trench or disposed off in sea quite away from the sea shore. Therefore, the site selected for such a plant should have adequate arrangement for the disposal of radioactive waste.

(*iii*) *Distance from populated areas.* The site selected for a nuclear power station should be quite away from the populated areas as there is a danger of presence of radioactivity in the atmosphere near the plant. However, as a precautionary measure, *a dome* is used in the plant which does not allow the radioactivity to spread by wind or underground waterways.

(*iv*) *Transportation facilities.* The site selected for a nuclear power station should have adequate facilities in order to transport the heavy equipment during erection and to facilitate the movement of the workers employed in the plant.

From the above mentioned factors it becomes apparent that ideal choice for a nuclear power station would be near sea or river and away from thickly populated areas.

Nuclear Power Station

Example 2.17. *An atomic power reactor can deliver 300 MW. If due to fission of each atom of* $_{92}U^{235}$ *, the energy released is 200 MeV, calculate the mass of uranium fissioned per hour.*

Solution.

Energy received from the reactor

$$= 300 \text{ MW} = 3 \times 10^8 \text{ W (or Js}^{-1})$$

$$\text{Energy received/hour} = (3 \times 10^8) \times 3600 = 108 \times 10^{10} \text{ J}$$

$$\text{Energy released/fission} = 200 \text{ MeV} = 200 \times 10^6 \times 1{\cdot}6 \times 10^{-19} \text{ J} = 3{\cdot}2 \times 10^{-11} \text{J}$$

Number of atoms fissioned per hour

$$= \frac{108 \times 10^{10}}{3{\cdot}2 \times 10^{-11}} = 33{\cdot}75 \times 10^{21}$$

Now 1 gram-atom (*i.e.*, 235g) has $6{\cdot}023 \times 10^{23}$ atoms.

∴ Mass of Uranium fissioned per hour

$$= \frac{235}{6{\cdot}023 \times 10^{23}} \times 33{\cdot}75 \times 10^{21} = \mathbf{13{\cdot}17g}$$

Example 2.18. *What is the power ouput of a* $_{92}U^{235}$ *reactor if it takes 30 days to use up 2 kg of fuel? Given that energy released per fission is 200 MeV and Avogadro's number* = $6{\cdot}023 \times 10^{26}$ *per kilomole.*

Solution.

$$\text{Number of atoms in 2 kg fuel} = \frac{2}{235} \times 6{\cdot}023 \times 10^{26} = 5{\cdot}12 \times 10^{24}$$

These atoms fission in 30 days. Therefore, the fission rate (*i.e.*, number of fissions per second)

$$= \frac{5{\cdot}12 \times 10^{24}}{30 \times 24 \times 60 \times 60} = 1{\cdot}975 \times 10^{18}$$

$$\text{Energy released per fission} = 200 \text{ MeV} = (200 \times 10^6) \times 1{\cdot}6 \times 10^{-19} = 3{\cdot}2 \times 10^{-11} \text{J}$$

∴ Energy released per second *i.e.*, power output *P* is

$$P = (3{\cdot}2 \times 10^{-11}) \times (1{\cdot}975 \times 10^{18}) \text{ W}$$

$$= 63{\cdot}2 \times 10^6 \text{ W} = \mathbf{63{\cdot}2 \text{ MW}}$$

2.16 Gas Turbine Power Plant

A generating station which employs gas turbine as the prime mover for the generation of electrical energy is known as a **gas turbine power plant**

In a gas turbine power plant, air is used as the working fluid. The air is compressed by the compressor and is led to the combustion chamber where heat is added to air, thus raising its temperature. Heat is added to the compressed air either by burning fuel in the chamber or by the use of air heaters. The hot and high pressure air from the combustion chamber is then passed to the gas turbine where it expands and does the mechanical work. The gas turbine drives the alternator which converts mechanical energy into electrical energy.

It may be mentioned here that compressor, gas turbine and the alternator are mounted on the same shaft so that a part of mechanical power of the turbine can be utilised for the operation of the compressor. Gas turbine power plants are being used as standby plants for hydro-electric stations, as a starting plant for driving auxiliaries in power plants etc.

Advantages

(*i*) It is simple in design as compared to steam power station since no boilers and their auxiliaries are required.

(*ii*) It is much smaller in size as compared to steam power station of the same capacity. This is expected since gas turbine power plant does not require boiler, feed water arrangement etc.

(*iii*) The initial and operating costs are much lower than that of equivalent steam power station.

(*iv*) It requires comparatively less water as no condenser is used.

(*v*) The maintenance charges are quite small.

(*vi*) Gas turbines are much simpler in construction and operation than steam turbines.

(*vii*) It can be started quickly form cold conditions.

(*viii*) There are no standby losses. However, in a steam power station, these losses occur because boiler is kept in operation even when the steam turbine is supplying no load.

Disadvantages

(*i*) There is a problem for starting the unit. It is because before starting the turbine, the compressor has to be operated for which power is required from some external source. However, once the unit starts, the external power is not needed as the turbine itself supplies the necessary power to the compressor.

(*ii*) Since a greater part of power developed by the turbine is used in driving the compressor, the net output is low.

(*iii*) The overall efficiency of such plants is low (about 20%) because the exhaust gases from the turbine contain sufficient heat.

(*iv*) The temperature of combustion chamber is quite high (3000°F) so that its life is comparatively reduced.

2.17 Schematic Arrangement of Gas Turbine Power Plant

The schematic arrangement of a gas turbine power plant is shown in Fig. 2.9. The main components of the plant are :

(*i*) Compressor (*ii*) Regenerator

(*iii*) Combustion chamber (*iv*) Gas turbine

(*v*) Alternator (*vi*) Starting motor

(*i*) **Compressor.** The compressor used in the plant is generally of rotatory type. The air at atmospheric pressure is drawn by the compressor *via* the filter which removes the dust from air. The rotatory blades of the compressor push the air between stationary blades to raise its pressure. Thus air at high pressure is available at the output of the compressor.

(*ii*) **Regenerator.** A regenerator is a device which recovers heat from the exhaust gases of the turbine. The exhaust is passed through the regenerator before wasting to atmosphere. A regenerator consists of a nest of tubes contained in a shell. The compressed air from the compressor passes through the tubes on its way to the combustion chamber. In this way, compressed air is heated by the hot exhaust gases.

(*iii*) **Combustion chamber.** The air at high pressure from the compressor is led to the combustion chamber *via* the regenerator. In the combustion chamber, heat* is added to the air by burning oil. The oil is injected through the burner into the chamber at high pressure to ensure atomisation of oil and its thorough mixing with air. The result is that the chamber attains a very high temperature (about 3000°F). The combustion gases are suitably cooled to 1300°F to 1500°F and then delivered to the gas turbine.

(*iv*) **Gas turbine.** The products of combustion consisting of a mixture of gases at high temperature and pressure are passed to the gas turbine. These gases in passing over the turbine blades expand and thus do the mechanical work. The temperature of the exhaust gases from the turbine is about 900°F.

---

* Only hot pressurised air makes it possible to convert heat into mechanical work. Heating air at atmospheric pressure generally does not make it permissible to convert heat into mehanical work.

2.18 Comparison of the Various Power Plants

The comparison of steam power plant, hydro-electric plant, diesel power plant and nuclear power plant is given below in the tabular form :

S.No.	Item	Steam Power Station	Hydro-electric Power Plant	Diesel Power Plant	Nuclear power Plant
1.	*Site*	Such plants are located at a place where ample supply of water and coal is available, transportation facilities are adequate	Such plants are located where large reservoirs can be obtained by constructing a dam *e.g.* in hilly areas.	Such plants can be located at any place because they require less space and small quantity of water.	These plants are located away from thickly populated areas to avoid radioactive pollution.
2.	*Initial cost*	Initial cost is lower than those of hydroelectric and nuclear power plants.	Initial cost is very high because of dam construction and excavation work.	Initial cost is less as compared to other plants.	Initial cost is highest because of huge investement on building a nuclear reactor.
3.	*Running cost*	Higher than hydroelectric and nuclear plant because of the requirement of huge amount of coal.	Practically nil because no fuel is required.	Highest among all plants because of high price of diesel.	Except the hydroelectric plant, it has the minimum running cost because small amount of fuel can produce relatively large amount of power.
4.	*Limit of source of power*	Coal is the source of power which has limited reserves all over the world.	Water is the source of power which is not dependable becuase of wide variations in the rainfall every year.	Diesel is the source of power which is not available in huge quantities due to limited reserves.	The source of power is the nuclear fuel which is available in sufficient quantity. It is because small amount of fuel can produce huge power.
5.	*Cost of fuel transportation*	Maximum because huge amount of coal is transported to the plant site.	Practically nil.	Higher than hydro and nuclear power plants	Minimum because small quantity of fuel is required.
6.	*Cleanliness and simplicity*	Least clean as atmosphere is polluted due to smoke.	Most simple and clean.	More clean than steam power and nuclear power plants.	Less cleaner than hydroelectric and diesel power plants.

S.No.	Item	Steam Power Station	Hydro-electric Power Plant	Diesel Power Plant	Nuclear power Plant
7.	*Overall efficiency*	Least efficient. Overall efficiency is about 25%.	Most efficient. Overall efficiency is about 85%.	More efficient than steam power station. Efficiency is about 35%.	More efficient than steam power station.
8.	*Starting*	Requires a lot of time for starting.	Can be started instantly.	Can be started quickly.	Can be started easily.
9.	*Space required*	These plants need sufficient space because of boilers and other auxiliaries.	Require very large area because of the reservoir.	Require less space.	These require minimum space as compared to any other plant of equivalent capacity.
10.	*Maintenance cost*	Quite high as skilled operating staff is required.	Quite low.	Less	Very high as highly trained personnel are required to handle the plant.
11.	*Transmission and distribution cost*	Quite low as these are generally located near the load centres.	Quite high as these are located quite away from the load centres.	Least as they are generally located at the centre of gravity of the load.	Quite low as these are located near load centres.
12.	*Standby losses*	Maximum as the boiler remains in operation even when the turbine is not working.	No standby losses.	Less standby losses.	Less.

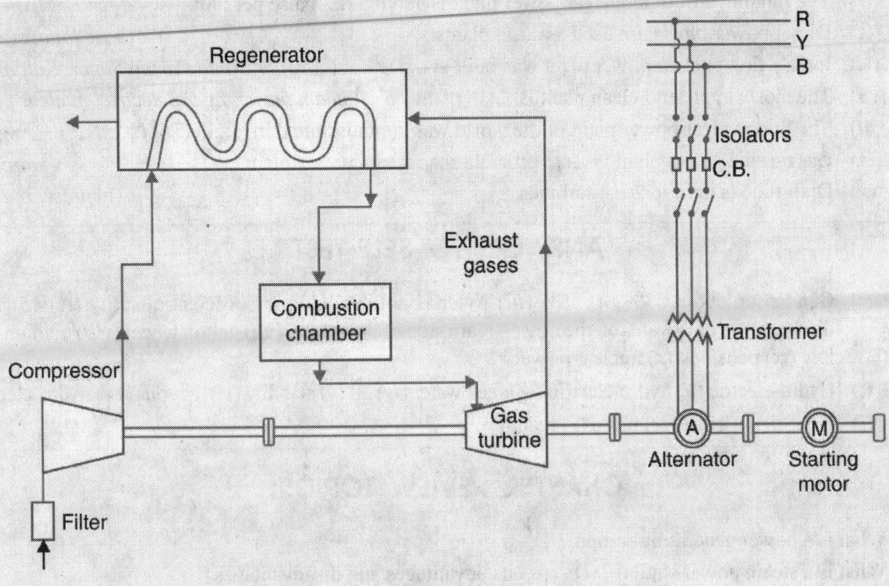

Schematic arrangement of gas turbine power plant.

Fig. 2.9

(*v*) **Alternator.** The gas turbine is coupled to the alternator. The alternator converts mechanical energy of the turbine into electrical energy. The output from the alternator is given to the bus-bars through transformer, circuit breakers and isolators.

(*vi*) **Starting motor.** Before starting the turbine, compressor has to be started. For this purpose, an electric motor is mounted on the same shaft as that of the turbine. The motor is energised by the batteries. Once the unit starts, a part of mechanical power of the turbine drives the compressor and there is no need of motor now.

SELF-TEST

1. **Fill in the blanks by inserting appropriate words/figures :**
   (*i*) The major heat loss in a steam power station occurs in ....
   (*ii*) The thermal efficiency of a steam power station is about ....
   (*iii*) Cooling towers are used where ....
   (*iv*) The running cost of medium power stations is about .... paise per unit.
   (*v*) In a hydro-electric plant, spillways are used ....
   (*vi*) The running cost of a hydro-electric plant is about .... paise per unit.
   (*vii*) For high head hydro-electric plants, the turbine used is ....
   (*viii*) Francis and Kaplan turbines are used for .... heads.
   (*ix*) Surge tank is provided for the protection of ....
   (*x*) Of all the plants, minimum quantity of fuel is required in .... plant.
2. **Pick up the correct words from the brackets and fill in the blanks :**
   (*i*) The cost of fuel transportation is minimum in .... plant.
   (*steam power, hydro-electric, nuclear power*)
   (*ii*) The cheapest plant in operation and maintenance is .... plant.
   (*diesel power, hydro-electric, steam power*)
   (*iii*) Economisers are used to heat .... (*air, feed water, steam*)

(iv) The running cost of a nuclear power plant is about .... paise per unit. (20, 48, 64)
(v) Diesel power plants are used as .... plants. (*base load, standby*)
(vi) India's first nuclear power plant was built at .... (*Tarapur, Rana Partap Sagar, Kalpakkam*)
(vii) The most simple and clean plant is .... plant (*steam power, hydro-electric, nuclear power*)
(viii) The first nuclear power plant in the world was commissioned in ....(*U.S.A., U.S.S.R., England*)
(ix) Gas turbine power plant is .... efficient than steam power plant, (*more, less*)
(x) Draft tube is used in .... turbines. (*impulse, reaction*)

ANSWERS TO SELF-TEST

1. (*i*) Condenser, about 53% (*ii*) 28% (*iii*) water is not available in sufficient quantity (*iv*) 15 (*v*) to discharge surplus water on the downstream side of dam (*vi*) 5 (*vii*) pelton wheel (*viii*) medium and low (*ix*) penstock (*x*) nuclear power.
2. (*i*) Hydro-electric (*ii*) hydro-electric (*iii*) feed water (*iv*) 20 (*v*) standby (*vi*) Tarapur (*vii*) hydro-electric (*viii*) U.S.S.R. in 1954 (*ix*) more (*x*) reaction.

CHAPTER REVIEW TOPICS

1. What is a power generating station ?
2. What is a steam power station ? Discuss its advantages and disadvantages.
3. Draw the schematic diagram of a modern steam power station and explain its operation.
4. Explain the important components of a steam power station.
5. What factors are taken into account while selecting the site for a steam power station ?
6. Discuss the merits and demerits of a hydro-electric plant.
7. Draw a neat schematic diagram of a hydro-electric plant and explain the functions of various components.
8. Explain the essential factors which influence the choice of site for a hydro-electric plant.
9. Explain the functions of the following :
   (*i*) dam (*ii*) spillways (*iii*) surge tank (*iv*) headworks (*v*) draft tube.
10. Draw the flow diagram of a diesel power station and discuss its operation.
11. Discuss the advantages and disadvantages of a diesel power station.
12. Draw the schematic diagram of a nuclear power station and discuss its operation.
13. Explain with a neat sketch the various parts of a nuclear reactor.
14. Discuss the factors for the choice of site for a nuclear power plant.
15. Explain the working of a gas turbine power plant with a schematic diagram.
16. Give the comparison of steam power plant, hydro-electric power plant, diesel power plant and nuclear power plant on the basis of operating cost, initial cost, efficiency, maintenance cost and availability of source of power.

DISCUSSION QUESTIONS

1. Why is the overall efficiency of a steam power station very low ?
2. Why is a condenser used in a steam power station ?
3. Why hydro-electric stations have high transmission and distribution costs ?
4. Why are nuclear power stations becoming very popular ?
5. Why hot gas at high pressure and not hot gas at atmospheric pressure is used in gas turbine power plants?
6. How do the various devices protect the penstock ?
7. Why cannot diesel power stations be employed to generate bulk power ?
8. Why is regenerator used in gas turbine power plant ?

CHAPTER

3

Variable Load on Power Stations

Introduction

The function of a power station is to deliver power to a large number of consumers. However, the power demands of different consumers vary in accordance with their activities. The result of this variation in demand is that load on a power station is never constant, rather it varies from time to time. Most of the complexities of modern power plant operation arise from the inherent variability of the load demanded by the users. Unfortunately, electrical power cannot be stored and, therefore, the power station must produce power as and when demanded to meet the requirements of the consumers. On one hand, the power engineer would like that the alternators in the power station should run at their rated capacity for maximum efficiency and on the other hand, the demands of the consumers have wide variations. This makes the design of a power station highly complex. In this chapter, we shall focus our attention on the problems of variable load on power stations.

3.1 Structure of Electric Power System

The function of an electric power system is to connect the power station to the consumers' loads by means of interconnected system of transmission and distribution networks. Therefore, an electric power system consists of three principal components : the *power station*, the *transmission lines* and the *distribution systems*. The transmission lines are the connecting link between the power station and the distribution systems. A distribution system connects all the individual loads in a given locality to the transmission lines.

Fig. 3.1. shows the one-line diagram of a very simple electric power system. The power station generally houses two or more 3-phase alternators operated in parallel. Due to technical reasons, the generation voltage is not very high ; it is usually 11 kV to 25kV. The power stations are located at favourable places, generally quite far away from the consumers. Owing to economical reasons, electric power has to be transmitted through transmission lines at a very high voltage. Therefore, voltage at the power station is stepped to a high value (say 220 kV) by the step-up transformer. The electric power is carried by the transmission lines at this high voltage. At the outskirts of the city, this voltage is reduced considerably (say 11 kV) by a step-down transformer (also called transmission sub-station). The feeders carry the power to distribution sub-station. Feeders should not be tapped for direct supply. At the distribution sub-station, the step-down transformer reduces the voltage to 400 V (11kV/0.4 kV transformer — 400 V line voltage and 230 V phase voltage). Distributors are used to supply power to various consumers. The following points may be noted :

(*i*) For the sake of simplicity, we show the 3-phase circuit by a single-line diagram.

(*ii*) The 3-phase alternators at the power station are designed to produce balanced voltages. Since a balanced 3-phase system is always solved as a single-phase circuit composed of one of the three lines and a neutral return, it is seldom necessary to show more than one phase and the neutral return when drawing a diagram of the circuit. Often the diagram is simplified further by omitting the neutral wire and by indicating the component parts by standard symbols rather than by their equivalent circuits. Such a simplified diagram of an electric power system is called one-line diagram.

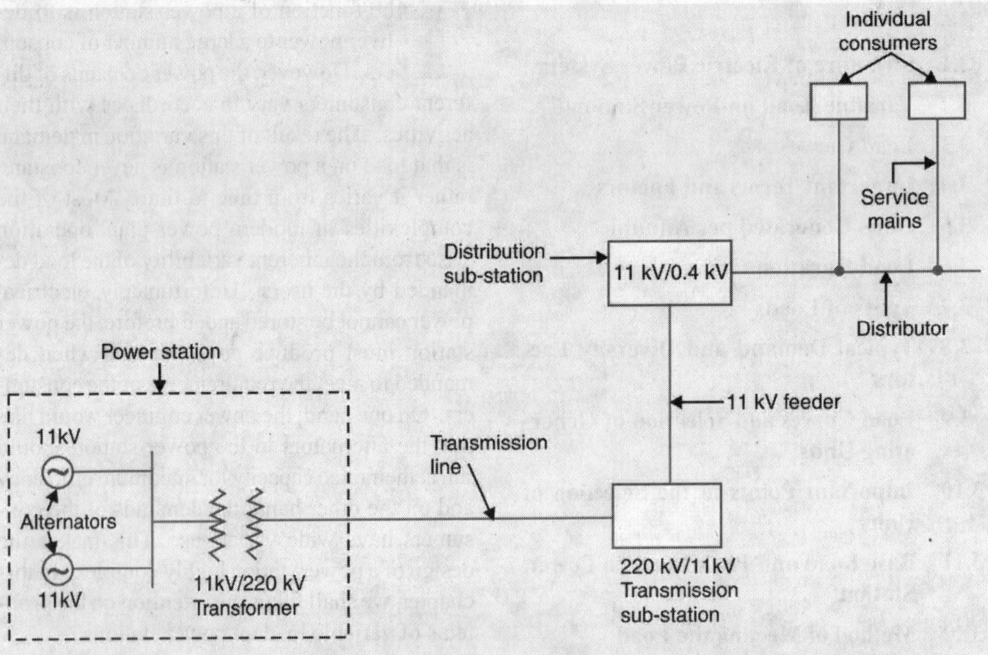

Fig. 3.1

(*iii*) In an actual electric power system, there are a number of protective devices (fuses, circuit breakers, relays etc.) and other equipment. The amount of information included in the single-line diagram depends on the purpose for which the diagram is intended. For example, in this chapter, we are to make load study. For this purpose, circuit breakers, relays etc. are unimportant and are omitted in the diagram shown in Fig. 3.1.

(*iv*) The power demanded by the consumers is supplied by the power station through the transmission and distribution networks. As the consumers' load demand changes, the power supply by the power station changes accordingly.

Transmission line

3.2 Variable Load on Power Station

The load on a power station varies from time to time due to uncertain demands of the consumers and is known as **variable load on the station.**

A power station is designed to meet the load requirements of the consumers. An ideal load on the station, from stand point of equipment needed and operating routine, would be one of constant magnitude and steady duration. However, such a steady load on the station is never realised in actual practice. The consumers require their small or large block of power in accordance with the demands of their activities. Thus the load demand of one consumer at any time may be different from that of the other consumer. The result is that load on the power station varies from time to time.

Effects of variable load. The variable load on a power station introduces many perplexities in its operation. Some of the important effects of variable load on a power station are :

(*i*) **Need of additional equipment.** The variable load on a power station necessitates to have additional equipment. By way of illustration, consider a steam power station. Air, coal and water are the raw materials for this plant. In order to produce variable power, the supply of these materials will be required to be varied correspondingly. For instance, if the power demand on the plant increases, it must be followed by the increased flow of coal, air and water to the boiler in order to meet the increased demand. Therefore, additional equipment has to be installed to accomplish this job. As a matter of fact, in a modern power plant, there is much equipment devoted entirely to adjust the rates of supply of raw materials in accordance with the power demand made on the plant.

(*ii*) **Increase in production cost.** The variable load on the plant increases the cost of the production of electrical energy. An alternator operates at maximum efficiency near its rated capacity. If a single alternator is used, it will have poor efficiency during periods of light loads on the plant. Therefore, in actual practice, a number of alternators of different capacities are installed so that most of the alternators can be operated at nearly full load capacity. However, the use of a number of generating units increases the initial cost per kW of the plant capacity as well as floor area required. This leads to the increase in production cost of energy.

3.3 Load Curves

The curve showing the variation of load on the power station with respect to (w.r.t) time is known as a **load curve.**

The load on a power station is never constant; it varies from time to time. These load variations during the whole day (*i.e.,* 24 hours) are recorded half-hourly or hourly and are plotted against time on the graph. The curve thus obtained is known as *daily load curve* as it shows the variations of load *w.r.t.* time during the day. Fig. 3.2. shows a typical daily load curve of a power station. It is clear that load on the power station is varying, being maximum at 6 P.M. in this case. It may be seen that load curve indicates at a glance the general character of the load that is being imposed on the plant. Such a clear representation cannot be obtained from tabulated figures.

The *monthly load curve* can be obtained from the daily load curves of that month. For this purpose, average* values of power over a month at different times of the day are calculated and then plotted on the graph. The monthly load curve is generally used to fix the rates of energy. The *yearly load curve* is obtained by considering the monthly load curves of that particular year. The yearly load curve is generally used to determine the annual load factor.

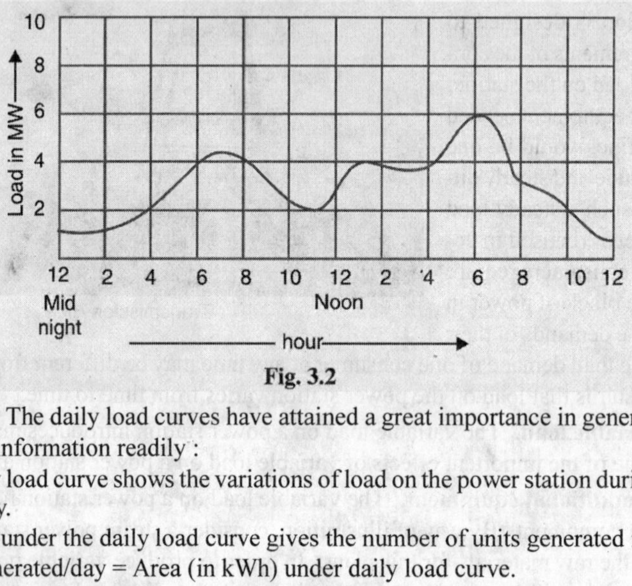

Fig. 3.2

Importance. The daily load curves have attained a great importance in generation as they supply the following information readily :

(*i*) The daily load curve shows the variations of load on the power station during different hours of the day.

(*ii*) The area under the daily load curve gives the number of units generated in the day.
Units generated/day = Area (in kWh) under daily load curve.

(*iii*) The highest point on the daily load curve represents the maximum demand on the station on that day.

(*iv*) The area under the daily load curve divided by the total number of hours gives the average load on the station in the day.

$$\text{Average load} = \frac{\text{Area (in kWh) under daily load curve}}{\text{24 hours}}$$

(*v*) The ratio of the area under the load curve to the total area of rectangle in which it is contained gives the load factor.

$$\text{Load factor} = \frac{\text{Average load}}{\text{Max. demand}} = \frac{\text{Average load} \times 24}{\text{Max. demand} \times 24}$$

$$= \frac{\text{Area (in kWh) under daily load curve}}{\text{Total area of rectangle in which the load curve is contained}}$$

---

* For instance, if we consider the load on power station at mid-night during the various days of the month, it may vary slightly. Then the average will give the load at mid-night on the monthly curve.

(*vi*) The load curve helps in selecting* the size and number of generating units.

(*vii*) The load curve helps in preparing the operation schedule** of the station.

3.4 Important Terms and Factors

The variable load problem has introduced the following terms and factors in power plant engineering:

(*i*) **Connected load.** *It is the sum of continuous ratings of all the equipments connected to supply system.*

A power station supplies load to thousands of consumers. Each consumer has certain equipment installed in his premises. The sum of the continuous ratings of all the equipments in the consumer's premises is the "connected load" of the consumer. For instance, if a consumer has connections of five 100-watt lamps and a power point of 500 watts, then connected load of the consumer is $5 \times 100 + 500 = 1000$ watts. The sum of the connected loads of all the consumers is the connected load to the power station.

(*ii*) **Maximum demand** : *It is the greatest demand of load on the power station during a given period.*

The load on the power station varies from time to time. The maximum of all the demands that have occurred during a given period (*say* a day) is the maximum demand. Thus referring back to the load curve of Fig. 3.2, the maximum demand on the power station during the day is 6 MW and it occurs at 6 P.M. Maximum demand is generally less than the connected load because all the consumers do not switch on their connected load to the system at a time. The knowledge of maximum demand is very important as it helps in determining the installed capacity of the station. The station must be capable of meeting the maximum demand.

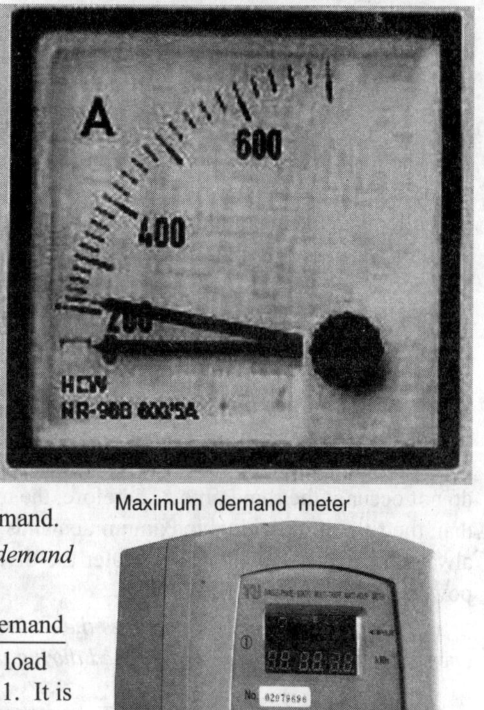

Maximum demand meter

(*iii*) **Demand factor.** *It is the ratio of maximum demand on the power station to its connected load i.e.,*

$$\text{Demand factor} = \frac{\text{Maximum demand}}{\text{Connected load}}$$

The value of demand factor is usually less than 1. It is expected because maximum demand on the power station is generally less than the connected load. If the maximum demand on the power station is 80 MW and the connected load is 100 MW, then demand factor = 80/100 = 0·8. The knowledge of demand factor is vital in determining the capacity of the plant equipment.

(*iv*) **Average load.** *The average of loads occurring on the power station in a given period (day or month or year) is known as* **average load** *or* **average demand.**

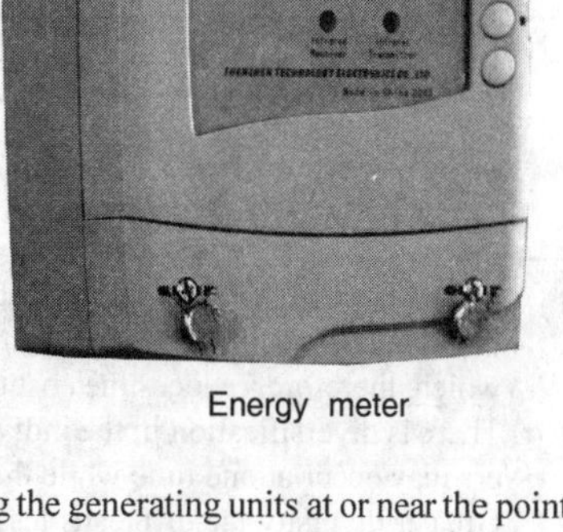
Energy meter

* It will be shown in Art. 3.9 that number and size of the generating units are selected to fit the load curve. This helps in operating the generating units at or near the point of maximum efficiency.

** It is the sequence and time for which the various generating units (*i.e.,* alternators) in the plant will be put in operation.

$$\text{Daily average load} = \frac{\text{No. of units (kWh) generated in a day}}{\text{24 hours}}$$

$$\text{Monthly average load} = \frac{\text{No. of units (kWh) generated in a month}}{\text{Number of hours in a month}}$$

$$\text{Yearly average load} = \frac{\text{No. of units (kWh) generated in a year}}{\text{8760 hours}}$$

(*v*) Load factor. *The ratio of average load to the maximum demand during a given period is known as* **load factor** *i.e.,*

$$\text{Load factor} = \frac{\text{Average load}}{\text{Max. demand}}$$

If the plant is in operation for T hours,

$$\text{Load factor} = \frac{\text{Average load} \times \text{T}}{\text{Max. demand} \times \text{T}}$$

$$= \frac{\text{Units generated in T hours}}{\text{Max. demand} \times \text{T hours}}$$

The load factor may be daily load factor, monthly load factor or annual load factor if the time period considered is a day or month or year. Load factor is always less than 1 because average load is smaller than the maximum demand. The load factor plays key role in determining the overall cost per unit generated. Higher the load factor of the power station, lesser* will be the cost per unit generated.

(*vi*) Diversity factor. *The ratio of the sum of individual maximum demands to the maximum demand on power station is known as* **diversity factor** *i.e.,*

$$\text{Diversity factor} = \frac{\text{Sum of individual max. demands}}{\text{Max. demand on power station}}$$

A power station supplies load to various types of consumers whose maximum demands generally do not occur at the same time. Therefore, the maximum demand on the power station is always less than the sum of individual maximum demands of the consumers. Obviously, diversity† factor will always be greater than 1. The greater the diversity factor, the lesser‡ is the cost of generation of power.

(*vii*) Plant capacity factor. *It is the ratio of actual energy produced to the maximum possible energy that could have been produced during a given period i.e.,*

$$\text{Plant capacity factor} = \frac{\text{Actual energy produced}}{\text{Max. energy that could have been produced}}$$

$$= \frac{\text{Average demand} \times \text{T}^{**}}{\text{Plant capacity} \times \text{T}}$$

$$= \frac{\text{Average demand}}{\text{Plant capacity}}$$

---

* It is because higher load factor factor means lesser maximum demand. The station capacity is so selected that it must meet the maximum demand. Now, lower maximum demand means lower capacity of the plant which, therefore, reduces the cost of the plant.

† There is diversification in the individual maximum demands *i.e.,* the maximum demand of some consumers may occur at one time while that of others at some other time. Hence, the name diversity factor

‡ Greater diversity factor means lesser maximum demand. This in turn means that lesser plant capcity is required. Thus, the capital investment on the plant is reduced.

** Suppose the period is *T* hours.

Thus if the considered period is one year,

$$\text{Annual plant capacity factor} = \frac{\text{Annual kWh output}}{\text{Plant capacity} \times 8760}$$

The plant capacity factor is an indication of the reserve capacity of the plant. A power station is so designed that it has some reserve capacity for meeting the increased load demand in future. Therefore, the installed capacity of the plant is always somewhat greater than the maximum demand on the plant.

$$\text{Reserve capacity} = \text{Plant capacity} - \text{Max. demand}$$

It is interesting to note that difference between load factor and plant capacity factor is an indication of reserve capacity. If the maximum demand on the plant is equal to the plant capacity, then load factor and plant capacity factor will have the same value. In such a case, the plant will have no reserve capacity.

(*viii*) **Plant use factor.** *It is ratio of kWh generated to the product of plant capacity and the number of hours for which the plant was in operation i.e.*

$$\text{Plant use factor} = \frac{\text{Station output in kWh}}{\text{Plant capacity} \times \text{Hours of use}}$$

Suppose a plant having installed capacity of 20 MW produces annual output of $7 \cdot 35 \times 10^6$ kWh and remains in operation for 2190 hours in a year. Then,

$$\text{Plant use factor} = \frac{7 \cdot 35 \times 10^6}{\left(20 \times 10^3\right) \times 2190} = 0 \cdot 167 = 16 \cdot 7\%$$

3.5 Units Generated per Annum

It is often required to find the kWh generated per annum from maximum demand and load factor. The procedure is as follows :

$$\text{Load factor} = \frac{\text{Average load}}{\text{Max. demand}}$$

$$\therefore \quad \text{Average load} = \text{Max. demand} \times \text{L.F.}$$

$$\text{Units generated/annum} = \text{Average load (in kW)} \times \text{Hours in a year}$$

$$= \text{Max. demand (in kW)} \times \text{L.F.} \times 8760$$

3.6 Load Duration Curve

When the load elements of a load curve are arranged in the order of descending magnitudes, the curve thus obtained is called a **load duration curve.**

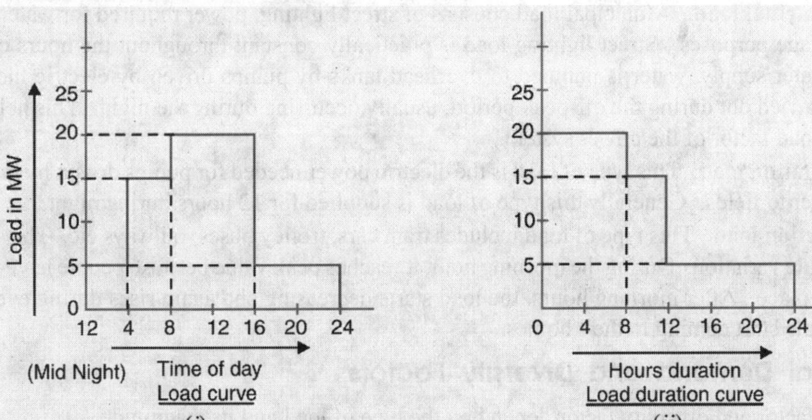

Fig. 3.3

The load duration curve is obtained from the same data as the load curve but the ordinates are arranged in the order of descending magnitudes. In other words, the maximum load is represented to the left and decreasing loads are represented to the right in the descending order. Hence the area under the load duration curve and the area under the load curve are equal. Fig. 3.3 (*i*) shows the daily load curve. The daily load duration curve can be readily obtained from it. It is clear from daily load curve [See Fig. 3.3. (*i*)], that load elements in order of descending magnitude are : 20 MW for 8 hours; 15 MW for 4 hours and 5 MW for 12 hours. Plotting these loads in order of descending magnitude, we get the daily load duration curve as shown in Fig. 3.3 (*ii*).

The following points may be noted about load duration curve :

(*i*) The load duration curve gives the data in a more presentable form. In other words, it readily shows the number of hours during which the given load has prevailed.

(*ii*) The area under the load duration curve is equal to that of the corresponding load curve. Obviously, area under daily load duration curve (in kWh) will give the units generated on that day.

(*iii*) The load duration curve can be extended to include any period of time. By laying out the abscissa from 0 hour to 8760 hours, the variation and distribution of demand for an entire year can be summarised in one curve. The curve thus obtained is called the *annual load duration curve.*

3.7 Types of Loads

A device which taps electrical energy from the electric power system is called a load on the system. The load may be resistive (*e.g.,* electric lamp), inductive (*e.g.,* induction motor), capacitive or some combination of them. The various types of loads on the power system are :

(*i*) **Domestic load.** Domestic load consists of lights, fans, refrigerators, heaters, television, small motors for pumping water etc. Most of the residential load occurs only for some hours during the day (*i.e.,* 24 hours) *e.g.,* lighting load occurs during night time and domestic appliance load occurs for only a few hours. For this reason, the load factor is low (10% to 12%).

(*ii*) **Commercial load.** Commercial load consists of lighting for shops, fans and electric appliances used in restaurants etc. This class of load occurs for more hours during the day as compared to the domestic load. The commercial load has seasonal variations due to the extensive use of air-conditioners and space heaters.

(*iii*) **Industrial load.** Industrial load consists of load demand by industries. The magnitude of industrial load depends upon the type of industry. Thus small scale industry requires load upto 25 kW, medium scale industry between 25kW and 100 kW and large-scale industry requires load above 500 kW. Industrial loads are generally not weather dependent.

(*iv*) **Municipal load.** Municipal load consists of street lighting, power required for water supply and drainage purposes. Street lighting load is practically constant throughout the hours of the night. For water supply, water is pumped to overhead tanks by pumps driven by electric motors. Pumping is carried out during the off-peak period, usually occurring during the night. This helps to improve the load factor of the power system.

(*v*) **Irrigation load.** This type of load is the electric power needed for pumps driven by motors to supply water to fields. Generally this type of load is supplied for 12 hours during night.

(*vi*) **Traction load.** This type of load includes tram cars, trolley buses, railways etc. This class of load has wide variation. During the morning hour, it reaches peak value because people have to go to their work place. After morning hours, the load starts decreasing and again rises during evening since the people start coming to their homes.

3.8 Typical Demand and Diversity Factors

The demand factor and diversity factor depend on the type of load and its magnitude.

TYPICAL DEMAND FACTORS

Type of consumer		Demand factor
Residence lighting	$\frac{1}{4}$ kW	1·00
	$\frac{1}{2}$ kW	0·60
	Over 1 kW	0·50
Commercial lighting	Restaurants	0·70
	Theatres	0·60
	Hotels	0·50
	Schools	0·55
	Small industry	0·60
	Store	0·70
General power service	0–10 H.P.	0·75
	10–20 H.P.	0·65
	20–100 H.P.	0·55
	Over 100 H.P.	0·50

TYPICAL DIVERSITY FACTORS

	Residential lighting	*Commercial lighting*	*General power supply*
Between consumers	3 – 4	1·5	1·5
Between transformers	1·3	1·3	1·3
Between feeders	1·2	1·2	1·2
Between substations	1·1	1·1	1·1

Illustration. Load and demand factors are always less than 1 while diversity factors are more than unity. High load and diversity factors are the desirable qualities of the power system. Indeed, these factors are used to predict the load. Fig. 3.4 shows a small part of electric power system where a distribution transformer is supplying power to the consumers. For simplicity, only three consumers *a*, *b*, and *c* are shown in the figure. The maximum demand of consumer *a* is the product of its connected load and the appropriate demand factor. Same is the case for consumers *b* and *c*. The maximum demand on the transformer is the sum of *a*, *b* and *c*'s maximum demands divided by the diversity factors between the consumers. Similarly, the maximum demand on the feeder is the sum of maximum demands on the distribution transformers connected to it divided by the diversity factor between transformers. Likewise diversification between feeders is recognised when obtaining substation maximum demands and substation diversification when predicting maximum load on the power station. Note that diversity factor is the sum of the individual maximum demands of the subdivisions of a system taken as they may occur during the daily cycle divided by the maximum simultaneous demand of the system. The "system" may be a group of consumers served by a certain transformer, a group of transformers served by a feeder etc. Since individual variations have diminishing effect as one goes

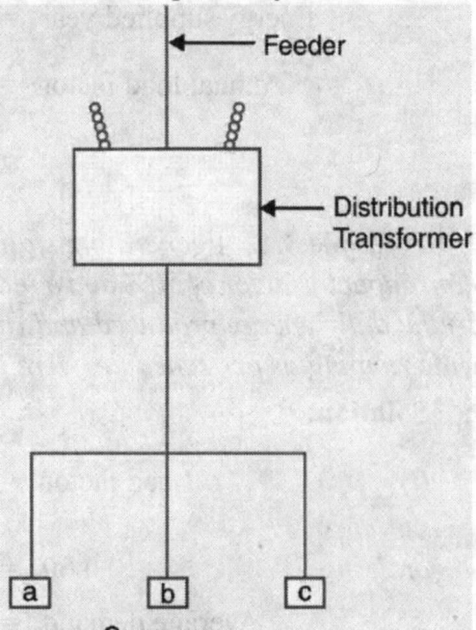

Fig. 3.4

farther from the ultimate consumer in making measurements, one should expect decreasing numerical values of diversity factor as the power plant end of the system is approached. This is clear from the above table showing diversity factors between different elements of the power system.

Example 3.1. *The maximum demand on a power station is 100 MW. If the annual load factor is 40%, calculate the total energy generated in a year.*

Solution.

$$\text{Energy generated/year} = \text{Max. demand} \times \text{L.F.} \times \text{Hours in a year}$$
$$= (100 \times 10^3) \times (0{\cdot}4) \times (24 \times 365) \text{ kWh}$$
$$= \mathbf{3504 \times 10^5 \text{ kWh}}$$

Example 3.2. *A generating station has a connected load of 43MW and a maximum demand of 20 MW; the units generated being 61·5 × 10⁶ per annum. Calculate (i) the demand factor and (ii) load factor.*

Solution.

(*i*) $$\text{Demand factor} = \frac{\text{Max. demand}}{\text{Connected load}} = \frac{20}{43} = \mathbf{0{\cdot}465}$$

(*ii*) $$\text{Average demand} = \frac{\text{Units generated / annum}}{\text{Hours in a year}} = \frac{61{\cdot}5 \times 10^6}{8760} = 7020 \text{ kW}$$

$$\therefore \quad \text{Load factor} = \frac{\text{Average demand}}{\text{Max. demand}} = \frac{7020}{20 \times 10^3} = \mathbf{0{\cdot}351 \text{ or } 35{\cdot}1\%}$$

Example 3.3. *A 100 MW power station delivers 100 MW for 2 hours, 50 MW for 6 hours and is shut down for the rest of each day. It is also shut down for maintenance for 45 days each year. Calculate its annual load factor.*

Solution.

Energy supplied for each working day

$$= (100 \times 2) + (50 \times 6) = 500 \text{ MWh}$$

$$\text{Station operates for} = 365 - 45 = 320 \text{ days in a year}$$

$$\therefore \quad \text{Energy supplied/year} = 500 \times 320 = 160{,}000 \text{ MWh}$$

$$\text{Annual load factor} = \frac{\text{MWh supplied per annum}}{\text{Max. demand in MW} \times \text{Working hours}} \times 100$$
$$= \frac{160{,}000}{(100) \times (320 \times 24)} \times 100 = \mathbf{20{\cdot}8\%}$$

Example 3.4. *A generating station has a maximum demand of 25MW, a load factor of 60%, a plant capacity factor of 50% and a plant use factor of 72%. Find (i) the reserve capacity of the plant (ii) the daily energy produced and (iii) maximum energy that could be produced daily if the plant while running as per schedule, were fully loaded.*

Solution.

(*i*) $$\text{Load factor} = \frac{\text{Average demand}}{\text{Maximum demand}}$$

or $$0{\cdot}60 = \frac{\text{Average demand}}{25}$$

$$\therefore \quad \text{Average demand} = 25 \times 0{\cdot}60 = 15 \text{ MW}$$

$$\text{Plant capacity factor} = \frac{\text{Average demand}}{\text{Plant capacity}}$$

$$\therefore \quad \text{Plant capacity} = \frac{\text{Average demand}}{\text{Plant capacity factor}} = \frac{15}{0{\cdot}5} = 30 \text{ MW}$$

$\therefore$ Reserve capacity of plant = Plant capacity – maximum demand

= 30 – 25 = **5 MW**

(*ii*) Daily energy produced = Average demand × 24

= 15 × 24 = **360 MWh**

(*iii*) Maximum energy that could be produced

$$= \frac{\text{Actual energy produced in a day}}{\text{Plant use factor}}$$

$$= \frac{360}{0 \cdot 72} = \textbf{500 MWh/day}$$

Example 3.5. *A diesel station supplies the following loads to various consumers :*

Industrial consumer = 1500 kW ; Commercial establishment = 750 kW

Domestic power = 100 kW; Domestic light = 450 kW

If the maximum demand on the station is 2500 kW and the number of kWh generated per year is 45×10^5, determine (i) the diversity factor and (ii) annual load factor.

Solution.

(*i*) $$\text{Diversity factor} = \frac{1500 + 750 + 100 + 450}{2500} = \mathbf{1{\cdot}12}$$

(*ii*) $$\text{Average demand} = \frac{\text{kWh generated / annum}}{\text{Hours in a year}} = 45 \times 10^5/8760 = 513{\cdot}7 \text{ kW}$$

$$\therefore \quad \text{Load factor} = \frac{\text{Average load}}{\text{Max. demand}} = \frac{513 \cdot 7}{2500} = 0{\cdot}205 = \mathbf{20{\cdot}5\%}$$

Example 3.6. *A power station has a maximum demand of 15000 kW. The annual load factor is 50% and plant capacity factor is 40%. Determine the reserve capacity of the plant.*

Solution.

Energy generated/annum = Max. demand × L.F. × Hours in a year

= (15000) × (0·5) × (8760) kWh

= $65{\cdot}7 \times 10^6$ kWh

$$\text{Plant capacity factor} = \frac{\text{Units generated / annum}}{\text{Plant capacity} \times \text{Hours in a year}}$$

$$\therefore \quad \text{Plant capacity} = \frac{65 \cdot 7 \times 10^6}{0 \cdot 4 \times 8760} = 18{,}750 \text{ kW}$$

Reserve capacity = Plant capacity – Max. demand

= 18,750 – 15000 = **3750 kW**

Example 3.7. *A power supply is having the following loads :*

Type of load	*Max. demand (kW)*	*Diversity of group*	*Demand factor*
Domestic	*1500*	*1·2*	*0·8*
Commercial	*2000*	*1·1*	*0·9*
Industrial	*10,000*	*1·25*	*1*

If the overall system diversity factor is 1·35, determine (i) the maximum demand and (ii) connected load of each type.

Solution.

(*i*) The sum of maximum demands of three types of loads is = 1500 + 2000 + 10,000 = 13,500 kW. As the system diversity factor is 1·35,

$\therefore$ Max. demand on supply system = 13,500/1·35 = **10,000 kW** (*ii*) Each type of load has its own diversity factor among its consumers.

Sum of max. demands of different domestic consumers

= Max. domestic demand × diversity factor

= 1500 × 1·2 = 1800 kW

∴ Connected domestic load = 1800/0·8 = **2250 kW**

Connected commercial load = 2000 × 1·1/0·9 = **2444 kW**

Connected industrial load = 10,000 × 1·25/1= **12,500 kW**

Example 3.8. *At the end of a power distribution system, a certain feeder supplies three distribution transformers, each one supplying a group of customers whose connected loads are as under:*

Transformer	*Load*	*Demand factor*	*Diversity of groups*
Transformer No. 1	*10 kW*	*0·65*	*1·5*
Transformer No. 2	*12 kW*	*0·6*	*3·5*
Transformer No. 3	*15 kW*	*0·7*	*1·5*

If the diversity factor among the transformers is 1·3, find the maximum load on the feeder.

Solution. Fig. 3.5 shows a feeder supplying three distribution transformers.

Sum of max. demands of customers on Transformer 1

= connected load × demand factor = 10 × 0·65 = 6·5 kW

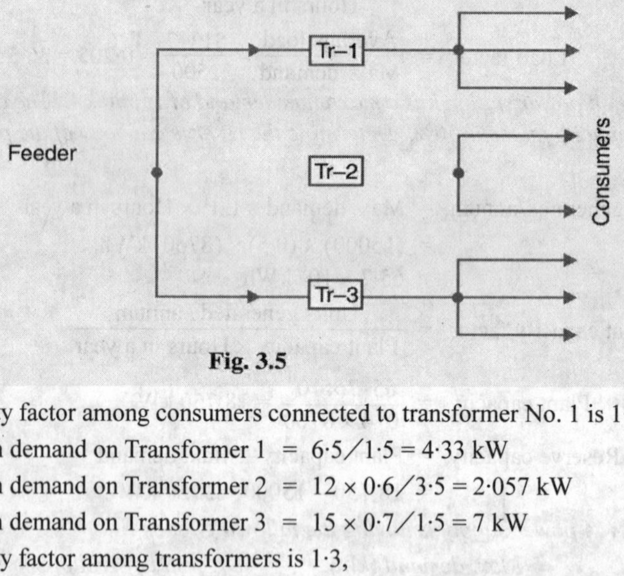

Fig. 3.5

As the diversity factor among consumers connected to transformer No. 1 is 1·5,

∴ Maximum demand on Transformer 1 = 6·5/1·5 = 4·33 kW

Maximum demand on Transformer 2 = 12 × 0·6/3·5 = 2·057 kW

Maximum demand on Transformer 3 = 15 × 0·7/1·5 = 7 kW

As the diversity factor among transformers is 1·3,

$$\therefore \quad \text{Maximum demand on feeder} = \frac{4\cdot33 + 2\cdot057 + 7}{1\cdot3} = \mathbf{10\cdot3\ kW}$$

Example 3.9. *It has been desired to install a diesel power station to supply power in a suburban area having the following particulars :*

(*i*) *1000 houses with average connected load of 1·5 kW in each house. The demand factor and diversity factor being 0·4 and 2·5 respectively.*

(*ii*) *10 factories having overall maximum demand of 90 kW.*

(*iii*) *7 tubewells of 7 kW each and operating together in the morning.*

The diversity factor among above three types of consumers is 1·2. What should be the minimum capacity of power station ?

Solution.

Sum of max. demands of houses $= (1{\cdot}5 \times 0{\cdot}4) \times 1000 = 600$ kW

Max. demand for domestic load $= 600/2{\cdot}5 = 240$ kW

Max. demand for factories $= 90$ kW

Max. demand for tubewells $= 7^* \times 7 = 49$ kW

The sum of maximum demands of three types of loads is = 240 + 90 + 49 = 379 kW. As the diversity factor among the three types of loads is 1·2,

$\therefore$ Max. demand on station $= 379/1{\cdot}2 = 316$ kW

$\therefore$ Minimum capacity of station requried = **316 kW**

Example 3.10. *A generating station has the following daily load cycle :*

Time (Hours)	*0—6*	*6—10*	*10—12*	*12—16*	*16—20*	*20—24*
Load (MW)	*40*	*50*	*60*	*50*	*70*	*40*

Draw the load curve and find (i) maximum demand (ii) units generated per day (iii) average load and (iv) load factor.

Solution. Daily curve is drawn by taking the load along *Y*-axis and time along *X*-axis. For the given load cycle, the load curve is shown in Fig. 3.6.

(*i*) It is clear from the load curve that maximum demand on the power station is 70 MW and occurs during the period 16—20 hours.

$\therefore$ Maximum demand = **70 MW**

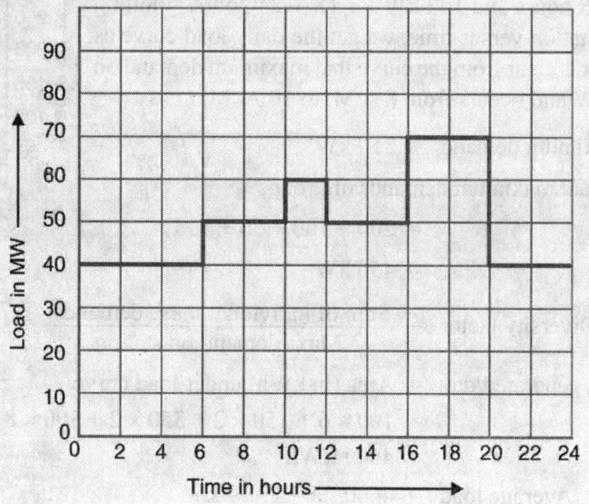

Fig. 3.6

(*ii*) Units generated/day = Area (in kWh) under the load curve

$= 10^3\,[40 \times 6 + 50 \times 4 + 60 \times 2 + 50 \times 4 + 70 \times 4 + 40 \times 4]$

$= 10^3\,[240 + 200 + 120 + 200 + 280 + 160]$ kWh

$= \mathbf{12 \times 10^5}$ **kWh**

(*iii*) Average load $= \dfrac{\text{Units generated / day}}{24 \text{ hours}} = \dfrac{12 \times 10^5}{24} =$ **50,000 kW**

(*iv*) Load factor $= \dfrac{\text{Average load}}{\text{Max. demand}} = \dfrac{50{,}000}{70 \times 10^3} = 0{\cdot}714 =$ **71·4%**

* Since the tubewells operate together, the diversity factor is 1.

Example 3.11. *A power station has to meet the following demand :*

Group A : 200 kW between 8 A.M. and 6 P.M.

Group B : 100 kW between 6 A.M. and 10 A.M.

Group C : 50 kW between 6 A.M. and 10 A.M.

Group D : 100 kW between 10 A.M. and 6 P.M. and then between 6 P.M. and 6 A.M.

Plot the daily load curve and determine (i) diversity factor (ii) units generated per day (iii) load factor.

Solution. The given load cycle can be tabulated as under :

Time (Hours)	0—6	6—8	8—10	10—18	18—24
Group A	—	—	200 kW	200 kW	—
Group B	—	100 kW	100 kW	—	—
Group C	—	50 kW	50 kW	—	—
Group D	100 kW	—	—	100 kW	100 kW
Total load on power station	100 kW	150 kW	350 kW	300 kW	100 kW

From this table, it is clear that total load on power station is 100 kW for 0—6 hours, 150 kW for 6—8 hours, 350 kW for 8—10 hours, 300 kW for 10—18 hours and 100 kW for 18—24 hours. Plotting the load on power station versus time, we get the daily load curve as shown in Fig. 3.7. It is clear from the curve that maximum demand on the station is 350 kW and occurs from 8 A.M. to 10 A. M. *i.e.,*

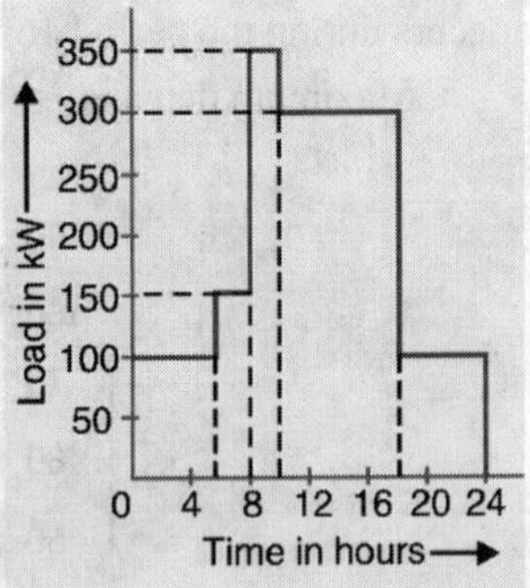

Fig. 3.7

$$\text{Maximum demand} = 350 \text{ kW}$$

Sum of individual maximum demands of groups

$$= 200 + 100 + 50 + 100$$

$$= 450 \text{ kW}$$

(*i*) $$\text{Diversity factor} = \frac{\text{Sum of individual max. demands}}{\text{Max. demand on station}} = 450/350 = \mathbf{1{\cdot}286}$$

(*ii*) $$\text{Units generated/day} = \text{Area (in kWh) under load curve}$$

$$= 100 \times 6 + 150 \times 2 + 350 \times 2 + 300 \times 8 + 100 \times 6$$

$$= \mathbf{4600 \ kWh}$$

(*iii*) $$\text{Average load} = 4600/24 = 191{\cdot}7 \text{ kW}$$

$$\therefore \quad \text{Load factor} = \frac{191{\cdot}7}{350} \times 100 = \mathbf{54{\cdot}8\%}$$

Example 3.12. *The daily demands of three consumers are given below :*

Time	*Consumer 1*	*Consumer 2*	*Consumer 3*
12 midnight to 8 A.M.	*No load*	*200 W*	*No load*
8 A.M. to 2 P.M.	*600 W*	*No load*	*200 W*
2 P.M. to 4 P.M.	*200 W*	*1000 W*	*1200 W*
4 P.M. to 10 P.M.	*800 W*	*No load*	*No load*
10 P.M. to midnight	*No load*	*200 W*	*200 W*

Plot the load curve and find (i) maximum demand of individual consumer (ii) load factor of individual consumer (iii) diversity factor and (iv) load factor of the station.

Solution. Fig. 3.8 shows the load curve.

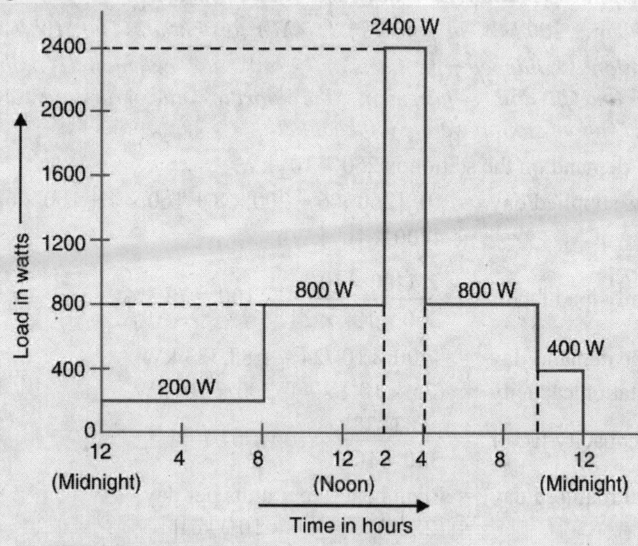

Fig. 3.8

(*i*) Max. demand of consumer 1 = **800 W**

Max. demand of consumer 2 = **1000 W**

Max. demand of consumer 3 = **1200 W**

(*ii*) L.F. of consumer 1 $= \dfrac{\text{Energy consumed / day}}{\text{Max. demand} \times \text{Hours in a day}} \times 100$

$$= \frac{600 \times 6 + 200 \times 2 + 800 \times 6}{800 \times 24} \times 100 = \mathbf{45{\cdot}8\%}$$

L.F. of consumer 2 $= \dfrac{200 \times 8 + 1000 \times 2 + 200 \times 2}{1000 \times 24} \times 100 = \mathbf{16{\cdot}7\%}$

L.F. of consumer 3 $= \dfrac{200 \times 6 + 1200 \times 2 + 200 \times 2}{1200 \times 24} \times 100 = \mathbf{13{\cdot}8\%}$

(*iii*) The simultaneous maximum demand on the station is 200 + 1000 + 1200 = 2400 W and occurs from 2 P.M. to 4 P.M.

∴ Diversity factor $= \dfrac{800 + 1000 + 1200}{2400} = \mathbf{1{\cdot}25}$

(*iv*) Station load factor $= \dfrac{\text{Total energy consumed / day}}{\text{Simultaneous max. demand} \times 24} \times 100$

$$= \frac{8800 + 4000 + 4000}{2400 \times 24} \times 100 = \mathbf{29{\cdot}1\%}$$

Example 3.13. *A daily load curve which exhibited a 15-minute peak of 3000 kW is drawn to scale of 1 cm = 2 hours and 1 cm = 1000 kW. The total area under the load curve is measured by planimeter and is found to be 12 cm^2. Calculate the load factor based on 15-min. peak.*

Solution.

1 cm^2 of load curve represents 1000 × 2 = 2000 kWh

Average demand $= \dfrac{2000 \times \text{Area of load curve}}{\text{Hours in a day}} = 2000 \times \dfrac{12}{24} = 1000 \text{ kW}$

$\therefore$ Load factor $= \frac{1000}{3000} \times 100 = \mathbf{33 \cdot 3\%}$

Example 3.14. *A power station has a daily load cycle as under :*

260 MW for 6 hours ; 200 MW for 8 hours : 160 MW for 4 hours, 100 MW for 6 hours.

If the power station is equipped with 4 sets of 75 MW each, calculate (i) daily load factor (ii) plant capacity factor and (iii) daily requirement if the calorific value of oil used were 10,000 kcal/kg and the average heat rate of station were 2860 kcal/kWh.

Solution. Max. demand on the station is 260×10^3 kW.

Units supplied/day $= 10^3\,[260 \times 6 + 200 \times 8 + 160 \times 4 + 100 \times 6]$

$= 4400 \times 10^3$ kWh

(*i*) Daily load factor $= \frac{4400 \times 10^3}{260 \times 10^3 \times 24} \times 100 = \mathbf{70 \cdot 5\%}$

(*ii*) Average demand/day $= 4400 \times 10^3/24 = 1{,}83{,}333$ kW

Station capacity $= (75 \times 10^3) \times 4 = 300 \times 10^3$ kW

$\therefore$ Plant capacity factor $= \frac{1{,}83{,}333}{300 \times 10^3} \times 100 = \mathbf{61 \cdot 1\ \%}$

(*iii*) Heat required/day = Plant heat rate × units per day

$= (2860) \times (4400 \times 10^3)$ kcal

Fuel required/day $= \frac{2860 \times 4400 \times 10^3}{10000} = 1258 \cdot 4 \times 10^3$ kg $=$ **1258·4 tons**

Example 3.15. *A power station has the following daily load cycle :*

Time in Hours	*6—8*	*8—12*	*12—16*	*16—20*	*20—24*	*24—6*
Load in MW	*20*	*40*	*60*	*20*	*50*	*20*

Plot the load curve and load duratoin curve. Also calculate the energy generated per day.

Solution. Fig. 3.9 (*i*) shows the daily load curve, whereas Fig. 3.9 (*ii*) shows the daily load duraton curve. It can be readily seen that area under the two load curves is the same. Note that load duration curve is drawn by arranging the loads in the order of descending magnitudes.

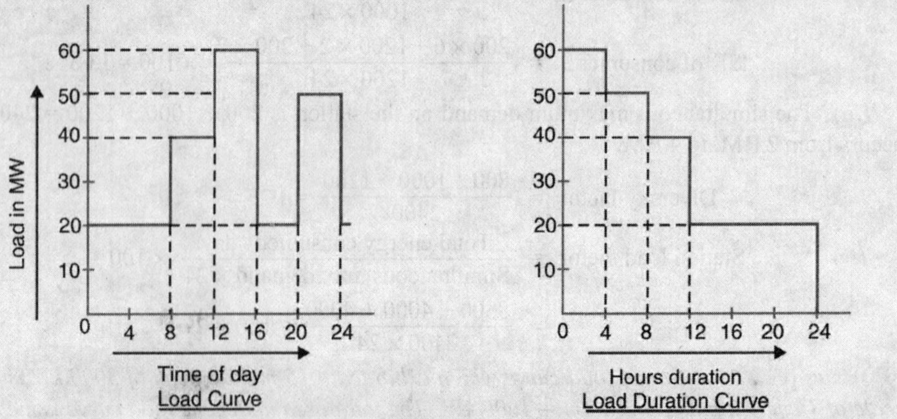

Fig. 3.9

Units generated/day = Area (in kWh) under daily load curve

$= 10^3\,[20 \times 8 + 40 \times 4 + 60 \times 4 + 20 \times 4 + 50 \times 4]$

$=$ **840×10^3 kWh**

Alternatively :

$$\text{Units generated/day} = \text{Area (in kWh) under daily load duration curve}$$
$$= 10^3 [60 \times 4 + 50 \times 4 + 40 \times 4 + 20 \times 12]$$
$$= \mathbf{840 \times 10^3\ kWh}$$

which is the same as above.

Example 3.16. *The annual load duration curve of a certain power station can be considered as a straight line from 20 MW to 4 MW. To meet this load, three turbine-generator units, two rated at 10 MW each and one rated at 5 MW are installed. Determine (i) installed capacity (ii) plant factor (iii) units generated per annum (iv) load factor and (v) utilisation factor.*

Solution. Fig. 3.10 shows the annual load duration curve of the power station.

(*i*) Installed capacity = 10 + 10 + 5 = **25 MW**

(*ii*) Referring to the load duration curve,

$$\text{Average demand} = \frac{1}{2}[20 + 4] = 12 \text{ MW}$$

$$\therefore \quad \text{Plant factor} = \frac{\text{Average demand}}{\text{Plant capacity}} = \frac{12}{25} = 0{\cdot}48 = \mathbf{48\%}$$

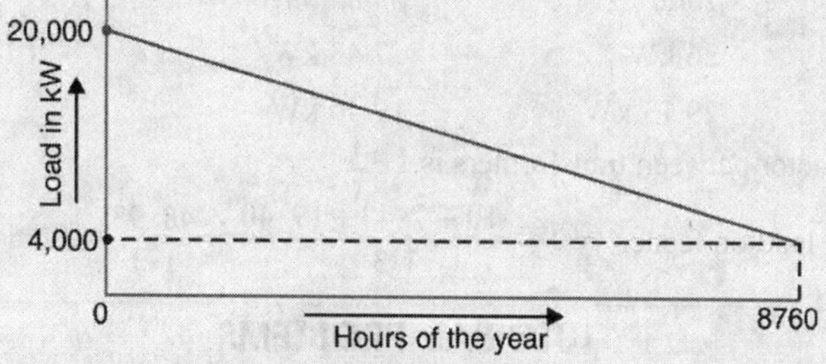

Fig. 3.10

(*iii*) Units generated/annum = Area (in kWh) under load duration curve

$$= \frac{1}{2}[4000 + 20{,}000] \times 8760 \text{ kWh} = \mathbf{105{\cdot}12 \times 10^6\ kWh}$$

(*iv*) $$\text{Load factor} = \frac{12{,}000}{20{,}000} \times 100 = \mathbf{60\%}$$

(*v*) $$\text{Utilisation factor} = \frac{\text{Max. demand}}{\text{Plant capacity}} = \frac{20{,}000}{25000} = 0{\cdot}8 = \mathbf{80\%.}$$

Example 3.17. *At the end of a power distribution system, a certain feeder supplies three distribution transformers, each one supplying a group of customers whose connected load are listed as follows :*

Transformer 1	*Transformer 2*	*Transformer 3*
General power service and lighting	*Residence lighting*	*Store lighting and power*
a : 10 H.P., 5kW	*e : 5 kW*	*j : 10 kW, 5 H.P.*
b : 7·5 H.P., 4kW	*f : 4 kW*	*k : 8 kW, 25 H.P.*
c : 15 H.P.	*g : 8 kW*	*l : 4 kW*
d : 5 H.P., 2 kW	*h : 15 kW*	
	i : 20 kW	

Use the factors given in Art. 3.8 and predict the maximum demand on the feeder. The H.P. load is motor load and assume an efficiency of 72%.

Solution. The individual maximum demands of the group of consumers connected to transformer 1are obtained with factors from the table on page 49.

a: $\left(10 \times \frac{0 \cdot 746}{0 \cdot 72}\right) \times 0{\cdot}65 + 5 \times 0{\cdot}60^* = 9{\cdot}74$ kW

b: $\left(7 \cdot 5 \times \frac{0 \cdot 746}{0 \cdot 72}\right) \times 0{\cdot}75 + 4 \times 0{\cdot}60 = 8{\cdot}23$ kW

c: $\left(15 \times \frac{0 \cdot 746}{0 \cdot 72}\right) \times 0{\cdot}65 = 10{\cdot}10$ kW

d: $\left(5 \times \frac{0 \cdot 746}{0 \cdot 72}\right) \times 0{\cdot}75 + 2 \times 0{\cdot}60 = 5{\cdot}09$ kW

Total $= 33{\cdot}16$ kW

The diversity factor between consumers of this type of service is 1·5 (From the table of article 3.8).

$\therefore$ Maximum demand on transformer 1 $= \frac{33 \cdot 16}{1 \cdot 5} = 22{\cdot}10$ kW

In a similar manner, the other transformer loads are determined to be

	Total	*Simultaneous*
Transformer 2	26 kW	7·43 kW
Transformer 3	29·13 kW	19·40 kW

The diversity factor between transformers is 1·3.

$\therefore$ Maximum load on feeder $= \frac{22 \cdot 10 + 7 \cdot 43 + 19 \cdot 40}{1 \cdot 3} = \frac{48 \cdot 93}{1 \cdot 3} = \mathbf{37{\cdot}64\ kW}$

TUTORIAL PROBLEMS

1. A generating station has a connected load of 40 MW and a maximum demand of 20 MW : the units generated being 60×10^6. Calculate (*i*) the demand factor (*ii*) the load factor. **[(*i*) 0·5 (*ii*) 34·25%]**
2. A 100 MW powers stations delivers 100 MW for 2 hours, 50 MW for 8 hours and is shut down for the rest of each day. It is also shut down for maintenance for 60 days each year. Calculate its annual load factor. **[21%]**
3. A power station is to supply four regions of loads whose peak values are 10,000 kW, 5000 kW, 8000 kW and 7000 kW. The diversity factor of the load at the station is 1.5 and the average annual load factor is 60%. Calculate the maximum demand on the station and annual energy supplied from the station. **[20,000 kW ; 105·12 × 10^6 kWh]**
4. A generating station supplies the following loads : 15000 kW, 12000 kW, 8500 kW, 6000 kW and 450 kW. The station has a maximum demand of 22000 kW. The annual load factor of the station is 48%. Calculate (*i*) the number of units supplied annually (*ii*) the diversity factor and (*iii*) the demand factor. **[(*i*) 925 × 10^5 kWh (*ii*) 52·4% (*iii*) 1·9]**
5. A generating station has a maximum demand of 20 MW, a load factor of 60%, a plant capacity factor of 48% and a plant use factor of 80% . Find :
   (*i*) the daily energy produced
   (*ii*) the reserve capacity of the plant

* Since demand factor for a particular load magnitude in not given in the table, it is reasonable to assume the average value *i.e.*

$$\text{Demand Factor} = \frac{0 \cdot 7 + 0 \cdot 5}{2} = \frac{1 \cdot 2}{2} = 0{\cdot}6$$

(iii) the maximum energy that could be produced daily if the plant was running all the time

(iv) the maximum energy that could be produced daily if the plant was running fully loaded and operating as per schedule. **[(i) 288 × 10^3 kWh (ii) 0 (iii) 4·80 × 10^3 kWh (iv) 600 × 10^3 kWh]**

6. A generating station has the following daily load cycle :

Time (hours)	0—6	6—10	10—12	12—16	16—20	20—24
Load (MW)	20	25	30	25	35	20

Draw the load curve and find

(i) maximum demand,

(ii) units generated per day,

(iii) average load,

(iv) load factor, **[(i) 35 MW (ii) 560 × 10^3 kWh (iii) 23333 kW (iv) 66·67%]**

7. A power station has to meet the following load demand :

Load *A*	50 kW	between	10 A.M. and 6 P.M.
Load *B*	30 kW	between	6 P.M. and 10 P.M.
Load *C*	20 kW	between	4 P.M. and 10 A.M.

Plot the daily load curve and determine (i) diversity factor (ii) units generated per day (iii) load factor. **[(i) 1·43 (ii) 880 kWh (iii) 52·38%]**

8. A substation supplies power by four feeders to its consumers. Feeder no. 1 supplies six consumers whose individual daily maximum demands are 70 kW, 90 kW, 20 kW, 50 kW, 10 kW and 20 kW while the maximum demand on the feeder is 200 kW. Feeder no. 2 supplies four consumers whose daily maximum demands are 60 kW, 40 kW, 70 kW and 30 kW, while the maximum demand on the feeder is 160 kW. Feeder nos. 3 and 4 have a daily maximum demand of 150 kW and 200 kW respectively while the maximum demand on the station is 600 kW.

Determine the diversity factors for feeder no. 1. feeder no. 2 and for the four feeders. **[1·3, 1·25, 1·183]**

9. A central station is supplying energy to a community through two substations. Each substation feeds four feeders. The maximum daily recorded demands are :

POWER STATION........ 12,000 KW

Substation *A* ...... 6000 kW	Sub-station *B* .... 9000 kW
Feeder 1 .... 1700 kW	Feeder 1 .... 2820 kW
Feeder 2 .... 1800 kW	Feeder 2 .... 1500 kW
Feeder 3 .... 2800 kW	Feeder 3 .... 4000 kW
Feeder 4 .... 600 kW	Feeder 4 .... 2900 kW

Calculate the diversity factor between (i) substations (ii) feeders on substation A and (iii) feeders on substation B. **[(i) 1·25 (ii) 1·15 (iii) 1·24]**

10. The yearly load duration curve of a certain power station can be approximated as a straight line ; the maximum and minimum loads being 80 MW and 40 MW respectively. To meet this load, three turbine-generator units, two rated at 20 MW each and one at 10 MW are installed. Determine (i) installed capacity (ii) plant factor (iii) kWh output per year (iv) load factor. **[(i) 50MW (ii) 48% (iii) 210 × 10^6 (iv) 60%]**

3.9 Load Curves and Selection of Generating Units

The load on a power station is seldom constant; it varies from time to time. Obviously, a single generating unit (*i.e.*, alternator) will not be an economical proposition to meet this varying load. It is because a single unit will have very poor* efficiency during the periods of light loads on the power station. Therefore, in actual practice, a number of generating units of different sizes are installed in a power station. The selection of the number and sizes of the units is decided from the annual load curve of the station. *The number and size of the units are selected in such a way that they correctly*

* The efficiency of a machine (alternator in this case) is maximum at nearly 75% of its rated capacity.

fit the station load curve. Once this underlying principle is adhered to, it becomes possible to operate the generating units at or near the point of maximum efficiency.

Illustration. The principle of selection of number and sizes of generating units with the help of load curve is illustrated in Fig. 3.11. In Fig. 3.11 (*i*), the annual load curve of the station is shown. It is clear form the curve that load on the station has wide variations ; the minimum load being somewhat near 50 kW and maximum load reaching the value of 500 kW. It hardly needs any mention that use of a single unit to meet this varying load will be highly uneconomical.

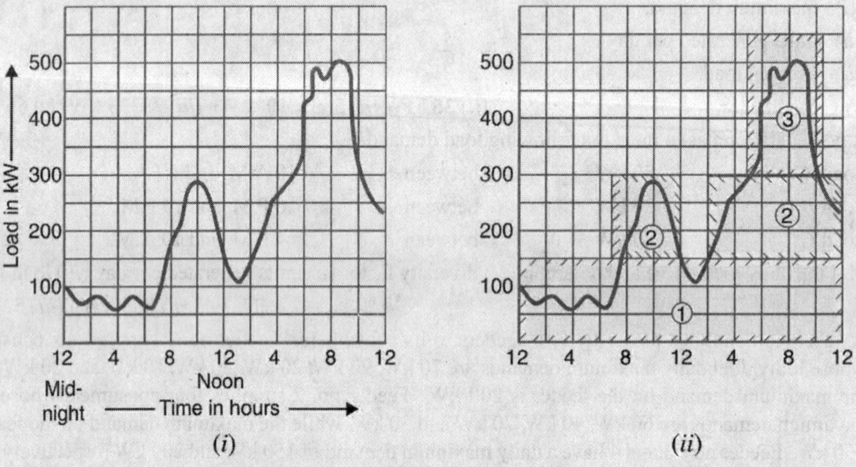

Fig. 3.11

As discussed earlier, the total plant capacity is divided into several generating units of different sizes to fit the load curve. This is illustrated in Fig. 3.11(*ii*) where the plant capacity is divided into three* units numbered as 1, 2 and 3. The cyan colour outline shows the units capacity being used. The three units employed have different capacities and are used according to the demand on the station. In this case, the operating schedule can be as under :

Time	*Units in operation*
From 12 midnight to 7 A.M.	Only unit no.1 is put in operation.
From 7 A.M. to 12.00 noon	Unit no. 2 is also started so that both units 1 and 2 are in operation.
From 12.00 noon to 2 P.M.	Unit no. 2 is stopped and only unit 1operates.
From 2 P.M. to 5 P.M.	Unit no. 2 is again started. Now units 1 and 2 are in operation.
From 5 P.M. to 10.30 P.M.	Units 1, 2 and 3 are put in operation.
From 10. 30 P.M. to 12.00 midnight	Units 1 and 2 are put in operation.

Thus by selecting the proper number and sizes of units, the generating units can be made to operate near maximum efficiency. This results in the overall reduction in the cost of production of electrical energy.

3.10 Important Points in the Selection of Units

While making the selection of number and sizes of the generating units, the following points should be kept in view :

(*i*) The number and sizes of the units should be so selected that they approximately fit the annual load curve of the station.

* It may be seen that the generating units can fit the load curve more closely if more units of smaller sizes are employed. However, using greater number of units increases the investment cost per kW of the capacity.

(*ii*) The units should be *preferably* of different capacities to meet the load requirements. Although use of identical units (*i.e.*, having same capacity) ensures saving* in cost, they often do not meet the load requirement.

(*iii*) The capacity of the plant should be made 15% to 20% more than the maximum demand to meet the future load requirements.

(*iv*) There should be a spare generating unit so that repairs and overhauling of the working units can be carried out.

(*v*) The tendency to select a large number of units of smaller capacity in order to fit the load curve very accurately should be avoided. It is because the investment cost per kW of capacity increases as the size of the units decreases.

Example 3.18. *A proposed station has the following daily load cycle :*

Time in hours	*6—8*	*8—11*	*11—16*	*16—19*	*19—22*	*22—24*	*24—6*
Load in MW	*20*	*40*	*50*	*35*	*70*	*40*	*20*

Draw the load curve and select suitable generator units from the 10,000, 20,000, 25,000, 30,000 kVA. Prepare the operation schedule for the machines selected and determine the load factor from the curve

Solution. The load curve of the power station can be drawn to some suitable scale as shown in Fig. 3.12.

$$\text{Units generated per day} = \text{Area (in kWh) under the load curve}$$
$$= 10^3 \, [20 \times 8 + 40 \times 3 + 50 \times 5 + 35 \times 3 + 70 \times 3 + 40 \times 2]$$
$$= 10^3 \, [160 + 120 + 250 + 105 + 210 + 80] \text{ kWh}$$
$$= 925 \times 10^3 \text{ kWh}$$

$$\text{Average load} = \frac{925 \times 10^3}{24} = 38541{\cdot}7 \text{ kW}$$

$$\text{Load factor} = \frac{38541 \cdot 7}{70 \times 10^3} \times 100 = \mathbf{55{\cdot}06\%}$$

Selection of number and sizes of units : Assuming power factor of the machines to be 0·8, the output of the generating units available will be 8, 16, 20 and 24 MW. There can be several possibilities. However, while selecting the size and number of units, it has to be borne in mind that (*i*) one set of highest capacity should be kept as standby unit (*ii*) the units should meet the maximum demand (70 MW in this case) on the station (*iii*) there should be overall economy.

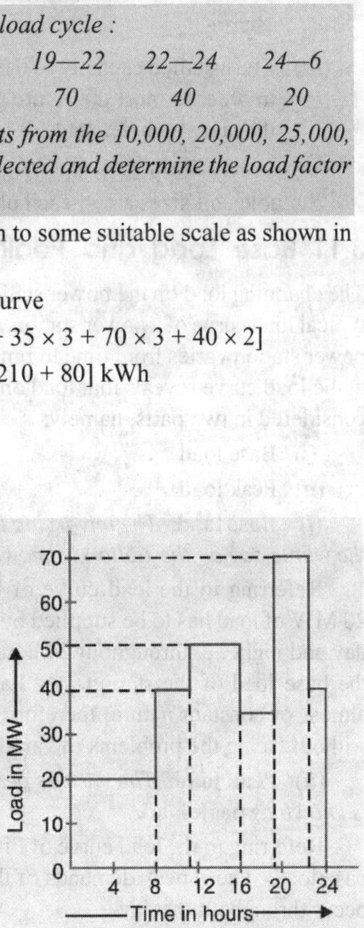

Fig. 3.12

Keeping in view the above facts, *4 sets of 24 MW each may be chosen.* Three sets will meet the maximum demand of 70 MW and one unit will serve as a standby unit.

Operational schedule. Referring to the load curve shown in Fig. 3.12, the operational schedule will be as under :

(*i*) Set No. 1 will run for 24 hours.

(*ii*) Set No. 2 will run from 8.00 hours to midnight.

(*iii*) Set No. 3 will run from 11.00 hours to 16 hours and again from 19 hours to 22 hours.

Example 3.19. *A generating station is to supply four regions of load whose peak loads are 10 MW, 5 MW, 8 MW and 7 MW. The diversity factor at the station is 1·5 and the average annual load factor is 60%. Calculate :*

---

* Due to duplication of sizes and dimensions of pipes, foundations etc.

(*i*) *the maximum demand on the station.*
(*ii*) *annual energy supplied by the station.*
(*iii*) *Suggest the installed capacity and the number of units.*

Solution.

(*i*) $$\text{Max. demand on station} = \frac{\text{Sum of max. demands of the regions}}{\text{Diversity factor}}$$
$$= (10 + 5 + 8 + 7)/1{\cdot}5 = \mathbf{20\ MW}$$

(*ii*) $$\text{Units generated/annum} = \text{Max. demand} \times \text{L.F.} \times \text{Hours in a year}$$
$$= (20 \times 10^3) \times (0{\cdot}6) \times (8760) \text{ kWh}$$
$$= \mathbf{105{\cdot}12 \times 10^6\ kWh}$$

(*iii*) The installed capacity of the station should be 15% to 20% more than the maximum demand in order to meet the future growth of load. Taking installed capacity to be 20% more than the maximum demand,

$$\text{Installed capacity} = 1{\cdot}2 \times \text{Max. demand} = 1{\cdot}2 \times 20 = \mathbf{24\ MW}$$

Suitable unit sizes are 4, each of 6 MW capacity.

3.11 Base Load and Peak Load on Power Station

The changing load on the power station makes its load curve of variable nature. Fig. 3.13. shows the typical load curve of a power station. It is clear that load on the power station varies from time to time. However, a close look at the load curve reveals that load on the power station can be considered in two parts, namely;

(*i*) Base load
(*ii*) Peak load

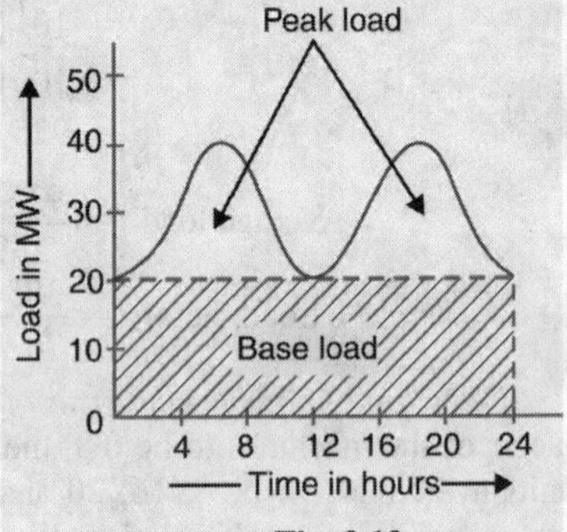

Fig. 3.13

(*i*) **Base load.** *The unvarying load which occurs almost the whole day on the station is known as* **base load.**

Referring to the load curve of Fig. 3.13, it is clear that 20 MW of load has to be supplied by the station at all times of day and night *i.e.* throughout 24 hours. Therefore, 20 MW is the base load of the station. As base load on the station is almost of constant nature, therefore, it can be suitably supplied (as discussed in the next Article) without facing the problems of variable load.

(*ii*) **Peak load.** *The various peak demands of load over and above the base load of the station is known as* **peak load.**

Referring to the load curve of Fig. 3.13, it is clear that there are peak demands of load excluding base load. These peak demands of the station generally form a small part of the total load and may occur throughout the day.

3.12 Method of Meeting the Load

The total load on a power station consists of two parts *viz.*, base load and peak load. In order to achieve overall economy, *the best method to meet load is to interconnect two different power stations*. The more efficient plant is used to supply the base load and is known *as base load power station*. The less efficient plant is used to supply the peak loads and is known as *peak load power station*. There is no hard and fast rule for selection of base load and peak load stations as it would depend upon the particular situation. For example, both hydro-electric and steam power stations are quite efficient and can be used as base load as well as peak load station to meet a particular load requirement.

Illustration. The interconnection of steam and hydro plants is a beautiful illustration to meet the load. When water is available in sufficient quantity as in summer and rainy season, the hydro-electric plant is used to carry the base load and the steam plant supplies the peak load as shown in Fig 3.14 (*i*).

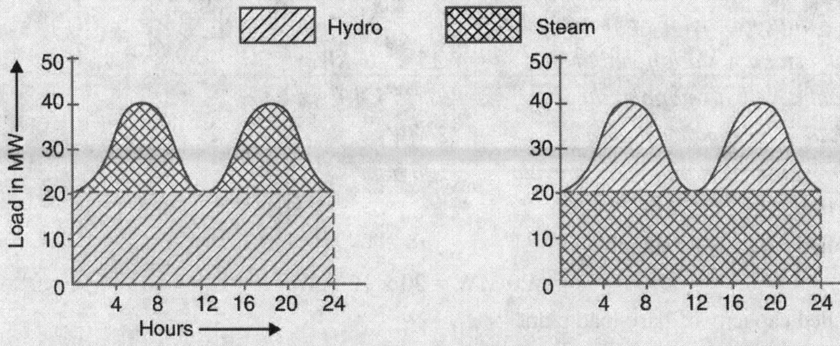

Load division on Hydro-steam system.

(*i*) High Flow day (*ii*) Low Flow day

Fig. 3.14

However, when the water is not available in sufficient quantity as in winter, the steam plant carries the base load, whereas the hydro-electric plant carries the peak load as shown in Fig. 3.14 (*ii*).

3.13 Interconnected Grid System

The connection of several generating stations in parallel is known as **interconnected grid system.**

The various problems facing the power engineers are considerably reduced by interconnecting different power stations in parallel. Although interconnection of station involves extra cost, yet considering the benefits derived from such an arrrangement, it is gaining much favour these days. Some of the advantages of interconnected system are listed below :

(*i*) *Exchange of peak loads* : An important advantage of interconnected system is that the peak load of the power station can be exchanged. If the load curve of a power station shows a peak demand that is greater than the rated capacity of the plant, then the excess load can be shared by other stations interconnected with it.

(*ii*) *Use of older plants* : The interconnected system makes it possible to use the older and less efficient plants to carry peak loads of short durations. Although such plants may be inadequate when used alone, yet they have sufficient capacity to carry short peaks of loads when interconnected with other modern plants. Therefore, interconnected system gives a direct key to the use of obsolete plants.

(*iii*) *Ensures economical operation* : The interconnected system makes the operation of concerned power stations quite economical. It is because sharing of load among the stations is arranged in such a way that more efficient stations work continuously throughouts the year at a high load factor and the less efficient plants work for peak load hours only.

(*iv*) *Increases diversity factor* : The load curves of different interconnected stations are generally different. The result is that the maximum demand on the system is much reduced as compared to the sum of individual maximum demands on different stations. In other words, the diversity factor of the system is improved, thereby increasing the effective capacity of the system.

(*v*) *Reduces plant reserve capacity* : Every power station is required to have a standby unit for emergencies. However, when several power stations are connected in parallel, the reserve capacity of the system is much reduced. This increases the efficiency of the system.

(*vi*) *Increases reliability of supply* : The interconnected system increases the reliability of supply. If a major breakdown occurs in one station, continuity of supply can be maintained by other healthy stations.

Example 3.20. *A base load station having a capacity of 18 MW and a standby station having a capacity of 20 MW share a common load. Find the annual load factors and plant capacity factors of two power stations from the following data :*

Annual standby station output	$= 7{\cdot}35 \times 10^6$ *kWh*
Annual base load station output	$= 101{\cdot}35 \times 10^6$ *kWh*
Peak load on standby station	*= 12 MW*
Hours of use by standby station/year	*= 2190 hours*

Solution.

Installed capacity of standby unit

$$= 20 \text{ MW} = 20 \times 10^3 \text{ kW}$$

Installed capacity of base load plant

$$= 18 \text{ MW} = 18 \times 10^3 \text{ kW}$$

Standby station

$$\text{Annual load factor} = \frac{\text{kWh generated / annum}}{\text{Max. demand} \times \text{Annual working hours}} \times 100$$

$$= \frac{7{\cdot}35 \times 10^6}{(12 \times 10^3) \times 2190} \times 100 = \mathbf{28\%}$$

$$\text{Annual plant capacity factor} = \frac{\text{kWh output / annum}}{\text{Installed capacity} \times \text{Hours in a year}} \times 100$$

$$= \frac{7{\cdot}35 \times 10^6}{(20 \times 10^3) \times 8760} \times 100 = \mathbf{4{\cdot}2\%}$$

Base load station. It is reasonable to assume that the maximum demand on the base load station is equal to the installed capacity (*i.e.*, 18 MW). It operates throughout the year *i.e.*, for 8760 hours.

$$\therefore \quad \text{Annual load factor} = \frac{101{\cdot}35 \times 10^6}{(18 \times 10^3) \times 8760} = \mathbf{64{\cdot}2\%}$$

As the base load station has no reserves above peak load and it is in continuous operation, therefore, its capacity factor is also **64.2%**.

Example 3.21. *The load duration curve for a typical heavy load being served by a combined hydro-steam system may be approximated by a straight line; maximum and minimum loads being 60,000 kW and 20,000 kW respectively. The hydro power available at the time of minimum regulated flow is just sufficient to take a peak load of 50,000 kWh per day. It is observed that it will be economical to pump water from tail race to the reservoir by utilising the steam power plant during the off-peak periods and thus running the station at 100% load factor. Determine the maximum capacity of each type of plant. Assume the efficiency of steam conversion to be 60%.*

Solution. *OCBA* represents the load duration curve for the combined system as shown in Fig. 3.15. The total maximum demand (*i.e.*, 60,000 kW) is represented by *OC,* whereas the minimum demand (*i.e.*, 20,000 kW) is represented by *OD*.

Let *OE* = Capacity of steam plant

EC = Capacity of hydro plant

Area *CHI* = The energy available from hydro plant in the low flow period.

$$\text{Area } FGB = \text{The off-peak* period energy available from steam plant}$$

Obviously, the energy of hydro plant represented by area *HEFI* and available from reservoir has been supplied by steam power plant represented by area *FGB*. As steam electric conversion is 60%,

$$\therefore \quad \text{Area } HEFI = 0{\cdot}6 \times \text{Area } FGB \quad \text{... (i)}$$

But
$$\text{Area } HEFI = \text{Area } CFE - \text{Area } CHI$$
$$= \frac{1}{2}xy - 50{,}000\dagger$$

Now
$$\text{Area } FGB = \frac{1}{2} \times FG \times GB = \frac{1}{2}(24 - x)(40{,}000 - y)$$

Putting the various values in exp. (*i*), we get,

$$\frac{1}{2}xy - 50{,}000 = 0{\cdot}6\left[\frac{1}{2}(24 - x)(40{,}000 - y)\right]$$

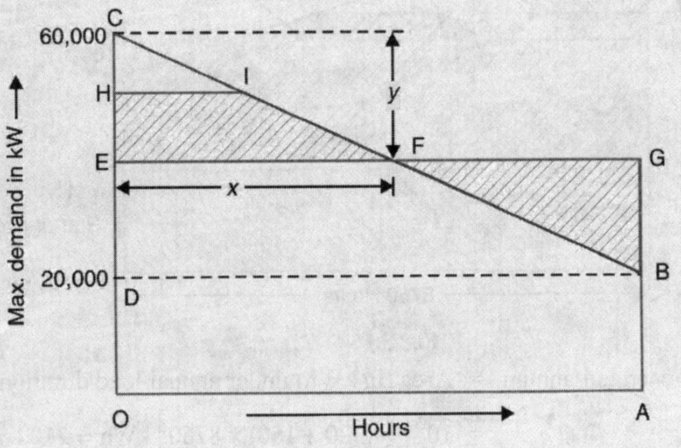

Fig. 3.15

or $\quad 0{\cdot}2\,xy + 12000\,x + 7{\cdot}2\,y - 3{,}38{,}000 = 0 \quad$ ... (*ii*)

Also from similar triangles *CEF* and *CDB*, we get,

$$\frac{y}{40{,}000} = \frac{x}{24}$$

$$\therefore \quad y = \frac{40{,}000\,x}{24} \quad \text{... (iii)}$$

Putting $y = 40{,}000\,x/24$ from exp. (*iii*) into exp. (*ii*), we get,

$$333\,x^2 + 24000\,x - 3{,}38{,}000 = 0$$

or
$$x^2 + 72x - 1015 = 0$$

$$\therefore \quad x = \frac{-72 \pm \sqrt{5184 + 4060}}{2} = \frac{-72 \pm 96}{24} = 12$$

$\therefore$ Capacity of the hydro plant is

$$y (= EC) = \frac{40{,}000 \times 12}{24} = \mathbf{20{,}000\ kW}$$

Capacity of steam plant $= 60{,}000 - 20{,}000 = \mathbf{40{,}000\ kW}$

Example 3.22. *The annual load duration curve for a typical heavy load being served by a steam station, a run-of-river station and a reservoir hydro-electric station is as shown in Fig. 3.16. The ratio of number of units supplied by these stations is as follows :*

* It is clear from load duration curve that the capacity of steam plant represented by area *FGB* is not being utilised efficiently. This steam energy can be used to pump water in tail race back to the reservoir.

† Because during minimum regulated flow, hydro energy supplied is 50,000 kWh.

Steam : Run-of-river : Reservoir : : 7 : 4 : 1

The run-of-river station is capable of generating power continuously and works as a base load station. The reservoir station works as a peak load station. Determine (i) the maximum demand of each station and (ii) load factor of each station.

Solution. *ODCA* is the annual load duration curve for the system as shown in Fig. 3.16. The energy supplied by the reservoir plant is represented by area *DFG* ; steam station by area *FGCBE* and run-of-river by area *OEBA*. The maximum and minimum loads on the system are 320 MW and 160 MW respectively.

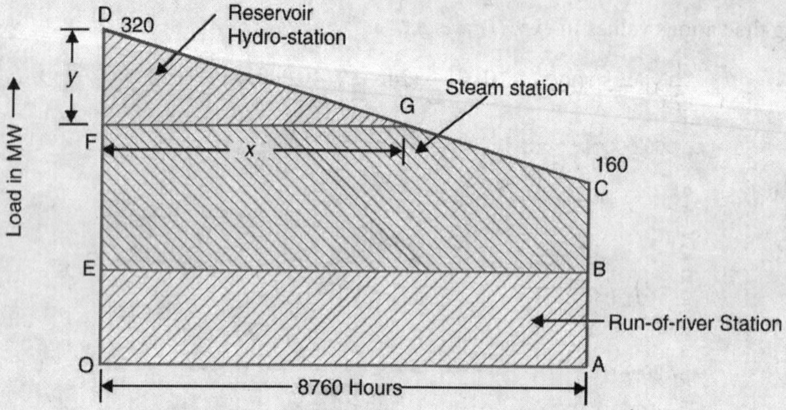

Fig. 3.16

$$\text{Units generated/annum} = \text{Area (in kWh) under annual load duration curve}$$

$$= 10^3\left[\frac{1}{2}(320+160)\times 8760\right]\text{kWh} = 2102\cdot4\times10^6\text{ kWh}$$

As the steam plant, run-of-river plant and hydro plant generate units in the ratio of 7 : 4 : 1, therefore, units generated by each plant are given by :

$$\text{Steam plant} = 2102\cdot4\times10^6\times7/12 = 1226\cdot4\times10^6\text{ kWh}$$

$$\text{Run-of-river plant} = 2102\cdot4\times10^6\times4/12 = 700\cdot8\times10^6\text{ kWh}$$

$$\text{Reservoir plant} = 2102\cdot4\times10^6\times1/12 = 175\cdot2\times10^6\text{ kWh}$$

(*i*) Maximum demand on run-of-river plant

$$= \frac{\text{Area } OEBA}{OA} = \frac{700\cdot8\times10^6}{8760} = \mathbf{80{,}000\ kW}$$

Suppose the maximum demand of reservoir plant is y MW and it operates for x hours (See Fig. 3.16).

Then, $$\frac{y}{160} = \frac{x}{8760} \quad\text{or}\quad x = \frac{8760\,y}{160}$$

Units generated per annum by reservoir plant

$$= \text{Area (in kWh) } DFG$$

$$= 10^3\left(\frac{1}{2}xy\right) = 10^3\left(\frac{1}{2}\times\frac{8760\,y}{160}y\right)$$

$$= \frac{y^2}{32}\times8{,}76{,}000$$

But the units generated by reservoir plant are $175\cdot2\times10^6$ kWh.

$$\therefore \quad \frac{y^2}{32}\times8{,}76{,}000 = 175\cdot2\times10^6$$

$$y^2 = 6400 \quad \text{or} \quad y = \sqrt{6400} = 80 \text{ MW}$$

∴ Maximum demand on reservoir station is

$$FD = \mathbf{80\ MW}$$

Maximum demand on steam station is

$$EF = 320 - 80 - 80 = \mathbf{160\ MW}$$

(*ii*) L.F. of run of river plant = **100* %**

$$\text{L.F. of reservoir plant} = \frac{\text{Units generated / annum}}{\text{Maximum demand} \times 8760} \times 100$$

$$= \frac{175 \cdot 2 \times 10^6}{(80 \times 10^3) \times 8760} \times 100 = \mathbf{25\%}$$

$$\text{L.F. of steam plant} = \frac{1226 \cdot 4 \times 10^6}{(160 \times 10^3) \times 8760} \times 100 = \mathbf{87.5\%}$$

SELF - TEST

1. Fill in the blanks by inserting appropriate words/figures :

(*i*) The area under the daily load curve gives ....
(*ii*) The connected load is generally .... than the maximum demand.
(*iii*) The value of demand factor is .... than 1.
(*iv*) The higher the load factor of a power station, the .... is the cost per unit generated.
(*v*) The value of diversity factor is .... than 1.
(*vi*) The lesser the diversity factor, the .... is the cost of generation of power.
(*vii*) A generating unit operates with maximum efficiency at about .... % of its rated capacity.
(*viii*) According to Indian Electricity Supply Act (1948), the capacity of the spare set should be ....
(*ix*) In an annual load curve, ... is taken along *Y*-axis and .... along *X*-axis.
(*x*) Base load occurs on the power station for .... hours in a day.

2. Pick up the correct words/figures from the brackets and fill in the blanks :

(*i*) Area under the daily load curve divided by 24 gives .... *(average load, maximum demand, units generated)*
(*ii*) The knowledge of diversity factor helps in determining .... *(average load, units generated, plant capacity)*
(*iii*) More efficient plants are used as .... *(base load stations, peak load stations)*
(*iv*) A diesel power plant is generally used as a .... *(base load station, peak load station)*
(*v*) In a hydro-steam system, steam power station carries the base load during .... *(high flow day, low flow day)*
(*vi*) In an interconnected grid system, the diversity factor of the whole system .... *(increases, decreases, remains constant)*
(*vii*) Installed capacity of a power station is .... then the maximum demand. *(less, more)*
(*viii*) Annual load factor is determined from .... load curve. *(daily, monthly, annual)*

ANSWERS TO SELF-TEST

1. (*i*) units generated in the day (*ii*) more (*iii*) less (*iv*) lesser (*v*) more (*vi*) greater (*vii*) 75% (*viii*) highest of all sets (*ix*) load, hours (*x*) 24.
2. (*i*) average load (*ii*) plant capacity (*iii*) base load stations (*iv*) peak load station (*v*) low flow day (*vi*) increases (*vii*) more (*viii*) annual.

* Since it operates continuously at rated capacity (*i.e.* it is a base load station).

CHAPTER REVIEW TOPICS

1. Why is the load on a power station variable ? What are the effects of variable load on the operation of the power station ?
2. What do you understand by the load curve ? What informations are conveyed by a load curve ?
3. Define and explain the importance of the following terms in generation :
   (*i*) connected load (*ii*) maximum demand (*iii*) demand factor (*iv*) average load.
4. Explain the terms load factor and divesity factor. How do these factors influence the cost of generation?
5. Explain how load curves help in the selection of size and number of generating units.
6. Discuss the important points to be taken into consideration while selecting the size and number of units.
7. What do you understand by (*i*) base load and (*ii*) peak load of a power station ?
8. Discuss the method of meeting the peak load of an electrified area.
9. Discuss the advantages of interconnected grid system.
10. Write short notes on the following :
    (*i*) load curves,
    (*ii*) load division on hydro-steam system,
    (*iii*) load factor,
    (*iv*) plant capacity factor,

DISCUSSION QUESTIONS

1. Why are load curves drawn ?
2. How will you improve the diversity factor of a power station ?
3. What is the importance of load factor ?
4. What is the importance of diversity factor ?
5. The values of demand factor and load factor are always less than 1. Why ?

CHAPTER

4

Economics of Power Generation

Introduction

A power station is required to deliver power to a large number of consumers to meet their requirements. While designing and building a power station, efforts should be made to achieve overall economy so that the per unit cost of production is as low as possible. This will enable the electric supply company to sell electrical energy at a profit and ensure reliable service. The problem of determining the cost of production of electrical energy is highly complex and poses a challenge to power engineers. There are several factors which influence the production cost such as cost of land and equipment, depreciation of equipment, interest on capital investment etc. Therefore, a careful study has to be made to calculate the cost of production. In this chapter, we shall focus our attention on the various aspects of economics of power generation.

4.1 Economics of Power Generation

The art of determining the per unit (i.e., one kWh) cost of production of electrical energy is known as **economics of power generation.**

The economics of power generation has assumed a great importance in this fast developing power plant engineering. A consumer will use electric power only if it is supplied at reasonable rate. Therefore, power engineers have to find convenient methods to produce electric power as cheap as possible so that consumers are tempted to use electrical methods. Before passing on to the subject further, it is desirable that the readers get themselves acquainted with the following terms much used in the economics of power generation :

(*i*) **Interest.** *The cost of use of money is known as* **interest.**

A power station is constructed by investing a huge capital. This money is generally borrowed from banks or other financial institutions and the supply company has to pay the annual interest on this amount. Even if company has spent out of its reserve funds, the interest must be still allowed for, since this amount could have earned interest if deposited in a bank. Therefore, while calculating the cost of production of electrical energy, the interest payable on the capital investment must be included. The rate of interest depends upon market position and other factors, and may vary from 4% to 8% per annum.

(*ii*) **Depreciation.** *The decrease in the value of the power plant equipment and building due to constant use is known as* **depreciation.**

If the power station equipment were to last for ever, then interest on the capital investment would have been the only charge to be made. However, in actual practice, every power station has a useful life ranging from fifty to sixty years. From the time the power station is installed, its equipment steadily deteriorates due to wear and tear so that there is a gradual reduction in the value of the plant. This reduction in the value of plant every year is known as *annual depreciation*. Due to depreciation, the plant has to be replaced by the new one after its useful life. Therefore, suitable amount must be set aside every year so that by the time the plant retires, the collected amount by way of depreciation equals the cost of replacement. It becomes obvious that while determining the cost of production, annual depreciation charges must be included. There are several methods of finding the annual depreciation charges and are discussed in Art. 4.4.

4.2 Cost of Electrical Energy

The total cost of electrical energy generated can be divided into three parts, namely ;

(*i*) Fixed cost ; (*ii*) Semi-fixed cost ; (*iii*) Running or operating cost.

(*i*) **Fixed cost.** *It is the cost which is independent of maximum demand and units generated.*

The fixed cost is due to the *annual cost of central organisation*, *interest on capital cost of land* and *salaries of high officials*. The annual expenditure on the central organisation and salaries of high officials is fixed since it has to be met whether the plant has high or low maximum demand or it generates less or more units. Further, the capital investment on the land is fixed and hence the amount of interest is also fixed.

(*ii*) **Semi-fixed cost.** *It is the cost which depends upon maximum demand but is independent of units generated.*

The semi-fixed cost is directly proportional to the maximum demand on power station and is on account of *annual interest and depreciation on capital investment of building and equipment, taxes, salaries of management and clerical staff*. The maximum demand on the power station determines its size and cost of installation. The greater the maximum demand on a power station, the greater is its size and cost of installation. Further, the taxes and clerical staff depend upon the size of the plant and hence upon maximum demand.

(*iii*) **Running cost.** *It is the cost which depends only upon the number of units generated.*

The running cost is on account of *annual cost of fuel, lubricating oil, maintenance, repairs* and *salaries of operating staff*. Since these charges depend upon the energy output, the running cost is directly proportional to the number of units generated by the station. In other words, if the power station generates more units, it will have higher running cost and *vice-versa*.

4.3 Expressions for Cost of Electrical Energy

The overall annual cost of electrical energy generated by a power station can be expressed in two forms *viz three part form* and *two part form*.

(*i*) **Three part form.** In this method, the overall annual cost of electrical energy generated is divided into three parts *viz* fixed cost, semi-fixed cost and running cost *i.e.*

Total annual cost of energy = Fixed cost + Semi-fixed cost + Running cost

= Constant + Proportional to max. demand + Proportional to kWh generated.

= Rs $(a + b \text{ kW} + c \text{ kWh})$

where a = annual fixed cost independent of maximum demand and energy output. It is on account of the costs mentioned in Art. 4.2.

b = constant which when multiplied by maximum kW demand on the station gives the annual semi-fixed cost.

c = a constant which when multiplied by kWh output per annum gives the annual running cost.

(*ii*) **Two part form.** It is sometimes convenient to give the annual cost of energy in two part form. In this case, the annual cost of energy is divided into two parts *viz.,* a fixed sum per kW of maximum demand *plus* a running charge per unit of energy. The expression for the annual cost of energy then becomes :

Total annual cost of energy = Rs. $(A \text{ kW} + B \text{ kWh})$

where A = a constant which when multiplied by maximum kW demand on the station gives the annual cost of the first part.

B = a constant which when multiplied by the annual kWh generated gives the annual running cost.

It is interesting to see here that two-part form is a simplification of three-part form. A little reflection shows that constant "*a*" of the three part form has been merged in fixed sum per kW maximum demand (*i.e.* constant *A*) in the two-part form.

4.4 Methods of Determining Depreciation

There is reduction in the value of the equipment and other property of the plant every year due to depreciation. Therefore, a suitable amount (known as *depreciation charge*) must be set aside annually so that by the time the life span of the plant is over, the collected amount equals the cost of replacement of the plant.

The following are the commonly used methods for determining the annual depreciation charge :

(*i*) Straight line method ;

(*ii*) Diminishing value method ;

(*iii*) Sinking fund method.

(*i*) **Straight line method.** In this method, a constant depreciation charge is made every year on the basis of total depreciation and the useful life of the property. Obviously, annual depreciation charge will be equal to the total depreciation divided by the useful life of the property. Thus, if the initial cost of equipment is Rs 1,00,000 and its scrap value is Rs 10,000 after a useful life of 20 years, then,

$$\text{Annual depreciation charge} = \frac{\text{Total depreciation}}{\text{Useful life}} = \frac{1{,}00{,}000 - 10{,}000}{20} = \text{Rs } 4{,}500$$

In general, the annual depreciation charge on the straight line method may be expressed as :

$$\text{Annual depreciation charge} = \frac{P - S}{n}$$

where P = Initial cost of equipment

n = Useful life of equipment in years

S = Scrap or salvage value after the useful life of the plant.

The straight line method is extremely simple and is easy to apply as the annual depreciation charge can be readily calculated from the total depreciation and useful life of the equipment. Fig. 4.1 shows the graphical representation of the method. It is clear that initial value P of the equipment reduces uniformly, through depreciation, to the scrap value S in the useful life of the equipment. The depreciation curve (PA) follows a straight line path, indicating constant annual depreciation charge. However, this method suffers from two defects. Firstly, the assumption of constant depreciation charge every year is not correct. Secondly, it does not account for the interest which may be drawn during accumulation.

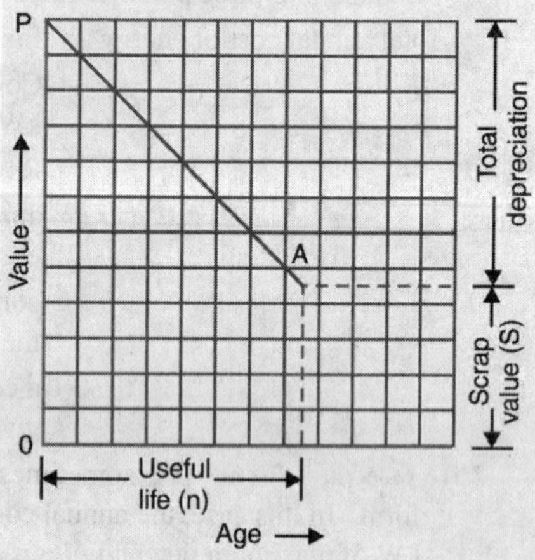

Fig. 4.1

(*ii*) **Diminishing value method.** In this method, depreciation charge is made every year at a fixed rate on the diminished value of the equipment. In other words, depreciation charge is first applied to the initial cost of equipment and then to its diminished value. As an example, suppose the initial cost of equipment is Rs 10,000 and its scrap value after the useful life is zero. If the annual rate of depreciation is 10%, then depreciation charge for the first year will be $0{\cdot}1 \times 10{,}000$ = Rs 1,000. The value of the equipment is diminished by Rs 1,000 and becomes Rs 9,000. For the second year, the depreciation charge will be made on the diminished value (*i.e.* Rs 9,000) and becomes $0{\cdot}1 \times 9{,}000$ = Rs 900. The value of the equipment now becomes 9000 − 900 = Rs 8100. For the third year, the depreciation charge will be $0{\cdot}1 \times 8100$ = Rs 810 and so on.

Mathematical treatment

Let P = Capital cost of equipment

n = Useful life of equipment in years

S = Scrap value after useful life

Suppose the annual unit* depreciation is x. It is desired to find the value of x in terms of P, n and S.

Value of equipment after one year

$$= P - Px = P(1 - x)$$

Value of equipment after 2 years

$$= \text{Diminished value} - \text{Annual depreciation}$$
$$= [P - Px] - [(P - Px)x]$$
$$= P - Px - Px + Px^2$$
$$= P(x^2 - 2x + 1)$$
$$= P(1 - x)^2$$

∴ Value of equipment after n years

$$= P(1 - x)^n$$

* If annual depreciation is 10%, then we can say that annual unit depreciation is 0·1.

But the value of equipment after n years (*i.e.*, useful life) is equal to the scrap value S.

$$\therefore \quad S = P(1-x)^n$$

or $$(1-x)^n = S/P$$

or $$1 - x = (S/P)^{1/n}$$

or $$x = 1 - (S/P)^{1/n} \quad ...(i)$$

From exp. (*i*), the annual depreciation can be easily found. Thus depreciation to be made for the first year is given by :

$$\text{Depreciation for the first year} = xP$$
$$= P[1 - (S/P)^{1/n}]$$

Similarly, annual depreciation charge for the subsequent years can be calculated.

This method is more rational than the straight line method. Fig. 4.2 shows the graphical representation of diminishing value method. The initial value P of the equipment reduces, through depreciation, to the scrap value S over the useful life of the equipment. The depreciation curve follows the path PA. It is clear from the curve that depreciation charges are heavy in the early years but decrease to a low value in the later years. This method has two drawbacks. Firstly, low depreciation charges are made in the late years when the maintenance and repair charges are quite heavy. Secondly, the depreciation charge is independent of the rate of interest which it may draw during accumulation. Such interest moneys, if earned, are to be treated as income.

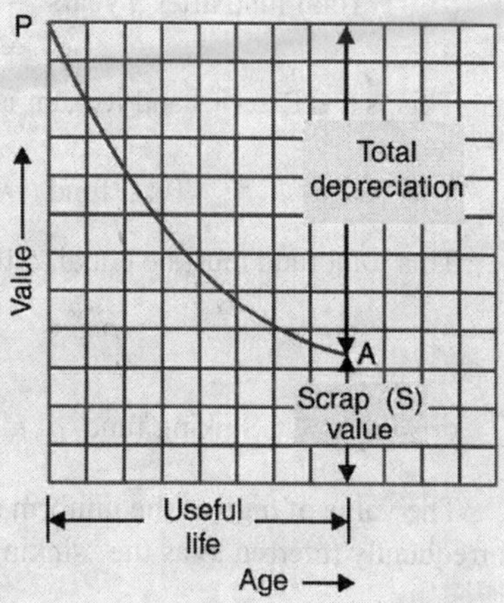

Fig. 4.2

(*iii*) Sinking fund method. In this method, a fixed depreciation charge is made every year and interest compounded on it annually. The constant depreciation charge is such that total of annual instalments plus the interest accumulations equal to the cost of replacement of equipment after its useful life.

Let
P = Initial value of equipment
n = Useful life of equipment in years
S = Scrap value after useful life
r = Annual rate of interest expressed as a decimal

$$\text{Cost of replacement} = P - S$$

Let us suppose that an amount of q is set aside as depreciation charge every year and interest compounded on it so that an amount of $P - S$ is available after n years. An amount q at annual interest rate of r will become *$q(1+r)^n$ at the end of n years.

Now, the amount q deposited at the end of first year will earn compound interest for $n-1$ years and shall become $q(1+r)^{n-1}$ *i.e.*,

Amount q deposited at the end of first year becomes

$$= q(1+r)^{n-1}$$

* This can be easily proved.

At the end of first year, amount is $= q + rq = q(1+r)$

At the end of second year, amount is $= (q+rq) + r(q+rq) = q + rq + rq + r^2q$

Similarly, at the end of n years, amount is $= q(1+r)^n$

Amount q deposited at the end of 2nd year becomes

$$= q\,(1+r)^{n-2}$$

Amount q deposited at the end of 3rd year becomes

$$= q\,(1+r)^{n-3}$$

Similarly amount q deposited at the end of $n-1$ year becomes

$$= q\,(1+r)^{n-(n-1)}$$
$$= q\,(1+r)$$

$$\therefore \quad \text{Total fund after } n \text{ years} = q\,(1+r)^{n-1} + q\,(1+r)^{n-2} + ... + q\,(1+r)$$
$$= q\,[(1+r)^{n-1} + (1+r)^{n-2} + ... + (1+r)]$$

This is a G.P. series and its sum is given by :

$$\text{Total fund} = \frac{q(1+r)^n - 1}{r}$$

This total fund must be equal to the cost of replacement of equipment *i.e.*, $P - S$.

$$\therefore \quad P - S = q\frac{(1+r)^n - 1}{r}$$

$$\text{or} \quad \text{Sinking fund, } q = (P-S)\left[\frac{r}{(1+r)^n - 1}\right] \quad ...(i)$$

The value of q gives the uniform annual depreciation charge. The paraenthetical term in eq. (*i*) is frequently referred to as the "sinking fund factor".

$$\therefore \quad \text{Sinking fund factor} = \frac{r}{(1+r)^n - 1}$$

Though this method does not find very frequent application in practical depreciation accounting, it is the fundamental method in making economy studies.

Example 4.1. *A transformer costing Rs 90,000 has a useful life of 20 years. Determine the annual depreciation charge using straight line method. Assume the salvage value of the equipment to be Rs 10,000.*

Solution :

Initial cost of transformer, P = Rs 90,000

Useful life, n = 20 years

Salvage value, S = Rs 10,000

Using straight line method,

$$\text{Annual depreciation charge} = \frac{P-S}{n} = \text{Rs}\ \frac{90{,}000 - 10{,}000}{20} = \textbf{Rs 4000}$$

Example 4.2. *A distribution transformer costs Rs 2,00,000 and has a useful life of 20 years. If the salvage value is Rs 10,000 and rate of annual compound interest is 8%, calculate the amount to be saved annually for replacement of the transformer after the end of 20 years by sinking fund method.*

Solution :

Initial cost of transformer, P = Rs 2,00,000

Salvage value of transformer, S = Rs 10,000

Useful life, n = 20 years

Annual interest rate, r = 8% = 0·08

Annual payment for sinking fund,

$$q = (P - S)\left[\frac{r}{(1+r)^n - 1}\right]$$

$$= (2{,}00{,}000 - 10{,}000)\left[\frac{0\cdot 08}{(1+0\cdot 08)^{20} - 1}\right]$$

$$= 1{,}90{,}000\left[\frac{0\cdot 08}{4\cdot 66 - 1}\right] = \textbf{Rs 4153}$$

Example 4.3. *The equipment in a power station costs Rs 15,60,000 and has a salvage value of Rs 60,000 at the end of 25 years. Determine the depreciated value of the equipment at the end of 20 years on the following methods :*

(*i*) Straight line method ;

(*ii*) Diminishing value method ;

(*iii*) Sinking fund method at 5% compound interest annually.

Solution :

Initial cost of equipment, P = Rs 15,60,000

Salvage value of equipment, S = Rs 60,000

Useful life, n = 25 years

(*i*) *Straight line method*

$$\text{Annual depreciation} = \frac{P-S}{n} = \text{Rs } \frac{15{,}60{,}000 - 60{,}000}{25} = \text{Rs } 60{,}000$$

Value of equipment after 20 years

$$= P - \text{Annual depreciation} \times 20$$

$$= 15{,}60{,}000 - 60{,}000 \times 20 = \textbf{Rs 3,60,000}$$

(*ii*) *Diminishing value method*

$$\text{Annual unit depreciation, } x = 1 - (S/P)^{1/n}$$

$$= 1 - \left(\frac{60{,}000}{15{,}60{,}000}\right)^{1/25} = 1 - 0\cdot 878 = 0\cdot 122$$

Value of equipment after 20 years

$$= P(1-x)^{20}$$

$$= 15{,}60{,}000\,(1 - 0\cdot 122)^{20} = \textbf{Rs 1,15,615}$$

(*iii*) *Sinking fund method*

Rate of interest, r = 5% = 0·05

Annual deposit in the sinking fund is

$$q = (P - S)\left[\frac{r}{(1+r)^n - 1}\right]$$

$$= (15{,}60{,}000 - 60{,}000)\left[\frac{0\cdot 05}{(1+0\cdot 05)^{25} - 1}\right]$$

$$= \text{Rs } 31{,}433$$

∴ Sinking fund at the end of 20 years

$$= q\frac{(1+r)^{20} - 1}{r} = 31{,}433\,\frac{(1+0\cdot 05)^{20} - 1}{0\cdot 05} = \text{Rs } 10{,}39{,}362$$

Value of plant after 20 years = Rs (15,60,000 – 10,39,362) = **Rs 5,20,638**

4.5 Importance of High Load Factor

The load factor plays a vital role in determining the cost of energy. Some important advantages of high load factor are listed below :

(*i*) *Reduces cost per unit generated* : A high load factor reduces the overall cost per unit generated. The higher the load factor, the lower is the generation cost. It is because higher load factor means that for a given maximum demand, the number of units generated is more. This reduces the cost of generation.

(*ii*) *Reduces variable load problems* : A high load factor reduces the variable load problems on the power station. A higher load factor means comparatively less variations in the load demands at various times. This avoids the frequent use of regulating devices installed to meet the variable load on the station.

Example 4.4. *A generating station has a maximum demand of 50,000 kW. Calculate the cost per unit generated from the following data :*

Capital cost = Rs 95 × 10^6 *;* *Annual load factor = 40%*

Annual cost of fuel and oil = Rs 9 × 10^6 *;* *Taxes, wages and salaries etc. = Rs 7·5 × 10*^{6}

Interest and depreciation = 12%

Solution :

$$\text{Units generated/annum} = \text{Max. demand} \times \text{L.F.} \times \text{Hours in a year}$$

$$= (50{,}000) \times (0{\cdot}4) \times (8760)\ \text{kWh} = 17{\cdot}52 \times 10^7\ \text{kWh}$$

Annual fixed charges

$$\text{Annual interest and depreciation} = 12\%\ \text{of capital cost}$$

$$= \text{Rs}\ 0{\cdot}12 \times 95 \times 10^6 = \text{Rs}\ 11{\cdot}4 \times 10^6$$

Annual Running Charges

$$\text{Total annual running charges} = \text{Annual cost of fuel and oil + Taxes, wages etc.}$$

$$= \text{Rs}\ (9 \times 10^6 + 7{\cdot}5 \times 10^6) = \text{Rs}\ 16{\cdot}5 \times 10^6$$

$$\text{Total annual charges} = \text{Rs}\ (11{\cdot}4 \times 10^6 + 16{\cdot}5 \times 10^6) = \text{Rs}\ 27{\cdot}9 \times 10^6$$

$$\therefore \quad \text{Cost per unit} = \text{Rs}\ \frac{27{\cdot}9 \times 10^6}{17{\cdot}52 \times 10^7} = \text{Re}\ 0{\cdot}16 = \textbf{16 paise}$$

Example 4.5. *A generating station has an installed capacity of 50,000 kW and delivers 220 × 10*^{6} *units per annum. If the annual fixed charges are Rs 160 per kW installed capacity and running charges are 4 paise per kWh, determine the cost per unit generated.*

Solution :

$$\text{Annual fixed charges} = 160 \times \text{Plant capacity}$$

$$= \text{Rs}\ 160 \times 50{,}000 = \text{Rs}\ 80 \times 10^5$$

$$\text{Annual running charges} = \text{Rs}\ 0{\cdot}04 \times 220 \times 10^6 = \text{Rs}\ 88 \times 10^5$$

$$\text{Total annual charges} = \text{Rs}\ (80 \times 10^5 + 88 \times 10^5) = \text{Rs}\ 168 \times 10^5$$

$$\text{Cost per unit} = \text{Rs}\ \frac{168 \times 10^5}{220 \times 10^6} = \text{Re}\ 0{\cdot}0764 = \textbf{7·64 paise}$$

Example 4.6. *A generating plant has a maximum capacity of 100 kW and costs Rs 1,60,000. The annual fixed charges are 12% consisting of 5% interserst, 5% depreciation and 2% taxes. Find the fixed charges per kWh if the load factor is (i) 100% and (ii) 50%.*

Solution :

$$\text{Maximum demand} = 100\ \text{kW}$$

$$\text{Annual fixed charges} = \text{Rs}\ 0{\cdot}12 \times 1{,}60{,}000 = \text{Rs}\ 19{,}200$$

(*i*) *When load factor is 100%*

$$\text{Units generated/annum} = \text{Max. demand} \times \text{L.F.} \times \text{Hours in a year}$$
$$= 100 \times 1 \times 8760 = 8{,}76{,}000 \text{ kWh}$$
$$\text{Fixed charges/kWh} = \text{Rs } \frac{19{,}200}{8{,}76{,}000} = \text{Rs } 0{\cdot}0219 = \mathbf{2{\cdot}19 \text{ paise}}$$

(*ii*) *When load factor is 50%*

$$\text{Units generated/annum} = 100 \times 0{\cdot}5 \times 8760 = 4{,}38{,}000 \text{ kWh}$$
$$\text{Fixed charges/kWh} = \text{Rs } \frac{19{,}200}{4{,}38{,}000} = \text{Re } 0{\cdot}0438 = \mathbf{4{\cdot}38 \text{ paise}}$$

It is interesting to note that by decreasing the load factor from 100% to 50%, the fixed charges/kWh have increased two-fold. Incidentally, this illustrates the utility of high load factor.

Example 4.7. *Estimate the generating cost per kWh delivered from a generating station from the following data :*

Plant capacity = 50 MW ; Annual load factor = 40%

Capital cost = 1·2 crores ; annual cost of wages, taxation etc. = Rs 4 lakhs ; cost of fuel, lubrication, maintenance etc. = 1·0 paise/kWh generated. Interest 5% per annum, depreciation 6% per annum of initial value.

Solution : The maximum demand on the station may be assumed equal to the plant capacity *i.e.*, 50 MW.

Annual fixed charges

$$\text{Interest and depreciation} = \text{Rs } 120 \times 10^5 \times (5 + 6)/100 = \text{Rs } 13{\cdot}2 \times 10^5$$
$$\text{Wages and taxation} = \text{Rs } 4 \times 10^5$$
$$\text{Total annual fixed charges} = \text{Rs } (13{\cdot}2 \times 10^5 + 4 \times 10^5) = \text{Rs } 17{\cdot}2 \times 10^5$$

Annual running charges

$$\text{Units generated/annum} = \text{Max. demand} \times \text{L.F.} \times \text{Hours in a year}$$
$$= (50 \times 10^3) \times (0{\cdot}4) \times (8760) \text{ kWh}$$
$$= 1752 \times 10^5 \text{ kWh}$$
$$\text{Cost of fuel, lubrication etc.} = \text{Rs } 1752 \times 10^5 \times 0{\cdot}01 = \text{Rs } 17{\cdot}52 \times 10^5$$
$$\text{Total annual charges} = \text{Rs } (17{\cdot}2 \times 10^5 + 17{\cdot}52 \times 10^5) = \text{Rs } 34{\cdot}72 \times 10^5$$
$$\therefore \quad \text{Cost per kWh} = \text{Rs} \frac{34{\cdot}72 \times 10^5}{1752 \times 10^5} = \text{Re } 0{\cdot}02 = \mathbf{2 \text{ paise}}$$

Example 4.8. *A generating station has the following data :*

Installed capacity = 300 MW ; Capacity factor = 50% ; Annual load factor = 60%

Annual cost of fuel, oil etc. = Rs 9×10^7 *; capital cost = Rs* 10^9 *; annual interest and depreciation = 10%. Calculate* (*i*) *the minimum reserve capacity of the station and* (*ii*) *the cost per kWh generated.*

Solution :

(*i*)
$$\text{Capacity factor, C.F.} = \frac{\text{Average demand}}{\text{Installed capacity}} \quad \text{...(i)}$$
$$\text{Load factor, L.F.} = \frac{\text{Average demand}}{\text{Max. demand}} \quad \text{...(ii)}$$

Dividing (*i*) by (*ii*), we get,

$$\frac{\text{C.F.}}{\text{L.F.}} = \frac{\text{Max. demand}}{\text{Installed capacity}}$$

or $\text{Max. demand} = \text{Installed capacity} \times \frac{\text{C.F.}}{\text{L.F.}} = 300 \times \frac{0{\cdot}5}{0{\cdot}6} = 250 \text{ MW}$

$\therefore$ $\text{Reserve capacity} = 300 - 250 = \mathbf{50\ MW}$

(*ii*) $\text{Units generated/annum} = \text{Max. demand} \times \text{L.F.} \times \text{Hours in a year}$

$= (250 \times 10^3) \times (0{\cdot}6) \times 8760 \text{ kWh} = 1314 \times 10^6 \text{ kWh}$

$\text{Annual fixed charges} = \text{Annual interest and depreciation}$

$= \text{Rs } 0{\cdot}1 \times 10^9 = \text{Rs } 10^8$

$\text{Annual running charges} = \text{Rs } 9 \times 10^7$

$\therefore$ $\text{Total annual charges} = \text{Rs } (10^8 + 9 \times 10^7) = \text{Rs } 19 \times 10^7$

$\therefore$ $\text{Cost per kWh} = \text{Rs } \frac{19 \times 10^7}{1314 \times 10^6} = \text{Re } 0{\cdot}14 = \mathbf{14\ paise}$

Example 4.9. *The capital cost of a hydro-power station of 50 MW capacity is Rs 1,000 per kW. The annual depreciation charges are 10% of the capital cost. A royalty of Re 1 per kW per year and Re 0·01 per kWh generated is to be paid for using the river water for generation of power. The maximum demand on the power station is 40 MW and annual load factor is 60%. Annual cost of salaries, maintenance charges etc. is Rs 7,00,000. If 20% of this expense is also chargeable as fixed charges, calculate the generation cost in two part form.*

Solution :

$\text{Units generated/annum} = (40 \times 10^3) \times (0{\cdot}6) \times 8760 = 210{\cdot}24 \times 10^6 \text{ kWh}$

$\text{Capital cost of plant} = \text{Rs } 50 \times 10^3 \times 1000 = \text{Rs } 50 \times 10^6$

Annual fixed charges

$\text{Depreciation} = \text{Rs } 0{\cdot}1 \times 50 \times 10^6 = \text{Rs } 5 \times 10^6$

$\text{Salaries, maintenance etc.} = \text{Rs } 0{\cdot}2 \times 7{,}00{,}000 = \text{Rs } 1{\cdot}4 \times 10^5$

$\text{Total annual fixed charges} = \text{Rs } (5 \times 10^6 + 1{\cdot}4 \times 10^5) = \text{Rs } 51{\cdot}4 \times 10^5$

$\text{Cost per kW} = \text{Cost per kW due to fixed charges} + \text{Royalty}$

$= \text{Rs } \frac{51{\cdot}4 \times 10^5}{40 \times 10^3} + \text{Re } 1 = \text{Rs } 128{\cdot}5 + \text{Re } 1 = \text{Rs } 129{\cdot}5$

Annual running charges

$\text{Salaries, maintenance etc.} = \text{Rs } 0{\cdot}8 \times 7{,}00{,}000 = \text{Rs } 5{\cdot}6 \times 10^5$

$\text{Cost per kWh} = \text{Cost/kWh due to running charges} + \text{Royalty}$

$= \text{Rs } \frac{5{\cdot}6 \times 10^5}{210{\cdot}24 \times 10^6} + \text{Re } 0{\cdot}01$

$= \text{Re } 0{\cdot}0027 + \text{Re } 0{\cdot}01 = \text{Re } 0{\cdot}0127$

$\therefore$ Total generation cost in two part form is given by ;

$$\mathbf{Rs\ (129{\cdot}5 \times kW + 0{\cdot}0127 \times kWh)}$$

Example 4.10. *The annual working cost of a power station is represented by the formula Rs (a + b kW + c kWh) where the various terms have their usual meaning. Determine the values of a, b and c for a 60 MW station operating at annual load factor of 50% from the following data :*

(*i*) *capital cost of building and equipment is Rs 5 × 10*^{6}

(*ii*) *the annual cost of fuel, oil, taxation and wages of operating staff is Rs 9,00,000*

(*iii*) *the interest and depreciation on building and equipment are 10% per annum*

(*iv*) *annual cost of organisation and interest on cost of site etc. is Rs 5,00,000.*

Solution :

$$\text{Units generated/annum} = \text{Max. demand} \times \text{L.F.} \times \text{Hours in a year}$$
$$= (60 \times 10^3) \times (0{\cdot}5) \times (8760) \text{ kWh} = 262{\cdot}8 \times 10^6 \text{ kWh}$$
$$\text{Annual operating cost} = \text{Rs } (a + b \times \text{kW} + c \times \text{kWh})$$

where
$$a = \text{annual fixed cost}$$
$$b \times \text{kW} = \text{annual semi-fixed cost}$$
$$c \times \text{kWh} = \text{annual running cost}$$

Annual fixed cost. The annual fixed cost is due to the annual cost of organisation and interest on the cost of site.

$$\therefore \quad a = \textbf{Rs 5,00,000}$$

Annual semi-fixed cost. This is on account of annual interest and depreciation on building and equipment.

$$\text{Annual semi-fixed cost} = \text{Rs } 0{\cdot}1 \times 5 \times 10^6 = \text{Rs } 5{,}00{,}000$$

But annual semi-fixed cost is equal to $b \times$ kW where b is the cost per kW of maximum demand.

$$\therefore \quad b \times 60 \times 10^3 = \text{Rs } 5{,}00{,}000$$

or
$$b = \text{Rs } \frac{5{,}00{,}000}{60 \times 10^3} = \textbf{Rs 8·34}$$

Annual running cost. This is due to the annual cost of fuel, oil, taxation, salaries of operating staff.

$$\therefore \quad c \times \text{kWh generated} = \text{Rs } 9{,}00{,}000$$

or
$$c \times 262{\cdot}8 \times 10^6 = \text{Rs } 9{,}00{,}000$$
$$\therefore \quad c = \text{Rs } \frac{9{,}00{,}000}{262{\cdot}8 \times 10^6} = \textbf{Re 0·0034}$$

Example 4.11. *A hydro-electric plant costs Rs 3000 per kW of installed capacity. The total annual charges consist of 5% as interest ; depreciation at 2%, operation and maintenance at 2% and insurance, rent etc. 1·5%. Determine a suitable two-part tariff if the losses in transmission and distribution are 12·5% and diversity of load is 1·25. Assume that maximum demand on the station is 80% of the capacity and annual load factor is 40%. What is the overall cost of generation per kWh?*

Solution : Let the installed capacity of the station be 100 kW.

$$\text{Maximum demand} = 100 \times 0{\cdot}8 = 80 \text{ kW}$$
$$\text{Average demand} = 80 \times 0{\cdot}4 = 32 \text{ kW}$$
$$\text{Capital cost of plant} = \text{Rs } 100 \times 3000 = \text{Rs } 3 \times 10^5$$

Annual fixed charges. The annual fixed charges are due to interest (5%) and depreciation (2%).

$$\therefore \quad \text{Annual fixed charges} = \text{Rs } 3 \times 10^5 \times (5 + 2)/100 = \text{Rs } 21000$$
$$\text{Aggregate of max. demand} = 80 \times 1{\cdot}25 = 100 \text{ kW}$$
$$\therefore \quad \text{Annual fixed charges} = \text{Rs } 21000/100 = \text{Rs } 210 \text{ per kW of max. demand}$$

Annual running charges. The annual running charges are due to operation and maintenance (2%) and insurance, rent (1·5%) etc.

$$\text{Annual running charges} = \text{Rs } 3 \times 10^5 \times (2 + 1{\cdot}5)/100 = \text{Rs } 10{,}500$$
$$\text{Units generated/annum} = \text{Average demand} \times \text{Hours in a year}$$
$$= 32 \times 8760 = 2{,}80{,}320 \text{ kWh}$$
$$\text{Units reaching the consumer} = 2{,}80{,}320 \times 0{\cdot}875 = 2{,}45{,}280 \text{ kWh}$$
$$\therefore \quad \text{Annual running charge} = \text{Rs } \frac{10{,}500}{2{,}45{,}280} = \text{Re } 0{\cdot}043 \text{ per kWh}$$

The generation cost in two-part form is

$$\text{Rs } (210 \times \text{kW} + 0{\cdot}043 \times \text{kWh})$$

$$\text{Total annual charges} = \text{Rs } (21{,}000 + 10{,}500) = \text{Rs } 31{,}500$$

$$\text{Cost per kWh} = \text{Rs } \frac{31{,}500}{2{,}45{,}280} = \text{Re } 0{\cdot}128 = \mathbf{12{\cdot}8 \text{ paise}}$$

Example 4.12. *Compare the annual cost of supplying a factory load having a maximum demand of 1 MW and a load factor of 50% by energy obtained from (i) a private oil engine generating plant and (ii) public supply.*

(i) Private oil engine generating unit :

Capital cost = Rs 12 × 10^5 ; Cost of repair and maintenance = Rs 0·005 per kWh generated

Cost of fuel = Rs 1600 per 1000 kg ; Interest and depreciation = 10% per annum

Fuel consumption = 0·3 kg/kWh generated ; Wages = Rs 50,000 per annum

(ii) Public supply company :

Rs 150 per kW of maximum demand plus 15 paise per kWh

Solution :

$$\text{Units generated/annum} = (1000) \times (0{\cdot}5) \times 8760 = 438 \times 10^4 \text{ kWh}$$

(*i*) Private oil engine generating plant

$$\text{Annual fuel consumption} = 0{\cdot}3 \times 438 \times 10^4 = 13{\cdot}14 \times 10^5 \text{ kg}$$

$$\text{Annual cost of fuel} = \text{Rs } 13{\cdot}14 \times 10^5 \times 1600/1000 = \text{Rs } 21{,}02{,}400$$

$$\text{Annual cost of repair and maintenance} = \text{Rs } 0{\cdot}005 \times 438 \times 10^4 = \text{Rs } 21{,}900$$

$$\text{Annual wages} = \text{Rs } 50{,}000$$

$$\text{Annual interest and depreciation} = \text{Rs } 0{\cdot}1 \times 12 \times 10^5 = \text{Rs } 1{,}20{,}000$$

$$\therefore \quad \text{Total annual charges} = \text{Rs } (21{,}02{,}400 + 21{,}900 + 50{,}000 + 1{,}20{,}000)$$

$$= \mathbf{Rs\ 22{,}94{,}300}$$

(*ii*) Public supply

$$\text{Annual fixed charges} = \text{Rs } 150 \times 1000 = \text{Rs } 1{,}50{,}000$$

$$\text{Annual running charges} = \text{Rs } 0{\cdot}15 \times 438 \times 10^4 = \text{Rs } 6{,}57{,}000$$

$$\text{Total annual charges} = \text{Rs } (1{,}50{,}000 + 6{,}57{,}000) = \mathbf{Rs\ 8{,}07{,}000}$$

Example 4.13. *A power station having a maximum demand of 100 MW has a load factor of 30% and is to be supplied by one of the following schemes :*

(i) a steam station in conjunction with a hydro-electric station, the latter supplying 100 × 10^6 kWh per annum with a maximum output of 40 MW.

(ii) a steam station capable of supplying the whole load.

(iii) a hydro-station capable of supplying the whole load.

Compare the overall cost per kWh generated, assuming the following data :

	Steam	*Hydro*
(a) Capital cost/kW installed	*Rs 1250*	*Rs 2500*
(b) Interest and depreciation on capital investment	*12%*	*10%*
(c) Operating cost/kWh	*5 paise*	*1·5 paise*
(d) Transmission cost/kWh	*negligible*	*0·2 paise*

Solution :

$$\text{Units generated/annum} = \text{Max. demand} \times \text{L.F.} \times \text{Hours in a year}$$

$$= (100 \times 10^3) \times (0{\cdot}3) \times (8760) = 262{\cdot}8 \times 10^6 \text{ kWh}$$

(i) Steam station in conjunction with hydro station

Units supplied by hydro-station $= 100 \times 10^6$ kWh

$\therefore$ Units supplied by steam station $= (262{\cdot}8 - 100) \times 10^6 = 162{\cdot}8 \times 10^6$ kWh

Since the maximum output of hydro station is 40 MW, the balance 100 – 40 = 60 MW shall be supplied by steam station.

(a) *Steam Station*

$$\text{Capital Cost} = \text{Rs } 60 \times 10^3 \times 1250 = \text{Rs } 75 \times 10^6$$

$$\text{Annual interest and depreciation} = \text{Rs } 0{\cdot}12 \times 75 \times 10^6 = \text{Rs } 9 \times 10^6$$

$$\text{Operating Cost} = \text{Rs } 0{\cdot}05 \times 162{\cdot}8 \times 10^6 = \text{Rs } 8{\cdot}14 \times 10^6$$

$$\text{Transmission cost} = \text{negligible}$$

$$\therefore \quad \text{Total annual cost} = \text{Rs } (9 + 8{\cdot}14) \times 10^6 = \text{Rs } 17{\cdot}14 \times 10^6$$

(b) *Hydro station*

$$\text{Capital Cost} = \text{Rs } 2500 \times 40 \times 10^3 = \text{Rs } 100 \times 10^6$$

$$\text{Annual interest and depreciation} = \text{Rs } 0{\cdot}1 \times 100 \times 10^6 = \text{Rs } 10 \times 10^6$$

$$\text{Operating cost} = \text{Rs } 0{\cdot}015 \times 100 \times 10^6 = \text{Rs } 1{\cdot}5 \times 10^6$$

$$\text{Transmission cost} = \text{Rs } 0{\cdot}002 \times 100 \times 10^6 = \text{Rs } 0{\cdot}2 \times 10^6$$

$$\text{Total annual cost} = \text{Rs } (10 + 1{\cdot}5 + 0{\cdot}2) \times 10^6 = \text{Rs } 11{\cdot}7 \times 10^6$$

Total annual charges for both steam and hydro stations

$$= \text{Rs } (17{\cdot}14 + 11{\cdot}7) \times 10^6 = \text{Rs } 28{\cdot}84 \times 10^6$$

$$\therefore \quad \text{Overall cost/kWh} = \text{Rs } \frac{28{\cdot}84 \times 10^6}{262{\cdot}8 \times 10^6} = \text{Re } 0{\cdot}1097 = \mathbf{10{\cdot}97 \text{ paise}}$$

(ii) Steam station

$$\text{Capital cost} = \text{Rs } 1250 \times 100 \times 10^3 = \text{Rs } 125 \times 10^6$$

$$\text{Annual interest and depreciation} = \text{Rs } 0{\cdot}12 \times 125 \times 10^6 = \text{Rs } 15 \times 10^6$$

$$\text{Fixed charges/kWh} = \text{Rs } \frac{15 \times 10^6}{262{\cdot}8 \times 10^6} = \text{Re } 0{\cdot}0571 = 5{\cdot}71 \text{ paise}$$

$$\text{Operating cost/kWh} = 5 \text{ paise}$$

$$\text{Transmission cost/kWh} = \text{negligible}$$

$$\therefore \quad \text{Overall cost/kWh} = 5{\cdot}71 + 5 = \mathbf{10{\cdot}71 \text{ paise}}$$

(iii) Hydro station

$$\text{Capital cost} = \text{Rs } 2500 \times 100 \times 10^3 = \text{Rs } 250 \times 10^6$$

$$\text{Annual interest and depreciation} = \text{Rs } 0{\cdot}1 \times 250 \times 10^6 = \text{Rs } 25 \times 10^6$$

$$\therefore \quad \text{Fixed charges/kWh} = \text{Rs } \frac{25 \times 10^6}{262{\cdot}8 \times 10^6} = \text{Re } 0{\cdot}0951 = 9{\cdot}51 \text{ paise}$$

Adding the operating cost/unit and transmission cost per unit, we get,

$$\text{Overall cost/kWh} = 9{\cdot}51 + 1{\cdot}5 + 0{\cdot}2 = \mathbf{11{\cdot}21 \text{ paise}}$$

Example 4.14. *A load having a maximum value of 150 MW can be supplied either by a hydro-electric station or steam power plant. The costs are as follows :*

Plant	*Capital cost per kW installed*	*Operating cost per kWh*	*Interest*
Steam Plant	*Rs 1600*	*Re 0·06*	*7%*
Hydro Plant	*Rs 3000*	*Re 0·03*	*7%*

Calculate the minimum load factor above which the hydro-electric plant will be more economical.

Solution :

$$\text{Maximum demand} = 150 \text{ MW} = 150 \times 10^3 \text{ kW}$$

Let the total number of units generated per annum be x.

Steam plant

$$\text{Capital cost} = \text{Rs } 1600 \times 150 \times 10^3 = \text{Rs } 240 \times 10^6$$

$$\text{Annual interest} = \text{Rs } 0{\cdot}07 \times 240 \times 10^6 = \text{Rs } 16{\cdot}8 \times 10^6$$

$$\text{Fixed cost/unit} = \text{Rs } \frac{16{\cdot}8 \times 10^6}{x}$$

$$\text{Running cost/unit} = \text{Re } 0{\cdot}06 \text{ (given)}$$

$$\therefore \quad \text{Total cost/unit} = \text{Rs}\left(\frac{16{\cdot}8 \times 10^6}{x} + 0{\cdot}06\right) \quad \text{...(i)}$$

Hydro plant

$$\text{Capital cost} = \text{Rs } 3000 \times 150 \times 10^3 = \text{Rs } 450 \times 10^6$$

$$\text{Annual interest} = 0{\cdot}07 \times 450 \times 10^6 = \text{Rs } 31{\cdot}5 \times 10^6$$

$$\text{Fixed cost/unit} = \text{Re } \frac{31{\cdot}5 \times 10^6}{x}$$

$$\text{Running cost/unit} = \text{Re } 0{\cdot}03 \text{ (given)}$$

$$\therefore \quad \text{Total cost/unit} = \text{Rs}\left(\frac{31{\cdot}5 \times 10^6}{x} + 0{\cdot}03\right) \quad \text{...(ii)}$$

The overall cost per unit of steam plant will be equal to hydro plant if exp. (*i*) = exp (*ii*) *i.e.*,

$$\frac{16{\cdot}8 \times 10^6}{x} + 0{\cdot}06 = \frac{31{\cdot}5 \times 10^6}{x} + 0{\cdot}03$$

or $$16{\cdot}8 \times 10^6 + 0{\cdot}06\,x = 31{\cdot}5 \times 10^6 + 0{\cdot}03\,x$$

$$\therefore \quad x = \frac{14{\cdot}7 \times 10^6}{0{\cdot}03} = 490 \times 10^6 \text{ kWh}$$

It follows, therefore, that if the units generated per annum are more than 490×10^6, the hydro plant will be more economical.

$$\therefore \quad \text{Load factor} = \frac{490 \times 10^6}{(150 \times 10^3) \times 8760} \times 100 = \mathbf{37{\cdot}3\%}$$

Therefore, the minimum load factor above which the hydro plant will be economical is 37·3%.

Example 4.15. *A particular area can be supplied either by hydro station or steam station. The following data is available :*

	Hydro	*Steam*
Capital cost/kW	*Rs 2100*	*Rs 1200*
Running cost/kWh	*3·2 paise*	*5 paise*
Interest and depreciation	*7·5%*	*9%*
Reserve capacity	*33%*	*25%*

(*i*) *At what load factor would the overall cost be the same in both cases ?*

(*ii*) *What would be the cost of generating 40×10^6 units at this load factor ?*

Solution : Let x kW be the maximum demand. Let y be the annual load factor at which cost/unit of steam and hydro stations is the same.

$$\therefore \quad \text{Units generated/annum} = x \times y \times 8760 = 8760\,xy \text{ kWh}$$

(*i*) The installed capacity of steam station will be $1{\cdot}25x$ kW (keeping 25% as reserve capacity), whereas the installed capacity of hydro station would be $1{\cdot}33x$ kW (keeping 33% as reserve capacity).

Steam station

$$\text{Capital cost} = \text{Rs } 1200 \times 1{\cdot}25x = \text{Rs } 1500x$$

$$\text{Interest and depreciation} = \text{Rs } 0{\cdot}09 \times 1500x = \text{Rs } 135x$$

$$\text{Running cost/annum} = \text{Rs } 0{\cdot}05 \times 8760xy = \text{Rs } 438xy$$

$$\therefore \quad \text{Overall cost/kWh} = \text{Re } \frac{(135x + 438\,xy)}{8760\,xy} \qquad \text{...}(i)$$

Hydro station

$$\text{Capital cost} = \text{Rs } 2100 \times 1{\cdot}33x = \text{Rs } 2793x$$

$$\text{Interest and depreciation} = \text{Rs } 0{\cdot}075 \times 2793x = \text{Rs } 210x$$

$$\text{Running cost/kWh} = \text{Rs } 0{\cdot}032 \times 8760xy = \text{Rs } 280xy$$

$$\therefore \quad \text{Overall cost/kWh} = \text{Re } \frac{(210x + 280\,xy)}{8760\,xy} \qquad \text{...}(ii)$$

As the overall cost per unit is the same in each case, therefore, equating exps. (*i*) and (*ii*), we get,

$$\frac{(135x + 438\,xy)}{8760\,xy} = \frac{(210x + 280\,xy)}{8760\,xy}$$

or $$75\,x = 158\,xy$$

$\therefore$ $$\text{Load factor, } y = 75\,x/158\,x = 0{\cdot}4746 = \mathbf{47{\cdot}46\%}$$

(*ii*) $$\text{Units generated/annum} = 8760\,xy$$

or $$40 \times 10^6 = 8760 \times x \times 0{\cdot}4746$$

$\therefore$ $$\text{Max. demand, } x = \frac{40 \times 10^6}{8760 \times 0{\cdot}4746} = 9{\cdot}62 \times 10^3 \text{ kW}$$

$\therefore$ $$\text{Cost of generation} = \text{Rs } (135\,x + 438\,xy)$$

$$= \text{Rs } (135 \times 9{\cdot}62 \times 10^3 + 438 \times 9{\cdot}62 \times 10^3 \times 0{\cdot}4746)$$

$$= \text{Rs } (1298{\cdot}7 \times 10^3 + 2000 \times 10^3) = \mathbf{Rs\ 3298{\cdot}7 \times 10^3}$$

Example 4.16. *The load duration curve of a system for the whole year of 8760 hours is as shown in Fig. 4.3. The system is supplied by two stations A and B having the following annual costs:*

Station A = Rs (75,000 + 80 × kW + 0·02 × kWh)

Station B = Rs (50,000 + 50 × kW + 0·03 × kWh)

Determine the installed capacity required for each station and for how many hours per year peak load station should be operated to give the minimum cost per unit generated.

Solution : Fig. 4.3 shows the annual load duration curve of the system. As station *A* has the lower operating cost, it should be used as the base load station. On the other hand, station *B* should be used as the peak load station.

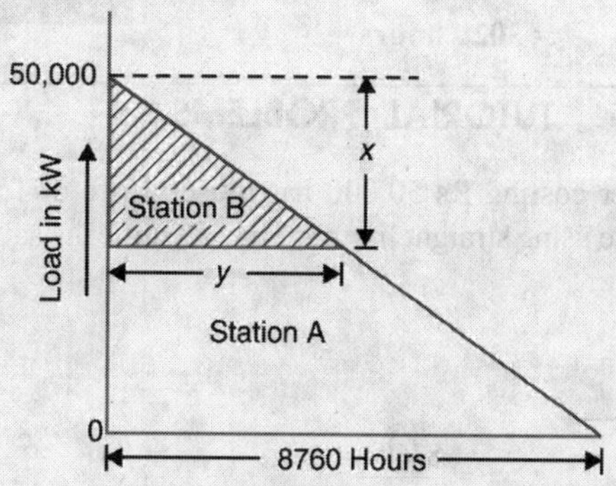

Fig. 4.3

Let x = Installed capacity of station B in kW

y = Hours of operation of station B

$\therefore$ Installed capacity of station A = $(50,000 - x)$ kW

$$\text{Units generated/annum by station } B = \frac{1}{2}xy = \frac{1}{2} \times x \times \frac{8760 * x}{50,000} = 0{\cdot}0876\, x^2$$

$$\text{Unit generated/annum by station } A = \frac{1}{2} \times 50,000 \times 8760 - 0{\cdot}0876\, x^2$$

$$= 219 \times 10^6 - 0{\cdot}0876\, x^2$$

$$\text{Annual cost of station } B,\ C_B = \text{Rs } (50,000 + 50\,x + 0{\cdot}03 \times 0{\cdot}0876\, x^2)$$

$$= \text{Rs } (50,000 + 50\,x + 0{\cdot}00262\, x^2)$$

$$\text{Annual cost of station } A,\ C_A = \text{Rs } (75,000 + 80\,(50,000 - x) + 0{\cdot}02\,(219 \times 10^6 - 0{\cdot}0876\, x^2)$$

$$= \text{Rs } (8{\cdot}455 \times 10^6 - 80\,x - 0{\cdot}00175\, x^2)$$

$\therefore$ Total annual operating cost of stations A and B

$$C = C_A + C_B$$

$$= \text{Rs } (50,000 + 50\,x + 0{\cdot}00262\, x^2) + (8{\cdot}455 \times 10^6 - 80\,x - 0{\cdot}00175\, x^2)$$

$$= \text{Rs } (85,05,000 - 30\,x + 0{\cdot}00087\, x^2)$$

For minimum annual operating cost, $\dfrac{dC}{dx} = 0$

$$\therefore \quad \frac{dC}{dx} = 0 - 30 + 2 \times 0{\cdot}00087\, x$$

or $$0 = -30 + 0{\cdot}00174\, x$$

or $$x = \frac{30}{0 \cdot 00174} = 17,241 \text{ kW}$$

$\therefore$ Capacity of station B = **17,241 kW**

Capacity of station A = 50,000 − 17,241

= **32,758 kW**

No. of hours of operation of station B is

$$y = \frac{8760x}{50,000} = \frac{8760 \times 17,241}{50,000}$$

= **3020 hours**

Steam Power Station

TUTORIAL PROBLEMS

1. A distribution transformer costing Rs 50,000 has a useful life of 15 years. Determine the annual depreciation charge using straight line method. Assume the salvage value of the equipment to be Rs 5,000.

[Rs 3,000]

---

* $\because \quad \dfrac{y}{8760} = \dfrac{x}{50,000} \qquad \therefore \quad y = \dfrac{8760\,x}{50,000}$

2. The power generation equipment of a power station cost Rs 15,75,000 and has a useful life of 25 years. If the salvage value of the equipment is Rs 75,000 and annual interest rate is 5%, determine annual amount to be saved by sinking fund method. **[Rs 31,400]**
3. A 500 kVA distribution transformer costs Rs 50,000 and has a useful life of 20 years. If the salvage value is Rs 5,000 and annual compound interest rate is 8%, determine the value of the transformer at the end of 10 years using sinking fund method. **[Rs 35,700]**
4. A generating station has a maximum demand of 10 MW. Calculate the cost per unit generated from the following data :

   Annual load factor = 35%
   Capital cost = Rs 12,50,000
   Annual cost of fuel and oil = Rs 8,00,000
   Taxes, wages and salaries = Rs 7,00,000
   Interest and depreciation = 10%
5. From the following data, estimate the cost per kWh for the generating station :

   Plant capacity = 50 MW
   Annual load factor = 40%
   Capital cost = Rs 12×10^6
   Annual cost of wages, taxes etc = Rs 4,00,000 ; cost of fuel, lubrication, maintenance = Rs 17,52,000 ; Annual interest and depreciation = 10% of initial value. **[1·913 paise per kWh]**
6. In a particular area both steam and hydro stations are equally possible. It has been estimated that capital costs and running costs of these two types will be as under :

Plant	*Capital cost/kW*	*Running cost/kWh*	*Interest*
Hydro	Rs 3000	3 paise	5%
Steam	Rs 2000	10 paise	5%

   If the expected average load factor is 40%, which is economical to operate, steam or hydro ? **[Hydro]**
7. The energy cost of a 100 MW steam station working at 40% load factor comes out to be 12 paise/kWh of energy generated. What will be the cost of energy generated if the load factor is improved to 60% ? The fuel cost of the power station due to increased generation increase the annual generation cost by 5%. **[8·4 paise]**
8. A load having a peak value of 50 MW and 45% load factor is to be supplied by a steam power station in conjunction with a hydro-power station, the latter supplying 60 million units of energy per annum with a maximum output of 20 MW.

   (*i*) Capital cost — Steam : Rs 2000 per kW installed; Hydro : Rs 3000 per kW installed
   (*ii*) Cost of fuel — Coal Rs 150 per metric ton ; consumption being 0·5 kg per unit generated
   (*iii*) Maintenance charges — Steam : 0·7 paise per unit; Hydro : 0·5 paise per unit
   (*iv*) Interest and depreciation — Steam : 10% per annum; Hydro : 9% per annum

   Calculate the average cost of energy. **[11·6 paise]**

SEIF-TEST

1. **Fill in the blanks by inserting appropriate words/figures.**
   (*i*) Depreciation is the .... in value of equipment due to ....
   (*ii*) The cost of electrical energy can be divided into three parts *viz.,* .... and ....
   (*iii*) The number of units generated will be more if the load factor is ....
   (*iv*) Semi-fixed cost is directly proportional to .... on power station.
   (*v*) The running cost is directly proportional to ....
   (*vi*) In the diminishing value method, depreciation charges are heavy in .... years.
   (*vii*) The annual deposit is .... in sinking fund method as compared to straight line method.
2. **Pick up the correct words/figures from the brackets and fill in the blanks.**
   (*i*) Fixed cost of electrical energy .... maximum demand. (*depends upon, does not depend upon*)
   (*ii*) For the same maximum demand, if load factor is decreased, the cost of energy is .... (*increased, decreased, not affected*)
   (*iii*) Average load is .... if the load factor increases. (*increased, decreased*)
   (*iv*) The annual, ... cost is due to the annual cost of fuel, oil, taxation, wages and salaries to the operating staff. (*running, fixed*)

ANSWERS TO SELF-TEST

1. (*i*) decrease, wear and tear, (*ii*) fixed, semi-fixed, running cost, (*iii*) more, (*iv*) maximum demand
   (*v*) units generated, (*vi*) early, (*vii*) smaller.
2. (*i*) does not depend upon, (*ii*) increased, (*iii*) increased, (*iv*) running.

CHAPTER REVIEW TOPICS

1. Explain the terms interest and depreciation as applied to economics of power generation.
2. Discuss the different classifications of costs of electrical energy.
3. Give the basis for expressing the cost of electrical energy as $a + b$ kW + c kWh and explain the factors on which a, b and c depend.
4. Discuss the various methods of determining the depreciation of the equipment.
5. Enlist the effects of high load factor on the operation of power plants.
6. Write short notes on the following :
   (*i*) Advantages of high load factor.
   (*ii*) Sinking fund method of depreciation.
   (*iii*) Three-part form of cost of electrical energy.

DISCUSSION QUESTIONS

1. What is the importance of interest on capital investment in calculating the cost of electrical energy ?
2. What is the significance of depreciation in the economics of power generation ?
3. Why is fixed cost independent of maximum demand and units generated ?
4. How does high load factor reduce the variable load problems on the power station ?

CHAPTER

5

Tariff

Introduction

The electrical energy produced by a power station is delivered to a large number of consumers. The consumers can be persuaded to use electrical energy if it is sold at reasonable rates. The tariff *i.e.*, the rate at which electrical energy is sold naturally becomes attention inviting for electric supply company. The supply company has to ensure that the tariff is such that it not only recovers the total cost of producing electrical energy but also earns profit on the capital investment. However, the profit must be marginal particularly for a country like India where electric supply companies come under public sector and are always subject to criticism. In this chapter, we shall deal with various types of tariff with special references to their advantages and disadvantages.

5.1 Tariff

The rate at which electrical energy is supplied to a consumer is known as **tariff**.

Although tariff should include the total cost of producing and supplying electrical energy plus the profit, yet it cannot be the same for all types of consumers. It is because the cost of producing electrical energy depends to a considerable extent upon the magnitude of electrical energy consumed by the user and his load conditions. Therefore, in all fairness, due consideration has to be given to different types of consumers (*e.g.*, industrial, domestic and commercial) while fixing the tariff. This makes the problem of suitable rate making highly complicated.

Objectives of tariff. Like other commodities, electrical energy is also sold at such a rate so that it not only returns the cost but also earns reasonable profit. Therefore, a tariff should include the following items :

(*i*) Recovery of cost of producing electrical energy at the power station.
(*ii*) Recovery of cost on the capital investment in transmission and distribution systems.
(*iii*) Recovery of cost of operation and maintenance of supply of electrical energy *e.g.*, metering equipment, billing etc.
(*iv*) A suitable profit on the capital investment.

5.2 Desirable Characteristics of a Tariff

A tariff must have the following desirable characteristics :

(*i*) *Proper return* : The tariff should be such that it ensures the proper return from each consumer. In other words, the total receipts from the consumers must be equal to the cost of producing and supplying electrical energy plus reasonable profit. This will enable the electric supply company to ensure continuous and reliable service to the consumers.

(*ii*) *Fairness* : The tariff must be fair so that different types of consumers are satisfied with the rate of charge of electrical energy. Thus a big consumer should be charged at a lower rate than a small consumer. It is because increased energy consumption spreads the fixed charges over a greater number of units, thus reducing the overall cost of producing electrical energy. Similarly, a consumer whose load conditions do not deviate much from the ideal (*i.e.*, non-variable) should be charged at a lower* rate than the one whose load conditions change appreciably from the ideal.

(*iii*) *Simplicity* : The tariff should be simple so that an ordinary consumer can easily understand it. A complicated tariff may cause an opposition from the public which is generally distrustful of supply companies.

(*iv*) *Reasonable profit* : The profit element in the tariff should be reasonable. An electric supply company is a public utility company and generally enjoys the benefits of monopoly. Therefore, the investment is relatively safe due to non-competition in the market. This calls for the profit to be restricted to 8% or so per annum.

(*v*) *Attractive* : The tariff should be attractive so that a large number of consumers are encouraged to use electrical energy. Efforts should be made to fix the tariff in such a way so that consumers can pay easily.

5.3 Types of Tariff

There are several types of tariff. However, the following are the commonly used types of tariff :

1. **Simple tariff.** *When there is a fixed rate per unit of energy consumed, it is called a* **simple tariff** *or* **uniform rate tariff.**

In this type of tariff, the price charged per unit is constant *i.e.*, it does not vary with increase or decrease in number of units consumed. The consumption of electrical energy at the consumer's terminals is recorded by means of an energy meter. This is the simplest of all tariffs and is readily understood by the consumers.

---

* The cost of producing electrical energy is not same for all consumers but increases with the increasing departure of consumer's load conditions from the ideal (*i.e.*, constant load).

Disadvantages

(*i*) There is no discrimination between different types of consumers since every consumer has to pay equitably for the fixed* charges.

(*ii*) The cost per unit delivered is high.

(*iii*) It does not encourage the use of electricity.

2. Flat rate tariff. *When different types of consumers are charged at different uniform per unit rates, it is called a* **flat rate tariff**.

In this type of tariff, the consumers are grouped into different classes and each class of consumers is charged at a different uniform rate. For instance, the flat rate per kWh for lighting load may be 60 paise, whereas it may be slightly less† (say 55 paise per kWh) for power load. The different classes of consumers are made taking into account their diversity and load factors. The advantage of such a tariff is that it is more fair to different types of consumers and is quite simple in calculations.

Disadvantages

(*i*) Since the flat rate tariff varies according to the way the supply is used, separate meters are required for lighting load, power load etc. This makes the application of such a tariff expensive and complicated.

(*ii*) A particular class of consumers is charged at the same rate irrespective of the magnitude of energy consumed. However, a big consumer should be charged at a lower rate as in his case the fixed charges per unit are reduced.

3. Block rate tariff. *When a given block of energy is charged at a specified rate and the succeeding blocks of energy are charged at progressively reduced rates, it is called a* **block rate tariff**.

In block rate tariff, the energy consumption is divided into blocks and the price per unit is fixed in each block. The price per unit in the first block is the highest** and it is progressively reduced for the succeeding blocks of energy. For example, the first 30 units may be charged at the rate of 60 paise per unit; the next 25 units at the rate of 55 paise per unit and the remaining additional units may be charged at the rate of 30 paise per unit.

The advantage of such a tariff is that the consumer gets an incentive to consume more electrical energy. This increases the load factor of the system and hence the cost of generation is reduced. However, its principal defect is that it lacks a measure of the consumer's demand. This type of tariff is being used for majority of residential and small commercial consumers.

4. Two-part tariff. *When the rate of electrical energy is charged on the basis of maximum demand of the consumer and the units consumed, it is called a* **two-part tariff**.

In two-part tariff, the total charge to be made from the consumer is split into two components *viz.*, fixed charges and running charges. The fixed charges depend upon the maximum demand of the consumer while the running charges depend upon the number of units consumed by the consumer. Thus, the consumer is charged at a certain amount per kW of maximum†† demand plus a certain amount per kWh of energy consumed *i.e.*,

---

* The total cost of electrical energy consists of *fixed charges* and *running charges*. The greater the number of units consumed, the lesser the fixed charges per unit. Therefore, a consumer who consumes more units must pay less fixed charges per unit.

† The flat rate for power load is always less than lighting load. It is because power load is much more than the lighting load and, therefore, improves the load factor of the system to a great extent.

** Generally, fixed charges are merged into the running charges for the first and second blocks of energy so that price per unit for these blocks is high.

†† The maximum demand of consumer is generally assessed on the basis of rateable value of the premises or on the number of rooms or on the connected load.

$$\text{Total charges} = \text{Rs} (b \times \text{kW} + c \times \text{kWh})$$

where, b = charge per kW of maximum demand

c = charge per kWh of energy consumed

This type of tariff is mostly applicable to industrial consumers who have appreciable maximum demand.

Advantages

(*i*) It is easily understood by the consumers.

(*ii*) It recovers the fixed charges which depend upon the maximum demand of the consumer but are independent of the units consumed.

Disadvantages

(*i*) The consumer has to pay the fixed charges irrespective of the fact whether he has consumed or not consumed the electrical energy.

(*ii*) There is always error in assessing the maximum demand of the consumer.

5. Maximum demand tariff. It is similar to two-part tariff with the only difference that the maximum demand is actually measured by installing maximum demand meter in the premises of the consumer. This removes the objection of two-part tariff where the maximum demand is assessed merely on the basis of the rateable value. This type of tariff is mostly applied to big consumers. However, it is not suitable for a small consumer (*e.g.*, residential consumer) as a separate maximum demand meter is required.

6. Power factor tariff. *The tariff in which power factor of the consumer's load is taken into consideration is known as* **power factor tariff.**

In an a.c. system, power factor plays an important role. A low* power factor increases the rating of station equipment and line losses. Therefore, a consumer having low power factor must be penalised. The following are the important types of power factor tariff :

(*i*) *kVA maximum demand tariff* : It is a modified form of two-part tariff. In this case, the fixed charges are made on the basis of maximum demand in kVA and *not* in kW. As kVA is inversely proportional to power factor, therefore, a consumer having low power factor has to contribute more towards the fixed charges. This type of tariff has the advantage that it encourages the consumers to operate their appliances and machinery at improved power factor.

(*ii*) *Sliding scale tariff* : This is also know as average power factor tariff. In this case, an average power factor, say 0·8 lagging, is taken as the reference. If the power factor of the consumer falls below this factor, suitable additional charges are made. On the other hand, if the power factor is above the reference, a discount is allowed to the consumer.

(*iii*) *kW and kVAR tariff* : In this type, both active power (kW) and reactive power (kVAR) supplied are charged separately. A consumer having low power factor will draw more reactive power and hence shall have to pay more charges.

7. Three-part tariff. *When the total charge to be made from the consumer is split into three parts viz., fixed charge, semi-fixed charge and running charge, it is known as a* **three-part tariff.** *i.e.*,

$$\text{Total charge} = \text{Rs} (a + b \times \text{kW} + c \times \text{kWh})$$

where a = fixed charge made during each billing period. It includes interest and depreciation on the cost of secondary distribution and labour cost of collecting revenues,

b = charge per kW of maximum demand,

c = charge per kWh of energy consumed.

* See chapter on power factor improvement.

It may be seen that by adding fixed charge or consumer's charge (*i.e.*, *a*) to two-part tariff, it becomes three-part tariff. The principal objection of this type of tariff is that the charges are split into three components. This type of tariff is generally applied to big consumers.

Power Factor Improvement

Example 5.1. *A consumer has a maximum demand of 200 kW at 40% load factor. If the tariff is Rs. 100 per kW of maximum demand plus 10 paise per kWh, find the overall cost per kWh.*

Solution.

$$\text{Units consumed/year} = \text{Max. demand} \times \text{L.F.} \times \text{Hours in a year}$$
$$= (200) \times (0{\cdot}4) \times 8760 = 7{,}00{,}800 \text{ kWh}$$
$$\text{Annual charges} = \text{Annual M.D. charges} + \text{Annual energy charges}$$
$$= \text{Rs} (100 \times 200 + 0{\cdot}1 \times 7{,}00{,}800)$$
$$= \text{Rs } 90{,}080$$
$$\therefore \quad \text{Overall cost/kWh} = \text{Rs} \frac{90{,}080}{7{,}00{,}800} = \text{Re } 0{\cdot}1285 = \mathbf{12{\cdot}85 \text{ paise}}$$

Example 5.2. *The maximum demand of a consumer is 20 A at 220 V and his total energy consumption is 8760 kWh. If the energy is charged at the rate of 20 paise per unit for 500 hours use of the maximum demand per annum plus 10 paise per unit for additional units, calculate : (i) annual bill (ii) equivalent flat rate.*

Solution.

Assume the load factor and power factor to be unity.

$$\therefore \quad \text{Maximum demand} = \frac{220 \times 20 \times 1}{1000} = 4 \cdot 4 \text{ kW}$$

(*i*) $\text{Units consumed in 500 hrs} = 4{\cdot}4 \times 500 = 2200 \text{ kWh}$

$$\text{Charges for 2200 kWh} = \text{Rs } 0{\cdot}2 \times 2200 = \text{Rs } 440$$
$$\text{Remaining units} = 8760 - 2200 = 6560 \text{ kWh}$$

$$\text{Charges for 6560 kWh} = \text{Rs } 0{\cdot}1 \times 6560 = \text{Rs } 656$$

$$\therefore \quad \text{Total annual bill} = \text{Rs } (440 + 656) = \textbf{Rs. 1096}$$

(ii) $$\text{Equivalent flat rate} = \text{Rs } \frac{1096}{8760} = \text{Re } 0 \cdot 125 = \textbf{12.5 paise}$$

Example 5.3. *The following two tariffs are offered :*

(a) Rs 100 plus 15 paise per unit ;

(b) A flat rate of 30 paise per unit ;

At what consumption is first tariff economical ?

Solution.

Let x be the number of units at which charges due to both tariffs become equal. Then,

$$100 + 0{\cdot}15x = 0{\cdot}3x$$

or $$0{\cdot}15x = 100$$

$$\therefore \quad x = 100/0{\cdot}15 = \textbf{666·67 units}$$

Therefore, tariff (*a*) is economical if consumption is more than 666·67 units.

Example 5.4. *A supply is offered on the basis of fixed charges of Rs 30 per annum plus 3 paise per unit or alternatively, at the rate of 6 paise per unit for the first 400 units per annum and 5 paise per unit for all the additional units. Find the number of units taken per annum for which the cost under the two tariffs becomes the same.*

Solution. Let x (> 400) be the number of units taken per annum for which the annual charges due to both tariffs become equal.

$$\text{Annual charges due to first tariff} = \text{Rs } (30 + 0{\cdot}03\,x)$$

$$\text{Annual charges due to second tariff} = \text{Rs } [(0{\cdot}06 \times 400) + (x - 400) \times 0{\cdot}05]$$

$$= \text{Rs } (4 + 0{\cdot}05\,x)$$

As the charges in both cases are equal,

$$\therefore \quad 30 + 0{\cdot}03\,x = 4 + 0{\cdot}05\,x$$

or $$x = \frac{30 - 4}{0 \cdot 05 - 0 \cdot 03} = \textbf{1300 kWh}$$

Example 5.5. *An electric supply company having a maximum load of 50 MW generates 18 × 10^7 units per annum and the supply consumers have an aggregate demand of 75 MW. The annual expenses including capital charges are :*

For fuel = Rs 90 lakhs

Fixed charges concerning generation = Rs 28 lakhs

Fixed charges concerning transmission and distribution = Rs 32 lakhs

Assuming 90% of the fuel cost is essential to running charges and the loss in transmission and distribution as 15% of kWh generated, deduce a two part tariff to find the actual cost of supply to the consumers.

Solution.

Annual fixed charges

$$\text{For generation} = \text{Rs } 28 \times 10^5$$

$$\text{For transmission and distribution} = \text{Rs } 32 \times 10^5$$

$$\text{For fuel (10\% only)} = \text{Rs } 0{\cdot}1 \times 90 \times 10^5 = \text{Rs } 9 \times 10^5$$

$$\text{Total annual fixed charge} = \text{Rs } (28 + 32 + 9) \times 10^5 = \text{Rs } 69 \times 10^5$$

This cost has to be spread over the aggregate maximum demand of all the consumers *i.e.*, 75 MW.

$\therefore$ Cost per kW of maximum demand $= \text{Rs}\dfrac{69\times10^5}{75\times10^3} = \textbf{Rs. 92}$

Annual running charges.

Cost of fuel (90%) $= \text{Rs } 0{\cdot}9\times90\times10^5 = \text{Rs } 81\times10^5$

Units delivered to consumers $=$ 85% of units generated

$= 0{\cdot}85\times18\times10^7 = 15{\cdot}3\times10^7$ kWh

This cost is to be spread over the units delivered to the consumers.

$\therefore$ Cost/kWh $= \text{Rs}\dfrac{81\times10^5}{15\cdot3\times10^7} = \text{Re } 0\cdot053 = \textbf{5.3 paise}$

$\therefore$ Tariff is Rs 92 per kW of maximum demand plus 5·3 paise per kWh.

Example 5.6. *A generating station has a maximum demand of 75 MW and a yearly load factor of 40%. Generating costs inclusive of station capital costs are Rs. 60 per annum per kW demand plus 4 paise per kWh transmitted. The annual capital charges for transmission system are Rs 20,00,000 and for distribution system Rs 15,00,000 ; the respective diversity factors being 1·2 and 1·25. The efficiency of transmission system is 90% and that of the distribution system inclusive of substation losses is 85%. Find the yearly cost per kW demand and cost per kWh supplied :*

(*i*) *at the substation* (*ii*) *at the consumers premises.*

Solution.

Maximum demand $=$ 75 MW $=$ 75,000 kW

Annual load factor $=$ 40% $=$ 0·4

(*i*) **Cost at substation.** The cost per kW of maximum demand is to be determined from the total annual fixed charges associated with the supply of energy at the substation. The cost per kWh shall be determined from the running charges.

(*a*) **Annual fixed charges**

Generation cost $= \text{Rs } 60\times75\times10^3 = \text{Rs } 4{\cdot}5\times10^6$

Transmission cost $= \text{Rs } 2\times10^6$

Total annual fixed charges at the substation

$= \text{Rs } (4{\cdot}5+2)\times10^6 = \text{Rs } 6{\cdot}5\times10^6$

Aggregate of all maximum demands by the various substations

$=$ Max. demand on generating station $\times$ Diversity factor

$= (75\times10^3)\times1{\cdot}2 = 90\times10^3$ kW

The total annual fixed charges have to be spread over the aggregate maximum demands by various substations *i.e.*, 90×10^3 kW.

Annual cost per kW of maximum demand

$= \text{Rs}\dfrac{6\cdot5\times10^6}{90\times10^3} = \textbf{Rs. 72.22}$

(*b*) **Running Charges.** It is given that cost of 1 kWh transmitted to substation is 4 paise. As the transmission efficiency is 90%, therefore, for every kWh transmitted, 0·9 kWh reaches the substation.

$\therefore$ Cost/kWh at substation $= 4/0{\cdot}9 = \textbf{4·45 paise}$

Hence at sub-station, the cost is **Rs 72·22** per annum per kW maximum demand plus **4·45 paise** per kWh.

(*ii*) Cost at consumer's premises. The total annual fixed charges at consumer's premises is the sum of annual fixed charges at substation (*i.e.* Rs $6{\cdot}5 \times 10^6$) and annual fixed charge for distribution (*i.e.*, Rs $1{\cdot}5 \times 10^6$).

$\therefore$ Total annual fixed charges at consumer's premises

$$= \text{Rs}\,(6{\cdot}5 + 1{\cdot}5) \times 10^6 = \text{Rs}\,8 \times 10^6$$

Aggregate of maximum demands of all consumers

$$= \text{Max. demand on Substation} \times \text{Diversity factor}$$
$$= (90 \times 10^3) \times 1{\cdot}25 = 112{\cdot}5 \times 10^3\ \text{kW}$$

$\therefore$ Annual cost per kW of maximum demand

$$= \text{Rs}\,\frac{8 \times 10^6}{112 \cdot 5 \times 10^3} = \textbf{Rs. 71.11}$$

As the distribution efficiency is 85%, therefore, for each kWh delivered from substation, only 0·85 kWh reaches the consumer's premises.

$\therefore$ Cost per kWh at consumer's premises

$$= \frac{\text{Cost per kWh at substation}}{0 \cdot 85} = \frac{4 \cdot 45}{0 \cdot 85} = \textbf{5.23 paise}$$

Hence at consumer's premises, the cost is **Rs. 71·11** per annum per kW maximum demand plus **5·23 paise** per kWh.

Example 5.7. *Determine the load factor at which the cost of supplying a unit of electricity from a Diesel and from a steam station is the same if the annual fixed and running charges are as follows :*

Station	*Fixed charges*	*Running charges*
Diesel	*Rs 300 per kW*	*25 paise/kWh*
Steam	*Rs 1200 per kW*	*6·25 paise/kWh*

Solution. Suppose energy supplied in one year is 100 units *i.e.*, 100 kWh. Let L be the load factor at which the cost of supplying a unit of electricity is the same for diesel and steam station.

Diesel Station.

$$\text{Average power} = \frac{100\ \text{kWh}}{8760\ \text{hrs}} = 0{\cdot}0114\ \text{kW}$$

$$\text{Maximum demand} = \frac{0 \cdot 0114}{L}\ \text{kW}$$

$$\text{Fixed charges} = \text{Rs}\,300 \times \frac{0 \cdot 0114}{L} = \text{Rs}\,\frac{3 \cdot 42}{L}$$

$$\text{Running charges} = \text{Rs}\,100 \times 0{\cdot}25 = \text{Rs}\,25$$

$\therefore$ Fixed and running charges for 100 kWh

$$= \text{Rs}\left(\frac{3 \cdot 42}{L} + 25\right) \qquad \text{...}(i)$$

Steam station.

$$\text{Fixed charges} = \text{Rs}\,1200 \times \frac{0 \cdot 0114}{L} = \text{Rs}\,\frac{13 \cdot 68}{L}$$

$$\text{Running charges} = \text{Rs}\,100 \times 0{\cdot}0625 = \text{Rs}\,6{\cdot}25$$

$\therefore$ Fixed and running charges for 100 kWh

$$= \text{Rs}\left(\frac{13 \cdot 68}{L} + 6 \cdot 25\right) \qquad \text{...}(ii)$$

As the two charges are same, therefore, equating exps. (*i*) and (*ii*), we get,

$$\frac{3 \cdot 42}{L} + 25 = \frac{13 \cdot 68}{L} + 6 \cdot 25$$

or

$$\frac{10 \cdot 26}{L} = 18 \cdot 75$$

$$\therefore \quad L = 10{\cdot}26/18{\cdot}75 = 0{\cdot}5472 = \mathbf{54{\cdot}72\%}$$

TUTORIAL PROBLEMS

1. A consumer has a maximum demand of 100 MW at 60% load factor. If the tariff is Rs 20 per kW of maximum demand plus 1 paise per kWh, find the overall cost per kWh. **[1·38 paise]**
2. The maximum demand of a consumer is 25A at 220 V and his total energy consumption is 9750 kWh. If energy is charged at the rate of 20 paise per kWh for 500 hours use of maximum demand plus 5 paise per unit for all additional units, estimate his annual bill and the equivalent flat rate. **[Rs 900 ; 9·2 paise]**
3. A consumer has an annual consumption of 2×10^5 units. The tariff is Rs 50 per kW of maximum demand plus 10 paise per kWh.

   (*i*) Find the annual bill and the overall cost per kWh if the load factor is 35%.

   (*ii*) What is the overall cost per kWh if the consumption were reduced by 25% with the same load factor ?

   (*iii*) What is the overall cost per kWh if the load factor were 25% with the same consumption as in (*i*) ?

   [(*i*) Rs 23,400 ; 11·7 paise (*ii*) 11·7 paise (*iii*) 12·28 paise]
4. Daily load of an industry is 200 kW for first one hour, 150 kW for next seven hours, 50 kW for next eight hours and 1 kW for remaining time. If tariff in force is Rs. 100 per kW of maximum demand per annum plus 5 paise per kWh, find the annual bill. **[Rs 50,258·5]**
5. A consumer requires one million units per year and his annual load factor is 50%. The tariff in force is Rs. 120 per kW per annum plus 5 paise per unit consumed. Estimate the saving in his energy costs if he improves the load factor to 100%. **[Rs 13,692]**
6. An industrial undertaking has a connected load of 100 kW. The maximum demand is 80 kW. On an average, each machine works for 60 per cent time. Find the yearly expenditure on the electricity if the tariff is

   Rs 10,000 + Rs 1000 per kW of maximum demand per year + Re 1 per kWh. **[Rs 615600]**

Example 5.8. *Calculate annual bill of a consumer whose maximum demand is 100 kW, p.f. = 0·8 lagging and load factor = 60%. The tariff used is Rs 75 per kVA of maximum demand plus 15 paise per kWh consumed.*

Solution.

$$\begin{aligned}
\text{Units consumed/year} &= \text{Max. demand} \times \text{L.F.} \times \text{Hours in a year} \\
&= (100) \times (0{\cdot}6) \times (8760)\ \text{kWh} \\
&= 5{\cdot}256 \times 10^5\ \text{kWh} \\
\text{Max. demand in kVA} &= 100/\text{p.f.} = 100/0{\cdot}8 = 125 \\
\text{Annual bill} &= \text{Max. demand charges} + \text{Energy charges} \\
&= \text{Rs}\ 75 \times 125 + \text{Rs}\ 0{\cdot}15 \times 5{\cdot}256 \times 10^5 \\
&= \text{Rs}\ 9375 + \text{Rs}\ 78{,}840 = \mathbf{Rs\ 88{,}215}
\end{aligned}$$

Example 5.9. *A factory has a maximum load of 240 kW at 0·8 p.f. lagging with an annual consumption of 50,000 units. The tariff is Rs 50 per kVA of maximum demand plus 10 paise per unit. Calculate the flat rate of energy consumption. What will be annual saving if p.f. is raised to unity?*

Solution.

Maximum demand in kVA at a p.f. of 0·8

$$= 240/0{\cdot}8 = 300$$

$\therefore$ Annual bill = Demand charges + Energy charges

$$= \text{Rs } 50 \times 300 + \text{Rs } 0{\cdot}1 \times 50{,}000$$

$$= \text{Rs } 15{,}000 + \text{Rs } 5000 = \text{Rs } 20{,}000$$

$$\therefore \quad \text{Flat rate/unit} = \text{Rs } \frac{20{,}000}{50{,}000} = \text{Rs } 0{\cdot}40 = \textbf{40 paise}$$

When p.f. is raised to unity, the maximum demand in kVA

$$= 240/1 = 240$$

$$\text{Annual bill} = \text{Rs } 50 \times 240 + \text{Rs } 0{\cdot}1 \times 50{,}000$$

$$= \text{Rs } 12{,}000 + \text{Rs } 5{,}000 = \text{Rs } 17{,}000$$

$$\text{Annual saving} = \text{Rs } (20{,}000 - 17{,}000) = \textbf{Rs 3000}$$

Example 5.10. *The monthly readings of a consumer's meter are as follows :*

Maximum demand = 50 kW

Energy consumed = 36,000 kWh

Reactive energy = 23,400 kVAR

If the tariff is Rs 80 per kW of maximum demand plus 8 paise per unit plus 0·5 paise per unit for each 1% of power factor below 86%, calculate the monthly bill of the consumer.

Solution.

$$\text{Average load} = \frac{36{,}000}{24 \times 30} = 50 \text{ kW}$$

$$\text{Average reactive power} = \frac{23{,}400}{24 \times 30} = 32{\cdot}5 \text{ kVAR}$$

Suppose ϕ is the power factor angle.

$$\therefore \quad \tan\phi = \frac{\text{kVAR}}{\text{Active power}} = \frac{32{\cdot}5}{50} = 0{\cdot}65$$

or $\phi = \tan^{-1}(0{\cdot}65) = 33{\cdot}02^\circ$

$\therefore$ Power factor, $\cos\phi = \cos 33{\cdot}02^\circ = 0{\cdot}8384$

$$\text{Power factor surcharge} = \text{Rs } \frac{36{,}000 \times 0{\cdot}5}{100}(86 - 83{\cdot}84) = \text{Rs } 388{\cdot}8$$

$$\text{Monthly bill} = \text{Rs } (80 \times 50 + 0{\cdot}08 \times 36{,}000 + 388{\cdot}8)$$

$$= \text{Rs } (4000 + 2880 + 388{\cdot}8) = \textbf{Rs 7268·8}$$

Example 5.11. *The tariff in force is Rs 150 per kVA of maximum demand and 8 paise per unit consumed. If the load factor is 30%, find the overall cost per unit at (i) unity p. f. and (ii) 0·7 p. f.*

Solution. Suppose the maximum demand is 1 kVA.

(*i*) *When p.f. is unity*

$$\text{Max. demand charge/unit} = \frac{150 \times 100}{8760 \times 0{\cdot}30} = 5{\cdot}7 \text{ paise}$$

$$\text{Energy charge/unit} = 8 \text{ paise}$$

$$\text{Overall cost/unit} = 5{\cdot}7 + 8 = \textbf{13·7 paise}$$

(*ii*) *When p.f. is 0·7*

$$\text{Max. demand charge/unit} = \frac{150 \times 100}{8760 \times 0{\cdot}30 \times 0{\cdot}7} = 8{\cdot}15 \text{ paise}$$

$$\text{Energy charge/unit} = 8 \text{ paise}$$

$$\text{Overall cost/unit} = 8{\cdot}15 + 8 = \textbf{16·15 paise}$$

Example 5.12. *Two systems of tariff are available for a factory working 8 hours a day for 300 working days in a year.*

(*i*) *High-voltage supply at 5 paise per unit plus Rs 4·50 per month per kVA of maximum demand.*

(*ii*) *Low-voltage supply at Rs 5 per month per kVA of maximum demand plus 5·5 paise per unit.*

The factory has an average load of 200 kW at 0·8 p.f. and a maximum demand of 250 kW at the same p.f. The high voltage equipment costs Rs 50 per kVA and the losses can be taken as 4%. Interest and depreciation charges are 12%. Calculate the difference in the annual costs between the two systems.

Solution.

(*i*) **High voltage supply**

Max. demand in kVA = 250/0·8 = 312·5

As the losses in *h.v.* equipment are 4%, therefore, capacity of *h.v.* equipment

= 312·5/0·96 = 325·5 kVA

Capital investment on *h.v.* equipment

= Rs 50 × 325·5 = Rs 16,275

Annual interest and depreciation = Rs 16,275 × 0·12 = Rs 1953

Annual charge due to maximum kVA demand

= Rs 325·5 × 4·5 × 12 = Rs 17,577

$$\text{Units consumed/year} = \frac{200 \times 8 \times 300}{0 \cdot 96} = 5 \times 10^5 \text{ kWh}$$

Annual charge due to kWh consumption

= Rs 0·05 × 5 × 10^5 = Rs 25,000

Total annual cost = Rs (1953 + 17,577 + 25,000) = Rs 44,530

(*ii*) **Low voltage supply.** There is no loss of energy as no equipment is used.

Max. demand in kVA = 250/0·8 = 312·5

Annual charge due to maximum kVA demand

= Rs 312·5 × 5 × 12 = Rs 18,750

Units consumed/year = 200 × 8 × 300 = 48 × 10^4 kWh

Annual charge due to kWh consumption

= Rs 0·055 × 48 × 10^4 = Rs 26,400

Total annual cost = Rs (18,750 + 26,400) = Rs 45,150

Difference in the annual costs of two systems

= Rs (45,150 – 44,530) = **Rs 620**

Hence, high-voltage supply is cheaper than low-voltage supply by Rs 620.

Example 5.13. *A generating station has two 1000 kW diesel-generator sets. The load is estimated to reach a maximum demand of 2500 kW after two years with an increase of 5.5 ×10^6 units over the present value. To meet this demand, the following two alternatives are available :*

(*i*) *Purchasing one more set of 1000 kW at Rs 400 per kW. The annual interest and depreciation of the new set are 10% of the capital investment. The cost of generation for the station is Rs 75 per kW maximum demand plus 5 paise per kWh.*

(*ii*) *Purchasing bulk power from a grid supply at Rs 120 per kW maximum demand plus 3 paise per kWh.*

Find which alternative in cheaper and by how much ?

Solution.

In order to determine the cheaper alternative, we shall find the annual cost in each case.

(*i*) **Purchasing diesel set**

$$\text{Capital cost of set} = \text{Rs } 400 \times 1000 = \text{Rs } 4{,}00{,}000$$

Annual interest and depreciation on capital investment

$$= \text{Rs } 4{,}00{,}000 \times 0{\cdot}1 = \text{Rs } 40{,}000$$

The present capacity of the station is 2000 kW and the expected maximum demand after two years is 2500 kW. Therefore, extra power to be generated is

$$= 2500 - 2000 = 500 \text{ kW}$$

Annual charge due to extra kW max. demand

$$= \text{Rs } 500 \times 75 = \text{Rs } 37{,}500$$

Annual charge due to extra kWh consumption

$$= \text{Rs } 0{\cdot}05 \times 5{\cdot}5 \times 10^6 = \text{Rs } 2{,}75{,}000$$

$$\text{Total annual cost} = \text{Rs } (40{,}000 + 37{,}500 + 2{,}75{,}000)$$

$$= \text{Rs } 3{,}52{,}500$$

(*ii*) **Purchasing from grid supply**

Annual charge due to extra kW max. demand

$$= \text{Rs } 500 \times 120 = \text{Rs } 60{,}000$$

Annual charge due to extra kWh consumption

$$= \text{Rs } 0{\cdot}03 \times 5{\cdot}5 \times 10^6 = \text{Rs } 1{,}65{,}000$$

$$\text{Total annual cost} = \text{Rs } (60{,}000 + 1{,}65{,}000) = \text{Rs } 2{,}25{,}000$$

Hence alternative (*ii*) is cheaper by 3,52,500 – 2,25,000 = **Rs 1,27,500 per annum**

Example 5.14. *A supply company offers the following alternate tariffs for supply to a factory :*

(*i*) *H.V. supply at Rs 70 per kVA per annum plus 3 paise per kWh.*

(*ii*) *L.V. supply at Rs 65 per kVA per annum plus 4 paise per kWh.*

The cost of transformers and switchgears for H.V. supply is Rs 50 per kVA and full transformation losses are 2%. The annual fixed charges on the capital cost of H.V. plant are 15%. If the factory runs for 6 hours a day, find the number of days above which the factory should be run so that the H.V. supply is cheaper.

Solution.

Let x = Factory load in kW

y = No. of working days above which H.V. supply is cheaper

(*i*) **H. V. Supply.** Assume the power factor of the load to be unity. As the transformation losses are 2%,

∴ Rating of transformer and switchgear = $x/0{\cdot}98$ kVA

$$\text{Energy consumed per annum} = (x/0{\cdot}98) \times y \times 6 = 6{\cdot}12\, xy \text{ kWh}$$

Annual fixed charges of H. V. supply due to kVA demand

$$= \text{Rs } 70 \times x/0{\cdot}98 = \text{Rs. } 71{\cdot}42x$$

$$\text{Cost of transformer and switchgear} = \text{Rs } 50 \times x/0{\cdot}98 = \text{Rs } 51x$$

Annual fixed charges of transformer and switchgear

$$= 15\% \text{ cost of transformer and switchgear}$$

$$= 0{\cdot}15 \times 51\,x = \text{Rs } 7{\cdot}65\,x$$

$$\text{Total annual fixed charges of H. V. supply} = \text{Rs } (71{\cdot}42\,x + 7{\cdot}65\,x) = \text{Rs } 79{\cdot}07\,x$$

$$\text{Total annual running charges of H. V. supply} = \text{Rs } 6{\cdot}12\,xy \times 0{\cdot}03 = \text{Rs } 0{\cdot}1836\,xy$$

Total annual charges of H.V. supply $= \text{Rs}\,(79{\cdot}07\,x + 0{\cdot}1836\,xy)$ ...(*i*)

(*ii*) **L. V. Supply**

$$\text{Energy consumed per annum} = x \times y \times 6 = 6\,xy\,\text{kWh}$$

$$\text{Annual fixed charges of L. V. supply} = \text{Rs}\,65\,x$$

$$\text{Annual running charges of L. V. supply} = \text{Rs}\,0{\cdot}04 \times 6\,xy = \text{Rs}\,0{\cdot}24\,xy$$

$$\text{Total annual charges of L. V. supply} = \text{Rs}\,(65\,x + 0{\cdot}24\,xy) \quad ...(ii)$$

The two tariffs will give equal annual cost if the factory is run for *y* days. Therefore, equating exp. (*i*) and exp. (*ii*), we get,

$$79{\cdot}07\,x + 0{\cdot}1836\,xy = 65\,x + 0{\cdot}24\,xy$$

or
$$14{\cdot}07\,x = 0{\cdot}057\,xy$$

or
$$y = \frac{14{\cdot}07}{0{\cdot}057} = \mathbf{247\ days}$$

i.e., if the factory is run for more than 247 days, then H. V. supply will be cheaper.

TUTORIAL PROBLEMS

1. An industrial consumer has a maximum demand of 120 kW and maintains a load factor of 80%. The tariff in force is Rs. 60 per kVA of maximum demand plus 8 paise per unit. If the average p.f. is 0·8 lagging, calculate the total energy consumed per annum and the annual bill. **[8,40,960 kWh ; Rs 76,276·8]**
2. A customer is offered power at Rs 50 per annum per kVA of maximum demand plus 5 paise per unit. He proposes to install a motor to carry his estimated maximum demand of 300 b.h.p. (metric). The motor available has a power factor of 0·83 at full load. How many units will be required at 30% load factor and what will be the annual bill ? The motor efficiency is 90%. **[6,44,307 ; Rs 46,985·35]**
3. A factory has a maximum load of 300 kW at 0·72 p.f. lagging with an annual consumption of 40,000 units. The tariff in force is Rs 4·5 per kVA of maximum demand plus 2 paise per unit. Calculate the flat rate of energy consumption. What will be the annual saving if p.f. is raised to unity ? **[4·69 paise ; Rs 525]**
4. The monthly readings of a consumer's meter are under ;

   Maximum demand = 60 kW

   Energy consumed = 24,000 kWh

   Reactive energy = 15,600 kVAR

   If the tariff is Rs 20 per kW of maximum demand plus 3 paise per unit plus 0·1 paise per unit for each 1% power factor below 85%, calculate the monthly bill of the consumer. **[Rs 1960·4]**
5. Compare the annual cost of power supply to a factory having a maximum demand of 500 kW and a load factor of 40% by having the supply from :

   (*i*) the factory's own diesel generating plant.

   (*ii*) a public supply.

   With regards to (*i*), the capital cost of factory's own generating plant is Rs 8 lakhs, cost of fuel oil is Rs 200 per ton, fuel consumption 0·65 lbs per kWh. Capital charges, cost of repairs and maintenance, interest and depreciation 15% of the total capital cost. Salaries and wages of the operating staff are Rs 15,000 per year. With regards to (*ii*), the tariff is Rs 150 per kW per annum of maximum demand plus 2·5 paise per kWh. Which of the two alternatives is favourable for the operation of the factory ?

   [(*i*) 13·5 paise/unit (*ii*) 6·8 paise/unit]
6. An industrial load can be supplied from the following alternative tariffs :

   (*i*) High voltage supply at Rs 65 per kW per annum plus 3 paise per kWh.

   (*ii*) Low voltage supply at Rs 65 per kW per annum plus 3·3 paise per kWh.

   The high voltage equipment costs Rs 50 per kW and the losses can be taken as 3%. Interest and depreciation charges are 15% per annum. If there are 40 working weeks in a year, find working hours per week above which high voltage supply is cheaper. **[55·42 hours/week]**

7. A supply company offers the following alternative tariffs :
   (*i*) Standing charges of Rs 75 per annum plus 3 paise/kWh.
   (*ii*) first 300 kWh at 20 paise/kWh ; and additional energy at 5 paise/kWh.
   If the annual consumption is 1800 kWh, which tariff is more economical and by how much ?
   [Tariff (*i*) is economical by Rs 6 per annum]
8. A factory has a maximum demand of 500 kW, the load factor being 60% during working hours. The following two tariffs are available :
   (*i*) Rs 8 per kW of maximum demand plus 3 paise per kWh.
   (*ii*) a flat rate of Re 0·1/kWh.
   Determine the working hours per week above which tariff (*i*) will be cheaper. **[44 hours/week]**

SELF-TEST

1. **Fill in the blanks by inserting appropriate words/figures :**
   (*i*) The flat rate for power load is generally .... than the lighting load.
   (*ii*) In block rate tariff, the rate of energy in first one or two blocks is ...... because .... charges are merged into ...... charges.
   (*iii*) The block rate tariff is mostly applicable to .... consumers.
   (*iv*) A big consumer is charged at a lower rate than a small consumer because ......
   (*v*) Maximum demand tariff is not applied to domestic consumers because ......
2. **Pick up the correct words/figures from brackets and fill in the blanks :**
   (*i*) A consumer whose load conditions do not deviate from ideal one should be charged at ..... rate than the one whose load conditions change appreciably. (*lower, higher*)
   (*ii*) A consumer who consumes more electrical energy should pay ...... fixed charges per unit. (*less, more*)
   (*iii*) The ideal tariff for any type of consumer is .... tariff. (*two-part, three-part*)
   (*iv*) The maximum kVA demand of the consumer is .... proportional to power factor. (*inversely, directly*)

ANSWERS TO SELF-TEST

1. (*i*) lower (*ii*) high, fixed, running (*iii*) domestic (*iv*) it improves the load factor (*v*) their maximum demand is small.
2. (*i*) lower (*ii*) less (*iii*) three-part (*iv*) inversely.

CHAPTER REVIEW TOPICS

1. What do you understand by tariff ? Discuss the objectives of tariff.
2. Describe the desirable characteristics of a tariff.
3. Describe some of the important types of tariff commonly used.
4. Write short notes on the following :
   (*i*) Two-part tariff.
   (*ii*) Power factor tariff.
   (*iii*) Three-part tariff.

DISCUSSION QUESTIONS

1. Why is tariff for power load less than the lighting load ?
2. What is the effect of power factor on the cost of generation ?
3. Can the load factor of the system be 100% ?
4. What is the importance of power factor tariff ?

CHAPTER

6

Power Factor Improvement

Introduction

The electrical energy is almost exclusively generated, transmitted and distributed in the form of alternating current. Therefore, the question of power factor immediately comes into picture. Most of the loads (*e.g.* induction motors, arc lamps) are inductive in nature and hence have low lagging power factor. The low power factor is highly undesirable as it causes an increase in current, resulting in additional losses of active power in all the elements of power system from power station generator down to the utilisation devices. In order to ensure most favourable conditions for a supply system from engineering and economical standpoint, it is important to have power factor as close to unity as possible. In this chapter, we shall discuss the various methods of power factor improvement.

6.1 Power Factor

The cosine of angle between voltage and current in an a.c. circuit is known as **power factor.**

In an a.c. circuit, there is generally a phase difference ϕ between voltage and current. The term $\cos\phi$ is called the power factor of the circuit. If the circuit is inductive, the current lags behind the voltage and the power factor is referred to as lagging. However, in a capacitive circuit, current leads the voltage and power factor is said to be leading.

Consider an inductive circuit taking a lagging current I from supply voltage V; the angle of lag being ϕ. The phasor diagram of the circuit is shown in Fig. 6.1. The circuit current I can be resolved into two perpendicular components, namely ;

(a) $I \cos \phi$ in phase with V

(b) $I \sin \phi$ 90° out of phase with V

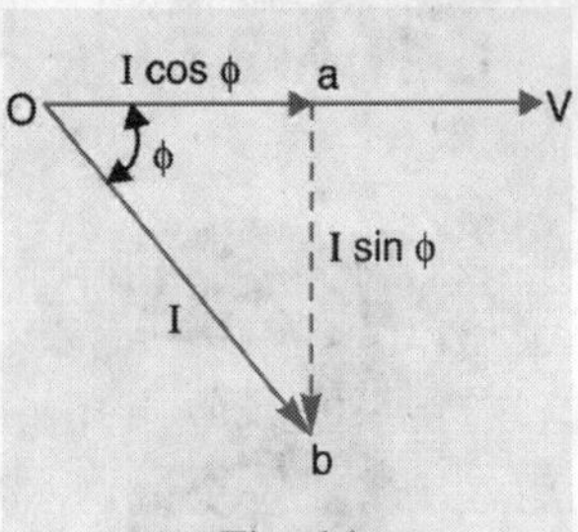

Fig. 6.1

The component $I \cos \phi$ is known as active or wattful component, whereas component $I \sin \phi$ is called the reactive or wattless component. The reactive component is a measure of the power factor. If the reactive component is small, the phase angle ϕ is small and hence power factor $\cos \phi$ will be high. Therefore, a circuit having small reactive current (*i.e.*, $I \sin \phi$) will have high power factor and *vice-versa*. It may be noted that value of power factor can never be more than unity.

(i) It is a usual practice to attach the word 'lagging' or 'leading' with the numerical value of power factor to signify whether the current lags or leads the voltage. Thus if the circuit has a p.f. of 0·5 and the current lags the voltage, we generally write p.f. as 0·5 lagging.

(ii) Sometimes power factor is expressed as a percentage. Thus 0·8 lagging power factor may be expressed as 80% lagging.

6.2 Power Triangle

The analysis of power factor can also be made in terms of power drawn by the a.c. circuit. If each side of the current triangle *oab* of Fig. 6.1 is multiplied by voltage V, then we get the power triangle OAB shown in Fig. 6.2 where

OA = $VI \cos \phi$ and represents the *active power* in watts or kW

AB = $VI \sin \phi$ and represents the *reactive power* in VAR or kVAR

OB = VI and represents the *apparent power* in VA or kVA

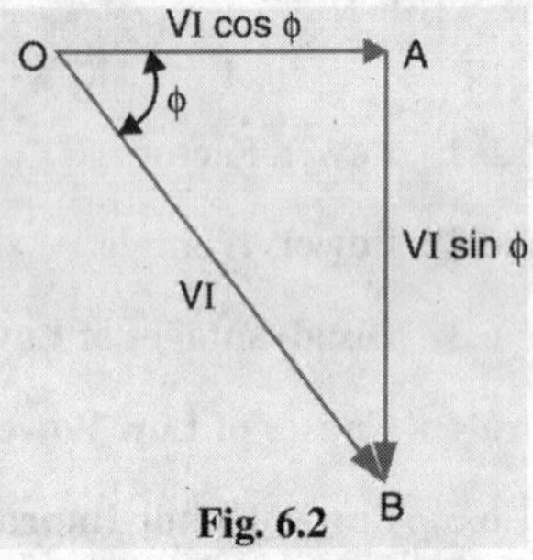

Fig. 6.2

The following points may be noted form the power triangle :

(i) The apparent power in an a.c. circuit has two components *viz*., active and reactive power at right angles to each other.

$$OB^2 = OA^2 + AB^2$$

or $$(\text{apparent power})^2 = (\text{active power})^2 + (\text{reactive power})^2$$

or $$(\text{kVA})^2 = (\text{kW})^2 + (\text{kVAR})^2$$

(ii) Power factor, $$\cos \phi = \frac{OA}{OB} = \frac{\text{active power}}{\text{apparent power}} = \frac{\text{kW}}{\text{kVA}}$$

Thus the power factor of a circuit may also be defined as the ratio of active power to the apparent power. This is a perfectly general definition and can be applied to all cases, whatever be the waveform.

(iii) The lagging* reactive power is responsible for the low power factor. It is clear from the power triangle that smaller the reactive power component, the higher is the power factor of the circuit.

$$\text{kVAR} = \text{kVA} \sin \phi = \frac{\text{kW}}{\cos \phi} \sin \phi$$

$$\therefore \quad \text{kVAR} = \text{kW} \tan \phi$$

* If the current lags behind the voltage, the reactive power drawn is known as lagging reactive power. However, if the circuit current leads the voltage, the reactive power is known as leading reactive power.

(*iv*) For leading currents, the power triangle becomes reversed. This fact provides a key to the power factor improvement. If a device taking leading reactive power (*e.g.* capacitor) is connected in parallel with the load, then the lagging reactive power of the load will be partly neutralised, thus improving the power factor of the load.

(*v*) The power factor of a circuit can be defined in one of the following three ways :

(*a*) Power factor $= \cos\phi$ = cosine of angle between V and I

(*b*) Power factor $= \dfrac{R}{Z} = \dfrac{\text{Resistance}}{\text{Impedance}}$

(*c*) Power factor $= \dfrac{VI\cos\phi}{VI} = \dfrac{\text{Active power}}{\text{Apparent Power}}$

(*vi*) The reactive power is neither consumed in the circuit nor it does any useful work. It merely flows back and forth in both directions in the circuit. A wattmeter does not measure reactive power.

Illustration. Let us illustrate the power relations in an a.c. circuit with an example. Suppose a circuit draws a current of 10 A at a voltage of 200 V and its p.f. is 0·8 lagging. Then,

$$\text{Apparent power} = VI = 200 \times 10 = 2000 \text{ VA}$$

$$\text{Active power} = VI\cos\phi = 200 \times 10 \times 0{\cdot}8 = 1600 \text{ W}$$

$$\text{Reactive power} = VI\sin\phi = 200 \times 10 \times 0{\cdot}6 = 1200 \text{ VAR}$$

The circuit receives an apparent power of 2000 VA and is able to convert only 1600 watts into active power. The reactive power is 1200 VAR and does no useful work. It merely flows into and out of the circuit periodically. In fact, reactive power is a liability on the source because the source has to supply the additional current (*i.e.*, $I\sin\phi$).

6.3 Disadvantages of Low Power Factor

The power factor plays an importance role in a.c. circuits since power consumed depends upon this factor.

$$P = V_L I_L \cos\phi \qquad \text{(For single phase supply)}$$

$$\therefore \quad I_L = \frac{P}{V_L\cos\phi} \qquad ...(i)$$

$$P = \sqrt{3}\, V_L I_L \cos\phi \qquad \text{(For 3 phase supply)}$$

$$\therefore \quad I_L = \frac{P}{\sqrt{3}\, V_L\cos\phi} \qquad ...(ii)$$

It is clear from above that for fixed power and voltage, the load current is inversely proportional to the power factor. Lower the power factor, higher is the load current and *vice-versa*. A power factor less than unity results in the following disadvantages :

(*i*) **Large kVA rating of equipment.** The electrical machinery (*e.g.*, alternators, transformers, switchgear) is always rated in *kVA.

Now, $$\text{kVA} = \frac{\text{kW}}{\cos\phi}$$

It is clear that kVA rating of the equipment is inversely proportional to power factor. The smaller the power factor, the larger is the kVA rating. Therefore, at low power factor, the kVA rating of the equipment has to be made more, making the equipment larger and expensive.

(*ii*) **Greater conductor size.** To transmit or distribute a fixed amount of power at constant voltage, the conductor will have to carry more current at low power factor. This necessitates

* The electrical machinery is rated in kVA because the power factor of the load is not known when the machinery is manufactured in the factory.

large conductor size. For example, take the case of a single phase a.c. motor having an input of 10 kW on full load, the terminal voltage being 250 V. At unity p.f., the input full load current would be 10,000/250 = 40 A. At 0·8 p.f; the kVA input would be 10/0·8 = 12·5 and the current input 12,500/250 = 50 A. If the motor is worked at a low power factor of 0·8, the cross-sectional area of the supply cables and motor conductors would have to be based upon a current of 50 A instead of 40 A which would be required at unity power factor.

(*iii*) **Large copper losses.** The large current at low power factor causes more I^2R losses in all the elements of the supply system. This results in poor efficiency.

(*iv*) **Poor voltage regulation.** The large current at low lagging power factor causes greater voltage drops in alternators, transformers, transmission lines and distributors. This results in the decreased voltage available at the supply end, thus impairing the performance of utilisation devices. In order to keep the receiving end voltage within permissible limits, extra equipment (*i.e.*, voltage regulators) is required.

(*v*) **Reduced handling capacity of system.** The lagging power factor reduces the handling capacity of all the elements of the system. It is because the reactive component of current prevents the full utilisation of installed capacity.

The above discussion leads to the conclusion that low power factor is an objectionable feature in the supply system

6.4 Causes of Low Power Factor

Low power factor is undesirable from economic point of view. Normally, the power factor of the whole load on the supply system in lower than 0·8. The following are the causes of low power factor:

(*i*) Most of the a.c. motors are of induction type (1ϕ and 3ϕ induction motors) which have low lagging power factor. These motors work at a power factor which is extremely small on light load (0·2 to 0·3) and rises to 0·8 or 0·9 at full load.

(*ii*) Arc lamps, electric discharge lamps and industrial heating furnaces operate at low lagging power factor.

(*iii*) The load on the power system is varying ; being high during morning and evening and low at other times. During low load period, supply voltage is increased which increases the magnetisation current. This results in the decreased power factor.

6.5 Power Factor Improvement

The low power factor is mainly due to the fact that most of the power loads are inductive and, therefore, take lagging currents. In order to improve the power factor, some device taking leading power should be connected in parallel with the load. One of such devices can be a capacitor. The capacitor draws a leading current and partly or completely neutralises the lagging reactive component of load current. This raises the power factor of the load.

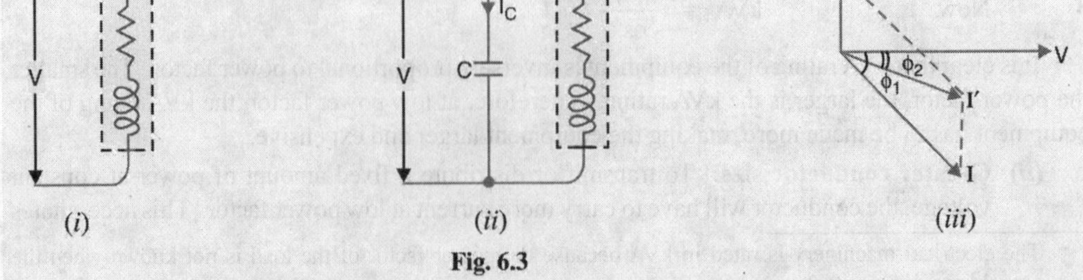

Fig. 6.3

Illustration. To illustrate the power factor improvement by a capacitor, consider a single *phase load taking lagging current I at a power factor $\cos \phi_1$ as shown in Fig. 6.3.

The capacitor C is connected in parallel with the load. The capacitor draws current I_C which leads the supply voltage by 90°. The resulting line current I' is the phasor sum of I and I_C and its angle of lag is ϕ_2 as shown in the phasor diagram of Fig. 6.3. (*iii*). It is clear that ϕ_2 is less than ϕ_1, so that $\cos \phi_2$ is greater than $\cos \phi_1$. Hence, the power factor of the load is improved. The following points are worth noting :

(*i*) The circuit current I' after p.f. correction is less than the original circuit current I.

(*ii*) The active or wattful component remains the same before and after p.f. correction because only the lagging reactive component is reduced by the capacitor.

$$\therefore \quad I \cos \phi_1 = I' \cos \phi_2$$

(*iii*) The lagging reactive component is reduced after p.f. improvement and is equal to the difference between lagging reactive component of load ($I \sin \phi_1$) and capacitor current (I_C) *i.e.*,

$$I' \sin \phi_2 = I \sin \phi_1 - I_C$$

(*iv*) As $\quad I \cos \phi_1 = I' \cos \phi_2$

$\therefore \quad VI \cos \phi_1 = VI' \cos \phi_2$ [Multiplying by V]

Therefore, active power (kW) remains unchanged due to power factor improvement.

(*v*) $\quad I' \sin \phi_2 = I \sin \phi_1 - I_C$

$\therefore \quad VI' \sin \phi_2 = VI \sin \phi_1 - VI_C$ [Multiplying by V]

i.e., Net kVAR *after* p.f. correction = Lagging kVAR *before* p.f. correction – leading kVAR of equipment

6.6 Power Factor Improvement Equipment

Normally, the power factor of the whole load on a large generating station is in the region of 0·8 to 0·9. However, sometimes it is lower and in such cases it is generally desirable to take special steps to improve the power factor. This can be achieved by the following equipment :

1. Static capacitors. 2. Synchronous condenser. 3. Phase advancers.

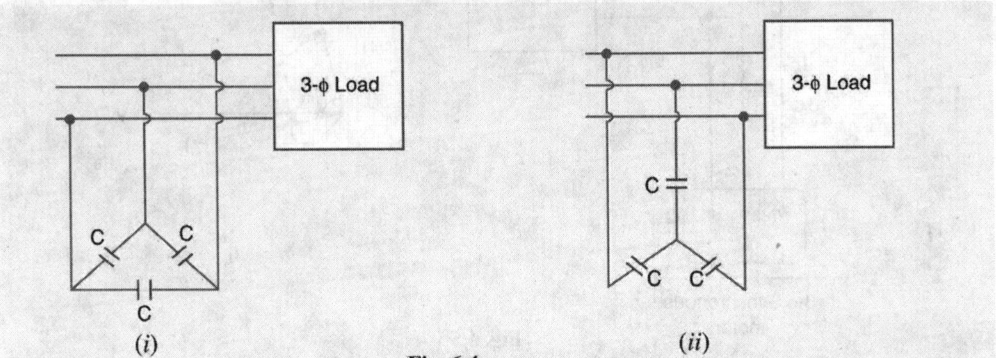

Fig. 6.4

1. Static capacitor. The power factor can be improved by connecting capacitors in parallel with the equipment operating at lagging power factor. The capacitor (generally known as static**

* The treatment can be used for 3-phase balanced loads *e.g.*, 3-ϕ induction motor. In a balanced 3-ϕ load, analysis of one phase leads to the desired results.

** To distinguish from the so called *synchronous condenser* which is a synchrnous motor running at no load and taking leading current.

capacitor) draws a leading current and partly or completely neutralises the lagging reactive component of load current. This raises the power factor of the load. For three-phase loads, the capacitors can be connected in delta or star as shown in Fig. 6.4. Static capacitors are invariably used for power factor improvement in factories.

Advantages

(*i*) They have low losses.

(*ii*) They require little maintenance as there are no rotating parts.

(*iii*) They can be easily installed as they are light and require no foundation.

(*iv*) They can work under ordinary atmospheric conditions.

Disadvantages

(*i*) They have short service life ranging from 8 to 10 years.

(*ii*) They are easily damaged if the voltage exceeds the rated value.

(*iii*) Once the capacitors are damaged, their repair is uneconomical.

2. Synchronous condenser. A synchronous motor takes a leading current when over-excited and, therefore, behaves as a capacitor. An over-excited synchronous motor running on no load is known as *synchronous condenser.* When such a machine is connected in parallel with the supply, it takes a leading current which partly neutralises the lagging reactive component of the load. Thus the power factor is improved.

Fig 6.5 shows the power factor improvement by synchronous condenser method. The 3ϕ load takes current I_L at low lagging power factor cos ϕ_L. The synchronous condenser takes a current I_m which leads the voltage by an angle ϕ_m*. The resultant current I is the phasor sum of I_m and I_L and lags behind the voltage by an angle ϕ. It is clear that ϕ is less than ϕ_L so that cos ϕ is greater than cos ϕ_L. Thus the power factor is increased from cos ϕ_L to cos ϕ. Synchronous condensers are generally used at major bulk supply substations for power factor improvement.

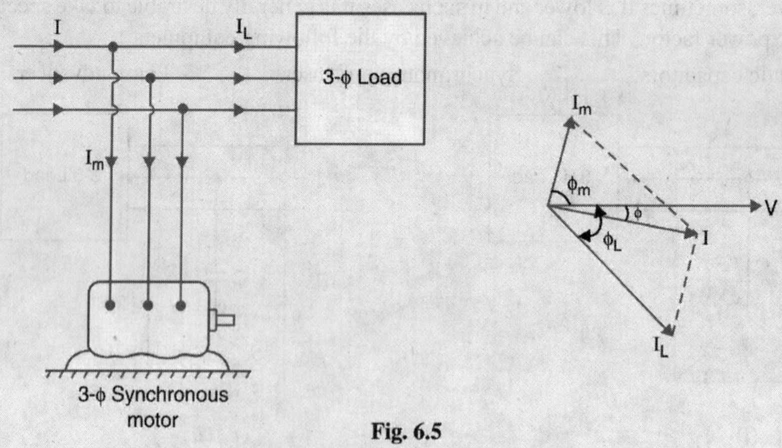

Fig. 6.5

Advantages

(*i*) By varying the field excitation, the magnitude of current drawn by the motor can be changed by any amount. This helps in achieving stepless † control of power factor.

* If the motor is ideal *i.e.,* there are no losses, then $\phi_m = 90°$. However, in actual practice, losses do occur in the motor even at no load. Therefore, the currents I_m leads the voltage by an angle less than 90°.

† The *p.f.* improvement with capacitors can only be done in steps by switching on the capacitors in various groupings. However, with synchronous motor, any amount of capacitive reactance can be provided by changing the field excitation.

(*ii*) The motor windings have high thermal stability to short circuit currents.

(*iii*) The faults can be removed easily.

Disadvantages

(*i*) There are considerable losses in the motor.

(*ii*) The maintenance cost is high.

(*iii*) It produces noise.

(*iv*) Except in sizes above 500 kVA, the cost is greater than that of static capacitors of the same rating.

(*v*) As a synchronous motor has no self-starting torque, therefore, an auxiliary equipment has to be provided for this purpose.

Note. The reactive power taken by a synchronous motor depends upon two factors, the d.c. field excitation and the mechanical load delivered by the motor. Maximum leading power is taken by a synchronous motor with maximum excitation and zero load.

Synchronous Condenser

Static Capacitor

3. **Phase advancers.** Phase advancers are used to improve the power factor of induction motors. The low power factor of an induction motor is due to the fact that its stator winding draws exciting current which lags behind the supply voltage by 90°. If the exciting ampere turns can be provided from some other a.c. source, then the stator winding will be relieved of exciting current and the power factor of the motor can be improved. This job is accomplished by the phase advancer which is simply an a.c. exciter. The phase advancer is mounted on the same shaft as the main motor and is connected in the rotor circuit of the motor. It provides exciting ampere turns to the rotor circuit at slip frequency. By providing more ampere turns than required, the induction motor can be made to operate on leading power factor like an over-excited synchronous motor.

Phase advancers have two principal advantages. Firstly, as the exciting ampere turns are supplied at slip frequency, therefore, lagging kVAR drawn by the motor are considerably reduced. Secondly, phase advancer can be conveniently used where the use of synchronous motors is unadmissible. However, the major disadvantage of phase advancers is that they are not economical for motors below 200 H.P.

6.7 Calculations of Power Factor Correction

Consider an inductive load taking a lagging current I at a power factor cos ϕ_1. In order to improve the power factor of this circuit, the remedy is to connect such an equipment in parallel with the load which takes a leading reactive component and partly cancels the lagging reactive component of the load. Fig. 6.6 (*i*) shows a capacitor connected across the load. The capacitor takes a current I_C which leads the supply voltage V by 90°. The current I_C partly cancels the lagging reactive component of the load current as shown in the phasor diagram in Fig. 6.6 (*ii*). The resultant circuit current becomes I' and its angle of lag is ϕ_2. It is clear that ϕ_2 is less than ϕ_1 so that new p.f. cos ϕ_2 is more than the previous p.f. cos ϕ_1.

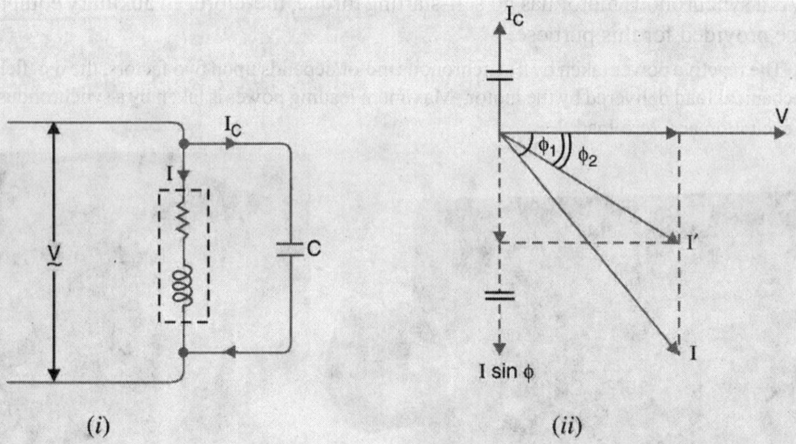

Fig. 6.6

From the phasor diagram, it is clear that after p.f. correction, the lagging reactive component of the load is reduced to $I'\sin\phi_2$.

Obviously, $I'\sin\phi_2 = I\sin\phi_1 - I_C$

or $I_C = I\sin\phi_1 - I'\sin\phi_2$

∴ Capacitance of capacitor to improve p.f. from cos ϕ_1 to cos ϕ_2

$$= \frac{I_C}{\omega V} \qquad \left(\because X_C = \frac{V}{I_C} = \frac{1}{\omega C}\right)$$

Power triangle. The power factor correction can also be illustrated from power triangle. Thus referring to Fig. 6.7, the power triangle *OAB* is for the power factor cos ϕ_1, whereas power triangle *OAC* is for the improved power factor cos ϕ_2. It may be seen that active power (*OA*) does not change with power factor improvement. However, the lagging kVAR of the load is reduced by the p.f. correction equipment, thus improving the p.f. to cos ϕ_2.

Leading kVAR supplied by p.f. correction equipment

$$= BC = AB - AC$$
$$= \text{kVAR}_1 - \text{kVAR}_2$$
$$= OA(\tan\phi_1 - \tan\phi_2)$$
$$= \text{kW}(\tan\phi_1 - \tan\phi_2)$$

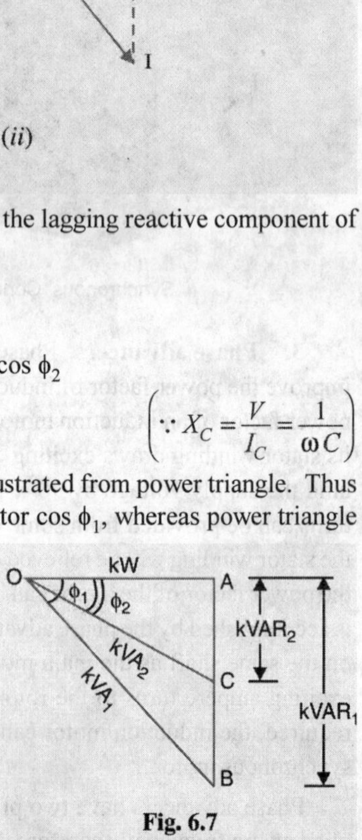

Fig. 6.7

Knowing the leading kVAR supplied by the p.f. correction equipment, the desired results can be obtained.

Example 6.1 *An alternator is supplying a load of 300 kW at a p.f. of 0·6 lagging. If the power factor is raised to unity, how many more kilowatts can alternator supply for the same kVA loading ?*

Solution :

$$kVA = \frac{kW}{\cos\phi} = \frac{300}{0{\cdot}6} = 500 \text{ kVA}$$

$$\text{kW at } 0{\cdot}6 \text{ p.f.} = 300 \text{ kW}$$

$$\text{kW at 1 p.f.} = 500 \times 1 = 500 \text{ kW}$$

$\therefore$ Increased power supplied by the alternator

$$= 500 - 300 = \mathbf{200\ kW}$$

Note the importance of power factor improvement. When the p.f. of the alternator is unity, the 500 kVA are also 500 kW and the engine driving the alternator has to be capable of developing this power together with the losses in the alternator. But when the power factor of the load is 0·6, the power is only 300 kW. Therefore, the engine is developing only 300 kW, though the alternator is supplying its rated output of 500 kVA.

Example 6.2 *A single phase motor connected to 400 V, 50 Hz supply takes 31·7A at a power factor of 0·7 lagging. Calculate the capacitance required in parallel with the motor to raise the power factor to 0·9 lagging.*

Solution : The circuit and phasor diagrams are shown in Figs. 6.8 and 6.9 respectively. Here motor M is taking a current I_M of 31·7A. The current I_C taken by the capacitor must be such that when combined with I_M, the resultant current I lags the voltage by an angle ϕ where $\cos\phi = 0{\cdot}9$.

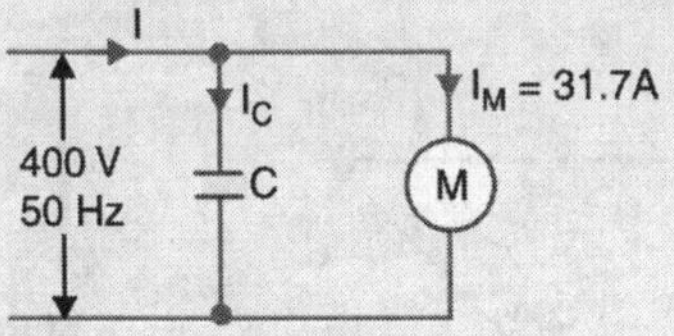

Fig. 6.8

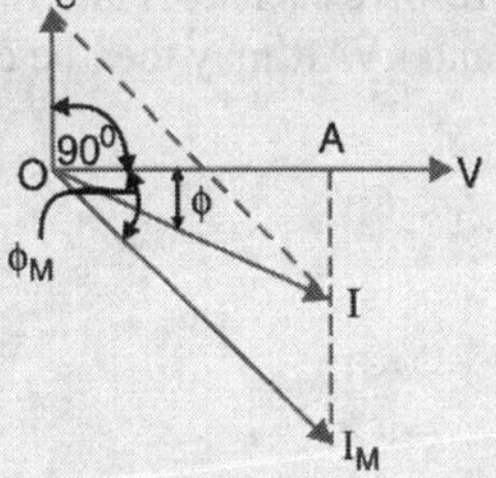

Fig. 6.9

Referring to the phasor diagram in Fig. 6.9,

$$\text{Active component of } I_M = I_M \cos\phi_M = 31{\cdot}7 \times 0{\cdot}7 = 22{\cdot}19\text{A}$$

$$\text{Active component of } I = I\cos\phi = I \times 0{\cdot}9$$

These components are represented by OA in Fig. 6.9.

$$\therefore \quad I = \frac{22{\cdot}19}{0{\cdot}9} = 24{\cdot}65\text{A}$$

$$\text{Reactive component of } I_M = I_M \sin\phi_M = 31{\cdot}7 \times 0{\cdot}714^* = 22{\cdot}6\text{A}$$

$$\text{Reactive component of } I = I\sin\phi = 24{\cdot}65\sqrt{1-(0{\cdot}9)^2}$$

$$= 24{\cdot}65 \times 0{\cdot}436 = 10{\cdot}75 \text{ A}$$

It is clear from Fig. 6.9 that :

$$I_C = \text{Reactive component of } I_M - \text{Reactive component of } I$$

$$= 22{\cdot}6 - 10{\cdot}75 = 11{\cdot}85\text{A}$$

But

$$I_C = \frac{V}{X_C} = V \times 2\pi f C$$

or

$$11{\cdot}85 = 400 \times 2\pi \times 50 \times C$$

$$\therefore \quad C = 94{\cdot}3 \times 10^{-6} \text{F} = \mathbf{94{\cdot}3\ \mu F}$$

* $\sin\phi_M = \sqrt{1-\cos^2\phi_M} = \sqrt{1-(0{\cdot}7)^2} = 0{\cdot}714$

Note the effect of connecting a 94·3 μF capacitor in parallel with the motor. The current taken from the supply is reduced from 31·7 A to 24·65 A without altering the current or power taken by the motor. This enables an economy to be affected in the size of generating plant and in the cross-sectional area of the conductors.

Example 6.3 *A single phase a.c. generator supplies the following loads :*

(i) *Lighting load of 20 kW at unity power factor.*

(ii) *Induction motor load of 100 kW at p.f. 0·707 lagging.*

(iii) *Synchronous motor load of 50 kW at p.f. 0·9 leading.*

Calculate the total kW and kVA delivered by the generator and the power factor at which it works.

Solution : Using the suffixes 1, 2 and 3 to indicate the different loads, we have,

$$kVA_1 = \frac{kW_1}{\cos\phi_1} = \frac{20}{1} = 20 \text{ kVA}$$

$$kVA_2 = \frac{kW_2}{\cos\phi_2} = \frac{100}{0\cdot707} = 141\cdot4 \text{ kVA}$$

$$kVA_3 = \frac{kW_3}{\cos\phi_3} = \frac{50}{0\cdot9} = 55\cdot6 \text{ kVA}$$

These loads are represented in Fig. 6.10. The three kVAs' are not in phase. In order to find the total kVA, we resolve each kVA into rectangular components – kW and kVAR as shown in Fig. 6.10. The total kW and kVAR may then be combined to obtain total kVA.

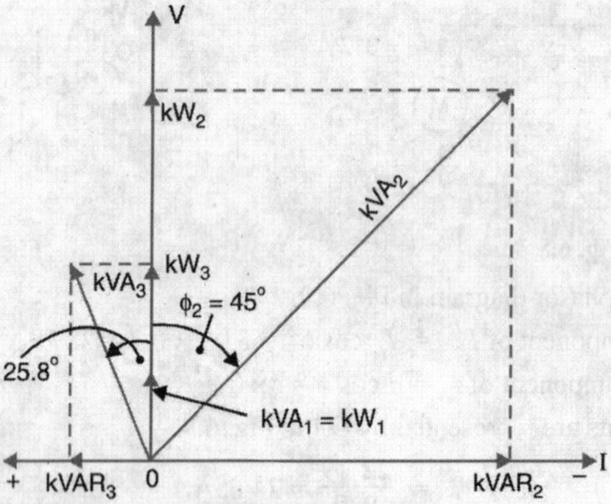

Fig. 6.10

$$kVAR_1 = kVA_1 \sin\phi_1 = 20 \times 0 = 0$$

$$kVAR_2 = kVA_2 \sin\phi_2 = -141\cdot4 \times 0\cdot707 = -100 \text{ kVAR}$$

$$kVAR_3 = kVA_3 \sin\phi_3 = +55\cdot6 \times 0\cdot436 = +24\cdot3 \text{ kVAR}$$

Note that $kVAR_2$ and $kVAR_3$ are in opposite directions ; $kVAR_2$ being a lagging while $kVAR_3$ being a leading kVAR.

$$\text{Total kW} = 20 + 100 + 50 = \mathbf{170\ kW}$$

$$\text{Total kVAR} = 0 - 100 + 24\cdot3 = -75\cdot7 \text{ kVAR}$$

$$\text{Total kVA} = \sqrt{(kW)^2 + (kVAR)^2} = \sqrt{(170)^2 + (75\cdot7)^2} = \mathbf{186\ kVA}$$

$$\text{Power factor} = \frac{\text{Total kW}}{\text{Total kVA}} = \frac{170}{186} = \mathbf{0\cdot914\ lagging}$$

The power factor must be lagging since the resultant kVAR is lagging.

Example 6.4 *A 3-phase, 5 kW induction motor has a p.f. of 0·75 lagging. A bank of capacitors is connected in delta across the supply terminals and p.f. raised to 0·9 lagging. Determine the kVAR rating of the capacitors connected in each phase.*

Solution :

Original p.f., $\cos\phi_1$ = 0·75 lag ; Motor input, P = 5 kW

Final p.f., $\cos\phi_2$ = 0·9 lag ; Efficiency, η = 100 % (assumed)

$\phi_1 = \cos^{-1}(0{\cdot}75) = 41{\cdot}41^\circ$; $\tan\phi_1 = \tan 41{\cdot}41^\circ = 0{\cdot}8819$

$\phi_2 = \cos^{-1}(0{\cdot}9) = 25{\cdot}84^\circ$; $\tan\phi_2 = \tan 25{\cdot}84^\circ = 0{\cdot}4843$

Leading kVAR taken by the condenser bank

$$= P(\tan\phi_1 - \tan\phi_2)$$

$$= 5\,(0{\cdot}8819 - 0{\cdot}4843) = 1{\cdot}99 \text{ kVAR}$$

∴ Rating of capacitors connected in each phase

$$= 1{\cdot}99/3 = \mathbf{0{\cdot}663\ kVAR}$$

Example 6.5 *A 3-phase, 50 Hz, 400 V motor develops 100 H.P. (74·6 kW), the power factor being 0·75 lagging and efficiency 93%. A bank of capacitors is connected in delta across the supply terminals and power factor raised to 0·95 lagging. Each of the capacitance units is built of 4 similar 100 V capacitors. Determine the capacitance of each capacitor.*

Solution :

Original p.f., $\cos\phi_1$ = 0·75 lag ; Final p.f., $\cos\phi_2$ = 0·95 lag

Motor input, P = output/η = 74·6/0·93 = 80 kW

$\phi_1 = \cos^{-1}(0{\cdot}75) = 41{\cdot}41^\circ$

$\tan\phi_1 = \tan 41{\cdot}41^\circ = 0{\cdot}8819$

$\phi_2 = \cos^{-1}(0{\cdot}95) = 18{\cdot}19^\circ$

$\tan\phi_2 = \tan 18{\cdot}19^\circ = 0{\cdot}3288$

Leading kVAR taken by the condenser bank

$$= P(\tan\phi_1 - \tan\phi_2)$$

$$= 80\,(0{\cdot}8819 - 0{\cdot}3288) = 44{\cdot}25 \text{ kVAR}$$

Leading kVAR taken by each of three sets

$$= 44{\cdot}25/3 = 14{\cdot}75 \text{ kVAR} \quad \ldots (i)$$

Fig. 6.11 shows the delta* connected condenser bank. Let C farad be the capacitance of 4 capacitors in each phase.

Phase current of capacitor is

$$I_{CP} = V_{ph}/X_C = 2\pi f C V_{ph}$$

$$= 2\pi \times 50 \times C \times 400$$

$$= 1{,}25{,}600\ C \text{ amperes}$$

$$\text{kVAR/phase} = \frac{V_{ph}\, I_{CP}}{1000}$$

$$= \frac{400 \times 1{,}25{,}600\ C}{1000}$$

$$= 50240\ C \quad \ldots (ii)$$

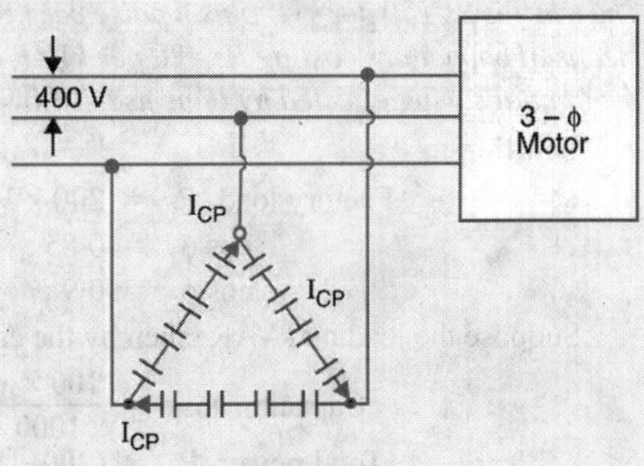

Fig. 6.11

* In practice, capacitors are always connected in delta since the capacitance of the capacitor required is one-third of that required for star connection.

Equating exps. (*i*) and (*ii*), we get,

$$50240\,C = 14{\cdot}75$$

$$\therefore \quad C = 14{\cdot}75/50{,}240 = 293{\cdot}4 \times 10^{-6}\ \text{F} = 293{\cdot}4\ \mu\text{F}$$

Since it is the combined capacitance of four equal capacitors joined in series,

$\therefore$ Capacitance of each capacitor = 4 × 293·4 = **1173·6 μF**

Example 6.6. *The load on an installation is 800 kW, 0·8 lagging p.f. which works for 3000 hours per annum. The tariff is Rs 100 per kVA plus 20 paise per kWh. If the power factor is improved to 0·9 lagging by means of loss-free capacitors costing Rs 60 per kVAR, calculate the annual saving effected. Allow 10% per annum for interest and depreciation on capacitors.*

Solution.

Load, P = 800 kW

$\cos\phi_1 = 0{\cdot}8$; $\tan\phi_1 = \tan(\cos^{-1} 0{\cdot}8) = 0{\cdot}75$

$\cos\phi_2 = 0{\cdot}9$; $\tan\phi_2 = \tan(\cos^{-1} 0{\cdot}9) = 0{\cdot}4843$

Leading kVAR taken by the capacitors

$$= P(\tan\phi_1 - \tan\phi_2) = 800\,(0{\cdot}75 - 0{\cdot}4843) = 212{\cdot}56$$

Annual cost before p.f. correction

Max. kVA demand = 800/0·8 = 1000

kVA demand charges = Rs 100 × 1000 = Rs 1,00,000

Units consumed/year = 800 × 3000 = 24,00,000 kWh

Energy charges/year = Rs 0·2 × 24,00,000 = Rs 4,80,000

Total annual cost = Rs (1,00,000 + 4,80,000) = Rs 5,80,000

Annual cost after p.f. correction

Max. kVA demand = 800/0·9 = 888·89

kVA demand charges = Rs 100 × 888·89 = Rs 88,889

Energy charges = Same as before *i.e.*, Rs 4,80,000

Capital cost of capacitors = Rs 60 × 212·56 = Rs 12,750

Annual interest and depreciation = Rs 0·1 × 12750 = Rs 1275

Total annual cost = Rs (88,889 + 4,80,000 + 1275) = Rs 5,70,164

$\therefore$ Annual saving = Rs (5,80,000 – 5,70,164) = **Rs 9836**

Example 6.7. *A factory takes a load of 200 kW at 0·85 p.f. lagging for 2500 hours per annum. The traiff is Rs 150 per kVA plus 5 paise per kWh consumed. If the p.f. is improved to 0·9 lagging by means of capacitors costing Rs 420 per kVAR and having a power loss of 100 W per kVA, calculate the annual saving effected by their use. Allow 10% per annum for interest and depreciation.*

Solution :

Factory load, P_1 = 200 kW

$\cos\phi_1 = 0{\cdot}85$; $\tan\phi_1 = 0{\cdot}62$

$\cos\phi_2 = 0{\cdot}9$; $\tan\phi_2 = 0{\cdot}4843$

Suppose the leading kVAR taken by the capacitors is x.

$$\therefore \quad \text{Capacitor loss} = \frac{100 \times x}{1000} = 0{\cdot}1\,x\ \text{kW}$$

Total power, $P_2 = (200 + 0{\cdot}1x)$ kW

Leading kVAR taken by the capacitors is

$$x = P_1 \tan\phi_1 - P_2 \tan\phi_2$$

$$= 200 \times 0{\cdot}62 - (200 + 0{\cdot}1x) \times 0{\cdot}4843$$

or $x = 124 - 96{\cdot}86 - 0{\cdot}04843\,x$

$\therefore$ $x = 27{\cdot}14/1{\cdot}04843 = 25{\cdot}89$ kVAR

Annual cost before p.f. improvement

Max. kVA demand = 200/0.85 = 235.3

kVA demand charges = Rs 150 × 235·3 = Rs 35,295

Units consumed/year = 200 × 2500 = 5,00,000 kWh

Energy charges = Rs 0·05 × 5,00,000 = Rs 25,000

Total annual cost = Rs (35,295 + 25,000) = Rs 60,295

Annual cost after p.f. improvement

Max. kVA demand = 200/0·9 = 222·2

kVA demand charges = Rs 150 × 222·2 = Rs 33,330

Energy charges = same as before *i.e.*, Rs 25,000

Annual interest and depreciation = Rs 420 × 25·89 × 0·1 = Rs 1087

Annual energy loss in capacitors = $0{\cdot}1\,x \times 2500 = 0{\cdot}1 \times 25{\cdot}89 \times 2500 = 6472$ kWh

Annual cost of losses occurring in capacitors

= Rs 0·05 × 6472 = Rs 323

$\therefore$ Total annual cost = Rs (33,330 + 25,000 + 1087 + 323) = Rs 59,740

Annual saving = Rs (60,295 – 59,740) = **Rs 555**

Example 6.8. *A factory operates at 0·8 p.f. lagging and has a monthly demand of 750 kVA. The monthly power rate is Rs 8·50 per kVA. To improve the power factor, 250 kVA capacitors are installed in which there is negligible power loss. The installed cost of equipment is Rs 20,000 and fixed charges are estimated at 10% per year. Calculate the annual saving effected by the use of capacitors.*

Solution.

Monthly demand is 750 kVA.

$\cos\phi = 0{\cdot}8\,;\ \sin\phi = \sin(\cos^{-1} 0{\cdot}8) = 0{\cdot}6$

kW component of demand = kVA × cos ϕ = 750 × 0·8 = 600

kVAR component of demand = kVA × sin ϕ = 750 × 0·6 = 450

Leading kVAR taken by the capacitors is 250 kVAR. Therefore, net kVAR after p.f. improvement is 450 – 250 = 200.

$\therefore$ kVA after p.f. improvement = $\sqrt{(600)^2 + (200)^2} = 632{\cdot}45$

Reduction in kVA = 750 – 632·45 = 117·5

Monthly saving on kVA charges = Rs 8·5 × 117·5 = Rs 998·75

Yearly saving on kVA charges = Rs 998·75 × 12 = Rs 11,985

Fixed charges/year = Rs 0·1 × 20,000 = Rs 2000

Net annual saving = Rs (11,985 – 2000) = **Rs 9,985**

Example 6.9. *A synchronous motor improves the power factor of a load of 200 kW from 0.8 lagging to 0.9 lagging. Simultaneously the motor carries a load of 80 kW. Find (i) the leading kVAR taken by the motor (ii) kVA rating of the motor and (iii) power factor at which the motor operates.*

Solution.

Load, $P_1 = 200$ kW ; Motor load, $P_2 = 80$ kW

p.f. of load, $\cos\phi_1 = 0{\cdot}8$ lag

p.f. of combined load, $\cos\phi_2 = 0{\cdot}9$ lag

Combined load, $P = P_1 + P_2 = 200 + 80 = 280$ kW

In Fig. 6.12, Δ *OAB* is the power triangle for load, Δ *ODC* for combined load and Δ *BEC* for the motor.

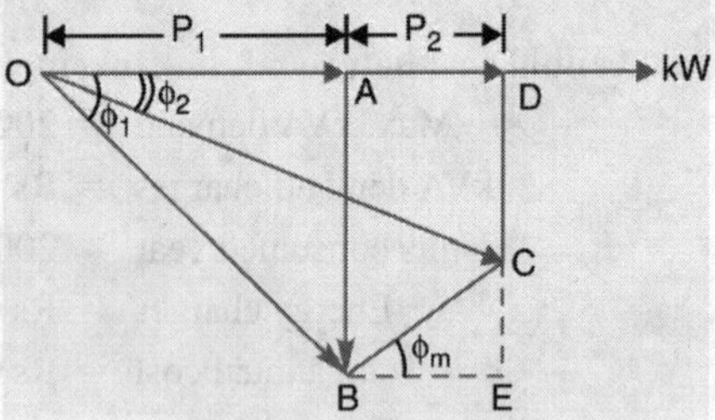

Fig. 6.12

(*i*) Leading kVAR taken by the motor

$= CE = DE - DC = AB - DC$

$[\because AB = DE]$

$= P_1 \tan \phi_1 - P^* \tan \phi_2$

$= 200 \tan(\cos^{-1} 0{\cdot}8) - 280 \tan(\cos^{-1} 0{\cdot}9)$

$= 200 \times 0{\cdot}75 - 280 \times 0{\cdot}4843$

$=$ **14·4 kVAR**

(*ii*) kVA rating of the motor $= BC = \sqrt{(BE)^2 + (EC)^2} = \sqrt{(80)^2 + (14{\cdot}4)^2} =$ **81·28 kVA**

(*iii*) p.f. of motor, $\cos \phi_m = \dfrac{\text{Motor kW}}{\text{Motor kVA}} = \dfrac{80}{81{\cdot}28} =$ **0·984 leading**

Example 6.10. *A factory load consists of the following :*

(*i*) *an induction motor of 50 H.P. (37·3 kW) with 0·8 p.f. and efficiency 0·85.*

(*ii*) *a synchronous motor of 25 H.P. (18·65 kW) with 0·9 p.f. leading and efficiency 0·9.*

(*iii*) *lighting load of 10 kW at unity p.f.*

Find the annual electrical charges if the tariff is Rs 60 per kVA of maximum demand per annum plus 5 paise per kWh ; assuming the load to be steady for 2000 hours in a year.

Solution.

Input power to induction motor $= 37{\cdot}3/0{\cdot}85 = 43{\cdot}88$ kW

Lagging kVAR taken by induction motor $= 43{\cdot}88 \tan(\cos^{-1} 0{\cdot}8) = 32{\cdot}91$

Input power to synchronous motor

$= 18{\cdot}65/0{\cdot}9 = 20{\cdot}72$ kW

Leading kVAR taken by synchronous motor

$= 20{\cdot}72 \tan(\cos^{-1} 0{\cdot}9) = 10$

Since lighting load works at unity p.f., its lagging kVAR = 0.

Net lagging kVAR $= 32{\cdot}91 - 10 = 22{\cdot}91$

Total active power $= 43{\cdot}88 + 20{\cdot}72 + 10 = 74{\cdot}6$ kW

Total kVA $= \sqrt{(74{\cdot}6)^2 + (22{\cdot}91)^2} = 78$

Annual kVA demand charges = Rs 60 × 78 = Rs 4,680

Energy consumed/year = 74·6 × 2000 = 1,49,200 kWh

Annual Energy charges = Rs 0·05 × 1,49,200 = Rs 7,460

Total annual bill = kVA demand charges + Energy charges

= Rs (4680 + 7460) = **Rs 12,140**

Example 6.11. *A supply system feeds the following loads* (*i*) *a lighting load of 500 kW* (*ii*) *a load of 400 kW at a p.f. of 0·707 lagging* (*iii*) *a load of 800 kW at a p.f. of 0·8 leading* (*iv*) *a load of 500 kW at p.f. 0·6 lagging* (*v*) *a synchronous motor driving a 540 kW d.c. generator and having an overall efficiency of 90%. Calculate the power factor of synchronous motor so that the station power factor may become unity.*

---

* In right angled triangle *OAB*, $AB = P_1 \tan \phi_1$

In right angled triangle *ODC*, $DC = OD \tan \phi_2 = (P_1 + P_2) \tan \phi_2 = P \tan \phi_2$

Solution. The lighting load works at unity p.f. and, therefore, its lagging kVAR is zero. The lagging kVAR are taken by the loads (*ii*) and (*iv*), whereas loads (*iii*) and (*v*) take the leading kVAR. For station power factor to be unity, the total lagging kVAR must be neutralised by the total leading kVAR. We know that kVAR = kW tan ϕ.

$\therefore$ Total lagging kVAR taken by loads (*ii*) and (*iv*)

$$= 400 \tan (\cos^{-1} 0{\cdot}707) + 500 \tan (\cos^{-1} 0{\cdot}6)$$
$$= 400 \times 1 + 500 \times 1{\cdot}33 = 1065$$

Leading kVAR taken by the load (*iii*)

$$= 800 \tan (\cos^{-1} 0{\cdot}8) = 800 \times 0{\cdot}75 = 600$$

$\therefore$ Leading kVAR to be taken by synchronous motor

$$= 1065 - 600 = 465 \text{ kVAR}$$

$$\text{Motor input} = \text{output/efficiency} = 540/0{\cdot}9 = 600 \text{ kW}$$

If ϕ is the phase angle of synchronous motor, then,

$$\tan \phi = \text{kVAR/kW} = 465/600 = 0{\cdot}775$$

$$\therefore \quad \phi = \tan^{-1} 0{\cdot}775 = 37{\cdot}77^{\circ}$$

$\therefore$ p.f. of synchronous motor $= \cos \phi = \cos 37{\cdot}77^{\circ} =$ **0·79 leading**

Therefore, in order that the station power factor may become unity, the synchronous motor should be operated at a p.f. of 0·79 leading.

Example 6.12. *An industrial load consists of* (*i*) *a synchronous motor of 100 metric h.p.* (*ii*) *induction motors aggregating 200 metric h.p., 0·707 power factor lagging and 82% efficiency and* (*iii*) *lighting load aggregating 30 kW.*

The tariff is Rs 100 per annum per kVA maximum demand plus 6 paise per kWh. Find the annual saving in cost if the synchronous motor operates at 0·8 p.f. leading, 93% efficiency instead of 0·8 p.f. lagging at 93% efficiency.

Solution. The annual power bill will be calculated under two conditions *viz.*, (*a*) when synchronous motor runs with lagging p.f. and (*b*) when synchronous motor runs with a leading p.f.

(*a*) When synchronous motor runs at p.f. 0·8 lagging. We shall find the combined kW and then calculate total kVA maximum demand using the relation :

$$\text{kVA} = \sqrt{(\text{kW})^2 + (\text{kVAR})^2}$$

$$\text{Input to synchronous motor} = \frac{100 \times 735{\cdot}5}{0{\cdot}93 \times 1000} = 79 \text{ kW}$$

*Lagging kVAR taken by the synchronous motor

$$= 79 \tan (\cos^{-1} 0{\cdot}8) = 79 \times 0{\cdot}75 = 59{\cdot}25 \text{ kVAR}$$

$$\text{Input to induction motors} = \frac{200 \times 735{\cdot}5}{0{\cdot}82 \times 1000} = 179.4 \text{ kW}$$

Lagging kVAR taken by induction motors

$$= 179{\cdot}4 \tan (\cos^{-1} 0{\cdot}707) = 179{\cdot}4 \times 1 = 179{\cdot}4 \text{ kVAR}$$

Since lighting load works at unity p.f., its lagging kVAR is zero.

$$\therefore \quad \text{Total lagging kVAR} = 59{\cdot}25 + 179{\cdot}4 = 238{\cdot}65 \text{ kVAR}$$

$$\text{Total active power} = 79 + 179{\cdot}4 + 30 = 288{\cdot}4 \text{ kW}$$

$$\text{Total kVA} = \sqrt{(238.65)^2 + (288{\cdot}4)^2} = 374{\cdot}4 \text{ kVA}$$

$$\text{Annual kVA demand charges} = \text{Rs } 100 \times 374{\cdot}4 = \text{Rs } 37{,}440$$

* Since the synchronous motor in this case runs at lagging p.f., it takes lagging kVAR.

Energy consumed/year = $288{\cdot}4 \times 8760 = 25{,}26384$ kWh

Annual energy charges = Rs $0{\cdot}06 \times 25{,}26{,}384$ = Rs 1,51,583

Total annual bill = Rs (37,440 + 1,51,583) = Rs 1,89,023

(*b*) When synchronous motor runs at p.f. 0·8 leading. As the synchronous motor runs at leading p.f. of 0·8 (instead of 0·8 p.f. lagging), therefore, it takes now 59·25 leading kVAR. The lagging kVAR taken by induction motors are the same as before *i.e.,* 179·4.

∴ Net lagging kVAR = $179{\cdot}4 - 59{\cdot}25 = 120{\cdot}15$

Total active power = Same as before *i.e.*, 288·4 kW

∴ Total kVA = $\sqrt{(120.15)^2 + (288 \cdot 4)^2} = 312{\cdot}4$

Annual kVA demand charges = Rs $100 \times 312{\cdot}4$ = Rs 31,240

Annual energy charges = Same as before *i.e.*, Rs 1,51,583

Total annual bill = Rs (31,240 + 1,51,583) = Rs 1,82,823

∴ Annual saving = Rs (1,89,023 − 1,82,823) = **Rs 6200**

TUTORIAL PROBLEMS

1. What should be the kVA rating of a capacitor which would raise the power factor of load of 100 kW from 0·5 lagging to 0·9 lagging ? **[125 kVA]**
2. A 3-phase, 50 Hz, 3300 V star connected induction motor develops 250 H.P. (186·5 kW), the power factor being 0·707 lagging and the efficiency 0·86. Three capacitors in delta are connected across the supply terminals and power factor raised to 0·9 lagging. Calculate :
   (*i*) the kVAR rating of the capacitor bank.
   (*ii*) the capacitance of each unit. **[(*i*) 111·8 kVAR (*ii*) 10·9 μF]**
3. A 3-phase, 50 Hz, 3000 V motor develops 600 H.P. (447·6 kW), the power factor being 0·75 lagging and the efficiency 0·93. A bank of capacitors is connected in delta across the supply terminals and power factor raised to 0·95 lagging. Each of the capacitance units is built of five similar 600-V capacitors. Determine the capacitance of each capacitor. **[156 μF]**
4. A factory takes a load of 800 kW at 0·8 p.f. (lagging) for 3000 hours per annum and buys energy on tariff of Rs 100 per kVA plus 10 paise per kWh. If the power factor is improved to 0·9 lagging by means of capacitors costing Rs 60 per kVAR and having a power loss of 100 W per kVA, calculate the annual saving effected by their use. Allow 10% per annum for interest and depreciation on the capacitors. **[Rs 3972]**
5. A station supplies 250 kVA at a lagging power factor of 0·8. A synchronous motor is connected in parallel with the load. If the combined load is 250 kW with a lagging p.f. of 0.9, determine :
   (*i*) the leading kVAR taken by the motor.
   (*ii*) kVA rating of the motor.
   (*iii*) p.f. at which the motor operates. **[(*i*) 28·9 kVAR (*ii*) 57·75 kVA (*iii*) 0·866 lead]**
6. A generating station supplies power to the following :
   (*i*) a lighting load of 100 kW;
   (*ii*) an induction motor 800 h.p. (596·8 kW) p.f. 0·8 lagging, efficiency 92%;
   (*iii*) a rotary converter giving 150 A at 400 V at an efficiency of 0·95.
   What must be the power factor of the rotary convertor in order that power factor of the supply station may become unity ? **[0·128 leading]**
7. A 3-phase, 400 V synchronous motor having a power consumption of 50 kW is connected in parallel with an induction motor which takes 200 kW at a power factor of 0·8 lagging.
   (*i*) Calculate the current drawn from the mains when the power factor of the synchronous motor is unity.

(ii) At what power factor should the synchronous motor operate so that the current drawn from the mains is minimum. ? **[(i) 421 A (ii) 0·316 leading]**

8. A factory load consists of the following :
   (i) an induction motor of 150 h.p. (111·9 kW) with 0·7 p.f. lagging and 80% efficiency ;
   (ii) a synchronous motor of 100 h.p. (74·6 kW) with 0·85 p.f. leading at 90% efficiency ;
   (iii) a lighting load of 50 kW.

   Find the annual electric charges if the tariff is Rs 100 per annum per kVA maximum demand plus 7 paise per kWh ; assuming the load to be steady throughout the year. **[Rs 1,96,070]**

9. A 3-phase synchronous motor having a mechanical load (including losses) of 122 kW is connected in parallel with a load of 510 kW at 0·8 p.f. lagging. The excitation of the motor is adjusted so that the kVA input to the motor becomes 140 kVA. Determine the new power factor of the whole system. **[0·8956 lagging]**

10. A 3-phase synchronous motor is connected in parallel with a load of 700 kW at 0·7 power factor lagging and its excitation is adjusted till it raises the total p.f. to 0.9 lagging. Mechanical load on the motor including losses is 150 kW. Find the power factor of the synchronous motor. **[0·444 leading]**

6.8 Importance of Power Factor Improvement

The improvement of power factor is very important for both consumers and generating stations as discussed below :

(i) *For consumers.* A consumer* has to pay electricity charges for his maximum demand in kVA plus the units consumed. If the consumer imporves the power factor, then there is a reduction† in his maximum kVA demand and consequently there will be annual saving due to maximum demand charges. Although power factor improvement involves extra annual expenditure on account of p.f. correction equipment, yet improvement of p.f. to a *proper value* results in the net annual saving for the consumer.

(ii) *For generating stations.* A generating station is as much concerned with power factor improvement as the consumer. The generators in a power station are rated in kVA but the useful output depends upon kW output. As station output is kW = kVA × cos ϕ, therefore, number of units supplied by it depends upon the power factor. The greater the power factor of the generating station, the higher is the kWh it delivers to the system. This leads to the conclusion that improved power factor increases the earning capacity of the power station.

6.9 Most Economical Power Factor

If a consumer improves the power factor, there is reduction in his maximum kVA demand and hence there will be annual saving over the maximum demand charges. However, when power factor is improved, it involves capital investment on the power factor correction equipment. The consumer will incur expenditure every year in the shape of annual interest and depreciation on the investment made over the p.f. correction equipment. Therefore, the *net annual saving* will be equal to the annual saving in maximum demand charges *minus* annual expenditure incurred on p.f. correction equipment.

The value to which the power factor should be improved so as to have maximum net annual saving is known as the **most economical power factor.**

Consider a consumer taking a peak load of P kW at a power factor of cos ϕ_1 and charged at a rate of Rs x per kVA of maximum demand per annum. Suppose the consumer improves the power factor

---

* This is not applicable to domestic consumers because the domestic load (*e.g.,* lighting load) has a p.f. very close to unity. Here, consumer means industrial and other big consumers.

† Max. demand in kVA = $\dfrac{\text{Peak kW}}{\cos\phi}$

If cos ϕ is more, maximum kVA demand will be less and *vice-versa.*

to cos ϕ_2 by installing p.f. correction equipment. Let expenditure incurred on the p.f. correction equipment be Rs y per kVAR per annum. The power triangle at the original p.f. cos ϕ_1 is OAB and for the improved p.f. cos ϕ_2, it is OAC [See Fig. 6.13].

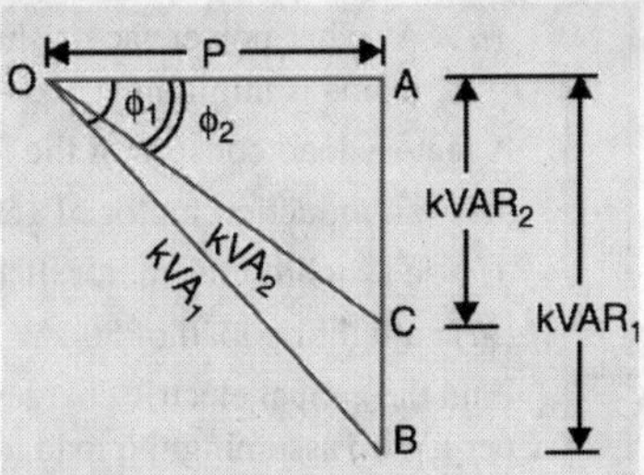

Fig. 6.13

kVA max. demand at cos ϕ_1, $kVA_1 = P/\cos\phi_1 = P\sec\phi_1$

kVA max. demand at cos ϕ_2, $kVA_2 = P/\cos\phi_2 = P\sec\phi_2$

Annual saving in maximum demand charges

$$= \text{Rs } x\,(kVA_1 - kVA_2)$$
$$= \text{Rs } x\,(P\sec\phi_1 - P\sec\phi_2)$$
$$= \text{Rs } x\,P\,(\sec\phi_1 - \sec\phi_2) \quad \text{...}(i)$$

Reactive power at cos ϕ_1, $kVAR_1 = P\tan\phi_1$

Reactive power at cos ϕ_2, $kVAR_2 = P\tan\phi_2$

Leading kVAR taken by p.f. correction equipment

$$= P\,(\tan\phi_1 - \tan\phi_2)$$

Annual cost of p.f. correction equipment

$$= \text{Rs } Py\,(\tan\phi_1 - \tan\phi_2) \quad \text{...}(ii)$$

$$\text{Net annual saving, } S = \text{exp. }(i) - \text{exp. }(ii)$$
$$= xP\,(\sec\phi_1 - \sec\phi_2) - yP\,(\tan\phi_1 - \tan\phi_2)$$

In this expression, only ϕ_2 is variable while all other quantities are fixed. Therefore, the net annual saving will be maximum if differentiation of above expression *w.r.t.* ϕ_2 is zero *i.e.*

$$\frac{d}{d\phi_2}(S) = 0$$

or $$\frac{d}{d\phi_2}[xP\,(\sec\phi_1 - \sec\phi_2) - yP\,(\tan\phi_1 - \tan\phi_2)] = 0$$

or $$\frac{d}{d\phi_2}(xP\sec\phi_1) - \frac{d}{d\phi_2}(xP\sec\phi_2) - \frac{d}{d\phi_2}(yP\tan\phi_1) + yP\frac{d}{d\phi_2}(\tan\phi_2) = 0$$

or $$0 - xP\sec\phi_2\tan\phi_2 - 0 + yP\sec^2\phi_2 = 0$$

or $$-x\tan\phi_2 + y\sec\phi_2 = 0$$

or $$\tan\phi_2 = \frac{y}{x}\sec\phi_2$$

or $$\sin\phi_2 = y/x$$

$\therefore$ $$\text{Most economical power factor, } \cos\phi_2 = \sqrt{1-\sin^2\phi_2} = \sqrt{1-(y/x)^2}$$

It may be noted that the most economical power factor (cos ϕ_2) depends upon the relative costs of supply and p.f. correction equipment but is independent of the original p.f. cos ϕ_1.

Example 6.13 *A factory which has a maximum demand of 175 kW at a power factor of 0·75 lagging is charged at Rs 72 per kVA per annum. If the phase advancing equipment costs Rs 120 per kVAR, find the most economical power factor at which the factory should operate. Interest and depreciation total 10% of the capital investment on the phase advancing equipment.*

Solution :

Power factor of the factory, cos ϕ_1 = 0·75 lagging

Max. demand charges, x = Rs 72 per kVA per annum

Expenditure on phase advancing equipment, y = Rs 120 × 0·1 = Rs 12* /kVAR/annum

* The total investment for producing 1 kVAR is Rs 120. The annual interest and depreciation is 10%. It means that an expenditure of Rs 120 × 10/100 = Rs 12 is incurred on 1 kVAR per annum.

∴ Most economical p.f. at which factory should operate is

$$\cos\phi_2 = \sqrt{1-(y/x)^2} = \sqrt{1-(12/72)^2} = \mathbf{0{\cdot}986\ lagging}$$

Example 6.14 *A consumer has an average demand of 400 kW at a p.f. of 0·8 lagging and annual load factor of 50%. The tariff is Rs 50 per kVA of maximum demand per annum plus 5 paise per kWh. If the power factor is improved to 0·95 lagging by installing phase advancing equipment, calculate :*

(*i*) *the capacity of the phase advancing equipment*

(*ii*) *the annual saving effected*

The phase advancing equipment costs Rs 100 per kVAR and the annual interest and depreciation together amount to 10%.

Solution :

Max. kW demand, $P = 400/0{\cdot}5 = 800$ kW

Original p.f., $\cos\phi_1 = 0{\cdot}8$ lag ; Final p.f., $\cos\phi_2 = 0{\cdot}95$ lag

$\phi_1 = \cos^{-1}(0{\cdot}8) = 36{\cdot}9^\circ$; $\tan\phi_1 = \tan 36{\cdot}9^\circ = 0{\cdot}75$

$\phi_2 = \cos^{-1}(0{\cdot}95) = 18{\cdot}2^\circ$; $\tan\phi_2 = \tan 18{\cdot}2^\circ = 0{\cdot}328$

(*i*) Leading kVAR taken by phase advancing equipment

$$= P(\tan\phi_1 - \tan\phi_2) = 800\,(0{\cdot}75 - 0{\cdot}328) = 337 \text{ kVAR}$$

∴ Capacity of phase advancing equipment should be **337 kVAR.**

(*ii*) Max. demand charges, x = Rs 50/kVA/annum

Expenditure on phase advancing equipment

y = Rs 0·1 × 100 = Rs 10/kVAR/annum

Max. kVA demand at 0·8 p.f. = 800/0·8 = 1000 kVA

Max. kVA demand at 0·95 p.f. = 800/0·95 = 842 kVA

Annual saving in maximum demand charges

= Rs 50 (1000 – 842) = Rs 7900

Annual expenditure on phase advancing equipment

= Rs (y × capacity of equipment)

= Rs 10 × 337 = 3370

∴ Net annual saving = Rs (7900 – 3370) = **Rs 4530**

Example 6.15 *A factory has an average demand of 50 kW and an annual load factor of 0·5. The power factor is 0·75 lagging. The tariff is Rs 100 per kVA of maximum demand per annum plus 5 paise per kWh. If loss free capacitors costing Rs 600 per kVAR are to be utilised, find the value of power factor at which maximum saving will result. The interest and depreciation together amount to 10%. Also determine the annual saving effected by improving the p.f. to this value.*

Solution :

Max. demand charge, x = Rs 100/kVA/annum

Expenditure on capacitors, y = Rs 0·1 × 600 = Rs 60/kVAR/annum

Most economical p.f., $\cos\phi_2 = \sqrt{1-(y/x)^2} = \sqrt{1-(60/100)^2} = \mathbf{0{\cdot}8\ lag}$

Max. kW demand = 50/0·5 = 100 kW

The maximum kVA demand at 0·75 p.f. is = 100/0·75 = 133·34 kVA, whereas it is = 100/0·8 = 125 kVA at 0·8 p.f.

∴ Annual saving = Rs 100 (133·34 – 125) = **Rs 834**

Example 6.16 *A factory takes a steady load of 200 kW at a lagging power factor of 0·8. The tariff is Rs 100 per kVA of maximum demand per annum plus 5 paise per kWh. The phase advancing plant costs Rs 500 per kVAR and the annual interest and depreciation together amount to 10%. Find:*

(i) the value to which the power factor be improved so that annual expenditure is minimum
(ii) the capacity of the phase advancing plant
(iii) the new bill for energy, assuming that the factory works for 5000 hours per annum.

Solution :

Peak load of factory, P = 200 kW
Original power factor, $\cos\phi_1$ = 0·8 lagging
Max. demand charges, x = Rs 100/kVA/annum
Charges on phase advancing plant, y = Rs 500 × 0·1
= Rs 50/kVAR/annum

(*i*) Most economical power factor, $\cos\phi_2 = \sqrt{1-(y/x)^2} = \sqrt{1-(50/100)^2}$ = **0·866 lagging**

(*ii*) Capacity of phase advancing plant $= P\,[\tan\phi_1 - \tan\phi_2]$
$= 200\,[\tan(\cos^{-1} 0{\cdot}8) - \tan(\cos^{-1} 0{\cdot}866)]$
$= 200\,[0{\cdot}75 - 0{\cdot}5774] =$ **34·52 kVAR**

(*iii*) Units consumed/year $= 200 \times 5000 = 10^6$ kWh
Annual energy charges = Rs $0{\cdot}05 \times 10^6$ = Rs 50,000
Annual cost of phase advancing plant = Rs y × Capacity of plant
= Rs 50 × 34·52 = Rs 1726
Max. demand charge = Rs $x \times P/\cos\phi_2$ = Rs 100 × 200/0·866 = Rs 23,094
Annual bill for energy = Rs (50,000 + 1726 + 23,094) = **Rs 74,820**

Example 6.17 *An industrial load takes 80,000 units in a year, the average power factor being 0·707 lagging. The recorded maximum demand is 500 kVA. The tariff is Rs 120 per kVA of maximum demand plus 2·5 paise per kWh. Calculate the annual cost of supply and find out the annual saving in cost by installing phase advancing plant costing Rs 50 per kVAR which raises the p.f. from 0·707 to 0·9 lagging. Allow 10% per year on the cost of phase advancing plant to cover all additional costs.*

Solution.

Energy consumed/year = 80,000 kWh
Maximum kVA demand = 500
Annual cost of supply = M.D. Charges + Energy charges
= Rs (120 × 500 + 0·025 × 80,000)
= Rs (60,000 + 2000) = **Rs 62,000**

$\cos\phi_1 = 0{\cdot}707$ lag ; $\cos\phi_2 = 0{\cdot}9$ lag

Max. kW demand at 0·707 p.f., P = 500 × 0·707 = 353·3 kW

Leading kVAR taken by phase advancing equipment
$= P\,[\tan\phi_1 - \tan\phi_2]$
$= 353{\cdot}3\,[\tan(\cos^{-1} 0{\cdot}707) - \tan(\cos^{-1} 0{\cdot}9)]$
$= 353{\cdot}3\,[1 - 0{\cdot}484] = 182{\cdot}3$ kVAR

Annual cost of phase advancing equipment
= Rs 182·3 × 50 × 0·1 = Rs 912

When p.f. is raised from 0·707 lag to 0·9 lag, new maximum kVA demand is = 353·3/0·9 = 392·6 kVA.

Reduction in kVA demand = 500 − 392·6 = 107·4

Annual saving in kVA charges = Rs 120 × 107·4 = Rs 12,888

As the units consumed remain the same, therefore, saving will be equal to saving in M.D. charges *minus* annual cost of phase advancing plant.

∴ Annual saving = Rs (12,882 − 912) = **Rs 11,976**

TUTORIAL PORBLEMS

1. A factory which has a maximum demand of 175 kW at a power factor of 0·75 lagging is charged at Rs 72 per kVA per annum. If the phase advancing equipment costs Rs 120 per kVAR, find the most economical power factor at which the factory should operate. Interest and depreciation total 10% of the capital investment on the phase advancing equipment. **[0·986 leading]**
2. A consumer has a steady load of 500 kW at a power factor of 0·8 lagging. The tariff in force is Rs 60 per kVA of maximum demand plus 5 paise per kWh. If the power factor is improved to 0·95 lagging by installing phase advancing equipment, calculate :
   (*i*) The capacity of the phase advancing equipment.
   (*ii*) The annual saving effected.
   The phase advancing equipment costs Rs 100 per kVAR and the annual interest and depreciation together amount to 10%. **[(*i*) 210·6 kVAR (*ii*) Rs. 3,815]**
3. A factory has an average demand of 320 kW and an annual load factor of 50%. The power factor is 0·8 lagging. The traiff is Rs 80 per annum per kVA of maximum demand plus 5 paise per kWh. If the loss free capacitors costing Rs 100 per kVAR are to be utilised, find the value of power factor at which maximum saving will result. The interest and depreciation together amount to 12%. Also determine the annual saving effected by improving the power factor to this value. **[0·988 lagging ; Rs 3040]**
4. What will be the kVA rating of a phase advancing plant if it improves p.f. from 0·8 lagging to 0·891 lagging ? The consumer load is 1000 kW and the current taken by the phase advancer leads the supply voltage at a p.f. of 0·1. **[230 kVA]**
5. A consumer takes a steady load of 300 kW at a lagging power factor of 0·7 for 3000 hours a year. The tariff is Rs 130 per kVA of maximum demand annually and 4 paise per kWh. The annual cost of phase advancing plant is Rs 13 per kVAR. Determine the annual saving if the power factor of the load is improved ? **[Rs 12929·8]**

6.10 Meeting the Increased kW Demand on Power Stations

The useful output of a power station is the kW output delivered by it to the supply system. Sometimes, a power station is required to deliver more kW to meet the increase in power demand. This can be achieved by either of the following two methods :

(*i*) By increasing the kVA capacity of the power station at the same power factor (say cos ϕ_1). Obviously, extra cost will be incurred to increase the kVA capacity of the station.

(*ii*) By improving the power factor of the station from cos ϕ_1 to cos ϕ_2 without increasing the kVA capacity of the station. This will also involve extra cost on account of power factor correction equipment.

Economical comparison of two methods. It is clear that each method of increasing kW capacity of the station involves extra cost. It is, therefore, desirable to make economical comparison of the two methods. Suppose a power station of rating *P* kVA is supplying load at p.f. of cos ϕ_1. Let us suppose that the new power demand can be met either by increasing the p.f. to cos ϕ_2 at *P* kVA *or* by increasing the kVA rating of the station at the original p.f. $\cos \phi_1$. The power* triangles for the whole situation are shown in Fig. 6.14.

(*i*) **Cost of increasing kVA capacity of station.** Referring to Fig. 6.14, the increase in kVA capacity of the station at $\cos \phi_1$ to meet the new demand is given by :

Increase in kVA capacity

$$= BD = \frac{BF}{\cos \phi_1} = \frac{AC}{\cos \phi_1} \quad (\because BF = AC)$$

$$= \frac{OC - OA}{\cos \phi_1}$$

$$= \frac{OE \cos \phi_2 - OB \cos \phi_1}{\cos \phi_1}$$

$$= \frac{P(\cos \phi_2 - \cos \phi_1)}{\cos \phi_1} \quad [\because OE = OB = P]$$

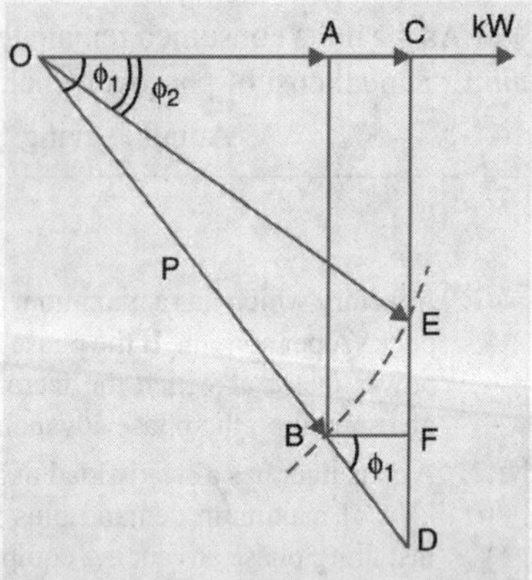

Fig. 6.14

If Rs x is the annual cost per kVA of the station, then,

Annual cost due to increase in kVA capacity

$$= \text{Rs } \frac{xP(\cos \phi_2 - \cos \phi_1)}{\cos \phi_1} \quad ...(i)$$

(*ii*) **Cost of p.f. correction equipment.** Referring to Fig. 6.14, the new demand in kW can be met by increasing the p.f. from $\cos \phi_1$ to $\cos \phi_2$ at the original kVA of the station. The leading kVAR to be taken by the p.f. correction equipment is given by *ED i.e.*

Leading kVAR taken by p.f. correction equipment

$$= ED = CD - CE$$

$$= OD \sin \phi_1 - OE \sin \phi_2$$

$$= \frac{OC}{\cos \phi_1} \sin \phi_1 - OE \sin \phi_2$$

$$= \frac{OE \cos \phi_2}{\cos \phi_1} \sin \phi_1 - OE \sin \phi_2$$

$$= OE (\tan \phi_1 \cos \phi_2 - \sin \phi_2)$$

$$= P (\tan \phi_1 \cos \phi_2 - \sin \phi_2)$$

If Rs. y is the annual cost per kVAR of the p.f. correction equipment, then,

Annual cost on p.f. correction equipment

$$= \text{Rs } y\,P (\tan \phi_1 \cos \phi_2 - \sin \phi_2) \quad ...(ii)$$

Different cases

(*a*) The p.f. correction equipment will be cheaper if

$$\text{exp. } (ii) < \text{exp. } (i)$$

or $$yP (\tan \phi_1 \cos \phi_2 - \sin \phi_2) < \frac{xP(\cos \phi_2 - \cos \phi_1)}{\cos \phi_1}$$

---

* Note the construction. Here Δ *OAB* is the power triangle for the station supplying *P* kVA at $\cos \phi_1$. The demand on the station is *OA* kW. The new demand is *OC kW*. This can be met :

(*i*) either by increasing the kVA demand of the station to *OD* at the same p.f. $\cos \phi_1$. Obviously, Δ OCD is the power triangle when the station is supplying *OC* kW at $\cos \phi_1$.

(*ii*) or by increasing the p.f. from $\cos \phi_1$ to $\cos \phi_2$ at same kVA *i.e., P* kVA. Obviously, *OB* = *OE*. Therefore, Δ *OCE* is the power triangle when the station is supplying *OC* kW at improved p.f. $\cos \phi_2$.

or $$y(\tan\phi_1\cos\phi_2 - \sin\phi_2) < x\,\frac{(\cos\phi_2 - \cos\phi_1)}{\cos\phi_1}$$

(*b*) The maximum annual cost per kVAR (*i.e.*, *y*) of p.f. correction equipment that would justify its installation is when

$$\text{exp. }(i) = \text{exp. }(ii)$$

or $$yP(\tan\phi_1\cos\phi_2 - \sin\phi_2) = \frac{xP(\cos\phi_2 - \cos\phi_1)}{\cos\phi_1}$$

or $$y\left(\frac{\sin\phi_1}{\cos\phi_1}\cos\phi_2 - \sin\phi_2\right) = \frac{x(\cos\phi_2 - \cos\phi_1)}{\cos\phi_1}$$

or $$y\left(\frac{\sin\phi_1\cos\phi_2 - \sin\phi_2\cos\phi_1}{\cos\phi_1}\right) = \frac{x(\cos\phi_2 - \cos\phi_1)}{\cos\phi_1}$$

or $$y\sin(\phi_1 - \phi_2) = x(\cos\phi_2 - \cos\phi_1)$$

$$\therefore \quad y = \frac{x(\cos\phi_2 - \cos\phi_1)}{\sin(\phi_1 - \phi_2)}$$

Example 6.18 *A power plant is working at its maximum kVA capacity with a lagging p.f. of 0·7. It is now required to increase its kW capacity to meet the demand of additional load. This can be done :*

(*i*) *by increasing the p.f. to 0·85 lagging by p.f. correction equipment*

or

(*ii*) *by installing additional generation plant costing Rs 800 per kVA.*

What is the maximum cost per kVA of p.f. correction equipment to make its use more economical than the additional plant ?

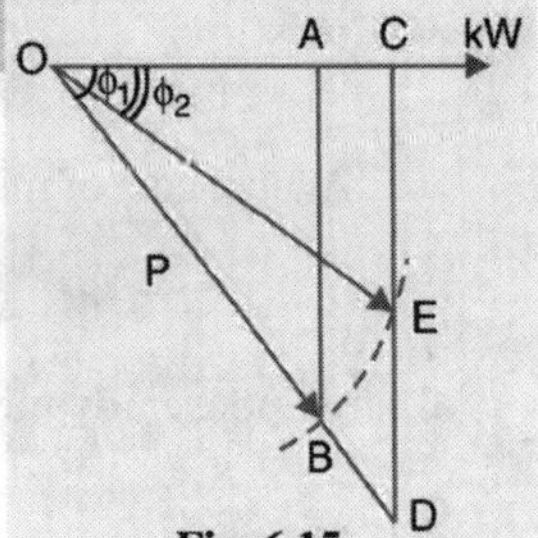

Fig. 6.15

Soloution. Let the initial capacity of the plant be *OB* kVA at a p.f. $\cos\phi_1$. Referring to Fig. 6.15, the new kW demand (*OC*) can be met by increasing the p.f. from 0·7 ($\cos\phi_1$) to 0·85 lagging ($\cos\phi_2$) at *OB* kVA *or* by increasing the capacity of the station to *OD* kVA at $\cos\phi_1$.

Cost of increasing plant capacity. Referring to Fig. 6.15, the increase in kVA capacity is *BD*.

Now $$OE\cos\phi_2 = OD\cos\phi_1$$

or $$OB\cos\phi_2 = OD\cos\phi_1 \quad (\because OE = OB)$$

$$\therefore \quad OD = OB \times \cos\phi_2/\cos\phi_1 = OB \times 0{\cdot}85/0{\cdot}7 = 1{\cdot}2143\ OB$$

Increase in the kVA capacity of the plant is

$$BD = OD - OB = 1{\cdot}2143 \times OB - OB = 0{\cdot}2143\ OB$$

$\therefore$ Total cost of increasing the plant capacity

$$= \text{Rs } 800 \times 0{\cdot}2143 \times OB$$
$$= \text{Rs } 171{\cdot}44 \times OB \qquad \ldots(i)$$

Cost of p.f. correction equipment.

$$\cos\phi_1 = 0{\cdot}7 \qquad \therefore \quad \sin\phi_1 = 0{\cdot}714$$
$$\cos\phi_2 = 0{\cdot}85 \qquad \therefore \quad \sin\phi_2 = 0{\cdot}527$$

Leading kVAR taken by p.f. correction equipment is

$$ED = CD - CE = OD\sin\phi_1 - OE\sin\phi_2$$
$$= 1{\cdot}2143 \times OB\sin\phi_1 - OB\sin\phi_2$$
$$= OB(1{\cdot}2143 \times 0{\cdot}714 - 0{\cdot}527) = 0{\cdot}34 \times OB$$

Let the cost per kVAR of the equipment be Rs y.

∴ Total cost of p.f. correction equipment

$$= \text{Rs } 0{\cdot}34 \times OB \times y \quad ...(ii)$$

The cost per kVAR of the equipment that would justify its installation is when exp. (*i*) = exp. (*ii*)

i.e.,

$$171{\cdot}44 \times OB = 0{\cdot}34 \times OB \times y$$

$$\therefore \quad y = \text{Rs } 171{\cdot}44/0{\cdot}34 = \text{Rs } 504{\cdot}2 \text{ per kVAR}$$

If the losses in p.f. correction equipment are neglected, then its kVAR = kVA. Therefore, the maximum cost per kVA of p.f. correction equipment that can be paid is **Rs 504·2.**

Example 6.19. *A system is working at its maximum kVA capacity with a lagging power factor 0·7. An anticipated increase of load can be met by one of the following two methods :*

(*i*) *By raising the p.f. of the system to 0·866 by installing phase advancing equipment.*

(*ii*) *By installing extra generating plant.*

If the total cost of generating plant is Rs 100 per kVA, estimate the limiting cost per kVA of phase advancing equipment to make its use more economical than the additional generating plant. Interest and depreciation charges may be assumed 10% in each case.

Solution. The original demand is OA and the increased demand is OC. Fig. 6.16 shows the two methods of meeting the increased kW demand (OC).

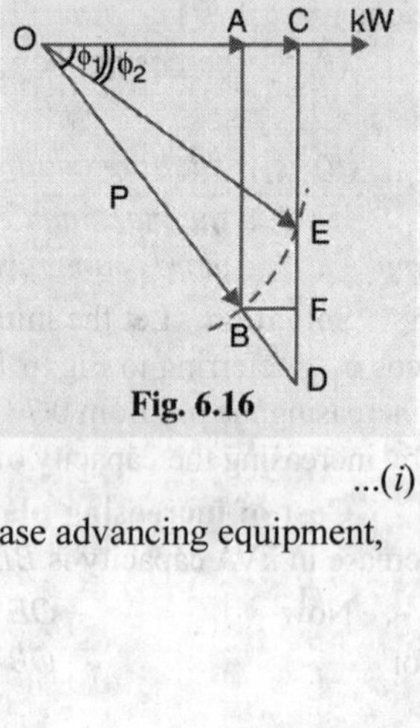

Fig. 6.16

Cost of increasing plant capacity

$$BD = OD - OB$$

$$= OB \times \frac{0{\cdot}866}{0{\cdot}70} - OB$$

$$= OB\,(1{\cdot}237 - 1)$$

$$= 0{\cdot}237 \times OB$$

∴ Annual cost of increasing the plant capacity

$$= \text{Rs } 10 \times 0{\cdot}237 \times OB$$

$$= \text{Rs. } 2{\cdot}37 \times OB \quad ...(i)$$

Cost of phase advancing equipment. Leading kVAR taken by phase advancing equipment,

$$ED = CD - CE$$

$$= OD^* \sin\phi_1 - OE \sin\phi_2$$

$$= 1{\cdot}237 \times OB \times \sin\phi_1 - OB \sin\phi_2$$

$$= OB\,(1{\cdot}237 \times 0{\cdot}174 - 0{\cdot}5) = 0{\cdot}383 \times OB$$

Let the cost per kVAR of the equipment be Rs y.

Annual cost of phase advancing equipment

$$= \text{Rs } 0{\cdot}1 \times y \times 0{\cdot}383 \times OB \quad ...(ii)$$

For economy, the two costs should be equal *i.e.*, exp. (*i*) = exp. (*ii*).

$$\therefore \quad 0{\cdot}1 \times y \times 0{\cdot}383 \times OB = 2{\cdot}37 \times OB$$

or

$$y = \text{Rs } \frac{2{\cdot}37}{0{\cdot}1 \times 0{\cdot}383} = \text{Rs } 61{\cdot}88$$

If the losses in the phase advancing equipment are neglected, then its kVAR = kVA. Hence, the maximum cost per kVA of phase advancing equipment that can be paid is **Rs 61·88**.

* $OD = OB + BD = OB + 0{\cdot}237 \times OB = 1{\cdot}237 \times OB$

TUTORIAL PROBLEMS

1. A system is working at its maximum capacity with a lagging power factor of 0·707. An anticipated increase in load can be met by (*i*) raising the power factor of the system to 0·87 lagging by the installation of phase advancers and (*ii*) by installing extra generating cables etc. to meet the increased power demand. The total cost of the latter method is Rs 110 per kVA. Estimate the limiting cost per kVA of the phase advancing plant which would justify the installation. **[Rs 76·26 per kVAR]**
2. For increasing the kW capacity of a power station working at 0·7 lagging power factor, the necessary increase in power can be obtained by raising power factor to 0·9 lagging or by installing additional plant. What is the maximum cost per kVA of power factor correction apparatus to make its use more economical than the additional plant at Rs 800 per kVA ? **[Rs 474 per kVA]**
3. An electrical system is working at its maximum kVA capacity with a lagging p.f. of 0·8. An anticipated increase of load can be met either by raising the p.f. of the system to 0·95 lagging by the installation of phase advancing plant or by erecting an extra generating plant and the required accessories. The total cost of the latter method is Rs 80 per kVA. Determine the economic limit cost per kVA of the phase advancing plant. Interest and depreciation may be assumed 12% in either case. **[Rs 37.50 per kVA]**

SELF-TEST

1. **Fill in the blanks by inserting appropriate words/figures.**
   (*i*) The power factor of an a.c. circuit is given by .... power divided by .... power.
   (*ii*) The lagging power factor is due to .... power drawn by the circuit.
   (*iii*) Power factor can be improved by installing such a device in parallel with load which takes ....
   (*iv*) The major reason for low lagging power factor of supply system is due to the use of .... motors.
   (*v*) An over-excited synchronous motor on no load is known as ....
2. **Pick up the correct words/figures from the brackets and fill in the blanks.**
   (*i*) The smaller the lagging reactive power drawn by a circuit. the .... is its power factor. (*smaller, greater*)
   (*ii*) The maximum value of power factor can be .... (*1, 0·5, 0·9*)
   (*iii*) KVAR = ... tan ϕ (*kW, KVA*)
   (*iv*) By improving the power factor of the system, the kilowatts delivered by the generating station are .... (*decreased, increased, not changed*)
   (*v*) The most economical power factor for a consumer is generally .... (*0·95 lagging, unity, 0·6 lagging*)

ANSWER TO SELF-TEST

1. (*i*) active, apparent, (*ii*) lagging reactive (*iii*) leading reactive power, (*iv*) induction (*v*) synchronous condenser.
2. (*i*) greater, (*ii*) 1, (*iii*) kW, (*iv*) increased, (*v*) 0·95 lagging.

CHAPTER REVIEW TOPICS

1. Why is there phase difference between voltage and current in an a.c. circuit ? Explain the concept of power factor.
2. Discuss the disadvantages of a low power factor.
3. Explain the causes of low power factor of the supply system.
4. Discuss the various methods for power factor improvement.
5. Derive an expression for the most economical value of power factor which may be attained by a consumer.

6. Show that the economical limit to which the power factor of a load can be raised is independent of the original value of power factor when the tariff consists of a fixed charge per kVA of maximum demand plus a flat rate per kWh.
7. Write short notes on the following :
   (*i*) Power factor improvement by synchronous condenser
   (*ii*) Importance of p.f. improvement
   (*iii*) Economics of p.f. improvement

DISCUSSION QUESTIONS

1. What is the importance of power factor in the supply system ?
2. Why is the power factor not more than unity ?
3. What is the effect of low power factor on the generating stations ?
4. Why is unity power factor not the most economical p.f. ?
5. Why a consumer having low power factor is charged at higher rates ?

CHAPTER

7

Supply Systems

Introduction

In early days, there was a little demand for electrical energy so that small power stations were built to supply lighting and heating loads. However, the widespread use of electrical energy by modern civilisation has necessitated to produce bulk electrical energy economically and efficiently. The increased demand of electrical energy can be met by building big power stations at favourable places where fuel (coal or gas) or water energy is available in abundance. This has shifted the site of power stations to places quite away from the consumers. The electrical energy produced at the power stations has to be supplied to the consumers. There is a large network of conductors between the power station and the consumers. This network can be broadly divided into two parts *viz.*, transmission and distribution. The purpose of this chapter is to focus attention on the various aspects of transmission of electric power.

7.1 Electric Supply System

The conveyance of electric power from a power station to consumers' premises is known as **electric supply system**.

An electric supply system consists of three principal components *viz.*, the power station, the transmission lines and the distribution system. Electric power is produced at the power stations which are located at favourable places, generally quite away from the consumers. It is then transmitted over large distances to load centres with the help of conductors known as transmission lines. Finally, it is distributed to a large number of small and big consumers through a distribution network.

The electric supply system can be broadly classified into (*i*) d.c. or a.c. system (*ii*) overhead or underground system. Now-a-days, 3-phase, 3-wire a.c. system is universally adopted for generation and transmission of electric power as an economical proposition. However, distribution of electric power is done by 3-phase, 4-wire a.c. system. The underground system is more expensive than the overhead system. Therefore, in our country, overhead system is *mostly adopted for transmission and distribution of electric power.

7.2 Typical a.c. Power Supply Scheme

The large network of conductors between the power station and the consumers can be broadly divided into two parts *viz.*, transmission system and distribution system. Each part can be further sub-divided into two—primary transmission and secondary transmission and primary distribution and secondary distribution. Fig. 7.1. shows the layout of a typical a.c. power supply scheme by a single line diagram. It may be noted that it is not necessary that all power schemes include all the stages shown in the figure. For example, in a certain power scheme, there may be no secondary transmission and in another case, the scheme may be so small that there is only distribution and no transmission.

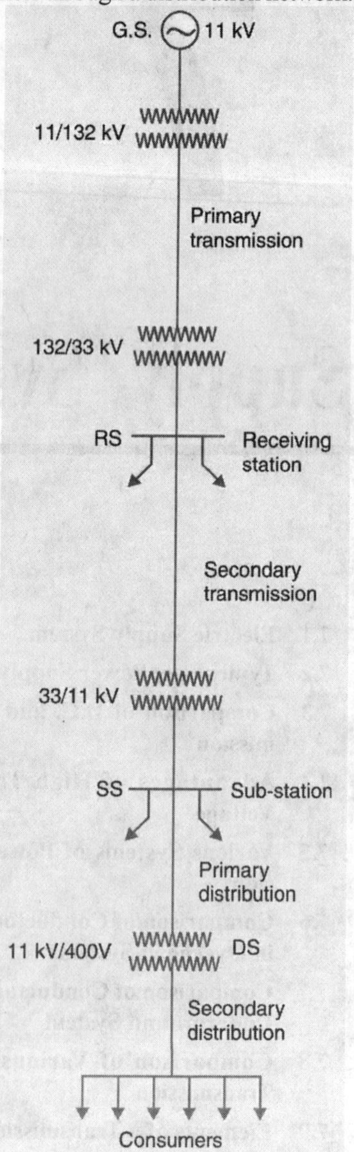

Fig. 7.1

(*i*) *Generating station* : In Fig 7.1, G.S. represents the generating station where electric power is produced by 3-phase alternators operating in parallel. The usual generation voltage is †11 kV. For economy in the transmission of electric power, the generation voltage (*i.e.*, 11 kV) is stepped upto 132 kV (or **more) at the generating station with the help of 3-phase transformers. The transmission of electric power at high voltages has several advantages including the saving of conductor material and high transmission efficiency. It may appear advisable to use the highest possible voltage for transmission of electric power to save conductor material and have other advantages. But there is a limit to which this voltage can be increased. It is because increase in transmission voltage introduces insulation problems as

---

* In certain densely populated cities, the underground system is being employed for distribution. This is to eliminate the danger to human life which would be present with overhead system and to avoid ugly appearance and inconvenience of pole lines running down the main thorough fares.

† It may be 6·6 kV or even 33 kV in certain cases.

** Depending upon the length of transmission line and the amount of power to be transmitted.

well as the cost of switchgear and transformer equipment is increased. Therefore, the choice of proper transmission voltage is essentially a question of economics. Generally the primary transmission is carried at 66 kV, 132 kV, 220 kV or 400 kV.

(*ii*) *Primary transmission.* The electric power at 132 kV is transmitted by 3-phase, 3-wire overhead system to the outskirts of the city. This forms the primary transmission.

(*iii*) *Secondary transmission.* The primary transmission line terminates at the receiving station (*RS*) which usually lies at the outskirts of the city. At the receiving station, the voltage is reduced to 33kV by step-down transformers. From this station, electric power is transmitted at 33kV by 3-phase, 3-wire overhead system to various sub-stations (*SS*) located at the strategic points in the city. This forms the secondary transmission.

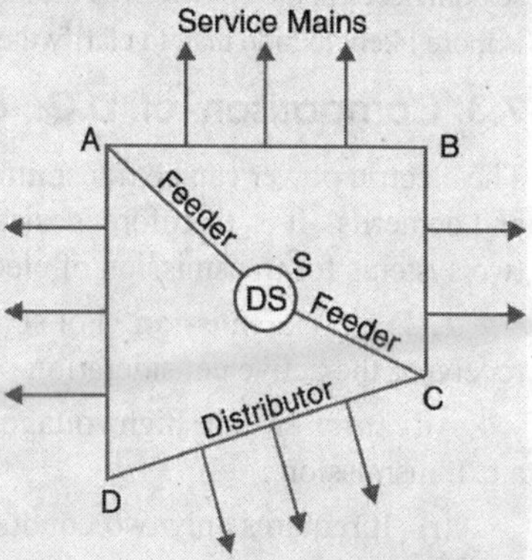

Fig. 7.2

(*iv*) *Primary distribution.* The secondary transmission line terminates at the sub-station (*SS*) where voltage is reduced from 33 kV to 11kV, 3-phase, 3-wire. The 11 kV lines run along the important road sides of the city. This forms the primary distribution. It may be noted that big consumers (having demand more than 50 kW) are generally supplied power at 11 kV for further handling with their own sub-stations.

(*v*) *Secondary distribution.* The electric power from primary distribution line (11 kV) is delivered to distribution sub-stations (*DS*). These sub-stations are located near the consumers' localities and step down the voltage to 400 V, 3-phase, 4-wire for secondary distribution. The voltage between any two phases is 400 V and between any phase and neutral is 230 V. The single-phase residential lighting load is connected between any one phase and neutral, whereas 3-phase, 400 V motor load is connected across 3-phase lines directly.

It may be worthwhile to mention here that secondary distribution system consists of *feeders, distributors and service mains*. Fig. 7.2 shows the elements of low voltage distribution system. Feeders (*SC* or *SA*) radiating from the distribution sub-station (*DS*) supply power to the distributors (*AB*, *BC*, *CD* and *AD*). No consumer is given direct connection from the feeders. Instead, the consumers are connected to the distributors through their service mains.

Power Transformer

Note. A practical power system has a large number of auxiliary equipments (*e.g.*, fuses, circuit breakers, voltage control devices etc.). However, such equipments are not shown in Fig. 7.1. It is because the amount of information included in the diagram depends on the purpose for which the diagram is intended. Here our purpose is to display general lay out of the power system. Therefore, the location of circuit breakers, relays etc., is unimportant.

Further, the structure of power system is shown by a single line diagram. The complete 3-phase circuit is seldom necessary to convey even the most detailed information about the system. In fact, the complete diagram is more likely to hide than to clarify the information we are seeking from the system viewpoint.

7.3 Comparison of D.C. and A.C. Transmission

The electric power can be transmitted either by means of d.c. or a.c. Each system has its own merits and demerits. It is, therefore, desirable to discuss the technical advantages and disadvantages of the two systems for transmission of electric power.

1. D.C. transmission. For some years past, the transmission of electric power by d.c. has been receiving the active consideration of engineers due to its numerous advantages.

Advantages. The high voltage d.c. transmission has the following advantages over high voltage a.c. transmission :

(*i*) It requires only two conductors as compared to three for a.c. transmission.

(*ii*) There is no inductance, capacitance, phase displacement and surge problems in d.c. transmission.

(*iii*) Due to the absence of inductance, the voltage drop in a d.c. transmission line is less than the a.c. line for the same load and sending end voltage. For this reason, a d.c. transmission line has better voltage regulation.

(*iv*) There is no skin effect in a d.c. system. Therefore, entire cross-section of the line conductor is utilised.

(*v*) For the same working voltage, the potential stress on the insulation is less in case of d.c. system than that in a.c. system. Therefore, a d.c. line requires less insulation.

(*vi*) A d.c. line has less corona loss and reduced interference with communication circuits.

(*vii*) The high voltage d.c. transmission is free from the dielectric losses, particularly in the case of cables.

(*viii*) In d.c. transmission, there are no stability problems and synchronising difficulties.

Disadvantages

(*i*) Electric power cannot be generated at high d.c. voltage due to commutation problems.

(*ii*) The d.c. voltage cannot be stepped up for transmission of power at high voltages.

(*iii*) The d.c. switches and circuit breakers have their own limitations.

2. A.C. transmission. Now-a-days, electrical energy is almost exclusively generated, transmitted and distributed in the form of a.c.

Advantages

(*i*) The power can be generated at high voltages.

(*ii*) The maintenance of a.c. sub-stations is easy and cheaper.

(*iii*) The a.c. voltage can be stepped up or stepped down by transformers with ease and efficiency. This permits to transmit power at high voltages and distribute it at safe potentials.

Disadvantages

(*i*) An a.c. line requires more copper than a d.c. line.

(*ii*) The construction of a.c. transmission line is more complicated than a d.c. transmission line.

(*iii*) Due to skin effect in the a.c. system, the effective resistance of the line is increased.

(*iv*) An a.c. line has capacitance. Therefore, there is a continuous loss of power due to charging current even when the line is open.

Conclusion. From the above comparison, it is clear that high voltage d.c. transmission is superior to high voltage a.c. transmission. Although at present, transmission of electric power is carried by a.c., there is an increasing interest in d.c. transmission. The introduction of mercury arc rectifiers and thyratrons have made it possible to convert a.c. into d.c. and *vice-versa* easily and efficiently. Such devices can operate upto 30 MW at 400 kV in single units. The present day trend is towards a.c. for generation and distribution and high voltage d.c. for transmission.

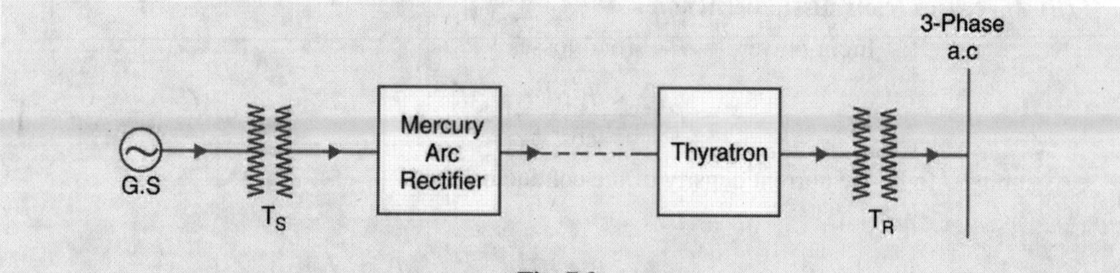

Fig. 7.3

Fig. 7.3 shows the single line diagram of high voltage d.c. transmission. The electric power is generated as a.c. and is stepped up to high voltage by the sending end transformer T_S. The a.c. power at high voltage is fed to the mercury arc rectifiers which convert a.c. into d.c. The transmission of electric power is carried at high d.c. voltage. At the receiving end, d.c. is converted into a.c. with the help of thyratrons. The a.c. supply is stepped down to low voltage by receiving end transformer T_R for distribution.

7.4 Advantages of High Transmission Voltage

The transmission of electric power is carried at high voltages due to the following reasons :

(i) **Reduces volume of conductor material.** Consider the transmission of electric power by a three-phase line.

Let P = power transmitted in watts

V = line voltage in volts

$\cos\phi$ = power factor of the load

l = length of the line in metres

R = resistance per conductor in ohms

ρ = resistivity of conductor material

a = area of X-section of conductor

$$\text{Load current, } I = \frac{P}{\sqrt{3}\,V\cos\phi}$$

$$\text{Resistance/conductor, } R = \rho l/a$$

$$\text{Total power loss, } W = 3I^2R = 3\left(\frac{P}{\sqrt{3}\,V\cos\phi}\right)^2 \times \frac{\rho l}{a}$$

$$= \frac{P^2\rho l}{V^2\cos^2\phi\, a}$$

$$\therefore \quad \text{Area of X-section, } a = \frac{P^2\rho l}{W V^2\cos^2\phi}$$

Total volume of conductor material required

$$= 3\,a\,l = 3\left(\frac{P^2\rho l}{W V^2\cos^2\phi}\right) l$$

$$= \frac{3P^2 \rho l^2}{W V^2 \cos^2 \phi} \quad ...(i)$$

It is clear from exp. (*i*) that for given values of P, l, ρ and W, the volume of conductor material required is inversely proportional to the square of transmission voltage and power factor. In other words, the greater the transmission voltage, the lesser is the conductor material required.

(*ii*) Increases transmission efficiency

$$\text{Input power} = P + \text{Total losses}$$

$$= P + \frac{P^2 \rho l}{V^2 \cos^2 \phi \, a}$$

Assuming J to be the current density of the conductor, then,

$$a = I/J$$

$$\therefore \quad \text{Input power} = P + \frac{P^2 \rho l J}{V^2 \cos^2 \phi \, I} = P + \frac{P^2 \rho l J}{V^2 \cos^2 \phi} \times \frac{1}{I}$$

$$= P + \frac{P^2 \rho l J}{V^2 \cos^2 \phi} \times \frac{{}^*\sqrt{3}\, V \cos \phi}{P}$$

$$= P + \frac{\sqrt{3}\, P J \rho l}{V \cos \phi} = P\left[1 + \frac{\sqrt{3}\, J \rho l}{V \cos \phi}\right]$$

$$\text{Transmission efficiency} = \frac{\text{Output power}}{\text{Input power}} = \frac{P}{P\left[1 + \dfrac{\sqrt{3}\, J \rho l}{V \cos \phi}\right]} = \frac{1}{\left[1 + \dfrac{\sqrt{3}\, J \rho l}{V \cos \phi}\right]}$$

$$= {}^{**}\left[1 - \frac{\sqrt{3}\, J \rho l}{V \cos \phi}\right] \text{ approx.} \quad ...(ii)$$

As J, ρ and l are constants, therefore, transmission efficiency increases when the line voltage is increased.

(*iii*) Decreases percentage line drop

$$\text{Line drop} = I R = I \times \frac{\rho l}{a}$$

$$= I \times \rho l \times J/I = \rho l J \quad [\because a = I/J]$$

$$\%\text{age line drop} = \frac{J \rho l}{V} \times 100 \quad ...(iii)$$

As J, ρ and l are constants, therefore, percentage line drop decreases when the transmission voltage increases.

Limitations of high transmission voltage. From the above discussion, it might appear advisable to use the highest possible voltage for transmission of power in a bid to save conductor material. However, it must be realised that high transmission voltage results in

(*i*) the increased cost of insulating the conductors

(*ii*) the increased cost of transformers, switchgear and other terminal apparatus.

Therefore, there is a limit to the higher transmission voltage which can be economically employed in a particular case. This limit is reached when the saving in cost of conductor material due to

---

* $I = \dfrac{P}{\sqrt{3} V \cos \phi}$

** Binomial theorem.

higher voltage is offset by the increased cost of insulation, transformer, switchgear etc. Hence, the choice of proper transmission voltage is essentially a question of economics. The reader may find further discussion on this topic later in this chapter.

7.5 Various Systems of Power Transmission

It has already been pointed out that for transmission of electric power, 3-phase, 3-wire a.c. system is universally adopted. However, other systems can also be used for transmission under special circumstances. The different possible systems of transmission are :

1. **D.C. system**
   (*i*) D.C. two-wire.
   (*ii*) D.C. two-wire with mid-point earthed.
   (*iii*) D.C. three-wire.
2. **Single-phase A.C. system**
   (*i*) Single-phase two-wire.
   (*ii*) Single-phase two-wire with mid-point earthed.
   (*iii*) Single-phase three-wire.
3. **Two-phase A.C. system**
   (*i*) Two-phase four-wire.
   (*ii*) Two-phase three wire.
4. **Three-phase A.C. system**
   (*i*) Three-phase three-wire.
   (*ii*) Three-phase four-wire.

From the above possible systems of power transmission, it is difficult to say which is the best system unless and until some method of comparison is adopted. Now, the cost of conductor material is one of the most important charges in a system. Obviously, the best system for transmission of power is that for which the volume of conductor material required is minimum. Therefore, the volume of conductor material required forms the basis of comparison between different systems.

While comparing the amount of conductor material required in various systems, the proper comparison shall be on the basis of equal maximum stress on the *dielectric. There are two cases :

(*i*) *When transmission is by overhead system.* In the overhead system, the maximum disruptive stress** exists between the conductor and the earth. Therefore, the comparison of the system in this case has to be made on the basis of maximum voltage between conductor and earth.

(*ii*) *When transmission is by underground system.* In the underground system, the chief stress on the insulation is between conductors. Therefore, the comparison of the systems in this case should be made on the basis of maximum potential difference between conductors.

7.6 Comparison of Conductor Material in Overhead System

In comparing the relative amounts of conductor material necessary for different systems of transmission, similar conditions will be assumed in each case *viz.,*

---

* In long transmission lines, the voltage is only limited by the problem of insulating the conductors against disruptive discharge. Therefore, comparison should be on the basis of equality of maximum potential difference *i.e.,* equal maximum stress on the dielectric.

** In overhead system, insulation between conductors whether at the supports or intermediate points is always provided by suitably spacing the conductors. Therefore, electric dischrage cannot occur between conductors. However, the insulation has to be provided between the conductor and supporting structure. Therefore, maximum stress is between conductor and earth.

(*i*) same power (*P watts*) transmitted by each system.

(*ii*) the distance (*l metres*) over which power is transmitted remains the same.

(*iii*) the line losses (*W watts*) are the same in each case.

(*iv*) the maximum voltage between any conductor and earth (V_m) is the same in each case.

1. Two-wire d.c. system with one conductor earthed

In the 2-wire d.c. system, one is the outgoing or positive wire and the other is the return or negative wire as shown in Fig. 7.4. The load is connected between the two wires.

Max. voltage between conductors = V_m

Power to be transmitted = P

∴ Load current, $I_1 = P/V_m$

If R_1 is the resistance of each line conductor, then,

$$R_1 = \rho\, l/a_1$$

where a_1 is the area of X-section of the conductor.

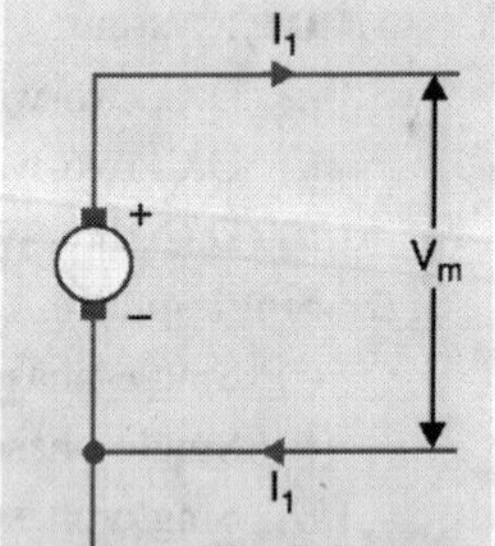

Fig. 7.4

$$\text{Line losses,}\quad W = 2 I_1^2 R_1 = 2\left(\frac{P}{V_m}\right)^2 \rho \frac{l}{a_1}$$

$$\therefore\quad \text{Area of X-section,}\ a_1 = \frac{2P^2 \rho\, l}{W V_m^2}$$

Volume of conductor material required

$$= 2a_1 l = 2\left(\frac{2P^2\rho\, l}{W V_m^2}\right) l = \frac{4P^2 \rho\, l^2}{W V_m^2}$$

It is a usual practice to make this system as the basis for comparison with other systems. Therefore, volume of conductor material required in this system shall be taken as the basic quantity *i.e.*

$$\frac{4P^2\rho\, l^2}{W V_m^2} = K \text{ (say)}$$

2. Two-wire d.c. system with mid-point earthed. Fig. 7.5 shows the two-wire d.c. system with mid-point earthed. The maximum voltage between any conductor and earth is V_m so that maximum voltage between conductors is $2V_m$.

Load current, $I_2 = P/2V_m$

Let a_2 be the area of X-section of the conductor.

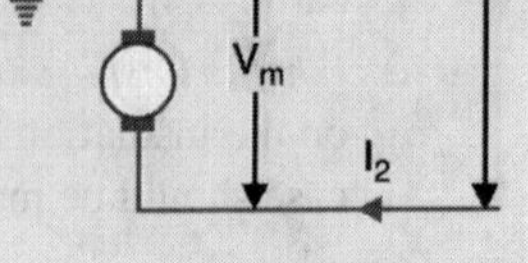

Fig. 7.5

$$\text{Line losses,}\ W = 2I_2^2 R_2 = 2\left(\frac{P}{2V_m}\right)^2 \times \frac{\rho l}{a_2}$$

$$[\because R_2 = \rho\, l/a_2]$$

$$\therefore\quad W = \frac{P^2 \rho\, l}{2a_2 V_m^2}$$

$$\therefore\quad \text{Area of X-section,}\ a_2 = \frac{P^2\rho\, l}{2W V_m^2}$$

∴ Volume of conductor material required

$$= 2a_2 l = 2\left(\frac{P^2\rho\, l}{2W V_m^2}\right) l = \frac{P^2\rho\, l^2}{W V_m^2}$$

$$= \frac{K}{4} \qquad \left[\because K = \frac{4P^2\rho l^2}{W\,V_m^2}\right]$$

Hence, the volume of conductor material required in this system is *one-fourth* of that required in a two-wire d.c. system with one conductor earthed.

3. Three-wire d.c. system. In a 3-wire d.c. system, there are two outers and a middle or neutral wire which is earthed at the generator end as shown in Fig. 7.6. If the load is balanced, the current in the neutral wire is zero. Assuming balanced loads,

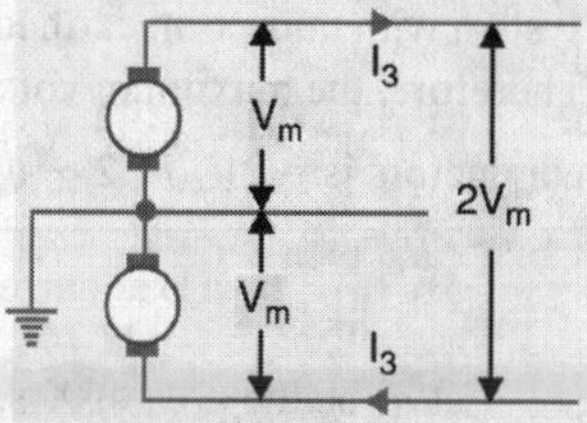

Fig. 7.6

$$\text{Load current, } I_3 = P/2V_m$$

Let a_3 be the area of X-section of each outer wire.

$$\text{Line losses, } W = 2I_3^2\,R_3 = 2\left(\frac{P}{2\,V_m}\right)^2 \times \rho\,\frac{l}{a_3} = \frac{P^2\rho\,l}{2\,V_m^2\,a_3}$$

$$\therefore \quad \text{Area of X-section, } a_3 = \frac{P^2\rho\,l}{2W\,V_m^2}$$

Assuming the area of X-section of neutral wire to be half that of the outer wire,

Volume of conductor material required

$$= 2{\cdot}5\,a_3\,l = 2{\cdot}5\left(\frac{P^2\rho\,l}{2\,W\,V_m^2}\right)l = \frac{2{\cdot}5}{2}\left(\frac{P^2\rho\,l^2}{W\,V_m^2}\right)$$

$$= \frac{5}{16}K \qquad \left[\because K = \frac{4P^2\rho l^2}{W\,V_m^2}\right]$$

Hence the volume of conductor material required in this system is 5/16th of what is required for a 2-wire d.c. system with one conductor earthed.

4. Single phase 2-wire a.c. system with one conductor earthed. Fig. 7.7 shows a single phase 2-wire a.c. system with one conductor earthed. The maximum voltage between conductors is V_m so that r.m.s. value of voltage between them is $V_m/\sqrt{2}$. Assuming the load power factor to be cos φ,

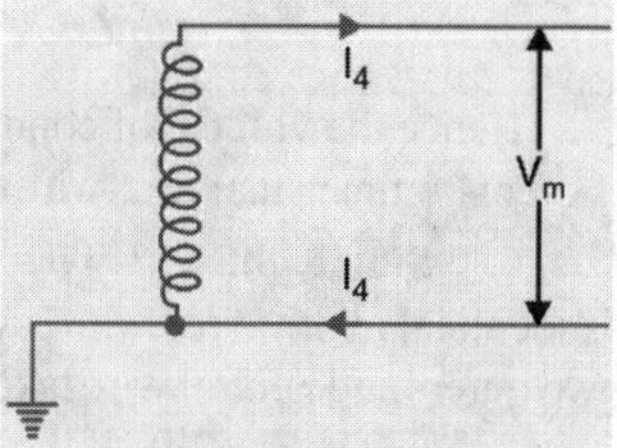

Fig. 7.7

$$\text{Load current, } I_4 = \frac{P}{(V_m/\sqrt{2})\cos\phi} = \frac{\sqrt{2}\,P}{V_m\cos\phi}$$

Let a_4 be the area of X-section of the conductor.

$$\therefore \quad \text{Line losses, } W = 2\,I_4^2\,R_4 = 2\left(\frac{\sqrt{2}\,P}{V_m\cos\phi}\right)^2 \times \frac{\rho\,l}{a_4} = \frac{4\,P^2\rho\,l}{\cos^2\phi\,V_m^2\,a_4}$$

$$\therefore \quad \text{Area of X-section, } a_4 = \frac{4\,P^2\rho\,l}{\cos^2\phi\,WV_m^2}$$

Volume of conductor material required

$$= 2\,a_4\,l = 2\left(\frac{4\,P^2\rho\,l}{V_m^2\,W\cos^2\phi}\right)l$$

$$= \frac{2}{\cos^2\phi} \times \frac{4\,P^2\rho\,l^2}{W\,V_m^2}$$

$$= \frac{2\,K}{\cos^2\phi} \qquad \left[\because K = \frac{4\,P^2\rho\,l^2}{W\,V_m^2}\right]$$

Hence, the volume of conductor material required in this system is $2/\cos^2\phi$ times that of 2-wire d.c. system with the one conductor earthed.

5. Single phase 2-wire system with mid-point earthed. Fig. 7.8 shows a single phase a.c. system with mid-point earthed. The two wires possess equal and opposite voltages to earth (*i.e.*, V_m). Therefore, the maximum voltage between the two wires is $2V_m$. The r.m.s. value of voltage between conductors is $= 2V_m/\sqrt{2} = \sqrt{2}V_m$. Assuming the power factor of the load to be $\cos\phi$,

$$\text{Load current, } I_5 = \frac{P}{\sqrt{2}\,V_m\cos\phi}$$

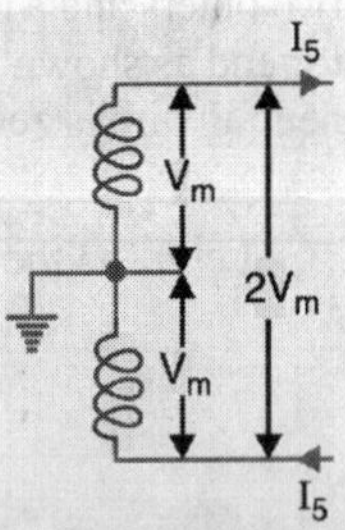

Fig. 7.8

Let a_5 be the area of X-section of the conductor.

$$\text{Line losses, } W = 2\,I_5^2 R_5 = 2\left(\frac{P}{\sqrt{2}\,V_m\cos\phi}\right)^2 R_5$$

$$\therefore\quad W = \frac{P^2\rho\, l}{a_5 V_m^2\cos^2\phi}$$

$$\therefore\quad \text{Area of X-section, } a_5 = \frac{P^2\rho\, l}{W V_m^2\cos^2\phi}$$

Volume of conductor material required

$$= 2\,a_5\, l = 2\left(\frac{P^2\rho\, l}{W V_m^2\cos^2\phi}\right) l = \frac{2\,P^2\rho\, l^2}{W V_m^2\cos^2\phi}$$

$$= \frac{2}{\cos^2\phi}\times\frac{P^2\rho l^2}{W V_m^2}$$

$$= \frac{K}{2\cos^2\phi} \qquad \left[\because K = \frac{4\,P^2\rho\, l^2}{W V_m^2}\right]$$

Hence the volume of conductor material required in this system is $1/2\cos^2\phi$ times that of 2-wire d.c. system with one conductor earthed.

6. Single phase, 3-wire system. The single phase 3-wire system is identical in principle with 3-wire d.c. system. The system consists of two outers and neutral wire taken from the mid-point of the phase winding as shown in Fig. 7.9. If the load is balanced, the current through the neutral wire is zero. Assuming balanced load,

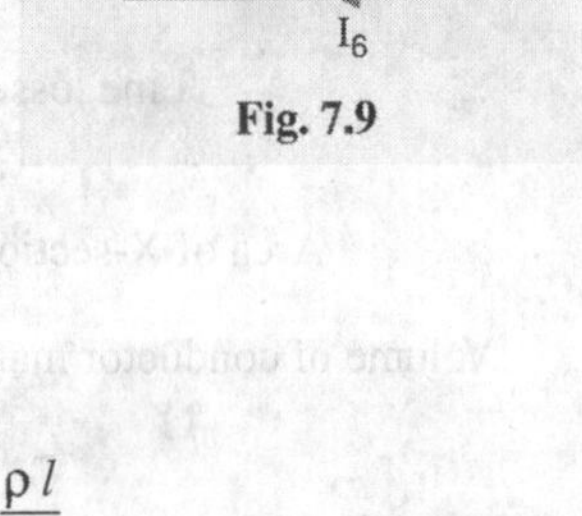

Fig. 7.9

Max. voltage between conductors $= 2\,V_m$

R.M.S. value of voltage between conductors $= 2V_m/\sqrt{2} = \sqrt{2}V_m$

If the p.f of the load is $\cos\phi$, then,

$$\text{Load current, } I_6 = \frac{P}{\sqrt{2}\,V_m\cos\phi}$$

Let a_6 be the area of X-section of each outer conductor.

$$\text{Line losses, } W = 2\,I_6^2 R_6 = 2\left(\frac{P}{\sqrt{2}\,V_m\cos\phi}\right)^2 \times\frac{\rho\, l}{a_6}$$

$$= \frac{P^2\rho\, l}{a_6 V_m^2\cos^2\phi}$$

$$\therefore\quad \text{Area of X-section, } a_6 = \frac{P^2\,\rho\, l}{W V_m^2\cos^2\phi}$$

Assuming the area of X-section of neutral wire to be half that of the outer wire,

Volume of conductor material required

$$= 2{\cdot}5\, a_6\, l = 2{\cdot}5 \left(\frac{P^2 \rho l}{W V_m^2 \cos^2 \phi} \right) l = \frac{2 \cdot 5\, P^2 \rho l^2}{W V_m^2 \cos^2 \phi}$$

$$= \frac{2 \cdot 5}{\cos^2 \phi} \times \frac{P^2 \rho l^2}{W V_m^2}$$

$$= \frac{5K}{8 \cos^2 \phi} \qquad \left[\because K = \frac{4P^2 \rho l^2}{W V_m^2} \right]$$

Hence, the volume of conductor material required in this system is 5/8 $\cos^2 \phi$ times that required in a 2-wire d.c. system with one conductor earthed.

7. **Two phase, 4-wire a.c. system.** As shown in Fig. 7.10, the four wires are taken from the ends of the two-phase windings and the mid-points of the two windings are connected together. This system can be considered as two independent single phase systems, each transmitting one half of the total power.

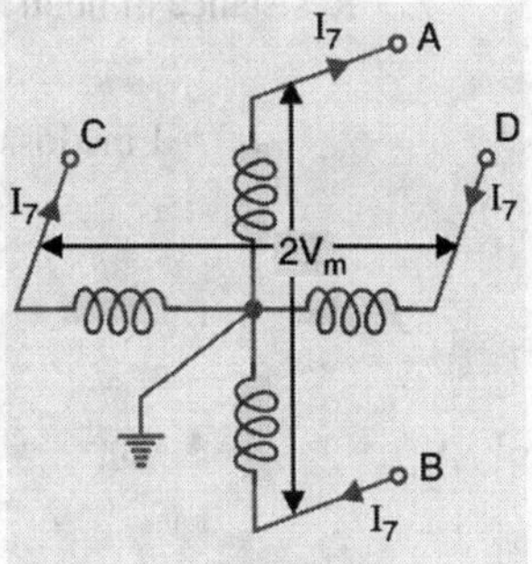

Fig. 7.10

Max. voltage between outers A and $B = 2V_m$

R.M.S. value of voltage $= 2V_m / \sqrt{2} = \sqrt{2}\, V_m$

Power supplied per phase (*i.e.*, by outers A and B) $= P/2$

Assuming p.f. of the load to be $\cos \phi$,

$$\text{Load current, } I_7 = \frac{P/2}{\sqrt{2}\, V_m \cos \phi} = \frac{P}{2\sqrt{2}\, V_m \cos \phi}$$

Let a_7 be the area of X-section of one conductor.

$$\text{Line losses, } W = 4\, I_7^2 R_7 = 4 \left(\frac{P}{2\sqrt{2}\, V_m \cos \phi} \right)^2 \times \frac{\rho l}{a_7}$$

$$\therefore \quad W = \frac{P^2 \rho l}{2 a_7 V_m^2 \cos^2 \phi}$$

$$\therefore \quad \text{Area of X-section, } a_7 = \frac{P^2 \rho l}{2\, W V_m^2 \cos^2 \phi}$$

$\therefore$ Volume of conductor material required

$$= 4\, a_7\, l$$

$$= 4 \left(\frac{P^2 \rho l}{2\, W V_m^2 \cos^2 \phi} \right) l = \frac{4P^2 \rho l^2}{2\, W V_m^2 \cos^2 \phi}$$

$$= \frac{1}{2 \cos^2 \phi} \times \frac{4P^2 \rho l^2}{W V_m^2}$$

$$= \frac{K}{2 \cos^2 \phi} \qquad \left[\because K = \frac{4P^2 \rho l^2}{W V_m^2} \right]$$

Hence, the volume of conductor material required for this system is $1/2 \cos^2 \phi$ times that of 2-wire d.c. system with one conductor earthed.

8. **Two-phase, 3-wire system.** Fig. 7.11 shows two-phase, 3-wire a.c. system. The third or neutral wire is taken from the junction of two-phase windings whose voltages are in quadrature with each other. Obviously, each phase transmits one half of the total power. The R.M.S. voltage between

outgoing conductor and neutral

$$= V_m/\sqrt{2}$$

$\therefore$ Current in each outer, $I_8 = \dfrac{P/2}{\dfrac{V_m}{\sqrt{2}}\cos\phi} = \dfrac{P}{\sqrt{2}\,V_m\cos\phi}$

Current in neutral* wire $= \sqrt{I_8^2 + I_8^2} = \sqrt{2}\,I_8$

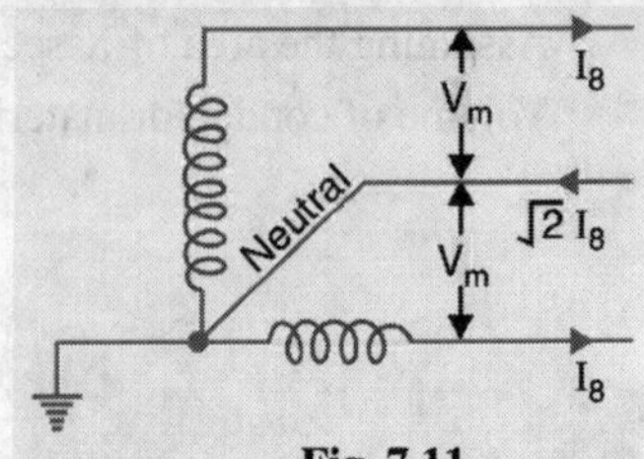

Fig. 7.11

Assuming the current density to be constant, the area of X-section of the neutral wire will be $\sqrt{2}$ ** times that of either of the outers.

$\therefore$ Resistance of neutral wire $= \dfrac{R_8}{\sqrt{2}} = \dfrac{\rho\, l}{\sqrt{2}\, a_8}$

Line losses, $W = 2\,I_8^2\,R_8 + \left(\sqrt{2}\,I_8\right)^2 \dfrac{R_8}{\sqrt{2}} = I_8^2\,R_8\left(2+\sqrt{2}\right)$

$$= \left(\frac{P}{\sqrt{2}\,V_m\cos\phi}\right)^2 \times \frac{\rho\, l}{a_8}\left(2+\sqrt{2}\right)$$

$$\therefore \quad W = \frac{P^2 \rho\, l}{2\,a_8\,V_m^2\cos^2\phi}\left(2+\sqrt{2}\right)$$

$\therefore$ Area of X-section, $a_8 = \dfrac{P^2\,\rho\, l}{2W\,V_m^2\cos^2\phi}\left(2+\sqrt{2}\right)$

Volume of conductor material required

$$= 2\,a_8\,l + \sqrt{2}\,a_8\,l = a_8\,l\left(2+\sqrt{2}\right)$$

$$= \frac{P^2\,\rho\, l^2}{2W\,V_m^2\cos^2\phi}\left(2+\sqrt{2}\right)^2$$

$$= \frac{1\cdot457}{\cos^2\phi}K \qquad \left[\because K = \frac{4P^2\,\rho\, l^2}{W\,V_m^2}\right]$$

Hence, the volume of conductor material required for this system is $1\cdot457/\cos^2\phi$ times that of 2-wire d.c. system with one conductor earthed.

9. 3-Phase, 3-wire system. This system is almost universally adopted for transmission of electric power. The 3-phase, 3-wire system may be star connected or delta connected. Fig. 7.12 shows 3-phase, 3-wire star† connected system. The neutral point N is earthed.

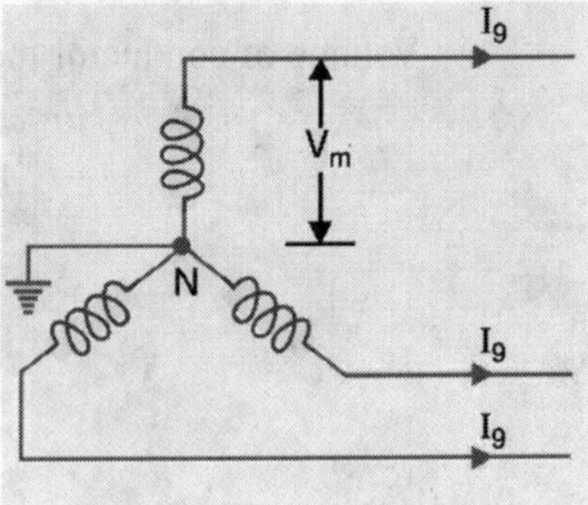

Fig. 7.12

R.M.S. voltage per phase $= V_m/\sqrt{2}$

Power transmitted per phase $= P/3$

* Current in the neutral wire is the phasor sum of currents in the outer wires. Now, the currents in the outers are in quadrature (*i.e.*, 90° apart) with each other.

** Since the neutral wire carries $\sqrt{2}$ times the current in each of the outers, its X-section must be increased in the same ratio to maintain the same current density.

† The same result will be obtained if Δ-connected system is considered.

Load current per phase, $I_9 = \dfrac{P/3}{(V_m/\sqrt{2}\cos\phi)} = \dfrac{\sqrt{2}\,P}{3\,V_m\cos\phi}$

Let a_9 be the area of X-section of each conductor.

$$\text{Line losses, } W = 3\,I_9^2\,R_9 = 3\left(\frac{\sqrt{2}\,P}{3\,V_m\cos\phi}\right)^2\frac{\rho\,l}{a_9} = \frac{2\,P^2\,\rho\,l}{3\,a_9\,V_m^2\cos^2\phi}$$

$$\therefore \quad \text{Area of X-section, } a_9 = \frac{2\,P^2\,\rho\,l}{3\,W\,V_m^2\cos^2\phi}$$

Volume of conductor material required

$$= 3\,a_9\,l = 3\left(\frac{2\,P^2\,\rho\,l}{3\,W\,V_m^2\cos^2\phi}\right)l = \frac{2\,P^2\,\rho\,l^2}{W\,V_m^2\cos^2\phi}$$

$$= \frac{0{\cdot}5\,K}{\cos^2\phi} \qquad \left[\because K = \frac{4P^2\,\rho\,l^2}{W\,V_m^2}\right]$$

Hence, the volume of conductor material required for this system is $0{\cdot}5/\cos^2\phi$ times that required for 2-wire d.c. system with one conductor earthed.

10. 3-phase, 4-wire system. In this case, 4th or neutral wire is taken from the neutral point as shown in Fig. 7.13. The area of X-section of the neutral wire is generally one-half that of the line conductor. If the loads are balanced, then current through the neutral wire is zero. Assuming balanced loads and p.f. of the load as $\cos\phi$,

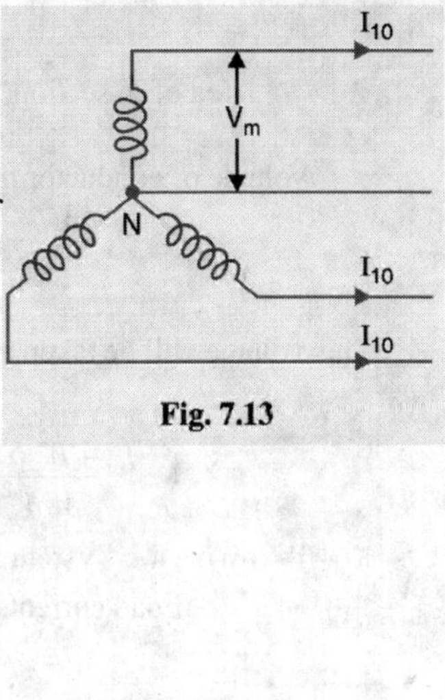

Fig. 7.13

$$\text{Line losses, } W = \text{Same as in 3 phase, 3-wire}$$

$$= \frac{2\,P^2\,\rho\,l}{3\,a_{10}\,V_m^2\cos^2\phi}$$

$$\therefore \quad \text{Area of X-section, } a_{10} = \frac{2\,P^2\,\rho\,l}{3\,W\,V_m^2\cos^2\phi}$$

As the area of X-section of neutral wire is one-half that of any line conductor,

$\therefore$ Volume of conductor material required

$$= 3{\cdot}5\,a_{10}\,l = 3{\cdot}5\left(\frac{2\,P^2\,\rho\,l}{3\,W\,V_m^2\cos^2\phi}\right)\times l$$

$$= \frac{7\,P^2\,\rho\,l^2}{3\,W\,V_m^2\cos^2\phi} = \frac{7}{3\cos^2\phi}\times\frac{P^2\rho\,l^2}{W\,V_m^2}$$

$$= \frac{7\,K}{12\cos^2\phi} \qquad \left[\because K = \frac{4P^2\,\rho\,l^2}{W\,V_m^2}\right]$$

Hence, the volume of conductor material required for this system is $7/12\cos^2\phi$ times that required for 2-wire d.c. system with one conductor earthed.

7.7 Comparison of Conductor Material in Underground System

In an underground transmission using multi-core* belted type cables, the chief stress on the insulation is usually between conductors. Under such situations, comparison is made on the basis of maximum voltage between conductors. Again assumptions made are :

(*i*) The same power (*P* watts) is transmitted by each system.

(*ii*) The distance (*l* metres) over which power is transmitted remains the same.

(*iii*) The line losses (*W* watts) are the same in each case.

(*iv*) The maximum voltage between conductors (V_m) is the same in each case.

1. Two-wire d.c. system. If V_m denotes the maximum potential difference between the conductors, it will also be the working voltage in this case.

$$\text{Load current, } I_1 = P/V_m$$

$$\text{Line losses, } W = 2I_1^2 R_1 = 2\left(\frac{P}{V_m}\right)^2 \frac{\rho l}{a_1}$$

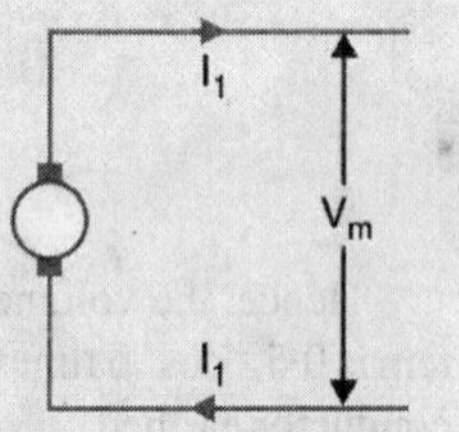

Fig. 7.14

$$\therefore \quad W = \frac{2P^2 \rho l}{a_1 V_m^2}$$

$$\therefore \quad \text{Area of X-section, } a_1 = \frac{2P^2 \rho l}{W V_m^2}$$

∴ Volume of conductor material required

$$= 2\, a_1\, l = 2\left(\frac{2P^2 \rho l}{W V_m^2}\right) l = \frac{4 P^2 \rho l^2}{W V_m^2} = K \text{ (say)}$$

This volume will be taken as the basic quantity and comparison shall be made for other systems *i.e.,*

$$\frac{4 P^2 \rho l^2}{W V_m^2} = K$$

2. Two-wire d.c. system with mid point earthed

$$\text{Load current, } I_2 = P/V_m$$

$$\text{Line losses, } W = 2I_2^2 R_2 = 2\left(\frac{P}{V_m}\right)^2 \rho \frac{l}{a_2}$$

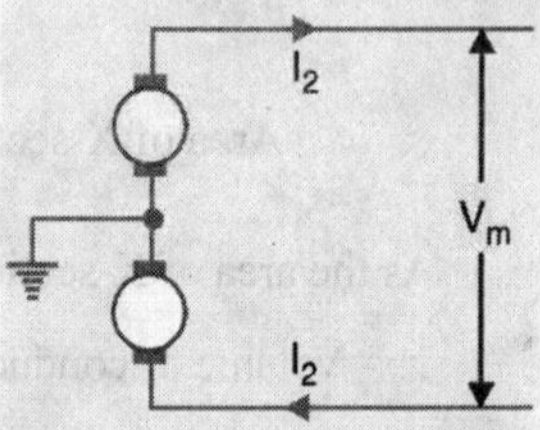

Fig. 7.15

$$\therefore \quad W = \frac{2P^2 \rho l}{V_m^2 a_2}$$

$$\therefore \quad \text{Area of X-section, } a_2 = \frac{2P^2 \rho l}{W V_m^2}$$

Volume of conductor material required

$$= 2\, a_2\, l = 2\left(\frac{2P^2 \rho l}{W V_m^2}\right) l = \frac{4P^2 \rho l^2}{W V_m^2} = K$$

---

* When the underground transmission is by single-core or multi-core cables of S.L. type, the stress is from conductor to earth. Under such conditions, maximum voltage between conductor and earth forms basis of comparison of various systems.

Hence, the volume of conductor material required in this system is the same as that for 2-wire d.c. system.

3. **Three wire d.c. system.** The maximum voltage between the outers is V_m. Assuming balanced load, the current through the neutral wire will be zero.

Load current, $I_3 = P/V_m$

Line losses, $W = 2\,I_3^2\,R_3 = 2\left(\frac{P}{V_m}\right)^2 \rho\,\frac{l}{a_3}$

$$\therefore \quad W = \frac{2P^2\,\rho\,l}{V_m^2\,a_3}$$

$$\therefore \quad \text{Area of } X\text{-section, } a_3 = \frac{2P^2\,\rho\,l}{W\,V_m^2}$$

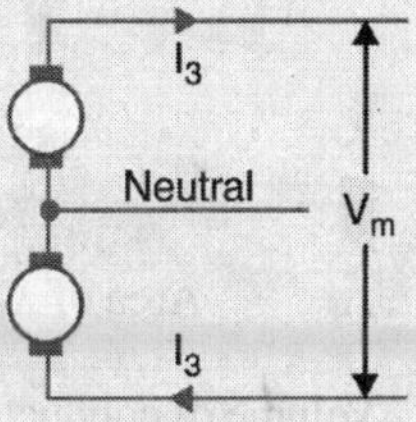

Fig. 7.16

Assuming the area of *X*-section of neutral wire to be half of that of either outers,

Volume of conductor material required

$$= 2{\cdot}5\,a_3 l = 2{\cdot}5\left(\frac{2P^2\,\rho\,l}{W\,V_m^2}\right)l = \frac{5P^2\,\rho\,l^2}{W\,V_m^2}$$

$$= 1{\cdot}25\,K \qquad \left[\because K = \frac{4P^2\,\rho l^2}{W\,V_m^2}\right]$$

Hence, the volume of conductor material required in the system is 1·25 times that required for 2-wire d.c. system.

4. **Single phase, 2-wire a.c. system.** As the maximum voltage between conductors is V_m, therefore, r.m.s. value of voltage is $V_m/\sqrt{2}$. Assuming the p.f. of the load to be cos ϕ,

Load current, $I_4 = \frac{P}{V_m/\sqrt{2}\cos\phi} = \frac{\sqrt{2}P}{V_m\cos\phi}$

Line losses, $W = 2I_4^2 R_4$

$$= 2\left(\frac{\sqrt{2}P}{V_m\cos\phi}\right)^2 \rho\frac{l}{a_4} = \frac{4P^2\,\rho\,l}{a_4\,V_m^2\cos^2\phi}$$

$$\therefore \quad \text{Area of } X\text{-section, } a_4 = \frac{4P^2\,\rho\,l}{W\,V_m^2\cos^2\phi}$$

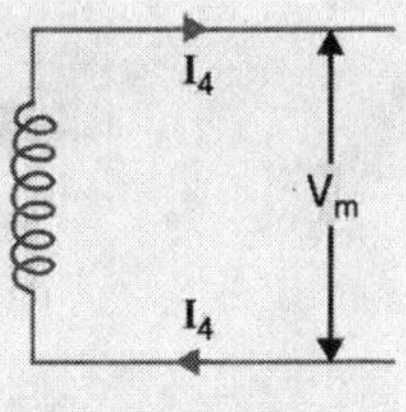

Fig. 7.17

Volume of conductor material required

$$= 2a_4\,l = 2\left(\frac{4P^2\,\rho\,l}{W\,V_m^2\cos^2\phi}\right)l = \frac{8P^2\,\rho\,l^2}{W\,V_m^2\cos^2\phi}$$

$$= \frac{2}{\cos^2\phi} \times \frac{4P^2\,\rho\,l^2}{W\,V_m^2}$$

$$= \frac{2\,K}{\cos^2\phi} \qquad \left[\because K = \frac{4P^2\,\rho l^2}{W\,V_m^2}\right]$$

Hence, the volume of conductor material required for this system is $2/\cos^2\phi$ times that required in a 2-wire d.c. system.

5. **Single phase, 2-wire system with mid-point earthed.** The maximum value of voltage between the outers is V_m. Therefore, the r.m.s. value of voltage is $V_m/\sqrt{2}$. If the p.f.of the load is cos ϕ,

Load current, $I_5 = \dfrac{\sqrt{2}P}{V_m \cos\phi}$

Line losses, $W = 2\, I_5^2\, R_5 = 2\left(\dfrac{\sqrt{2}P}{V_m \cos\phi}\right)^2 \rho \dfrac{l}{a_5}$

$\therefore \quad W = \dfrac{4P^2 \rho\, l}{a_5 V_m^2 \cos^2\phi}$

$\therefore \quad$ Area of X-section, $a_5 = \dfrac{4P^2 \rho\, l}{W V_m^2 \cos^2\phi}$

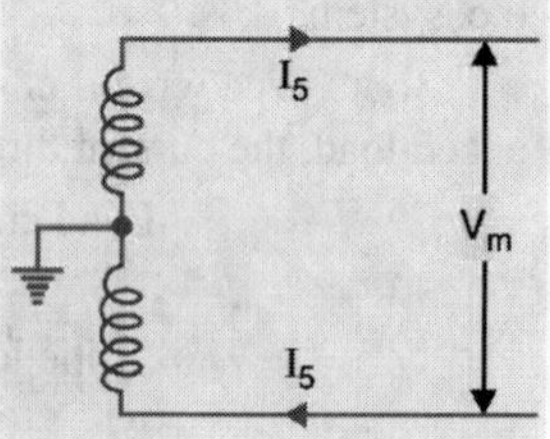

Fig. 7.18

Volume of conductor material required

$$= 2\, a_5\, l = 2\left(\frac{4P^2 \rho\, l}{W V_m^2 \cos^2\phi}\right) l = \frac{8P^2 \rho\, l^2}{W V_m^2 \cos^2\phi}$$

$$= \frac{2}{\cos^2\phi} \times \frac{4P^2 \rho\, l^2}{W V_m^2}$$

$$= \frac{2\,K}{\cos^2\phi} \qquad \left[\because K = \frac{4P^2 \rho\, l^2}{W V_m^2}\right]$$

Hence, the volume of conductor material required in this system is $2/\cos^2\phi$ times that required in 2-wire d.c. system.

6. Single phase, 3-wire system. If the load is considered balanced, the system reduces to a single phase 2-wire except that a neutral wire is provided in addition. Assuming the area of X-section of the neutral wire to be half of either of the outers,

Volume of conductor material required

$$= 2{\cdot}5\ {}^*a_4\, l$$

$$= 2{\cdot}5\left(\frac{4\,P^2 \rho\, l}{W V_m^2 \cos^2\phi}\right) l$$

$$= \frac{10\,P^2 \rho\, l^2}{W V_m^2 \cos^2\phi}$$

$$= \frac{2\cdot 5}{\cos^2\phi} \times \frac{4P^2 \rho\, l^2}{W V_m^2}$$

$$= \frac{2{\cdot}5\,K}{\cos^2\phi} \qquad \left[\because K = \frac{4P^2 \rho\, l^2}{W V_m^2}\right]$$

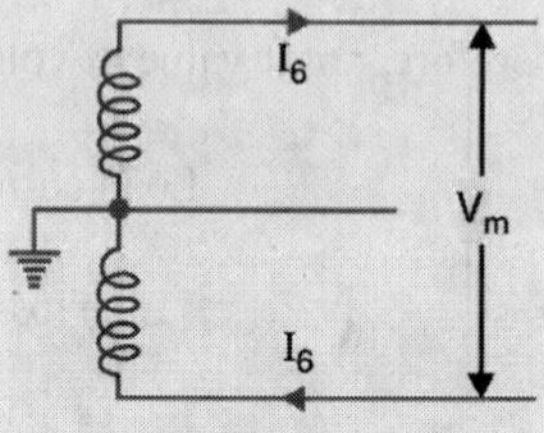

Fig. 7.19

Hence, the volume of conductor material required in this system is $2{\cdot}5/\cos^2\phi$ times that required in 2-wire d.c. system.

7. Two-phase, 4-wire system. This system can be considered as two independent single phase systems, each transmitting one-half of the total power. It is clear form Fig. 7.20 that voltage across outers (*AB* or *CD*) is twice that of single phase 2-wire (refer back to Fig. 7.17). Therefore, current (I_7) in each conductor will be half that in single-phase 2-wire system. Consequently, area of X-section of each conductor is also half but as there are four wires, so volume of conductor

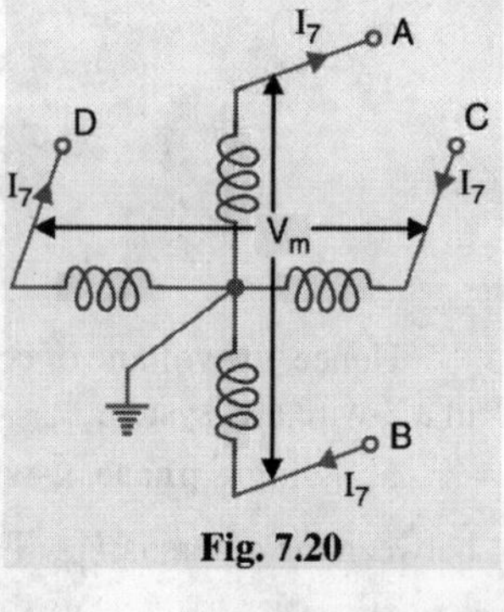

Fig. 7.20

* Area of X-section of conductor for single phase 2-wire system.

material used is the same as in a single phase, 2 wire system *i.e.*

Volume of conductor material required.

$$= \frac{2K}{\cos^2 \phi}$$

Hence, volume of conductor material required in this system is $2/\cos^2 \phi$ times that required in 2-wire d.c. system.

8. Two-phase, 3-wire system. Fig. 7.21 shows two phase, 3-wire a.c. system. Let us suppose that maximum voltage between the outers is V_m. Then maximum voltage between either outer and neutral wire is $V_m/\sqrt{2}$ *.

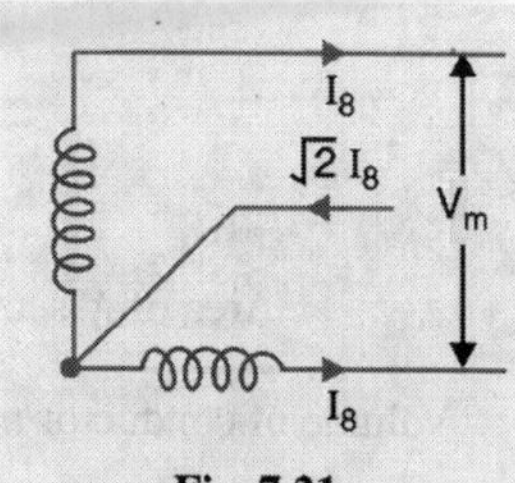

Fig. 7.21

R.M.S. voltage between outer and neutral wire

$$= \frac{V_m/\sqrt{2}}{\sqrt{2}} = \frac{V_m}{2}$$

Current in each outer, $I_8 = \dfrac{P/2}{V_m/2 \cos \phi} = \dfrac{P}{V_m \cos \phi}$

Current in neutral wire $= \sqrt{I_8^2 + I_8^2} = \sqrt{2} I_8$

Assuming the current density to be constant, the area of X-section of neutral wire will be $\sqrt{2}$ times that of either of the outers.

$\therefore$ Resistance of neutral wire $= \dfrac{R_8}{\sqrt{2}} = \dfrac{\rho l}{\sqrt{2}\, a_8}$

$$\text{Line losses, } W = 2I_8^2 R_8 + (\sqrt{2} I_8)^2 \frac{R_8}{\sqrt{2}} = I_8^2 R_8 (2 + \sqrt{2})$$

$$= \left(\frac{P}{V_m \cos \phi}\right)^2 \times \rho \frac{l}{a_8} (2 + \sqrt{2})$$

$$\therefore \quad W = \frac{P^2 \rho l}{a_8 V_m^2 \cos^2 \phi} (2 + \sqrt{2})$$

$$\therefore \quad \text{Area of X-section, } a_8 = \frac{P^2 \rho l}{W V_m^2 \cos^2 \phi} (2 + \sqrt{2})$$

Volume of conductor material required

$$= 2 a_8 l + \sqrt{2} a_8 l = a_8 l (2 + \sqrt{2})$$

$$= \frac{P^2 \rho l^2}{W V_m^2 \cos^2 \phi} (2 + \sqrt{2})^2$$

$$= \frac{2{\cdot}194\, K}{\cos^2 \phi} \qquad \left[\because K = \frac{4 P^2 \rho l^2}{W V_m^2}\right]$$

Hence, the volume of conductor material required in this system is $2{\cdot}914/\cos^2 \phi$ times that required in 2-wire d.c. system.

9. 3-Phase, 3-wire system. Suppose that the maximum value of voltage between the conductors is V_m. Then maximum voltage between each phase and neutral is $V_m/\sqrt{3}$. Therefore, r.m.s. value of voltage per phase

* The voltages in two phase windings are 90° out of phase.

$$= \frac{V_m}{\sqrt{3}} \times \frac{1}{\sqrt{2}} = \frac{V_m}{\sqrt{6}}$$

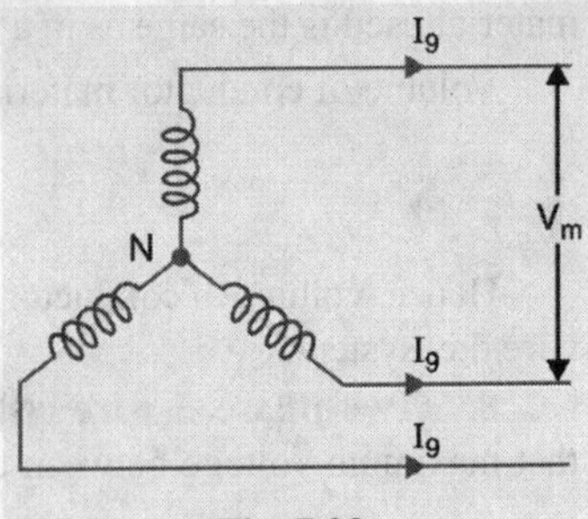

Fig. 7.22

Power transmitted per phase $= P/3$

$\therefore$ Load current per phase, $I_9 = \dfrac{P/3}{V_m/\sqrt{6}\cos\phi} = \dfrac{\sqrt{6}\,P}{3\,V_m\cos\phi}$

Line losses, $W = 3\,I_9^2\,R_9$

$$= 3\left(\frac{\sqrt{6}\,P}{3V_m\cos\phi}\right)^2 \times \rho\frac{l}{a_9}$$

$$\therefore \quad W = \frac{2P^2\rho\,l}{a_9 V_m^2\cos^2\phi}$$

$\therefore$ Area of X-section, $a_9 = \dfrac{2P^2\,\rho l}{W\,V_m^2\cos^2\phi}$

Volume of conductor material required

$$= 3\,a_9\,l = 3\left(\frac{2P^2\rho\,l}{W\,V_m^2\cos^2\phi}\right) l = \frac{6P^2\rho\,l^2}{W\,V_m^2\cos^2\phi}$$

$$= \frac{1\cdot 5}{\cos^2\phi} \times \frac{4P^2\rho\,l^2}{W\,V_m^2}$$

$$= \frac{1\cdot 5\,K}{\cos^2\phi} \qquad \left[\because K = \frac{4\,P^2\,\rho\,l^2}{W\,V_m^2}\right]$$

Hence, the volume of conductor material required in this system is $1\cdot 5/\cos^2\phi$ times that required in 2-wire d.c. system.

Underground transmission system.

10. 3-phase, 4-wire system. Fig. 7.23 shows the 3-phase, 4-wire system. If the loads are balanced, then neutral wire carries no current. Consequently, the system reduces to a 3-phase, 3-wire system except that there is additional neutral wire. Assuming the area of X-section of the neutral wire to be half that of line conductor,

Volume of conductor material required

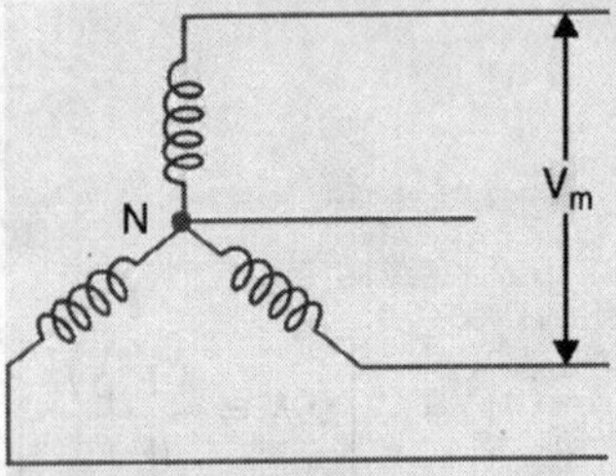

Fig. 7.23

$$= 3\cdot 5\,a_9\,l$$

$$= 3\cdot 5\left(\frac{2P^2\rho\,l}{W\,V_m^2\cos^2\phi}\right) l$$

$$= \frac{7P^2\rho\,l^2}{W\,V_m^2\cos^2\phi} = \frac{7}{\cos^2\phi} \times \frac{P^2\,\rho\,l^2}{W\,V_m^2}$$

$$= \frac{1\cdot 75\,K}{\cos^2\phi} \qquad \left[\because K = \frac{4\,P^2\,\rho\,l^2}{W\,V_m^2}\right]$$

Hence the volume of conductor material required in this system is $1\cdot 75/\cos^2\phi$ times that required in 2-wire d.c. system.

7.8 Comparison of Various Systems of Transmission

Below is given the table which shows the ratio of conductor-material in any system compared with that in the corresponding 2-wire d.c. system. Cos ϕ is the power factor in an a.c. system.

System	Same maximum voltage to earth	Same maximum voltage between conductors
1. D.C. system		
(*i*) Two-wire	1	1
(*ii*) Two-wire mid-point earthed	0·25	1
(*iii*) 3-wire	0·3125	1·25
2. Single phase system		
(*i*) 2 wire	$2/\cos^2\phi$	$2/\cos^2\phi$
(*ii*) 2-wire with mid-point earthed	$\dfrac{0\cdot 5}{\cos^2\phi}$	$\dfrac{2}{\cos^2\phi}$
(*iii*) 3-wire	$\dfrac{0\cdot 625}{\cos^2\phi}$	$\dfrac{2\cdot 5}{\cos^2\phi}$
3. Two-phase system		
(*i*) 2-phase, 4-wire	$\dfrac{0\cdot 5}{\cos^2\phi}$	$\dfrac{2}{\cos^2\phi}$
(*ii*) 2-phase, 3-wire	$\dfrac{1\cdot 457}{\cos^2\phi}$	$\dfrac{2\cdot 914}{\cos^2\phi}$
4. Three-phase system		
(*i*) 3-phase, 3-wire	$\dfrac{0\cdot 5}{\cos^2\phi}$	$\dfrac{1\cdot 5}{\cos^2\phi}$
(*ii*) 3-phase, 4-wire	$\dfrac{0\cdot 583}{\cos^2\phi}$	$\dfrac{1\cdot 75}{\cos^2\phi}$

The following points may be noted from the above table :

(*i*) There is a great saving in conductor material if d.c. system is adopted for transmission of electric power. However, due to technical difficulties, d.c. system is not used for transmission.

(*ii*) Considering the a.c. system, the 3-phase a.c. system is most suitable for transmission due to two reasons. Firstly, there is considerable saving in conductor material. Secondly, this system is convenient and efficient.

Example 7.1 *What is the percentage saving in feeder copper if the line voltage in a 2-wire d.c. system is raised from 200 volts to 400 volts for the same power transmitted over the same distance and having the same power loss ?*

Solution. Fig. 7.24 (*i*) shows 200 volts system, whereas Fig. 7.24 (*ii*) shows 400 volts system. Let P be the power delivered and W be power loss in both cases. Let v_1 and a_1 be the volume and area of X-section for 200 V system and v_2 and a_2 for that of 400 V system.

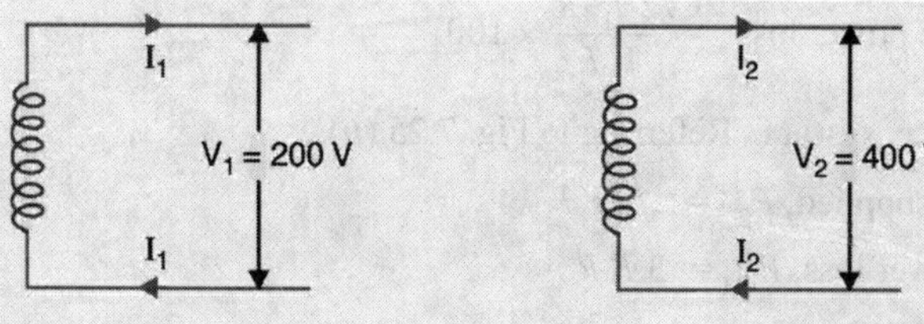

Fig. 7.24

Now, $P = V_1 I_1 = 200\, I_1$ ...(*i*)

And $P = V_2 I_2 = 400\, I_2$ ...(*ii*)

As same power is delivered in both cases,

$\therefore \quad 200 I_1 = 400\, I_2$ or $I_2 = (200/400)\, I_1 = 0{\cdot}5 I_1$

Power loss in 200 *V* system, $W_1 = 2 I_1^{\,2} R_1$

Power loss in 400 *V* system, $W_2 = 2 I_2^{\,2} R_2 = 2(0{\cdot}5\, I_1)^2 R_2 = 0{\cdot}5 I_1^{\,2} R_2$

As power loss in the two cases is the same,

$\therefore \quad W_1 = W_2$

or $\quad 2\, I_1^2 R_1 = 0{\cdot}5\, I_1^2 R_2$

or $\quad R_2/R_1 = 2/0{\cdot}5 = 4$

or $\quad a_1/a_2 = 4$

or $\quad \upsilon_1/\upsilon_2 = 4$

or $\quad \upsilon_2/\upsilon_1 = 1/4 = 0{\cdot}25$

$$\therefore \quad \text{\% age saving in feeder copper} = \frac{\upsilon_1 - \upsilon_2}{\upsilon_1} \times 100 = \left(\frac{\upsilon_1}{\upsilon_1} - \frac{\upsilon_2}{\upsilon_1}\right) \times 100$$

$$= (1 - 0{\cdot}25) \times 100 = 75\%$$

Example 7.2 *A d.c. 2-wire system is to be converted into a.c. 3-phase, 3-wire system by the addition of a third conductor of the same cross-section as the two existing conductors. Calculate the percentage additional load which can now be supplied if the voltage between wires and the percentage loss in the line remain unchanged. Assume a balanced load of unity power factor.*

Solution. Fig. 7.25 (*i*) shows the 2-wire d.c. system, whereas Fig. 7.25 (*ii*) shows the 3-phase, 3-wire system. Suppose *V* is the voltage between conductors for the two cases. Let *R* be the resistance per conductor in each case.

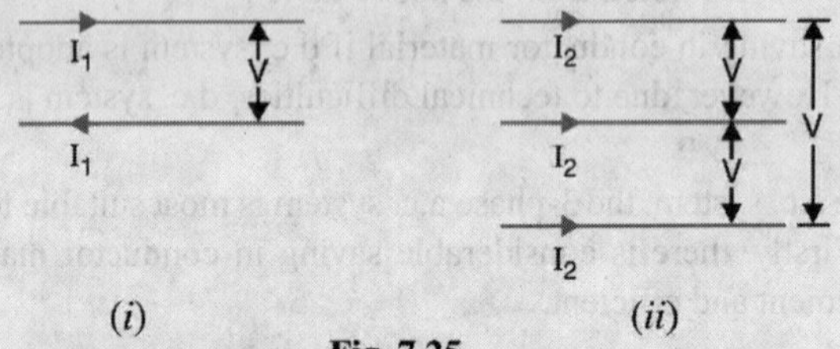

Fig. 7.25

Two-wire d.c. system. Referring to Fig. 7.25 (*i*),

Power supplied, $P_1 = V I_1$

Power loss, $W_1 = 2\, I_1^2 R$

$$\text{Percentage power loss} = \frac{2\, I_1^2\, R}{V\, I_1} \times 100 \quad \text{...(}i\text{)}$$

3-phase, 3-wire a.c. system. Referring to Fig. 7.25 (*ii*),

Power supplied, $P_2 = \sqrt{3}\, V I_2$

Power loss, $W_2 = 3\, I_2^2 R$

$$\text{Percentage power loss} = \frac{3\, I_2^2\, R}{\sqrt{3}\, V\, I_2} \times 100 \quad \text{...(}ii\text{)}$$

As the percentage power loss in the two cases is the same,

$$\therefore \quad \text{exp. } (i) = \text{exp. } (ii)$$

or $$\frac{2 I_1^2 R}{V I_1} \times 100 = \frac{3 I_2^2 R}{\sqrt{3} V I_2} \times 100$$

or $$2 I_1 = \sqrt{3} I_2$$

or $$I_2 = \frac{2}{\sqrt{3}} I_1$$

Now, $$\frac{P_2}{P_1} = \frac{\sqrt{3} V I_2}{V I_1} = \frac{\sqrt{3} V \times \frac{2}{\sqrt{3}} I_1}{V I_1} = 2$$

$$\therefore \quad P_2 = 2 P_1$$

i.e. additional power which can be supplied at unity p.f. by 3-phase, 3-wire a.c. system = **100%**.

Example 7.3. *A d.c. 3-wire system is to be converted into a 3-phase, 4-wire system by adding a fourth wire equal in X-section to each outer of the d.c. system. If the percentage power loss and voltage at the consumer's terminals are to be the same in the two cases, find the extra power at unity power factor that can be supplied by the a.c. system. Assume loads to be balanced.*

Solution. Fig. 7.26 (*i*) shows the 3-wire d.c. system, whereas Fig. 7.26 (*ii*) shows 3-phase, 4-wire system. Suppose that V is consumer's terminal voltage (*i.e.*, between conductor and neutral) in the two cases. Let R be the resistance per conductor in each case.

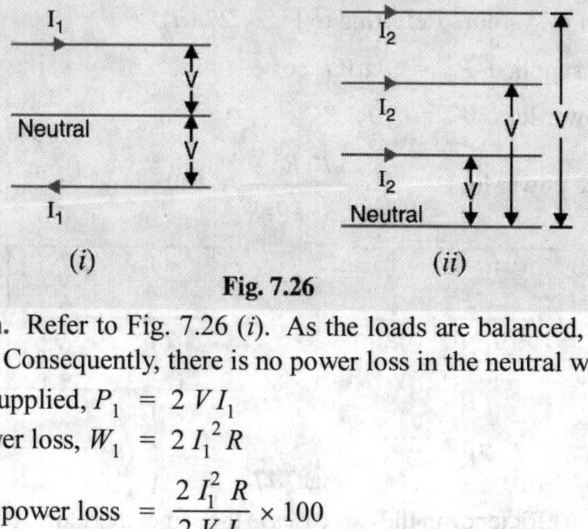

Fig. 7.26

3-wire d.c. system. Refer to Fig. 7.26 (*i*). As the loads are balanced, therefore, neutral wire carries no current. Consequently, there is no power loss in the neutral wire.

$$\text{Power supplied, } P_1 = 2 V I_1$$

$$\text{Power loss, } W_1 = 2 I_1^2 R$$

$$\text{Percentage power loss} = \frac{2 I_1^2 R}{2 V I_1} \times 100 \qquad ...(i)$$

3-phase, 4-wire a.c. system. Refer to Fig. 7.26 (*ii*). Since the loads are balanced, the neutral wire carries no current and hence there is no power loss in it.

$$\text{Power supplied, } P_2 = {}^*3 V I_2 \qquad [\because \cos\phi = 1]$$

$$\text{Power loss, } W_2 = 3 I_2^2 R$$

$$\text{Percentage power loss} = \frac{3 I_2^2 R}{3 V I_2} \times 100 \qquad ...(ii)$$

As the percentage power loss in the two cases is the same, therefore, exp. (*i*) is equal to exp. (*ii*)

i.e.,

* Power per phase = $V I_2$

$\therefore$ Power in 3-phases = $3 V I_2$

$$\frac{2 I_1^2 R}{2 V I_1} \times 100 = \frac{3 I_2^2 R}{3 V I_2} \times 100$$

or $$I_1 = I_2$$

Now $$\frac{P_2}{P_1} = \frac{3 V I_2}{2 V I_1} = \frac{3 V I_1}{2 V I_1} = 1{\cdot}5$$

$\therefore$ $$P_2 = 1{\cdot}5\, P_1$$

i.e., extra power that can be supplied at unity power factor by 3-phase, 4-wire a.c. system = **50%.**

Example 7.4. *A single phase a.c. system supplies a load of 200 kW and if this system is converted to 3-phase, 3-wire a.c. system by running a third similar conductor, calculate the 3-phase load that can now be supplied if the voltage between the conductors is the same. Assume the power factor and transmission efficiency to be the same in the two cases.*

Solution. Fig. 7.27 (*i*) shows the single phase 2-wire a.c. system, whereas Fig. 7.27 (*ii*) shows 3-phase, 3-wire system. Suppose that V is the voltage between the conductors in the two cases. Let R be the resistance per conductor and cos ϕ the power factor in each case.

Single phase 2-wire system. Referring to Fig. 7.27 (*i*),

$$\text{Power supplied, } P_1 = V I_1 \cos\phi$$

$$\text{Power loss, } W_1 = 2 I_1^2 R$$

$$\text{\% age power loss} = \frac{2 I_1^2 R}{V I_1 \cos\phi} \times 100 \quad ...(i)$$

3-phase, 3-wire a.c. system. Referring to Fig. 7.27 (*ii*),

$$\text{Power supplied, } P_2 = \sqrt{3}\, V I_2 \cos\phi$$

$$\text{Power loss, } W_2 = 3 I_2^2 R$$

$$\text{\% age power loss} = \frac{3 I_2^2 R}{\sqrt{3}\, V I_2 \cos\phi} \times 100 \quad ...(ii)$$

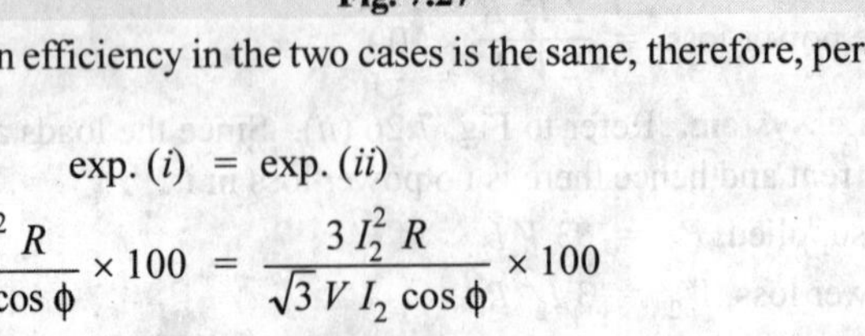

Fig. 7.27

As the transmission efficiency in the two cases is the same, therefore, percentage power loss will also be the same *i.e.*,

$$\text{exp. } (i) = \text{exp. } (ii)$$

or $$\frac{2 I_1^2 R}{V I_1 \cos\phi} \times 100 = \frac{3 I_2^2 R}{\sqrt{3}\, V I_2 \cos\phi} \times 100$$

or $$2 I_1 = \sqrt{3}\, I_2$$

or $$I_2 = \frac{2}{\sqrt{3}} I_1$$

Now, $$\frac{P_2}{P_1} = \frac{\sqrt{3}\, V I_2 \cos\phi}{V I_1 \cos\phi} = \frac{\sqrt{3}\, V \frac{2}{\sqrt{3}} I_1 \cos\phi}{V I_1 \cos\phi} = 2$$

As the power supplied by single phase, 2-wire (*i.e.,* P_1) is 200 kW,

∴ Power supplied by 3-phase, 3-wire a.c. system is

$$P_2 = 2P_1 = 2 \times 200 = \mathbf{400\ kW}$$

It may be seen that 3-phase, 3-wire system can supply 100% additional load.

Example 7.5. *A 50 km long transmission line supplies a load of 5 MVA at 0·8 p.f. lagging at 33 kV. The efficiency of transmission is 90%. Calculate the volume of aluminium conductor required for the line when (i) single phase, 2-wire system is used (ii) 3-phase, 3-wire system is used. The specific resistance of aluminium is 2·85 × 10^{-8} Ω m.*

Solution.

$$\text{Power transmitted} = \text{MVA} \times \cos\phi = 5 \times 0{\cdot}8 = 4\ \text{MW} = 4 \times 10^6\ \text{W}$$

$$\text{Line loss, } W = 10\%\text{ of power transmitted} = (10/100) \times 4 \times 10^6 = 4 \times 10^5\ \text{W}$$

$$\text{Length of line, } l = 50\ \text{km} = 50 \times 10^3\ \text{m}$$

(*i*) Single phase, 2-wire system

$$\text{Apparent power} = V I_1$$

$$\therefore \quad I_1 = \frac{\text{Apparent power}}{V} = \frac{5 \times 10^6}{33 \times 10^3} = 151{\cdot}5\ \text{A}$$

Suppose a_1 is the area of cross-section of aluminium conductor.

$$\text{Line loss, } W = 2 I_1^2 R_1 = 2 I_1^2 \left(\rho \frac{l}{a_1}\right)$$

$$\therefore \quad \text{Area of X-section, } a_1 = \frac{2 I_1^2 \rho l}{W} = \frac{2 \times (151 \cdot 5)^2 \times (2 \cdot 85 \times 10^{-8})\, 50 \times 10^3}{4 \times 10^5}$$

$$= 1{\cdot}635 \times 10^{-4}\ \text{m}^2$$

$$\text{Volume of conductor required} = 2\, a_1\, l = 2 \times (1{\cdot}635 \times 10^{-4}) \times 50 \times 10^3 = \mathbf{16{\cdot}35\ m^3}$$

(*ii*) 3-phase, 3-wire system

$$\text{Line current, } I_2 = \frac{\text{Apparent power}}{\sqrt{3}\, V} = \frac{5 \times 10^6}{\sqrt{3} \times 33 \times 10^3} = 87{\cdot}5\ \text{A}$$

Suppose a_2 is the area of cross-section of the conductor in this case.

$$\text{Line loss, } W = 3 I_2^2 R_2 = 3 I_2^2 \left(\rho \frac{l}{a_2}\right)$$

$$\therefore \quad \text{Area of X-section, } a_2 = \frac{3 I_2^2 \rho l}{W} = \frac{3 \times (87 \cdot 5)^2 \times (2 \cdot 85 \times 10^{-8}) \times 50 \times 10^3}{4 \times 10^5}$$

$$= 0{\cdot}818 \times 10^{-4}\ \text{m}^2$$

$$\text{Volume of conductor required} = 3\, a_2\, l = 3 \times (0{\cdot}818 \times 10^{-4}) \times 50 \times 10^3 = \mathbf{12{\cdot}27\ m^3}$$

Note that volume of conductor (and hence weight) required is less in case of 3-phase, 3-wire system.

Example 7.6. *A sub-station supplies power at 11 kV, 0·8 p.f. lagging to a consumer through a single phase transmission line having total resistance (both go and return) of 0.15 Ω. The voltage drop in the line is 15%. If the same power is to be supplied to the same consumer by two wire d.c. system by a new line having a total resistance of 0·05 Ω and if the allowable voltage drop is 25%, calculate the d.c. supply voltage.*

Solution :

Single phase system

Let I_1 be the load current. Then,

$$\text{Voltage drop} = I_1R_1 = I_1 \times 0{\cdot}15 \text{ volts} \quad ...(i)$$

$$\text{Also voltage drop} = \frac{15}{100} \times 11000 = 1650 \text{ volts} \quad ...(ii)$$

From eqs. (*i*) and (*ii*), $I_1 \times 0{\cdot}15 = 1650$

$$\therefore \quad I_1 = \frac{1650}{0{\cdot}15} = 11000 \text{ A}$$

$$\text{Power received by consumer} = \text{Apparent power} \times \cos\phi$$

$$= \frac{(11{,}000 \times 11{,}000) \times 0{\cdot}8}{1000} \text{ kW} = 9{\cdot}68 \times 10^4 \text{ kW}$$

D.C. two-wire system. The power to be supplied by d.c. 2-wire system is $9{\cdot}68 \times 10^4$ kW = $9{\cdot}68 \times 10^7$ W. Let V volts be the supply voltage.

$$\therefore \quad \text{Load current, } I_2 = \frac{9{\cdot}68 \times 10^7}{V}$$

$$\text{Voltage drop} = I_2 R_2 = \left(\frac{9{\cdot}68 \times 10^7}{V}\right) \times 0{\cdot}05 \quad ...(iii)$$

$$\text{Allowable voltage drop} = \frac{25}{100} \times \text{V} = 0{\cdot}25 \text{ V} \quad ...(iv)$$

$$\therefore \quad \frac{9{\cdot}68 \times 10^7 \times 0{\cdot}05}{V} = 0{\cdot}25 \text{ V}$$

$$or \quad V^2 = \frac{9{\cdot}68 \times 10^7 \times 0{\cdot}05}{0{\cdot}25} = 1936 \times 10^4$$

$$\therefore \quad V = \sqrt{1936 \times 10^4} = \mathbf{4400\ V}$$

TUTORIAL PROBLEMS

1. What is the percentage saving in copper feeder if the line voltage in a 2-wire d.c. system is raised from 220 V to 500 V for the same power transmitted over the same distance and having the same power loss? **[80·64%]**
2. A single phase load of 5 MW is transmitted by a pair of overhead conductors. If a third conductor of the same cross-section and material be added and 3-phase supply be thus substituted for the original single phase, calculate the 3-phase load which can now be transmitted if the voltage between the conductors and the percentage loss in the lines remains unchanged. **[10 MW]**
3. Electric power of 50 MW is to be transmitted over a 132 KV, 3-phase, 3-wire transmission line. The length of the line is 300 km and the efficiency of transmission is 85%. Aluminium is used for conductor material which has resistivity of 3×10^{-9} Ωm. Calculate the volume of conductor material required for a power factor of 0·8 lagging. **[242 m^3]**
4. A 3-phase, 4-wire system is used for lighting. Compare the amount of copper required with that needed for a 2-wire d.c. system with the same lamp voltage. Assume the same losses and balanced load. The neutral is one-half the cross-section of one of the respective outers. **[Copper for 3φ, 4 wire : Copper for 2-wire d.c. = 0·292 : 1]**
5. 30,000 kW at power factor 0·8 lagging is being transmitted over a 220 kV, three-phase transmission line. The length of the line is 275 km and the efficiency of transmission is 90%. Calculate the weight of copper required. Also calculate the weight of copper had the power been transmitted over a single-phase transmission line for the same line voltage and losses. Assume that the resistance of 1 km long conductor and 1 sq. cm is 0·173 Ω and specific gravity of copper is 8·9. **[338 × 10^3 kg : 450·67 × 10^3 kg]**

7.9 Elements of a Transmission Line

For reasons associated with economy, transmission of electric power is done at high voltage by 3-phase, 3-wire overhead system. The principal elements of a high-voltage transmission line are :

(i) *Conductors,* usually three for a single-circuit line and six for a double-circuit line. The usual material is aluminium reinforced with steel.

(ii) *Step-up and step-down transformers,* at the sending and receiving ends respectively. The use of transformers permits power to be transmitted at high efficiency.

(iii) *Line insulators,* which mechanically support the line conductors and isolate them electrically from the ground.

(iv) *Support,* which are generally steel towers and provide support to the conductors.

(v) *Protective devices,* such as ground wires, lightning arrestors, circuit breakers, relays etc. They ensure the satisfactory service of the transmission line.

(vi) *Voltage regulating devices,* which maintain the voltage at the receiving end within permissible limits.

All these elements will be discussed in detail in the subsequent chapters.

7.10 Economics of Power Transmission

While designing any scheme of power transmission, the engineer must have before him the commercial aspect of the work entrusted to him. He must design the various parts of transmission scheme in a way that maximum economy is achieved. The economic design and layout of a complete power transmission scheme is outside the scope of this book. However, the following two fundamental economic principles which closely influence the electrical design of a transmission line will be discussed :

(i) Economic choice of conductor size
(ii) Economic choice of transmission voltage

7.11 Economic Choice of Conductor Size

The cost of conductor material is generally a very considerable part of the total cost of a transmission line. Therefore, the determination of proper size of conductor for the line is of vital importance. The most economical area of conductor is that for which the total annual cost of transmission line is minimum*. This is known as *Kelvin's Law* after Lord Kelvin who first stated it in 1881. The total annual cost of transmission line can be divided broadly into two parts *viz.*, annual charge on capital outlay and annual cost of energy wasted in the conductor.

(i) *Annual charge on capital outlay.* This is on account of interest and depreciation on the capital cost of complete installation of transmission line. In case of overhead system, it will be the annual interest and depreciation on the capital cost of conductors, supports and insulators and the cost of their erection. Now, for an overhead line, insulator cost is constant, the conductor cost is proportional to the area of X-section and the cost of supports and their erection is partly constant and partly proportional to area of X-section of the conductor. Therefore, annual charge on an overhead† transmission line can be expressed as :

$$\text{Annual charge} = P_1 + P_2\, a \qquad ...(i)$$

* The question of voltage regulation is unimportant in a transmission line. Generally, the X-sectional area of the conductor is decided on the basis of minimum annual cost.

† **Underground system.** A similar relationship exists for underground system. In this system, the annual charge is on account of interest and depreciation on the cost of conductors, insulation and the cost of laying the cables. Now, the cost of insulation is constant and the cost of conductor is proportional to area of X-section of conductor.

$$\therefore \quad \text{Annual charge} = P_1 + P_2\, a$$

where P_1 and P_2 are constants and a is the area of X-section of the conductor.

(*ii*) *Annual cost of energy wasted.* This is on account of energy lost mainly‡ in the conductor due to I^2R losses. Assuming a constant current in the conductor throughout the year, the energy lost in the conductor is proportional to resistance. As resistance is inversely proportional to the area of X-section of the conductor, therefore, the energy lost in the conductor is inversely proportional to area of X-section. Thus, the annual cost of energy wasted in an overhead transmission line can be expressed as :

$$\text{Annual cost of energy wasted} = P_3/a \qquad ...(ii)$$

where P_3 is a constant.

$$\text{Total annual cost, } C = \text{exp. } (i) + \text{exp. } (ii)$$
$$= (P_1 + P_2\, a) + P_3/a$$
$$\therefore \quad C = P_1 + P_2\, a + P_3/a \qquad ...(iii)$$

In exp. (*iii*), only area of X-section a is variable. Therefore, the total annual cost of transmission line will be minimum if differentiation of C w.r.t. a is zero *i.e.*

$$\frac{d}{da}(C) = 0$$

$$\text{or} \quad \frac{d}{da}(P_1 + P_2\, a + P_3/a) = 0$$

$$\text{or} \quad P_2 - \frac{P_3}{a^2} = 0$$

$$\text{or} \quad P_2 = P_3/a^2$$

$$\text{or} \quad P_2\, a = \frac{P_3}{a}$$

i.e. Variable part of annual charge = Annual cost of energy wasted

Therefore Kelvin's Law can also be stated in an another way *i.e. the most economical area of conductor is that for which the variable part* of annual charge is equal to the cost of energy losses per year.*

Graphical illustration of Kelvin's law. Kelvin's law can also be illustrated graphically by plotting annual cost against X-sectional area 'a' of the conductor as shown in Fig. 7.28. In the diagram, the straight line (1) shows the relation between the annual charge (*i.e.*, $P_1 + P_2a$) and the area of X-section a of the conductor. Similarly, the rectangular hyperbola (2) gives the relation between annual cost of energy wasted and X-sectional area a. By adding the ordinates of curves (1) and (2), the curve (3) is obtained. This latter curve shows the relation between total annual cost ($P_1 + P_2a + P_3/a$) of transmission line and area of X-section a. The lowest point on the curve (*i.e.*, point P) represents the most economical area of X-section.

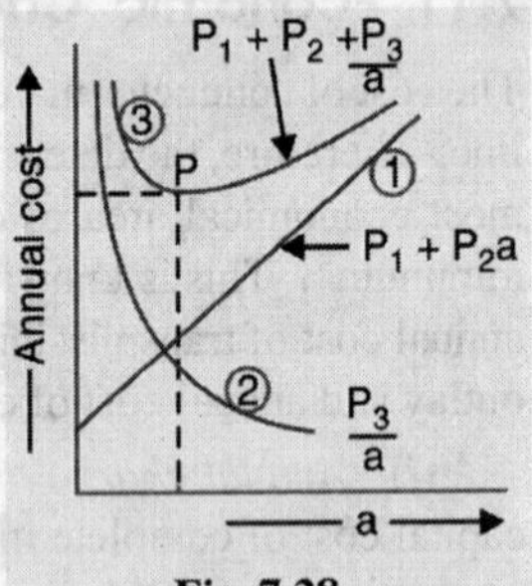

Fig. 7.28

Limitations of Kelvin's law. Although theoretically Kelvin's law holds good, there is often considerable difficulty in applying it to a proposed scheme of power transmission. In practice, the limitations of this law are :

(*i*) It is not easy to estimate the energy loss in the line without actual load curves, which are not available at the time of estimation.

(*ii*) The assumption that annual cost on account of interest and depreciation on the capital outlay is in the form $P_1 + P_2a$ is strictly speaking not true. For instance, in cables neither the cost of cable dielectric and sheath nor the cost of laying vary in this manner.

‡ There may be some losses in the insulating material but these are quite small and may be neglected.

* That part of annual charge (*i.e.*, $P_2\, a$) which is proportional to the area of X-section of conductor.

(*iii*) This law does not take into account several physical factors like safe current density, mechanical strength, corona loss etc.

(*iv*) The conductor size determined by this law may not always be practicable one because it may be too small for the safe carrying of necessary current.

(*v*) Interest and depreciation on the capital outlay cannot be determined accurately.

Example 7.7. *A 2-conductor cable 1 km long is required to supply a constant current of 200 A throughout the year. The cost of cable including installation is Rs. (20 a + 20) per metre where 'a' is the area of X-section of the conductor in cm². The cost of energy is 5P per kWh and interest and depreciation charges amount to 10%. Calculate the most economical conductor size. Assume resistivity of conductor material to be 1·73 μ Ω cm.*

Solution.

$$\text{Resistance of one conductor} = \frac{\rho l}{a} = \frac{1\cdot 73 \times 10^{-6} \times 10^{5}}{a} = \frac{0\cdot 173}{a}\ \Omega$$

$$\text{Energy lost per annum} = \frac{2I^2 Rt}{1000}\text{kWh}$$

$$= \frac{2 \times (200)^2 \times 0\cdot 173 \times 8760}{1000 \times a} = \frac{1{,}21{,}238\cdot 4}{a}\text{ kWh}$$

$$\text{Annual cost of energy lost} = \text{Cost per kWh} \times \text{Annual energy loss}$$

$$= \text{Rs } \frac{5}{100} \times \frac{1{,}21{,}238\cdot 4}{a}$$

$$= \text{Rs } 6062/a \qquad ...(i)$$

The capital cost (variable) of the cable is given to be Rs 20 *a* per metre. Therefore, for 1 km length of the cable, the capital cost (variable) is Rs. 20 *a* × 1000 = Rs. 20,000 *a*.

Variable annual charge = Annual interest and depreciation on capital cost (variable) of cable

$$= \text{Rs } 0\cdot 1 \times 20{,}000\ a$$

$$= \text{Rs } 2000\ a \qquad ...(ii)$$

According to Kelvin's law, for most economical X-section of the conductor,

$$\text{Variable annual charge} = \text{Annual cost of energy lost}$$

or

$$2000\ a = 6062/a$$

$$\therefore \quad a = \sqrt{\frac{6062}{2000}} = \mathbf{1\cdot 74\ cm^2}$$

Example 7.8 *The cost of a 3-phase overhead transmission line is Rs (25000 a + 2500) per km where 'a' is the area of X-section of each conductor in cm². The line is supplying a load of 5 MW at 33kV and 0·8 p.f. lagging assumed to be constant throughout the year. Energy costs 4P per kWh and interest and depreciation total 10% per annum. Find the most economical size of the conductor. Given that specific resistance of conductor material is 10^{-6} Ω cm.*

Solution.

$$\text{Resistance of each conductor, } R = \frac{\rho l}{a} = \frac{10^{-6} \times 10^{5}}{a} = \frac{0\cdot 1}{a}\ \Omega$$

$$\text{Line current, } I = \frac{P}{\sqrt{3}V\cos\phi} = \frac{5 \times 10^6}{\sqrt{3} \times 33 \times 10^3 \times 0\cdot 8} = 109\cdot 35\text{ A}$$

$$\text{Energy lost per annum} = \frac{3I^2 Rt}{1000}\text{ kWh} = \frac{3 \times (109\cdot 35)^2 \times 0\cdot 1 \times 8760}{1000 \times a} = \frac{31{,}424}{a}\text{ kWh}$$

$$\text{Annual cost of energy lost} = \text{Rs } 0{\cdot}04 \times 31{,}424/a = \text{Rs } \frac{1256{\cdot}96}{a}$$

The capital cost (variable) of the cable is given to be Rs 25000 *a* per km length of the line.

$\therefore$ Variable annual charge = 10% of capital cost (variable) of line

$$= \text{Rs } 0{\cdot}1 \times 25{,}000a = \text{Rs } 2{,}500\ a$$

According to Kelvin's law, for most economical *X*-section of the conductor,

Variable annual charge = Annual cost of energy lost

$$2500\ a = \frac{1256{\cdot}96}{a}$$

or
$$a = \sqrt{\frac{1256{\cdot}96}{2500}} = \mathbf{0{\cdot}71\ cm^2}$$

Example 7.9. *A 2-wire feeder carries a constant current of 250 A throughout the year. The portion of capital cost which is proportional to area of X-section is Rs 5 per kg of copper conductor. The interest and depreciation total 10% per annum and the cost of energy is 5P per kWh. Find the most economical area of X-section of the conductor. Given that the density of copper is $8{\cdot}93 gm/cm^3$ and its specific resistance is $1{\cdot}73 \times 10^{-8}\ \Omega\ m$.*

Solution. Consider 1 metre length of the feeder. Let *a* be the most economical area of X-section of each conductor in m^2.

$$\text{Resistance of each conductor, } R = \frac{\rho\, l}{a} = \frac{1{\cdot}73 \times 10^{-8} \times 1}{a} = \frac{1{\cdot}73 \times 10^{-8}}{a}\ \Omega$$

$$\text{Energy lost per annum} = \frac{2I^2 R\, t}{1000}\ \text{kWh} = \frac{2 \times (250)^2 \times 1{\cdot}73 \times 10^{-8} \times 8760}{1000 \times a}$$

$$= \frac{18{,}94{,}350}{a} \times 10^{-8}\ \text{kWh}$$

$$\text{Annual cost of energy lost} = \text{Rs } \frac{5}{100} \times \frac{18{,}94{,}350 \times 10^{-8}}{a} = \text{Rs } \frac{94{,}717{\cdot}5}{a} \times 10^{-8}$$

$$\text{Mass of 1 metre feeder} = 2\,(\text{Volume} \times \text{density}) = 2 \times a \times 1 \times 8{\cdot}93 \times 10^3\ \text{kg}$$

$$= 17{\cdot}86 \times 10^3 a\ \text{kg}$$

$$\text{Capital cost (variable)} = \text{Rs } 5 \times 17{\cdot}86 \times 10^3\ a = \text{Rs } 89{\cdot}3 \times 10^3 a$$

Variable Annual charge = 10% of capital cost (variable)

$$= 0{\cdot}1 \times 89{\cdot}3 \times 10^3 a = \text{Rs } 8930\ a$$

For most economical area of X-section,

Variable annual charge = Annual cost of energy lost

or
$$\text{Rs } 8930\ a = \frac{94{,}717{\cdot}5}{a} \times 10^{-8}$$

$\therefore$
$$a = \sqrt{\frac{94{,}717{\cdot}5 \times 10^{-8}}{8930}} = 3{\cdot}25 \times 10^{-4} m^2 = \mathbf{3{\cdot}25\ cm^2}$$

Example 7.10. *Determine the most economical cross-section for a 3-phase transmission line, 1 km long to supply at a constant voltage of 110 kV for the following daily load cycle :*

6 hours	*20 MW*	*at p.f. 0·8 lagging*
12 hours	*5 MW*	*at p.f. 0·8 lagging*
6 hours	*6 MW*	*at p.f. 0·8 lagging*

The line is used for 365 days yearly. The cost per km of line including erection is Rs (9000 + 6000 a) where 'a' is the area of X-section of conductor in cm^2. The annual rate of interest and *depreciation is 10% and the energy costs 6P per kWh. The resistance per km of each conductor is 0·176/a.*

Solution.

Resistance per km of each conductor is

$$R = \frac{0 \cdot 176}{a}\ \Omega$$

$$\text{Line voltage, } V = 110\text{ k}V = 110 \times 10^3 \text{ V}$$

The load currents at various loads are :

$$\text{At 20 } MW, I_1 = \frac{20 \times 10^6}{\sqrt{3} \times 110 \times 10^3 \times 0 \cdot 8} = 131 \cdot 2 \text{ A}$$

$$\text{At 5 } MW, I_2 = \frac{5 \times 10^6}{\sqrt{3} \times 110 \times 10^3 \times 0 \cdot 8} = 32 \cdot 8 \text{ A}$$

$$\text{At 6 } MW, I_3 = \frac{6 \times 10^6}{\sqrt{3} \times 110 \times 10^3 \times 0 \cdot 8} = 39 \cdot 36 \text{ A}$$

Energy loss per day in 3-phase line

$$= 3 \times \frac{0 \cdot 176}{a} \times \frac{1}{1000} [(131 \cdot 2)^2 \times 6 + (32 \cdot 8)^2 \times 12 + (39 \cdot 36)^2 \times 6]$$

$$= \frac{0 \cdot 528}{1000\ a} [1{,}03{,}280 \cdot 64 + 12{,}910 \cdot 08 + 9{,}295 \cdot 26] \text{ kWh}$$

$$= \frac{66 \cdot 26}{a} \text{ kWh}$$

$$\text{Energy lost per annum} = \frac{66 \cdot 26}{a} \times 365 = \frac{24{,}184 \cdot 9}{a} \text{ kWh}$$

$$\text{Annual cost of energy} = \text{Rs } \frac{6}{100} \times \frac{24{,}184 \cdot 9}{a} = \frac{1451 \cdot 09}{a}$$

Variable annual charge = 10% of capital cost (variable) of line

$$= \text{Rs } 0 \cdot 1 \times 6000\ a = \text{Rs } 600\ a$$

According to Kelvin's law, for most economical X-section of the conductor,

Variable annual charge = Annual cost of energy

or
$$600\ a = \frac{1451 \cdot 09}{a}$$

$$\therefore \quad a = \sqrt{\frac{1451 \cdot 09}{600}} = \mathbf{1 \cdot 56\ cm^2}$$

TUTORIAL PROBLEMS

1. Determine the best current density in A/mm^2 for *a* 3-ϕ overhead line if the line is in use for 2000 hours per year and if the conductor costing Rs 3·0 per kg has a specific resistance of 1·73 Ω m and weighs 6200 kg/m^3. Cost of energy is 10 P/unit. Interest and depreciation is 12% of conductor cost. **[0·705 A/mm^2]**
2. Determine the most economical size of a 3-phase line which supplies the following loads at 10 kV :
   (*i*) 100 kW at 0·8 p.f. (lag) for 10 hours
   (*ii*) 500 kW at 0·9 p.f. (lag) for 8 hours
   (*iii*) 100 kW at unity p.f. for 6 hours.
   The above gives the daily load cycle. The cost per km of the completely erected line is Rs (8000 *a* + 1500) where *a* is the area of cross-section of each conductor. The combined interest and depreciation is 10% per annum of capital cost. Cost of energy losses is 5 paise per kWh. Resistivity of conductor material = 1·72 × 10^{-6} Ω cm. **[0·844 cm^2]**

3. If the cost of an overhead line is Rs 2000 A (where A is the cross-section in cm^2) and if the interest and depreciation charges of the line are 8%, estimate the most economical current density to use for a transmission requiring full load current for 60% of the year. The cost of generating electrical energy is 5 paise/kWh. The resistance of the conductor one km long and 1 cm^2 X-section is 0·18 Ω. [41·12 A/cm^2]

7.12 Economic Choice of Transmission Voltage

It has been shown earlier in the chapter that if transmission voltage is increased, the volume of conductor material required is reduced. This decreases the expenditure on the conductor material. It may appear advisable to use the highest possible transmission voltage in order to reduce the expenditure on conductors to a minimum. However, it may be remembered that as the transmission voltage is increased, the cost of insulating the conductors, cost of transformers, switchgear and other terminal apparatus also increases. Therefore, for every transmission line, there is optimum transmission voltage, beyond which there is nothing to be gained in the matter of economy. The transmission voltage for which the cost of conductors, cost of insulators, transformers, switchgear and other terminal apparatus is minimum is called *economical transmission voltage.*

The method of finding the economical transmission voltage is as follows. Power to be transmitted, generation voltage and length of transmission line are assumed to be known. We choose some standard transmission voltage and work out the following costs :

(*i*) *Transformers,* at the generating and receiving ends of transmission line. For a given power, this cost increases slowly with the increase in transmission voltage.

(*ii*) *Switchgear.* This cost also increases with the increase in transmission voltage.

(*iii*) *Lightning arrestor.* This cost increases rapidly with the increase in transmission voltage.

(*iv*) *Insulation and supports.* This cost increases sharply with the increase in transmission voltage.

(*v*) *Conductor.* This cost decreases with the increase in transmission voltage.

The sum of all above costs gives the total cost of transmission for the voltage considered. Similar calculations are made for other transmission voltages. Then, a curve is drawn for total cost of transmission against voltage as shown in Fig. 7.29. The lowest point (P) on the curve gives the economical transmission voltage. Thus, in the present case, OA is the optimum transmission voltage. This method of finding the economical transmission voltage is rarely used in practice as different costs cannot be determined with a fair degree of accuracy.

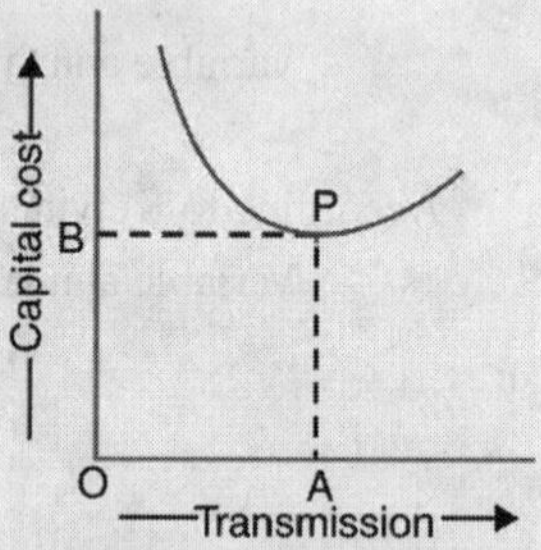

Fig. 7.29

The present day trend is to follow certain empirical formulae for finding the economical transmission voltage. Thus, according to American practice, the economic voltage between lines in a 3-phase a.c. system is

$$V = 5{\cdot}5\sqrt{0{\cdot}62\, l + \frac{3P}{150}}$$

where V = line voltage in kV

P = maximum kW per phase to be delivered to single circuit

l = distance of transmission line in km

It may be noted here that in the above formula, power to be transmitted and distance of transmission line have been taken into account. It is because both these factors influence the economic voltage of a transmission line. This can be easily explained. If the distance of transmission line is increased, the cost of terminal apparatus is decreased, resulting in higher economic transmission voltage. Also if power to be transmitted is large, large generating and transforming units can be employed. This reduces the cost per kW of the terminal station equipment.

7.13 Requirements of Satisfactory Electric Supply

The electric power system in India is 3-phase a.c. operating at a frequency of 50 Hz. The power station delivers power to consumers through its transmission and distribution systems. The power delivered must be characterised by constant or nearly constant voltage, dependability of service, balanced voltage, efficiency so as to give minimum annual cost, sinusoidal waveform and freedom from inductive interference with telephone lines.

(i) **Voltage regulation.** A voltage variation has a large effect upon the operation of both power machinery and lights. A motor is designed to have its best characteristics at the rated voltage and consequently a voltage that is too high or too low will result in a decrease in efficiency. If the fluctuations in the voltage are sudden, these may cause the tripping of circuit breakers and consequent interruptions to service. Usually the voltage at the generator terminals, where this is done, in some cases the voltage variations at the load may be made sufficiently small by keeping the resistance and reactance of the lines and feeders low.

(ii) **Dependability.** One important requirement of electric supply is to furnish uninterrupted service. The losses which an industrial consumer sustains due to the failure of electric power supply are usually vastly greater than the actual value of the power that he would use during this period. It is on account of the expense of idle workmen and machines and other overhead charges. Interruptions to service cause irritation and are sometimes positively dangerous to life and property. For example, failure of power in hospitals, in crowded theatres and stores may lead to very grave consequences. Therefore, it is the duty of electric supply company to keep the power system going and to furnish uninterrupted service.

(iii) **Balanced voltage.** It is very important that the polyphase voltage should be balanced. If an unbalanced polyphase voltage is supplied to a consumer operating synchronous or induction motors, it will result in a decrease in the efficiency of his machinery and also a decrease in its maximum power output. Motors called upon to deliver full load when their terminal voltages are unbalanced are liable to considerable damage due to overheating. One method of maintaining balance of voltage is by having balanced loads connected to the circuit.

(iv) **Efficiency.** The efficiency of a transmission system is not of much importance in itself. The important economic feature of the design being the layout of the system as a whole so as to perform the requisite function of generating and delivering power with a minimum overall annual cost. The annual cost can be minimised to a considerable extent by taking care of power factor of the system. It is because losses in the lines and machinery are largely determined by power factor. Therefore, it is important that consumers having loads of low power factor should be penalised by being charged at a higher rate per kWh than those who take power at high power factors. Loads of low power factor also require greater generator capacity than those of high power factor (for the same amount of power) and produce larger voltage drops in the lines and transformers.

(v) **Frequency.** The frequency of the supply system must be maintained constant. It is because a change in frequency would change the motor speed, thus interfering with the manufacturing operations.

(vi) **Sinusoidal waveform.** The alternating voltage supplied to the consumers should have a sine waveform. It is because any harmonics which might be present would have detrimental effect upon the efficiency and maximum power output of the connected machinery. Harmonics may be avoided by using generators of good design and by avoidance of high flux densities in transformers.

(vii) **Freedom from inductive interference.** Power lines running parallel to telephone lines produce electrostatic and electromagnetic field disturbances. These fields tend to cause objectionable noises and hums in the apparatus connected to communication circuits. Inductive interference with telephone lines may be avoided by limiting as much as possible the amount of zero-sequence and harmonic current and by the proper transposition of both power lines and telephone lines.

SELF - TEST

1. **Fill in the blanks by inserting appropriate words/figures.**
   (*i*) In India .... system is adopted for transmission of electric power.
   (*ii*) .... voltage is used for power transmission as a matter of economy.
   (*iii*) The distribution system comprises of three elements *viz.*, ... and ....
   (*iv*) D.C. transmission is .... to a.c. transmission.
   (*v*) The higher the transmission voltage, the .... is the conductor material required.
   (*vi*) The choice of proper transmission voltage is essentially a question of ....
   (*vii*) In overhead system, the comparison of various systems is made on the basis of maximum voltage between ....
   (*viii*) The economic size of conductor is determined by ....
   (*ix*) In a transmission system, the cost of conductor is proportional to .... of conductor.
   (*x*) The economic transmission voltage is one for which the transmission cost is ....
2. **Pick up the correct words/figures from brackets and fill in the blanks.**
   (*i*) Primary transmission is done by 3-phase ....wire a.c. system. (*3, 4*)
   (*ii*) The .... distribution is done by 3-phase, 4-wire a.c. system. (*primary, secondary*)
   (*iii*) The greater the power to be transmitted, the .... is the economic transmission voltage. (*smaller, larger*)
   (*iv*) The annual charge of a transmission line can be expressed as $P_1 + P_2$ .... (a, $1/a$)
   (*v*) The economic transmission voltage .... the distance of transmission. (*depends upon, does not depend upon*)

ANSWERS TO SELF-TEST

1. (*i*) 3-phase, 3-wire (*ii*) high (*iii*) feeders, distributors, service mains (*iv*) superior (*v*) lesser (*vi*) economics (*vii*) conductor and earth (*viii*) Kelvin's law (*ix*) area (*x*) minimum.
2. (*i*) 3 (*ii*) secondary (*iii*) larger (*iv*) a (*v*) depends upon.

CHAPTER REVIEW TOPICS

1. What is electric power supply system ? Draw a single line diagram of a typical a.c power supply scheme.
2. What are the advantages and disadvantages of d.c. transmission over a.c. transmission ?
3. Discuss the advantages of high transmission voltage.
4. Compare the volume of conductor material required for a d.c. 3-wire system and 3-phase, 3-wire system on the basis of equal maximum potential difference between one conductor and earth. Make suitable assumptions.
5. Compare the volume of conductor materiel required in d.c. single phase and three-phase a.c. system.
6. State and prove Kelvin's law for size of conductor for transmission. Discuss its limitations.
7. How will you determine the economic transmission voltage ?

DISCUSSION QUESTIONS

1. What is the need of primary distribution in an electric supply scheme ?
2. The present trend is towards a.c for generation and distribution and d.c. for transmission. Discuss the reasons for it.
3. In an overhead system, the basis of comparison is the maximum voltage between conductor and ground. Why ?
4. Kelvin's law does not give the exact economical size of conductor. Give reasons in support of your answer.

CHAPTER

8

Mechanical Design of Overhead Lines

Introduction

Electric power can be transmitted or distributed either by means of underground cables or by overhead lines. The underground* cables are rarely used for power transmission due to two main reasons. Firstly, power is generally transmitted over long distances to load centres. Obviously, the installation costs for underground transmission will be very heavy. Secondly, electric power has to be transmitted at high voltages for economic reasons. It is very difficult to provide proper insulation† to the cables to withstand such higher pressures. Therefore, as a rule, power trasmission over long distances is carried out by using overhead lines. With the growth in power demand and consequent rise in voltage levels, power transmission by overhead lines has assumed considerable importance.

* The undergound system is much more expensive than overhead system. Therefore, it has limited use for distribution in congested areas where safety and good appearances are the main considerations.

† In overhead lines, bare conductors are used and air acts as the insulation. The necessary insulation between the conductors can be provided by adjusting the spacing between them.

An overhead line is subjected to uncertain weather conditions and other external interferences. This calls for the use of proper mechanical factors of saftey in order to ensure the continuity of operation in the line. In general, the strength of the line should be such so as to provide against the worst *probable* weather conditions. In this chapter, we shall focus our attention on the various aspects of mechanical design of overhead lines.

8.1 Main Components of Overhead Lines

An overhead line may be used to transmit or distribute electric power. The succcessful operation of an overhead line depends to a great extent upon the mechanical design of the line. While constructing an overhead line, it should be ensured that mechanical strength of the line is such so as to provide against the most *probable* weather conditions. In general, the main components of an overhead line are:

(*i*) *Conductors* which carry electric power from the sending end station to the receiving end station.

(*ii*) *Supports* which may be poles or towers and keep the conductors at a suitable level above the ground.

(*iii*) *Insulators* which are attached to supports and insulate the conductors from the ground.

(*iv*) *Cross arms* which provide support to the insulators.

(*v*) *Miscellaneous items* such as phase plates, danger plates, lightning arrestors, anti-climbing wires etc.

The continuity of operation in the overhead line depends upon the judicious choice of above components. Therefore, it is profitable to have detailed discussion on them.

8.2 Conductor MaterialS

The conductor is one of the important items as most of the capital outlay is invested for it. Therefore, proper choice of material and size of the conductor is of considerable importance. The conductor material used for transmission and distribution of electric power should have the following properties :

(*i*) high electrical conductivity.

(*ii*) high tensile strength in order to withstand mechanical stresses.

(*iii*) low cost so that it can be used for long distances.

(*iv*) low specific gravity so that weight per unit volume is small.

All above requirements are not found in a single material. Therefore, while selecting a conductor material for a particular case, a compromise is made between the cost and the required electrical and mechanical properties.

Commonly used conductor materials. The most commonly used conductor materials for overhead lines are *copper, aluminium, steel-cored aluminium, galvanised steel* and *cadmium copper*. The choice of a particular material will depend upon the cost, the required electrical and mechanical properties and the local conditions.

All conductors used for overhead lines are preferably stranded* in order to increase the flexibility. In stranded conductors, there is generally one central wire and round this, successive layers of wires containing 6, 12, 18, 24 ...... wires. Thus, if there are n layers, the total number of individual wires is $3n(n + 1) + 1$. In the manufacture of stranded conductors, the consecutive layers of wires are twisted or spiralled in opposite directions so that layers are bound together.

1. Copper. Copper is an ideal material for overhead lines owing to its high electrical conductivity and greater tensile strength. It is always used in the hard drawn form as stranded conductor.

* Solid wires are only used when area of X-section is small. If solid wires are used for larger X-section and longer spans, continuous vibrations and swinging would produce mechanical fatigue and they would fracture at the points of support.

Although hard drawing decreases the electrical conductivity slightly yet it increases the tensile strength considerably.

Copper has high current density *i.e.*, the current carrying capacity of copper per unit of X-sectional area is quite large. This leads to two advantages. Firstly, smaller X-sectional area of conductor is required and secondly, the area offered by the conductor to wind loads is reduced. Moreover, this metal is quite homogeneous, durable and has high scrap value.

There is hardly any doubt that copper is an ideal material for transmission and distribution of electric power. However, due to its higher cost and non-availability, it is rarely used for these purposes. Now-a-days the trend is to use aluminium in place of copper.

2. Aluminium. Aluminium is cheap and light as compared to copper but it has much smaller conductivity and tensile strength. The relative comparison of the two materials is briefed below :

(*i*) The conductivity of aluminium is 60% that of copper. The smaller conductivity of aluminium means that for any particular transmission efficiency, the X-sectional area of conductor must be larger in aluminium than in copper. For the same resistance, the diameter of aluminium conductor is about 1·26 times the diameter of copper conductor.

The increased X-section of aluminium exposes a greater surface to wind pressure and, therefore, supporting towers must be designed for greater transverse strength. This often requires the use of higher towers with consequence of greater sag.

(*ii*) The specific gravity of aluminium (2·71 gm/cc) is lower than that of copper (8·9 gm/cc). Therefore, an aluminium conductor has almost one-half the weight of equivalent copper conductor. For this reason, the supporting structures for aluminium need not be made so strong as that of copper conductor.

(*iii*) Aluminium conductor being light, is liable to greater swings and hence larger cross-arms are required.

(*iv*) Due to lower tensile strength and higher co-efficient of linear expansion of aluminium, the sag is greater in aluminium conductors.

Considering the combined properties of cost, conductivity, tensile strength, weight etc., aluminium has an edge over copper. Therefore, it is being widely used as a conductor material. It is particularly profitable to use aluminium for heavy-current transmission where the conductor size is large and its cost forms a major proportion of the total cost of complete installation.

3. Steel cored aluminium. Due to low tensile strength, aluminium conductors produce greater sag. This prohibits their use for larger spans and makes them unsuitable for long distance transmission. In order to increase the tensile strength, the aluminium conductor is reinforced with a core of galvanised steel wires. The *composite conductor thus obtained is known as *steel cored aluminium* and is abbreviated as A.C.S.R. (aluminium conductor steel reinforced).

Steel-cored aluminium conductor consists of central core of †galvanised steel wires surrounded by a number of aluminium strands. Usually, diameter of both steel and aluminium wires is the same. The X-section of the two metals are generally in the ratio of 1 : 6 but can be modified to 1 : 4 in order to get more tensile strength for the conductor. Fig. 8.1 shows steel cored aluminium conductor having one steel wire surrounded by six wires of aluminium. The result of this composite conductor is that steel core takes greater percentage of

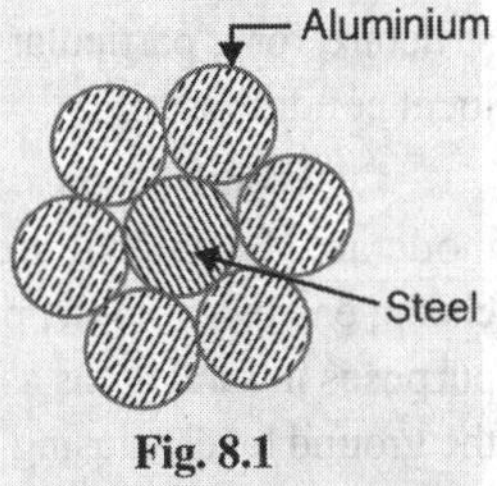

Fig. 8.1

* The reader may think that reinforcement with steel increases the weight but actually the weight of composite conductor is 25% less as compared with equivalent copper conductor.

† The galvanised steel is used in order to prevent rusting and electrolytic corrosion.

mechanical strength while aluminium strands carry the bulk of current. The steel cored aluminium conductors have the following advantages :

(*i*) The reinforcement with steel increases the tensile strength but at the same time keeps the composite conductor light. Therefore, steel cored aluminium conductors will produce smaller sag and hence longer spans can be used.

(*ii*) Due to smaller sag with steel cored aluminium conductors, towers of smaller heights can be used.

4. **Galvanised steel.** Steel has very high tensile strength. Therefore, galvanised steel conductors can be used for extremely long spans or for short line sections exposed to abnormally high stresses due to climatic conditions. They have been found very suitable in rural areas where cheapness is the main consideration. Due to poor conductivity and high resistance of steel, such conductors are not suitable for transmitting large power over a long distance. However, they can be used to advantage for transmitting a small power over a small distance where the size of the copper conductor desirable from economic considerations would be too small and thus unsuitable for use because of poor mechanical strength.

5. **Cadmium copper.** The conductor material now being employed in certain cases is copper alloyed with cadmium. An addition of 1% or 2% cadmium to copper increases the tensile strength by about 50% and the conductivity is only reduced by 15% below that of pure copper. Therefore, cadmium copper conductor can be useful for exceptionally long spans. However, due to high cost of cadmium, such conductors will be economical only for lines of small X-section *i.e.*, where the cost of conductor material is comparatively small compared with the cost of supports.

8.3 Line Supports

The supporting structures for overhead line conductors are various types of poles and towers called *line supports*. In general, the line supports should have the following properties :

(*i*) High mechanical strength to withstand the weight of conductors and wind loads etc.

(*ii*) Light in weight without the loss of mechanical strength.

(*iii*) Cheap in cost and economical to maintain.

(*iv*) Longer life.

(*v*) Easy accessibility of conductors for maintenance.

The line supports used for transmission and distribution of electric power are of various types including *wooden poles, steel poles, R.C.C. poles* and *lattice steel towers*. The choice of supporting structure for a particular case depends upon the line span, X-sectional area, line voltage, cost and local conditions.

1. **Wooden poles.** These are made of seasoned wood (sal or chir) and are suitable for lines of moderate X-sectional area and of relatively shorter spans, say upto 50 metres. Such supports are cheap, easily available, provide insulating properties and, therefore, are widely used for distribution purposes in rural areas as an economical proposition. The wooden poles generally tend to rot below the ground level, causing foundation failure. In order to prevent this, the portion of the pole below the ground level is impregnated with preservative compounds like *creosote oil*. Double pole structures of the '*A*' or '*H*' type are often used (See Fig. 8.2) to obtain a higher transverse strength than could be economically provided by means of single poles.

The main objections to wooden supports are : (*i*) tendency to rot below the ground level (*ii*) comparatively smaller life (20-25 years) (*iii*) cannot be used for voltages higher than 20 kV (*iv*) less mechanical strength and (*v*) require periodical inspection.

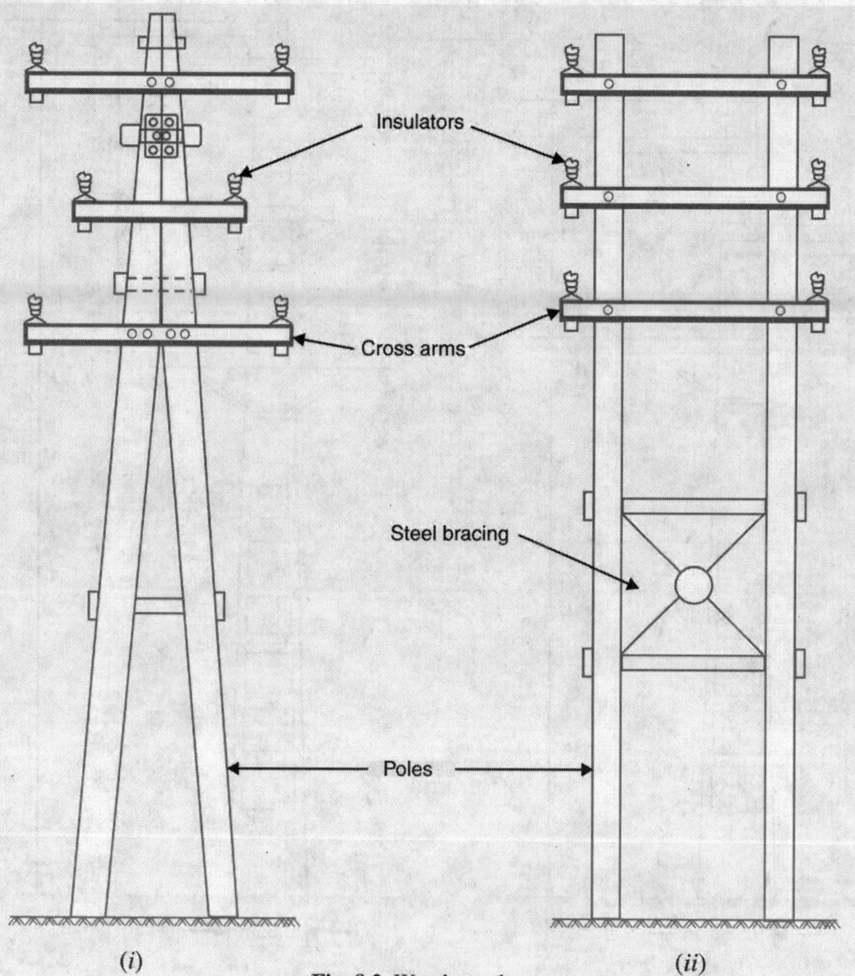

Fig. 8.2 Wooden poles

2. **Steel poles.** The steel poles are often used as a substitute for wooden poles. They possess greater mechanical strength, longer life and permit longer spans to be used. Such poles are generally used for distribution purposes in the cities. This type of supports need to be galvanised or painted in order to prolong its life. The steel poles are of three types *viz.,* (*i*) rail poles (*ii*) tubular poles and (*iii*) rolled steel joints.

3. **RCC poles.** The reinforced concrete poles have become very popular as line supports in recent years. They have greater mechanical strength, longer life and permit longer spans than steel poles. Moreover, they give good outlook, require little maintenance and have good insulating properties. Fig. 8.3 shows R.C.C. poles for single and double circuit. The holes in the poles facilitate the climbing of poles and at the same time reduce the weight of line supports.

The main difficulty with the use of these poles is the high cost of transport owing to their heavy weight. Therefore, such poles are often manufactured at the site in order to avoid heavy cost of transportation.

4. **Steel towers.** In practice, wooden, steel and reinforced concrete poles are used for distribution purposes at low voltages, say upto 11 kV. However, for long distance transmission at higher voltage, steel towers are invariably employed. Steel towers have greater mechanical strength, longer

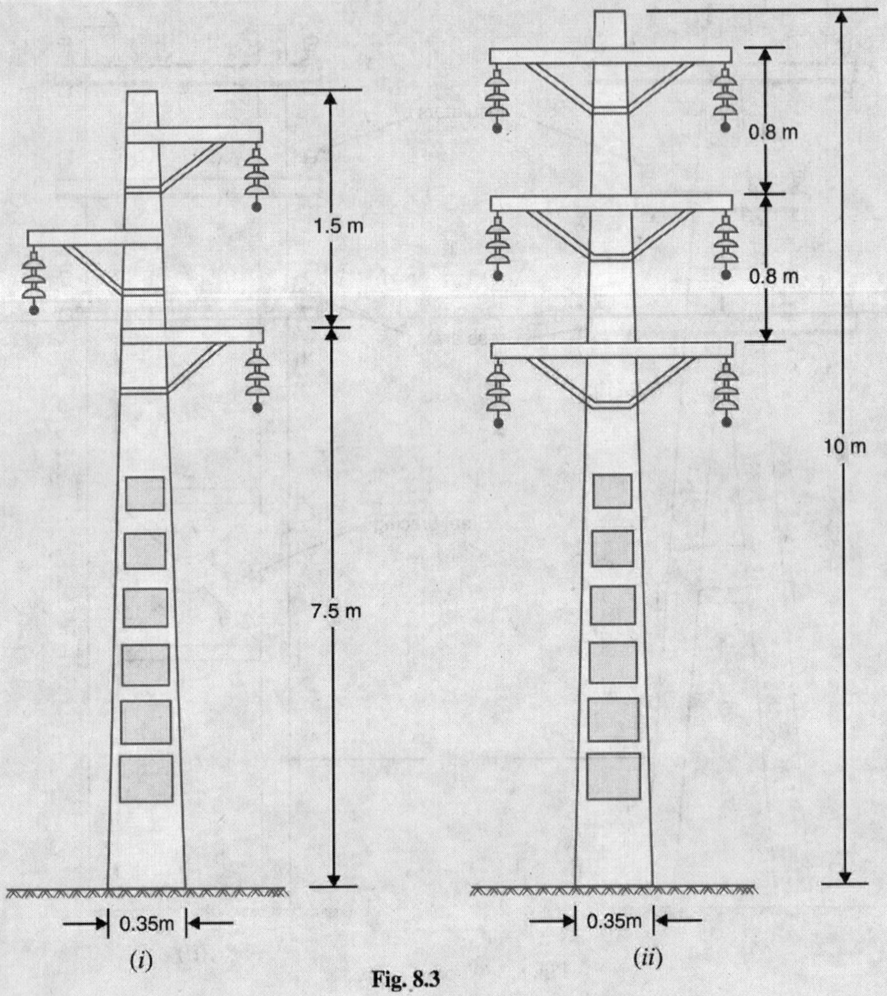

Fig. 8.3

life, can withstand most severe climatic conditions and permit the use of longer spans. The risk of interrupted serivce due to broken or punctured insulation is considerably reduced owing to longer spans. Tower footings are usually grounded by driving rods into the earth. This minimises the lightning troubles as each tower acts as a lightning conductor.

Fig. 8.4 (*i*) shows a single circuit tower. However, at a moderate additional cost, double circuit tower can be provided as shown in Fig. 8.4 (*ii*). The double circuit has the advantage that it ensures continuity of supply. It case there is breakdown of one circuit, the continuity of supply can be maintained by the other circuit.

8.4 Insulators

The overhead line conductors should be supported on the poles or towers in such a way that currents from conductors do not flow to earth through supports *i.e.*, line conductors must be properly insulated from supports. This is achieved by securing line conductors to supports with the help of *insulators*. The insulators provide necessary insulation between line conductors and supports and thus prevent any leakage current from conductors to earth. In general, the insulators should have the following desirable properties :

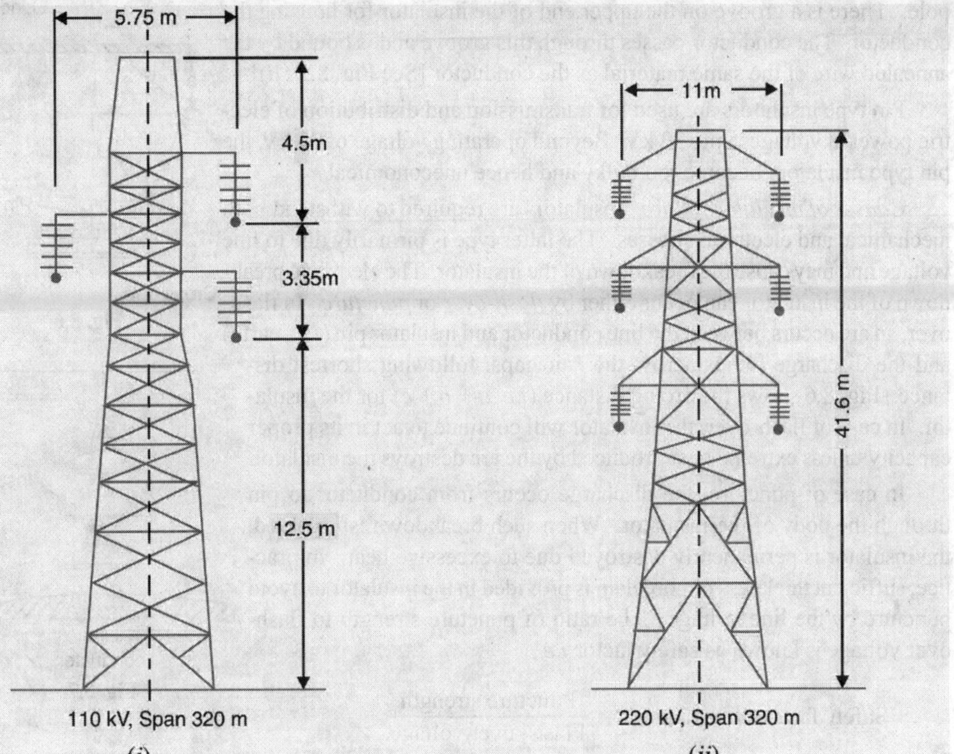

Fig. 8.4 Steel Towers

(*i*) High mechanical strength in order to withstand conductor load, wind load etc.

(*ii*) High electrical resistance of insulator material in order to avoid leakage currents to earth.

(*iii*) High relative permittivity of insulator material in order that dielectric strength is high.

(*iv*) The insulator material should be non-porous, free from impurities and cracks otherwise the permittivity will be lowered.

(*v*) High ratio of puncture strength to flashover.

The most commonly used material for insulators of overhead line is *porcelain* but glass, steatite and special composition materials are also used to a limited extent. Porcelain is produced by firing at a high temperature a mixture of kaolin, feldspar and quartz. It is stronger mechanically than glass, gives less trouble from leakage and is less effected by changes of temperature.

8.5 Types of Insulators

The successful operation of an overhead line depends to a considerable extent upon the proper selection of insulators. There are several types of insulators but the most commonly used are pin type, suspension type, strain insulator and shackle insulator.

1. Pin type insulators. The part section of a pin type insulator is shown in Fig. 8.5 (*i*). As the name suggests, the pin type insulator is secured to the cross-arm on the

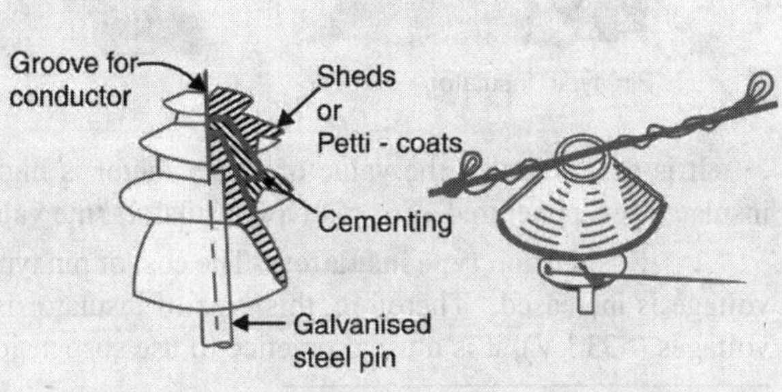

Fig. 8.5. Pin-type insulator

pole. There is a groove on the upper end of the insulator for housing the conductor. The conductor passes through this groove and is bound by the annealed wire of the same material as the conductor [See Fig. 8.5 (*ii*)].

Pin type insulators are used for transmission and distribution of electric power at voltages upto 33 kV. Beyond operating voltage of 33 kV, the pin type insulators become too bulky and hence uneconomical.

Causes of insulator failure. Insulators are required to withstand both mechanical and electrical stresses. The latter type is pirmarily due to line voltage and may cause the breakdown of the insulator. The electrical breakdown of the insulator can occur either by *flash-over* or *puncture.* In flash-over, an arc occurs between the line conductor and insulator pin (*i.e.*, earth) and the discharge jumps across the *air gaps, following shortest distance. Fig. 8.6 shows the arcing distance (*i.e.* $a + b + c$) for the insulator. In case of flash-over, the insulator will continue to act in its proper capacity unless extreme heat produced by the arc destroys the insulator.

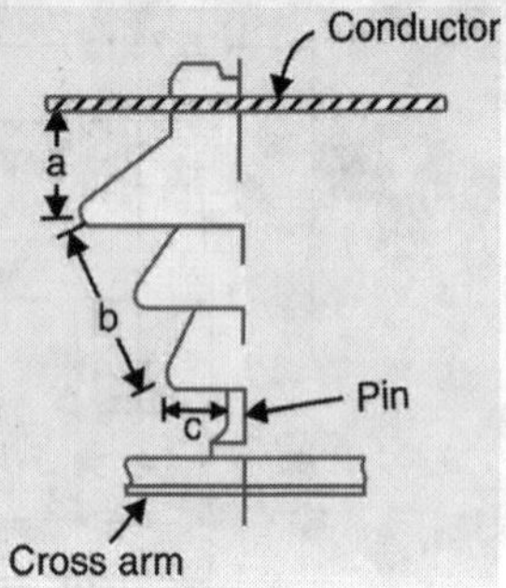

Fig. 8.6

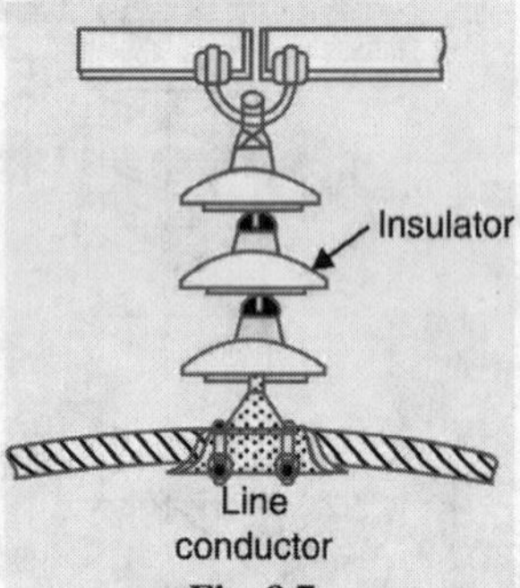

Fig. 8.7

In case of puncture, the discharge occurs from conductor to pin through the body of the insulator. When such breakdown is involved, the insulator is permanently destroyed due to excessive heat. In practice, sufficient thickness of porcelain is provided in the insulator to avoid puncture by the line voltage. The ratio of puncture strength to flash-over voltage is known as safety factor *i.e.,*

$$\text{Safety factor of insulator} = \frac{\text{Puncture strength}}{\text{Flash - over voltage}}$$

Pin type insulator

Suspension insulator

It is desirable that the value of safety factor is high so that flash-over takes place before the insulator gets punctured. For pin type insulators, the value of safety factor is about 10.

2 **Suspension type insulators.** The cost of pin type insulator increases rapidly as the working voltage is increased. Therefore, this type of insulator is not economical beyond 33 kV. For high voltages (>33 kV), it is a usual practice to use suspension type insulators shown in Fig. 8.7. They

* The insulator is generally dry and its surfaces have proper insulating properties. Therefore, arc can only occur through air gap between conductor and insulator pin.

consist of a number of porcelain discs connected in series by metal links in the form of a string. The conductor is suspended at the bottom end of this string while the other end of the string is secured to the cross-arm of the tower. Each unit or disc is designed for low voltage, say 11 kV. The number of discs in series would obviously depend upon the working voltage. For instance, if the working voltage is 66 kV, then six discs in series will be provided on the string.

Advantages

(*i*) Suspension type insulators are cheaper than pin type insulators for voltages beyond 33 kV.

(*ii*) Each unit or disc of suspension type insulator is designed for low voltage,usually 11 kV. Depending upon the working voltage, the desired number of discs can be connected in series.

(*iii*) If any one disc is damaged, the whole string does not become useless because the damaged disc can be replaced by the sound one.

(*iv*) The suspension arrangement provides greater flexibility to the line. The connection at the cross arm is such that insulator string is free to swing in any direction and can take up the position where mechanical stresses are minimum.

(*v*) In case of increased demand on the transmission line, it is found more satisfactory to supply the greater demand by raising the line voltage than to provide another set of conductors. The additional insulation required for the raised voltage can be easily obtained in the suspension arrangement by adding the desired number of discs.

(*vi*) The suspension type insulators are generally used with steel towers. As the conductors run below the earthed cross-arm of the tower, therefore, this arrangement provides partial protection from lightning.

3. **Strain insulators.** When there is a dead end of the line or there is corner or sharp curve, the line is subjected to greater tension. In order to relieve the line of excessive tension, strain insulators are used. For low voltage lines (< 11 kV), shackle insulators are used as strain insulators. However, for high voltage transmission lines, strain insulator consists of an assembly of suspension insulators as shown in Fig. 8.8. The discs of strain insulators are used in the vertical plane. When the tension in lines is exceedingly high, as at long river spans, two or more strings are used in parallel.

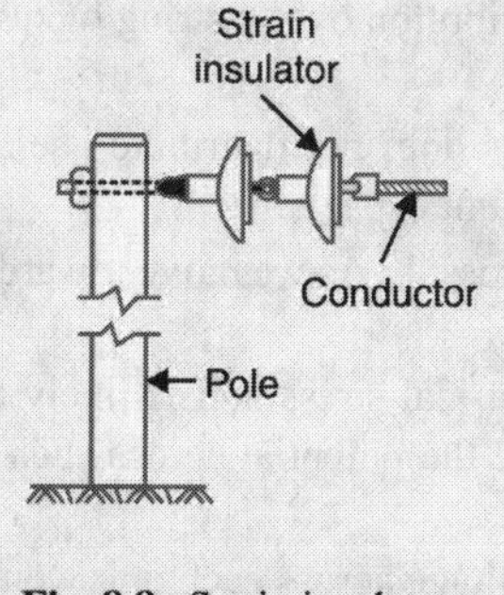

Fig. 8.8. Strain insulator.

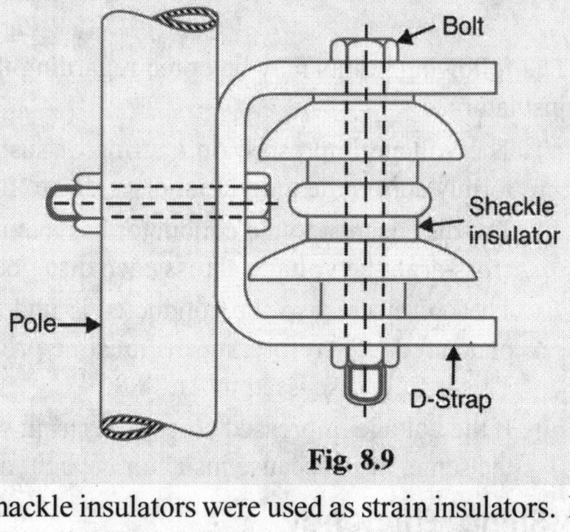

Fig. 8.9

4. **Shackle insulators.** In early days, the shackle insulators were used as strain insulators. But now a days, they are frequently used for low voltage distribution lines. Such insulators can be used either in a horizontal position or in a vertical position. They can be directly fixed to the pole with a bolt or to the cross arm. Fig. 8.9 shows a shackle insulator fixed to the pole. The conductor in the groove is fixed with a soft binding wire.

8.6 Potential Distribution over Suspension Insulator String

A string of suspension insulators consists of a number of porcelain discs connected in series through metallic links. Fig. 8.10 (*i*) shows 3-disc string of suspension insulators. The porcelain portion of each disc is inbetween two metal links. Therefore, each disc forms a capacitor C as shown in Fig. 8.10 (*ii*). This is known as *mutual capacitance* or *self-capacitance*. If there were mutual capacitance alone, then charging current would have been the same through all the discs and consequently voltage across each unit would have been the same *i.e.*, $V/3$ as shown in Fig. 8.10 (*ii*). However, in actual practice, capacitance also exists between metal fitting of each disc and tower or earth. This is known as *shunt capacitance* C_1. Due to shunt capacitance, charging current is not the same through all the discs of the string [See Fig. 8.10 (*iii*)]. Therefore, voltage across each disc will be different. Obviously, the disc nearest to the line conductor will have the maximum* voltage. Thus referring to Fig. 8.10 (*iii*), V_3 will be much more than V_2 or V_1.

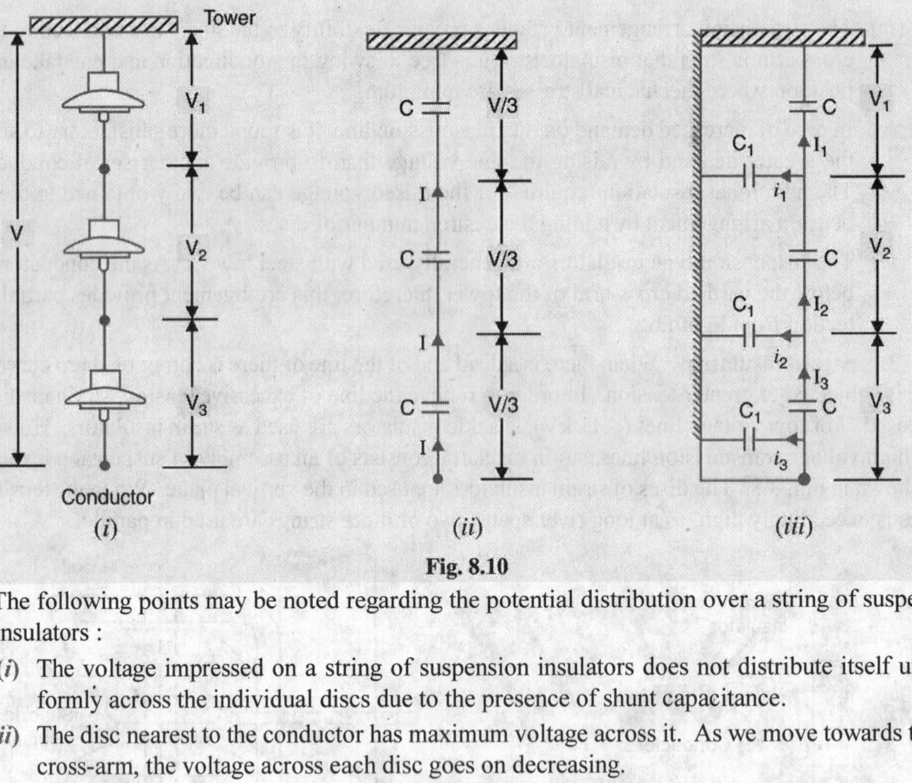

Fig. 8.10

The following points may be noted regarding the potential distribution over a string of suspension insulators :

(*i*) The voltage impressed on a string of suspension insulators does not distribute itself uniformly across the individual discs due to the presence of shunt capacitance.

(*ii*) The disc nearest to the conductor has maximum voltage across it. As we move towards the cross-arm, the voltage across each disc goes on decreasing.

(*iii*) The unit nearest to the conductor is under maximum electrical stress and is likely to be punctured. Therefore, means must be provided to equalise the potential across each unit. This is fully discussed in Art. 8.8.

(*iv*) If the voltage impressed across the string were d.c., then voltage across each unit would be the same. It is because insulator capacitances are ineffective for d.c.

8.7 String Efficiency

As stated above, the voltage applied across the string of suspension insulators is not uniformly distributed across various units or discs. The disc nearest to the conductor has much higher potential than the other discs. This unequal potential distribution is undesirable and is usually expressed in

* Because charging current through the string has the maximum value at the disc nearest to the conductor.

terms of string efficiency.

The ratio of voltage across the whole string to the product of number of discs and the voltage across the disc nearest to the conductor is known as **string efficiency** *i.e.,*

$$\text{String efficiency} = \frac{\text{Voltage across the string}}{n \times \text{Voltage across disc nearest to conductor}}$$

where n = number of discs in the string.

String efficiency is an important consideration since it decides the potential distribution along the string. The greater the string efficiency, the more uniform is the voltage distribution. Thus 100% string efficiency is an ideal case for which the volatge across each disc will be exactly the same. Although it is impossible to achieve 100% string efficiency, yet efforts should be made to improve it as close to this value as possible.

Mathematical expression. Fig. 8.11 shows the equivalent circuit for a 3-disc string. Let us suppose that self capacitance of each disc is C. Let us further assume that shunt capacitance C_1 is some fraction K of self-capacitance *i.e.,* $C_1 = KC$. Starting from the cross-arm or tower, the voltage across each unit is V_1, V_2 and V_3 respectively as shown.

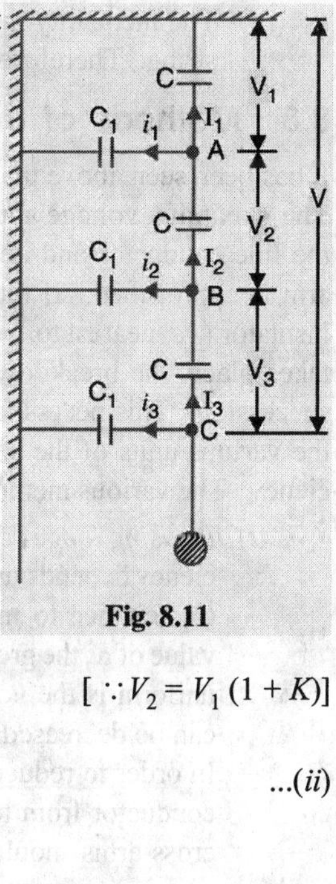

Fig. 8.11

Applying Kirchhoff's current law to node A, we get,

$$I_2 = I_1 + i_1$$

or $$V_2 \omega C^* = V_1 \omega C + V_1 \omega C_1$$

or $$V_2 \omega C = V_1 \omega C + V_1 \omega K C$$

$\therefore$ $$V_2 = V_1 (1 + K) \quad \ldots(i)$$

Applying Kirchhoff's current law to node B, we get,

$$I_3 = I_2 + i_2$$

or $$V_3 \, \omega C = V_2 \omega C + (V_1 + V_2) \, \omega \, C_1 \dagger$$

or $$V_3 \, \omega C = V_2 \omega C + (V_1 + V_2) \, \omega K C$$

or $$V_3 = V_2 + (V_1 + V_2) K$$

$$= K V_1 + V_2 (1 + K)$$

$$= K V_1 + V_1 (1 + K)^2 \quad [\because V_2 = V_1 (1 + K)]$$

$$= V_1 [K + (1 + K)^2]$$

$\therefore$ $$V_3 = V_1 [1 + 3K + K^2] \quad \ldots(ii)$$

Voltage between conductor and earth (*i.e.,* tower) is

$$V = V_1 + V_2 + V_3$$

$$= V_1 + V_1(1 + K) + V_1 (1 + 3K + K^2)$$

$$= V_1 (3 + 4K + K^2)$$

$\therefore$ $$V = V_1 (1 + K)(3 + K) \quad \ldots(iii)$$

From expressions (*i*), (*ii*) and (*iii*), we get,

$$\frac{V_1}{1} = \frac{V_2}{1 + K} = \frac{V_3}{1 + 3K + K^2} = \frac{V}{(1 + K)(3 + K)} \quad \ldots(iv)$$

$\therefore$ Voltage across top unit, $$V_1 = \frac{V}{(1 + K)(3 + K)}$$

---

* Note that current through capacitor = $\dfrac{\text{Voltage}}{\text{Capacitive reactance}}$

† Voltage across second shunt capacitance C_1 from top = $V_1 + V_2$. It is because one point of it is connected to B and the other point to the tower.

Voltage across second unit from top, $V_2 = V_1(1 + K)$

Voltage across third unit from top, $V_3 = V_1(1 + 3K + K^2)$

$$\text{\%age String efficiency} = \frac{\text{Voltage across string}}{n \times \text{Voltage across disc nearest to conductor}} \times 100$$

$$= \frac{V}{3 \times V_3} \times 100$$

The following points may be noted from the above mathematical analysis :

(*i*) If $K = 0{\cdot}2$ (Say), then from exp. (*iv*), we get, $V_2 = 1{\cdot}2\ V_1$ and $V_3 = 1{\cdot}64\ V_1$. This clearly shows that disc nearest to the conductor has maximum voltage across it; the voltage across other discs decreasing progressively as the cross-arm in approached.

(*ii*) The greater the value of $K (= C_1/C)$, the more non-uniform is the potential across the discs and lesser is the string efficiency.

(*iii*) The inequality in voltage distribution increases with the increase of number of discs in the string. Therefore, shorter string has more efficiency than the larger one.

8.8 Methods of Improving String Efficiency

It has been seen above that potential distribution in a string of suspension insulators is not uniform. The maximum voltage appears across the insulator nearest to the line conductor and decreases progressively as the cross-arm is approached. If the insulation of the highest stressed insulator (*i.e.* nearest to conductor) breaks down or flash over takes place, the breakdown of other units will take place in succession. This necessitates to equalise the potential across the various units of the string *i.e.* to improve the string efficiency. The various methods for this purpose are :

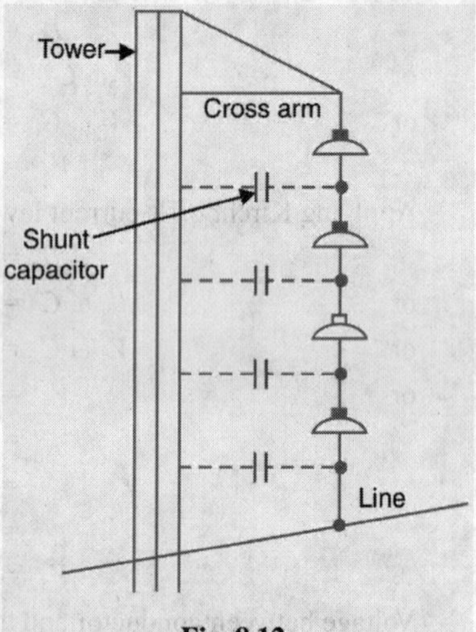

Fig. 8.12

(*i*) *By using longer cross-arms.* The value of string efficiency depends upon the value of K *i.e.*, ratio of shunt capacitance to mutual capacitance. The lesser the value of K, the greater is the string efficiency and more uniform is the voltage distribution. The value of K can be decreased by reducing the shunt capacitance. In order to reduce shunt capacitance, the distance of conductor from tower must be increased *i.e.*, longer cross-arms should be used. However, limitations of cost and strength of tower do not allow the use of very long cross-arms. In practice, $K = 0{\cdot}1$ is the limit that can be achieved by this method.

(*ii*) *By grading the insulators.* In this method, insulators of different dimensions are so chosen that each has a different capacitance. The insulators are capacitance graded *i.e.* they are assembled in the string in such a way that the top unit has the minimum capacitance, increasing progressively as the bottom unit (*i.e.*, nearest to conductor) is reached. Since voltage is inversely proportional to capacitance, this method tends to equalise the potential distribution across the units in the string. This method has the disadvantage that a large number of different-sized insulators are required. However, good results can be obtained by using standard insulators for most of the string and larger units for that near to the line conductor.

(*iii*) *By using a guard ring.* The potential across each unit in a string can be equalised by using a guard ring which is a metal ring electrically connected to the conductor and surrounding the bottom insulator as shown in the Fig. 8.13. The guard ring introduces capacitance between metal fittings and the line conductor. The guard ring is contoured in such a way that shunt capacitance currents i_1, i_2 etc. are equal to metal fitting line capacitance currents i'_1, i'_2 etc. The result is that same charging current I flows through each unit of string. Consequently, there will be uniform potential distribution across the units.

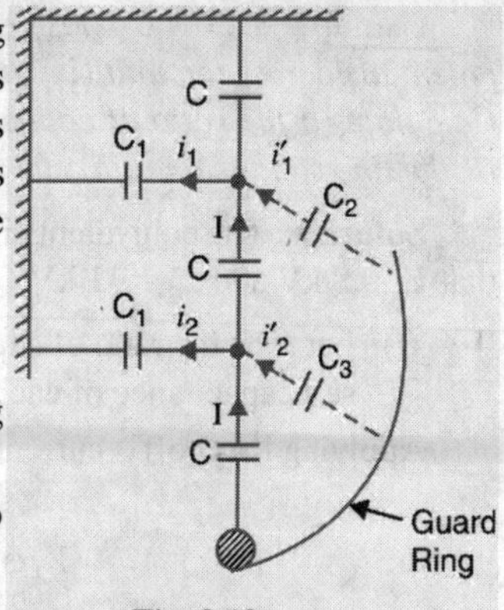

Fig. 8.13

8.9 Important Points

While solving problems relating to string efficiency, the following points must be kept in mind:

(*i*) The maximum voltage appears across the disc nearest to the conductror (*i.e.,* line conductor).

(*ii*) The voltage across the string is equal to phase voltage *i.e.,*

Voltage across string = Voltage between line and earth = Phase Voltage

(*iii*) Line Voltage = $\sqrt{3}$ × Voltage across string

Example 8.1. *In a 33 kV overhead line, there are three units in the string of insulators. If the capacitance between each insulator pin and earth is 11% of self-capacitance of each insulator, find (i) the distribution of voltage over 3 insulators and (ii) string efficiency.*

Solution. Fig. 8.14. shows the equivalent circuit of string insulators. Let V_1, V_2 and V_3 be the voltage across top, middle and bottom unit respectively. If C is the self-capacitance of each unit, then KC will be the shunt capacitance.

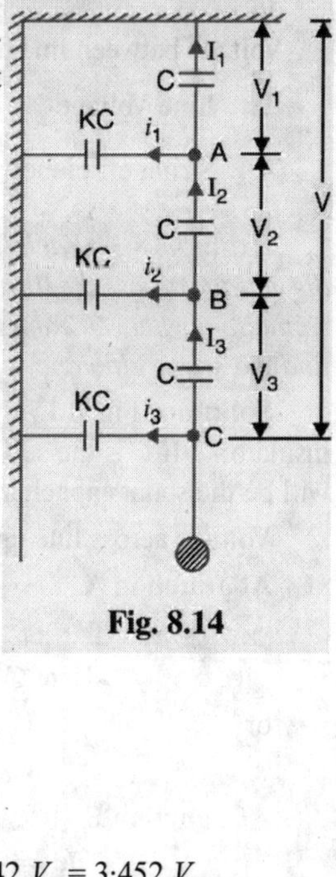

Fig. 8.14

$$K = \frac{\text{Shunt Capacitance}}{\text{Self - capacitance}} = 0.11$$

$$\text{Voltage across string, } V = 33/\sqrt{3} = 19{\cdot}05 \text{ kV}$$

At Junction A

$$I_2 = I_1 + i_1$$

or $$V_2 \omega C = V_1 \omega C + V_1 K \omega C$$

or $$V_2 = V_1 (1 + K) = V_1 (1 + 0{\cdot}11)$$

or $$V_2 = 1{\cdot}11\ V_1 \qquad \text{...}(i)$$

At Junction B

$$I_3 = I_2 + i_2$$

or $$V_3 \omega C = V_2 \omega C + (V_1 + V_2) K \omega C$$

or $$V_3 = V_2 + (V_1 + V_2) K$$

$$= 1{\cdot}11 V_1 + (V_1 + 1{\cdot}11\ V_1)\, 0{\cdot}11$$

$\therefore$ $$V_3 = 1{\cdot}342\ V_1$$

(*i*) Voltage across the whole string is

$$V = V_1 + V_2 + V_3 = V_1 + 1{\cdot}11\ V_1 + 1{\cdot}342\ V_1 = 3{\cdot}452\ V_1$$

or $$19{\cdot}05 = 3{\cdot}452\ V_1$$

$\therefore$ Voltage across top unit, V_1 = $19{\cdot}05/3{\cdot}452 =$ **5·52 kV**

Voltage across middle unit, V_2 = $1{\cdot}11\ V_1 = 1{\cdot}11 \times 5{\cdot}52 =$ **6.13 kV**

Voltage across bottom unit, V_3 = $1{\cdot}342\ V_1 = 1{\cdot}342 \times 5{\cdot}52 =$ **7.4 kV**

(*ii*) $$\text{String efficiency} = \frac{\text{Voltage across string}}{\text{No. of insulators} \times V_3} \times 100 = \frac{19{\cdot}05}{3 \times 7{\cdot}4} \times 100 = \mathbf{85.8\%}$$

Example 8.2. *A 3-phase transmission line is being supported by three disc insulators. The potentials across top unit (i.e., near to the tower) and middle unit are 8 kV and 11 kV respectively. Calculate (i) the ratio of capacitance between pin and earth to the self-capacitance of each unit (ii)the line voltage and (iii) string efficiency.*

Solution. The equivalent circuit of string insulators is the same as shown in Fig. 8.14. It is given that $V_1 = 8$ kV and $V_2 = 11$ kV.

(*i*) Let K be the ratio of capacitance between pin and earth to self capacitance. If C farad is the self capacitance of each unit, then capacitance between pin and earth = KC.

Applying Kirchoff's current law to Junction A,

$$I_2 = I_1 + i_1$$

or $$V_2 \omega C = V_1 \omega C + V_1 K \omega C$$

or $$V_2 = V_1 (1 + K)$$

$\therefore$ $$K = \frac{V_2 - V_1}{V_1} = \frac{11 - 8}{8} = 0{\cdot}375$$

(*ii*) Applying Kirchoff's current law to Junction B,

$$I_3 = I_2 + i_2$$

or $$V_3 \omega C = V_2 \omega C + (V_1 + V_2) K \omega C$$

or $$V_3 = V_2 + (V_1 + V_2) K = 11 + (8 + 11) \times 0{\cdot}375 = 18{\cdot}12 \text{ kV}$$

Voltage between line and earth $= V_1 + V_2 + V_3 = 8 + 11 + 18{\cdot}12 = 37{\cdot}12$ kV

$\therefore$ Line Voltage $= \sqrt{3} \times 37{\cdot}12 = \mathbf{64{\cdot}28\ kV}$

(*iii*) String efficiency $$= \frac{\text{Voltage across string}}{\text{No. of insulators} \times V_3} \times 100 = \frac{37{\cdot}12}{3 \times 18{\cdot}12} \times 100 = \mathbf{68.28\%}$$

Example 8.3. *Each line of a 3-phase system is suspended by a string of 3 similar insulators. If the voltage across the line unit is 17·5 kV, calculate the line to neutral voltage. Assume that the shunt capacitance betwen each insulator and earth is 1/8th of the capacitance of the insulator itself. Also find the string efficiency.*

Solution. Fig. 8·15 shows the equivalent circuit of string insulators. If C is the self capacitance of each unit, then KC will be the shunt capacitance where $K = 1/8 = 0{\cdot}125$.

Voltage across line unit, $V_3 = 17{\cdot}5$ kV

At Junction A

$$I_2 = I_1 + i_1$$

$$V_2 \omega C = V_1 \omega C + V_1 K \omega C$$

or $$V_2 = V_1 (1 + K) = V_1 (1 + 0.125)$$

$\therefore$ $$V_2 = 1{\cdot}125\, V_1$$

At Junction B

$$I_3 = I_2 + i_2$$

or $$V_3 \omega C = V_2 \omega C + (V_1 + V_2) K \omega C$$

or $$V_3 = V_2 + (V_1 + V_2) K$$

$$= 1{\cdot}125\, V_1 + (V_1 + 1{\cdot}125\, V_1) \times 0.125$$

$\therefore$ $$V_3 = 1{\cdot}39\, V_1$$

Voltage across top unit, $V_1 = V_3/1{\cdot}39 = 17{\cdot}5/1{\cdot}39$

$= 12{\cdot}59$ kV

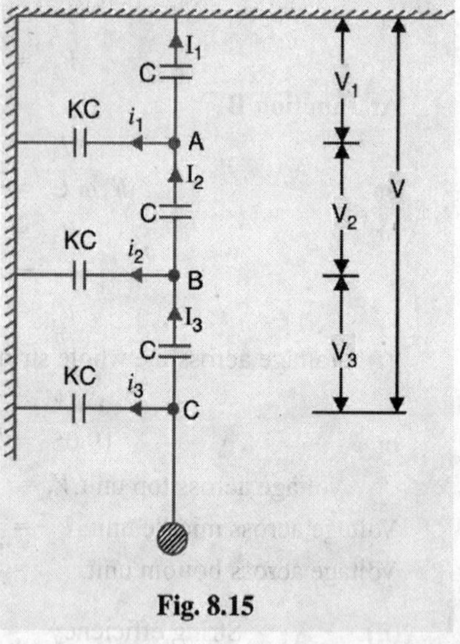

Fig. 8.15

Voltage across middle unit, $V_2 = 1 \cdot 125\ V_1 = 1 \cdot 125 \times 12 \cdot 59 = 14 \cdot 16$ kV

$\therefore$ Voltage between line and earth (*i.e.*, line to neutral)

$$= V_1 + V_2 + V_3 = 12 \cdot 59 + 14 \cdot 16 + 17 \cdot 5 = \mathbf{44.25\ kV}$$

$$\text{String efficiency} = \frac{44 \cdot 25}{3 \times 17 \cdot 5} \times 100 = \mathbf{84 \cdot 28\%}$$

Example 8.4. *The three bus-bar conductors in an outdoor substation are supported by units of post type insulators. Each unit consists of a stack of 3 pin type insulators fixed one on the top of the other. The voltage across the lowest insulator is 13·1 kV and that across the next unit is 11 kV. Find the bus-bar voltage of the station.*

Solution. The equivalent circuit of insulators is the same as shown in Fig. 8.15. It is given that $V_3 = 13 \cdot 1$ kV and $V_2 = 11$ kV. Let K be the ratio of shunt capacitance to self capacitance of each unit. Applying Kirchhoff's current law to Junctions A and B, we can easily derive the following equations (See example 8.3) :

$$V_2 = V_1 (1 + K)$$

or $$V_1 = \frac{V_2}{1 + K} \quad \text{...(i)}$$

and $$V_3 = V_2 + (V_1 + V_2)\, K \quad \text{...(ii)}$$

Putting the value of $V_1 = V_2/(1 + K)$ in eq. (*ii*), we get,

$$V_3 = V_2 + \left[\frac{V_2}{1 + K} + V_2\right] K$$

or $$V_3 (1 + K) = V_2 (1 + K) + [V_2 + V_2 (1 + K)]\, K$$

$$= V_2 [(1 + K) + K + (K + K^2)]$$

$$= V_2 (1 + 3K + K^2)$$

$\therefore$ $$13 \cdot 1 (1 + K) = 11[1 + 3K + K^2]$$

or $$11K^2 + 19 \cdot 9\, K - 2 \cdot 1 = 0$$

Solving this equation, we get, $K = 0.1$.

$\therefore$ $$V_1 = \frac{V_2}{1 + K} = \frac{11}{1 + 0 \cdot 1} = 10 \text{ kV}$$

Voltage between line and earth $= V_1 + V_2 + V_3 = 10 + 11 + 13 \cdot 1 = 34 \cdot 1$ kV

$\therefore$ Voltage between bus-bars (*i.e.*, line voltage)

$$= 34 \cdot 1 \times \sqrt{3} = \mathbf{59\ kV}$$

Example 8.5. *An insulator string consists of three units, each having a safe working voltage of 15 kV. The ratio of self-capacitance to shunt capacitance of each unit is 8 : 1. Find the maximum safe working voltage of the string. Also find the string efficiency.*

Solution. The equivalent circuit of string insulators is the same as shown in Fig. 8.15. The maximum voltage will appear across the lowest unit in the string.

$\therefore$ $$V_3 = 15 \text{ kV} \; ; \quad K = 1/8 = 0 \cdot 125$$

Applying Kirchhoff's current law to junction A, we get,

$$V_2 = V_1 (1 + K)$$

or $$V_1 = V_2/(1 + K) = V_2/(1 + 0 \cdot 125) = 0 \cdot 89\ V_2 \quad \text{...(i)}$$

Applying Kirchhoff's current law to Junction B, we get,

$$V_3 = V_2 + (V_1 + V_2)\, K = V_2 + (0.89\ V_2 + V_2) \times 0.125$$

$\therefore \quad V_3 = 1.236\ V_2$ ...(ii)

$\therefore$ Voltage across middle unit, $V_2 = V_3/1.236 = 15/1{\cdot}236 = 12{\cdot}13$ kV

Voltage across top unit, $V_1 = 0{\cdot}89\ V_2 = 0{\cdot}89 \times 12{\cdot}13 = 10{\cdot}79$ kV

Voltage across the String $= V_1 + V_2 + V_3 = 10{\cdot}79 + 12{\cdot}13 + 15 =$ **37·92 kV**

String efficiency $= \dfrac{37{\cdot}92}{3 \times 15} \times 100 =$ **84·26 %**

Example 8.6. *A string of 4 insulators has a self-capacitance equal to 10 times the pin to earth capacitance. Find (i) the voltage distribution across various units expressed as a percentage of total voltage across the string and (ii) string efficiency.*

Solution. When the number of insulators in a string exceeds 3, the nodal equation method becomes laborious. Under such circumstances, there is a simple method to solve the problem. In this method*, shunt capacitance (C_1) and self capacitance (C) of each insulator are represented by their equivalent reactances. As it is only the ratio of capacitances which determines the voltage distribution, therefore, the problem can be simplified by assigning unity value to X_C *i.e.,* assuming $X_C = 1\ \Omega$. If ratio of $C/C_1 = 10$, then we have $X_C = 1\ \Omega$ and $X_{C1} = 10\ \Omega$.

(*i*) Suppose $X_C = 1\ \Omega$. As the ratio of self-capacitance to shunt capacitance (*i.e.,* C/C_1) is 10, therefore, $X_{C1} = 10\ \Omega$ as shown in Fig. 8·16 (*i*). Suppose that potential V across the string is such that 1 A current flows in the top insulator. Now the potential across each insulator can be easily determined. Thus :

Voltage across top unit, $V_1 = 1\ \Omega \times 1\ \text{A} = 1$ volt

Voltage across **2nd unit, $V_2 = 1\ \Omega \times 1{\cdot}1\ \text{A} = 1{\cdot}1$ volts

Voltage across †3rd unit, $V_3 = 1\ \Omega \times 1{\cdot}31\ \text{A} = 1{\cdot}31$ volts

Voltage across 4th unit, $V_4 = 1\ \Omega \times 1{\cdot}65\ \text{A} = 1{\cdot}65$ volts

Voltage obtained across the string, $V = 1 + 1{\cdot}1 + 1{\cdot}31 + 1{\cdot}65 = 5{\cdot}06$ volts

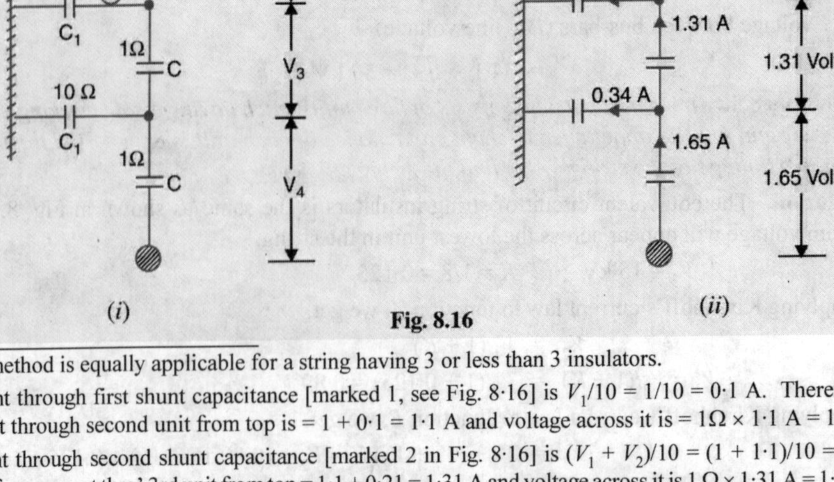

Fig. 8.16

* This method is equally applicable for a string having 3 or less than 3 insulators.

** Current through first shunt capacitance [marked 1, see Fig. 8·16] is $V_1/10 = 1/10 = 0{\cdot}1$ A. Therefore, the current through second unit from top is $= 1 + 0{\cdot}1 = 1{\cdot}1$ A and voltage across it is $= 1\Omega \times 1{\cdot}1\ \text{A} = 1{\cdot}1$ volts.

† Current through second shunt capacitance [marked 2 in Fig. 8·16] is $(V_1 + V_2)/10 = (1 + 1{\cdot}1)/10 = 0{\cdot}21$ A. Therefore, current thro' 3rd unit from top $= 1{\cdot}1 + 0{\cdot}21 = 1{\cdot}31$ A and voltage across it is $1\ \Omega \times 1{\cdot}31\ \text{A} = 1{\cdot}31$ volts.

The voltage across each unit expressed as a percentage of V (*i.e.*, 5·06 volts) becomes :

Top unit $= (1/5{\cdot}06) \times 100 = \mathbf{19{\cdot}76\%}$

Second from top $= (1{\cdot}1/5{\cdot}06) \times 100 = \mathbf{21{\cdot}74\%}$

Third from top $= (1{\cdot}31/5{\cdot}06) \times 100 = \mathbf{25{\cdot}9\%}$

Fourth from top $= (1{\cdot}65/5{\cdot}06) \times 100 = \mathbf{32{\cdot}6\%}$

(*ii*) String efficiency $= \dfrac{V}{4 \times V_4} \times 100 = \dfrac{5{\cdot}06}{4 \times 1{\cdot}65} \times 100 = \mathbf{76{\cdot}6\%}$

Example 8.7. *A string of 5 insulators is connected across a 100 kV line. If the capacitance of each disc to earth is 0·1 of the capacitance of the insulator, calculate (i) the distribution of voltage on the insulator discs and (ii) the string efficiency.*

Solution. Suppose $X_C = 1\ \Omega$. As the ratio of self capacitance to shunt capacitance is 10, therefore, $X_{C1} = 10\ \Omega$ as shown in Fig. 8·17 (*i*). Suppose that potential V across the string is such that 1A current flows in the top insulator. Then potential across each insulator will be as shown in Fig. 8·17 (*ii*).

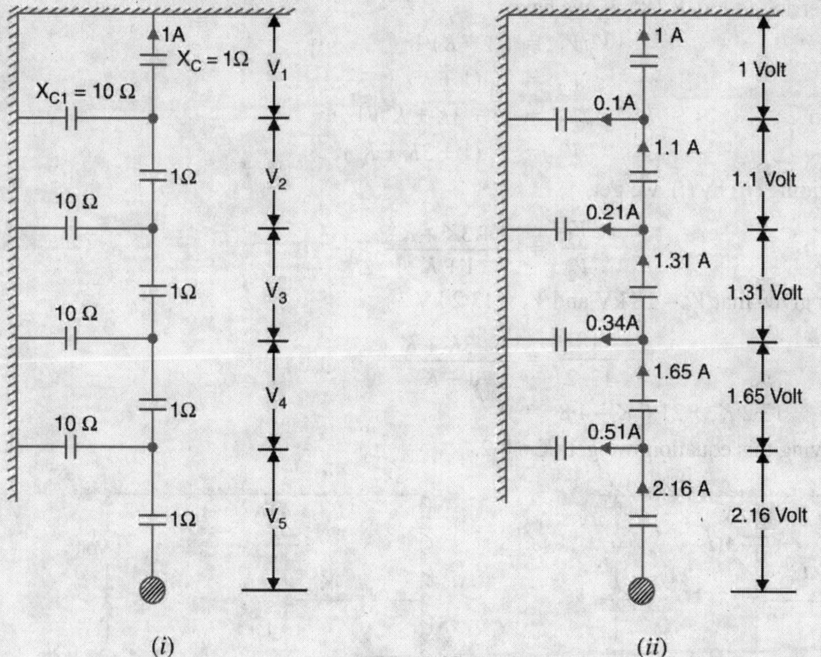

Fig. 8.17

The value obtained for $V = 1 + 1{\cdot}1 + 1{\cdot}31 + 1{\cdot}65 + 2{\cdot}16 = 7{\cdot}22$ volts and starting from top, the percentage of V (*i.e.*, 7·22 volts) across various units are :

*13·8 %, 15·2 %, 18·2 %, 22·8 % and 30%

Voltage across string $= 100/\sqrt{3} = 57{\cdot}7$ kV

(*i*) Voltage across top insulator, $V_1 = 0{\cdot}138 \times 57{\cdot}7 = \mathbf{7{\cdot}96\ kV}$

Voltage across 2nd from top, $V_2 = 0{\cdot}152 \times 57{\cdot}7 = \mathbf{8{\cdot}77\ kV}$

---

* % age of V (*i.e.*, 7·22 volts) across top unit $= \dfrac{1}{7{\cdot}22} \times 100 = 13{\cdot}8\%$

% age of V across 2nd from top $= \dfrac{1{\cdot}1}{7{\cdot}22} \times 100 = 15{\cdot}2\%$

Voltage across 3rd from top, $V_3 = 0{\cdot}182 \times 57{\cdot}7 =$ **10·5 kV**

Voltage across 4th from top, $V_4 = 0{\cdot}228 \times 57{\cdot}7 =$ **13·16 kV**

Voltage across 5th from top, $V_5 = 0{\cdot}3 \times 57{\cdot}7 =$ **17·3 kV**

(*ii*) String efficiency $= \dfrac{57{\cdot}7}{5 \times 17{\cdot}3} \times 100 =$ **66·7%**

Example 8.8. *Each conductor of a 3-phase high-voltage transmission line is suspended by a string of 4 suspension type disc insulators. If the potential difference across the second unit from top is 13·2 kV and across the third from top is 18 kV, determine the voltage between conductors.*

Solution. Suppose $X_C = 1\ \Omega$. If K is the ratio of shunt-capacitance to self-capacitance, then $X_{C1} = 1/K$ ohms as shown in Fig. 8.18 (*i*). Suppose voltage across string is such that current in top insulator disc is 1 A. Then voltage across each insulator can be easily determined [see Fig. 8.18 (*ii*)]. Thus the voltage across first shunt capacitance from top is 1 volt and its reactance is $1/K$ ohms. Therefore, current through it is K ampere. Hence current through second insulator from top is $(1 + K)$ amperes and voltage across it is $(1 + K) \times 1 = (1 + K)$ volts.

Referring to Fig. 8.18 (*ii*), we have,

$$V_2/V_1 = (1 + K)/1$$

or $$V_2 = V_1 (1 + K) \qquad ...(i)$$

Also $$V_3/V_1 = (1 + 3K + K^2)/1$$

$\therefore$ $$V_3 = V_1 (1 + 3K + K^2) \qquad ...(ii)$$

Dividing (*ii*) by (*i*), we get,

$$\frac{V_3}{V_2} = \frac{1 + 3K + K^2}{1 + K}$$

It is given that $V_3 = 18$ kV and $V_2 = 13{\cdot}2$ kV

$\therefore$ $$\frac{18}{13{\cdot}2} = \frac{1 + 3K + K^2}{1 + K}$$

or $$13{\cdot}2\,K^2 + 21{\cdot}6\,K - 4{\cdot}8 = 0$$

Solving this equation, we get, $K = 0{\cdot}2$.

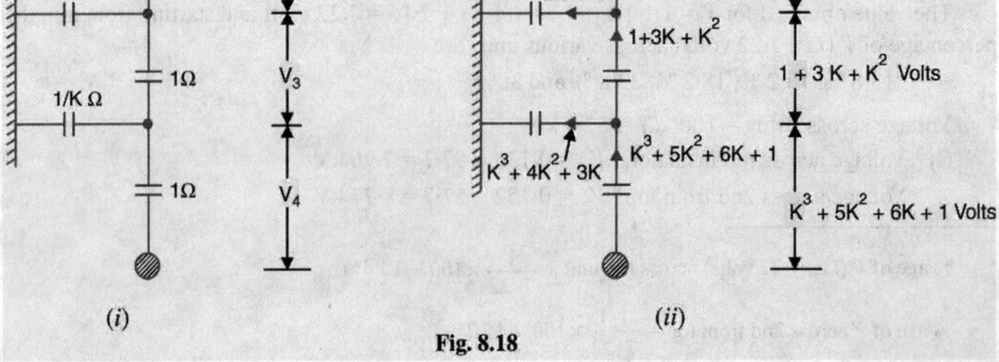

Fig. 8.18

$\therefore \quad V_1 = V_2/(1 + K) = 13{\cdot}2/1{\cdot}2 = 11$ kV

$V_4 = V_1 (1 + K^3 + 5K^2 + 6K) = 11\,(1 + 0{\cdot}008 + 0{\cdot}2 + 1{\cdot}2) = 26{\cdot}49$kV

Voltage between line and earth (*i.e.*, phase voltage)

$= V_1 + V_2 + V_3 + V_4$

$= 11 + 13{\cdot}2 + 18 + 26{\cdot}49 = 68{\cdot}69$ kV

Voltage between conductors (*i.e.*, line voltage)

$= 68{\cdot}69 \times \sqrt{3} =$ **119 kV**

Example 8.9. *A string of four insulators has a self-capacitance equal to 5 times pin to earth capacitance. Find (i) the voltage distribution across various units as a percentage of total voltage across the string and (ii) string efficiency.*

Solution. The ratio of self-capacitance (C) to pin-earth capacitance (C_1) is $C/C_1 = 5$. Suppose $X_C = 1\ \Omega$. Then $X_{C1} = 5\ \Omega$. Suppose the voltage V across string is such that current in the top insulator is 1A as shown in Fig. 8.19 (*i*). The potential across various insulators will be as shown in Fig. 8.19 (*ii*).

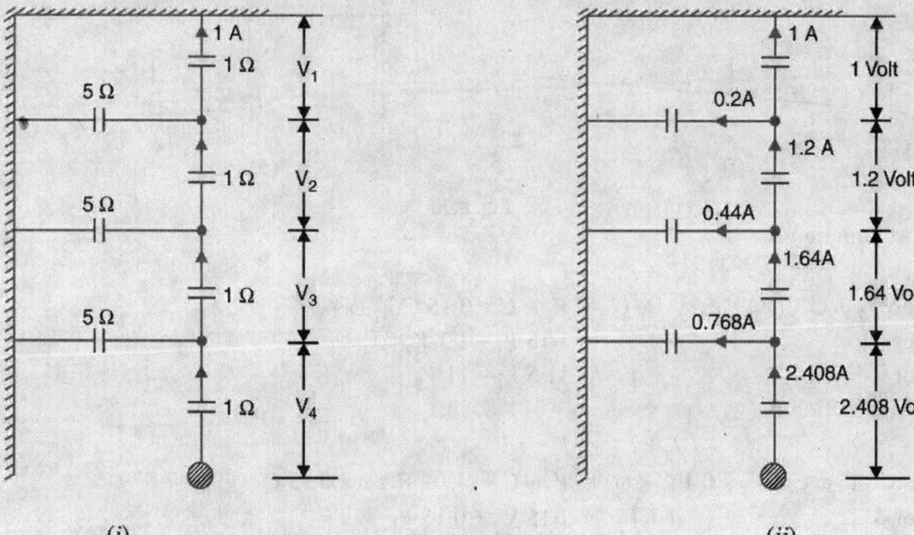

Fig. 8.19

The voltage obtained across the string is given by ;

$V = 1 + 1{\cdot}2 + 1{\cdot}64 + 2{\cdot}408 = 6{\cdot}248$ volts

(*i*) The voltage across each unit expressed as a percentage of V (*i.e.*, 6·248 volts) is given by :

Top Unit $= (1/6{\cdot}248) \times 100 =$ **16%**

Second from top $= (1{\cdot}2/6{\cdot}248) \times 100 =$ **19·2%**

Third from top $= (1{\cdot}64/6{\cdot}248) \times 100 =$ **26·3%**

Fourth from top $= (2{\cdot}408/6{\cdot}248) \times 100 =$ **38·5%**

(*ii*) String efficiency $= \dfrac{6\cdot 248}{4 \times 2\cdot 408} \times 100 =$ **64·86%**

Example 8.10. *The self capacitance of each unit in a string of three suspension insulators is C. The shunting capacitance of the connecting metal work of each insulator to earth is 0·15 C while for line it is 0·1 C. Calculate (i) the voltage across each insulator as a percentage of the line voltage to earth and (ii) string efficiency.*

Solution. In an actual string of insulators, three capacitances exist *viz.,* self-capacitance of each insulator, shunt capacitance and capacitance of each unit to line as shown in Fig. 8.20 (*i*). However, capacitance of each unit to line is very small and is usually neglected. Fig. 8.20 (*ii*) shows the equivalent circuit of string insulators.

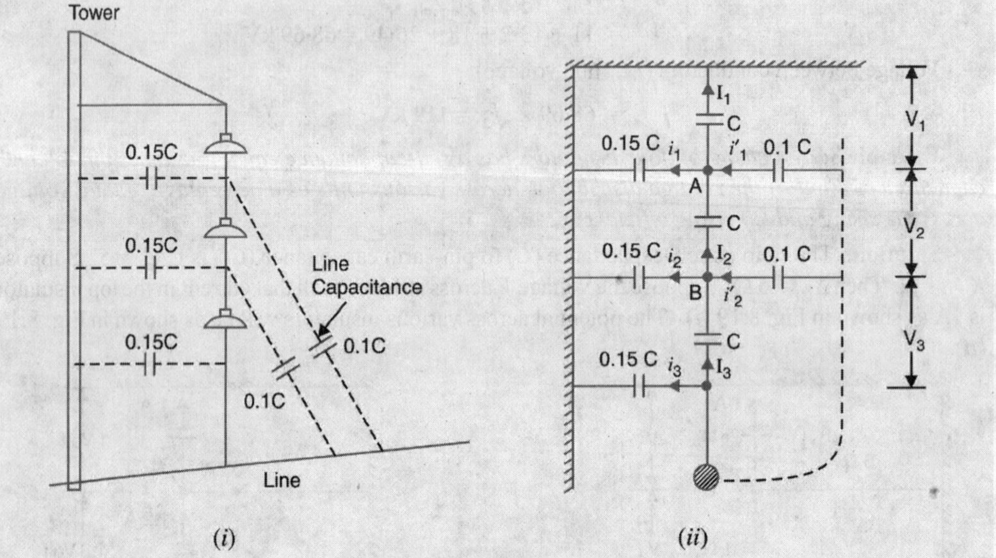

Fig. 8.20

At Junction A

$$I_2 + i'_1 = I_1 + i_1$$

or $V_2 \omega C + (V_2 + V_3)\, 0{\cdot}1 \omega C = V_1 \omega C + 0{\cdot}15\, C\, V_1\, \omega$

or $0{\cdot}1\, V_3 = 1{\cdot}15\, V_1 - 1{\cdot}1\, V_2$

or $V_3 = 11{\cdot}5\, V_1 - 11\, V_2$ ...(*i*)

At Junction B

$$I_3 + i'_2 = I_2 + i_2$$

or $V_3 \omega C + V_3 \times 0{\cdot}1\, C \times \omega = V_2 \omega C + (V_1 + V_2)\, \omega \times 0{\cdot}15\, C$

or $1{\cdot}1\, V_3 = 1{\cdot}15\, V_2 + 0{\cdot}15\, V_1$ ...(*ii*)

Putting the value of V_3 from exp (*i*). into exp. (*ii*), we get,

$$1{\cdot}1\,(11{\cdot}5\, V_1 - 11\, V_2) = 1{\cdot}15\, V_2 + 0{\cdot}15\, V_1$$

or $13{\cdot}25\, V_2 = 12{\cdot}5\, V_1$

or $V_2 = \dfrac{12{\cdot}5}{13{\cdot}25} V_1$ ...(*iii*)

Putting the value of V_2 from exp. (*iii*) into exp. (*i*), we get,

$$V_3 = 11{\cdot}5\, V_1 - 11\left(\frac{12{\cdot}5\, V_1}{13{\cdot}25}\right) = \left(\frac{14{\cdot}8}{13{\cdot}25}\right) V_1$$

Now voltage between conductor and earth is

$$V = V_1 + V_2 + V_3 = V_1\left(1 + \frac{12{\cdot}5}{13{\cdot}25} + \frac{14{\cdot}8}{13{\cdot}25}\right) = \left(\frac{40{\cdot}55\, V_1}{13{\cdot}25}\right) \text{volts}$$

$\therefore$ $V_1 = 13{\cdot}25\, V/40{\cdot}55 = 0{\cdot}326\, V$ volts

$V_2 = 12{\cdot}5 \times 0{\cdot}326\, V/13{\cdot}25 = 0{\cdot}307\, V$ volts

$V_3 = 14{\cdot}8 \times 0{\cdot}326\, V/13{\cdot}25 = 0{\cdot}364\, V$ volts (*i*) The voltage across each unit expressed as a percentage of V becomes:

Top unit $= V_1 \times 100/V = 0{\cdot}326 \times 100 = \mathbf{32{\cdot}6\%}$

Second from top $= V_2 \times 100/V = 0{\cdot}307 \times 100 = \mathbf{30{\cdot}7\%}$

Third from top $= V_3 \times 100/V = 0{\cdot}364 \times 100 = \mathbf{36{\cdot}4\%}$

(*ii*) String efficiency $= \dfrac{V}{3 \times 0{\cdot}364\, V} \times 100 = \mathbf{91{\cdot}5\ \%}$

Example 8.11. *Each line of a 3-phase system is suspended by a string of 3 indentical insulators of self-capacitance C farad. The shunt capacitance of connecting metal work of each insulator is 0·2 C to earth and 0·1 C to line. Calculate the string efficiency of the system if a guard ring increases the capacitance to the line of metal work of the lowest insulator to 0·3 C.*

Solution. The capacitance between each unit and line is artificially increased by using a guard ring as shown in Fig. 8.21. This arrangement tends to equalise the potential across various units and hence leads to improved string efficiency. It is given that with the use of guard ring, capacitance of the insulator link-pin to the line of the lowest unit is increased from 0·1 C to 0·3 C.

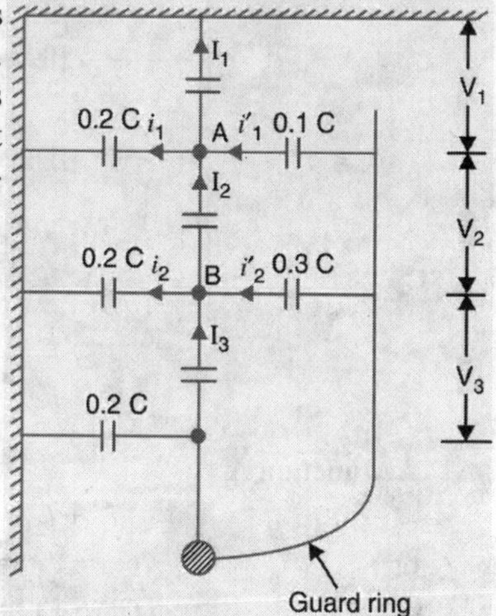

Fig. 8.21

At Junction A

$$I_2 + i'_1 = I_1 + i_1$$

or $$V_2\, \omega\, C + (V_2 + V_3)\, \omega \times 0{\cdot}1\, C = V_1\, \omega\, C + V_1 \times 0{\cdot}2\, C\, \omega$$

$$V_3 = 12\, V_1 - 11\, V_2 \qquad ...(i)$$

At Junction B

$$I_3 + i'_2 = I_2 + i_2$$

or $$V_3\, \omega\, C + V_3 \times 0{\cdot}3\, C \times \omega = V_2\, \omega\, C + (V_1 + V_2)\, \omega \times 0{\cdot}2\, C$$

or $$1{\cdot}3\, V_3 = 1{\cdot}2\, V_2 + 0{\cdot}2\, V_1 \qquad ...(ii)$$

Substituting the value of V_3 from exp. (*i*) into exp. (*ii*), we get,

$$1{\cdot}3\,(12\, V_1 - 11 V_2) = 1{\cdot}2\, V_2 + 0{\cdot}2\, V_1$$

or $$15{\cdot}5\, V_2 = 15{\cdot}4\, V_1$$

$\therefore$ $$V_2 = 15{\cdot}4\, V_1/15{\cdot}5 = 0{\cdot}993\, V_1 \qquad ...(iii)$$

Substituting the value of V_2 from exp. (*iii*) into exp. (*i*), we get,

$$V_3 = 12\, V_1 - 11 \times 0{\cdot}993\, V_1 = 1{\cdot}077\, V_1$$

Voltage between conductor and earth (*i.e.* phase voltage)

$$= V_1 + V_2 + V_3 = V_1 + 0{\cdot}993\, V_1 + 1{\cdot}077\, V_1 = 3{\cdot}07\, V_1$$

String efficiency $= \dfrac{3{\cdot}07\, V_1}{3 \times 1{\cdot}077\, V_1} \times 100 = \mathbf{95\%}$

Example 8.12. *It is required to grade a string having seven suspension insulators. If the pin to earth capacitance are all equal to C, determine the line to pin capacitance that would give the same voltage across each insulator of the string.*

Solution. Let $C_1, C_2 ... C_6$ respectively be the required line to pin capacitances of the units as shown in Fig. 8.22. As the voltage across each insulator has to be the same, therefore,

$$I_1 = I_2 = I_3 = I_4 = I_5 = I_6 = I_7$$

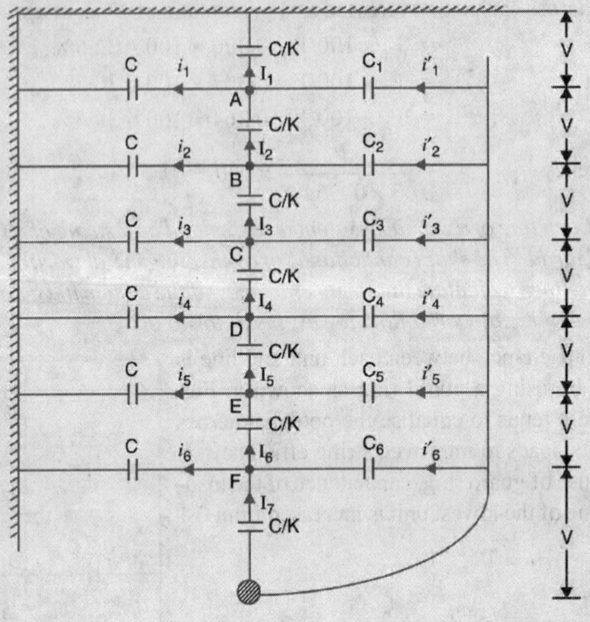

Fig. 8.22

At Junction A

$$i_1' + I_2 = i_1 + I_1$$

or $\quad i'_1 = i_1 \qquad (\because I_1 = I_2)$

or $\quad \omega C_1 (6V) = \omega CV \qquad (\because$ Voltage across $C_1 = 6V)$

$\therefore \quad C_1 = C/6 = \mathbf{0{\cdot}167\ C}$

At Junction B

$$i_2' = i_2$$

or $\quad \omega C_2 (5V) = \omega C (2V)$

$\therefore \quad C_2 = \frac{2C}{5} = \mathbf{0{\cdot}4\ C}$

At Junction C

$$i_3' = i_3$$

or $\quad \omega C_3 (4V) = \omega C (3V)$

$\therefore \quad C_3 = 3C/4 = \mathbf{0{\cdot}75\ C}$

At Junction D

$$i_4' = i_4$$

or $\quad \omega C_4 (3V) = \omega C (4V)$

$\therefore \quad C_4 = 4C/3 = \mathbf{1{\cdot}33\ C}$

At Junction E

$$i_5' = i_5$$

or $\quad \omega C_5 (2V) = \omega C (5V)$

$\therefore \quad C_5 = 5C/2 = \mathbf{2{\cdot}5\ C}$

At Junction F

$$i_6' = i_6$$

or $\quad \omega C_6 V = \omega C (6V)$

$\therefore \quad C_6 = \mathbf{6\ C}$

TUTORIAL PROBLEMS

1. In a 3-phase overhead system, each line is suspended by a string of 3 insulators. The voltage across the top unit (*i.e.* near the tower) and middle unit are 10 kV and 11 kV respectively. Calculate (*i*) the ratio of shunt capacitance to self capacitance of each insulator, (*ii*) the string efficiency and (*iii*) line voltage. **[(*i*) 0·1 (*ii*) 86·76% (*iii*) 59 kV]**

2. Each line of a 3-phase system is suspended by a string of 3 similar insulators. If the voltage across the line unit is 17·5 kV, calculate the line to neutral voltage and string efficiency. Assume that shunt capacitance between each insulator and earthed metal work of tower to be 1/10th of the capacitance of the insulator. **[52 kV, 86·67%]**
3. The three bus-bar conductors in an outdoor sub-station are supplied by units of post insulators. Each unit consists of a stack of 3-pin insulators fixed one on the top of the other. The voltage across the lowest insulator is 8·45 kV and that across the next is 7·25 kV. Find the bus-bar voltage of the station. **[38·8 kV]**
4. A string of suspension insulators consists of three units. The capacitance between each link pin and earth is one-sixth of the self-capacitance of each unit. If the maximum voltage per unit is not to exceed 35 kV, determine the maximum voltage that the string can withstand. Also calculate the string efficiency. **[84·7 kV; 80·67%]**
5. A string of 4 insulators has self-capacitance equal to 4 times the pin-to-earth capacitance. Calculate (*i*) the voltage distribution across various units as a percentage of total voltage across the string and (*ii*) string efficiency. **[(*i*) 14·5%, 18·1%, 26·2% and 40·9% (*ii*) 61·2 %]**
6. A string of four suspension insulators is connected across a 285 kV line. The self-capacitance of each unit is equal to 5 times pin to earth capacitance. Calculate :
   (*i*) the potential difference across each unit, (*ii*) the string efficiency.
   [(*i*) 27·65 kV, 33·04 kV, 43·85 kV, 60 kV (*ii*) 68·5%]
7. Each of three insulators forming a string has self-capacitance of "*C*" farad. The shunt capacitance of each cap of insulator is 0·25 C to earth and 0·15 C to line. Calculate the voltage distribution across each insulator as a pecentage of line voltage to earth and the string efficiency. **[31·7%, 29·4%, 38·9%; 85·7%]**
8. Each of the three insulators forming a string has a self capacitance of *C* farad. The shunt capacitance of each insulator is 0·2 C to earth and 0·1 C to line. A guard-ring increases the capacitance of line of the metal work of the lowest insulator to 0·3 C. Calculate the string efficiency of the arrangement :
   (*i*) with the guard ring, (*ii*) without guard ring. **[(*i*) 95% (*ii*) 86·13%]**
9. A three-phase overhead transmission line is being supported by three-disc suspension insulators; the potentials across the first and second insulator from the top are 8 kV and 11 kV respectively. Calcualte (*i*) the line voltage (*ii*) the ratio of capacitance between pin and earth to self capacitance of each unit (*iii*) the string efficiency. **[(*i*) 64·28 V (*ii*) 0·375 (*iii*) 68·28%]**
10. A 3-phase overhead transmission line is supported on 4-disc suspension insulators. The voltage across the second and third discs are 13·2 kV and 18 kV respectively. Calculate the line voltage and mention the nearest standard voltage. **[118·75 kV; 120 kV]**

8.10 Corona

When an alternating potential difference is applied across two conductors whose spacing is large as compared to their diameters, there is no apparent change in the condition of atmospheric air surrounding the wires if the applied voltage is low. However, when the applied voltage exceeds a certain value, called *critical disruptive voltage,* the conductors are surrounded by a faint violet glow called corona.

The phenomenon of corona is accompanied by a hissing sound, production of ozone, power loss and radio interference. The higher the voltage is raised, the larger and higher the luminous envelope becomes, and greater are the sound, the power loss and the radio noise. If the applied voltage is increased to breakdown value, a flash-over will occur between the conductors due to the breakdown of air insulation.

The phenomenon of violet glow, hissing noise and production of ozone gas in an overhead transmission line is known as **corona.**

If the conductors are polished and smooth, the corona glow will be uniform throughout the length of the conductors, otherwise the rough points will appear brighter. With d.c. voltage, there is difference in the appearance of the two wires. The positive wire has uniform glow about it, while the negative conductor has spotty glow.

Theory of corona formation. Some ionisation is always present in air due to cosmic rays, ultra-violet radiations and radioactivity. Therefore, under normal conditions, the air around the conductors contains some ionised particles (*i.e.,* free electrons and +ve ions) and neutral molecules. When p.d. is applied between the conductors, potential gradient is set up in the air which will have maximum value at the conductor surfaces. Under the influence of potential gradient, the existing free electrons acquire greater velocities. The greater the applied voltage, the greater the potential gradient and more is the velocity of free electrons.

When the potential gradient at the conductor surface reaches about 30 kV per cm (max. value), the velocity acquired by the free electrons is sufficient to strike a neutral molecule with enough force to dislodge one or more electrons from it. This produces another ion and one or more free electrons, which is turn are accelerated until they collide with other neutral molecules, thus producing other ions. Thus, the process of ionisation is cummulative. The result of this ionisation is that either corona is formed or spark takes place between the conductors.

8.11 Factors Affecting Corona

The phenomenon of corona is affected by the physical state of the atmosphere as well as by the conditions of the line. The following are the factors upon which corona depends :

(*i*) *Atmosphere.* As corona is formed due to ionsiation of air surrounding the conductors, therefore, it is affected by the physical state of atmosphere. In the stormy weather, the number of ions is more than normal and as such corona occurs at much less voltage as compared with fair weather.

(*ii*) *Conductor size.* The corona effect depends upon the shape and conditions of the conductors. The rough and irregular surface will give rise to more corona because unevenness of the surface decreases the value of breakdown voltage. Thus a stranded conductor has irregular surface and hence gives rise to more corona that a solid conductor.

(*iii*) *Spacing between conductors.* If the spacing between the conductors is made very large as compared to their diameters, there may not be any corona effect. It is because larger distance between conductors reduces the electro-static stresses at the conductor surface, thus avoiding corona formation.

(*iv*) *Line voltage.* The line voltage greatly affects corona. If it is low, there is no change in the condition of air surrounding the conductors and hence no corona is formed. However, if the line voltage has such a value that electrostatic stresses developed at the conductor surface make the air around the conductor conducting, then corona is formed.

8.12 Important Terms

The phenomenon of corona plays an important role in the design of an overhead transmission line. Therefore, it is profitable to consider the following terms much used in the analysis of corona effects:

(*i*) **Critical disruptive voltage.** *It is the minimum phase-neutral voltage at which corona occurs.*

Consider two conductors of radii r cm and spaced d cm apart. If V is the phase-neutral potential, then potential gradient at the conductor surface is given by:

$$g = \frac{V}{r \log_e \dfrac{d}{r}} \text{ volts / cm}$$

In order that corona is formed, the value of g must be made equal to the breakdown strength of air. The breakdown strength of air at 76 cm pressure and temperature of 25ºC is 30 kV/cm (*max*) or

21·2 kV/cm (*r.m.s.*) and is denoted by g_o. If V_c is the phase-neutral potential required under these conditions, then,

$$g_o = \frac{V_c}{r \log_e \frac{d}{r}}$$

where g_o = breakdown strength of air at 76 cm of mercury and 25°C
= 30 kV/cm (*max*) or 21·2 kV/cm (*r.m.s.*)

∴ Critical disruptive voltage, $V_c = g_o \, r \log_e \frac{d}{r}$

The above expression for disruptive voltage is under standard conditions *i.e.,* at 76 cm of Hg and 25°C. However, if these conditions vary, the air density also changes, thus altering the value of g_o. The value of g_o is directly proportional to air density. Thus the breakdown strength of air at a barometric pressure of b cm of mercury and temperature of t°C becomes $\delta \, g_o$ where

$$\delta = \text{air density factor} = \frac{3 \cdot 92 b}{273 + t}$$

Under standard conditions, the value of $\delta = 1$.

∴ Critical disruptive voltage, $V_c = g_o \, \delta \, r \log_e \frac{d}{r}$

Correction must also be made for the surface condition of the conductor. This is accounted for by multiplying the above expression by irregularity factor m_o.

∴ Critical disruptive voltage, $V_c = m_o \, g_o \, \delta \, r \log_e \frac{d}{r}$ kV/phase

where m_o = 1 for polished conductors
= 0·98 to 0·92 for dirty conductors
= 0·87 to 0·8 for stranded conductors

(*ii*) Visual critical voltage. *It is the minimum phase-neutral voltage at which corona glow appears all along the line conductors.*

It has been seen that in case of parallel conductors, the corona glow does not begin at the disruptive voltage V_c but at a higher voltage V_v, called *visual critical voltage.* The phase-neutral effective value of visual critical voltage is given by the following empirical formula :

$$V_v = m_v \, g_o \, \delta \, r \left(1 + \frac{0 \cdot 3}{\sqrt{\delta \, r}}\right) \log_e \frac{d}{r} \text{ kV/phase}$$

where m_v is another irregularity factor having a value of 1·0 for polished conductors and 0·72 to 0·82 for rough conductors.

(*iii*) Power loss due to corona. Formation of corona is always accompanied by energy loss which is dissipated in the form of light, heat, sound and chemical action. When disruptive voltage is exceeded, the power loss due to corona is given by :

$$P = 242 \cdot 2 \left(\frac{f + 25}{\delta}\right) \sqrt{\frac{r}{d}} \, (V - V_c)^2 \times 10^{-5} \text{ kW / km / phase}$$

where f = supply frequency in Hz
V = phase-neutral voltage (*r.m.s.*)
V_c = disruptive voltage (*r.m.s.*) per phase

8.13 Advantages and Disadvantages of Corona

Corona has many advantages and disadvantages. In the correct design of a high voltage overhead line, a balance should be struck between the advantages and disadvantages.

Advantages

(*i*) Due to corona formation, the air surrounding the conductor becomes conducting and hence virtual diameter of the conductor is increased. The increased diameter reduces the electrostatic stresses between the conductors.

(*ii*) Corona reduces the effects of transients produced by surges.

Disadvantages

(*i*) Corona is accompanied by a loss of energy. This affects the transmission efficiency of the line.

(*ii*) Ozone is produced by corona and may cause corrosion of the conductor due to chemical action.

(*iii*) The current drawn by the line due to corona is non-sinusoidal and hence non-sinusoidal voltage drop occurs in the line. This may cause inductive interference with neighbouring communication lines.

8.14 Methods of Reducing Corona Effect

It has been seen that intense corona effects are observed at a working voltage of 33 kV or above. Therefore, careful design should be made to avoid corona on the sub-stations or bus-bars rated for 33 kV and higher voltages otherwise highly ionised air may cause flash-over in the insulators or between the phases, causing considerable damage to the equipment. The corona effects can be reduced by the following methods :

(*i*) *By increasing conductor size.* By increasing conductor size, the voltage at which corona occurs is raised and hence corona effects are considerably reduced. This is one of the reasons that *ACSR* conductors which have a larger cross-sectional area are used in transmission lines.

(*ii*) *By increasing conductor spacing.* By increasing the spacing between conductors, the voltage at which corona occurs is raised and hence corona effects can be eliminated. However, spacing cannot be increased too much otherwise the cost of supporting structure (*e.g.*, bigger cross arms and supports) may increase to a considerable extent.

Example 8.13. *A 3-phase line has conductors 2 cm in diameter spaced equilaterally 1 m apart. If the dielectric strength of air is 30 kV (max) per cm, find the disruptive critical voltage for the line. Take air density factor $\delta = 0{\cdot}952$ and irregularity factor $m_o = 0{\cdot}9$.*

Solution.

Conductor radius, $r = 2/2 = 1$ cm

Conductor spacing, $d = 1$ m $= 100$ cm

Dielectric strength of air, $g_o = 30$ kV/cm (*max.*) $= 21{\cdot}2$ kV (*r.m.s.*) per cm

Disruptive critical voltage, $V_c = m_o\, g_o\, \delta\, r \log_e (d/r)$ kV*/phase (*r.m.s.* value)

$= 0{\cdot}9 \times 21{\cdot}2 \times 0{\cdot}952 \times 1 \times \log_e 100/1 = 83{\cdot}64$ kV/phase

$\therefore$ Line voltage (*r.m.s.*) $= \sqrt{3} \times 83{\cdot}64 =$ **144·8 kV**

Example 8.14. *A 132 kV line with 1·956 cm dia. conductors is built so that corona takes place if the line voltage exceeds 210 kV (r.m.s.). If the value of potential gradient at which ionisation occurs can be taken as 30 kV per cm, find the spacing between the conductors.*

---

* As g_o is taken in kV/cm, therefore, V_c will be in kV.

Solution.

Assume the line is 3-phase.

Conductor radius, $r = 1{\cdot}956/2 = 0{\cdot}978$ cm

Dielectric strength of air, $g_o = 30/\sqrt{2} = 21{\cdot}2$ kV (*r.m.s.*) per cm

Disruptive voltage/phase, $V_c = 210/\sqrt{3} = 121{\cdot}25$ kV

Assume smooth conductors (*i.e.*, irregularity factor $m_o = 1$) and standard pressure and temperature for which air density factor $\delta = 1$. Let d cm be the spacing between the conductors.

∴ Disruptive voltage (*r.m.s.*) per phase is

$$V_c = m_o\, g_o\, \delta\, r \log_e (d/r) \text{ kV}$$

$$= 1 \times 21{\cdot}2 \times 1 \times 0{\cdot}978 \times \log_e (d/r)$$

or $$121{\cdot}25 = 20{\cdot}733 \log_e (d/r)$$

or $$\log_e \frac{d}{r} = \frac{121{\cdot}25}{20{\cdot}733} = 5{\cdot}848$$

or $$2{\cdot}3 \log_{10} d/r = 5{\cdot}848$$

or $$\log_{10} d/r = 5{\cdot}848/2{\cdot}3 = 2{\cdot}5426$$

or $$d/r = \text{Antilog } 2{\cdot}5426$$

or $$d/r = 348{\cdot}8$$

∴ Conductor spacing, $d = 348{\cdot}8 \times r = 348{\cdot}8 \times 0{\cdot}978 = \mathbf{341\ cm}$

Example 8.15. *A 3-phase, 220 kV, 50 Hz transmission line consists of 1·5 cm radius conductor spaced 2 metres apart in equilateral triangular formation. If the temperature is 40°C and atmospheric pressure is 76 cm, calculate the corona loss per km of the line. Take $m_o = 0{\cdot}85$.*

Solution.

As seen from Art. 8.12, the corona loss is given by :

$$P = \frac{242{\cdot}2}{\delta}(f + 25)\sqrt{\frac{r}{d}}\,(V - V_c)^2 \times 10^{-5} \text{ kW/km/phase}$$

Now, $$\delta = \frac{3{\cdot}92\, b}{273 + t} = \frac{3{\cdot}92 \times 76}{273 + 40} = 0{\cdot}952$$

Assuming $g_o = 21{\cdot}2$ kV/cm (*r.m.s.*)

∴ Critical disruptive voltage per phase is

$$V_c = m_o\, g_o\, \delta\, r \log_e d/r \text{ kV}$$

$$= 0{\cdot}85 \times 21{\cdot}2 \times 0{\cdot}952 \times 1{\cdot}5 \times \log_e 200/1{\cdot}5 = 125{\cdot}9 \text{ kV}$$

Supply voltage per phase, $V = 220/\sqrt{3} = 127$ kV

Substituting the above values, we have corona loss as:

$$P = \frac{242{\cdot}2}{0{\cdot}952}(50 + 25) \times \sqrt{\frac{1{\cdot}5}{200}} \times (127 - 125{\cdot}9)^2 \times 10^{-5} \text{ kW/phase/km}$$

$$= \frac{242{\cdot}2}{0{\cdot}952} \times 75 \times 0{\cdot}0866 \times 1{\cdot}21 \times 10^{-5} \text{ kW/km/phase}$$

$$= 0{\cdot}01999 \text{ kW/km/phase}$$

∴ Total corona loss per km for three phases

$$= 3 \times 0{\cdot}01999 \text{ kW} = \mathbf{0{\cdot}05998\ kW}$$

Example 8.16. *A certain 3-phase equilateral transmission line has a total corona loss of 53 kW at 106 kV and a loss of 98 kW at 110·9 kV. What is the disruptive critical voltage? What is the corona loss at 113 kV?*

Solution.

The power loss due to corona for 3 phases is given by :

$$P = 3\times\frac{242\cdot 2(f+25)}{\delta}\sqrt{\frac{r}{d}}(V-V_c)^2\times 10^{-5}\ \text{kW/km}$$

As f, δ, r and d are the same for the two cases,

$$\therefore \quad P \propto (V-V_c)^2$$

For first case, $P = 53$ kW and $V = 106/\sqrt{3} = 61{\cdot}2$ kV

For second case, $P = 98$ kW and $V = 110\cdot 9/\sqrt{3} = 64$ kV

$$\therefore \quad 53 \propto (61{\cdot}2 - V_c)^2 \quad ...(i)$$

and

$$98 \propto (64 - V_c)^2 \quad ...(ii)$$

Dividing [(*ii*)/(*i*)], we get,

$$\frac{98}{53} = \frac{(64-V_c)^2}{(61\cdot 2-V_c)^2}$$

or $\quad V_c = \mathbf{54\ kV}$

Let W kilowatt be the power loss at 113 kV.

$$\therefore \quad W \propto \left(\frac{113}{\sqrt{3}} - V_c\right)^2$$

$$\propto (65{\cdot}2 - 54)^2 \quad ...(iii)$$

Dividing [(*iii*)/(*i*)], we get,

$$\frac{W}{53} = \frac{(65\cdot 2-54)^2}{(61\cdot 2-54)^2}$$

$$\therefore \quad W = (11{\cdot}2/7{\cdot}2)^2\times 53 = \mathbf{128\ kW}$$

TUTORIAL PROBLEMS

1. Estimate the corona loss for a three-phase, 110 kV, 50 Hz, 150 km long transmission line consisting of three conductors each of 10 mm diameter and spaced 2·5 m apart in an equilateral triangle formation. The temperature of air is 30ºC and the atmospheric pressure is 750 mm of mercury. Take irregularity factor as 0·85. Ionisation of air may be assumed to take place at a maximum voltage gradient of 30 kV/cm. **[316·8 kW]**
2. Taking the dielectric strength of air to be 30 kV/cm, calculate the disruptive critical voltage for a 3-phase line with conductors of 1 cm radius and spaced symmetrically 4 m apart. **[220 kV line voltage]**
3. A 3-phase, 220 kV, 50 Hz transmission line consists of 1·2 cm radius conductors spaced 2 m at the corners of an equilateral triangle. Calculate the corona loss per km of the line. The condition of the wire is smoothly weathered and the weather is fair with temperature of 20ºC and barometric pressure of 72·2 cm of Hg. **[2·148 kW]**

8.15 Sag in Overhead Lines

While erecting an overhead line, it is very important that conductors are under safe tension. If the conductors are too much stretched between supports in a bid to save conductor material, the stress in the conductor may reach unsafe value and in certain cases the conductor may break due to excessive tension. In order to permit safe tension in the conductors, they are not fully stretched but are allowed to have a dip or sag.

The difference in level between points of supports and the lowest point on the conductor is called **sag.**

Fig. 8.23. (*i*) shows a conductor suspended between two equilevel supports *A* and *B*. The conductor is not fully stretched but is allowed to have a dip. The lowest point on the conductor is *O* and the sag is *S*. The following points may be noted :

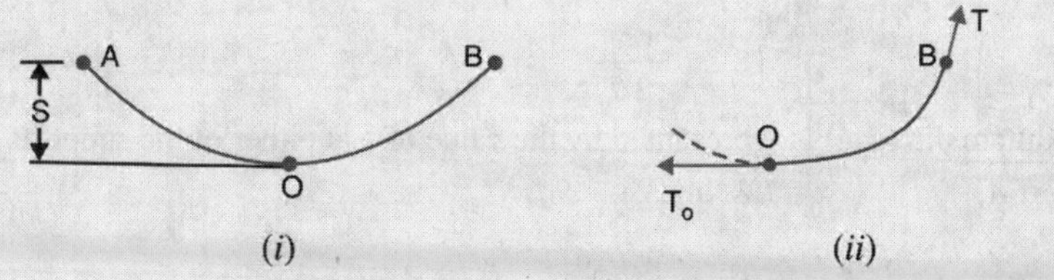

Fig. 8.23

(*i*) When the conductor is suspended between two supports at the same level, it takes the shape of catenary. However, if the sag is very small compared with the span, then sag-span curve is like a parabola.

(*ii*) The tension at any point on the conductor acts tangentially. Thus tension T_O at the lowest point *O* acts horizontally as shown in Fig. 8.23. (*ii*).

(*iii*) The horizontal component of tension is constant throughout the length of the wire.

(*iv*) The tension at supports is approximately equal to the horizontal tension acting at any point on the wire. Thus if *T* is the tension at the support *B*, then $T = T_O$.

Conductor sag and tension. This is an important consideration in the mechanical design of overhead lines. The conductor sag should be kept to a minimum in order to reduce the conductor material required and to avoid extra pole height for sufficient clearance above ground level. It is also desirable that tension in the conductor should be low to avoid the mechanical failure of conductor and to permit the use of less strong supports. However, low conductor tension and minimum sag are not possible. It is because low sag means a tight wire and high tension, whereas a low tension means a loosc wirc and increased sag. Therefore, in actual practice, a compromise in made between the two.

8.16 Calculation of Sag

In an overhead line, the sag should be so adjusted that tension in the conductors is within safe limits. The tension is governed by conductor weight, effects of wind, ice loading and temperature variations. It is a standard practice to keep conductor tension less than 50% of its ultimate tensile strength *i.e.*, minimum factor of safety in respect of conductor tension should be 2. We shall now calculate sag and tension of a conductor when (*i*) supports are at equal levels and (*ii*) supports are at unequal levels.

(*i*) **When supports are at equal levels.** Consider a conductor between two equilevel supports *A* and *B* with *O* as the lowest point as shown in Fig. 8.24. It can be proved that lowest point will be at the mid-span.

Let

l = Length of span

w = Weight per unit length of conductor

T = Tension in the conductor.

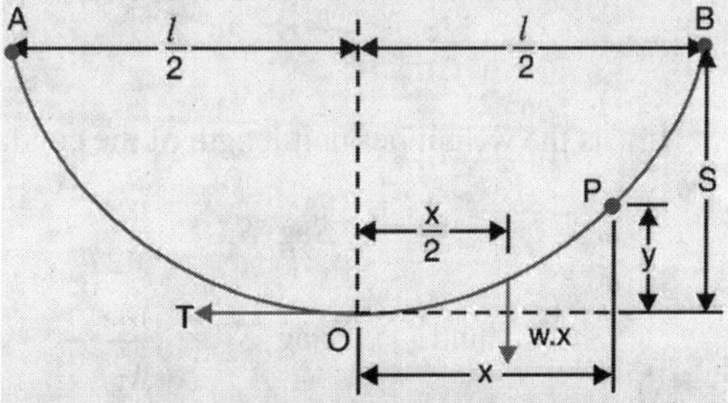

Fig. 8.24

Consider a point *P* on the conductor. Taking the lowest point *O* as the origin, let the co-ordinates of point *P* be *x* and *y*. Assuming that the curvature is so small that curved length is equal to its horizontal projection (*i.e.*, $OP = x$), the two forces acting on the portion *OP* of the conductor are :

(*a*) The weight wx of conductor acting at a distance $x/2$ from *O*.

(*b*) The tension *T* acting at *O*.

Equating the moments of above two forces about point O, we get,

$$T\,y = w\,x \times \frac{x}{2}$$

or
$$y = \frac{w\,x^2}{2\,T}$$

The maximum dip (sag) is represented by the value of y at either of the supports A and B.

At support A, $x = l/2$ and $y = S$

$$\therefore \quad \text{Sag, } S = \frac{w(l/2)^2}{2T} = \frac{w\,l^2}{8\,T}$$

(*ii*) **When supports are at unequal levels.** In hilly areas, we generally come across conductors suspended between supports at unequal levels. Fig. 8.25 shows a conductor suspended between two supports A and B which are at different levels. The lowest point on the conductor is O.

Let

l = Span length

h = Difference in levels between two supports

x_1 = Distance of support at lower level (*i.e., A*) from O

x_2 = Distance of support at higher level (*i.e. B*) from O

T = Tension in the conductor

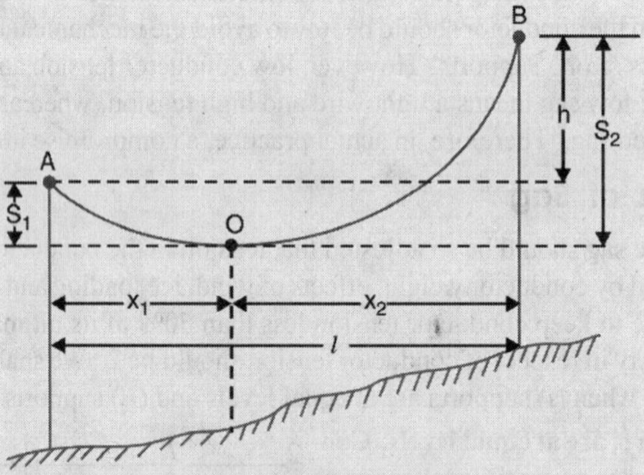

Fig. 8.25

If w is the weight per unit length of the conductor, then,

$$\text{Sag } S_1 = \frac{w\,x_1^{2*}}{2T}$$

and
$$\text{Sag } S_2 = \frac{w\,x_2^2}{2T}$$

Also
$$x_1 + x_2 = l \qquad \ldots(i)$$

$$^*\,y = \frac{w\,x^2}{2T}$$

At support A, $x = x_1$ and $y = S_1$.

$$\therefore \quad S_1 = \frac{w\,x_1^2}{2T}$$

Now $$S_2 - S_1 = \frac{w}{2T}[x_2^2 - x_1^2] = \frac{w}{2T}(x_2 + x_1)(x_2 - x_1)$$

$\therefore$ $$S_2 - S_1 = \frac{w\,l}{2T}(x_2 - x_1) \qquad [\because x_1 + x_2 = l]$$

But $$S_2 - S_1 = h$$

$\therefore$ $$h = \frac{w\,l}{2T}(x_2 - x_1)$$

or $$x_2 - x_1 = \frac{2\,Th}{w\,l} \qquad \text{...(ii)}$$

Solving exps. (*i*) and (*ii*), we get,

$$x_1 = \frac{l}{2} - \frac{T\,h}{w\,l}$$

$$x_2 = \frac{l}{2} + \frac{T\,h}{w\,l}$$

Having found x_1 and x_2, values of S_1 and S_2 can be easily calculated.

Effect of wind and ice loading. The above formulae for sag are true only in still air and at normal temperature when the conductor is acted by its weight only. However, in actual practice, a conductor may have ice coating and simultaneously subjected to wind pressure. The weight of ice acts vertically downwards *i.e.*, in the same direction as the weight of conductor. The force due to the wind is assumed to act horizontally *i.e.*, at right angle to the projected surface of the conductor. Hence, the total force on the conductor is the vector sum of horizontal and vertical forces as shown in Fig. 8.26 (*iii*).

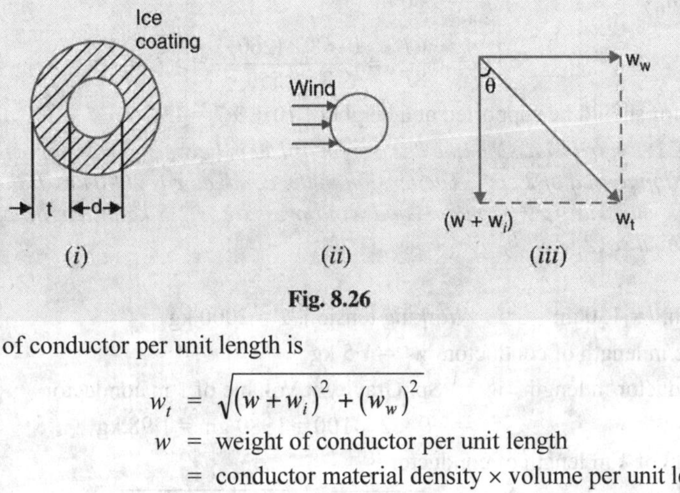

Fig. 8.26

Total weight of conductor per unit length is

$$w_t = \sqrt{(w + w_i)^2 + (w_w)^2}$$

where w = weight of conductor per unit length
= conductor material density × volume per unit length

w_i = weight of ice per unit length
= density of ice × volume of ice per unit length
= density of ice × $\frac{\pi}{4}[(d + 2t)^2 - d^2] \times 1$
= density of ice × $\pi\, t\,(d + t)$*

w_w = wind force per unit length
= wind pressure per unit area × projected area per unit length
= wind pressure × $[(d + 2t) \times 1]$

* Volume of ice per unit length = $\frac{\pi}{4}[(d + t)^2 - d^2] \times 1 = \frac{\pi}{4}[4dt + 4t^2] = \pi t\,(d + t)$

When the conductor has wind and ice loading also, the following points may be noted :

(*i*) The conductor sets itself in a plane at an angle θ to the vertical where

$$\tan\theta = \frac{w_w}{w + w_i}$$

(*ii*) The sag in the conductor is given by :

$$S = \frac{w_t\, l^2}{2T}$$

Hence S represents the slant sag in a direction making an angle θ to the vertical. *If no specific mention is made in the problem, then slant slag is calculated by using the above formula.*

(*iii*) The vertical sag = $S\cos\theta$

Example 8.17. *A 132 kV transmission line has the following data :*

Wt. of conductor = 680 kg/km ; *Length of span = 260 m*

Ultimate strength = 3100 kg ; *Safety factor = 2*

Calculate the height above ground at which the conductor should be supported. Ground clearance required is 10 metres.

Solution.

Wt. of conductor/metre run, $w = 680/1000 = 0{\cdot}68$ kg

Working tension, $T = \dfrac{\text{Ultimate strength}}{\text{Safety factor}} = \dfrac{3100}{2} = 1550$ kg

Span length, $l = 260$ m

$$\therefore \quad \text{Sag} = \frac{w\,l^2}{8T} = \frac{0{\cdot}68 \times (260)^2}{8 \times 1550} = 3{\cdot}7 \text{ m}$$

∴ Conductor should be supported at a height of 10 + 3·7 = **13·7 m**

Example 8.18. *A transmission line has a span of 150 m between level supports. The conductor has a cross-sectional area of 2 cm². The tension in the conductor is 2000 kg. If the specific gravity of the conductor material is 9·9 gm/cm³ and wind pressure is 1·5 kg/m length, calculate the sag. What is the vertical sag?*

Solution.

Span length, $l = 150$ m; Working tension, $T = 2000$ kg

Wind force/m length of conductor, $w_w = 1{\cdot}5$ kg

Wt. of conductor/m length, w = Sp. Gravity × Volume of 1 m conductor

= 9·9 × 2 × 100 = 1980 gm = 1·98 kg

Total weight of 1 m length of conductor is

$$w_t = \sqrt{w^2 + w_w^2} = \sqrt{(1{\cdot}98)^2 + (1{\cdot}5)^2} = 2{\cdot}48 \text{ kg}$$

$$\therefore \quad \text{Sag, } S = \frac{w_t\, l^2}{8T} = \frac{2{\cdot}48 \times (150)^2}{8 \times 2000} = \mathbf{3{\cdot}48\ m}$$

This is the value of slant sag in a direction making an angle θ with the vertical. Referring to Fig. 8.27, the value of θ is given by ;

$$\tan\theta = w_w/w = 1{\cdot}5/1{\cdot}98 = 0{\cdot}76$$

$$\therefore \quad \theta = \tan^{-1} 0{\cdot}76 = 37{\cdot}23°$$

∴ Vertical sag = $S\cos\theta$

= 3·48 × cos 37·23° = **2·77 m**

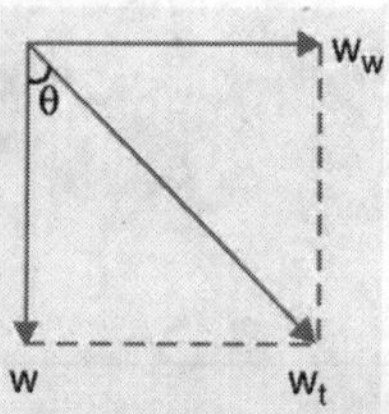

Fig. 8.27

Example 8.19. *A transmission line has a span of 200 metres between level supports. The conductor has a cross-sectional area of 1·29 cm², weighs 1170 kg/km and has a breaking stress of 4218 kg/cm². Calculate the sag for a safety factor of 5, allowing a wind pressure of 122 kg per square metre of projected area. What is the vertical sag?*

Solution.

Span length, l = 200 m

Wt. of conductor/m length, w = 1170/1000 = 1·17 kg

Working tension, *T = 4218 × 1·29/5 = 1088 kg

Diameter of conductor, $d = \sqrt{\dfrac{4 \times \text{area}}{\pi}} = \sqrt{\dfrac{4 \times 1 \cdot 29}{\pi}} = 1 \cdot 28$ cm

Wind force/m length, w_w = Pressure × projected area in m²

$= (122) \times (1 \cdot 28 \times 10^{-2} \times 1) = 1 \cdot 56$ kg

Total weight of conductor per metre length is

$$w_t = \sqrt{w^2 + w_w^2} = \sqrt{(1 \cdot 17)^2 + (1 \cdot 56)^2} = 1 \cdot 95 \text{ kg}$$

∴ Slant sag, $S = \dfrac{w_t\, l^2}{8T} = \dfrac{1 \cdot 95 \times (200)^2}{8 \times 1088} = \mathbf{8 \cdot 96\ m}$

The slant sag makes an angle θ with the vertical where value of θ is given by :

$$\theta = \tan^{-1}(w_w/w) = \tan^{-1}(1 \cdot 56/1 \cdot 17) = 53 \cdot 13°$$

∴ Vertical sag $= S \cos\theta = 8 \cdot 96 \times \cos 53 \cdot 13° = \mathbf{5 \cdot 37\ m}$

Example 8.20. *A transmission line has a span of 275 m between level supports. The conductor has an effective diameter of 1·96 cm and weighs 0·865 kg/m. Its ultimate strength is 8060 kg. If the conductor has ice coating of radial thickness 1·27 cm and is subjected to a wind pressure of 3·9 gm/cm² of projected area, calculate sag for a safety factor of 2. Weight of 1 c.c. of ice is 0·91 gm.*

Solution.

Span length, l = 275 m ; Wt. of conductor/m length, w = 0·865 kg

Conductor diameter, d = 1·96 cm ; Ice coating thickness, t = 1·27 cm

Working tension, T = 8060/2 = 4030 kg

Volume of ice per metre (*i.e.*, 100 cm) length of conductor

$= \pi t (d + t) \times 100$ cm³

$= \pi \times 1 \cdot 27 \times (1 \cdot 96 + 1 \cdot 27) \times 100 = 1288$ cm³

Weight of ice per metre length of conductor is

$$w_i = 0 \cdot 91 \times 1288 = 1172 \text{ gm} = 1 \cdot 172 \text{ kg}$$

Wind force/m length of conductor is

w_w = [Pressure] × [(d + 2t) × 100]

= [3·9] × (1·96 + 2 × 1·27) × 100 gm = 1755 gm = 1·755 kg

Total weight of conductor per metre length of conductor is

$$w_t = \sqrt{(w + w_i)^2 + (w_w)^2}$$

$$= \sqrt{(0 \cdot 865 + 1 \cdot 172)^2 + (1 \cdot 755)^2} = 2 \cdot 688 \text{ kg}$$

---

* Working stress $= \dfrac{\text{Ultimate Strength}}{\text{Safety factor}} = \dfrac{4218}{5}$

∴ Working Tension, T = Working stress × conductor area = 4218 × 1·29/5

$$\therefore \quad \text{Sag} = \frac{w_t\, l^2}{8T} = \frac{2\cdot688 \times (275)^2}{8 \times 4030} = \mathbf{6\cdot3\ m}$$

Example 8.21. *A transmission line has a span of 214 metres between level supports. The conductors have a cross-sectional area of 3·225 cm². Calculate the factor of safety under the following conditions :*

Vertical sag = 2·35 m ; *Wind pressure = 1·5 kg/m run*

Breaking stress = 2540 kg/cm² ; *Wt. of conductor = 1·125 kg/m run*

Solution.

Here, $l = 214$ m ; $w = 1\cdot125$ kg ; $w_w = 1\cdot5$ kg

Total weight of one metre length of conductor is

$$w_t = \sqrt{w^2 + w_w^2} = \sqrt{(1\cdot125)^2 + (1\cdot5)^2} = 1\cdot875 \text{ kg}$$

If f is the factor of safety, then,

$$\text{Working tension, } T = \frac{\text{Breaking stress} \times \text{conductor area}}{\text{safety factor}} = 2540 \times 3\cdot225/f = 8191/f \text{ kg}$$

$$\text{Slant Sag, } S = \frac{\text{Vertical sag}}{{}^*\cos\theta} = \frac{2\cdot35 \times 1\cdot875}{1\cdot125} = 3\cdot92 \text{ m}$$

Now $$S = \frac{w_t\, l^2}{8T}$$

or $$T = \frac{w_t\, l^2}{8\,S}$$

$$\therefore \quad \frac{8191}{f} = \frac{1\cdot875 \times (214)^2}{8 \times 3\cdot92}$$

or Safety factor, $$f = \frac{8191 \times 8 \times 3\cdot92}{1\cdot875 \times (214)^2} = \mathbf{3}$$

Example 8.22. *An overhead line has a span of 150 m between level supports. The conductor has a cross-sectional area of 2 cm². The ultimate strength is 5000 kg/cm² and safety factor is 5. The specific gravity of the material is 8·9 gm/cc. The wind pressure is 1·5 kg/m. Calculate the height of the conductor above the ground level at which it should be supported if a minimum clearance of 7 m is to be left between the ground and the conductor.*

Solution.

Span length, $l = 150$ m ; Wind force/m run, $w_w = 1\cdot5$ kg

Wt. of conductor/m run, w = conductor area × 100 cm × sp. gravity

= 2 × 100 × 8·9 = 1780 gm = 1·78 kg

Working tension, T = 5000 × 2/5 = 2000 kg

Total weight of one metre length of conductor is

$$w_t = \sqrt{w^2 + w_w^2} = \sqrt{(1\cdot78)^2 + (1\cdot5)^2} = 2\cdot33 \text{ kg}$$

$$\text{Slant sag, } S = \frac{w_t\, l^2}{8T} = \frac{2\cdot33 \times (150)^2}{8 \times 2000} = 3\cdot28 \text{ m}$$

$$\text{Vertical sag} = S\cos\theta = 3\cdot28 \times w/w_t = 3\cdot28 \times 1\cdot78/2\cdot33 = 2\cdot5 \text{ m}$$

Conductor should be supported at a height of 7 + 2·5 = **9·5 m**

---

* The slant sag makes an angle θ with the vertical.

$$\therefore \quad \cos\theta = w/w_t = 1\cdot125/1\cdot875$$

Example 8.23. *The towers of height 30 m and 90 m respectively support a transmission line conductor at water crossing. The horizontal distance betwen the towers is 500 m. If the tension in the conductor is 1600 kg, find the minimum clearance of the conductor and water and clearance mid-way between the supports. Weight of conductor is 1·5 kg/m. Bases of the towers can be considered to be at water level.*

Solution. Fig. 8.28 shows the conductor suspended between two supports *A* and *B* at different levels with *O* as the lowest point on the conductor.

Here, $l = 500$ m ; $w = 1{\cdot}5$ kg ; $T = 1600$ kg.

Difference in levels between supports, $h = 90 - 30 = 60$ m. Let the lowest point *O* of the conductor be at a distance x_1 from the support at lower level (*i.e.,* support *A*) and at a distance x_2 from the support at higher level (*i.e.,* support *B*).

Obviously, $$x_1 + x_2 = 500 \text{ m} \qquad ...(i)$$

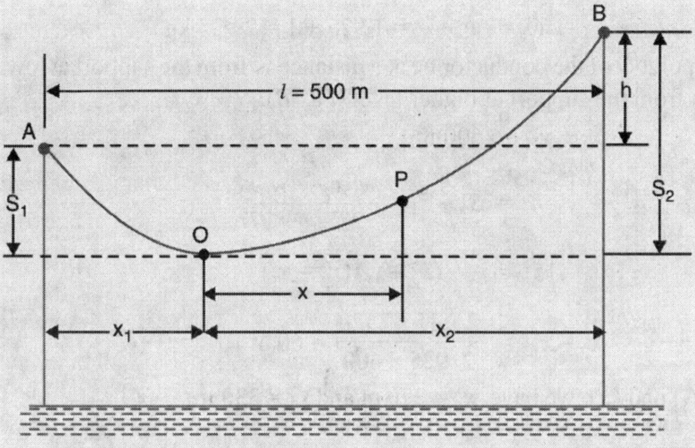

Fig. 8.28

Now $$\text{Sag } S_1 = \frac{w\,x_1^2}{2T} \quad \text{and} \quad \text{Sag } S_2 = \frac{w\,x_2^2}{2T}$$

$$\therefore \quad h = S_2 - S_1 = \frac{w\,x_2^2}{2T} - \frac{w\,x_1^2}{2T}$$

or $$60 = \frac{w}{2T}(x_2 + x_1)(x_2 - x_1)$$

$$\therefore \quad x_2 - x_1 = \frac{60 \times 2 \times 1600}{1{\cdot}5 \times 500} = 256 \text{ m} \qquad ...(ii)$$

Solving exps. (*i*) and (*ii*), we get, $x_1 = 122$ m; $x_2 = 378$ m

Now, $$S_1 = \frac{w\,x_1^2}{2T} = \frac{1{\cdot}5 \times (122)^2}{2 \times 1600} = 7 \text{ m}$$

Clearance of the lowest point *O* from water level

$$= 30 - 7 = \mathbf{23\ m}$$

Let the mid-point *P* be at a distance *x* from the lowest point *O*.

Clearly, $$x = 250 - x_1 = 250 - 122 = 128 \text{ m}$$

Sag at mid-point *P*, $$S_{mid} = \frac{w\,x^2}{2T} = \frac{1{\cdot}5 \times (128)^2}{2 \times 1600} = 7{\cdot}68 \text{ m}$$

Clearance of mid-point P from water level

$$= 23 + 7\cdot68 = \mathbf{30\cdot68\ m}$$

Example 8.24. *An overhead transmission line conductor having a parabolic configuration weighs 1·925 kg per metre of length. The area of X-section of the conductor is 2·2 cm² and the ultimate strength is 8000 kg/cm². The supports are 600 m apart having 15 m difference of levels. Calculate the sag from the taller of the two supports which must be allowed so that the factor of safety shall be 5. Assume that ice load is 1 kg per metre run and there is no wind pressure.*

Solution. Fig. 8.29. shows the conductor suspended between two supports at A and B at different levels with O as the lowest point on the conductor.

Here, $l = 600\text{ m}$; $w_i = 1\text{ kg}$; $h = 15\text{m}$

$w = 1\cdot925\text{ kg}$; $T = 8000 \times 2\cdot2/5 = 3520\text{ kg}$

Total weight of 1 m length of conductor is

$$w_t = w + w_i = 1\cdot925 + 1 = 2\cdot925\text{ kg}$$

Let the lowest point O of the conductor be at a distance x_1 from the support at lower level (*i.e., A*) and at a distance x_2 from the support at higher level (*i.e., B*).

Clearly, $x_1 + x_2 = 600\text{ m}$ ...(*i*)

Now, $$h = S_2 - S_1 = \frac{w_t\,x_2^2}{2T} - \frac{w_t\,x_1^2}{2T}$$

or $$15 = \frac{w_t}{2T}(x_2 + x_1)(x_2 - x_1)$$

$\therefore$ $$x_2 - x_1 = \frac{2 \times 15 \times 3520}{2\cdot925 \times 600} = 60\text{ m} \quad \text{...(ii)}$$

Solving exps. (*i*) and (*ii*), we have, $x_1 = 270$ m and $x_2 = 330$ m

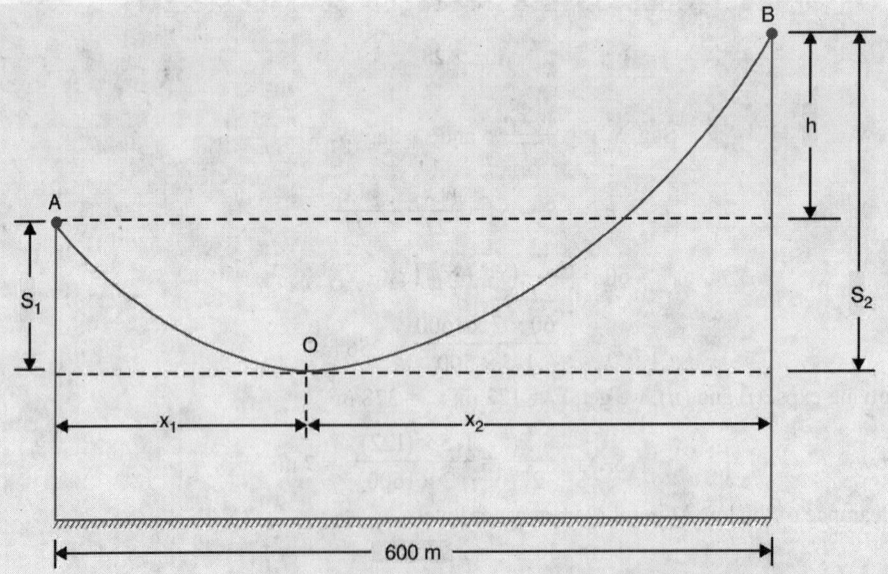

Fig. 8.29

Sag from the taller of the two towers is

$$S_2 = \frac{w_t\,x_2^2}{2T} = \frac{2\cdot925 \times (330)^2}{2 \times 3520} = \mathbf{45\cdot24\ m}$$

Example 8.25. *An overhead transmission line at a river crossing is supported from two towers at heights of 40 m and 90 m above water level, the horizontal distance between the towers being 400 m. If the maximum allowable tension is 2000 kg, find the clearance between the conductor and water at a point mid-way between the towers. Weight of conductor is 1 kg/m.*

Solution. Fig. 8.30 shows the whole arrangement.

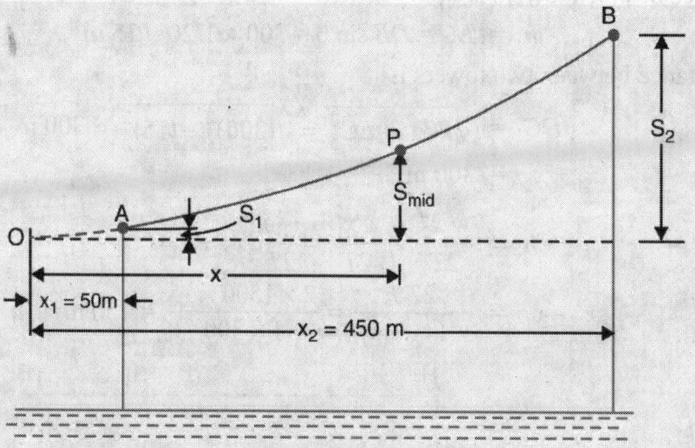

Fig. 8.30

Here, $h = 90 - 40 = 50$ m; $l = 400$ m

$T = 2000$ kg; $w = 1$ kg/m

Obviously, $x_1 + x_2 = 400$ m ...(*i*)

Now $$h = S_2 - S_1 = \frac{wx_2^2}{2T} - \frac{wx_1^2}{2T}$$

or $$50 = \frac{w}{2T}(x_2 + x_1)(x_2 - x_1)$$

$\therefore$ $$x_2 - x_1 = \frac{50 \times 2 \times 2000}{400} = 500 \text{ m} \qquad ...(ii)$$

Solving exps. (*i*) and (*ii*), we get, $x_2 = 450$ m and $x_1 = -50$ m

Now x_2 is the distance of higher support *B* from the lowest point *O* on the conductor, whereas x_1 is that of lower support *A*. As the span is 400 m, therefore, point *A* lies on the same side of *O* as *B* (see Fig. 8.30).

Horizontal distance of mid-point *P* from lowest point *O* is

x = Distance of *A* from $O + 400/2 = 50 + 200 = 250$ m

$\therefore$ Sag at point *P*, $$S_{mid} = \frac{w\,x^2}{2T} = \frac{1 \times (250)^2}{2 \times 2000} = 15{\cdot}6 \text{ m}$$

Now $$\text{Sag } S_2 = \frac{w\,x_2^2}{2T} = \frac{1 \times (450)^2}{2 \times 2000} = 50{\cdot}6 \text{ m}$$

Height of point *B* above mid-point *P*

$= S_2 - S_{mid} = 50{\cdot}6 - 15{\cdot}6 = 35$ m

$\therefore$ Clearance of mid-point *P* above water level

$= 90 - 35 =$ **55 m**

Example 8.26. *A transmission line over a hillside where the gradient is 1 : 20, is supported by two 22 m high towers with a distance of 300 m between them. The lowest conductor is fixed 2 m below the top of each tower. Find the clearance of the conductor from the ground. Given that conductor weighs 1 kg/m and the allowable tension is 1500 kg.*

Solution. The conductors are supported between towers *AD* and *BE* over a hillside having gradient of 1 : 20 as shown in Fig. 8.31. The lowest point on the conductor is *O* and $\sin\theta = 1/20$.

Effective height of each tower (*AD* or *BE*)

$$= 22 - 2 = 20 \text{ m}$$

Vertical distance between towers is

$$h = EC = DE \sin\theta = 300 \times 1/20 = 15 \text{ m}$$

Horizontal distance between two towers is

$$DC = \sqrt{DE^2 - EC^2} = \sqrt{(300)^2 - (15)^2} \simeq 300 \text{ m}$$

or $$x_1 + x_2 = 300 \text{ m} \quad ...(i)$$

Now $$h = \frac{w x_2^2}{2T} - \frac{w x_1^2}{2T} = \frac{w}{2T}(x_2 + x_1)(x_2 - x_1)$$

or $$x_2 - x_1 = \frac{2T h}{w(x_2 + x_1)} = \frac{2 \times 1500 \times 15}{1 \times 300} = 150 \text{ m} \quad ...(ii)$$

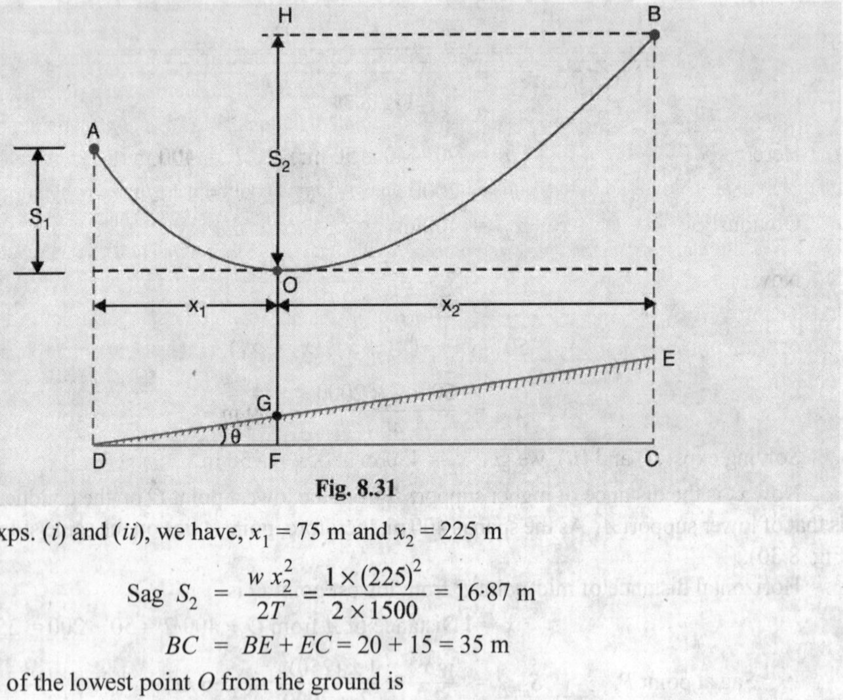

Fig. 8.31

Solving exps. (*i*) and (*ii*), we have, $x_1 = 75$ m and $x_2 = 225$ m

$$\text{Sag } S_2 = \frac{w x_2^2}{2T} = \frac{1 \times (225)^2}{2 \times 1500} = 16{\cdot}87 \text{ m}$$

Now $$BC = BE + EC = 20 + 15 = 35 \text{ m}$$

Clearance of the lowest point *O* from the ground is

$$OG = HF - S_2 - GF$$

$$= BC - S_2 - GF \quad (\because BC = HF)$$

$$[\text{Now } GF = x_1 \tan\theta = 75 \times 0{\cdot}05 = 3{\cdot}75 \text{ m}]$$

$$= 35 - 16{\cdot}87 - 3{\cdot}75 = \mathbf{14{\cdot}38 \text{ m}}$$

Example 8.27. *A transmission tower on a level ground gives a minimum clearance of 8 metres for its lowest conductor with a sag of 10 m for a span of 300 m. If the same tower is to be used over a slope of 1 in 15, find the minimum ground clearance obtained for the same span, same conductor and same weather conditions.*

Solution. On level ground

$$\text{Sag, } S = \frac{w l^2}{8T}$$

$$\therefore \quad \frac{w}{T} = \frac{8S}{l^2} = \frac{8 \times 10}{(300)^2} = \frac{8}{9 \times 10^3}$$

Height of tower = Sag + Clearance = 10 + 8 = 18 m

On sloping ground. The conductors are supported between towers *AD* and *BE* over a sloping ground having a gradient 1 in 15 as shown in Fig. 8.32. The height of each tower (*AD* or *BE*) is 18 m.

Vertical distance between the two towers is

$$h = EC = {}^*DE \sin\theta = 300 \times 1/15 = 20 \text{ m}$$

Now $$x_1 + x_2 = 300 \text{ m} \quad ...(i)$$

Also $$h = \frac{w x_2^2}{2T} - \frac{w x_1^2}{2T} = \frac{w}{2T}(x_2 + x_1)(x_2 - x_1)$$

$$\therefore \quad x_2 - x_1 = \frac{2\,T\,h}{w(x_2 + x_1)} = \frac{2 \times 9 \times 10^3 \times 20}{8 \times 300} = 150 \text{ m} \quad ...(ii)$$

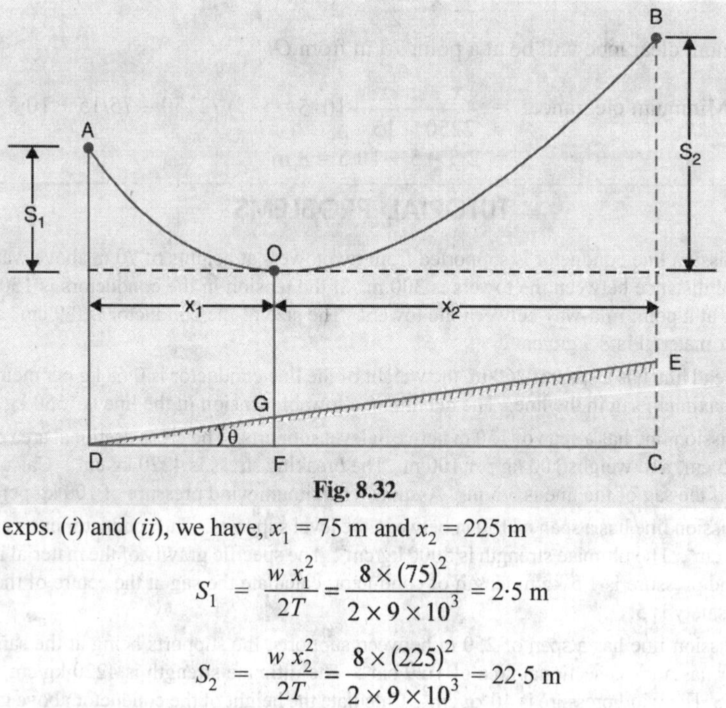

Fig. 8.32

Solving exps. (*i*) and (*ii*), we have, $x_1 = 75$ m and $x_2 = 225$ m

Now $$S_1 = \frac{w x_1^2}{2T} = \frac{8 \times (75)^2}{2 \times 9 \times 10^3} = 2{\cdot}5 \text{ m}$$

$$S_2 = \frac{w x_2^2}{2T} = \frac{8 \times (225)^2}{2 \times 9 \times 10^3} = 22{\cdot}5 \text{ m}$$

Clearance of point *O* from the ground is

$$OG = BC - S_2 - GF = 38 - 22{\cdot}5 - 5 = 10{\cdot}5 \text{ m}$$

$$[\because GF = x_1 \tan\theta = 75 \times 1/15 = 5\text{m}]$$

Since *O* is the origin, the equation of slope of ground is given by :

$$y = m\,x + A$$

Here $$m = 1/15 \text{ and } A = OG = -10{\cdot}5 \text{ m}$$

$$\therefore \quad y = \frac{x}{15} - 10{\cdot}5$$

∴ Clearance *C* from the ground at any point *x* is

* $DE \simeq DC = 300$ m

$$C = \text{Equation of conductor curve} - y = \left(\frac{w\,x^2}{2T}\right) - \left(\frac{x}{15} - 10\cdot5\right)$$

$$= \frac{8x^2}{2\times9\times10^3} - \left(\frac{x}{15} - 10\cdot5\right)$$

$$\therefore \quad C = \frac{x^2}{2250} - \frac{x}{15} + 10\cdot5$$

Clearance will be minimum when $dC/dx = 0$ *i.e.*,

$$\frac{d}{dx}\left[\frac{x^2}{2250} - \frac{x}{15} + 10\cdot5\right] = 0$$

or $$\frac{2x}{2250} - \frac{1}{15} = 0$$

or $$x = \frac{1}{15}\times\frac{2250}{2} = \mathbf{75\ m}$$

i.e., minimum clearance will be at a point 75 m from *O*.

$$\text{Minimum clearance} = \frac{x^2}{2250} - \frac{x}{15} + 10\cdot5 = (75)^2/2250 - 75/15 + 10\cdot5$$

$$= 2\cdot5 - 5 + 10\cdot5 = \mathbf{8\ m}$$

TUTORIAL PROBLEMS

1. A transmission line conductor is supported from two towers at heights of 70 m above water level. The horizontal distance between the towers is 300 m. If the tension in the conductors is 1500 kg, find the clearance at a point mid-way between the towers. The size of the conductor is 0·9 cm^2 and density of conductor material is 8·9 gm/cm^3. **[64 m]**
2. An overhead line has a span of 260 m, the weight of the line conductor is 0·68 kg per metre run. Calculate the maximum sag in the line. The maximum allowable tension in the line is 1550 kg. **[3·7 m]**
3. A transmission line has a span of 150 m between level supports. The cross-sectional area of the conductor is 1·25 cm^2 and weighs 100 kg per 100 m. The breaking stress is 4220 kg/cm^2. Calculate the factor of safety if the sag of the line is 3·5 m. Assume a maximum wind pressure of 100 kg per sq. metre. **[4]**
4. A transmission line has a span of 150 m between the level supports. The conductor has a cross-sectional area of 2 cm^2. The ultimate strength is 5000 kg/cm^2. The specific gravity of the material is 8·9 gm/cm^3. If the wind pressure is 1·5 kg/m length of conductor, calculate the sag at the centre of the conductor if factor of safety is 5. **[3·28 m]**
5. A transmission line has a span of 250 m between supports, the supports being at the same level. The conductor has a cross-sectional area of 1·29 cm^2. The ultimate strength is 4220 kg/cm^2 and factor of safety is 2. The wind pressure is 40 kg/cm^2. Calculate the height of the conductor above ground level at which it should be supported if a minimum clearance of 7m is to be kept between the ground and the conductor. **[10·24 m]**
6. A transmission line has a span of 150 m between level supports. The conductor has a cross-sectional area of 2 cm^2. The ultimate strength is 5000 kg/cm^2. The specific gravity of the material is 8·9 gm/cm^3. If the wind pressure is 1·5 kg/m length of the conductor, calculate the sag if factor of safety is 5. **[3·5 m]**
7. Two towers of height 40 m and 30 m respectively support a transmission line conductor at water crossing. The horizontal distance between the towers is 300 m. If the tension in the conductor is 1590 kg, find the clearance of the conductor at a point mid-way between the supports. Weight of conductor is 0·8 kg/m. Bases of the towers can be considered to be at the water level. **[59 m]**
8. An overhead transmission line at a river crossing is supported from two towers at heights of 50 m and 100 m above the water level. The horizontal distance between the towers is 400 m. If the maximum allowable tension is 1800 kg and the conductor weighs 1 kg/m, find the clearance between the conductor and water at a point mid-way between the supports. **[63·8 m]**

8.17 Some Mechanical Principles

Mechanical factors of safety to be used in transmission line design should depend to some extent on the importance of continuity of operation in the line under consideration. In general, the strength of the line should be such as to provide against the worst *probable* weather conditions. We now discuss some important points in the mechanical design of overhead transmission lines.

(*i*) **Tower height :** Tower height depends upon the length of span. With long spans, relatively few towers are required but they must be tall and correspondingly costly. It is not usually possible to determine the tower height and span length on the basis of direct construction costs because the lightning hazards increase greatly as the height of the conductors above ground is increased. This is one reason that horizontal spacing is favoured inspite of the wider right of way required.

(*ii*) **Conductor clearance to ground :** The conductor clearance to ground at the time of greatest sag should not be less than some specified distance (usually between 6 and 12 m), depending on the voltage, on the nature of the country and on the local laws. The greatest sag may occur on the hottest day of summer on account of the expansion of the wire or it may occur in winter owing to the formation of a heavy coating of ice on the wires. Special provisions must be made for melting ice from the power lines.

(*iii*) **Sag and tension :** When laying overhead transmission lines, it is necessary to allow a reasonable factor of safety in respect of the tension to which the conductor is subjected. The tension is governed by the effects of wind, ice loading and temperature variations. The relationship between tension and sag is dependent on the loading conditions and temperature variations. For example, the tension increases when the temperature decreases and there is a corresponding decrease in the sag. Icing-up of the line and wind loading will cause stretching of the conductor by an amount dependent on the line tension.

In planning the sag, tension and clearance to ground of a given span, a maximum stress is selected. It is then aimed to have this stress developed at the worst probable weather conditions (*i.e.* minimum expected temperature, maximum ice loading and maximum wind). Wind loading increases the sag in the direction of resultant loading but decreases the vertical component. Therefore, in clearance calculations, the effect of wind should not be included unless horizontal clearance is important.

(*iv*) **Stringing charts :** For use in the field work of stringing the conductors, temperature-sag and temperature-tension charts are plotted for the given conductor and loading conditions. Such curves are called stringing charts (see Fig. 8.33). These charts are very helpful while stringing overhead lines.

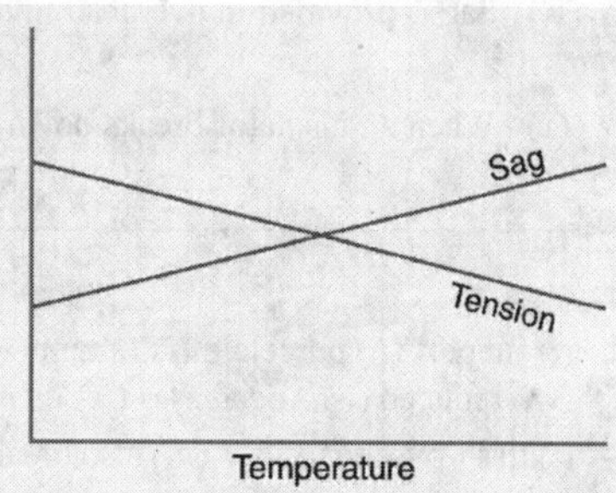

Fig. 8.33

(*v*) **Conductor spacing :** Spacing of conductors should be such so as to provide safety against flash-over when the wires are swinging in the wind. The proper spacing is a function of span length, voltage and weather conditions. The use of horizontal spacing eliminates the danger caused by unequal ice loading. Small wires or wires of light material are subjected to more swinging by the wind than heavy conductors. Therefore, light wires should be given greater spacings.

(*vi*) **Conductor vibration :** Wind exerts pressure on the exposed surface of the conductor. If the wind velocity is small, the swinging of conductors is harmless provided the clearance is sufficiently large so that conductors do not approach within the sparking distance of each other. A completely different type of vibration, called *dancing,* is caused by the action of fairly strong wind on a wire covered with ice, when the ice coating happens to take a form which makes a good air-foil section. Then the whole span may sail up like a kite until it reaches the limit of its slack, stops with a jerk and falls or sails back. The harmful effects of these vibrations occur at the clamps or supports where the conductor suffers fatigue and breaks eventually. In order to protect the conductors, dampers are used.

SELF-TEST

1. **Fill in the blanks by inserting appropriate words/figures.**
   (*i*) Cross-arms are used on poles or towers to provide .... to the insulators.
   (*ii*) The most commonly used material for insulators of overhead lines is ....
   (*iii*) The potential across the various discs of suspension string is different because of .... capacitance.
   (*iv*) In a string of suspension insulators, the maximum voltage appears across the unit ...... to the conductor.
   (*v*) If the string efficiency is 100%, it means that ....
   (*vi*) If shunt capacitance is reduced, then string efficiency is ....
   (*vii*) If the spacing between the conductors is increased, then corona effect is ....
   (*viii*) If sag in an overhead line increases, tension in the line ....
   (*ix*) By using a guard ring, string efficiency is ....
   (*x*) Shunt capacitance in suspension insulators can be decreased by increasing the distance of .... from ....
2. **Pick up the correct words/figures from the brackets and fill in the blanks.**
   (*i*) The insulator is so designed that it should fail only by .... (*flash-over, puncture*)
   (*ii*) Suspension type insulators are used for voltages beyond .... (*33 kV, 400 V, 11 kV*)
   (*iii*) In a string of suspension insulators, if the unit nearest to the conductor breaks down, then other units will .... (*also breakdown, remain intact*)
   (*iv*) A shorter string has .... string efficiency than a larger one. (*less, more*)
   (*v*) Corona effect is .... pronounced in stormy weather as compared to fair weather. (*more, less*)
   (*vi*) If the conductor size is increased, the corona effect is .... (*increased, decreased*)
   (*vii*) The longer the cross arm, the .... the string efficiency. (*greater, lesser*)
   (*viii*) The discs of the strain insulators are used in .... plane. (*vertical, horizontal*)
   (*ix*) Sag is provided in overhead lines so that .... (*Safe tension is not exceeded, repair can be done*)
   (*x*) When an insulator breaks down by puncture, it is .... damaged. (*permanently, only partially*)

ANSWERS

1. (*i*) support (*ii*) porcelain (*iii*) shunt (*iv*) nearest (*v*) potential across each disc is the same (*vi*) increased (*vii*) reduced (*viii*) decreases (*ix*) increased (*x*) conductor, tower.
2. (*i*) flash-over (*ii*) 33 kV (*iii*) also breakdown (*iv*) more (*v*) more (*vi*) decreased (*vii*) greater (*viii*) vertical (*ix*) safe tension is not exceeded (*x*) permanently.

CHAPTER REVIEW TOPICS

1. Name the important components of an overhead transmission line.
2. Discuss the various conductor materials used for overhead lines. What are their relative advantages and disadvantages ?
3. Discuss the various types of line supports.
4. Why are insulators used with overhead lines ? Discuss the desirable properties of insulators.

5. Discuss the advantages and disadvantages of (*i*) pin-type insulators (*ii*) suspension type insulators.
6. Explain how the electrical breakdown can occur in an insulator.
7. What is a strain insulator and where is it used ? Give a sketch to show its location.
8. Give reasons for unequal potential distribution over a string of suspension insulators.
9. Define and explain string efficiency. Can its value be equal to 100% ?
10. Show that in a string of suspension insulators, the disc nearest to the conductor has the highest voltage across it.
11. Explain various methods of improving string efficiency.
12. What is corona? What are the factors which affect corona ?
13. Discuss the advantages and disadvantages of corona.
14. Explain the following terms with reference to corona :
    (*i*) Critical disruptive voltage
    (*ii*) Visual critical voltage
    (*iii*) Power loss due to corona
15. Describe the various methods for reducing corona effect in an overhead transmission line.
16. What is a sag in overhead lines ? Discuss the disadvantages of providing too small or too large sag on a line.
17. Deduce an approximate expression for sag in overhead lines when
    (*i*) supports are at equal levels
    (*ii*) supports are at unequal levels.

DISCUSSION QUESTIONS

1. What is the need for stranding the conductors ?
2. Is sag a necessity or an evil ? Discuss.
3. String efficiency for a d.c. system is 100% ? Discuss.
4. Can string efficiency in an a.c. system be 100% ?
5. Why are suspension insulators preferred for high voltage power transmission ?
6. Give reasons for the following :
    (*i*) A.C.S.R. conductors are preferred for transmission and distribution lines.
    (*ii*) Conductors are not fully stretched between supports.

CHAPTER

9

Electrical Design of Overhead Lines

Introduction

It has already been discussed that transmission of electric power is done by 3-phase, 3-wire overhead lines. An a.c. transmission line has resistance, inductance and capacitance uniformly distributed along its length. These are known as constants or parameters of the line. The performance of a transmission line depends to a considerable extent upon these constants. For instance, these constants determine whether the efficiency and voltage regulation of the line will be good or poor. Therefore, a sound concept of these constants is necessary in order to make the electrical design of a transmission line a technical success. In this chapter, we shall focus our attention on the methods of calculating these constants for a given transmission line. Out of these three parameters of a transmission line, we shall pay greatest attention to inductance and capacitance. Resistance is certainly of equal importance but requires less explanation since it is not a function of conductor arrangement.

9.1 Constants of a Transmission Line

A transmission line has resistance, inductance and capacitance uniformly distributed along the whole length of the line. Before we pass on to the methods of finding these constants for a transmission line, it is profitable to understand them thoroughly.

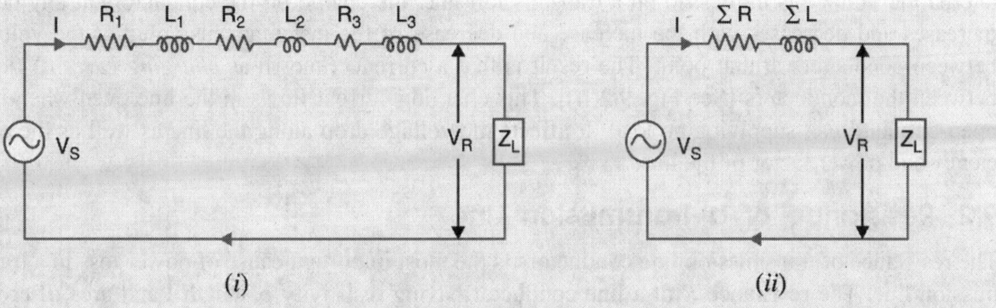

Fig. 9.1

(*i*) **Resistance.** It is the opposition of line conductors to current flow. The resistance is distributed uniformly along the whole length of the line as shown in Fig. 9.1 (*i*). However, the performance of a transmission line can be analysed conveniently if distributed resistance is considered as lumped as shown in Fig. 9.1(*ii*).

(*ii*) **Inductance.** When an alternating current flows through a conductor, a changing flux is set up which links the conductor. Due to these flux linkages, the conductor possesses inductance. Mathematically, inductance is defined as the flux linkages per ampere *i.e.,*

$$\text{Inductance,} \quad L = \frac{\psi}{I} \text{ henry}$$

where ψ = flux linkages in weber-turns

I = current in amperes

The inductance is also uniformly distributed along the length of the * line as show in Fig. 9.1(*i*). Again for the convenience of analysis, it can be taken to be lumped as shown in Fig. 9.1(*ii*).

(*iii*) **Capacitance.** We know that any two conductors separated by an insulating material constitute a capacitor. As any two conductors of an overhead transmission line are separated by air which acts as an insulation, therefore, capacitance exists between any two overhead line conductors. The capacitance between the conductors is the charge per unit potential difference *i.e.,*

$$\text{Capacitance,} \quad C = \frac{q}{v} \text{ farad}$$

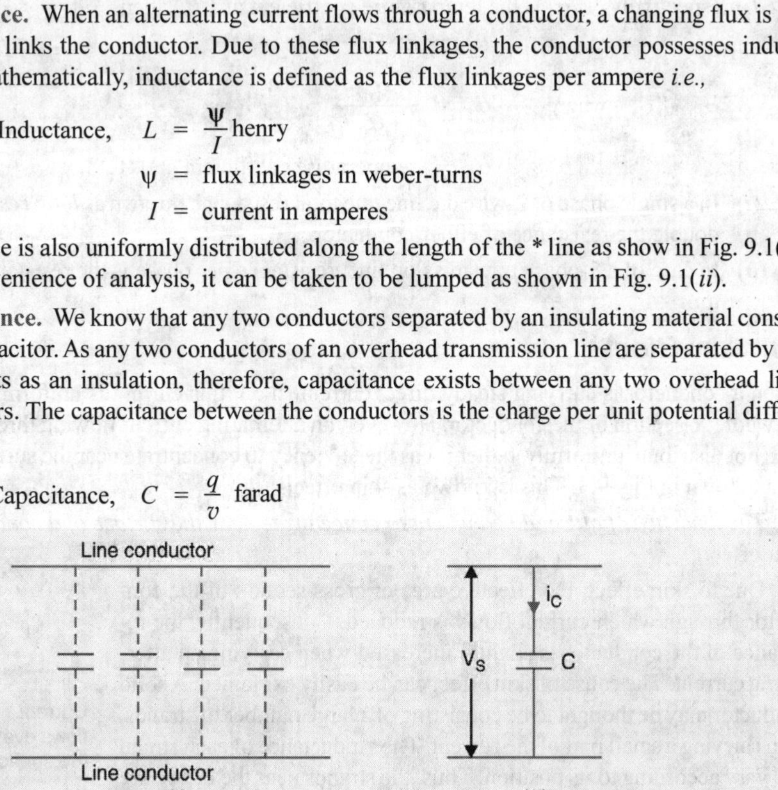

Fig. 9.2

* The two parallel conductors of a transmission line form a rectangular loop of one turn. The changing flux in the line links the loop and hence the line has inductance.

where q = charge on the line in coulomb

v = p.d. between the conductors in volts

The capacitance is uniformly distributed along the whole length of the line and may be regarded as a uniform series of capacitors connected between the conductors as shown in Fig. 9.2(*i*). When an alternating voltage is impressed on a transmission line, the charge on the conductors at any point increases and decreases with the increase and decrease of the instantaneous value of the voltage between conductors at that point. The result is that a current (known as *charging current*) flows between the conductors [See Fig. 9.2(*ii*)]. This charging current flows in the line even when it is open-circuited *i.e.,* supplying no load. It affects the voltage drop along the line as well as the efficiency and power factor of the line.

9.2 Resistance of a Transmission Line

The resistance of transmission line conductors is the most important cause of power loss in a transmission line. The resistance R of a line conductor having resistivity ρ, length l and area of cross-section a is given by ;

$$R = \rho \frac{l}{a}$$

The variation of resistance of metallic conductors with temperature is practically linear over the normal range of operation. Suppose R_1 and R_2 are the resistances of a conductor at t_1°C and t_2°C ($t_2 > t_1$) respectively. If α_1 is the temperature coefficient at t_1°C, then,

$$R_2 = R_1 \; [1 + \alpha_1 (t_2 - t_1)]$$

where $\alpha_1 = \dfrac{\alpha_0}{1 + \alpha_0 \, t_1}$

α_0 = temperature coefficient at 0° C

(*i*) In a single phase or 2-wire d.c line, the total resistance (known as *loop resistance*) is equal to double the resistance of either conductor.

(*ii*) In case of a 3-phase transmission line, resistance per phase is the resistance of one conductor.

9.3 Skin Effect

When a conductor is carrying steady direct current (d.c.), this current is uniformly distributed over the whole X-section of the conductor. However, an alternating current flowing through the conductor does not distribute uniformly, rather it has the tendency to concentrate near the surface of the conductor as shown in Fig. 9.3. This is known as skin effect.

The tendency of alternating current to concentrate near the surface of a conductor is known as **skin effect.**

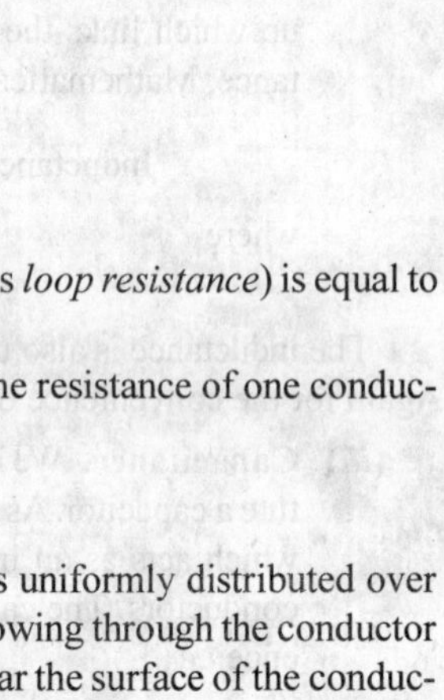

Fig. 9.3

Due to skin effect, the effective area of cross-section of the conductor through which current flows is reduced. Consequently, the resistance of the conductor is slightly increased when carrying an alternating current. The cause of skin effect can be easily explained. A solid conductor may be thought to be consisting of a large number of strands, each carrying a small part of the current. The *inductance of each strand will vary according to its position. Thus, the strands near the centre are surrounded by a greater magnetic flux and hence have larger inductance than that near the surface. The high reactance of inner strands

* For a direct current, inductance is zero and hence the current distributes uniformly over the entire X-section of the conductor.

causes the alternating current to flow near the surface of conductor. This crowding of current near the conductor surface is the skin effect. The skin effect depends upon the following factors :

(*i*) Nature of material

(*ii*) Diameter of wire – increases with the diameter of wire.

(*iii*) Frequency – increases with the increase in frequency.

(*iv*) Shape of wire – less for stranded conductor than the solid conductor.

It may be noted that skin effect is negligible when the supply frequency is low (< 50 Hz) and conductor diameter is small (< 1cm).

9.4 Flux Linkages

As stated earlier, the inductance of a circuit is defined as the flux linkages per unit current. Therefore, in order to find the inductance of a circuit, the determination of flux linkages is of primary importance. We shall discuss two important cases of flux linkages.

1. **Flux linkages due to a single current carrying conductor.** Consider a long straight cylindrical conductor of radirus r metres and carrying a current I amperes (r.m.s.) as shown in Fig. 9.4 (*i*). This current will set up magnetic field. The magnetic lines of force will exist inside the conductor as well as outside the conductor. Both these fluxes will contribute to the inductance of the conductor.

(*i*) ***Flux linkages due to internal flux.*** Refer to Fig. 9.4 (*ii*) where the X-section of the conductor is shown magnified for clarity. The magnetic field intensity at a point x metres from the centre is given by;

$$^*H_x = \frac{I_x}{2\pi x}$$

Assuming a uniform current density,

$$I_x = \frac{\pi x^2}{\pi r^2} I = \frac{x^2}{r^2} I$$

$$\therefore \quad H_x = \frac{x^2}{r^2} \times I \times \frac{1}{2\pi x} = \frac{x}{2\pi r^2} I \text{ AT/m}$$

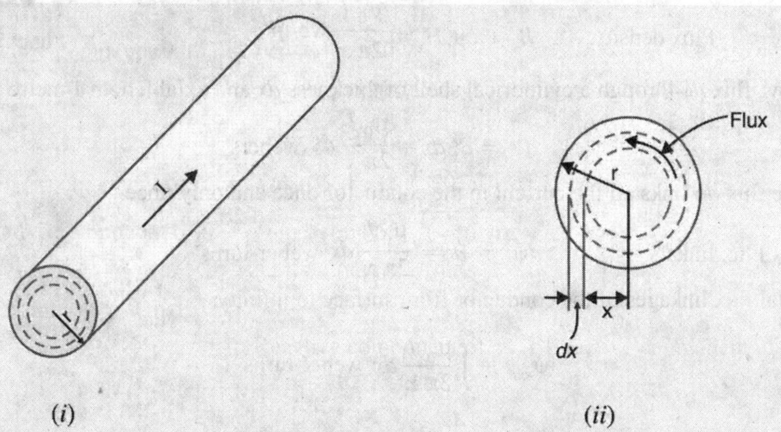

Fig. 9.4

* According to Ampere's law, m.m.f. (ampere-turns) around any closed path equals the current enclosed by the path. The current enclosed by the path is I_x and m.m.f. = $H_x \times 2\pi x$. $\therefore$ $H_x \times 2\pi x = I_x$.

If μ ($= \mu_0 \mu_r$) is the permeability of the conductor, then flux density at the considered point is given by;

$$B_x = \mu_0 \mu_r H_x \text{ wb/m}^2$$

$$= \frac{\mu_0 \mu_r x}{2\pi r^2} I = \frac{\mu_0 x I}{2\pi r^2} \text{ wb/m}^2 [\because \mu_r = 1 \text{ for non-magnetic material}]$$

Now, flux $d\phi$ through a cylindrical shell of radial thickness dx and axial length 1 m is given by;

$$d\phi = B_x \times 1 \times dx = \frac{\mu_0 x I}{2\pi r^2} dx \text{ weber}$$

This flux links with current $I_x \left(= \frac{I \pi x^2}{\pi r^2} \right)$ only. Therefore, flux linkages per metre length of the conductor is

$$d\psi = \frac{\pi x^2}{\pi r^2} d\phi = \frac{\mu_0 I x^3}{2\pi r^4} dx \text{ weber-turns}$$

Total flux linkages from centre upto the conductor surface is

$$\psi_{int} = \int_0^r \frac{\mu_0 I x^3}{2\pi r^4} dx$$

$$= \frac{\mu_0 I}{8\pi} \text{ weber-turns per metre length}$$

(ii) Flux linkages due to external flux. Now let us calculate the flux linkages of the conductor due to external flux. The external flux extends from the surface of the conductor to infinity. Referring to Fig. 9.5, the field intensity at a distance x metres (from centre) outside the conductor is given by ;

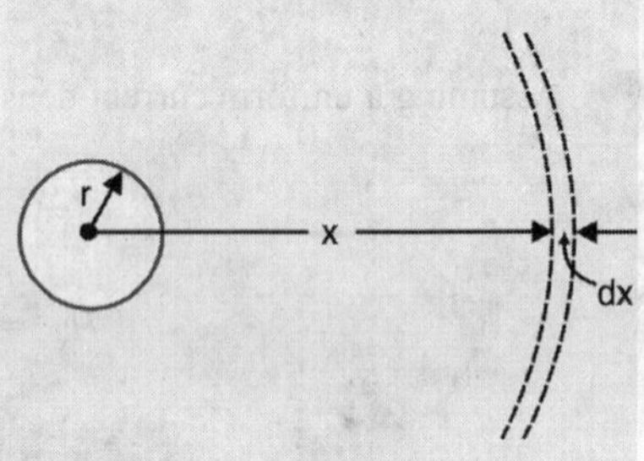

Fig. 9.5

$$H_x = \frac{I}{2\pi x} \text{ AT/m}$$

Flux density, $$B_x = \mu_0 H_x = \frac{\mu_0 I}{2\pi x} \text{ wb/m}^2$$

Now, flux $d\phi$ through a cylindrical shell of thickness dx and axial length 1 metre is

$$d\phi = B_x\, dx = \frac{\mu_0 I}{2\pi x} dx \text{ webers}$$

The flux $d\phi$ links all the current in the conductor once and only once.

$\therefore$ Flux linkages, $$d\psi = d\phi = \frac{\mu_0 I}{2\pi x} dx \text{ weber-turns}$$

Total flux linkages of the conductor from surface to infinity,

$$\psi_{ext} = \int_r^\infty \frac{\mu_0 I}{2\pi x} dx \text{ weber-turns}$$

$\therefore$ Overall flux linkages, $$\psi = \psi_{int} + \psi_{ext} = \frac{\mu_0 I}{8\pi} + \int_r^\infty \frac{\mu_0 I}{2\pi x} dx$$

$$\therefore \quad \psi = \frac{\mu_0 I}{2\pi} \left[\frac{1}{4} + \int_r^\infty \frac{dx}{x} \right] \text{ wb-turns/m length}$$

2. Flux linkages in parallel current-carrying conductors. We shall now determine the flux linkages in a group of parallel current carrying conductors. Fig. 9.6 shows the conductors A, B, C etc. carrying currents I_A, I_B, I_C etc. Let us consider the flux linkages with one conductor, say conductor A. There will be flux linkages with conductor A due to its own current as discussed previously. Also there will be flux linkages with this conductor due to the mutual inductance effects of I_B, I_C, I_D etc. We shall now determine the total flux linkages with conductor A.

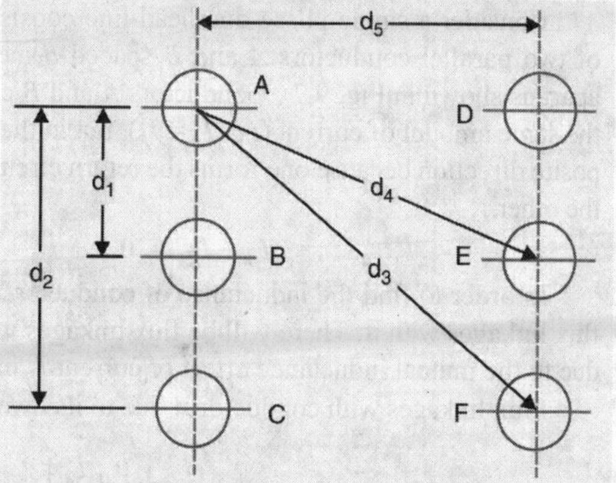

Fig. 9.6

Flux linkages with conductor A due to its own current

$$= \frac{\mu_0 I_A}{2\pi}\left[\frac{1}{4} + \int_r^\infty \frac{dx}{x}\right] \qquad ...(i)$$

Flux linkages with conductor A due to current I_B

$$= \frac{{}^*\mu_0 I_B}{2\pi}\int_{d_1}^\infty \frac{dx}{x} \qquad ...(ii)$$

Flux linkages with conductor A due to current I_C

$$= \frac{\mu_0 I_C}{2\pi}\int_{d_2}^\infty \frac{dx}{x} \qquad ...(iii)$$

∴ Total flux linkages with conductor A

$$= (i) + (ii) + (iii) + ...$$

$$= \frac{\mu_0 I_A}{2\pi}\left(\frac{1}{4} + \int_r^\infty \frac{dx}{x}\right) + \frac{\mu_0 I_B}{2\pi}\int_{d_1}^\infty \frac{dx}{x} + \frac{\mu_0 I_C}{2\pi}\int_{d_2}^\infty \frac{dx}{x} + ...$$

Similarly, flux linkages with other conductors can be determined. The above relation provides the basis for evaluating inductance of any circuit.

9.5 Inductance of a Single Phase Two-wire Line

A single phase line consists of two parallel conductors which form a rectangular loop of one turn. When an alternating current flows through such a loop, a changing magnetic flux is set up. The changing flux links the loop and hence the loop (or single phase line) possesses inductance. It may appear that inductance of a single phase line is negligible because it consists of a loop of one turn and the flux path is through air of high reluctance. But as the X -sectional area of the loop is very **large, even for a small flux density, the total flux linking the loop is quite large and hence the line has appreciable inductance.

* The conductor B carrying current I_B is at a distance d_1 from conductor A. Only the external flux due to I_B links with conductor A. This external flux due to I_B links with conductor A from d_1 to ∞ and hence the term

$$\frac{\mu_0 I_B}{2\pi}\int_{d_1}^\infty \frac{dx}{x}$$

** The conductors are spaced several metres and the length of the line is several kilometres. Therefore, the loop has a large X-sectional area.

Consider a single phase overhead line consisting of two parallel conductors A and B spaced d metres apart as shown in Fig. 9.7. Conductors A and B carry the same amount of current (*i.e.* $I_A = I_B$), but in the opposite direction because one forms the return circuit of the other.

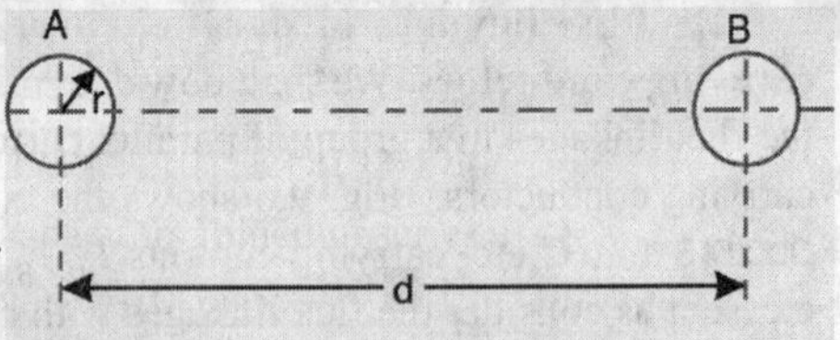

Fig. 9.7

$$\therefore \quad I_A + I_B = 0$$

In order to find the inductance of conductor A (or conductor B), we shall have to consider the flux linkages with it. There will be flux linkages with conductor A due to its own current I_A and also due to the mutual inductance effect of current I_B in the conductor B.

Flux linkages with conductor A due to its own current

$$= \frac{\mu_0 I_A}{2\pi}\left(\frac{1}{4} + \int_r^{\infty} \frac{dx}{x}\right) \quad \text{...(i)} \quad \text{[See Art. 9.4]}$$

Flux linkages with conductor A due to current I_B

$$= \frac{\mu_0 I_B}{2\pi}\int_d^{\infty} \frac{dx}{x} \quad \text{...(ii)}$$

Total flux linkages with conductor A is

$$\psi_A = \text{exp. }(i) + \text{exp }(ii)$$

$$= \frac{\mu_0 I_A}{2\pi}\left(\frac{1}{4} + \int_r^{\infty} \frac{dx}{x}\right) + \frac{\mu_0 I_B}{2\pi}\int_d^{\infty} \frac{dx}{x}$$

$$= \frac{\mu_0}{2\pi}\left[\left(\frac{1}{4} + \int_r^{\infty} \frac{dx}{x}\right) I_A + I_B \int_d^{\infty} \frac{dx}{x}\right]$$

$$= \frac{\mu_0}{2\pi}\left[\left(\frac{1}{4} + \log_e \infty - \log_e r\right) I_A + \left(\log_e \infty - \log_e d\right) I_B\right]$$

$$= \frac{\mu_0}{2\pi}\left[\left(\frac{I_A}{4} + \log_e \infty \left(I_A + I_B\right) - I_A \log_e r - I_B \log_e d\right)\right]$$

$$= \frac{\mu_0}{2\pi}\left[\frac{I_A}{4} - I_A \log_e r - I_B \log_e d\right] \qquad (\because I_A + I_B = 0)$$

Now, $\quad I_A + I_B = 0 \quad$ or $\quad -I_B = I_A$

$$\therefore \quad -I_B \log_e d = I_A \log_e d$$

$$\therefore \quad \psi_A = \frac{\mu_0}{2\pi}\left[\frac{I_A}{4} + I_A \log_e d - I_A \log_e r\right] \text{ wb-turns/m}$$

$$= \frac{\mu_0}{2\pi}\left[\frac{I_A}{4} + I_A \log_e \frac{d}{r}\right]$$

$$= \frac{\mu_0 I_A}{2\pi}\left[\frac{1}{4} + \log_e \frac{d}{r}\right] \text{ wb-turns/m}$$

$$\text{Inductance of conductor } A, L_A = \frac{\psi_A}{I_A}$$

$$= \frac{\mu_0}{2\pi}\left[\frac{1}{4} + \log_e \frac{d}{r}\right] \text{H/m} = \frac{4\pi \times 10^{-7}}{2\pi}\left[\frac{1}{4} + \log_e \frac{d}{r}\right] \text{H/m}$$

$\therefore \quad L_A = 10^{-7}\left[\frac{1}{2}+2\log_e \frac{d}{r}\right]$ H/m ...(*i*)

Loop inductance $= 2\,L_A$ H/m $= 10^{-7}\left[1+4\log_e \frac{d}{r}\right]$ H/m

$\therefore \quad$ Loop inductance $= 10^{-7}\left[1+4\log_e \frac{d}{r}\right]$ H/m ...(*ii*)

Note that eq. (*ii*) is the inductance of the two-wire line and is sometimes called loop inductance. However, inductance given by eq. (*i*) is the inductance per conductor and is equal to half the loop inductance.

Expression in alternate form. The expression for the inductance of a conductor can be put in a concise form.

$$L_A = 10^{-7}\left[\frac{1}{2}+2\log_e \frac{d}{r}\right] \text{ H/m}$$

$$= 2\times 10^{-7}\left[\frac{1}{4}+\log_e \frac{d}{r}\right]$$

$$= 2\times 10^{-7}\left[\log_e e^{1/4}+\log_e \frac{d}{r}\right]$$

$$\therefore \quad L_A = 2\times 10^{-7}\log_e \frac{d}{r\,e^{-1/4}}$$

If we put $r\,e^{-1/4} = r'$, then,

$$L_A = 2\times 10^{-7}\log_e \frac{d}{r'} \text{ H/m} \quad ...(iii)$$

The radius r' is that of a fictitious conductor assumed to have no internal flux but with the same inductance as the actual conductor of radius r. The quantity $e^{-1/4} = 0{\cdot}7788$ so that

$$r' = r\,e^{-1/4} = 0{\cdot}7788\,r$$

The term r' ($= r\,e^{-1/4}$) is called *geometric mean radius* (GMR) of the wire. Note that eq. (*iii*) gives the same value of inductance L_A as eq. (*i*). The difference is that eq. (*iii*) omits the term to account for internal flux but compensates for it by using an adjusted value of the radius of the conductor.

Loop inductance $= 2\,L_A = 2\times 2\times 10^{-7}\log_e \frac{d}{r'}$ H/m

Note that $r' = 0{\cdot}7788\,r$ is applicable to only solid round conductor.

9.6 Inductance of a 3-Phase Overhead Line

Fig. 9·8 shows the three conductors *A*, *B* and *C* of a 3-phase line carrying currents I_A, I_B and I_C respectively. Let d_1, d_2 and d_3 be the spacings between the conductors as shown. Let us further assume that the loads are balanced *i.e.* $I_A + I_B + I_C = 0$. Consider the flux linkages with conductor *A*. There will be flux linkages with conductor *A* due to its own current and also due to the mutual inductance effects of I_B and I_C.

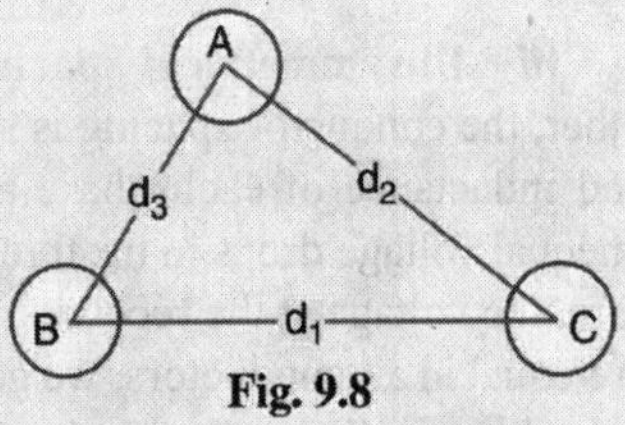

Fig. 9.8

Flux linkages with conductor *A* due to its own current

$$= \frac{\mu_0 I_A}{2\pi}\left(\frac{1}{4}+\int_r^{\infty}\frac{dx}{x}\right) \quad ...(i)$$

Flux linkages with conductor *A* due to current I_B

$$= \frac{\mu_0 I_B}{2\pi} \int_{d_3}^{\infty} \frac{dx}{x} \qquad ...(ii)$$

Flux linkages with conductor *A* due to current I_C

$$= \frac{\mu_0 I_C}{2\pi} \int_{d_2}^{\infty} \frac{dx}{x} \qquad ...(iii)$$

Total flux linkages with conductor *A* is

$$\psi_A = (i) + (ii) + (iii)$$

$$= \frac{\mu_0 I_A}{2\pi}\left(\frac{1}{4} + \int_r^{\infty} \frac{dx}{x}\right) + \frac{\mu_0 I_B}{2\pi}\int_{d_3}^{\infty} \frac{dx}{x} + \frac{\mu_0 I_C}{2\pi}\int_{d_2}^{\infty} \frac{dx}{x}$$

$$= \frac{\mu_0}{2\pi}\left[\left(\frac{1}{4} + \int_r^{\infty} \frac{dx}{x}\right) I_A + I_B \int_{d_3}^{\infty} \frac{dx}{x} + I_C \int_{d_2}^{\infty} \frac{dx}{x}\right]$$

$$= \frac{\mu_0}{2\pi}\left[\left(\frac{1}{4} - \log_e r\right) I_A - I_B \log_e d_3 - I_C \log_e d_2 + \log_e \infty \left(I_A + I_B + I_C\right)\right]$$

As $\qquad I_A + I_B + I_C = 0,$

$$\therefore \qquad \psi_A = \frac{\mu_0}{2\pi}\left[\left(\frac{1}{4} - \log_e r\right) I_A - I_B \log_e d_3 - I_C \log_e d_2\right]$$

(*i*) Symmetrical spacing. If the three conductors *A, B* and *C* are placed symmetrically at the corners of an equilateral triangle of side *d,* then, $d_1 = d_2 = d_3 = d$. Under such conditions, the flux linkages with conductor *A* become :

$$\psi_A = \frac{\mu_0}{2\pi}\left[\left(\frac{1}{4} - \log_e r\right) I_A - I_B \log_e d - I_C \log_e d\right]$$

$$= \frac{\mu_0}{2\pi}\left[\left(\frac{1}{4} - \log_e r\right) I_A - \left(I_B + I_C\right) \log_e d\right]$$

$$= \frac{\mu_0}{2\pi}\left[\left(\frac{1}{4} - \log_e r\right) I_A + I_A \log_e d\right] \qquad (\because I_B + I_C = -I_A)$$

$$= \frac{\mu_0 I_A}{2\pi}\left[\frac{1}{4} + \log_e \frac{d}{r}\right] \text{ werber-turns/m}$$

Inductance of conductor *A,* $\quad L_A = \frac{\psi_A}{I_A} \text{ H/m} = \frac{\mu_0}{2\pi}\left[\frac{1}{4} + \log_e \frac{d}{r}\right] \text{ H/m}$

$$= \frac{4\pi \times 10^{-7}}{2\pi}\left[\frac{1}{4} + \log_e \frac{d}{r}\right] \text{ H/m}$$

$$\therefore \qquad L_A = 10^{-7}\left[0 \cdot 5 + 2 \log_e \frac{d}{r}\right] \text{ H/m}$$

Derived in a similar way, the expressions for inductance are the same for conductors *B* and *C.*

(*ii*) Unsymmetrical spacing. When 3-phase line conductors are not equidistant from each other, the conductor spacing is said to be unsymmetrical. Under such conditions, the flux linkages and inductance of each phase are not the same. A different inductance in each phase results in unequal voltage drops in the three phases even if the currents in the conductors are balanced. Therefore, the voltage at the receiving end will not be the same for all phases. In order that voltage drops are equal in all conductors, we generally interchange the positions of the conductors at regular intervals along the line so that each conductor occupies the original position of every other conductor over an equal distance. Such an exchange of positions is known as *transposition.* Fig. 9.9 shows the transposed line. The phase conductors are designated as *A, B* and *C* and the positions occupied are numbered 1, 2 and 3. The effect of transposition is that each conductor has the same average inductance.

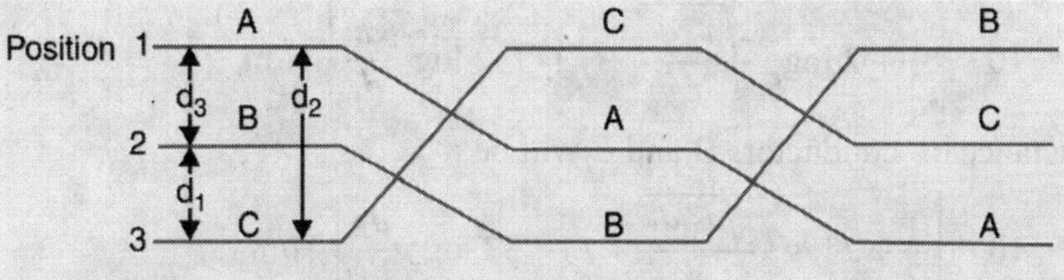

Fig. 9.9

Fig. 9.9 shows a 3-phase transposed line having unsymmetrical spacing. Let us assume that each of the three sections is 1 m in length. Let us further assume balanced conditions *i.e.,* $I_A + I_B + I_C = 0$. Let the line currents be :

$$I_A = I(1+j\,0)$$
$$I_B = I(-0{\cdot}5 - j\,0{\cdot}866)$$
$$I_C = I(-0{\cdot}5 + j\,0{\cdot}866)$$

As proved above, the total flux linkages per metre length of conductor *A* is

$$\psi_A = \frac{\mu_0}{2\pi}\left[\left(\frac{1}{4} - \log_e r\right)I_A - I_B \log_e d_3 - I_C \log_e d_2\right]$$

Putting the values of I_A, I_B and I_C, we get,

$$\psi_A = \frac{\mu_0}{2\pi}\left[\left(\frac{1}{4} - \log_e r\right)I - I(-0{\cdot}5 - j\,0{\cdot}866)\log_e d_3 - I(-0{\cdot}5 + j\,0{\cdot}866)\log_e d_2\right]$$

$$= \frac{\mu_0}{2\pi}\left[\frac{1}{4}I - I\log_e r + 0{\cdot}5\,I\log_e d_3 + j\,0{\cdot}866\log_e d_3 + 0{\cdot}5\,I\log_e d_2 - j\,0{\cdot}866\,I\log_e d_2\right]$$

$$= \frac{\mu_0}{2\pi}\left[\frac{1}{4}I - I\log_e r + 0{\cdot}5\,I(\log_e d_3 + \log_e d_2) + j\,0{\cdot}866\,I(\log_e d_3 - \log_e d_2)\right]$$

$$= \frac{\mu_0}{2\pi}\left[\frac{1}{4}I - I\log_e r + I^{*}\log_e \sqrt{d_2 d_3} + j\,0{\cdot}866\,I\log_e \frac{d_3}{d_2}\right]$$

$$= \frac{\mu_0}{2\pi}\left[\frac{1}{4}I + I\log_e \frac{\sqrt{d_2\,d_3}}{r} + j\,0{\cdot}866\,I\log_e \frac{d_3}{d_2}\right]$$

$$= \frac{\mu_0\,I}{2\pi}\left[\frac{1}{4} + \log_e \frac{\sqrt{d_2 d_3}}{r} + j\,0{\cdot}866\log_e \frac{d_3}{d_2}\right]$$

∴ Inductance of conductor *A* is

$$L_A = \frac{\psi_A}{I_A} = \frac{\psi_A}{I}$$

$$= \frac{\mu_0}{2\pi}\left[\frac{1}{4} + \log_e \frac{\sqrt{d_2\,d_3}}{r} + j\,0{\cdot}866\log_e \frac{d_3}{d_2}\right]$$

---

* $0{\cdot}5\,I(\log_e d_3 + \log_e d_2) = 0{\cdot}5\,I\log_e d_2 d_3 = I\log_e (d_2 d_3)^{0{\cdot}5} = I\log_e \sqrt{d_2 d_3}$

$$= \frac{4\pi \times 10^{-7}}{2\pi}\left[\frac{1}{4} + \log_e \frac{\sqrt{d_2\, d_3}}{r} + j\,0\cdot866 \log_e \frac{d_3}{d_2}\right] \text{ H/m}$$

$$= 10^{-7}\left[\frac{1}{2} + 2\log_e \frac{\sqrt{d_2\, d_3}}{r} + j\,1\cdot732 \log_e \frac{d_3}{d_2}\right] \text{ H/m}$$

Similarly inductance of conductors B and C will be :

$$L_B = 10^{-7}\left[\frac{1}{2} + 2\log_e \frac{\sqrt{d_3\, d_1}}{r} + j\,1\cdot732 \log_e \frac{d_1}{d_3}\right] \text{ H/m}$$

$$L_C = 10^{-7}\left[\frac{1}{2} + 2\log_e \frac{\sqrt{d_1\, d_2}}{r} + j\,1\cdot732 \log_e \frac{d_2}{d_1}\right] \text{ H/m}$$

Inducance of each line conductor

$$= \frac{1}{3}(L_A + L_B + L_C)$$

$$= {}^*\left[\frac{1}{2} + 2\log_e \frac{\sqrt[3]{d_1\, d_2\, d_3}}{r}\right] \times 10^{-7} \text{ H/m}$$

$$= \left[0\cdot5 + 2\log_e \frac{\sqrt[3]{d_1\, d_2\, d_3}}{r}\right] \times 10^{-7} \text{ H/m}$$

If we compare the formula of inductance of an unsymmetrically spaced transposed line with that of symmetrically spaced line, we find that inductance of each line conductor in the two cases will be equal if $d = \sqrt[3]{d_1\, d_2\, d_3}$. The distance d is known as *equivalent equilateral spacing* for unsymmetrically transposed line.

9.7 Concept of Self-GMD and Mutual-GMD

The use of *self geometrical mean distance* (abbreviated as self-GMD) and *mutual geometrical mean distance* (mutual-GMD) simplifies the inductance calculations, particularly relating to multiconductor arrangements. The symbols used for these are respectively D_s and D_m. We shall briefly discuss these terms.

(*i*) Self-GMD (D_s). In order to have concept of self-GMD (also sometimes called Geometrical mean radius ; GMR), consider the expression for inductance per conductor per metre already derived in Art. 9.5

$$\text{Inductance/conductor/m} = 2 \times 10^{-7}\left(\frac{1}{4} + \log_e \frac{d}{r}\right)$$

$$= 2 \times 10^{-7} \times \frac{1}{4} + 2 \times 10^{-7} \log_e \frac{d}{r} \quad ...(i)$$

In this expression, the term $2 \times 10^{-7} \times (1/4)$ is the inductance due to flux within the solid conductor. For many purposes, it is desirable to eliminate this term by the introduction of a concept called self-GMD or GMR. If we replace the original solid conductor by an equivalent hollow cylinder with extremely thin walls, the current is confined to the conductor surface and internal conductor flux linkage would be almost zero. Consequently, inductance due to internal flux would be zero and the term $2 \times 10^{-7} \times (1/4)$ shall be eliminated. The radius of this equivalent hollow cylinder must be sufficiently smaller than the physical radius of the conductor to allow room for enough additional flux

* On solving.

to compensate for the absence of internal flux linkage. It can be proved mathematically that for a **solid round conductor of radius** *r*, the self-GMD or GMR = 0·7788 *r*. Using self-GMD, the eq. (*i*) becomes :

$$\text{Inductance/conductor/m} = 2 \times 10^{-7} \log_e d/D_s{}^*$$

$$\text{where} \quad D_s = \text{GMR or self-GMD} = 0{\cdot}7788\, r$$

It may be noted that self-GMD of a conductor depends upon the size and shape of the conductor and is independent of the spacing between the conductors.

(*ii*) **Mutual-GMD.** The mutual-GMD is the geometrical mean of the distances form one conductor to the other and, therefore, must be between the largest and smallest such distance. In fact, mutual-GMD simply represents the equivalent geometrical spacing.

(*a*) The mutual-GMD between two conductors (assuming that spacing between conductors is large compared to the diameter of each conductor) is equal to the distance between their centres *i.e.*

$$D_m = \text{spacing between conductors} = d$$

(*b*) For a single circuit 3-φ line, the mutual-GMD is equal to the equivalent equilateral spacing *i.e.*, $(d_1\, d_2\, d_3)^{1/3}$.

$$D_m = (d_1\, d_2\, d_3)^{1/3}$$

(*c*) The principle of geometrical mean distances can be most profitably employed to 3-φ double circuit lines. Consider the conductor arrangement of the double circuit shown in Fig. 9·10. Suppose the radius of each conductor is *r*.

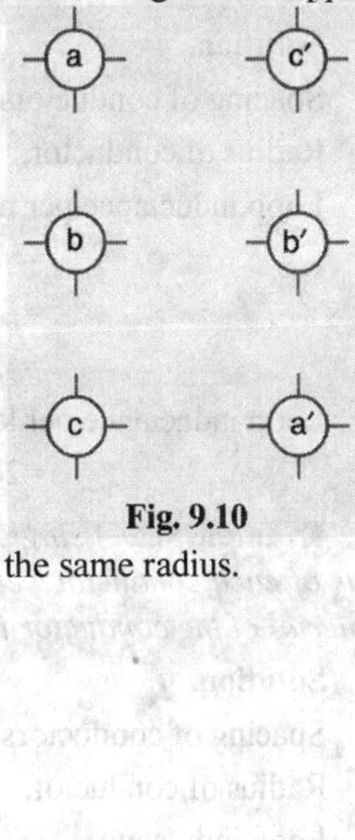

Fig. 9.10

Self-GMD of conductor = 0·7788 *r*

Self-GMD of combination *aa*′ is

$$D_{s1} = ({}^{**}D_{aa} \times D_{aa'} \times D_{a'a'} \times D_{a'a})^{1/4}$$

Self-GMD of combination *bb*′ is

$$D_{s2} = (D_{bb} \times D_{bb'} \times D_{b'b'} \times D_{b'b})^{1/4}$$

Self-GMD of combination *cc*′ is

$$D_{s3} = (D_{cc} \times D_{cc'} \times D_{c'c'} \times D_{c'c})^{1/4}$$

Equivalent self-GMD of one phase

$$D_s = (D_{s1} \times D_{s2} \times D_{s3})^{1/3}$$

The value of D_s is the same for all the phases as each conductor has the same radius.

Mutual-GMD between phases *A* and *B* is

$$D_{AB} = (D_{ab} \times D_{ab'} \times D_{a'b} \times D_{a'b'})^{1/4}$$

Mutual-GMD between phases *B* and *C* is

$$D_{BC} = (D_{bc} \times D_{bc'} \times D_{b'c} \times D_{b'c'})^{1/4}$$

Mutual-GMD between phases *C* and *A* is

$$D_{CA} = (D_{ca} \times D_{ca'} \times D_{c'a} \times D_{c'a'})^{1/4}$$

Equivalent mutual-GMD, $D_m = (D_{AB} \times D_{BC} \times D_{CA})^{1/3}$

It is worthwhile to note that mutual GMD depends only upon the spacing and is substantially independent of the exact size, shape and orientation of the conductor.

9.8 Inductance Formulas in Terms of GMD

The inductance formulas developed in the previous articles can be conveniently expressed in terms of geometrical mean distances.

* Basically, we have omitted the internal flux term while compensating for it by using an adjusted value for the radius of the conductor. Sometimes GMR is denoted by *r*′.

** D_{aa} or $D_{a'a'}$ means self-GMD of the conductor. $D_{aa'}$ means distance between *a* and *a*′.

(*i*) *Single phase line*

$$\text{Inductance/conductor/m} = 2 \times 10^{-7} \log_e \frac{D_m}{D_s}$$

where $D_s = 0{\cdot}7788\, r$ and D_m = Spacing between conductors = d

(*ii*) *Single circuit 3-φ line*

$$\text{Inductance/phase/m} = 2 \times 10^{-7} \log_e \frac{D_m}{D_s}$$

where $D_s = 0{\cdot}7788\, r$ and $D_m = (d_1\, d_2\, d_3)^{1/3}$

(*iii*) *Double circuit 3-φ line*

$$\text{Inductance/phase/m} = 2 \times 10^{-7} \log_e \frac{D_m}{D_s}$$

where $D_s = (D_{s1}\, D_{s2}\, D_{s3})^{1/3}$ and $D_m = (D_{AB} \times D_{BC} \times D_{CA})^{1/3}$

Example 9.1. *A single phase line has two parallel conductors 2 metres apart. The diameter of each conductor is 1·2 cm. Calculate the loop inductance per km of the line.*

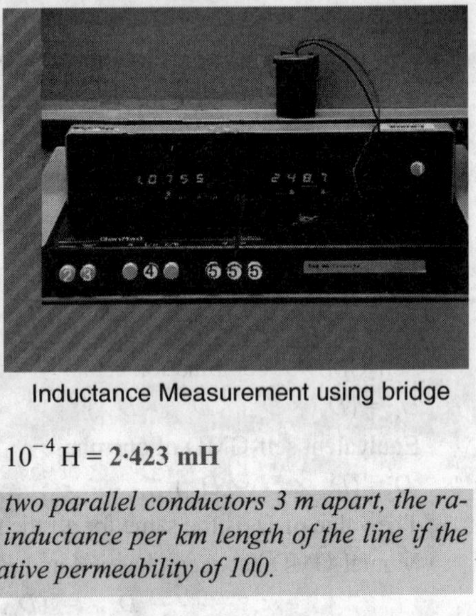

Inductance Measurement using bridge

Solution.

Spacing of conductors, $d = 2 \text{ m} = 200 \text{ cm}$

Radius of conductor, $r = 1{\cdot}2/2 = 0{\cdot}6 \text{ cm}$

Loop inductance per metre length of the line

$= 10^{-7}(1 + 4 \log_e d/r)$ H

$= 10^{-7}(1 + 4 \log_e 200/0{\cdot}6)$ H

$= 24{\cdot}23 \times 10^{-7}$ H

Loop inductance per km of the line

$= 24{\cdot}23 \times 10^{-7} \times 1000 = 24{\cdot}23 \times 10^{-4}$ H = **2·423 mH**

Example 9.2. *A single phase transmission line has two parallel conductors 3 m apart, the radius of each conductor being 1 cm. Calculate the loop inductance per km length of the line if the material of the conductor is (i) copper (ii) steel with relative permeability of 100.*

Solution.

Spacing of conductors, $d = 300$ cm

Radius of conductor, $r = 1$ cm

Loop inductance $= 10^{-7}(\mu_r + 4 \log_e d/r)$ H/m

(*i*) With copper conductors, $\mu_r = 1$

∴ Loop inductance/m $= 10^{-7}(1 + 4 \log_e d/r)$ H $= 10^{-7}(1 + 4 \log_e 300/1)$ H

$= 23{\cdot}8 \times 10^{-7}$ H

Loop inductance/km $= 23{\cdot}8 \times 10^{-7} \times 1000 = 2{\cdot}38 \times 10^{-3}$ H = **2·38 mH**

(*ii*) With steel conductors, $\mu_r = 100$

∴ Loop inductance/m $= 10^{-7}(100 + 4 \log_e 300/1)$ H $= 122{\cdot}8 \times 10^{-7}$ H

Loop inductance/km $= 122{\cdot}8 \times 10^{-7} \times 1000 = 12{\cdot}28 \times 10^{-3}$ H = **12·28 mH**

Example 9.3. *Find the inductance per km of a 3-phase transmission line using 1·24 cm diameter conductors when these are placed at the corners of an equilateral triangle of each side 2 m.*

Solution. Fig. 9·11 shows the three conductors of the three phase line placed at the corners of an equilateral triangle of each side 2 m. Here conductor spacing d = 2 m and conductor radius $r = 1{\cdot}24/2 = 0{\cdot}62$ cm.

$$\text{Inductance/phase/m} = 10^{-7}\,(0{\cdot}5 + 2\log_e d/r)\text{ H}$$
$$= 10^{-7}\,(0{\cdot}5 + 2\log_e 200/0{\cdot}62)\text{ H}$$
$$= 12 \times 10^{-7}\text{ H}$$
$$\text{Inductance/phase/km} = 12 \times 10^{-7} \times 1000$$
$$= 1{\cdot}2 \times 10^{-3}\text{ H} = \mathbf{1{\cdot}2\ mH}$$

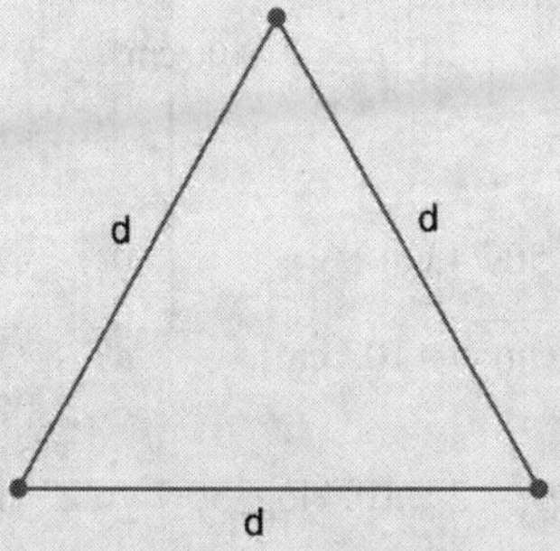

Fig. 9.11

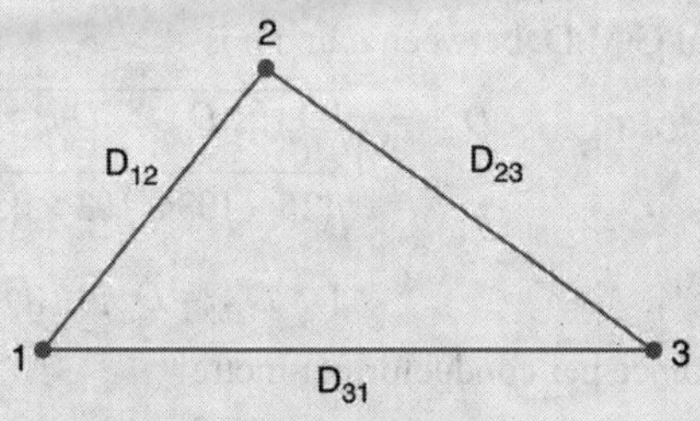

Fig. 9.12

Example 9.4. *The three conductors of a 3-phase line are arranged at the corners of a triangle of sides 2 m, 2·5 m and 4·5 m. Calculate the inductance per km of the line when the conductors are regularly transposed. The diameter of each conductor is 1·24 cm.*

Solution. Fig. 9·12 shows three conductors of a 3-phase line placed at the corners of a triangle of sides D_{12} = 2 m, D_{23} = 2·5 m and D_{31} = 4·5 m. The conductor radius $r = 1{\cdot}24/2 = 0{\cdot}62$ cm.

$$\text{Equivalent equilateral spacing, } D_{eq} = \sqrt[3]{D_{12} \times D_{23} \times D_{31}} = \sqrt[3]{2 \times 2{\cdot}5 \times 4{\cdot}5} = 2{\cdot}82\text{ m} = 282\text{ cm}$$
$$\text{Inductance/phase/m} = 10^{-7}(0{\cdot}5 + 2\log_e D_{eq}/r)\text{ H} = 10^{-7}(0{\cdot}5 + 2\log_e 282/0{\cdot}62)\text{ H}$$
$$= 12{\cdot}74 \times 10^{-7}\text{ H}$$
$$\text{Inductance/phase/km} = 12{\cdot}74 \times 10^{-7} \times 1000 = 1{\cdot}274 \times 10^{-3}\text{ H} = \mathbf{1{\cdot}274\ mH}$$

Example 9.5. *Calculate the inductance of each conductor in a 3-phase, 3-wire system when the conductors are arranged in a horizontal plane with spacing such that D_{31} = 4 m ; $D_{12} = D_{23}$ = 2m. The conductors are transposed and have a diameter of 2·5 cm.*

Solution. Fig. 9.13. shows the arrangement of the conductors of the 3phase line. The conductor radius $r = 2{\cdot}5/2 = 1{\cdot}25$ cm.

$$\text{Equivalent equilateral spacing, } D_{eq} = \sqrt[3]{D_{12} \times D_{23} \times D_{31}} = \sqrt[3]{2 \times 2 \times 4} = 2{\cdot}52\text{ m} = 252\text{ cm}$$
$$\text{Inductance/phase/m} = 10^{-7}\,(0{\cdot}5 + 2\log_e D_{eq}/r)\text{ H}$$
$$= 10^{-7}\,(0{\cdot}5 + 2\log_e 252/1{\cdot}25)\text{ H}$$
$$= 11{\cdot}1 \times 10^{-7}\text{H}$$
$$\text{Inductance/phase/km} = 11{\cdot}1 \times 10^{-7} \times 1000$$
$$= 1{\cdot}11 \times 10^{-3}\text{ H} = \mathbf{1{\cdot}11\ mH}$$

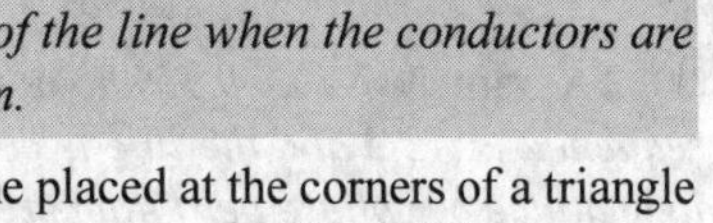

Fig. 9.13

Example 9.6. *Two conductors of a single phase line, each of 1 cm diameter, are arranged in a vertical plane with one conductor mounted 1 m above the other. A second identical line is mounted at the same height as the first and spaced horizontally 0·25 m apart from it. The two upper and the two lower conductors are connected in parallel. Determine the inductance per km of the resulting double circuit line.*

Solution. Fig. 9.14. shows the arrangement of double circuit single phase line. Conductors a, a' form one connection and conductors b, b' form the return connnection. The conductor radius, $r = 1/2 = 0{\cdot}5$ cm.

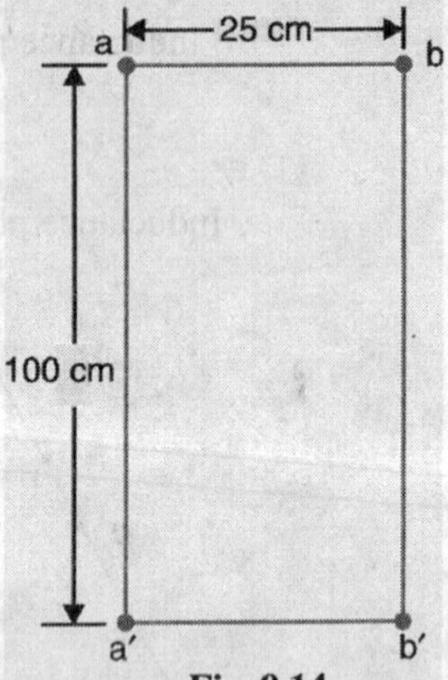

Fig. 9.14

G.M.R. of conductor $= 0{\cdot}7788\, r = 0{\cdot}7788 \times 0{\cdot}5 = 0{\cdot}389$ cm

Self G.M.D. of aa' combination is

$$D_s = \sqrt[4]{D_{aa} \times D_{aa'} \times D_{a'a'} \times D_{a'a}}$$

$$= \sqrt[4]{(0\cdot389 \times 100)^2} = 6{\cdot}23 \text{ cm}$$

Mutual G.M.D. between a and b is

$$D_m = \sqrt[4]{D_{ab} \times D_{ab'} \times D_{a'b} \times D_{a'b'}}$$

$$= \sqrt[4]{(25 \times 103 \times 103 \times 25)} = 50{\cdot}74 \text{ cm}$$

$$[\because D_{ab'} = D_{a'b} = \sqrt{25^2 + 100^2} = 103 \text{ cm}]$$

Inductance per conductor per metre

$$= 2 \times 10^{-7} \log_e D_m/D_s = 2 \times 10^{-7} \log_e 50{\cdot}74/6{\cdot}23 \text{ H}$$

$$= 0{\cdot}42 \times 10^{-6} \text{ H}$$

$\therefore$ Loop inductance per km of the line

$$= 2 \times 0{\cdot}42 \times 10^{-6} \times 1000 \text{ H} = \mathbf{0{\cdot}84 \text{ mH}}$$

Example 9.7. *Fig. 9.15 shows the spacings of a double circuit 3-phase overhead line. The phase sequence is ABC and the line is completely transposed. The conductor radius in 1·3 cm. Find the inductance per phase per kilometre.*

Solution.

G.M.R. of conductor $= 1{\cdot}3 \times 0{\cdot}7788 = 1{\cdot}01$ cm

Distance a to b' $= \sqrt{6^2 + 3^2} = 6{\cdot}7$ m

Distance a to a' $= \sqrt{6^2 + 6^2} = 8{\cdot}48$ m

Equivalent self G.M.D. of one phase is

$$D_s = \sqrt[3]{D_{s1} \times D_{s2} \times D_{s3}}$$

where D_{s1}, D_{s2} and D_{s3} represent the self-G.M.D. in positions 1, 2 and 3 respectively. Also D_s is the same for all the phases.

$$\text{Now } D_{s1} = \sqrt[4]{D_{aa} \times D_{aa'} \times D_{a'a'} \times D_{a'a}}$$

$$= \sqrt[4]{(1\cdot01 \times 10^{-2}) \times (8\cdot48) \times (1\cdot01 \times 10^{-2}) \times (8\cdot48)}$$

$$= 0.292 \text{ m} = D_{s3}$$

$$D_{s2} = \sqrt[4]{D_{bb} \times D_{bb'} \times D_{b'b'} \times D_{b'b}}$$

$$= \sqrt[4]{\left(1\cdot01 \times 10^{-2}\right) \times (6) \times \left(1\cdot01 \times 10^{-2}\right) \times (6)} = 0{\cdot}246 \text{ m}$$

$$D_s = \sqrt[3]{0\cdot292 \times 0\cdot246 \times 0\cdot292} = 0{\cdot}275 \text{ m}$$

Equivalent mutual G.M.D., $D_m = \sqrt[3]{D_{AB} \times D_{BC} \times D_{CA}}$

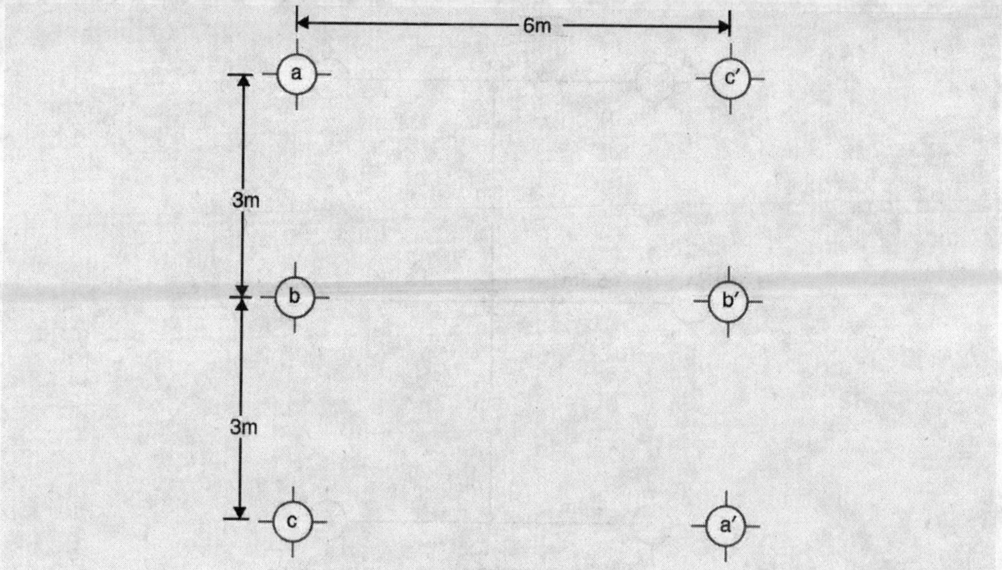

Fig. 9.15

where D_{AB}, D_{BC} and D_{CA} represent the mutual G.M.D. between phases *A* and *B*, *B* and *C* and *C* and *A* respectively.

Now
$$D_{AB} = \sqrt[4]{D_{ab} \times D_{ab'} \times D_{a'b} \times D_{a'b'}} = \sqrt[4]{3 \times 6{\cdot}7 \times 6{\cdot}7 \times 3}$$
$$= 4{\cdot}48 \text{ m} = D_{BC}$$
$$D_{CA} = \sqrt[4]{D_{ca} \times D_{ca'} \times D_{c'a} \times D_{c'a'}} = \sqrt[4]{6 \times 6 \times 6 \times 6} = 6 \text{ m}$$

$\therefore$
$$D_m = \sqrt[3]{4{\cdot}48 \times 4{\cdot}48 \times 6} = 4{\cdot}94 \text{ m}$$

$\therefore$ Inductance per phase per metre length
$$= 10^{-7} \times 2 \log_e D_m/D_s = 10^{-7} \times 2 \log_e 4{\cdot}94/0{\cdot}275$$
$$= 5{\cdot}7 \times 10^{-7} \text{ H}$$

Inductance /phase/km $= 5{\cdot}7 \times 10^{-7} \times 1000 = 0{\cdot}57 \times 10^{-3}$ H $=$ **0·57 mH**

Example 9.8. *Find the inductance per phase per km of double circuit 3-phase line shown in Fig. 9·16. The conductors are transposed and are of radius 0·75 cm each. The phase sequence is ABC.*

Solution.

G.M.R. of conductor $= 0{\cdot}75 \times 0{\cdot}7788 = 0{\cdot}584$ cm

Distance *a* to *b* $= \sqrt{3^2 + (0{\cdot}75)^2} = 3{\cdot}1$ m

Distance *a* to *b′* $= \sqrt{3^2 + (4{\cdot}75)^2} = 5{\cdot}62$ m

Distance *a* to *a′* $= \sqrt{6^2 + 4^2} = 7{\cdot}21$ m

Equivalent self G.M.D. of one phase is
$$D_s = \sqrt[3]{D_{s1} \times D_{s2} \times D_{s3}}$$

where
$$D_{s1} = \sqrt[4]{D_{aa} \times D_{aa'} \times D_{a'a'} \times D_{a'a}}$$

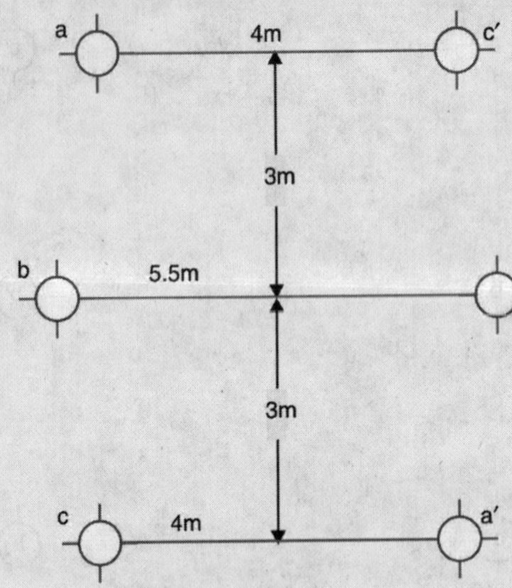

Fig. 9.16

$$= \sqrt[4]{(0 \cdot 584 \times 10^{-2}) \times (7 \cdot 21) \times (0 \cdot 584 \times 10^{-2}) \times (7 \cdot 21)}$$

$$= 0{\cdot}205 \text{ m} = D_{s3}$$

$$D_{s2} = \sqrt[4]{(D_{bb} \times D_{bb'} \times D_{b'b'} \times D_{b'b})}$$

$$= \sqrt[4]{(0 \cdot 584 \times 10^{-2}) \times (5 \cdot 5) \times (0 \cdot 584 \times 10^{-2}) \times 5 \cdot 5} = 0 \cdot 18 \text{ m}$$

$$\therefore \quad D_s = \sqrt[3]{0 \cdot 205 \times 0 \cdot 18 \times 0 \cdot 205} = 0{\cdot}195 \text{ m}$$

Equivalent mutual G.M.D. is

$$D_m = \sqrt[3]{D_{AB} \times D_{BC} \times D_{CA}}$$

where $D_{AB} = \sqrt[4]{D_{ab} \times D_{ab'} \times D_{a'b} \times D_{a'b'}} = \sqrt[4]{3 \cdot 1 \times 5 \cdot 62 \times 5 \cdot 62 \times 3 \cdot 1}$

$$= 4{\cdot}17 \text{ m} = D_{BC}$$

$$D_{CA} = \sqrt[4]{D_{ca} \times D_{ca'} \times D_{c'a} \times D_{c'a'}}$$

$$= \sqrt[4]{6 \times 4 \times 4 \times 6} = 4{\cdot}9 \text{ m}$$

$$\therefore \quad D_m = \sqrt[3]{4 \cdot 17 \times 4 \cdot 17 \times 4 \cdot 9} = 4{\cdot}4 \text{ m}$$

$\therefore$ Inductance/phase/m $= 10^{-7} \times 2 \log_e D_m/D_s = 10^{-7} \times 2 \log_e 4{\cdot}4/0{\cdot}195$ H

$= 6{\cdot}23 \times 10^{-7}$ H $= 0{\cdot}623 \times 10^{-3}$ mH

Inductance/phase/km $= 0{\cdot}623 \times 10^{-3} \times 1000 =$ **0·623 mH**

Example 9.9. *Calculate the inductance per phase per metre for a three-phase double-circuit line whose phase conductors have a radius of 5·3 cm with the horizontal conductor arrangement as shown in Fig. 9.17.*

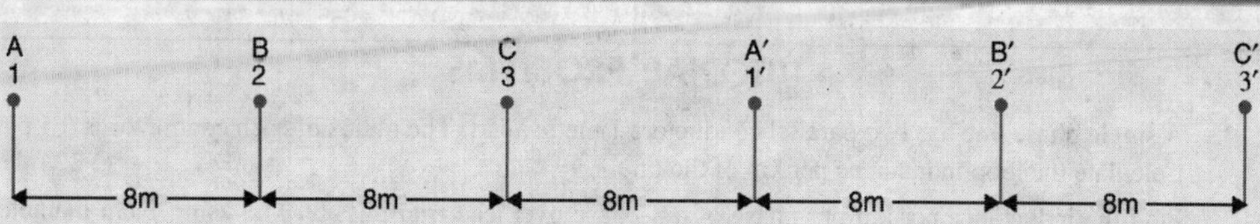

Fig. 9.17

Solution.

G.M.R. of conductor $= 0{\cdot}7788\, r = 0{\cdot}7788 \times 5{\cdot}3 \times 10^{-2} = 0{\cdot}0413$ m

Equivalent self-G.M.D. of one phase is

$$D_s = (D_{s1} \times D_{s2} \times D_{s3})^{1/3}$$

where
$$D_{s1} = (D_{AA} \times D_{AA'} \times D_{A'A'} \times D_{A'A})^{1/4} = (0{\cdot}0413 \times 24 \times 0{\cdot}0413 \times 24)^{1/4} = 0{\cdot}995 \text{ m}$$
$$D_{s2} = (D_{BB} \times D_{BB'} \times D_{B'B'} \times D_{B'B})^{1/4} = (0{\cdot}0413 \times 24 \times 0{\cdot}0413 \times 24)^{1/4} = 0{\cdot}995 \text{ m}$$

Similarly $D_{s3} = 0{\cdot}995$ m

$$\therefore \quad D_s = \sqrt[3]{0{\cdot}995 \times 0{\cdot}995 \times 0{\cdot}995} = 0{\cdot}995 \text{ m}$$

Equivalent mutual G.M.D. is

$$D_m = (D_{AB} \times D_{BC} \times D_{CA})^{1/3}$$

where
$$D_{AB} = (D_{AB} \times D_{AB'} \times D_{A'B} \times D_{A'B'})^{1/4} = (8 \times 32 \times 16 \times 8)^{1/4}$$
$$= 13{\cdot}45 = D_{BC}$$
$$D_{CA} = (D_{CA} \times D_{CA'} \times D_{C'A} \times D_{C'A'})^{1/4} = (16 \times 8 \times 40 \times 16)^{1/4}$$
$$= 16{\cdot}917 \text{ m}$$

$$\therefore \quad D_m = (13{\cdot}45 \times 13{\cdot}45 \times 16{\cdot}917)^{1/3} = 14{\cdot}518 \text{ m}$$

Inductance/phase/m $= 10^{-7} \times 2 \log_e D_m/D_s$ H/m

$$= 10^{-7} \times 2 \log_e \frac{14{\cdot}518}{0{\cdot}995} \text{ H/m}$$

$$= \mathbf{5{\cdot}36 \times 10^{-7} \text{ H/m}}$$

Example 9.10. *In a single phase line (See. Fig. 9.18), conductors a and a′ in parallel form one circuit while conductors b and b′ in parallel form the return path. Calculate the total inductance of the line per km assuming that current is equally shared by the two parallel conductors. Conductor diameter in 2·0 cm.*

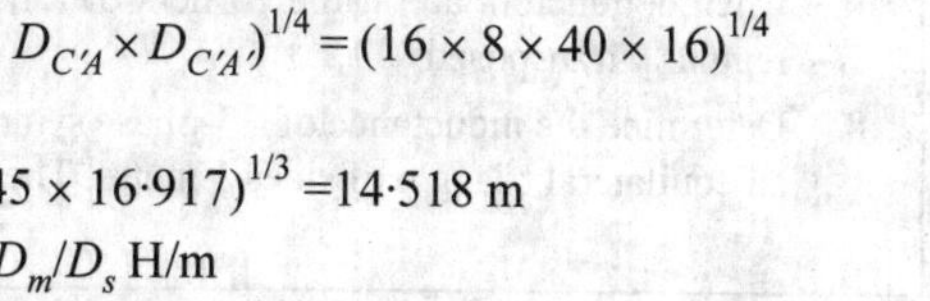

Fig. 9.18

Solution.

Loop inductance/km, $L = 4 \times 10^{-4} \log_e \dfrac{D_m}{D_s}$ H/km

Mutual G.M.D., $D_m = \sqrt[4]{D_{ab} \times D_{ab'} \times D_{a'b} \times D_{a'b'}}$

$$= \sqrt[4]{120 \times 140 \times 100 \times 120} = 119 \text{ cm}$$

Self G.M.D., $D_s = \sqrt[4]{D_{aa} \times D_{aa'} \times D_{a'a'} \times D_{a'a}}$

Here $D_{aa} = D_{a'a'} = 0{\cdot}7788$ cm; $D_{aa'} = D_{a'a} = 20$ cm

$$\therefore \quad D_s = \sqrt[4]{0{\cdot}7788 \times 0{\cdot}7788 \times 20 \times 20} = 3{\cdot}94 \text{ cm}$$

$$\therefore \quad L = 4 \times 10^{-4} \log_e \frac{119}{3{\cdot}94} = 1{\cdot}36 \times 10^{-3} \text{ H/km} = \mathbf{1{\cdot}36 \text{ mH/km}}$$

TUTORIAL PROBLEMS

1. A single phase line has two parallel conductors 1 metre apart. The radius of each conductor is 0·5 cm. Calculate the loop inductance per km of the line. **[2·22 mH]**
2. Find the inductance per km per phase of a 3-phase overhead transmission line using 2 cm diameter conductor when these are placed at the corners of an equilateral triangle of side 4 metres. **[1·25 mH]**
3. Find the loop inductance per km of a single phase overhead transmission line when conductors have relative permeability of (*i*) 1(*ii*) 100. Each conductor has a diameter of 1 cm and they are spaced 5 m apart. **[(*i*)1·02 mH (*ii*) 10·9 mH]**

   Hint. For a conductor of relative permeability μ_r (= 100 in the second case), loop inductance
   $$= (\mu_r + 4 \log_e d/r) \times 10^{-7} \text{ H/m}$$
4. A 20 km single phase line has two parallel conductors separated by 1·5 metres. The diameter of each conductor is 0·823 cm. If the conductor has a resistance of 0·311 Ω/km, find the loop impedance of this line at 50 Hz. **[19·86 Ω]**
5. The three conductors of a 3-phase line are arranged at the corners of a triangle of sides 4, 5 and 6 metres. Calculate inductance per km of the each conductor when conductors are regularly transposed. The diameter of each line conductor is 2 cm. **[1·285 mH]**
6. The three conductors of 3-phase overhead line are arranged in a horizontal plane with a spacing of 4 m between adjacent conductors. The diameter of each conductor is 2 cm. Determine the inductance per km per phase of the line assuming that the lines are transposed. **[1·3 mH]**
7. Determine the inductance per km of a 3-phase transmission line using 20 mm diameter conductors when conductors are at the corners of a triangle with spacing of 4,5 and 6 metres. Conductors are regularly transposed. **[1·29 mH/km/phase]**
8. Determine the inductance of a 3-phase symmetrical line whose conductors are placed at the corners of an equilateral triangle of sides 1 metre. The diameter of each conductor is 20 mm. **[0·971 mH/phase/km]**

9.9 Electric Potential

The electric potential at a point due to a charge is the work done in bringing a unit positive charge from infinity to that point. The concept of electric potential is extremely important for the determination of capacitance in a circuit since the latter is defined as the charge per unit potential. We shall now discuss in detail the electric potential due to some important conductor arrangements.

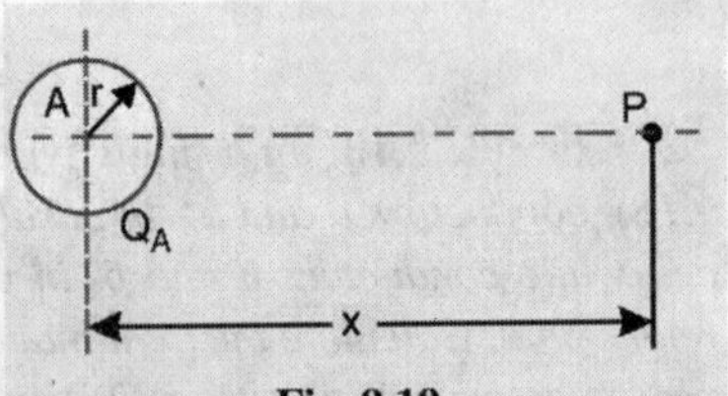

Fig. 9.19

(*i*) Potential at a charged single conductor. Consider a long straight cylindrical conductor *A* of radius *r* metres. Let the conductor operate at such a potential (V_A) that charge Q_A coulombs per metre exists on the conductor. It is desired to find the expression for V_A. The electric intensity *E* at a distance *x* from the centre of the conductor in air is given by:

$$E = \frac{Q_A}{2\pi x \varepsilon_0} \text{ volts/m}$$

where Q_A = charge per metre length

ε_0 = permittivity of free space

As *x* approaches infinity, the value of *E* approaches zero. Therefore, the potential difference between conductor *A* and infinity distant * neutral plane is given by :

* A plane where *E* and hence potential is zero.

$$\dagger V_A = \int_r^{\infty} \frac{Q_A}{2\pi x \varepsilon_0} dx = \frac{Q_A}{2\pi\varepsilon_0} \int_r^{\infty} \frac{dx}{x}$$

(ii) Potential at a conductor in a group of charged conductors. Consider a group of long straight conductors *A, B, C* etc. operating at potentials such that charges Q_A, Q_B, Q_C etc.coulomb per metre length exist on the respective conductors (see Fig. 9.20). Let us find the potential at *A* (*i.e.* V_A) in this arrangement. Potential at *A* due to its own charge (*i.e.* Q_A)

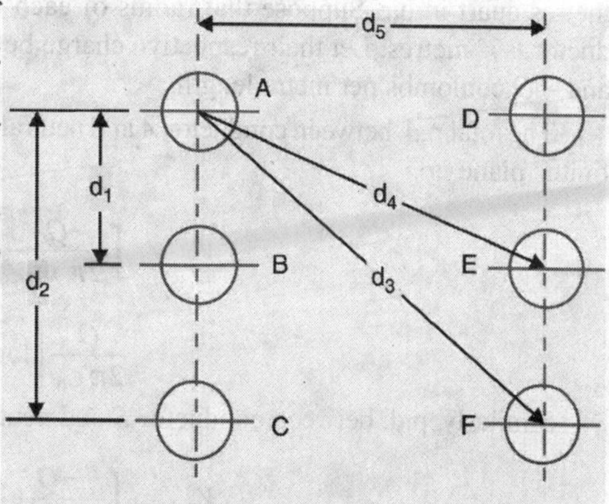

Fig. 9.20

$$= \int_r^{\infty} \frac{Q_A}{2\pi x \varepsilon_0} dx \quad ...(i)$$

Potential at conductor *A* due to charge Q_B

$$= \int_{d_1}^{\infty} \frac{* Q_B}{2\pi x \varepsilon_0} dx \quad ...(ii)$$

Potential at conductor *A* due to charge Q_C

$$= \int_{d_2}^{\infty} \frac{Q_C}{2\pi x \varepsilon_0} dx \quad ...(iii)$$

Overall potential difference between conductor *A* and infinite neutral plane is

$$V_A = (i) + (ii) + (iii) + ...$$

$$= \int_r^{\infty} \frac{Q_A}{2\pi x \varepsilon_0} dx + \int_{d_1}^{\infty} \frac{Q_B}{2\pi x \varepsilon_0} dx + \int_{d_2}^{\infty} \frac{Q_C}{2\pi x \varepsilon_0} dx + ...$$

$$= \frac{1}{2\pi \varepsilon_0} \Big[Q_A (\log_e \infty - \log_e r) + Q_B (\log_e \infty - \log_e d_1)$$

$$+ Q_C (\log_e \infty - \log_e d_2) + ... \Big]$$

$$= \frac{1}{2\pi \varepsilon_0} \Big[Q_A {}^{\dagger\dagger} \log_e \frac{1}{r} + Q_B \log_e \frac{1}{d_1} + Q_C \log_e \frac{1}{d_2}$$

$$+ \log_e \infty \,(Q_A + Q_B + Q_C) + ... \Big]$$

Assuming balanced condtions *i.e.*, $Q_A + Q_B + Q_C = 0$, we have,

$$V_A = \frac{1}{2\pi \varepsilon_0} \left[Q_A \log_e \frac{1}{r} + Q_B \log_e \frac{1}{d_1} + Q_C \log_e \frac{1}{d_2} + ... \right]$$

---

† Note the expression. Work is done in bringing a unit positive charge against *E* from infinity to conductor surface.

* Conductor *B* is d_1 metres away from conductor *A*. Therefore, the work done in bringing a unit positive charge (due to charge Q_B) from infinity to conductor *A* is

$$= \int_{d_1}^{\infty} \frac{Q_B}{2\pi x \varepsilon_0} dx$$

†† $-\log_e r = \log_e (r)^{-1} = \log_e \frac{1}{r}$

9.10 Capacitance of a Single Phase Two-wire Line

Consider a single phase overhead transmission line consisting of two parallel conductors A and B spaced d metres apart in air. Suppose that radius of each conductor is r metres. Let their respective charge be $+Q$ and $-Q$ coulombs per metre length.

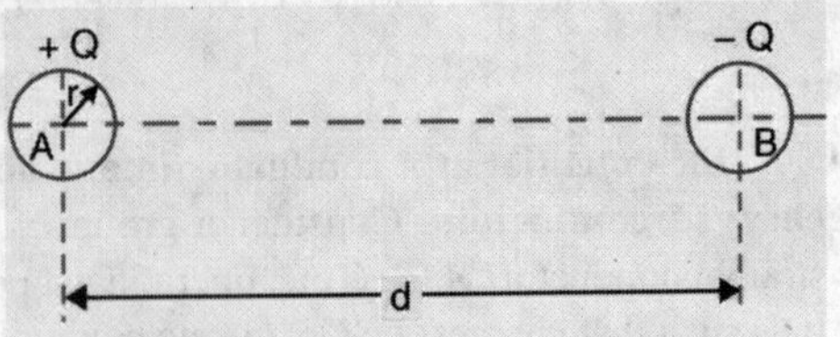

Fig. 9.21

The total p.d. between conductor A and neutral "infinite" plane is

$$V_A = \int_r^{\infty} \frac{Q}{2\pi x \varepsilon_0} dx + \int_d^{\infty} \frac{-Q}{2\pi x \varepsilon_0} dx$$

$$= \frac{Q}{2\pi \varepsilon_0}\left[\log_e \frac{\infty}{r} - \log_e \frac{\infty}{d}\right] \text{volts} = \frac{Q}{2\pi \varepsilon_0} \log_e \frac{d}{r} \text{ volts}$$

Similarly, p.d. between conductor B and neutral "infinite" plane is

$$V_B = \int_r^{\infty} \frac{-Q}{2\pi x \varepsilon_0} dx + \int_d^{\infty} \frac{Q}{2\pi x \varepsilon_0} dx$$

$$= \frac{-Q}{2\pi \varepsilon_0}\left[\log_e \frac{\infty}{r} - \log_e \frac{\infty}{d}\right] = \frac{-Q}{2\pi \varepsilon_0} \log_e \frac{d}{r} \text{ volts}$$

Both these potentials are *w.r.t.* the same neutral plane. Since the unlike charges attract each other, the potential difference between the conductors is

$$V_{AB} = 2V_A = \frac{2Q}{2\pi \varepsilon_0} \log_e \frac{d}{r} \text{ volts}$$

$\therefore$ Capacitance,

$$C_{AB} = Q/V_{AB} = \frac{Q}{\dfrac{2Q}{2\pi \varepsilon_0} \log_e \dfrac{d}{r}} \text{ F/m}$$

$$\therefore \quad C_{AB} = \frac{\pi \varepsilon_0}{\log_e \dfrac{d}{r}} \text{ F/m} \quad \text{...}(i)$$

Capacitance to neutral. Equation (i) gives the capacitance between the conductors of a two-wire line [See Fig. 9.22]. Often it is desired to know the capacitance between one of the conductors and a neutral point between them. Since potential of the mid-point between the conductors is zero, the potential difference between each conductor and the ground or neutral is half the potential difference between the conductors. Thus the *capacitance to ground* or capacitance to neutral for the two-wire line is *twice* the line-to-line capacitance (capacitance between conductors as shown in Fig 9.23).

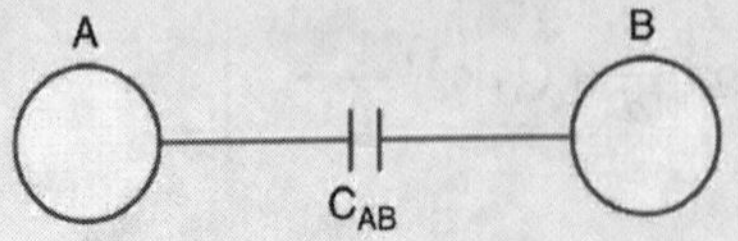

Fig. 9.22

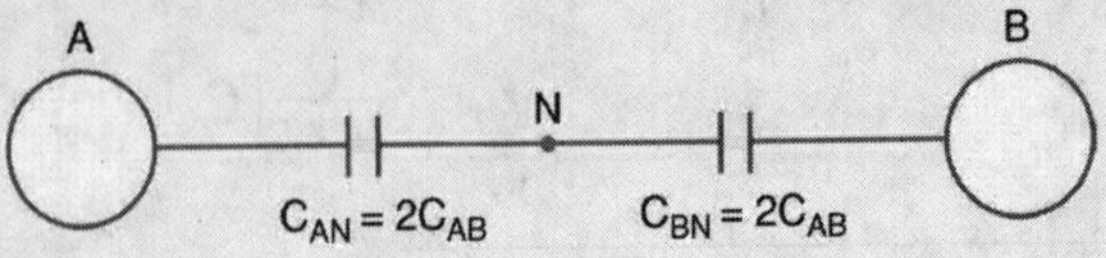

Fig. 9.23

$\therefore$ Capacitance to neutral, $C_N = C_{AN} = C_{BN} = 2C_{AB}$

$$\therefore \quad C_N = \frac{2\pi \varepsilon_0}{\log_e \dfrac{d}{r}} \text{ F/m} \quad \text{...}(ii)$$

The reader may compare eq. (ii) to the one for inductance. One difference between the equations for capacitance and inductance should be noted carefully. The radius in the equation for capacitance is the actual outside radius of the conductor and not the GMR of the conductor as in the inductance formula. Note that eq. (*ii*) applies only to a solid round conductor.

9.11 Capacitance of a 3-Phase Overhead Line

In a 3-phase transmission line, the capacitance of each conductor is considered instead of capacitance from conductor to conductor. Here, again two cases arise *viz.*, symmetrical spacing and unsymmetrical spacing.

(*i*) **Symmetrical Spacing.** Fig. 9.24 shows the three conductors A, B and C of the 3-phase overhead transmission line having charges Q_A, Q_B and Q_C per metre length respectively. Let the conductors be equidistant (d metres) from each other. We shall find the capacitance from line conductor to neutral in this symmetrically spaced line. Referring to Fig. 9.24, overall potential difference between conductor A and infinite neutral plane is given by (Refer to Art. 9.9);

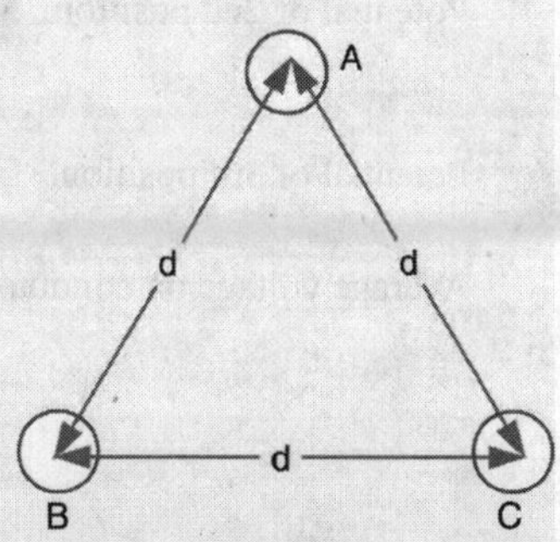

Fig. 9.24

$$V_A = \int_r^{\infty} \frac{Q_A}{2\,\pi x \varepsilon_0}\,dx + \int_d^{\infty} \frac{Q_B}{2\,\pi x \varepsilon_0}\,dx + \int_d^{\infty} \frac{Q_C}{2\,\pi x \varepsilon_0}\,dx$$

$$= \frac{1}{2\pi\,\varepsilon_0}\left[Q_A \log_e \frac{1}{r} + Q_B \log_e \frac{1}{d} + Q_C \log_e \frac{1}{d}\right]$$

$$= \frac{1}{2\pi\,\varepsilon_0}\left[Q_A \log_e \frac{1}{r} + (Q_B + Q_C) \log_e \frac{1}{d}\right]$$

Assuming balanced supply, we have, $Q_A + Q_B + Q_C = 0$

$$\therefore \quad Q_B + Q_C = -\,Q_A$$

$$\therefore \quad V_A = \frac{1}{2\pi\,\varepsilon_0}\left[Q_A \log_e \frac{1}{r} - Q_A \log_e \frac{1}{d}\right] = \frac{Q_A}{2\pi\varepsilon_0} \log_e \frac{d}{r} \text{ volts}$$

$\therefore$ Capacitance of conductor A w.r.t neutral,

$$C_A = \frac{Q_A}{V_A} = \frac{Q_A}{\dfrac{Q_A}{2\pi\,\varepsilon_0} \log_e \dfrac{d}{r}} \text{ F/m} = \frac{2\pi\,\varepsilon_0}{\log_e \dfrac{d}{r}} \text{ F/m}$$

$$\therefore \quad C_A = \frac{2\pi\,\varepsilon_0}{\log_e \dfrac{d}{r}} \text{ F/m}$$

Note that this equation is identical to capacitance to neutral for two-wire line. Derived in a similar manner, the expressions for capacitance are the same for conductors B and C.

(*ii*) **Unsymmetrical spacing.** Fig. 9.25 shows a 3-phase transposed line having unsymmetrical spacing. Let us assume balanced conditions *i.e.* $Q_A + Q_B + Q_C = 0$.

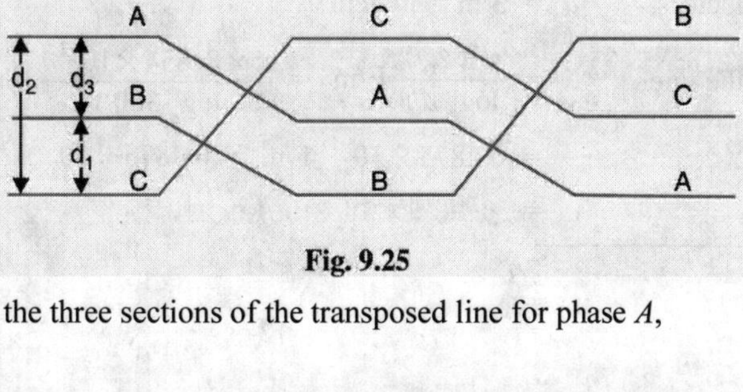

Fig. 9.25

Considering all the three sections of the transposed line for phase A,

Potential of 1st position, $V_1 = \frac{1}{2\pi\varepsilon_0}\left(Q_A \log_e \frac{1}{r} + Q_B \log_e \frac{1}{d_3} + Q_C \log_e \frac{1}{d_2}\right)$

Potential of 2nd position, $V_2 = \frac{1}{2\pi\varepsilon_0}\left(Q_A \log_e \frac{1}{r} + Q_B \log_e \frac{1}{d_1} + Q_C \log_e \frac{1}{d_3}\right)$

Potential of 3rd position, $V_3 = \frac{1}{2\pi\varepsilon_0}\left(Q_A \log_e \frac{1}{r} + Q_B \log_e \frac{1}{d_2} + Q_C \log_e \frac{1}{d_1}\right)$

Average voltage on condutor A is

$$V_A = \frac{1}{3}(V_1 + V_2 + V_3)$$

$$= \frac{1}{3 \times 2\pi\varepsilon_0}{}^{*}\left[Q_A \log_e \frac{1}{r^3} + (Q_B + Q_C)\log_e \frac{1}{d_1 d_2 d_3}\right]$$

As $Q_A + Q_B + Q_C = 0$, therefore, $Q_B + Q_C = - Q_A$

$$\therefore \quad V_A = \frac{1}{6\pi\varepsilon_0}\left[Q_A \log_e \frac{1}{r^3} - Q_A \log_e \frac{1}{d_1 d_2 d_3}\right]$$

$$= \frac{Q_A}{6\pi\varepsilon_0} \log_e \frac{d_1 d_2 d_3}{r^3}$$

$$= \frac{1}{3} \times \frac{Q_A}{2\pi\varepsilon_0} \log_e \frac{d_1 d_2 d_3}{r^3}$$

$$= \frac{Q_A}{2\pi\varepsilon_0} \log_e \left(\frac{d_1 d_2 d_3}{r^3}\right)^{1/3}$$

$$= \frac{Q_A}{2\pi\varepsilon_0} \log_e \frac{(d_1 d_2 d_3)^{1/3}}{r}$$

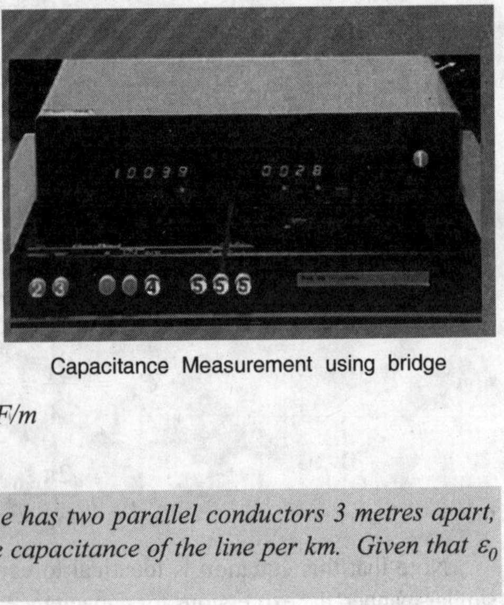

Capacitance Measurement using bridge

∴ Capacitance from conductor to neutral is

$$C_A = \frac{Q_A}{V_A} = \frac{2\pi\varepsilon_0}{\log_e \frac{\sqrt[3]{d_1 d_2 d_3}}{r}} \text{ F/m}$$

Example 9.11 *A single-phase transmission line has two parallel conductors 3 metres apart, radius of each conductor being 1 cm. Calculate the capacitance of the line per km. Given that $\varepsilon_0 = 8{\cdot}854 \times 10^{-12}$ F/m.*

Solution.

Conductor radius, $r = 1$ cm

Spacing of conductors, $d = 3$ m $= 300$ cm

Capacitance of the line $= \frac{\pi\varepsilon_0}{\log_e d/r}$ F/m $= \frac{\pi \times 8{\cdot}854 \times 10^{-12}}{\log_e 300/1}$ F/m

$= 0{\cdot}4875 \times 10^{-11}$ F/m $= 0{\cdot}4875 \times 10^{-8}$ F/km

$=$ **$0{\cdot}4875 \times 10^{-2}$ μF/km**

* On solving

Example 9.12. *A 3 phase overhead transmission line has its conductors arranged at the corners of an equilateral triangle of 2 m side. Calculate the capacitance of each line conductor per km. Given that diameter of each conductor is 1·25 cm.*

Solution.

Conductor radius, $r = 1{\cdot}25/2 = 0{\cdot}625$ cm

Spacing of conductors, $d = 2$ m $= 200$ cm

Capacitance of each line conductor

$$= \frac{2\,\pi\,\varepsilon_0}{\log_e d/r}\ \text{F/m} = \frac{2\,\pi \times 8{\cdot}854 \times 10^{-12}}{\log_e 200/0{\cdot}625}\ \text{F/m}$$

$$= 0{\cdot}0096 \times 10^{-9}\ \text{F/m} = 0{\cdot}0096 \times 10^{-6}\ \text{F/km} = \mathbf{0{\cdot}0096\ \mu F/km}$$

Example 9.13. *A 3-phase, 50 Hz, 66 kV overhead line conductors are placed in a horizontal plane as shown in Fig. 9.26. The conductor diameter is 1·25 cm. If the line length is 100 km, calculate (i) capacitance per phase, (ii) charging current per phase, assuming complete transposition of the line.*

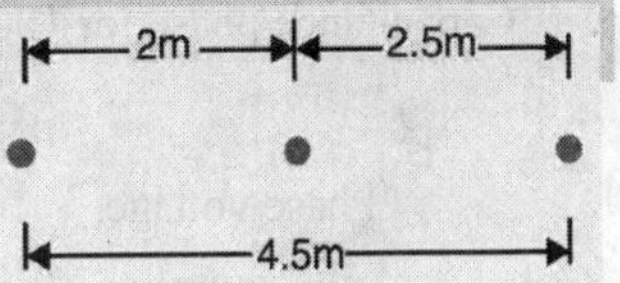

Fig. 9.26

Solution. Fig 9.26 shows the arrangement of conductors of the 3-phase line. The equivalent equilateral spacing is

$$d = \sqrt[3]{d_1 d_2 d_3} = \sqrt[3]{2 \times 2{\cdot}5 \times 4{\cdot}5} = 2{\cdot}82\ \text{m}$$

Conductor radius, $r = 1{\cdot}25/2 = 0{\cdot}625$ cm

Conductor spacing, $d = 2{\cdot}82$ m $= 282$ cm

(*i*) Line to neutral capacitance $= \dfrac{2\,\pi\,\varepsilon_0}{\log_e d/r}\ \text{F/m} = \dfrac{2\,\pi \times 8{\cdot}854 \times 10^{-12}}{\log_e 282/0{\cdot}625}\ \text{F/m}$

$$= 0{\cdot}0091 \times 10^{-9}\ \text{F/m} = 0{\cdot}0091 \times 10^{-6}\ \text{F/km} = 0{\cdot}0091\ \mu\text{F/km}$$

∴ Line to neutral capacitance for 100 km line is

$$C = 0{\cdot}0091 \times 100 = \mathbf{0{\cdot}91\ \mu F}$$

(*ii*) Charging current per phase is

$$I_C = \frac{V_{ph}}{X_C} = \frac{66{,}000}{\sqrt{3}} \times 2\pi\, f\, C$$

$$= \frac{66{,}000}{\sqrt{3}} \times 2\pi \times 50 \times 0{\cdot}91 \times 10^{-6} = \mathbf{10{\cdot}9\ A}$$

Example 9.14. *Calculate the capacitance of a 100 km long 3-phase, 50 Hz overhead transmission line consisting of 3 conductors, each of diameter 2 cm and spaced 2·5 m at the corners of an equilateral triangle.*

Solution.

Equilateral spacing, $d = 2{\cdot}5$ m $= 250$ cm

Radius of conductor, $r = 2/2 = 1$ cm

Capacitance of each conductor to netural

$$= \frac{2\,\pi\,\varepsilon_0}{\log_e d/r}\ \text{F/m} = \frac{2\,\pi \times 8{\cdot}85 \times 10^{-12}}{\log_e 250/1}\ \text{F/m}$$

$$= 10{\cdot}075 \times 10^{-12}\ \text{F/m} = 10{\cdot}075 \times 10^{-9}\ \text{F/km}$$

∴ Capacitance of 100 km line $= (10{\cdot}075 \times 10^{-9}) \times 100 = 1{\cdot}0075 \times 10^{-6}\ \text{F} = \mathbf{1{\cdot}0075\ \mu F/phase}$

Example 9.15. *A 3-phase, 50 Hz, 132 kV overhead line has conductors placed in a horizontal plane 4 m apart. Conductor diameter is 2 cm. If the line length is 100 km, calculate the charging current per phase assuming complete transposition.*

Solution. Fig 9.27 shows the conditions of the problem. The diameter of each conductor is 2 cm so that conductor radius $r = 2/2 = 1$ cm $= 1 \times 10^{-2}$ m.

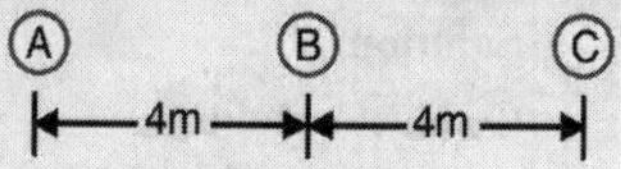

Fig. 9.27

Now $d_1 = AB = 4$m; $d_2 = BC = 4$ m; $d_3 = AC = 8$ m

$$\therefore \quad D_{eq} = \sqrt[3]{d_1 \times d_2 \times d_3} = \sqrt[3]{4 \times 4 \times 8} = 5{\cdot}04 \text{ m}$$

Capacitance of each conductor to neutral

$$= \frac{2\,\pi\,\varepsilon_0}{\log_e D_{eq}/r} \text{ F/m} = \frac{2\,\pi \times 8{\cdot}85 \times 10^{-12}}{\log_e 5{\cdot}04/1 \times 10^{-2}} \text{ F/m}$$

$$= 0{\cdot}00885 \times 10^{-6} \text{ F/km}$$

Capacitance/phase for 100 km line is

$$C_n = 0{\cdot}00885 \times 10^{-6} \times 100 = 0{\cdot}885 \times 10^{-6} \text{ F}$$

$$\text{Phase voltage,} \quad V_{ph} = \frac{\text{Line Voltage}}{\sqrt{3}} = \frac{132 \times 10^3}{\sqrt{3}} = 76210 \text{ V}$$

$\therefore$ Charging current/phase, $I_C = \omega C_n V_{ph} = (2\,\pi \times 50) \times (0{\cdot}885 \times 10^{-6}) \times 76210 =$ **21·18 A**

TUTORIAL PROBLEMS

1. A single phase transmission line has two parallel conductors 1·5 metres apart, the diameter of each conductor being 0·5 cm. Calculate line to neutral capacitance for a line 80 km long. **[3·48 μF]**
2. A 200 km, 3-phase transmission line has its conductors placed at the corners of an equilateral triangle of 2·5 m side. The radius of each conductor is 1 cm. Calculate :
   (*i*) line to neutral capacitance of the line,
   (*ii*) charging current per phase if the line is maintained at 66 kV, 50 Hz. **[(*i*) 2·02 μF (*ii*) 24·2 A]**
3. The three conductors *A*, *B* and *C* of a 3-ϕ line are arranged in a horizontal plane with $D_{AB} = 2$ m and $D_{BC} = 2{\cdot}5$ m. Find line-to-neutral capacitance per km if diameter of each conductor is 1·24 cm. The conductors are transposed at regular intervals. **[0·0091 μF/km]**
4. The three conductors of a 3-ϕ line are arranged at the corners of a right angled isosceles triangle. If each equal side of this triangle is 2 m, find line-to-neutral capacitance per km. Take the diameter of each conductor as 1·24 cm. The conductors are transposed at regular intervals. **[0·0094 μF/km]**
5. A 3-phase, 50 Hz, 132 kV overhead line has conductors placed in a horizontal plane 4·56 m apart. Conductor diameter is 22·4 mm. If the line length is 100 km, Calculate the charging current per phase, assuming complete transposition. **[21·345 A]**
6. Three conductors of a 3-phase overhead line are arranged in a horizontal plane 6 m apart. The diameter of each conductor is 1·24 cm. Find the capacitance of each conductor to neutral per 100 km of the line. **[0·785 μF]**

SELF - TEST

1. **Fill in the blanks by inserting appropriate words/figures.**
   (*i*) The power loss in an overhead transmission line is mainly due to ....
   (*ii*) If the length of a transmission line increases, its inductance is ....
   (*iii*) The d.c. resistance of a line conductor is .... than its a.c. resistance.
   (*iv*) If capacitance between two conductors of a 3-phase line is 4 μF, then capacitance of each conductor to neutral is ....

(v) If the length of the line is decreased, its capacitance is ....

(vi) Transposition of a 3-phase transmission line helps in ....

(vii) A neutral plane is one where .... is zero.

(viii) In a single phase overhead line, the neutral plane lies at ....

2. Pick up the correct words/figures from brackets and fill in the blanks

(i) If the supply frequency increases, then skin effect is .... [*increases, decreased*]

(ii) An overhead transmission line has appreciable inductance because the loop it forms has .... X-sectional area. [*large, small*]

(iii) If the spacing between the conductors is increased, the inductance of the line.......... [*increases, decreases*]

(iv) The skin effect is .... for stranded conductor than the solid conductor. [*less, more*]

(v) If the conductor diameter decreases, inductance of the line is .... [*increased, decreased*]

ANSWERS TO SELF-TEST

1. (i) Line conductor resistance (ii) increased (iii) less (iv) 8 μF (v) decreased (vi) equalising inductance and capacitance of the three phases (vii) electric intensity (viii) the centre of the distance between the conductors
2. (i) increased (ii) large (iii) increases (iv) less (v) increased

CHAPTER REVIEW TOPICS

1. What do you understand by the constants of an overhead transmission line ?
2. What is skin effect ? Why is it absent in the d.c. system ?
3. Find an expression for the flux linkages
   (i) due to a single current carrying conductor
   (ii) in parallel current carrying conductors
4. Derive an expression for the loop inductance of a single phase line.
5. Derive an expression for the inductance per phase for a 3-phase overhead transmission line when
   (i) conductors are symmetrically placed
   (ii) conductors are unsymmetrically placed but the line is completely transposed
6. What do you understand by electric potential ? Derive an expression for electric potential
   (i) at a charged single conductor
   (ii) at a conductor in a group of charged conductors
7. Derive an expression for the capacitance of a single phase overhead transmission line.
8. Deduce an expression for line to neutral capacitance for a 3-phase overhead transmission line when the conductors are
   (i) symmetrically placed
   (ii) unsymmetrically placed but transposed

DISCUSSION QUESTIONS

1. What is the effect of unsymmetrical spacing of conductors in a 3-phase transmission line ?
2. Will capacitance of a transmission line depend upon the ground effect ?
3. Why do we find line to neutral capacitance in a 3-phase system ?
4. How does skin effect vary with conductor material ?
5. What is proximity effect ?

CHAPTER

10

Performance of Transmission Lines

Introduction

The important considerations in the design and operation of a transmission line are the determination of voltage drop, line losses and efficiency of transmission. These values are greatly influenced by the line constants *R*, *L* and *C* of the transmission line. For instance, the voltage drop in the line depends upon the values of above three line constants. Similarly, the resistance of transmission line conductors is the most important cause of power loss in the line and determines the transmission efficiency. In this chapter, we shall develop formulas by which we can calculate voltage regulation, line losses and efficiency of transmission lines. These formulas are important for two principal reasons. Firstly, they provide an opportunity to understand the effects of the parameters of the line on bus voltages and the flow of power. Secondly, they help in developing an overall understanding of what is occuring on electric power system.

10.1 Classification of Overhead Transmission Lines

A transmission line has *three constants R, L and C distributed uniformly along the whole length of the line. The resistance and inductance form the series impedance. The capacitance existing between conductors for 1-phase line or from a conductor to neutral for a 3-phase line forms a shunt path throughout the length of the line. Therefore, capacitance effects introduce complications in transmission line calculations. Depending upon the manner in which capacitance is taken into account, the overhead transmission lines are classified as :

(i) *Short transmission lines.* When the length of an overhead transmission line is upto about 50 km and the line voltage is comparatively low (< 20 kV), it is usually considered as a short transmission line. Due to smaller length and lower voltage, the capacitance effects are small and hence can be neglected. Therefore, while studying the performance of a short transmisison line, only resistance and inductance of the line are taken into account.

(ii) *Medium transmission lines.* When the length of an overhead transmission line is about 50-150 km and the line voltage is moderatly high (>20 kV < 100 kV), it is considered as a medium transmission line. Due to sufficient length and voltage of the line, the capacitance effects are taken into account. For purposes of calculations, the distributed capacitance of the line is divided and lumped in the form of condensers shunted across the line at one or more points.

(iii) *Long transmission lines.* When the length of an overhead transmission line is more than 150 km and line voltage is very high (> 100 kV), it is considered as a long transmission line. For the treatment of such a line, the line constants are considered uniformly distributed over the whole length of the line and rigorous methods are employed for solution.

It may be emphasised here that exact solution of any tranmission line must consider the fact that the constants of the line are not lumped but are distributed unfiormly throughout the length of the line. However, reasonable accuracy can be obtained by considering these constants as lumped for short and medium transmission lines.

10.2 Important Terms

While studying the performance of a transmission line, it is desirable to determine its voltage regulation and transmission efficiency. We shall explain these two terms in turn.

(i) **Voltage regulation.** When a transmission line is carrying current, there is a voltage drop in the line due to resistance and inductance of the line. The result is that receiving end voltage (V_R) of the line is generally less than the sending end voltage (V_S). This voltage drop ($V_S - V_R$) in the line is expressed as a percentage of receiving end voltage V_R and is called voltage regulation.

*The difference in voltage at the receiving end of a transmission line **between conditions of no load and full load is called* **voltage regulation** and is expressed as a percentage of the receiving end voltage.

---

* There is also a fourth constant *i.e.*, shunt conductance. It represents the conductance between conductors or between conductor and ground and accounts for the leakage current at the insulators. It is very small in case of overhead lines and may be assumed zero.

** At no load, there is no drop in the line so that at no load, $V_R = V_S$. However, at full load, there is a voltage drop in the line so that receiving end voltage is V_R.

∴ Difference in voltage at receiving end between no load and full load

$$= V_S - V_R$$

Mathematically,

$$\% \text{ age Voltage regulation} = \frac{V_S - V_R}{V_R} \times 100$$

Obviously, it is desirable that the voltage regulation of a transmission line should be low *i.e.*, the increase in load current should make very little difference in the receiving end voltage.

(*ii*) **Transmission efficiency.** The power obtained at the receiving end of a transmission line is generally less than the sending end power due to losses in the line resistance.

The ratio of receiving end power to the sending end power of a transmission line is known as the **transmission efficiency** *of the line i.e.*

$$\% \text{ age Transmission efficiency, } \eta_T = \frac{\text{Receiving end power}}{\text{Sending end power}} \times 100$$

$$= \frac{V_R I_R \cos\phi_R}{V_S I_S \cos\phi_S} \times 100$$

where V_R, I_R and $\cos\phi_R$ are the receiving end voltage, current and power factor while V_S, I_S and $\cos\phi_S$ are the corresponding values at the sending end.

10.3 Performance of Single Phase Short Transmission Lines

As stated earlier, the effects of line capacitance are neglected for a short transmission line. Therefore, while studying the performance of such a line, only resistance and inductance of the line are taken into account. The equivalent circuit of a single phase short transmission line is shown in Fig. 10.1 (*i*). Here, the total line resistance and inductance are shown as concentrated or lumped instead of being distributed. The circuit is a simple a.c. series circuit.

Let
I = load current
R = loop resistance *i.e.*, resistance of both conductors
X_L = loop reactance
V_R = receiving end voltage
$\cos\phi_R$ = receiving end power factor (lagging)
V_S = sending end voltage
$\cos\phi_S$ = sending end power factor

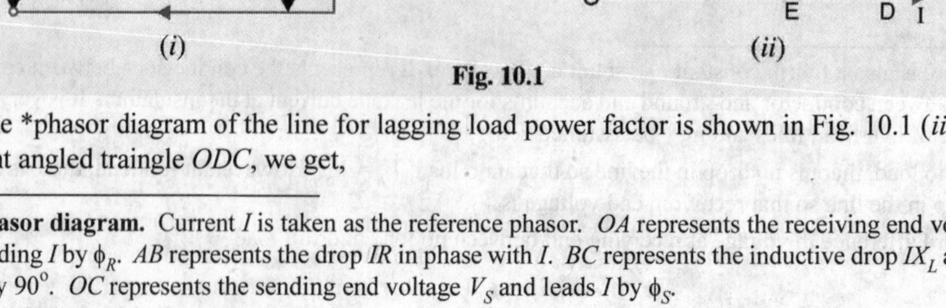

(*i*) (*ii*)

Fig. 10.1

The *phasor diagram of the line for lagging load power factor is shown in Fig. 10.1 (*ii*). From the right angled traingle *ODC*, we get,

* **Phasor diagram.** Current I is taken as the reference phasor. OA represents the receiving end voltage V_R leading I by ϕ_R. AB represents the drop IR in phase with I. BC represents the inductive drop IX_L and leads I by 90°. OC represents the sending end voltage V_S and leads I by ϕ_S.

$$(OC)^2 = (OD)^2 + (DC)^2$$

or
$$V_S^2 = (OE + ED)^2 + (DB + BC)^2$$
$$= (V_R \cos \phi_R + IR)^2 + (V_R \sin \phi_R + IX_L)^2$$

$$\therefore \quad V_S = \sqrt{(V_R \cos \phi_R + IR)^2 + (V_R \sin \phi_R + IX_L)^2}$$

(*i*) %age Voltage regulation $= \dfrac{V_S - V_R}{V_R} \times 100$

(*ii*) Sending end *p.f.*, $\cos \phi_S = \dfrac{OD}{OC} = \dfrac{V_R \cos \phi_R + IR}{V_S}$

(*iii*) Power delivered $= V_R I_R \cos \phi_R$

Line losses $= I^2R$

Power sent out $= V_R I_R \cos \phi_R + I^2 R$

%age Transmission efficiency $= \dfrac{\text{Power delivered}}{\text{Power sent out}} \times 100$

$$= \frac{V_R\, I_R \cos \phi_R}{V_R\, I_R \cos \phi_R + I^2 R} \times 100$$

An approximate expression for the sending end voltage V_S can be obtained as follows. Draw perpendicular from *B* and *C* on *OA* produced as shown in Fig. 10.2. Then *OC* is *nearly* equal to *OF i.e.*,

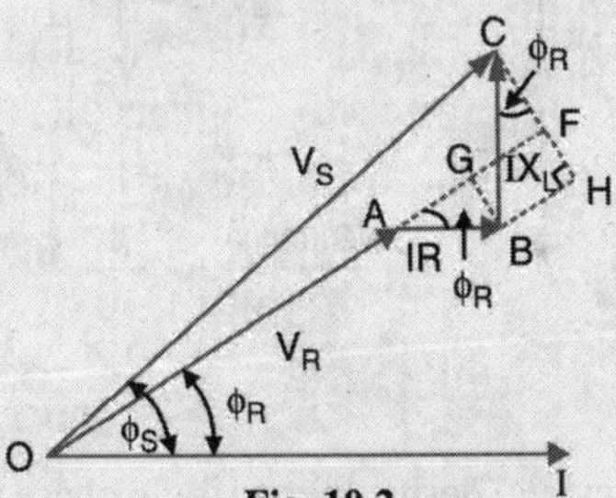

Fig. 10.2

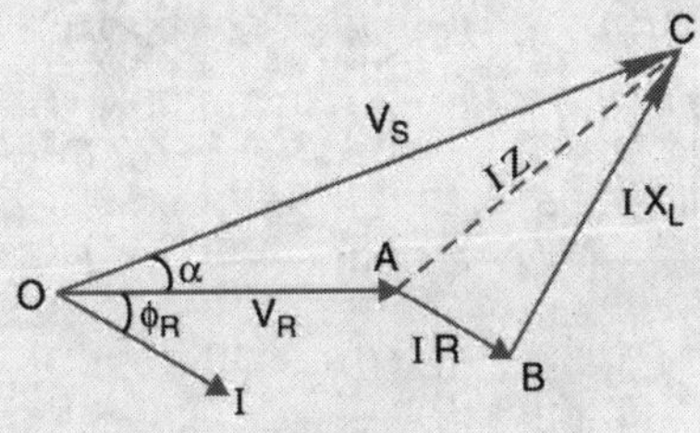

Fig. 10.3

$$OC = OF = OA + AF = OA + AG + GF$$
$$= OA + AG + BH$$

$$\therefore \quad V_S = V_R + I\,R \cos \phi_R + I\,X_L \sin \phi_R$$

Solution in complex notation. It is often convenient and profitable to make the line calculations in complex notation.

Taking $\vec{V_R}$ as the reference phasor, draw the phasor diagram as shown in Fig 10.3. It is clear that $\vec{V_S}$ is the phasor sum of $\vec{V_R}$ and $\vec{I}\,\vec{Z}$.

$$^*\vec{V_R} = V_R + j\,0$$
$$\vec{I} = \vec{I} \angle -\phi_R = I\,(\cos \phi_R - j \sin \phi_R)$$
$$\vec{Z} = R + jX_L$$
$$\therefore \quad \vec{V_S} = \vec{V_R} + \vec{I}\,\vec{Z}$$
$$= (V_R + j\,0) + I\,(\cos \phi_R - j \sin \phi_R)\,(R + j\,X_L)$$

---

* Phasors are shown by arrows and their magnitudes without arrow. Thus $\vec{V_R}$ is the receiving end voltage phasor, whereas V_R is its magnitude.

$$= (V_R + IR\cos\phi_R + IX_L\sin\phi_R) + j\,(IX_L\cos\phi_R - IR\sin\phi_R)$$

$$\therefore \quad V_S = \sqrt{(V_R + IR\cos\phi_R + IX_L\sin\phi_R)^2 + (IX_L\cos\phi_R - IR\sin\phi_R)^2}$$

The second term under the root is quite small and can be neglected with reasonable accuracy. Therefore, approximate expression for V_S becomes :

$$V_S = V_R + IR\cos\phi_R + IX_L\sin\phi_R$$

The following poins may be noted :

(*i*) The approximate formula for $V_S\,(= V_R + IR\cos\phi_R + IX_L\sin\phi_R)$ gives fairly correct results for lagging power factors. However, appreciable error is caused for leading power factors. Therefore, approximate expression for V_S should be used for lagging p.f. only.

(*ii*) The solution in complex notation is in more presentable form.

10.4 Three-Phase Short Transmission Lines

For reasons associated with economy, transmission of electric power is done by 3-phase system. This system may be regarded as consisting of three single phase units, each wire transmitting one-third of the total power. As a matter of convenience, we generally analyse 3-phase system by considering

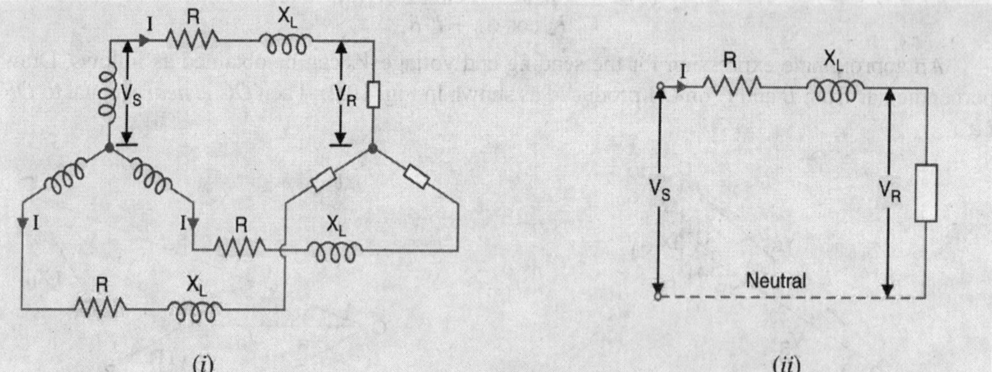

Fig. 10.4

*one phase only. Therefore, expression for regulation, efficiency etc. derived for a single phase line can also be applied to a 3-phase system. Since only one phase is considered, phase values of 3-phase system should be taken. Thus, V_S and V_R are the phase voltages, whereas R and X_L are the resistance and inductive reactance per phase respectively.

Fig. 10.4 (*i*) shows a *Y*-connected generator supplying a balanced *Y*-connected load through a transmission line. Each conductor has a resistance of $R\ \Omega$ and inductive reactance of $X_L\ \Omega$. Fig. 10.4 (*ii*) shows one phase separately. The calculations can now be made in the same way as for a single phase line.

10.5 Effect of Load p.f. on Regulation and Efficiency

The regulation and efficiency of a transmission line depend to a considerable extent upon the power factor of the load.

1. Effect on regulation. The expression for voltage regulation of a short transmission line is given by :

$$\%\text{age Voltage regulation} = \frac{IR\cos\phi_R + IX_L\sin\phi_R}{V_R} \times 100 \qquad \textit{(for lagging p.f.)}$$

---

* As similar conditions prevail in the three phases.

$$\text{\%age Voltage regulation} = \frac{I\,R\cos\phi_R - I\,X_L\sin\phi_R}{V_R} \times 100 \qquad \textit{(for leading p.f.)}$$

The following conclusions can be drawn from the above expressions :

(*i*) When the load p.f. is lagging or unity or such leading that $I\,R\cos\phi_R > I\,X_L\sin\phi_R$, then voltage regulation is positive *i.e.*, receiving end voltage V_R will be less than the sending end voltage V_S.

(*ii*) For a given V_R and I, the voltage regulation of the line increases with the decrease in p.f. for lagging loads.

(*iii*) When the load p.f. is leading to this extent that $I\,X_L\sin\phi_R > I\,R\cos\phi_R$, then voltage regulation is negative *i.e.* the receiving end voltage V_R is more than the sending end voltage V_S.

(*iv*) For a given V_R and I, the voltage regulation of the line decreases with the decrease in p.f. for leading loads.

2. Effect on transmission efficiency. The power delivered to the load depends upon the power factor.

$$P = V_R\,{}^{*}I\cos\phi_R \quad \text{(For 1-phase line)}$$

$$\therefore \quad I = \frac{P}{V_R\cos\phi_R}$$

$$P = 3\,V_R\,I\cos\phi_R \quad \text{(For 3-phase line)}$$

$$\therefore \quad I = \frac{P}{3V_R\cos\phi_R}$$

It is clear that in each case, for a given amount of power to be transmitted (P) and receiving end voltage (V_R), the load current I is inversely proportional to the load p.f. $\cos\phi_R$. Consequently, with the decrease in load p.f., the load current and hence the line losses are increased. This leads to the conclusion that transmission efficiency of a line decreases with the decrease in load p.f. and *vice-versa*,

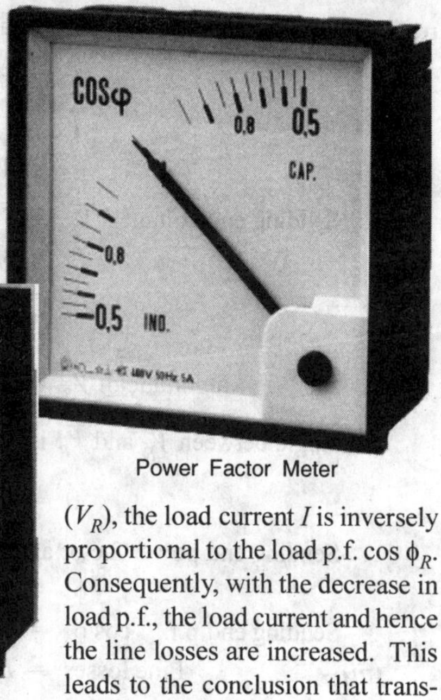

Power Factor Meter

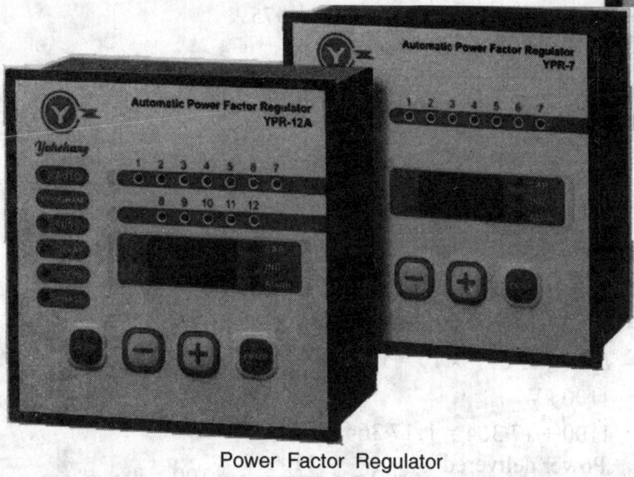

Power Factor Regulator

Example 10.1. *A single phase overhead transmission line delivers 1100 kW at 33 kV at 0·8 p.f. lagging. The total resistance and inductive reactance of the line are 10 Ω and 15 Ω respectively. Determine : (i) sending end voltage (ii) sending end power factor and (iii) transmission efficiency.*

Solution.

Load power factor, $\cos\phi_R = 0{\cdot}8$ lagging

Total line impedance, $\vec{Z} = R + j\,X_L = 10 + j\,15$

---

* $I_R = I_S = I$

Receiving end voltage, $V_R = 33$ kV = 33,000 V

$$\therefore \quad \text{Line current, } I = \frac{kW \times 10^3}{V_R \cos\phi_R} = \frac{1100 \times 10^3}{33{,}000 \times 0{\cdot}8} = 41{\cdot}67 \text{ A}$$

As $\cos\phi_R = 0{\cdot}8 \quad \therefore \quad \sin\phi_R = 0{\cdot}6$

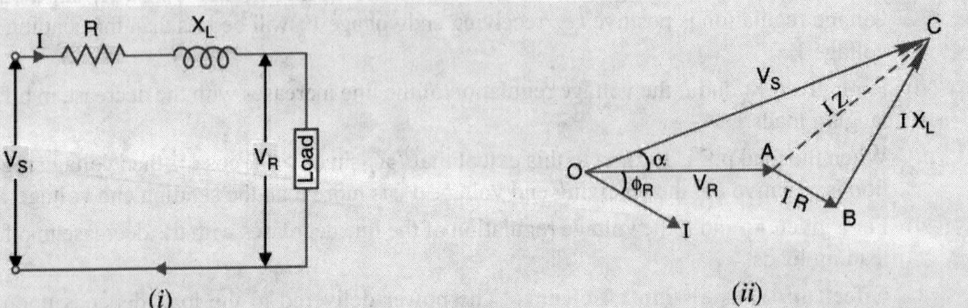

Fig. 10.5

The equivalent circuit and phasor diagram of the line are shown in Figs. 10.5 (*i*) and 10.5 (*ii*) respectively. Taking receiving end voltage $\vec{V_R}$ as the reference phasor,

$$\vec{V_R} = V_R + j\,0 = 33000 \text{ V}$$

$$\vec{I} = I(\cos\phi_R - j\sin\phi_R)$$

$$= 41{\cdot}67\,(0{\cdot}8 - j\,0{\cdot}6) = 33{\cdot}33 - j\,25$$

(*i*) Sending end voltage, $\vec{V_S} = \vec{V_R} + \vec{I}\,\vec{Z}$

$$= 33{,}000 + (33{\cdot}33 - j\,25{\cdot}0)\,(10 + j\,15)$$

$$= 33{,}000 + 333{\cdot}3 - j250 + j500 + 375$$

$$= 33{,}708{\cdot}3 + j\,250$$

$$\therefore \quad \text{Magnitude of } V_S = \sqrt{(33{,}708{\cdot}3)^2 + (250)^2} = \mathbf{33{,}709\ V}$$

(*ii*) Angle between $\vec{V_S}$ and $\vec{V_R}$ is

$$\alpha = \tan^{-1}\frac{250}{33{,}708{\cdot}3} = \tan^{-1} 0{\cdot}0074 = 0{\cdot}42^\circ$$

∴ Sending end power factor angle is

$$\phi_S = \phi_R + \alpha = 36{\cdot}87^\circ + 0{\cdot}42^\circ = 37{\cdot}29^\circ$$

∴ Sending end p.f., $\cos\phi_S = \cos 37{\cdot}29^\circ =$ **0·7956 lagging**

(*iii*) Line losses $= I^2R = (41{\cdot}67)^2 \times 10 = 17{,}364 \text{ W} = 17{\cdot}364 \text{ kW}$

Output delivered = 1100 kW

Power sent = 1100 + 17·364 = 1117·364 kW

$$\therefore \quad \text{Transmission efficiency} = \frac{\text{Power delivered}}{\text{Power sent}} \times 100 = \frac{1100}{1117{\cdot}364} \times 100 = \mathbf{98{\cdot}44\%}$$

Note. V_S and ϕ_S can also be calculated as follows :

$$V_S = V_R + I\,R\cos\phi_R + I\,X_L\sin\phi_R \text{ (approximately)}$$

$$= 33{,}000 + 41{\cdot}67 \times 10 \times 0{\cdot}8 + 41{\cdot}67 \times 15 \times 0{\cdot}6$$

$$= 33{,}000 + 333{\cdot}36 + 375{\cdot}03$$

$$= 33708{\cdot}39 \text{ V which is approximately the same as above}$$

$$\cos\phi_S = \frac{V_R\cos\phi_R + I\,R}{V_S} = \frac{33{,}000 \times 0{\cdot}8 + 41{\cdot}67 \times 10}{33{,}708{\cdot}39} = \frac{26{,}816{\cdot}7}{33{,}708{\cdot}39}$$

$$= 0{\cdot}7958$$

As stated earlicr, this method gives fairly correct results for lagging p.f. The reader will find that this method is used in the solution of some numericals.

Example 10.2. *What is the maximum length in km for a 1-phase transmission line having copper conductor of 0·775 cm² cross-section over which 200 kW at unity power factor and at 3300V are to be delivered ? The efficiecny of transmission is 90%. Take specific resistance as 1.725 μ Ω cm.*

Solution.

$$\text{Receiving end power} = 200 \text{ kW} = 2{,}00{,}000 \text{ W}$$

$$\text{Transmission efficiency} = 0{\cdot}9$$

$$\therefore \quad \text{Sending end power} = \frac{2{,}00{,}000}{0 \cdot 9} = 2{,}22{,}222 \text{ W}$$

$$\therefore \quad \text{Line losses} = 2{,}22{,}222 - 2{,}00{,}000 = 22{,}222 \text{ W}$$

$$\text{Line current, } I = \frac{200 \times 10^3}{3{,}300 \times 1} = 60{\cdot}6 \text{ A}$$

Let R Ω be the resistance of one conductor.

$$\text{Line losses} = 2 I^2 R$$

or
$$22{,}222 = 2\,(60{\cdot}6)^2 \times R$$

$$\therefore \quad R = \frac{22{,}222}{2 \times (60 \cdot 6)^2} = 3{\cdot}025 \ \Omega$$

Now,
$$R = \rho\, l/a$$

$$\therefore \quad l = \frac{Ra}{\rho} = \frac{3 \cdot 025 \times 0 \cdot 775}{1 \cdot 725 \times 10^{-6}} = 1{\cdot}36 \times 10^6 \text{ cm} = \mathbf{13{\cdot}6 \text{ km}}$$

Example 10.3. *An overhead 3-phase transmission line delivers 5000 kW at 22 kV at 0·8 p.f. lagging. The resistance and reactance of each conductor is 4 Ω and 6 Ω respectively. Determine : (i) sending end voltage (ii) percentage regulation (iii) transmission efficiency.*

Solution.

$$\text{Load power factor, } \cos\phi_R = 0{\cdot}8 \text{ lagging}$$

$$\text{Receiving end voltage/phase,}^* V_R = 22{,}000/\sqrt{3} = 12{,}700 \text{ V}$$

$$\text{Impedance/phase, } \quad \vec{Z} = 4 + j\,6$$

$$\text{Line current, } \quad I = \frac{5000 \times 10^3}{3 \times 12700 \times 0 \cdot 8} = 164 \text{ A}$$

As $\cos\phi_R = 0{\cdot}8 \quad \therefore \sin\phi_R = 0{\cdot}6$

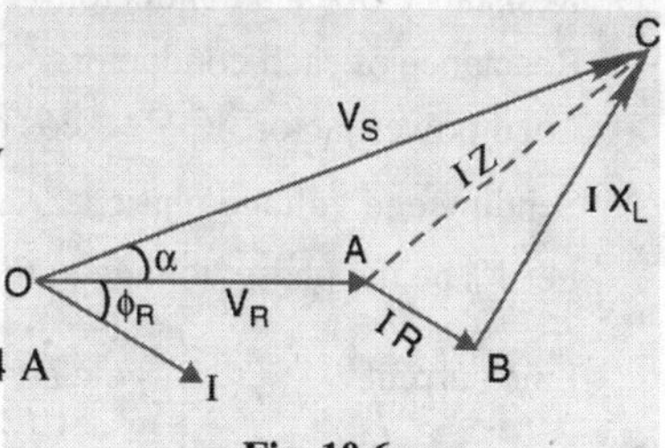

Fig. 10.6

Taking $\vec{V_R}$ as the reference phasor (see Fig. 10.6),

$$\vec{V_R} = V_R + j\,0 = 12700 \text{ V}$$

$$\vec{I} = I\,(\cos\phi_R - j\sin\phi_R) = 164\,(0{\cdot}8 - j\,0{\cdot}6) = 131{\cdot}2 - j\,98{\cdot}4$$

(*i*) Sending end voltage per phase is

$$\vec{V_S} = \vec{V_R} + \vec{I}\,\vec{Z} = 12700 + (131{\cdot}2 - j\,98{\cdot}4)\,(4 + j\,6)$$
$$= 12700 + 524{\cdot}8 + j\,787{\cdot}2 - j\,393{\cdot}6 + 590{\cdot}4$$
$$= 13815{\cdot}2 + j\,393{\cdot}6$$

$$\text{Magnitude of } V_S = \sqrt{(13815 \cdot 2)^2 + (393 \cdot 6)^2} = 13820{\cdot}8 \text{ V}$$

---

* If not mentioned in the problem, star-connection is understood.

$$\text{Line value of } V_S = \sqrt{3} \times 13820{\cdot}8 = 23938 \text{ V} = \mathbf{23{\cdot}938\ kV}$$

(ii) $$\%\text{ age Regulation} = \frac{V_S - V_R}{V_R} \times 100 = \frac{13820 \cdot 8 - 12700}{12700} \times 100 = \mathbf{8{\cdot}825\%}$$

(iii) $$\text{Line losses} = 3I^2R = 3 \times (164)^2 \times 4 = 3{,}22{,}752 \text{ W} = 322{\cdot}752 \text{ kW}$$

$$\therefore \quad \text{Transmission efficiency} = \frac{5000}{5000 + 322 \cdot 752} \times 100 = \mathbf{93{\cdot}94\%}$$

Example 10.4. *Estimate the distance over which a load of 15000 kW at a p.f. 0·8 lagging can be delivered by a 3-phase transmission line having conductors each of resistance 1 Ω per kilometre. The voltage at the receiving end is to be 132 kV and the loss in the transmission is to be 5%.*

Solution.

$$\text{Line current,} \quad I = \frac{\text{Power delivered}}{\sqrt{3} \times \text{line voltage} \times \text{power factor}} = \frac{15000 \times 10^3}{\sqrt{3} \times 132 \times 10^3 \times 0 \cdot 8} = 82 \text{ A}$$

$$\text{Line losses} = 5\% \text{ of power delivered} = 0{\cdot}05 \times 15000 = 750 \text{ kW}$$

Let R Ω be the resistance of one conductor.

$$\text{Line losses} = 3\,I^2R$$

or $$750 \times 10^3 = 3 \times (82)^2 \times R$$

$$\therefore \quad R = \frac{750 \times 10^3}{3 \times (82)^2} = 37{\cdot}18\ \Omega$$

Resistance of each conductor per km is 1 Ω (*given*).

$$\therefore \quad \text{Length of line} = \mathbf{37{\cdot}18\ km}$$

Example 10.5. *A 3-phase line delivers 3600 kW at a p.f. 0·8 lagging to a load. If the sending end voltage is 33 kV, determine (i) the receiving end voltage (ii) line current (iii) transmission efficiency. The resistance and reactance of each conductor are 5·31 Ω and 5·54 Ω respectively.*

Solution.

Resistance of each conductor, $R = 5{\cdot}31\ \Omega$

Reactance of each conductor, $X_L = 5{\cdot}54\ \Omega$

Load power factor, $\cos\phi_R = 0{\cdot}8$ (*lagging*)

Sending end voltage/phase, $V_S = 33{,}000/\sqrt{3} = 19{,}052$ V

Let V_R be the phase voltage at the receiving end.

$$\text{Line current,} \quad I = \frac{\text{Power delivered / phase}}{V_R \times \cos\phi_R} = \frac{1200 \times 10^3}{V_R \times 0 \cdot 8}$$

$$= \frac{150 \times 10^5}{V_R} \quad ...(i)$$

(*i*) Using approximate expression for V_S, we get,

$$V_S = V_R + I\,R\cos\phi_R + I\,X_L\sin\phi_R$$

or $$19{,}052 = V_R + \frac{15 \times 10^5}{V_R} \times 5{\cdot}31 \times 0{\cdot}8 + \frac{15 \times 10^5}{V_R} \times 5{\cdot}54 \times 0{\cdot}6$$

or $$V_R^2 - 19{,}052\,V_R + 1{,}13{,}58{,}000 = 0$$

Solving this equation, we get, $V_R = 18{,}435$ V

$$\therefore \quad \text{Line voltage at the receiving end} = \sqrt{3} \times 18{,}435 = 31{,}930 \text{ V} = \mathbf{31{\cdot}93\ kV}$$

(*ii*) Line current, $I = \frac{15 \times 10^5}{V_R} = \frac{15 \times 10^5}{18,435} = \mathbf{81 \cdot 36\ A}$

(*iii*) Line losses, $= 3 I^2 R = 3 \times (81 \cdot 36)^2 \times 5 \cdot 31 = 1,05,447\ W = 105 \cdot 447\ kW$

$\therefore$ Transmission efficiency $= \frac{3600}{3600 + 105 \cdot 447} \times 100 = \mathbf{97 \cdot 15\%}$

Example 10.6. *A short 3-ϕ transmission line with an impedance of $(6 + j\,8)\ \Omega$ per phase has sending and receiving end voltages of 120 kV and 110 kV respectively for some receiving end load at a p.f. of 0·9 lagging. Determine (i) power output and (ii) sending end power factor.*

Solution.

Resistance of each conductor, $R = 6\ \Omega$

Reactance of each conductor, $X_L = 8\ \Omega$

Load power factor, $\cos \phi_R = 0 \cdot 9$ lagging

Receiving end voltage/phase, $V_R = 110 \times 10^3/\sqrt{3} = 63508\ V$

Sending end voltage/phase, $V_S = 120 \times 10^3/\sqrt{3} = 69282\ V$

Let I be the load current. Using approximate expression for V_S, we get,

$$V_S = V_R + I R \cos \phi_R + I X_L \sin \phi_R$$

or $$69282 = 63508 + I \times 6 \times 0 \cdot 9 + I \times 8 \times 0 \cdot 435$$

or $$8 \cdot 88\, I = 5774$$

or $$I = 5774/8 \cdot 88 = 650 \cdot 2\ A$$

(*i*) Power output $= \frac{3\, V_R\, I \cos \phi_R}{1000}\ kW = \frac{3 \times 63508 \times 650 \cdot 2 \times 0 \cdot 9}{1000}$

$= \mathbf{1,11,490\ kW}$

(*ii*) Sending end p.f., $\cos \phi_S = \frac{V_R \cos \phi_R + I R}{V_S} = \frac{63508 \times 0 \cdot 9 + 650 \cdot 2 \times 6}{69282} = \mathbf{0 \cdot 88\ lag}$

Example 10.7. *An 11 kV, 3-phase transmission line has a resistance of 1·5 Ω and reactance of 4 Ω per phase. Calculate the percentage regulation and efficiency of the line when a total load of 5000 kVA at 0.8 lagging power factor is supplied at 11 kV at the distant end.*

Solution.

Resistance of each conductor, $R = 1 \cdot 5\ \Omega$

Reactance of each conductor, $X_L = 4\ \Omega$

Receiving end voltage/phase, $V_R = \frac{11 \times 10^3}{\sqrt{3}} = 6351\ V$

Load power factor, $\cos \phi_R = 0 \cdot 8$ lagging

Load current, $I = \frac{\text{Power delivered in kVA} \times 1000}{3 \times V_R}$

$= \frac{5000 \times 1000}{3 \times 6351} = 262 \cdot 43 A$

Using the approximate expression for V_S (sending end voltage per phase), we get,

$$V_S = V_R + I R \cos \phi_R + I X_L \sin \phi_R$$

$$= 6351 + 262 \cdot 43 \times 1 \cdot 5 \times 0 \cdot 8 + 262 \cdot 43 \times 4 \times 0 \cdot 6 = 7295 \cdot 8\ V$$

% regulation $= \frac{V_S - V_R}{V_R} \times 100 = \frac{7295 \cdot 8 - 6351}{6351} \times 100 = \mathbf{14 \cdot 88\%}$

Line losses $= 3 I^2 R = 3 \times (262 \cdot 43)^2 \times 1 \cdot 5 = 310 \times 10^3\ W = 310\ kW$

$$\text{Output power} = 5000 \times 0{\cdot}8 = 4000 \text{ kW}$$

$$\text{Input power} = \text{Ouput power + line losses} = 4000 + 310 = 4310 \text{ kW}$$

$$\text{Transmission efficiency} = \frac{\text{Output power}}{\text{Input power}} \times 100 = \frac{4000}{4310} \times 100 = \mathbf{92{\cdot}8\%}$$

Example 10.8. *A 3-phase, 50 Hz, 16 km long overhead line supplies 1000 kW at 11kV, 0·8 p.f. lagging. The line resistance is 0·03 Ω per phase per km and line inductance is 0·7 mH per phase per km. Calculate the sending end voltage, voltage regulation and efficiency of transmission.*

Solution.

Resistance of each conductor, $R = 0{\cdot}03 \times 16 = 0{\cdot}48\ \Omega$

Reactance of each conductor, $X_L = 2\pi f L \times 16 = 2\pi \times 50 \times 0{\cdot}7 \times 10^{-3} \times 16 = 3{\cdot}52\ \Omega$

Receiving end voltage/phase, $V_R = \dfrac{11 \times 10^3}{\sqrt{3}} = 6351 \text{ V}$

Load power factor, $\cos \phi_R = 0{\cdot}8$ lagging

$$\text{Line current, } I = \frac{1000 \times 10^3}{3 \times V_R \times \cos\phi} = \frac{1000 \times 10^3}{3 \times 6351 \times 0{\cdot}8} = 65{\cdot}6\text{A}$$

$$\text{Sending end voltage/phase, } V_S = V_R + I R \cos \phi_R + I X_L \sin \phi_R$$

$$= 6351 + 65{\cdot}6 \times 0{\cdot}48 \times 0{\cdot}8 + 65{\cdot}6 \times 3{\cdot}52 \times 0{\cdot}6 = 6515 \text{ V}$$

$$\therefore \quad \text{\%age Voltage regulation} = \frac{V_S - V_R}{V_R} \times 100 = \frac{6515 - 6351}{6351} \times 100 = \mathbf{2{\cdot}58\%}$$

$$\text{Line losses} = 3 I^2 R = 3 \times (65{\cdot}6)^2 \times 0{\cdot}48 = 6{\cdot}2 \times 10^3 \text{ W} = 6{\cdot}2 \text{ kW}$$

$$\text{Input power} = \text{Output power + Line losses} = 1000 + 6{\cdot}2 = 1006{\cdot}2 \text{ kW}$$

$$\therefore \quad \text{Transmission efficiency} = \frac{\text{Output power}}{\text{Input power}} \times 100 = \frac{1000}{1006{\cdot}2} \times 100 = \mathbf{99{\cdot}38\%}$$

Example 10.9. *A 3-phase load of 2000 kVA, 0·8 p.f. is supplied at 6·6 kV, 50 Hz by means of a 33 kV transmission line 20 km long and 33/6·6 kV step-down transfomer. The resistance and reactance of each conductor are 0·4 Ω and 0·5 Ω per km respectively. The resistance and reactance of transformer primary are 7.5 Ω and 13.2 Ω, while those of secondary are 0.35 Ω and 0.65 Ω respectively. Find the voltage necessary at the sending end of transmission line when 6.6 kV is maintained at the receiving end. Determine also the sending end power factor and transmission efficiency.*

Solution. Fig. 10.7 shows the single diagram of the transmission system. Here, the voltage drop will be due to the impedance of transmission line and also due to the impedance of transformer.

Resistance of each conductor $= 20 \times 0{\cdot}4 = 8\ \Omega$

Reactance of each conductor $= 20 \times 0{\cdot}5 = 10\ \Omega$

Let us transfer the impedance of transformer secondary to high tension side *i.e.*, 33 kV side.

Equivalent resistance of transformer referred to 33 kV side

$$= \text{Primary resistance} + 0{\cdot}35\,(33/6{\cdot}6)^2$$

$$= 7{\cdot}5 + 8{\cdot}75 = 16{\cdot}25\ \Omega$$

Equivalent reactance of transformer referred to 33 kV side

$$= \text{Primary reactance} + 0{\cdot}65\,(33/6{\cdot}6)^2$$

$$= 13{\cdot}2 + 16{\cdot}25 = 29{\cdot}45\ \Omega$$

Total resistance of line and transformer is

$$R = 8 + 16{\cdot}25 = 24{\cdot}25\ \Omega$$

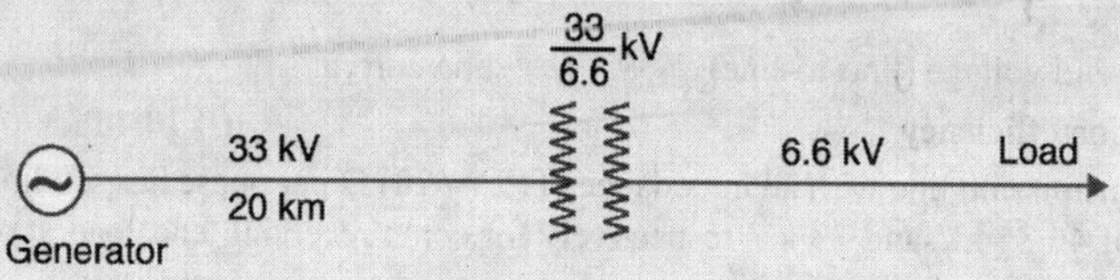

Fig. 10.7

Total reactance of line and transformer is

$$X_L = 10 + 29{\cdot}45 = 39{\cdot}45\ \Omega$$

Receiving end voltage per phase is

$$V_R = 33{,}000/\sqrt{3} = 19052\ \text{V}$$

Line current, $$I = \frac{2000 \times 10^3}{\sqrt{3} \times 33000} = 35\ \text{A}$$

Using the approximate expression for sending end voltage V_S per phase,

$$V_S = V_R + I R \cos\phi_R + I X_L \sin\phi_R$$
$$= 19052 + 35 \times 24{\cdot}25 \times 0{\cdot}8 + 35 \times 39{\cdot}45 \times 0{\cdot}6$$
$$= 19052 + 679 + 828 = 20559\ \text{V} = 20{\cdot}559\ \text{kV}$$

Sending end line voltage $= \sqrt{3} \times 20{\cdot}559\ \text{kV} = \mathbf{35{\cdot}6\ kV}$

Sending end p.f., $$\cos\phi_S = \frac{V_R \cos\phi_R + IR}{V_S} = \frac{19052 \times 0{\cdot}8 + 35 \times 24{\cdot}25}{20559} = \mathbf{0{\cdot}7826\ lag}$$

Line losses $$= \frac{3 I^2 R}{1000}\ \text{kW} = \frac{3 \times (35)^2 \times 24{\cdot}25}{1000} = 89{\cdot}12\ \text{kW}$$

Output power $= 2000\ \text{kVA} \times 0{\cdot}8 = 1600\ \text{kW}$

$\therefore$ Transmission efficiency $$= \frac{1600}{1600 + 89{\cdot}12} \times 100 = \mathbf{94{\cdot}72\%}$$

TUTORIAL PROBLEMS

1. A single phase overhead transmission line delivers 4000 kW at 11 kV at 0·8 p.f. lagging. If resistance and reactance per conductor are 0·15 Ω and 0·02 Ω respectively, calculate :
   (*i*) percentage regulation (*ii*) sending end power factor
   (*iii*) line losses **[(*i*) 19·83% (*ii*) 0·77 lag (*iii*) 620 kW]**
2. A single phase 11 kV line with a length of 15 km is to transmit 500 kVA. The inductive reactance of the line is 0·5 Ω/km and the resistance is 0·3 Ω/km. Calculate the efficiency and regulation of the line for 0·8 lagging power factor. **[97·74%, 3·34%]**
3. A load of 1000 kW at 0·8 p.f. lagging is received at the end of a 3-phase line 20 km long. The resistance and reactance of each conductor are 0·25 Ω and 0·28 Ω per km. If the receiving end line voltage is maintained at 11 kV, calculate :
   (*i*) sending end voltage (line-to-line) (*ii*) percentage regulation
   (*iii*) transmission efficiency **[(*i*) 11·84 kV (*ii*) 7·61% (*iii*) 94·32%]**
4. Estimate the distance over which a load of 15000 kW at 0·85 p.f. can be delivered by a 3-phase transmission line having conductors of steel-cored aluminium each of resistance 0·905 Ω/phase per kilometre. The voltage at the receiving end is to be 132 kV and the loss in transmission is to be 7·5% of the load. **[69·55 km]**
5. A 3-phase line 3 km long delivers 3000 kW at a p.f. 0·8 lagging to a load. The resistance and reactance per km of each conductor are 0·4 Ω and 0·3 Ω respectively. If the voltage at the supply end is maintained at 11 kV, calculate :

(*i*) receiving end voltage (line-to-line) (*ii*) line current
(*iii*) transmission efficiency. **[(*i*) 10.46 kV (*ii*) 207 A (*iii*) 95%]**

6. A short 3-ϕ transmission line with an impedance of $(5 + j\,20)\ \Omega$ per phase has sending end and receiving end voltages of 46·85 kV and 33 kV respectively for some receiving end load at a p.f.of 0·8 lagging. Determine :
   (*i*) power output
   (*ii*) sending end power factor **[(*i*) 22·86 kW (*ii*) 0·657 lag]**
7. A substation receives 6000 kVA at 6 kV, 0.8 p.f. lagging on low voltage side of a transformer from a generating station through a 3-phase cable system having resistance of 7 Ω and reactance of 2 Ω per phase. Identical 6600/33000 V transformers are installed at each end, 6600 V side being delta connected and 33000 V side star connected. The resistance and reactance of each transformer are 1 Ω and 9 Ω respectively, referred to *h.v.* side. Calculate the voltage at the generating station bus bars **[6778 V]**
8. A short 3-phase transmission line connected to a 33kV, 50 Hz generating station at the sending end is required to supply a load of 10 MW at 0·8 lagging power factor at 30 kV at the receiving end. If the minimum transmission efficiency is to be limited to 96%, estimate the per phase value of resistance and inductance of the line. **[2·4 Ω; 0·028 H]**
9. A single phase transmission line is delivering 500 kVA load at 2 kV. Its resistance is 0·2 Ω and inductive reactance is 0·4 Ω. Determine the voltage regulation if the load power factor is (*i*) 0·707 lagging (*ii*) 0·707 leading. **[(*i*) 5·3% (*ii*) −1·65%]**

10.6 Medium Transmission Lines

In short transmission line calculations, the effects of the line capacitance are neglected because such lines have smaller lengths and transmit power at relatively low voltages (< 20 kV). However, as the length and voltage of the line increase, the capacitance gradually becomes of greater importance. Since medium transmission lines have sufficient length (50-150 km) and usually operate at voltages greater than 20 kV, the effects of capacitance cannot be neglected. Therefore, in order to obtain reasonable accuracy in medium transmission line calculations, the line capacitance must be taken into consideration.

The capacitance is uniformly distributed over the entire length of the line. However, in order to make the calculations simple, the line capacitance is assumed to be lumped or concentrated in the form of capacitors shunted across the line at one or more points. Such a treatment of localising the line capacitance gives reasonably accurate results. The most commonly used methods (known as *localised capacitance methods*) for the solution of medium transmissions lines are :

(*i*) End condenser method (*ii*) Nominal *T* method (*iii*) Nominal π method.

Although the above methods are used for obtaining the performance calculations of medium lines, they can also be used for short lines if their line capacitance is given in a particular problem.

10.7 End Condenser Method

In this method, the capacitance of the line is lumped or concentrated at the receiving or load end as shown in Fig. 10.8. This method of localising the line capacitance at the load end overestimates the effects of capacitance. In Fig. 10.8, one phase of the 3-phase transmission line is shown as it is more convenient to work in phase instead of line-to-line values.

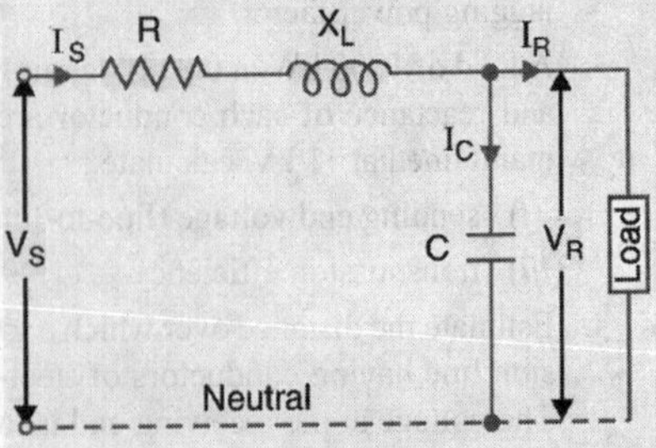

Fig. 10.8

Let I_R = load current per phase
R = resistance per phase
X_L = inductive reactance per phase
C = capacitance per phase
$\cos \phi_R$ = receiving end power factor (*lagging*)

V_S = sending end voltage per phase

The *phasor diagram for the circuit is shown in Fig 10.9. Taking the receiving end voltage $\vec{V_R}$ as the reference phasor, we have, $\vec{V_R} = V_R + j\,0$

Load current, $\vec{I_R} = I_R\,(\cos\phi_R - j\sin\phi_R)$

Capacitive current, $\vec{I_C} = j\,\vec{V_R}\,\omega\,C = j\,2\,\pi f C\,\vec{V_R}$

The sending end current $\vec{I_S}$ is the phasor sum of load current $\vec{I_R}$ and capacitive current $\vec{I_C}$ *i.e.*,

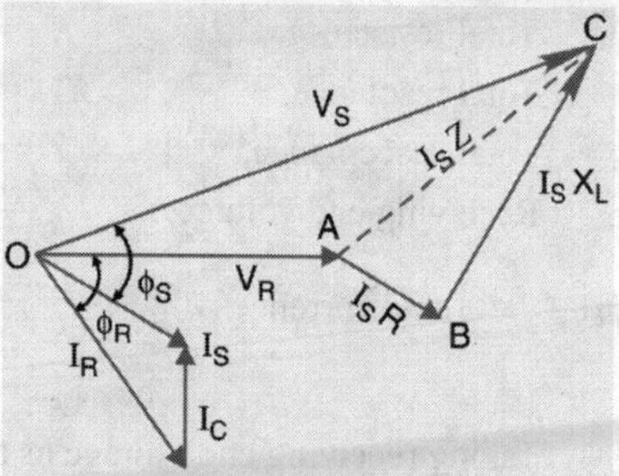

Fig. 10.9

$$\vec{I_S} = \vec{I_R} + \vec{I_C}$$
$$= I_R\,(\cos\phi_R - j\sin\phi_R) + j\,2\,\pi f C\,V_R$$
$$= I_R\cos\phi_R + j\,(-I_R\sin\phi_R + 2\,\pi f C V_R)$$

Voltage drop/phase $= \vec{I_S}\,\vec{Z} = \vec{I_S}\,(R + j X_L)$

Sending end voltage, $\vec{V_S} = \vec{V_R} + \vec{I_S}\,\vec{Z} = \vec{V_R} + \vec{I_S}\,(R + j X_L)$

Thus, the magnitude of sending end voltage V_S can be calculated.

$$\%\text{ Voltage regulation} = \frac{V_S - V_R}{V_R} \times 100$$

$$\%\text{ Voltage transmission efficiency} = \frac{\text{Power delivered / phase}}{\text{Power delivered / phase + losses / phase}} \times 100$$
$$= \frac{V_R\,I_R\cos\phi_R}{V_R\,I_R\cos\phi_R + I_S^2\,R} \times 100$$

Limitations. Although end condenser method for the solution of medium lines is simple to work out calculations, yet it has the following drawbacks :

(*i*) There is a considerable error (about 10%) in calculations because the distributed capacitance has been assumed to be lumped or concentrated.

(*ii*) This method overestimates the effects of line capacitance.

Example 10.10. *A (medium) single phase transmission line 100 km long has the following constants :*

Resistance/km = 0·25 Ω ; *Reactance/km = 0·8 Ω*

Susceptance/km = 14 × 10^{-6} siemen ; *Receiving end line voltage = 66,000 V*

Assuming that the total capacitance of the line is localised at the receiving end alone, determine (i) the sending end current (ii) the sending end voltage (iii) regulation and (iv) supply power factor. The line is delivering 15,000 kW at 0.8 power factor lagging. Draw the phasor diagram to illustrate your calculations.

Solution. Figs. 10.10 (*i*) and (*ii*) show the circuit diagram and phasor diagram of the line respectively.

---

* Note the construction of phasor diagram. The load current $\vec{I_R}$ lags behind $\vec{V_R}$ by ϕ_R. The capacitive current $\vec{I_C}$ leads $\vec{V_R}$ by 90° as shown. The phasor sum of $\vec{I_C}$ and $\vec{I_R}$ is the sending end current $\vec{I_S}$. The drop in the line resistance is $\vec{I_S}R$ (*AB*) in phase with $\vec{I_S}$ whereas inductive drop $\vec{I_S}X_L$ (*BC*) leads I_S by 90°. Therefore, *OC* represents the sending end voltage $\vec{V_S}$. The angle ϕ_S between the sending end voltage $\vec{V_S}$ and sending end current $\vec{I_S}$ determines the sending end power factor $\cos\phi_S$.

Total resistance, $R = 0{\cdot}25 \times 100 = 25\ \Omega$

Total reactance, $X_L = 0{\cdot}8 \times 100 = 80\ \Omega$

Total susceptance, $Y = 14 \times 10^{-6} \times 100 = 14 \times 10^{-4}\ S$

Receiving end voltage, $V_R = 66{,}000$ V

$\therefore$ Load current, $I_R = \dfrac{15{,}000 \times 10^3}{66{,}000 \times 0{\cdot}8} = 284$ A

$\cos\phi_R = 0{\cdot}8\ ; \qquad \sin\phi_R = 0{\cdot}6$

Taking receiving end voltage as the reference phasor [see Fig.10.10 (*ii*)], we have,

$$\vec{V_R} = V_R + j\,0 = 66{,}000\,V$$

Load current, $\vec{I_R} = I_R(\cos\phi_R - j\sin\phi_R) = 284\,(0{\cdot}8 - j\,0{\cdot}6) = 227 - j\,170$

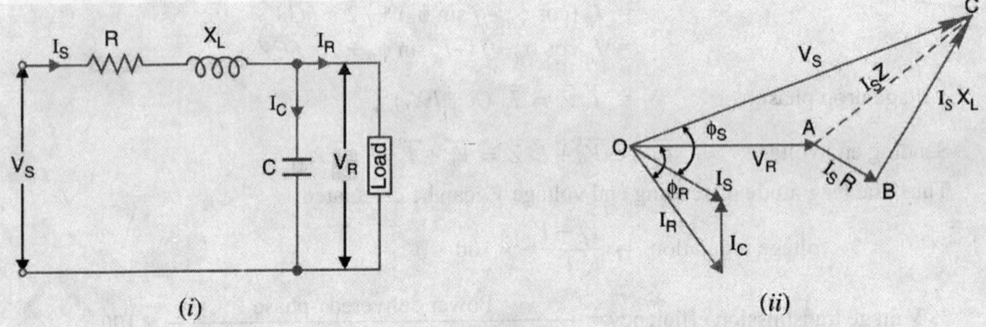

Fig. 10.10

Capacitive current, $\vec{I_C} = jY \times V_R = j\,14 \times 10^{-4} \times 66000 = j\,92$

(*i*) Sending end current, $\vec{I_S} = \vec{I_R} + \vec{I_C} = (227 - j\,170) + j\,92$

$= 227 - j\,78$ ... (*i*)

Magnitude of $I_S = \sqrt{(227)^2 + (78)^2} =$ **240 A**

(*ii*) Voltage drop $= \vec{I_S}\vec{Z} = \vec{I_S}(R + j\,X_L) = (227 - j\,78)\,(25 + j\,80)$

$= 5{,}675 + j\,18,160 - j\,1950 + 6240$

$= 11{,}915 + j\,16{,}210$

Sending end voltage, $\vec{V_S} = \vec{V_R} + \vec{I_S}\,\vec{Z} = 66{,}000 + 11{,}915 + j\,16{,}210$

$= 77{,}915 + j\,16{,}210$ ...(*ii*)

Magnitude of $V_S = \sqrt{(77915)^2 + (16210)^2} =$ **79583V**

(*iii*) % Voltage regulation $= \dfrac{V_S - V_R}{V_R} \times 100 = \dfrac{79{,}583 - 66{,}000}{66{,}000} \times 100 =$ **20·58%**

(*iv*) Referring to exp. (*i*), phase angle between $\vec{V_R}$ and $\vec{I_R}$ is :

$$\theta_1 = \tan^{-1} - 78/227 = \tan^{-1}(-\,0{\cdot}3436) = -\,18{\cdot}96^\circ$$

Referring to exp. (*ii*), phase angle between $\vec{V_R}$ and $\vec{V_S}$ is :

$$\theta_2 = \tan^{-1}\frac{16210}{77915} = \tan^{-1}(0\cdot 2036) = 11\cdot 50^\circ$$

$\therefore$ Supply power factor angle, $\phi_S = 18{\cdot}96^\circ + 11{\cdot}50^\circ = 30{\cdot}46^\circ$

$\therefore$ Supply p.f. $= \cos\phi_S = \cos 30{\cdot}46^\circ =$ **0·86 lag**

10.8 Nominal T Method

In this method, the whole line capacitance is assumed to be concentrated at the middle point of the line and half the line resistance and reactance are lumped on its either side as shown in Fig. 10.11. Therefore, in this arrangement, full charging current flows over half the line. In Fig. 10.11, one phase of 3-phase transmission line is shown as it is advantageous to work in phase instead of line-to-line values.

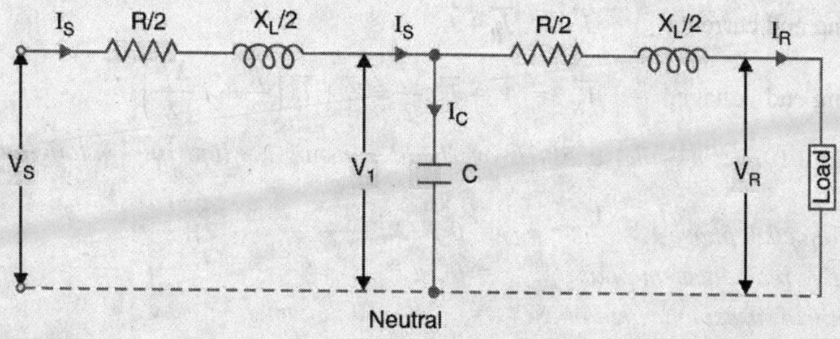

Fig. 10.11

Let I_R = load current per phase ; R = resistance per phase

X_L = inductive reactance per phase ; C = capacitance per phase

$\cos\phi_R$ = receiving end power factor (*lagging*) ; V_S = sending end voltage/phase

V_1 = voltage across capacitor C

The *phasor diagram for the circuit is shown in Fig. 10.12. Taking the receiving end voltage $\vec{V_R}$ as the reference phasor, we have,

Receiving end voltage, $\vec{V_R} = V_R + j\,0$

Load current, $\vec{I_R} = I_R(\cos\phi_R - j\sin\phi_R)$

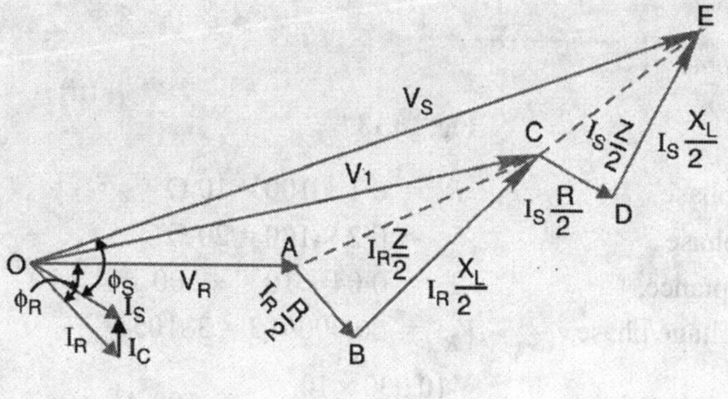

Fig. 10.12

* Note the construction of phasor diagram. $\vec{V_R}$ is taken as the reference phasor represented by OA. The load current $\vec{I_R}$ lags behind $\vec{V_R}$ by ϕ_R. The drop $AB = I_R R/2$ is in phase with $\vec{I_R}$ and $BC = I_R \cdot X_L/2$ leads $\vec{I_R}$ by 90°. The phasor OC represents the voltage $\vec{V_1}$ across condenser C. The capacitor current $\vec{I_C}$ leads $\vec{V_1}$ by 90° as shown. The phasor sum of $\vec{I_R}$ and $\vec{I_C}$ gives $\vec{I_S}$. Now $CD = I_S R/2$ is in phase with $\vec{I_S}$ while $DE = I_S X_L/2$ leads $\vec{I_S}$ by 90°. Then, OE represents the sending end voltage $\vec{V_S}$.

Voltage across C, $\vec{V_1} = \vec{V_R} + \vec{I_R}\,\vec{Z}/2$

$= V_R + I_R(\cos\phi_R - j\sin\phi_R)\left(\frac{R}{2} + j\frac{X_L}{2}\right)$

Capacitive current, $\vec{I_C} = j\omega C\,\vec{V_1} = j2\pi f C\,\vec{V_1}$

Sending end current, $\vec{I_S} = \vec{I_R} + \vec{I_C}$

Sending end voltage, $\vec{V_S} = \vec{V_1} + \vec{I_S}\,\frac{\vec{Z}}{2} = \vec{V_1} + \vec{I_S}\left(\frac{R}{2} + j\frac{X_L}{2}\right)$

Example 10.11. *A 3-phase, 50-Hz overhead transmission line 100 km long has the following constants :*

Resistance/km/phase $= 0.1\ \Omega$

Inductive reactance/km/phase $= 0{\cdot}2\ \Omega$

Capacitive susceptance/km/phase $= 0{\cdot}04 \times 10^{-4}$ *siemen*

Determine (i) the sending end current (ii) sending end voltage (iii) sending end power factor and (iv) transmission efficiency when supplying a balanced load of 10,000 kW at 66 kV, p.f. 0·8 lagging. Use nominal T method.

Solution. Figs. 10.13 (*i*) and 10.13 (*ii*) show the circuit diagram and phasor diagram of the line respectively.

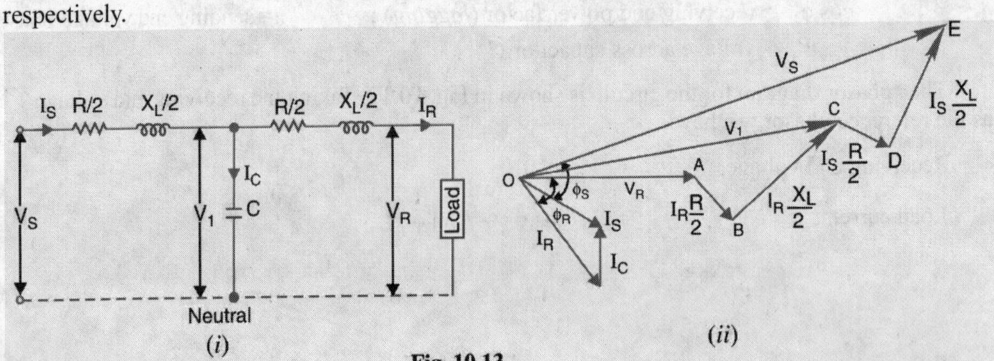

Fig. 10.13

Total resistance/phase, $R = 0{\cdot}1 \times 100 = 10\ \Omega$

Total reactance/phase. $X_L = 0{\cdot}2 \times 100 = 20\ \Omega$

Capacitive susceptance, $Y = 0{\cdot}04 \times 10^{-4} \times 100 = 4 \times 10^{-4}$ S

Receiving end voltage/phase, $V_R = 66{,}000/\sqrt{3} = 38105\ V$

Load current, $I_R = \dfrac{10{,}000 \times 10^3}{\sqrt{3} \times 66 \times 10^3 \times 0{\cdot}8} = 109$ A

$\cos\phi_R = 0{\cdot}8\ ;\ \ \sin\phi_R = 0{\cdot}6$

Impedance per phase, $\vec{Z} = R + jX_L = 10 + j20$

(*i*) Taking receiving end voltage as the reference phasor [see Fig. 10.13 (*ii*)], we have,

Receiving end voltage, $\vec{V_R} = V_R + j0 = 38{,}105\ V$

Load current, $\vec{I_R} = I_R(\cos\phi_R - j\sin\phi_R) = 109\,(0{\cdot}8 - j\,0{\cdot}6) = 87{\cdot}2 - j\,65{\cdot}4$

Voltage across C, $\vec{V_1} = \vec{V_R} + \vec{I_R}\,\vec{Z}/2 = 38,\ 105 + (87{\cdot}2\ - j\,65{\cdot}4)\,(5 + j\,10)$

$= 38{,}105 + 436 + j\,872 - j\,327 + 654 = 39{,}195 + j\,545$

Charging current, $\vec{I_C} = jY\vec{V_1} = j\,4 \times 10^{-4}(39{,}195 + j\,545) = -0\cdot218 + j\,15\cdot6$

Sending end current, $\vec{I_S} = \vec{I_R} + \vec{I_C} = (87\cdot2 - j\,65\cdot4) + (-0\cdot218 + j\,15\cdot6)$

$= 87\cdot0 - j\,49\cdot8 = 100 \angle -29°47' \text{ A}$

$\therefore$ Sending end current $=$ **100 A**

(*ii*) Sending end voltage, $\vec{V_S} = \vec{V_1} + \vec{I_S}\,\vec{Z}/2 = (39{,}195 + j\,545) + (87\cdot0 - j\,49\cdot8)(5 + j\,10)$

$= 39{,}195 + j\,545 + 434\cdot9 + j\,870 - j\,249 + 498$

$= 40128 + j\,1170 = 40145 \angle 1°40' \text{ V}$

$\therefore$ Line value of sending end voltage

$= 40145 \times \sqrt{3} = 69\,533 \text{ V} =$ **69·533 kV**

(*iii*) Referring to phasor diagram in Fig. 10.14,

θ_1 = angle between $\vec{V_R}$ and $\vec{V_S} = 1°40'$

θ_2 = angle between $\vec{V_R}$ and $\vec{I_S} = 29°\,47'$

$\therefore$ ϕ_S = angle between $\vec{V_S}$ and $\vec{I_S}$

$= \theta_1 + \theta_2 = 1°40' + 29°47' = 31°27'$

$\therefore$ Sending end power factor, $\cos\phi_S = \cos 31°27' =$ **0·853 lag**

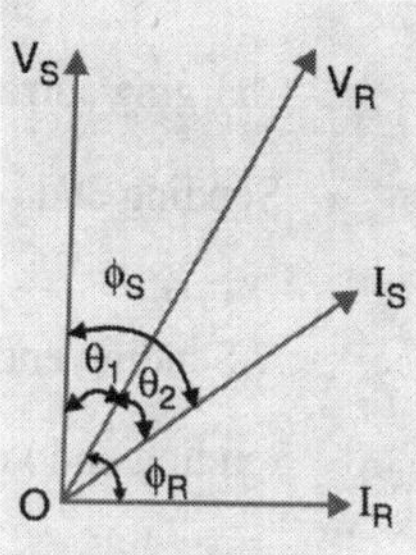

Fig. 10.14

(*iv*) Sending end power $= 3\,V_S I_S \cos\phi_S = 3 \times 40{,}145 \times 100 \times 0\cdot853$

$= 10273105 \text{ W} = 10273\cdot105 \text{ kW}$

Power delivered $= 10{,}000$ kW

$\therefore$ Transmission efficiency $= \dfrac{10{,}000}{10273\cdot105} \times 100 =$ **97·34%**

Example 10.12. *A 3-phase, 50 Hz transmission line 100 km long delivers 20 MW at 0·9 p.f. lagging and at 110 kV. The resistance and reactance of the line per phase per km are 0·2 Ω and 0·4 Ω respectively, while capacitance admittance is $2\cdot5 \times 10^{-6}$ siemen/km/phase. Calculate : (i) the current and voltage at the sending end (ii) efficiency of transmission. Use nominal T method.*

Solution. Figs. 10.15 (*i*) and 10.15 (*ii*) show the circuit diagram and phasor diagram respectively.

Total resistance/phase, $R = 0\cdot2 \times 100 = 20\ \Omega$

Total reactance/phase, $X_L = 0\cdot4 \times 100 = 40\ \Omega$

Total capacitance admittance/phase, $Y = 2\cdot5 \times 10^{-6} \times 100 = 2\cdot5 \times 10^{-4}$ S

Phase impedance, $\vec{Z} = 20 + j40$

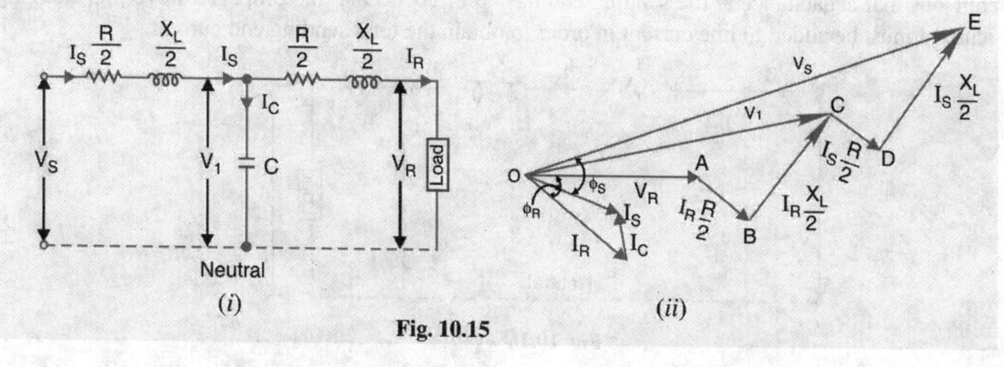

Fig. 10.15

Receiving end voltage/phase, $V_R = 110 \times 10^3/\sqrt{3} = 63508$ V

$$\text{Load current, } I_R = \frac{20 \times 10^6}{\sqrt{3} \times 110 \times 10^3 \times 0{\cdot}9} = 116.6 \text{ A}$$

$$\cos \phi_R = 0{\cdot}9 \; ; \; \sin \phi_R = 0{\cdot}435$$

(*i*) Taking receiving end voltage as the reference phasor [see phasor diagram 10.15 (*ii*)], we have,

$$\vec{V_R} = V_R + j0 = 63508 \text{ V}$$

Load current, $\vec{I_R} = I_R(\cos\phi_R - j\sin\phi_R) = 116{\cdot}6\,(0{\cdot}9 - j\,0{\cdot}435) = 105 - j50{\cdot}7$

Voltage across C, $\vec{V_1} = \vec{V_R} + \vec{I_R}\,\vec{Z}/2 = 63508 + (105 - j\,50{\cdot}7)(10 + j\,20)$

$= 63508 + (2064 + j1593) = 65572 + j1593$

Charging current, $\vec{I_C} = j\,Y\vec{V_1} = j\,2{\cdot}5 \times 10^{-4}\,(65572 + j1593) = -\,0{\cdot}4 + j\,16{\cdot}4$

Sending end current, $\vec{I_S} = \vec{I_R} + \vec{I_C} = (105 - j\,50{\cdot}7) + (-\,0{\cdot}4 + j16{\cdot}4)$

$= (104{\cdot}6 - j\,34{\cdot}3) = 110 \angle -18^\circ 9'$ A

∴ Sending end current $=$ **110 A**

Sending end voltage, $\vec{V_S} = \vec{V_1} + \vec{I_S}\,\vec{Z}/2$

$= (65572 + j1593) + (104{\cdot}6 - j34{\cdot}3)(10 + j20)$

$= 67304 + j\,3342$

∴ Magnitude of $V_S = \sqrt{(67304)^2 + (3342)^2} = 67387\ V$

∴ Line value of sending end voltage

$= 67387 \times \sqrt{3} = 116717 \text{ V} = $ **116·717 kV**

(*ii*) Total line losses for the three phases

$= 3\,I_S^2\,R/2 + 3I_R^2\,R/2$

$= 3 \times (110)^2 \times 10 + 3 \times (116{\cdot}6)^2 \times 10$

$= 0{\cdot}770 \times 10^6 \text{ W} = 0{\cdot}770 \text{ MW}$

∴ Transmission efficiency $= \dfrac{20}{20 + 0{\cdot}770} \times 100 = $ **96·29%**

10.9 Nominal π Method

In this method, capacitance of each conductor (*i.e.,* line to neutral) is divided into two halves; one half being lumped at the sending end and the other half at the receiving end as shown in Fig. 10.16. It is obvious that capacitance at the sending end has no effect on the line drop. However, its charging current must be added to line current in order to obtain the total sending end current.

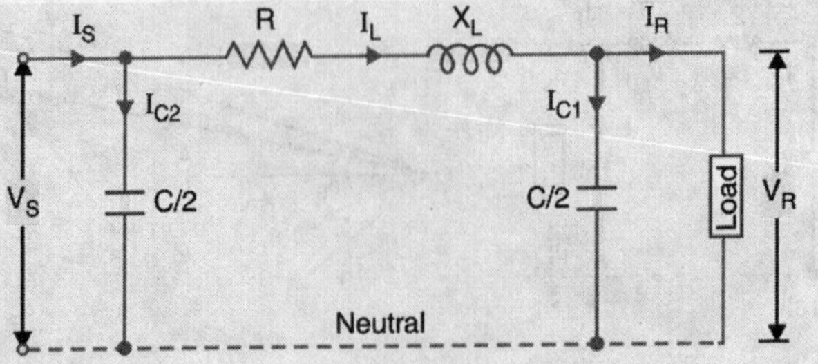

Fig. 10.16

Let

I_R = load current per phase
R = resistance per phase
X_L = inductive reactance per phase
C = capacitance per phase
$\cos \phi_R$ = receiving end power factor (*lagging*)
V_S = sending end voltage per phase

The *phasor diagram for the circuit is shown in Fig. 10.17. Taking the receiving end voltage as the reference phasor, we have,

$$\vec{V_R} = V_R + j\,0$$

Load current, $\vec{I_R} = I_R (\cos \phi_R - j \sin \phi_R)$

Charging current at load end is

$$\vec{I_{C1}} = j\,\omega\,(C/2)\,\vec{V_R} = j\,\pi\,f\,C\,\vec{V_R}$$

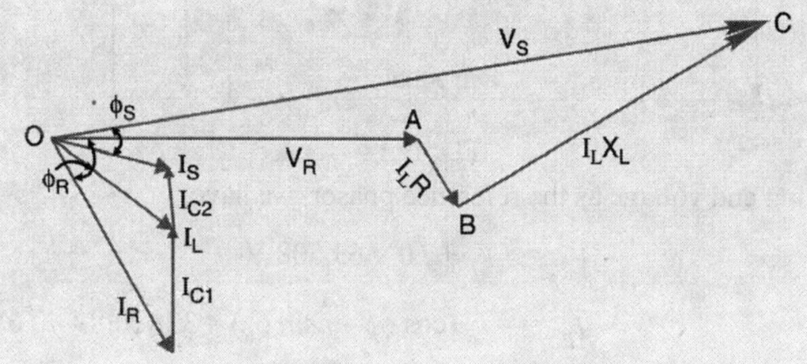

Fig. 10.17

Line current, $\vec{I_L} = \vec{I_R} + \vec{I_{C1}}$

Sending end voltage, $\vec{V_S} = \vec{V_R} + \vec{I_L}\,\vec{Z} = \vec{V_R} + \vec{I_L}\,(R + jX_L)$

Charging current at the sending end is

$$\vec{I_{C2}} = j\,\omega\,(C/2)\,\vec{V_S} = j\,\pi\,f\,C\,\vec{V_S}$$

∴ Sending end current, $\vec{I_S} = \vec{I_L} + \vec{I_{C2}}$

Example 10.13 *A 3-phase, 50Hz, 150 km line has a resistance, inductive reactance and capacitive shunt admittance of 0·1 Ω, 0·5 Ω and 3 × 10⁻⁶ S per km per phase. If the line delivers 50 MW at 110 kV and 0·8 p.f. lagging, determine the sending end voltage and current. Assume a nominal π circuit for the line.*

---

* Note the construction of phasor diagram. $\vec{V_R}$ is taken as the reference phasor represented by OA. The current $\vec{I_R}$ lags behind $\vec{V_R}$ by ϕ_R. The charging current $\vec{I_{C1}}$ leads $\vec{V_R}$ by 90°. The line current $\vec{I_L}$ is the phasor sum of $\vec{I_R}$ and $\vec{I_{C1}}$. The drop $AB = I_L R$ is in phase with $\vec{I_L}$ whereas drop $BC = I_L X_L$ leads $\vec{I_L}$ by 90°. Then OC represents the sending end voltage $\vec{V_S}$. The charging current $\vec{I_{C2}}$ leads $\vec{V_S}$ by 90°. Therefore, sending end current $\vec{I_S}$ is the phasor sum of the $\vec{I_{C2}}$ and $\vec{I_L}$. The angle ϕ_S between sending end voltage V_S and sending end current I_S determines the sending end p.f. $\cos \phi_S$.

Solution. Fig. 10.18 shows the circuit diagram for the line.

Total resistance/phase, $R = 0{\cdot}1 \times 150 = 15\ \Omega$

Total reactance/phase, $X_L = 0{\cdot}5 \times 150 = 75\ \Omega$

Capacitive admittance/phase, $Y = 3 \times 10^{-6} \times 150 = 45 \times 10^{-5}$ S

Receiving end voltage/phase, $V_R = 110 \times 10^3/\sqrt{3} = 63{,}508$ V

Load current, $I_R = \dfrac{50 \times 10^6}{\sqrt{3} \times 110 \times 10^3 \times 0{\cdot}8} = 328$ A

$\cos \phi_R = 0{\cdot}8\ ;\ \sin \phi_R = 0{\cdot}6$

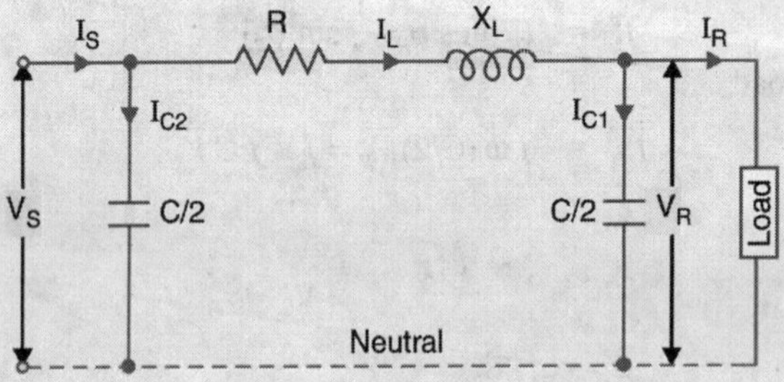

Fig. 10.18

Taking receiving end voltage as the reference phasor, we have,

$$\vec{V_R} = V_R + j\,0 = 63{,}508 \text{ V}$$

Load current, $\vec{I_R} = I_R(\cos \phi_R - j \sin \phi_R) = 328\,(0{\cdot}8 - j0{\cdot}6) = 262{\cdot}4 - j196{\cdot}8$

Charging current at the load end is

$$\vec{I_{C1}} = \vec{V_R}\, j\frac{Y}{2} = 63{,}508 \times j\frac{45 \times 10^{-5}}{2} = j\,14{\cdot}3$$

Line current, $\vec{I_L} = \vec{I_R} + \vec{I_{C1}} = (262{\cdot}4 - j\,196{\cdot}8) + j\,14{\cdot}3 = 262{\cdot}4 - j\,182{\cdot}5$

Sending end voltage, $\vec{V_S} = \vec{V_R} + \vec{I_L}\,\vec{Z} = \vec{V_R} + \vec{I_L}\,(R + j\,X_L)$

$= 63{,}508 + (262{\cdot}4 - j\,182{\cdot}5)\,(15 + j\,75)$

$= 63{,}508 + 3936 + j\,19{,}680 - j\,2737{\cdot}5 + 13{,}687$

$= 81{,}131 + j\,16{,}942{\cdot}5 = 82{,}881\ \angle\ 11°47'$ V

$\therefore$ Line to line sending end voltage $= 82{,}881 \times \sqrt{3} = 1{,}43{,}550$ V $=$ **143·55 kV**

Charging current at the sending end is

$$I_{C2} = j\vec{V_S}\,Y/2 = (81{,}131 + j\,16{,}942 \cdot 5)\, j\,\frac{45 \times 10^{-5}}{2}$$

$= -\,3{\cdot}81 + j\,18{\cdot}25$

Sending end current, $\vec{I_S} = \vec{I_L} + \vec{I_{C_2}} = (262{\cdot}4 - j\,182{\cdot}5) + (-\,3{\cdot}81 + j\,18{\cdot}25)$

$= 258{\cdot}6 - j\,164{\cdot}25 = 306{\cdot}4\ \angle - 32{\cdot}4°$ A

$\therefore$ Sending end current $=$ **306·4 A**

Example 10·14. *A 100-km long, 3-phase, 50-Hz transmission line has following line constants:*

Resistance/phase/km = 0·1 Ω

Reactance/phase/km = 0·5 Ω

Susceptance/phase/km = 10 × 10⁻⁶ S

If the line supplies load of 20 MW at 0·9 p.f. lagging at 66 kV at the receiving end, calculate by nominal π method :

(i) sending end power factor *(ii) regulation*

(iii) transmission efficiency

Solution. Fig. 10·19 shows the circuit diagram for the line.

Total resistance/phase, $R = 0{\cdot}1 \times 100 = 10\ \Omega$

Total reactance/phase, $X_L = 0{\cdot}5 \times 100 = 50\ \Omega$

Susceptance/phase, $Y = 10 \times 10^{-6} \times 100 = 10 \times 10^{-4}$ S

Receiving end voltage/phase, $V_R = 66 \times 10^3/\sqrt{3} = 38105$ V

Load current, $$I_R = \frac{20 \times 10^6}{\sqrt{3} \times 66 \times 10^3 \times 0{\cdot}9} = 195 \text{ A}$$

$$\cos \phi_R = 0{\cdot}9 \quad ; \quad \sin \phi_R = 0{\cdot}435$$

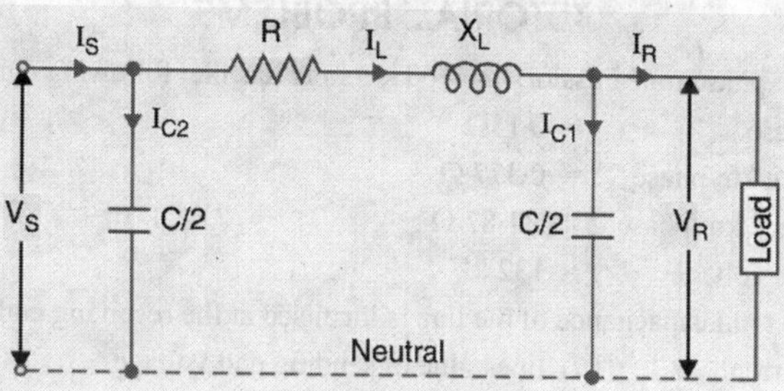

Fig. 10.19

Taking receiving end voltage as the reference phasor, we have,

$$\vec{V_R} = V_R + j0 = 38105 \text{ V}$$

Load current, $$\vec{I_R} = I_R(\cos \phi_R - j \sin \phi_R) = 195\,(0{\cdot}9 - j\,0{\cdot}435) = 176 - j\,85$$

Charging current at the receiving end is

$$\overrightarrow{I_{C1}} = \vec{V_R}\, j\frac{Y}{2} = 38105 \times j\,\frac{10 \times 10^{-4}}{2} = j\,19$$

Line current, $$\vec{I_L} = \vec{I_R} + \overrightarrow{I_{C1}} = (176 - j\,85) + j\,19 = 176 - j\,66$$

Sending end voltage , $$\vec{V_S} = \vec{V_R} + \vec{I_L}\,\vec{Z} = \vec{V_R} + \vec{I_L}\,(R + j\,X_L)$$

$$= 38{,}105 + (176 - j\,66)\,(10 + j\,50)$$

$$= 38{,}105 + (5060 + j\,8140)$$

$$= 43{,}165 + j\,8140 = 43{,}925 \angle 10{\cdot}65^\circ \text{ V}$$

Sending end line to line voltage $= 43,925 \times \sqrt{3} = 76 \times 10^3 \text{ V} = 76$ kV

Charging current at the sending end is

$$\overrightarrow{I_{C2}} = \vec{V_S}\, jY/2 = (43{,}165 + j\,8140)\, j\,\frac{10 \times 10^{-4}}{2}$$

$$= -\,4{\cdot}0 + j\,21{\cdot}6$$

∴ Sending end current, $$\vec{I_S} = \vec{I_L} + \overrightarrow{I_{C2}} = (176 - j\,66) + (-\,4{\cdot}0 + j\,21{\cdot}6)$$

$$= 172 - j\,44{\cdot}4 = 177{\cdot}6 \angle -14{\cdot}5^\circ \text{ A}$$

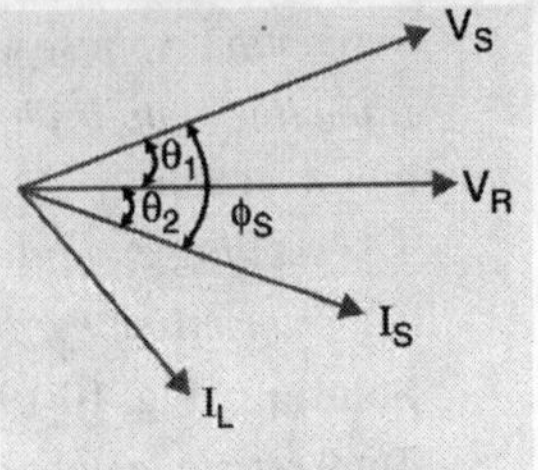

Fig. 10.20

(*i*) Referring to phasor diagram in Fig. 10·20,

θ_1 = angle between $\vec{V_R}$ and $\vec{V_S}$ = 10·65°

θ_2 = angle between $\vec{V_R}$ and $\vec{I_S}$ = − 14·5°

∴ ϕ_S = angle between $\vec{V_S}$ and $\vec{I_S}$ = $\theta_2 + \theta_1$

= 14·5° + 10·65° = 25·15°

∴ Sending end p.f., cos ϕ_S = cos 25·15° = **0·905 lag**

(*ii*) % Voltage regulation $= \dfrac{V_S - V_R}{V_R} \times 100 = \dfrac{43925 - 38105}{38105} \times 100$ = **15·27 %**

(*iii*) Sending end power $= 3\, V_S I_S \cos \phi_S$ = 3 × 43925 × 177·6 × 0·905

= 21·18 × 10^6 W = 21·18 MW

Transmission efficiency = (20/21·18) × 100 = **94 %**

TUTORIAL PROBLEMS

1. A (medium) single phase transmission line 100 km long has the following constants :

Resistance/km/phase = 0·15Ω
Inductive reactance/km/phase = 0·377 Ω
Capacitive reactance/km/phase = 31·87 Ω
Receiving end line voltage = 132 kV

Assuming that the total capacitance of the line is localised at the receiving end alone, determine :

(*i*) sending end current (*ii*) line value of sending end voltage
(*iii*) regulation (*iv*) sending end power factor

The line is delivering 72 MW at 0·8 p.f. lagging.

[(*i*) 377·3 A (*ii*) 155·7 kV (*iii*) 17·9% (*iv*) 0·774 lag]

2. A 3-phase, 50 Hz overhead transmission line has the following constants :

Resistance/phase = 9·6 Ω
Inductance/phase = 0·097 mH
Capacitance/phase = 0·765 μF

If the line is supplying a balanced load of 24,000 kVA 0·8 p.f. lagging at 66 kV, calculate :

(*i*) sending end current (*ii*) line value of sending end voltage
(*iii*) sending end power factor (*iv*) percentage regulation
(*v*) transmission efficiency **[(*i*) 204 A (*ii*) 75 kV (*iii*) 0·814 lag (*iv*) 13·63 % (*v*) 93·7%]**

3. A 3-phase, 50 Hz, overhead transmission line delivers 10 MW at 0·8 p.f. lagging and at 66 kV. The resistance and inductive reactance of the line per phase are 10 Ω and 20 Ω respectively while capacitance admittance is 4 × 10^{-4} siemen. Calculate :

(*i*) the sending end current (*ii*) sending end voltage (line-to-line)
(*iii*) sending end power factor (*iv*) transmission efficiency

Use nominal *T* method. **[(*i*) 100 A (*ii*) 69·8 kV (*iii*) 0·852 (*iv*) 97·5%]**

4. A 3-phase, 50 Hz, 100 km transmission line has the following constants ;

Resistance/phase/km = 0·1 Ω
Reactance/phase/km = 0·5 Ω
Susceptance/phase/km = 10^{-5} siemen

If the line supplies a load of 20 MW at 0·9 p.f. lagging at 66 kV at the receiving end, calculate by using nominal π method :

(*i*) sending end current (*ii*) line value of sending end voltage (*iii*) sending end power factor (*iv*) regulation

[(*i*) 177·6 A (*ii*) 76kV (*iii*) 0·905 lag (*iv*) 15·15%]

5. A 3-phase overhead transmission line has the following constants :

Resistance/phase = 10 Ω

Inductive reactance/phase = 35 Ω

Capacitive admittance/phase = 3×10^{-4} siemen

If the line supplied a balanced load of 40,000 kVA at 110 kV and 0·8 p.f. lagging, calculate :

(*i*) sending end power factor (*ii*) percentage regulation

(*iii*) transmission efficiency **[(*i*) 0·798 lag (*ii*) 10% (*iii*) 96·38%]**

6. A 3-phase, 50 Hz overhead transmission line, 100 km long, 110 kV between the lines at the receiving end has the following constants :

Resistance per km per phase = 0·153 Ω

Inductance per km per phase = 1·21 mH

Capacitance per km per phase = 0·00958 μF

The line supplies a load of 20,000 kW at 0·9 power factor lagging. Calculate using nominal π representation, the sending end voltage, current, power factor, regulation and the efficiency of the line. Neglect leakage. **[115·645 kV (line voltage) : 109 ∠ – 16·68° A ; 0·923 lag ; 5·13 %; 97·21 %]**

10.10 Long Transmission Lines

It is well known that line constants of the transmission line are uniformly distributed over the entire length of the line. However, reasonable accuracy can be obtained in line calculations for short and medium lines by considering these constants as lumped. If such an assumption of lumped constants is applied to long transmission lines (having length excess of about 150 km), it is found that serious errors are introduced in the performance calculations. Therefore, in order to obtain fair degree of accuracy in the performance calculations of long lines, the line constants are considered as uniformly distributed throughout the length of the line. Rigorous mathematical treatment is required for the solution of such lines.

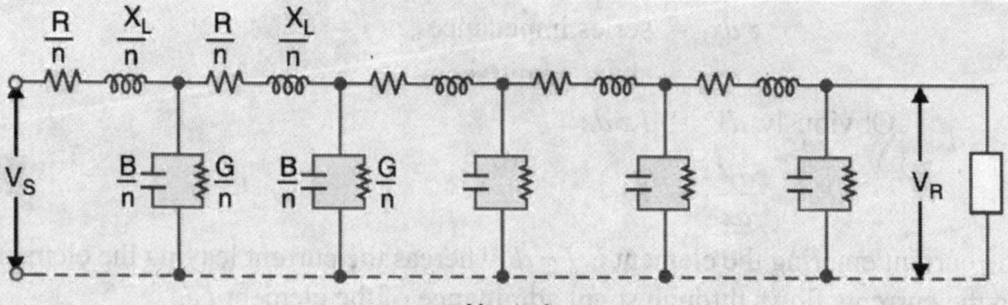

Fig. 10.21

Fig. 10.21 shows the equivalent circuit of a 3-phase long transmission line on a phase-neutral basis. The whole line length is divided into n sections, each section having line constants $\frac{1}{n}th$ of those for the whole line. The following points may by noted :

(*i*) The line constants are uniformly distributed over the entire length of line as is actually the case.

(*ii*) The resistance and inductive reactance are the series elements.

(*iii*) The leakage susceptance (B) and leakage conductance (G) are shunt elements. The leakage susceptance is due to the fact that capacitance exists between line and neutral. The leakage conductance takes into account the energy losses occurring through leakage over the insulators or due to corona effect between conductors. Admittance = $\sqrt{G^2 + B^2}$.

(*iv*) The leakage current through shunt admittance is maximum at the sending end of the line and decreases continuously as the receiving end of the circuit is approached at which point its value is zero.

10.11 Analysis of Long Transmission Line (*Rigorous method*)

Fig. 10.22 shows one phase and neutral connection of a 3-phase line with impedance and shunt admittance of the line uniformly distributed.

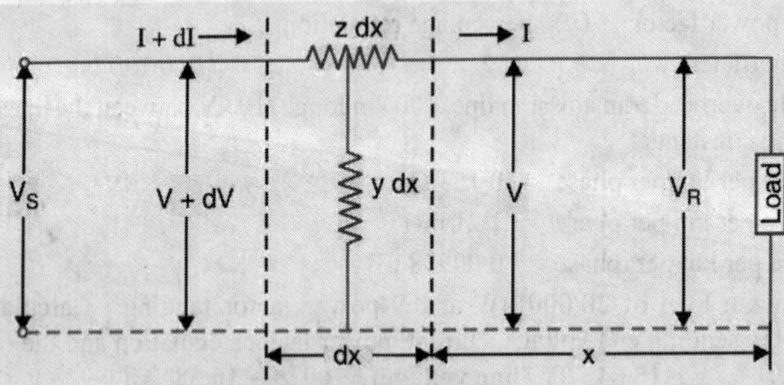

Fig. 10.22

Consider a small element in the line of length *dx* situated at a distance *x* from the receiving end.

Let z = series impedance of the line per unit length

y = shunt admittance of the line per unit length

V = voltage at the end of element towards receiving end

$V + dV$ = voltage at the end of element towards sending end

$I + dI$ = current entering the element dx

I = current leaving the element dx

Then for the small element *dx*,

$z\,dx$ = series impedance

$y\,dx$ = shunt admittance

Obviously, $dV = I\,z\,dx$

or $$\frac{dV}{dx} = I\,z \quad \text{...(i)}$$

Now, the current entering the element is $I + dI$ whereas the current leaving the element is I. The difference in the currents flows through shunt admittance of the element *i.e.*,

dI = Current through shunt admittance of element = $V\,y\,dx$

or $$\frac{dI}{dx} = V\,y \quad \text{...(ii)}$$

Differentiating eq. (*i*) w.r.t. *x*, we get,

$$\frac{d^2V}{dx^2} = z\,\frac{dI}{dx} = z\,(V\,y) \qquad \left[\because \frac{dI}{dx} = V\,y \textit{ from } \text{exp. (ii)}\right]$$

or $$\frac{d^2V}{dx^2} = y\,z\,V \quad \text{... (iii)}$$

The solution of this differential equation is

$$V = k_1 \cosh\left(x\sqrt{y\,z}\right) + k_2 \sinh\left(x\sqrt{y\,z}\right) \quad \text{...(iv)}$$

Differentiating exp. (*iv*) *w.r.t.* *x*, we have,

$$\frac{dV}{dx} = k_1 \sqrt{yz} \sinh (x\sqrt{yz}) + k_2 \sqrt{yz} \cosh (x\sqrt{yz})$$

But $$\frac{dV}{dx} = Iz \qquad \text{[from exp. (i)]}$$

$$\therefore \quad Iz = k_1 \sqrt{yz} \sinh (x\sqrt{yz}) + k_2 \sqrt{zy} \cosh (x\sqrt{yz})$$

or $$I = \sqrt{\frac{y}{z}}\left[k_1 \sinh (x\sqrt{yz}) + k_2 \cosh \left(x\sqrt{yz}\right)\right] \qquad ...(v)$$

Equations (*iv*) and (*v*) give the expressions for V and I in the form of unknown constants k_1 and k_2. The values of k_1 and k_2 can be found by applying end conditions as under :

$$\text{At } x = 0, \quad V = V_R \text{ and } I = I_R$$

Putting these values in eq. (*iv*), we have,

$$V_R = k_1 \cosh 0 + k_2 \sinh 0 = k_1 + 0$$

$$\therefore \quad V_R = k_1$$

Similarly, putting $x = 0$, $V = V_R$ and $I = I_R$ in eq. (*v*), we have,

$$I_R = \sqrt{\frac{y}{z}}\left[k_1 \sinh 0 + k_2 \cosh 0\right] = \sqrt{\frac{y}{z}}\left[0 + k_2\right]$$

$$\therefore \quad k_2 = \sqrt{\frac{z}{y}}\, I_R$$

Substituting the values of k_1 and k_2 in eqs. (*iv*) and (*v*), we get,

$$V = V_R \cosh (x\sqrt{yz}) + \sqrt{\frac{z}{y}}\, I_R \sinh (x\sqrt{yz})$$

and $$I = \sqrt{\frac{y}{z}}\, V_R \sinh (x\sqrt{yz}) + I_R \cosh (x\sqrt{yz})$$

The sending end voltage (V_S) and sending end current (I_S) are obtained by putting $x = l$ in the above equations *i.e.*,

$$V_S = V_R \cosh (l\sqrt{yz}) + \sqrt{\frac{z}{y}}\, I_R \sinh (l\sqrt{yz})$$

$$I_S = \sqrt{\frac{y}{z}}\, V_R \sinh (l\sqrt{yz}) + I_R \cosh (l\sqrt{yz})$$

Now, $$l\sqrt{yz} = \sqrt{l\,y \cdot l\,z} = \sqrt{YZ}$$

and $$\sqrt{\frac{y}{z}} = \sqrt{\frac{y\,l}{z\,l}} = \sqrt{\frac{Y}{Z}}$$

where Y = total shunt admittance of the line

Z = total series impedance of the line

Therefore, expressions for V_S and I_S become :

$$\mathbf{V}_S = \mathbf{V}_R \cosh \sqrt{\mathbf{YZ}} + \mathbf{I}_R \sqrt{\frac{\mathbf{Z}}{\mathbf{Y}}} \sinh \sqrt{\mathbf{YZ}}$$

$$\mathbf{I}_S = \mathbf{V}_R \sqrt{\frac{\mathbf{Y}}{\mathbf{Z}}} \sinh \sqrt{\mathbf{YZ}} + \mathbf{I}_R \cosh \sqrt{\mathbf{YZ}}$$

It is helpful to expand hyperbolic sine and cosine in terms of their power series.

$$\cosh \sqrt{YZ} = \left(1 + \frac{ZY}{2} + \frac{Z^2Y^2}{24} + \ldots\ldots\right)$$

$$\sinh \sqrt{YZ} = \left(\sqrt{YZ} + \frac{(YZ)^{3/2}}{6} + \ldots\ldots \right)$$

Example 10.15. *A 3-φ transmission line 200 km long has the following constants :*

Resistance/phase/km $= 0{\cdot}16\ \Omega$

Reactance/phase/km $= 0{\cdot}25\ \Omega$

Shunt admittance/phase/km $= 1{\cdot}5 \times 10^{-6}$ *S*

Calculate by rigorous method the sending end voltage and current when the line is delivering a load of 20 MW at 0·8 p.f. lagging. The receiving end voltage is kept constant at 110 kV.

Solution :

Total resistance/phase, $R = 0{\cdot}16 \times 200 = 32\ \Omega$

Total reactance/phase, $X_L = 0{\cdot}25 \times 200 = 50\ \Omega$

Total shunt admittance/phase, $Y = j\,1{\cdot}5 \times 10^{-6} \times 200 = 0{\cdot}0003 \angle 90^\circ$

Series Impedance/phase, $Z = R + j\,X_L = 32 + j\,50 = 59{\cdot}4 \angle 58^\circ$

The sending end voltage V_S per phase is given by :

$$V_S = V_R \cosh \sqrt{YZ} + I_R \sqrt{\frac{Z}{Y}} \sinh \sqrt{ZY} \quad \ldots(i)$$

Now $\sqrt{ZY} = \sqrt{59 \cdot 4 \angle 58^\circ \times 0 \cdot 0003 \angle 90^\circ} = 0 \cdot 133 \angle 74^\circ$

$$ZY = 0{\cdot}0178 \angle 148^\circ$$

$$Z^2Y^2 = 0{\cdot}00032 \angle 296^\circ$$

$$\sqrt{\frac{Z}{Y}} = \sqrt{\frac{59 \cdot 4 \angle 58^\circ}{0 \cdot 0003 \angle 90^\circ}} = 445 \angle -16^\circ$$

$$\sqrt{\frac{Y}{Z}} = \sqrt{\frac{0 \cdot 0003 \angle 90^\circ}{59 \cdot 4 \angle 58^\circ}} = 0 \cdot 00224 \angle 16^\circ$$

$\therefore$

$$\cosh \sqrt{YZ} = 1 + \frac{ZY}{2} + \frac{Z^2 Y^2}{24} \text{ approximately}$$

$$= 1 + \frac{0 \cdot 0178}{2} \angle 148^\circ + \frac{0 \cdot 00032}{24} \angle 296^\circ$$

$$= 1 + 0{\cdot}0089 \angle 148^\circ + 0{\cdot}0000133 \angle 296^\circ$$

$$= 1 + 0{\cdot}0089\,(-0{\cdot}848 + j\,0{\cdot}529) + 0{\cdot}0000133\,(0{\cdot}438 - j\,0{\cdot}9)$$

$$= 0{\cdot}992 + j\,0{\cdot}00469 = 0{\cdot}992 \angle 0{\cdot}26^\circ$$

$$\sinh \sqrt{YZ} = \sqrt{YZ} + \frac{(YZ)^{3/2}}{6} \text{ approximately}$$

$$= 0{\cdot}133 \angle 74^\circ + \frac{0 \cdot 0024 \angle 222^\circ}{6}$$

$$= 0{\cdot}133 \angle 74^\circ + 0{\cdot}0004 \angle 222^\circ$$

$$= 0{\cdot}133\,(0{\cdot}275 + j\,0{\cdot}961) + 0{\cdot}0004\,(-0{\cdot}743 - j\,0{\cdot}67)$$

$$= 0{\cdot}0362 + j\,0{\cdot}1275 = 0{\cdot}1325 \angle 74^\circ 6'$$

Receiving end voltage per phase is

$$V_R = 110 \times 10^3/\sqrt{3} = 63508 \text{ V}$$

Receiving end current, $I_R = \dfrac{20 \times 10^6}{\sqrt{3} \times 110 \times 10^3 \times 0 \cdot 8} = 131\ A$

Putting the various values in exp (*i*), we get,

$$\begin{aligned} V_S &= 63508 \times 0{\cdot}992 \angle 0{\cdot}26^\circ + 131 \times 445 \angle -16^\circ 0' \times 0{\cdot}1325 \angle 74^\circ 6' \\ &= 63000 \angle 0{\cdot}26^\circ + 7724 \angle 58^\circ 6' \\ &= 63000\,(0{\cdot}999 + j\,0{\cdot}0045) + 7724\,(0{\cdot}5284 + j\,0{\cdot}8489) \\ &= 67018 + j\,6840 = 67366 \angle 5^\circ 50' \text{ V} \end{aligned}$$

Sending end line-to-line voltage $= 67366 \times \sqrt{3} = 116{\cdot}67 \times 10^3$ V = **116·67 kV**

The sending end current I_S is given by :

$$I_S = V_R \sqrt{\frac{Y}{Z}} \sinh \sqrt{Y Z} + I_R \cosh \sqrt{Y Z}$$

Putting the various values, we get,

$$\begin{aligned} I_S &= 63508 \times 0{\cdot}00224 \angle 16^\circ \times 0{\cdot}1325 \angle 74^\circ 6' + 131 \times 0{\cdot}992 \angle 0{\cdot}26^\circ \\ &= 18{\cdot}85 \angle 90^\circ 6' + 130 \angle 0{\cdot}26^\circ \\ &= 18{\cdot}85\,(-0{\cdot}0017 + j\,0{\cdot}999) + 130\,(0{\cdot}999 + j\,0{\cdot}0045) \\ &= 129{\cdot}83 + j\,19{\cdot}42 = 131{\cdot}1 \angle 8^\circ \text{ A} \end{aligned}$$

∴ Sending end current = **131·1 A**

TUTORIAL PROBLEMS

1. A 3-phase overhead transmission line has a total series impedance per phase of 200 ∠80° ohms and a total shunt admittance of 0·0013∠90° siemen per phase. The line delivers a load of 80 MW at 0·8 p.f. lagging and 220 kV between the lines. Determine the sending end line voltage and current by rigorous method. **[263·574 kV ; 187·5 A]**
2. A 3-phase transmission line, 160 km long, has the following constants :
   Resistance/phase/km = 0·2 Ω
   Reactance/phase/km = 0·3127 Ω
   Shunt admittance/phase/km = $1{\cdot}875 \times 10^{-6}$ S
   Determine the sending end voltage and current by rigorous method when the line is delivering a load of 25 MVA at 0·8 p.f. lagging. The receiving end voltage is kept constant at 110 kV. **[116·67 kV ; 131·1 A]**

10.12 Generalised Circuit Constants of a Transmission Line

In any four terminal *network, the input voltage and input current can be expressed in terms of output voltage and output current. Incidentally, a transmission line is a 4-terminal network ; two input terminals where power enters the network and two output terminals where power leaves the network.

Therefore, the input voltage $\left(\vec{V_S}\right)$ and input current $\left(\vec{I_S}\right)$ of a 3-phase transmission line can be expressed as :

$$\vec{V_S} = \vec{A}\,\vec{V_R} + \vec{B}\,\vec{I_R}$$
$$\vec{I_S} = \vec{C}\,\vec{V_R} + \vec{D}\,\vec{I_R}$$

where $\vec{V_S}$ = sending end voltage per phase

$\vec{I_S}$ = sending end current

$\vec{V_R}$ = receiving end voltage per phase

$\vec{I_R}$ = receiving end current

* The network should be *passive* (containing no source of e.m.f.), *linear* (impedances independent of current flowing) and *bilateral* (impedances independent of direction of current flowing). This condition is fully met in transmission lines.

and $\vec{A}, \vec{B}, \vec{C}$ and $\vec{D}$ (generally complex numbers) are the constants known as *generalised circuit constants* of the transmission line. The values of these constants depned upon the particular method adopted for solving a transmission line. Once the values of these constants are known, performance calculations of the line can be easily worked out. The following points may be kept in mind :

(*i*) The constants $\vec{A}, \vec{B}, \vec{C}$ and $\vec{D}$ are generally complex numbers.

(*ii*) The constants $\vec{A}$ and $\vec{D}$ are dimensionless whereas the dimensions of $\vec{B}$ and $\vec{C}$ are ohms and siemen respectively.

(*iii*) For a given transmisson line,

$$\vec{A} = \vec{D}$$

(*iv*) For a given transmission line,

$$\vec{A}\vec{D} - \vec{B}\vec{C} = 1$$

We shall establish the correctness of above characteristics of generalised circuit constants in the following discussion.

10.13 Determination of Generalised Constants for Transmission Lines

As stated previously, the sending end voltage ($\vec{V_S}$) and sending end current ($\vec{I_S}$) of a transmission line can be expressed as :

$$\vec{V_S} = \vec{A}\,\vec{V_R} + \vec{B}\,\vec{I_R} \quad ...(i)$$

$$\vec{I_S} = \vec{C}\,\vec{V_R} + \vec{D}\,\vec{I_R} \quad ...(ii)$$

We shall now determine the values of these constants for different types of transmission lines.

(*i*) **Short lines.** In short transmission lines, the effect of line capacitance is neglected. Therefore, the line is considered to have series impedance. Fig. 10.23 shows the circuit of a 3-phase transmission line on a single phase basis.

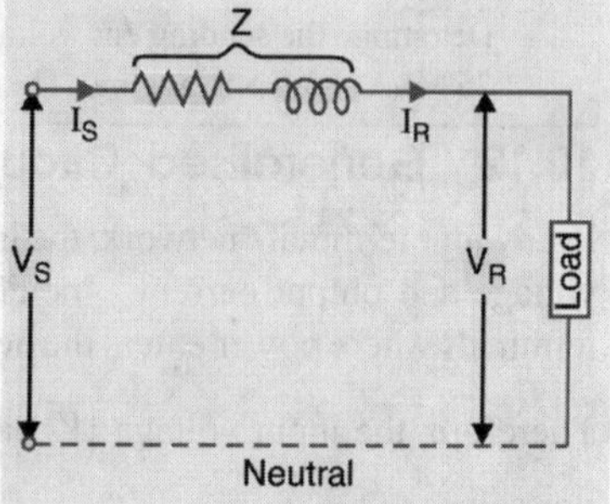

Fig. 10.23

Here, $$\vec{I_S} = \vec{I_R} \quad ...(iii)$$

and $$\vec{V_S} = \vec{V_R} + \vec{I_R}\vec{Z} \quad ...(iv)$$

Comparing these with eqs. (*i*) and (*ii*), we have,

$$\vec{A} = 1\,; \quad \vec{B} = \vec{Z}, \quad \vec{C} = 0 \quad \text{and} \quad \vec{D} = 1$$

Incidentally ; $\vec{A} = \vec{D}$

and $$\vec{A}\vec{D} - \vec{B}\vec{C} = 1 \times 1 - \vec{Z} \times 0 = 1$$

(*ii*) **Medium lines – Nominal T method.** In this method, the whole line to neutral capacitance is assumed to be concentrated at the middle point of the line and half the line resistance and reactance are lumped on either side as shown in Fig. 10.24.

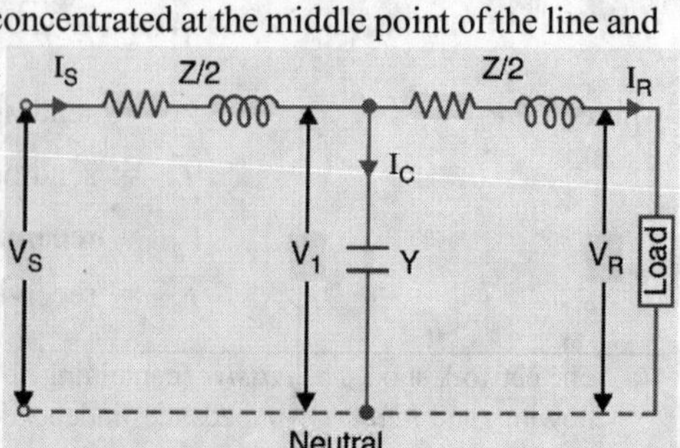

Fig. 10.24

Here, $$\vec{V_S} = \vec{V_1} + \vec{I_S}\,\vec{Z}/2 \quad ...(v)$$

and $$\vec{V_1} = \vec{V_R} + \vec{I_R}\,\vec{Z}/2$$

Now, $$\vec{I_C} = \vec{I_S} - \vec{I_R}$$

$$= \vec{V_1}\,\vec{Y} \quad \text{where } Y = \text{shunt admittance}$$

$$= \vec{Y}\left(\vec{V_R} + \frac{\vec{I_R}\,\vec{Z}}{2}\right)$$

$$\therefore \quad \vec{I_S} = \vec{I_R} + \vec{Y}\,\vec{V_R} + \vec{Y}\,\frac{\vec{I_R}\,\vec{Z}}{2}$$

$$= \vec{Y}\,\vec{V_R} + \vec{I_R}\left(1 + \frac{\vec{Y}\,\vec{Z}}{2}\right) \qquad \ldots(vi)$$

Substituting the value of V_1 in eq. (*v*), we get,

$$\vec{V_S} = \vec{V_R} + \frac{\vec{I_R}\,\vec{Z}}{2} + \frac{\vec{I_S}\,\vec{Z}}{2}$$

Substituing the value of I_S, we get,

$$\vec{V_S} = \left(1 + \frac{\vec{Y}\,\vec{Z}}{2}\right)\vec{V_R} + \left(\vec{Z} + \frac{\vec{Y}\,\vec{Z}^2}{4}\right)\vec{I_R} \qquad \ldots(vii)$$

Comparing eqs. (*vii*) and (*vi*) with those of (*i*) and (*ii*), we have,

$$\vec{A} = \vec{D} = 1 + \frac{\vec{Y}\,\vec{Z}}{2}\,; \quad \vec{B} = \vec{Z}\left(1 + \frac{\vec{Y}\,\vec{Z}}{4}\right); \; \vec{C} = \vec{Y}$$

Incidentally : $$\vec{A}\,\vec{D} - \vec{B}\,\vec{C} = \left(1 + \frac{YZ}{2}\right)^2 - Z\left(1 + \frac{YZ}{4}\right)Y$$

$$= 1 + \frac{Y^2Z^2}{4} + YZ - ZY - \frac{Z^2Y^2}{4} = 1$$

(*iii*) Medium lines—Nominal π method. In this method, line-to-neutral capacitance is divided into two halves ; one half being concentrated at the load end and the other half at the sending end as shown in Fig. 10.25.

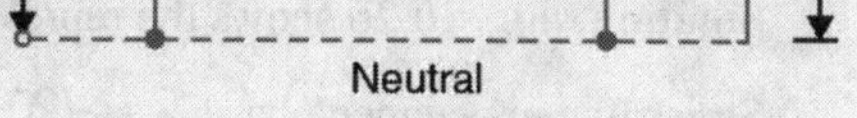

Fig. 10.25

Here, $\vec{Z} = R + jX_L$ = series impedenace/phase

$\vec{Y} = j\,\omega\,C$ = shunt admittance

$$\vec{I_S} = \vec{I_L} + \vec{I_{C2}}$$

or $$\vec{I_S} = \vec{I_L} + \vec{V_S}\,\vec{Y}/2 \qquad \ldots(viii)$$

Also $$\vec{I_L} = \vec{I_R} + \vec{I_{C1}} = \vec{I_R} + \vec{V_R}\,\vec{Y}/2 \qquad \ldots(ix)$$

Now $$\vec{V_S} = \vec{V_R} + \vec{I_L}\,\vec{Z} = \vec{V_R} + \left(\vec{I_R} + \vec{V_R}\,\vec{Y}/2\right)\vec{Z}$$ (Putting the value of $\vec{I_L}$)

$$\therefore \quad \vec{V_S} = \vec{V_R}\left(1 + \frac{\vec{Y}\,\vec{Z}}{2}\right) + \vec{I_R}\,\vec{Z} \qquad \ldots(x)$$

Also $$\vec{I_S} = \vec{I_L} + \vec{V_S}\,\vec{Y}/2 = \left(\vec{I_R} + \vec{V_R}\,\vec{Y}/2\right) + \vec{V_S}\,\vec{Y}/2$$

(Putting the value of $\vec{I_L}$)

Putting the value of $\vec{V_S}$ from eq. (*x*), we get,

$$\vec{I_S} = \vec{I_R} + \vec{V_R}\frac{\vec{Y}}{2} + \frac{\vec{Y}}{2}\left\{\vec{V_R}\left(1 + \frac{\vec{Y}\,\vec{Z}}{2}\right) + \vec{I_R}\,\vec{Z}\right\}$$

$$= \vec{I_R} + \vec{V_R}\frac{\vec{Y}}{2} + \frac{\vec{V_R}\,\vec{Y}}{2} + \frac{\vec{V_R}\,\vec{Y}^2\,\vec{Z}}{4} + \frac{\vec{Y}\,\vec{I_R}\,\vec{Z}}{2}$$

$$= \vec{I_R}\left(1+\frac{\vec{Y}\vec{Z}}{2}\right) + \vec{V_R}\,\vec{Y}\left(1+\frac{\vec{Y}\vec{Z}}{4}\right) \qquad ...(xi)$$

Comparing equations (*x*) and (*xi*) with those of (*i*) and (*ii*), we get,

$$\vec{A} = \vec{D} = \left(1+\frac{\vec{Y}\vec{Z}}{2}\right); \quad \vec{B} = \vec{Z}; \quad \vec{C} = \vec{Y}\left(1+\frac{\vec{Y}\vec{Z}}{4}\right)$$

Also
$$\vec{A}\vec{D} - \vec{B}\vec{C} = \left(1+\frac{YZ}{2}\right)^2 - ZY\left(1+\frac{YZ}{4}\right)$$

$$= 1 + \frac{Y^2Z^2}{4} + YZ - ZY - \frac{Z^2Y^2}{4} = 1$$

(*iv*) **Long lines—Rigorous method.** By rigorous method, the sending end voltage and current of a long transmission line are given by :

$$V_S = V_R \cosh\sqrt{YZ} + I_R\sqrt{\frac{Z}{Y}}\sinh\sqrt{YZ}$$

$$I_S = V_R\sqrt{\frac{Y}{Z}}\sinh\sqrt{YZ} + I_R\cosh\sqrt{YZ}$$

Comparing these equations with those of (*i*) and (*ii*), we get,

$$\vec{A} = \vec{D} = \cosh\sqrt{YZ}; \quad \vec{B} = \sqrt{\frac{Z}{Y}}\sinh\sqrt{YZ}; \quad \vec{C} = \sqrt{\frac{Y}{Z}}\sinh\sqrt{YZ}$$

Incidentally

$$\vec{A}\vec{D} - \vec{B}\vec{C} = \cosh\sqrt{YZ}\times\cosh\sqrt{YZ} - \sqrt{\frac{Z}{Y}}\sinh\sqrt{YZ}\times\sqrt{\frac{Y}{Z}}\sinh\sqrt{YZ}$$

$$= \cosh^2\sqrt{YZ} - \sinh^2\sqrt{YZ} = 1$$

Example 10.16. *A balanced 3-phase load of 30 MW is supplied at 132 kV, 50 Hz and 0·85 p.f. lagging by means of a transmission line. The series impedance of a single conductor is (20 + j52) ohms and the total phase-neutral admittance is 315 × 10⁻⁶ siemen. Using nominal T method, determine: (i) the A, B, C and D constants of the line (ii) sending end voltage (iii) regulation of the line.*

Solution. Fig. 10.26 shows the representation of 3-phase line on the single phase basis.

Series line impedance/phase, $\vec{Z} = (20 + j\,52)\ \Omega$

Shunt admittance/phase, $\vec{Y} = j\,315 \times 10^{-6}$ S

(*i*) **Generalised constants of line.** For nominal *T* method, various constants have the values as under :

$$\vec{A} = \vec{D} = 1 + \vec{Z}\,\vec{Y}/2 = 1 + \frac{20 + j\,52}{2} \times j\,315 \times 10^{-6}$$

$$= 0{\cdot}992 + j\,0{\cdot}00315 = \mathbf{0{\cdot}992\ \angle 0{\cdot}18^\circ}$$

$$\vec{B} = \vec{Z}\left(1+\frac{\vec{Z}\vec{Y}}{4}\right) = (20 + j\,52)\left[1 + \frac{(20 + j\,52)\,j\,315\times 10^{-6}}{4}\right]$$

$$= 19{\cdot}84 + j\,51{\cdot}82 = \mathbf{55{\cdot}5\ \angle 69^\circ}$$

$$\vec{C} = \vec{Y} = \mathbf{0{\cdot}000315\ \angle 90^\circ}$$

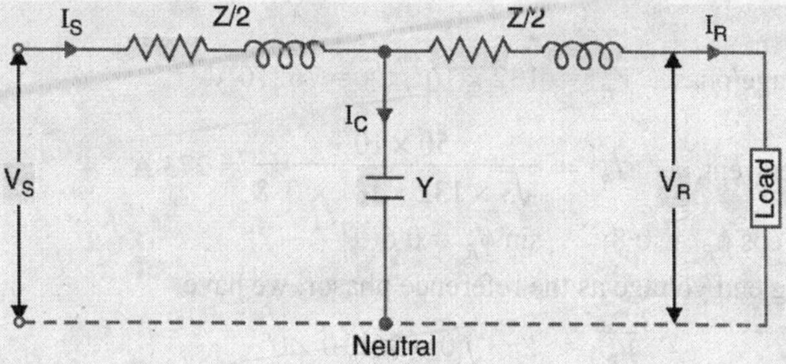

Fig. 10.26

***(ii)* Sending end voltage.**

Receiving end voltage/phase, $V_R = 132 \times 10^3/\sqrt{3} = 76210$ V

Receiving end current, $I_R = \dfrac{30 \times 10^6}{\sqrt{3} \times 132 \times 10^3 \times 0{\cdot}85} = 154$ A

$\cos \phi_R = 0{\cdot}85\,; \quad \sin \phi_R = 0{\cdot}53$

Taking receiving end voltage as the reference phasor, we have,

$$\vec{V_R} = V_R + j0 = 76210 \text{ V}$$

$$\vec{I_R} = I_R (\cos \phi_R - j \sin \phi_R) = 154\,(0{\cdot}85 - j\,0{\cdot}53) = 131 - j\,81{\cdot}62$$

Sending end voltage per phase is

$$\vec{V_S} = \vec{A}\,\vec{V_R} + \vec{B}\,\vec{I_R}$$

$$= (0{\cdot}992 + j\,0{\cdot}0032)\,76210 + (19{\cdot}84 + j\,51{\cdot}82)\,(131 - j\,81{\cdot}62)$$

$$= 82{,}428 + j\,5413$$

∴ Magnitude of sending end voltage is

$$V_S = \sqrt{(82{,}428)^2 + (5413)^2} = 82{\cdot}6 \times 10^3 \text{V} = 82{\cdot}6 \text{ kV}$$

∴ Sending end line-to-line voltage

$$= 82{\cdot}6 \times \sqrt{3} = \mathbf{143\ kV}$$

***(iii)* Regulation.** Regulation is defined as the change in voltage at the receiving end when full-load is thrown off.

Now, $\vec{V_S} = \vec{A}\,\vec{V_R} + \vec{B}\,\vec{I_R}$

At no load, $\vec{I_R} = 0$

∴ $\vec{V_S} = \vec{A}\,\vec{V_{R0}}$

where $\vec{V_{R0}}$ = voltage at receiving end at no load

or $\vec{V_{R0}} = \vec{V_S}/\vec{A}$

or $V_{R0} = V_S/A$ (in magnitude)

∴ $$\% \text{ Regulation} = \frac{(V_S/A - V_R)}{V_R} \times 100 = \frac{(82{\cdot}6/0{\cdot}992) - 76{\cdot}21}{76{\cdot}21} \times 100 = \mathbf{9{\cdot}25\%}$$

Example 10.17. *A 132 kV, 50 Hx, 3-phase transmission line delivers a load of 50 MW at 0·8 p.f. lagging at the receiving end. The generalised constants of the transmission line are :*

$A = D = 0{\cdot}95\ \angle 1{\cdot}4^\circ$; $B = 96\ \angle 78^\circ$; $C = 0{\cdot}0015\ \angle 90^\circ$

Find the regulation of the line and charging current. Use Nominal-T method.

Solution.

Receiving end voltage/phase, $V_R = 132 \times 10^3/\sqrt{3} = 76210$ V

Receiving end current, $I_R = \dfrac{50 \times 10^6}{\sqrt{3} \times 132 \times 10^3 \times 0{\cdot}8} = 273$ A

$\cos \phi_R = 0{\cdot}8; \quad \sin \phi_R = 0{\cdot}6$

Taking receiving end voltage as the reference phasor, we have,

$$\vec{V_R} = V_R + j\,0 = 76210 \angle 0^\circ$$

$$\vec{I_R} = I_R \angle -\phi_R = 273 \angle -36{\cdot}9^\circ$$

Sending end voltage per phase is

$$\begin{aligned}
\vec{V_S} &= \vec{A}\,\vec{V_R} + \vec{B}\,\vec{I_R} \\
&= 0{\cdot}95 \angle 1{\cdot}4^\circ \times 76210 \angle 0^\circ + 96 \angle 78^\circ \times 273 \angle -36{\cdot}9^\circ \\
&= 72400 \angle 1{\cdot}4^\circ + 26208 \angle 41{\cdot}1^\circ \\
&= 72400 (\cos 1{\cdot}4^\circ + j \sin 1{\cdot}4^\circ) + 26208 (\cos 41{\cdot}1^\circ + j \sin 41{\cdot}1^\circ) \\
&= 72400 (0{\cdot}9997 + j\,0{\cdot}0244) + 26208 (0{\cdot}7536 + j\,0{\cdot}6574) \\
&= (72378 + j\,1767) + (19750 + j\,17229) \\
&= 92128 + j\,18996 = 94066 \angle 11{\cdot}65^\circ \text{ V}
\end{aligned}$$

Sending end current,

$$\begin{aligned}
\vec{I_S} &= \vec{C}\,\vec{V_R} + \vec{D}\,\vec{I_R} \\
&= 0{\cdot}0015 \angle 90^\circ \times 76210 \angle 0^\circ + 0{\cdot}95 \angle 1{\cdot}4^\circ \times 273 \angle -36{\cdot}9^\circ \\
&= 114 \angle 90^\circ + 260 \angle -35{\cdot}5^\circ \\
&= 114 (\cos 90^\circ + j \sin 90^\circ) + 260 (\cos 35{\cdot}5^\circ - j \sin 35{\cdot}5^\circ) \\
&= 114 (0 + j) + 260 (0{\cdot}814 - j\,0{\cdot}58) \\
&= j\,114 + 211 - j\,150 = 211 - j\,36
\end{aligned}$$

Charging current,

$$\begin{aligned}
\vec{I_C} &= \vec{I_S} - \vec{I_R} = (211 - j36) - 273 \angle -36{\cdot}9^\circ \\
&= (211 - j\,36) - (218 - j\,164) = -7 + j\,128 = \mathbf{128{\cdot}2 \angle 93{\cdot}1^\circ \text{ A}}
\end{aligned}$$

$$\% \text{ Regulation} = \frac{(V_S/A) - V_R}{V_R} \times 100 = \frac{94066/0{\cdot}95 - 76210}{76210} \times 100 = \mathbf{30\%}$$

Example 10.18. *Find the following for a single circuit transmission line delivering a load of 50 MVA at 110 kV and p.f. 0·8 lagging :*

(i) sending end voltage (ii) sending end current (iii) sending end power (iv) efficiency of transmission. Given A = D = 0·98 ∠3°; B = 110 ∠75° ohm ; C = 0·0005 ∠80° siemen.

Solution.

Receiving end voltage/phase,

$$V_R = \frac{110}{\sqrt{3}} = 63{\cdot}5 \text{ kV}$$

Receiving end current, $I_R = \dfrac{50 \times 10^6}{\sqrt{3} \times 110 \times 10^3} = 262{\cdot}4$ A

Taking receiving end voltage as the reference phasor, we have,

$$\vec{V_R} = (63500 + j0)$$

$$\vec{I_R} = 262{\cdot}4 \angle -\cos^{-1} 0{\cdot}8 = 262{\cdot}4\,(0{\cdot}8 - j\,0{\cdot}6) = (210 - j\,157{\cdot}5)\text{ A}$$

(*i*) Now sending-end voltage per phase is

$$\vec{V_S} = \vec{A}\,\vec{V_R} + \vec{B}\,\vec{I_R}$$

Here $\vec{A}\,\vec{V_R} = 0{\cdot}98 \angle 3^\circ \times 63500 \angle 0^\circ = 62230 \angle 3^\circ = (62145 + j\,3260)\text{ V}$

and $\vec{B}\,\vec{I_R} = 110 \angle 75^\circ \times 262{\cdot}4 \angle -36{\cdot}86^\circ$

$$= 28865 \angle 38{\cdot}14^\circ = (22702 + j\,17826)\text{ V}$$

$\therefore$ $\vec{V_S} = (62145 + j\,3260) + (22702 + j\,17826)$

$$= 84847 + j\,21086 = 87427 \angle 14^\circ\text{ V}$$

$\therefore$ Magnitude of sending-end voltage/phase = **87427 V**

(*ii*) Sending-end current is given by ;

$$\vec{I_S} = \vec{C}\,\vec{V_R} + \vec{D}\,\vec{I_R}$$

Here $\vec{C}\,\vec{V_R} = 0{\cdot}0005 \angle 80^\circ \times 63500 \angle 0^\circ = 31{\cdot}75 \angle 80^\circ = (5{\cdot}5 + j\,31{\cdot}3)\text{ A}$

and $\vec{D}\,\vec{I_R} = 0{\cdot}98 \angle 3^\circ \times 262{\cdot}4 \angle -36{\cdot}86^\circ$

$$= 257{\cdot}15 \angle -33{\cdot}8^\circ = (213{\cdot}5 - j\,143{\cdot}3)\text{ A}$$

$\therefore$ $\vec{I_S} = (5{\cdot}5 + j\,31{\cdot}3) + (213{\cdot}5 - j\,143{\cdot}3)$

$$= 219 - j\,112 = 246 \angle -27^\circ\text{ A}$$

$\therefore$ Magnitude of sending-end current = **246 A**

(*iii*) Sending-end power $= 3\,V_S I_S \cos\phi_S$

Here $V_S = 87427$ V ; $I_S = 246$ A ; $\cos\phi_S = \cos(-27^\circ - 14^\circ)$

$\therefore$ Sending-end power $= 3 \times 87427 \times 246 \times \cos(-27^\circ - 14^\circ)$

$$= 48{\cdot}6 \times 10^6\text{ W} = \mathbf{48{\cdot}6\ MW}$$

(*iv*) Receiving end power $= 50 \times 0{\cdot}8 = 40$ MW

Transmission efficiency, $\eta = \dfrac{40}{48{\cdot}6} \times 100 = \mathbf{82{\cdot}3\%}$

TUTORIAL PROBLEMS

1. A 150 km, 3-ϕ, 110 kV, 50 Hz transmission line transmits a load of 40,000 kW at 0·8 p.f. lagging at the receiving end. Resistance/km/phase = 0·15Ω ; reactance/km/phase = 0·6 Ω; susceptance/km/phase = 10^{-5} S. Determine (*i*) the *A*, *B*, *C* and *D* constants of the line (*ii*) regulation of the line.

[(*i*) A = D = 0·968 ∠1° ; B = 92·8 ∠7·5°Ω ; C = 0·00145 ∠90·5° S (*ii*) 33·5%]

2. A balanced load of 30 MW is supplied at 132 kV, 50 Hz and 0·85 p.f. lagging by means of a transmission line. The series impedance of a single conductor is (20 + j52) ohms and the total phases-neutral admittance is 315 microsiemens. Shunt leakage may be neglected. Using the nominal *T* approximation, calculate the line voltage at the sending end of the line. If the load is removed and the sending end voltage remains constant, find the percentage rise in voltage at the receiving end.

[143 kV; 9%]

3. Calculate *A*, *B*, *C* and *D* constants of a 3-phase, 50 Hz transmission line 160 km long having the following distributed parameters :

$R = 0{\cdot}15\ \Omega/\text{km}$; $L = 1{\cdot}20 \times 10^{-3}$ H/km ; $C = 8 \times 10^{-9}$ F/km ; $G = 0$

[A = D = 0·988 ∠0·3° ; B = 64·2 ∠ 68·3° Ω ; C = 0·4 × 10^{-3} ∠90·2° S]

SELF-TEST

1. **Fill in the blanks by inserting appropriate words/figures.**
   (*i*) In short transmission lines, the effects of .... are neglected.
   (*ii*) .... of transmission lines, is the most important cause of power loss in the line.
   (*iii*) In the analysis of 3-phase transmission line, only .... is considered.
   (*iv*) For a given V_R and *I*, the regulation of the line .... with the decrease in p.f. for lagging loads.
   (*v*) If the p.f. of the load decreases, the line losses ....
   (*vi*) In medium transmission lines, effects of .... are taken into account.
   (*vii*) The rigorous solution of transmission lines takes into account the .... nature of line constants.
   (*viii*) In any transmission line, $AD - BC =$ ....
   (*ix*) In a transmission line, generalised constants .... and .... are equal.
   (*x*) The dimensions of constants *B* and *C* are respectively .... and ....
2. **Pick up the correct words/figures from the brackets and fill in the blanks.**
   (*i*) The line constants of a transmission line are .... [*uniformly distributed, lumped*]
   (*ii*) The length of a short transmission line is upto about .... [*50 km, 120 km, 200 km*]
   (*iii*) The capacitance of a transmission line is a .... element. [*series, shunt*]
   (*iv*) It is desirable that voltage regulation of a transmission line should be .... [*low, high*]
   (*v*) When the regulation is positive, then receiving and voltage (V_R) is .... than sending and voltage (V_S). [*more, less*]
   (*vi*) The shunt admittance of a transmission line is 3 microsiemens. Its complex notation will be .... siemen. [$3 \times 10^{-6} \angle 90°$, $3 \times 10^{-6} \angle 0°$]
   (*vii*) The exact solution of any transmission line must consider the fact that line constants are .... [*uniformly distributed, lumped*]
   (*viii*) The generalised constants *A* and *D* of the transmission line have .... [*no dimensions, dimensions of ohm*]
   (*ix*) $30 \angle 10° \times 60 \angle 20° =$ .... [$2 \angle 2°$, $1800 \angle 30°$, $1800 \angle 2°$]
   (*x*) $\sqrt{9 \angle 90° \times 4 \angle 10°} =$ .... [$6 \angle 50°$, $6 \angle 80°$, $6 \angle 10°$]

ANSWERS TO SELF-TEST

1. (*i*) capacitance (*ii*) resistance (*iii*) one phase (*iv*) increases (*v*) increase (*vi*) capacitance (*vii*) distributed (*viii*) 1 (*ix*) *A* and *D* (*x*) ohm, siemen
2. (*i*) uniformly distributed (*ii*) 50 km (*iii*) shunt (*iv*) low (*v*) less (*vi*) $3 \times 10^{-6} \angle 90°$ (*vii*) uniformly distributed (*viii*) no dimensions (*ix*) $1800 \angle 30°$ (*x*) $6 \angle 50°$

CHAPTER REVIEW TOPICS

1. What is the purpose of an overhead transmission line ? How are these lines classified ?
2. Discuss the terms voltage regulation and transmission efficiency as applied to transmission line.
3. Deduce an expression for voltage regulation of a short transmission line, giving the vector diagram.
4. What is the effect of load power factor on regulation and efficiency of a transmission line ?
5. What do you understand by medium transmission lines ? How capacitance effects are taken into account in such lines ?
6. Show how regulation and transmission efficiency are determined for medium lines using
   (*i*) end condensor method
   (*ii*) nominal *T* method
   (*iii*) nominal π method

   Illustrate your answer with suitable vector diagrams.

7. What do you understand by long transmission lines ? How capacitance effects are taken into account in such lines ?
8. Using rigorous method, derive expressions for sending end voltage and current for a long transmission line.
9. What do you understand by generalised circuit constants of a transmission line ? What is their importance?
10. Evaluate the generalised circuit constants for
    (*i*) short transmission line
    (*ii*) medium line — nominal *T* method
    (*iii*) medium line — nominal π method

DISCUSSION QUESTIONS

1. What is the justification in neglecting line capacitance in short transmission lines ?
2. What are the drawbacks of localised capacitance methods ?
3. A long transmission line is open circuited at the receiving end. Will there be any current in the line at the sending end ? Explain your answer.
4. Why is leakage conductance negligible in overhead lines ? What about underground system ?
5. Why do we analyse a 3-phase transmission line on single phase basis ?

CHAPTER

11

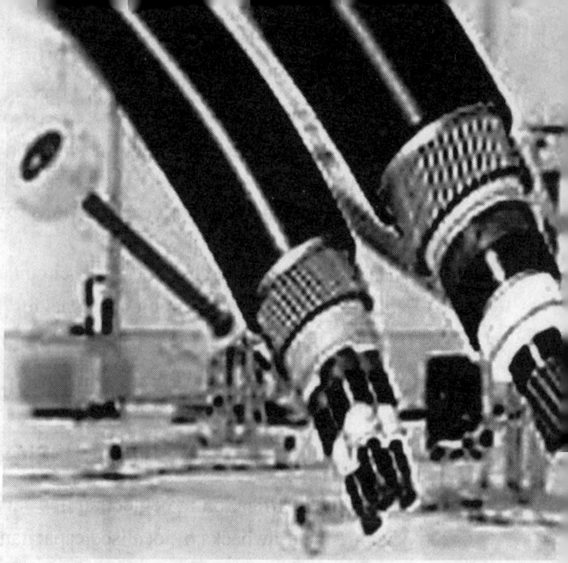

Underground Cables

Introduction

Electric power can be transmitted or distributed either by overhead system or by underground cables. The underground cables have serveral advantages such as less liable to damage through storms or lightning, low maintenance cost, less chances of faults, smaller voltage drop and better general appearance. However, their major drawback is that they have greater installation cost and introduce insulation problems at high voltages compared with the equivalent overhead system. For this reason, underground cables are employed where it is impracticable to use overhead lines. Such locations may be thickly populated areas where municipal authorities prohibit overhead lines for reasons of safety, or around plants and substations or where maintenance conditions do not permit the use of overhead construction.

The chief use of underground cables for many years has been for distribution of electric power in congested urban areas at comparatively low or moderate voltages. However, recent improvements in the design and manufacture have led to the development of cables suitable for use at high voltages. This has made it possible to employ underground cables for transmission of electric power for short or moderate distances. In this chapter, we shall focus our attention on the various aspects of underground cables and their increasing use in power system.

11.1 Underground Cables

*An **underground cable** essentially consists of one or more conductors covered with suitable insulation and surrounded by a protecting cover.*

Although several types of cables are available, the type of cable to be used will depend upon the working voltage and service requirements. In general, a cable must fulfil the following necessary requirements :

(*i*) The conductor used in cables should be tinned stranded copper or aluminium of high conductivity. Stranding is done so that conductor may become flexible and carry more current.

(*ii*) The conductor size should be such that the cable carries the desired load current without overheating and causes voltage drop within permissible limits.

(*iii*) The cable must have proper thickness of insulation in order to give high degree of safety and reliability at the voltage for which it is designed.

(*iv*) The cable must be provided with suitable mechanical protection so that it may withstand the rough use in laying it.

(*v*) The materials used in the manufacture of cables should be such that there is complete chemical and physical stability throughout.

11.2 Construction of Cables

Fig. 11.1 shows the general construction of a 3-conductor cable. The various parts are :

(*i*) *Cores or Conductors.* A cable may have one or more than one core (conductor) depending upon the type of service for which it is intended. For instance, the 3-conductor cable shown in Fig. 11.1 is used for 3-phase service. The conductors are made of tinned copper or aluminium and are usually stranded in order to provide flexibility to the cable.

(*ii*) *Insulatian.* Each core or conductor is provided with a suitable thickness of insulation, the thickness of layer depending upon the voltage to be withstood by the cable. The commonly used materials for insulation are impregnated paper, varnished cambric or rubber mineral compound.

(*iii*) *Metallic sheath.* In order to protect the cable from moisture, gases or other damaging liquids (acids or alkalies) in the soil and atmosphere, a metallic sheath of lead or aluminium is provided over the insulation as shown in Fig. 11.1

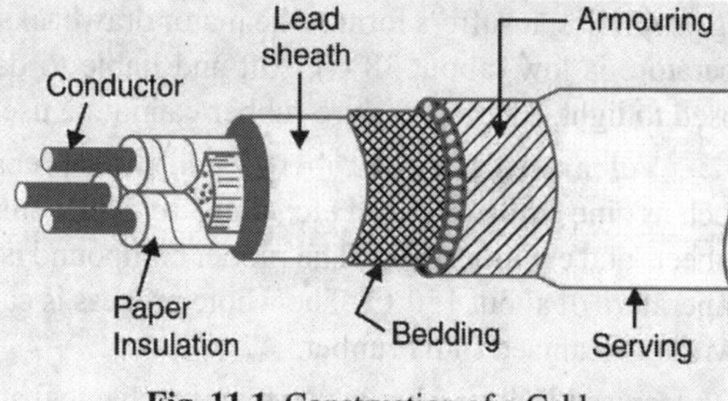

Fig. 11.1 Construction of a Cable

(*iv*) *Bedding.* Over the metallic sheath is applied a layer of bedding which consists of a fibrous material like jute or hessian tape. The purpose of bedding is to protect the metallic sheath against corrosion and from mechanical injury due to armouring.

(*v*) *Armouring.* Over the bedding, armouring is provided which consists of one or two layers of galvanised steel wire or steel tape. Its purpose is to protect the cable from mechanical injury while laying it and during the course of handling. Armouring may not be done in the case of some cables.

(*vi*) *Serving.* In order to protect armouring from atmospheric conditions, a layer of fibrous material (like jute) similar to bedding is provided over the armouring. This is known as *serving*.

It may not be out of place to mention here that bedding, armouring and serving are only applied to the cables for the protection of conductor insulation and to protect the metallic sheath from mechanical injury.

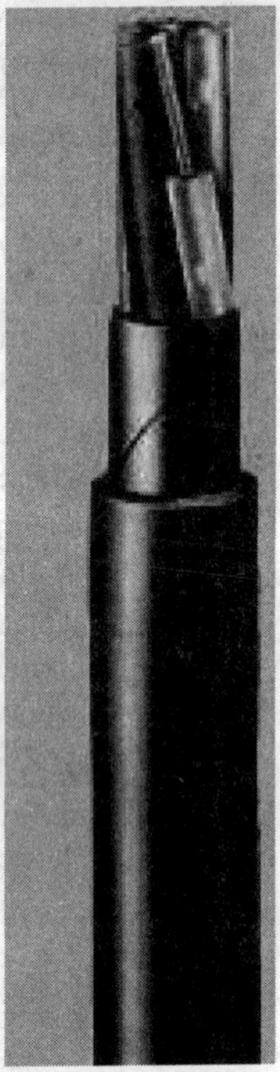

Underground cable

11.3 Insulating Materials for Cables

The satisfactory operation of a cable depends to a great extent upon the characteristics of insulation used. Therefore, the proper choice of insulating material for cables is of considerable importance. In general, the insulating materials used in cables should have the following properties :

(*i*) High insulation resistance to avoid leakage current.

(*ii*) High dielectric strength to avoid electrical breakdown of the cable.

(*iii*) High mechanical strength to withstand the mechanical handling of cables.

(*iv*) Non-hygroscopic *i.e.,* it should not absorb moisture from air or soil. The moisture tends to decrease the insulation resistance and hastens the breakdown of the cable. In case the insulating material is hygroscopic, it must be enclosed in a waterproof covering like lead sheath.

(*v*) Non-inflammable.

(*vi*) Low cost so as to make the underground system a viable proposition.

(*vii*) Unaffected by acids and alkalies to avoid any chemical action.

No one insulating material possesses all the above mentioned properties. Therefore, the type of insulating material to be used depends upon the purpose for which the cable is required and the quality of insulation to be aimed at. The principal insulating materials used in cables are rubber, vulcanised India rubber, impregnated paper, varnished cambric and polyvinyl chloride.

1. **Rubber.** Rubber may be obtained from milky sap of tropical trees or it may be produced from oil products. It has relative permittivity varying between 2 and 3, dielectric strength is about 30 kV/mm and resistivity of insulation is $10^{17}\Omega$ cm. Although pure rubber has reasonably high insulating properties, it suffers form some major drawbacks *viz.,* readily absorbs moisture, maximum safe temperature is low (about 38ºC), soft and liable to damage due to rough handling and ages when exposed to light. Therefore, pure rubber cannot be used as an insulating material.

2. **Vulcanised India Rubber (*V.I.R.*).** It is prepared by mixing pure rubber with mineral matter such as zine oxide, red lead etc., and 3 to 5% of sulphur. The compound so formed is rolled into thin sheets and cut into strips. The rubber compound is then applied to the conductor and is heated to a temperature of about 150ºC. The whole process is called *vulcanisation* and the product obtained is known as vulcanised India rubber.

Vulcanised India rubber has greater mechanical strength, durability and wear resistant property than pure rubber. Its main drawback is that sulphur reacts very quickly with copper and for this reason, cables using *VIR* insulation have tinned copper conductor. The *VIR* insulation is generally used for low and moderate voltage cables.

3. **Impregnated paper.** It consists of chemically pulped paper made from wood chippings and impregnated with some compound such as paraffinic or napthenic material. This type of insulation has almost superseded the rubber insulation. It is because it has the advantages of low cost, low capacitance, high dielectric strength and high insulation resistance. The only disadvantage is that paper is hygroscopic and even if it is impregnated with suitable compound, it absorbs moisture and thus lowers the insulation resistance of the cable. For this reason, paper insulated cables are always

provided with some protective covering and are never left unsealed. If it is required to be left unused on the site during laying, its ends are temporarily covered with wax or tar.

Since the paper insulated cables have the tendency to absorb moisture, they are used where the cable route has a *few joints. For instance, they can be profitably used for distribution at low voltages in congested areas where the joints are generally provided only at the terminal apparatus. However, for smaller installations, where the lenghts are small and joints are required at a number of places, *VIR* cables will be cheaper and durable than paper insulated cables.

4. Varnished cambric. It is a cotton cloth impregnated and coated with varnish. This type of insulation is also known as *empire tape.* The cambric is lapped on to the conductor in the form of a tape and its surfaces are coated with petroleum jelly compound to allow for the sliding of one turn over another as the cable is bent. As the varnished cambric is hygroscopic, therefore, such cables are always provided with metallic sheath. Its dielectric strength is about 4 kV/mm and permittivity is 2.5 to 3.8.

5. Polyvinyl chloride (*PVC*). This insulating material is a synthetic compound. It is obtained from the polymerisation of acetylene and is in the form of white powder. For obtaining this material as a cable insulation, it is compounded with certain materials known as plasticizers which are liquids with high boiling point. The plasticizer forms a gell and renders the material plastic over the desired range of temperature.

Polyvinyl chloride has high insulation resistance, good dielectric strength and mechanical toughness over a wide range of temperatures. It is inert to oxygen and almost inert to many alkalies and acids. Therefore, this type of insulation is preferred over *VIR* in extreme enviormental conditions such as in cement factory or chemical factory. As the mechanical properties (*i.e.*, elasticity etc.) of *PVC* are not so good as those of rubber, therefore, *PVC* insulated cables are generally used for low and medium domestic lights and power installations.

11.4 Classification of Cables

Cables for underground service may be classified in two ways according to (*i*) the type of insulating material used in their manufacture (*ii*) the voltage for which they are manufactured. However, the latter method of classification is generally preferred, according to which cables can be divided into the following groups :

(*i*) Low-tension (L.T.) cables — upto 1000 V
(*ii*) High-tension (H.T.) cables — upto 11,000 V
(*iii*) Super-tension (S.T.) cables — from 22 kV to 33 kV
(*iv*) Extra high-tension (E.H.T.) cables — from 33 kV to 66 kV
(*v*) Extra super voltage cables — beyond 132 kV

A cable may have one or more than one core depending upon the type of service for which it is intended. It may be (*i*) single-core (*ii*) two-core (*iii*) three-core (*iv*) four-core etc. For a 3-phase service, either 3-single-core cables or three-core cable can be used depending upon the operating voltage and load demand.

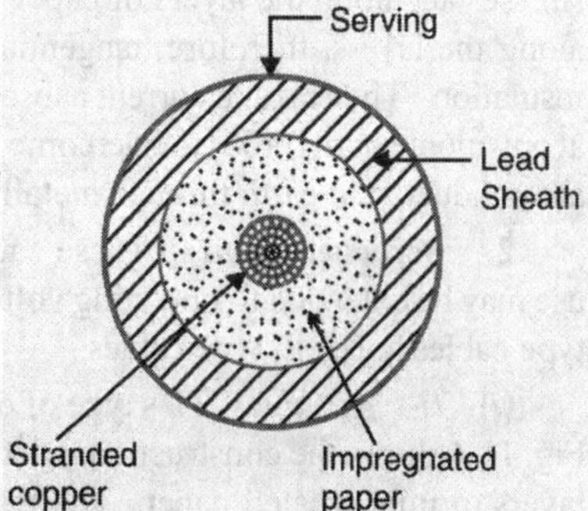

Fig. 11.2

Fig. 11.2 shows the constructional details of a single-core low tension cable. The cable has ordinary construction because the stresses developed in the cable for low voltages (upto 6600 V) are generally small. It consists of one circular core of tinned stranded copper (or aluminium) insulated by layers of

* Special precautions have to be taken to preclude moisture at joints. If the number of joints is more, the installation cost increases rapidly and prohibits the use of paper insulated cables.

impregnated paper. The insulation is surrounded by a lead sheath which prevents the entry of moisture into the inner parts. In order to protect the lead sheath from corrosion, an overall serving of compounded fibrous material (jute etc.) is provided. Single-core cables are not usually armoured in order to avoid excessive sheath losses. The principal advantages of single-core cables are simple construction and availability of larger copper section.

11.5 Cables for 3-Phase Service

In practice, underground cables are generally required to deliver 3-phase power. For the purpose, either three-core cable or *three single core cables may be used. For voltages upto 66 kV, 3-core cable (*i.e.,* multi-core construction) is preferred due to economic reasons. However, for voltages beyond 66 kV, 3-core-cables become too large and unwieldy and, therefore, single-core cables are used. The following types of cables are generally used for 3-phase service :

1. Belted cables — upto 11 kV
2. Screened cables — from 22 kV to 66 kV
3. Pressure cables — beyond 66 kV.

1. **Belted cables.** These cables are used for voltages upto 11kV but in extraordinary cases, their use may be extended upto 22kV. Fig. 11.3 shows the constructional details of a 3-core belted cable. The cores are insulated from each other by layers of impregnated paper. Another layer of impregnated paper tape, called *paper belt* is wound round the grouped insulated cores. The gap between the insulated cores is filled with fibrous insulating material (jute etc.) so as to give circular cross-section to the cable. The cores are generally stranded and may be of non-circular shape to make better use of available space. The belt is covered with lead sheath to protect the cable against ingress of moisture and mechanical injury. The lead sheath is covered with one or more layers of armouring with an outer serving (not shown in the figure).

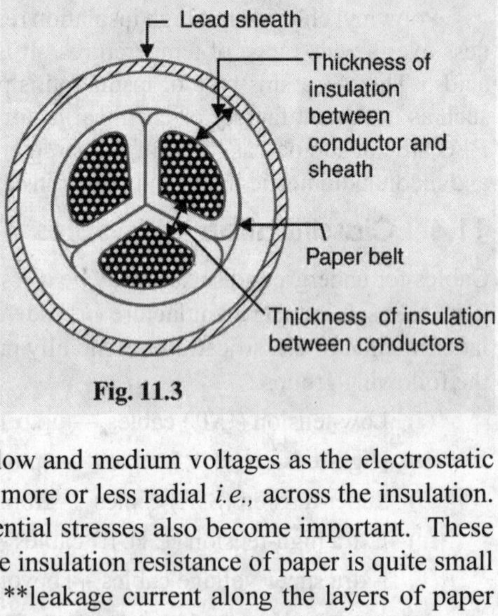

Fig. 11.3

The belted type construction is suitable only for low and medium voltages as the electrostatic stresses developed in the cables for these voltages are more or less radial *i.e.,* across the insulation. However, for high voltages (beyond 22 kV), the tangential stresses also become important. These stresses act along the layers of paper insulation. As the insulation resistance of paper is quite small along the layers, therefore, tangential stresses set up **leakage current along the layers of paper insulation. The leakage current causes local heating, resulting in the risk of breakdown of insulation at any moment. In order to overcome this difficulty, *screened cables* are used where leakage currents are conducted to earth through metallic screens.

2. **Screened cables.** These cables are meant for use upto 33 kV, but in particular cases their use may be extended to operating voltages upto 66 kV. Two principal types of screened cables are H-type cables and S.L. type cables.

(*i*) *H-type cables.* This type of cable was first designed by H. Hochstadter and hence the name. Fig. 11.4 shows the constructional details of a typical 3-core, *H*-type cable. Each core is insulated by layers of impregnated paper. The insulation on each core is covered with a metallic screen which usually consists of a perforated aluminium foil. The cores are laid in such a way that metallic screens

* Separate single-core cable for each phase.

** It is infact a leakage current but should not be confused with the capacitance current.

make contact with one another. An additional conducting belt (copper woven fabric tape) is wrapped round the three cores. The cable has no insulating belt but lead sheath, bedding, armouring and serving follow as usual. It is easy to see that each core screen is in electrical contact with the conducting belt and the lead sheath. As all the four screens (3 core screens and one conducting belt) and the lead sheath are at †earth potential, therefore, the electrical stresses are purely radial and consequently dielectric losses are reduced.

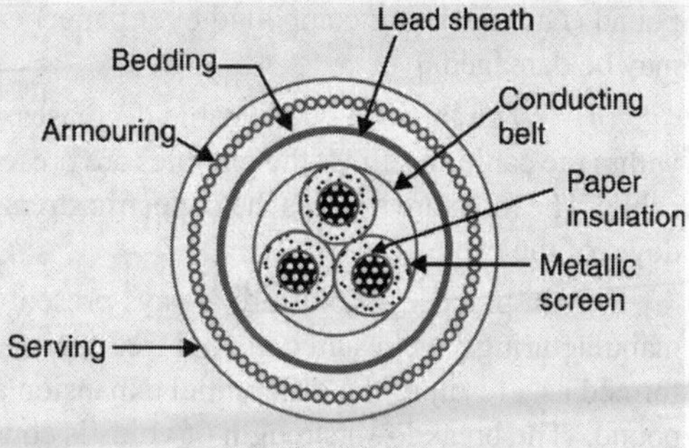

Fig. 11.4

Two principal advantages are claimed for *H*-type cables. Firstly, the perforations in the metallic screens assist in the complete impregnation of the cable with the compound and thus the possibility of air pockets or voids (vacuous spaces) in the dielectric is eliminated. The voids if present tend to reduce the breakdown strength of the cable and may cause considerable damage to the paper insulation. Secondly, the metallic screens increase the heat dissipating power of the cable.

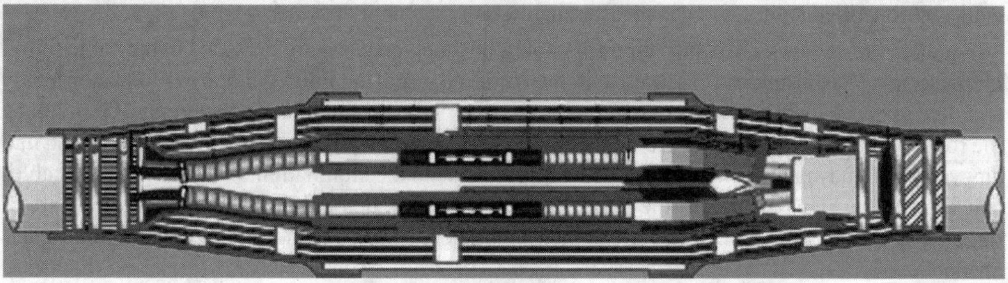

H-Type Cables

(*ii*) *S.L. type cables.* Fig. 11.5 shows the constructional details of a 3-core *S.L. (separate lead) type cable. It is basically *H*-type cable but the screen round each core insulation is covered by its own lead sheath. There is no overall lead sheath but only armouring and serving are provided. The S.L. type cables have two main advantages over *H*-type cables. Firstly, the separate sheaths minimise the possibility of core-to-core breakdown. Secondly, bending of cables becomes easy due to the elimination of overall lead sheath. However, the disadvantage is that the three lead sheaths of S.L. cable are much thinner than the single sheath of *H*-cable and, therefore, call for greater care in manufacture.

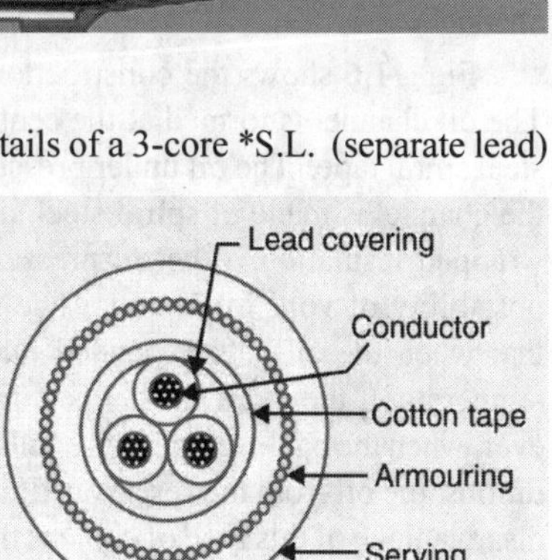

Fig. 11.5

Limitations of solid type cables. All the cables of above construction are referred to as solid type cables because solid insulation is used and no gas or oil circulates in the cable sheath. The voltage limit for solid type cables is 66 kV due to the following reasons :

(*a*) As a solid cable carries the load, its conductor temperature increases and the cable com-

---

† The four screens and lead sheath are in electrical contact and lead sheath is at earth potential.

* In this arrangement, each core is separately lead sheathed and hence the name S.L. cable.

pound (*i.e.,* insulating compound over paper) expands. This action stretches the lead sheath which may be damaged.

(*b*) When the load on the cable decreases, the conductor cools and a partial vacuum is formed within the cable sheath. If the pinholes are present in the lead sheath, moist air may be drawn into the cable. The moisture reduces the dielectric strength of insulation and may eventually cause the breakdown of the cable.

(*c*) In practice, †voids are always present in the insulation of a cable. Modern techniques of manufacturing have resulted in void free cables. However, under operating conditions, the voids are formed as a result of the differential expansion and contraction of the sheath and impregnated compound. The breakdown strength of voids is considerably less than that of the insulation. If the void is small enough, the electrostatic stress across it may cause its breakdown. The voids nearest to the conductor are the first to break down, the chemical and thermal effects of ionisation causing permanent damage to the paper insulation.

3. **Pressure cables** For voltages beyond 66 kV, solid type cables are unreliable because there is a danger of breakdown of insulation due to the presence of voids. When the operating voltages are greater than 66 kV, *pressure cables* are used. In such cables, voids are eliminated by increasing the pressure of compound and for this reason they are called pressure cables. Two types of pressure cables *viz* oil-filled cables and gas pressure cables are commonly used.

(*i*) *Oil-filled cables.* In such types of cables, channels or ducts are provided in the cable for oil circulation. The oil under pressure (it is the same oil used for impregnation) is kept constantly supplied to the channel by means of external reservoirs placed at suitable distances (say 500 m) along the route of the cable. Oil under pressure compresses the layers of paper insulation and is forced into any voids that may have formed between the layers. Due to the elimination of voids, oil-filled cables can be used for higher voltages, the range being from 66 kV upto 230 kV. Oil-filled cables are of three types *viz.,* single-core conductor channel, single-core sheath channel and three-core filler-space channels.

Fig. 11.6 shows the constructional details of a single-core conductor channel, oil filled cable. The oil channel is formed at the centre by stranding the conductor wire around a hollow cylindrical steel spiral tape. The oil under pressure is supplied to the channel by means of external reservoir. As the channel is made of spiral steel tape, it allows the oil to percolate between copper strands to the wrapped insulation. The oil pressure compresses the layers of paper insulation and prevents the possibility of void formation. The system is so designed that when the oil gets expanded due to increase in cable temperature, the extra oil collects in the reservoir. However, when the cable temperature falls during light load conditions, the oil from the reservoir flows to the channel. The disadvantage of this type of cable is that the channel is at the middle of the cable and is at full voltage *w.r.t.* earth, so that a very complicated system of joints is necessary.

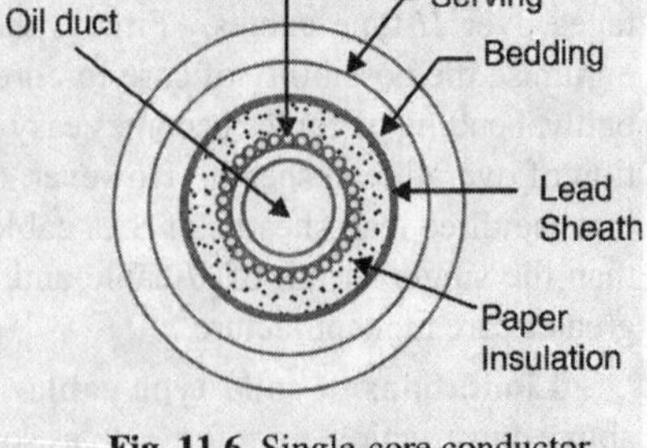

Fig. 11.6 Single-core conductor channel, oil-filled cable

Fig. 11.7 shows the constructional details of a single-core sheath channel oil-filled cable. In this type of cable, the conductor is solid similar to that of solid cable and is paper insulated. However, oil ducts are provided in the metallic sheath as shown. In the 3-core oil-filler cable shown in Fig. 11.8, the oil ducts are located in the filler spaces. These channels are composed of perforated metal-ribbon tubing and are at earth potential.

† Voids are unintentional spaces in the insulation of cable filled with air or gas, usually at low pressure.

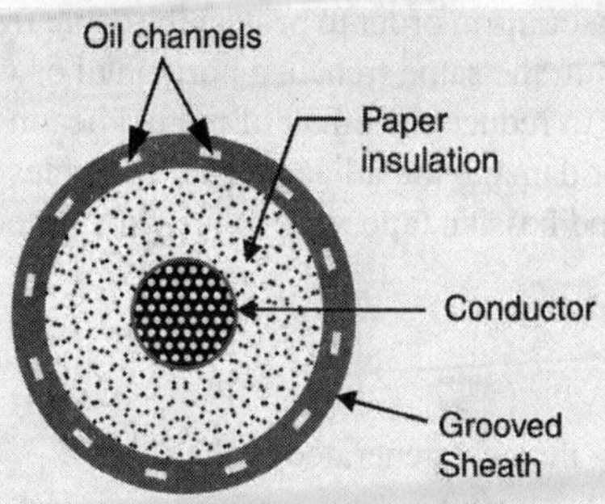

Fig. 11.7

Oil-filled cables

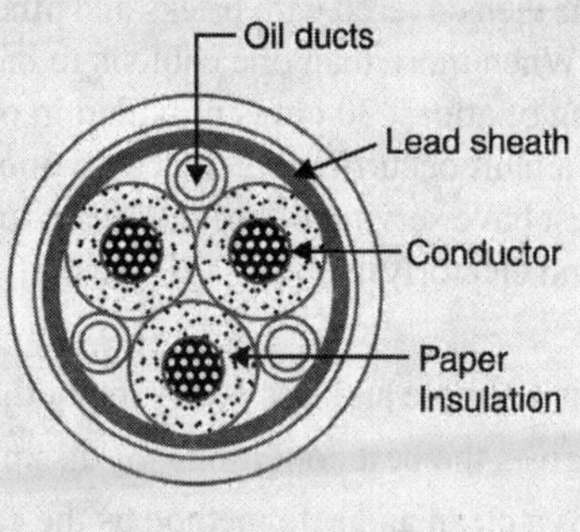

Fig. 11.8

The oil-filled cables have three principal advantages. Firstly, formation of voids and ionisation are avoided. Secondly, allowable temperature range and dielectric strength are increased. Thirdly, if there is leakage, the defect in the lead sheath is at once indicated and the possibility of earth faults is decreased. However, their major disadvantages are the high initial cost and complicated system of laying.

(ii) *Gas pressure cables.* The voltage required to set up ionisation inside a void increases as the pressure is increased. Therefore, if ordinary cable is subjected to a sufficiently high pressure, the ionisation can be altogether eliminated. At the same time, the increased pressure produces radial compression which tends to close any voids. This is the underlying principle of gas pressure cables.

Fig. 11.9 shows the section of external pressure cable designed by Hochstadter, Vogal and Bowden. The construction of the cable is similar to that of an ordinary solid type except that it is of triangular shape and thickness of lead sheath is 75% that of solid cable. The triangular section reduces the weight and gives low thermal resistance but the main reason for triangular shape is that the lead sheath acts as a pressure membrane. The sheath is protected by a thin metal tape. The cable is laid in a gas-tight steel pipe. The pipe is filled with dry nitrogen gas at 12 to 15 atmospheres. The gas pressure produces radial compression and closes the voids that may have formed between the layers of paper insulation. Such cables can carry more load current and operate at higher voltages than a normal cable. Moreover, maintenance cost is small and the nitrogen gas helps in quenching any flame. However, it has the disadvantage that the overall cost is very high.

Fig. 11.9

11.6 Laying of Underground Cables

The reliability of underground cable network depends to a considerable extent upon the proper laying and attachment of fittings *i.e.,* cable end boxes, joints, branch connectors etc. There are three main methods of laying underground cables *viz.,* direct laying, draw-in system and the solid system.

1. Direct laying. This method of laying underground cables is simple and cheap and is much favoured in modern practice. In this method, a trench of about 1·5 metres deep and 45 cm wide is dug. The trench is covered with a layer of fine sand (of about 10 cm thickness) and the cable is laid over this sand bed. The sand prevents the entry of moisture from the ground and thus protects the cable from decay. After the cable has been laid in the trench, it is covered with another layer of sand of about 10 cm thickness.

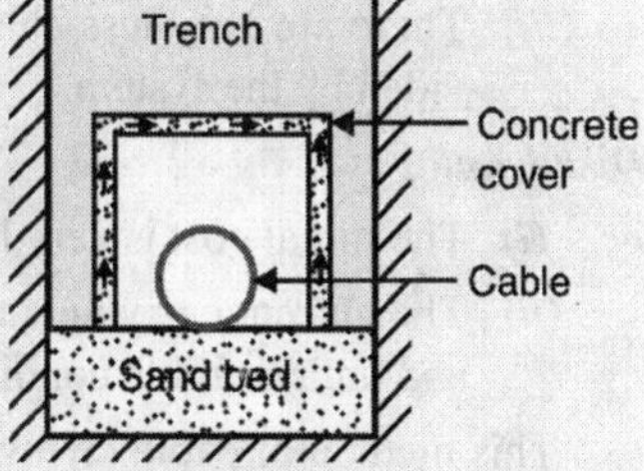

Fig. 11.10

The trench is then covered with bricks and other materials in order to protect the cable from mechanical injury. When more than one cable is to be laid in the same trench, a horizontal or vertical inter-axial spacing of atleast 30 cm is provided in order to reduce the effect of mutual heating and also to ensure that a fault occurring on one cable does not damage the adjacent cable. Cables to be laid in this way must have serving of bituminised paper and hessian tape so as to provide protection against corrosion and electorlysis.

Advantages

(*i*) It is a simple and less costly method.

(*ii*) It gives the best conditions for dissipating the heat generated in the cables.

(*iii*) It is a clean and safe method as the cable is invisible and free from external disturbances.

Disadvantages

(*i*) The extension of load is possible only by a completely new excavation which may cost as much as the original work.

(*ii*) The alterations in the cable netwok cannot be made easily.

(*iii*) The maintenance cost is very high.

(*iv*) Localisation of fault is difficult.

(*v*) It cannot be used in congested areas where excavation is expensive and inconvenient.

This method of laying cables is used in open areas where excavation can be done conveniently and at low cost.

2. Draw-in system. In this method, conduit or duct of glazed stone or cast iron or concrete are laid in the ground with manholes at suitable positions along the cable route. The cables are then pulled into position from manholes. Fig. 11.11 shows section through four-way underground duct line. Three of the ducts carry transmission cables and the fourth duct carries relay protection connection, pilot wires. Care must be taken that where the duct line changes direction ; depths, dips and offsets be made with a very long radius or it will be difficult to pull a large cable between the manholes. The distance between the manholes should not be too long so as to simplify the pulling in of the cables. The cables to be laid in this way need not be armoured but must be provided with serving of hessian and jute in order to protect them when being pulled into the ducts.

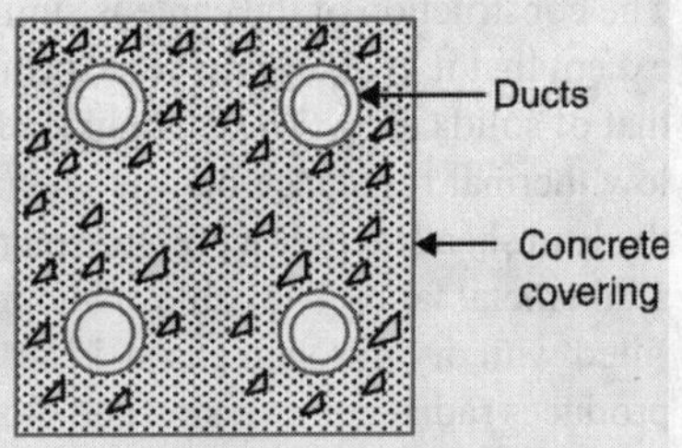

Fig. 11.11

Advantages

(*i*) Repairs, alterations or additions to the cable network can be made without opening the ground.

(*ii*) As the cables are not armoured, therefore, joints become simpler and maintenance cost is reduced considerably.

(*iii*) There are very less chances of fault occurrence due to strong mechanical protection provided by the system.

Disadvantages

(*i*) The initial cost is very high.

(*ii*) The current carrying capacity of the cables is reduced due to the close grouping of cables and unfavourable conditions for dissipation of heat.

This method of cable laying is suitable for congested areas where excavation is expensive and inconvenient, for once the conduits have been laid, repairs or alterations can be made without opening the ground. This method is generally used for short length cable routes such as in workshops, road crossings where frequent digging is costlier or impossible.

3. **Solid system.** In this method of laying, the cable is laid in open pipes or troughs dug out in earth along the cable route. The troughing is of cast iron, stoneware, asphalt or treated wood. After the cable is laid in position, the troughing is filled with a bituminous or asphaltic compound and covered over. Cables laid in this manner are usually plain lead covered because troughing affords good mechanical protection.

Disadvantages

(*i*) It is more expensive than direct laid system.

(*ii*) It requires skilled labour and favourable weather conditions.

(*iii*) Due to poor heat dissipation facilities, the current carrying capacity of the cable is reduced.

In view of these disadvantages, this method of laying underground cables is rarely used now-a-days.

11.7 Insulation Resistance of a Single-Core Cable

The cable conductor is provided with a suitable thickness of insulating material in order to prevent leakage current. The path for leakage current is radial through the insulation. The opposition offered by insulation to leakage current is known as insulation resistance of the cable. For satisfactory operation, the insulation resistance of the cable should be very high.

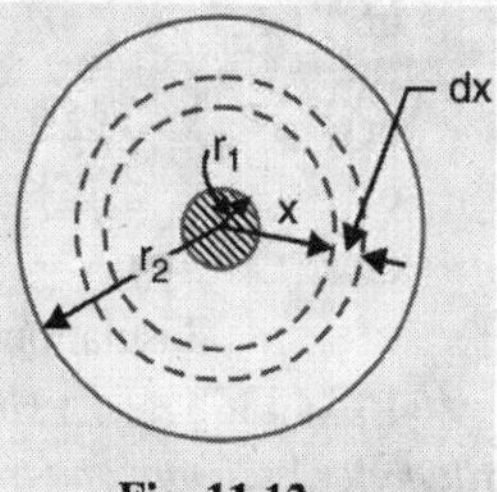

Fig. 11.12

Consider a single-core cable of conductor radius r_1 and internal sheath radius r_2 as shown in Fig. 11.12. Let l be the length of the cable and ρ be the resistivity of the insulation.

Consider a very small layer of insulation of thickness dx at a radius x. The length through which leakage current tends to flow is dx and the area of X-section offered to this flow is $2\pi x l$.

∴ Insulation resistance of considered layer

$$= \rho \frac{dx}{2\pi x l}$$

Insulation resistance of the whole cable is

$$R = \int_{r_1}^{r_2} \rho \frac{dx}{2\pi x l} = \frac{\rho}{2\pi l} \int_{r_1}^{r_2} \frac{1}{x} dx$$

$$\therefore \quad R = \frac{\rho}{2\pi l} \log_e \frac{r_2}{r_1}$$

This shows that insulation resistance of a cable is inversely proportional to its length. In other words, if the cable length increases, its insulation resistance decreases and *vice-versa.*

Example 11.1. *A single-core cable has a conductor diameter of 1cm and insulation thickness of 0·4 cm. If the specific resistance of insulation is 5 × 10^{14} Ω-cm, calculate the insulation resistance for a 2 km length of the cable.*

Solution

Conductor radius, $r_1 = 1/2 = 0{\cdot}5$ cm

Length of cable, $l = 2$ km $= 2000$ m

Resistivity of insulation, $\rho = 5 \times 10^{14}$ Ω-cm $= 5 \times 10^{12}$ Ω-m

Internal sheath radius, $r_2 = 0{\cdot}5 + 0{\cdot}4 = 0{\cdot}9$ cm

$\therefore$ Insulation resistance of cable is

$$R = \frac{\rho}{2\pi l}\log_e\frac{r_2}{r_1} = \frac{5\times10^{12}}{2\pi\times2000}\log_e\frac{0\cdot9}{0\cdot5}$$

$$= 0\cdot234\times10^9\ \Omega = \mathbf{234\ M\Omega}$$

Example 11.2. *The insulation resistance of a single-core cable is 495 MΩ per km. If the core diameter is 2·5 cm and resistivity of insulation is 4·5 × 10*^{14} *Ω-cm, find the insulation thickness.*

Solution.

Length of cable, $l = 1\text{ km} = 1000\text{ m}$

Cable insulation resistance, $R = 495\text{ M}\Omega = 495\times10^6\Omega$

Conductor radius, $r_1 = 2\cdot5/2 = 1\cdot25\text{ cm}$

Resistivity of insulation, $\rho = 4\cdot5\times10^{14}\ \Omega\text{-cm} = 4\cdot5\times10^{12}\Omega\text{m}$

Let r_2 cm be the internal sheath radius.

Now, $$R = \frac{\rho}{2\pi l}\log_e\frac{r_2}{r_1}$$

or $$\log_e\frac{r_2}{r_1} = \frac{2\pi l R}{\rho} = \frac{2\pi\times1000\times495\times10^6}{4\cdot5\times10^{12}} = 0\cdot69$$

or $$2\cdot3\log_{10} r_2/r_1 = 0\cdot69$$

or $$r_2/r_1 = \text{Antilog } 0\cdot69/2\cdot3 = 2$$

or $$r_2 = 2\,r_1 = 2\times1\cdot25 = 2\cdot5\text{ cm}$$

$\therefore$ Insulation thickness $= r_2 - r_1 = 2\cdot5 - 1\cdot25 = \mathbf{1\cdot25\ cm}$

Example 11.3. *A single core cable 5 km long has an insulation resistance of 0·4 MΩ. The core diameter is 20 mm and the diameter of the cable over the insulation is 50 mm. Calculate the resistivity of the insulating material.*

Solution.

Length of cable, $l = 5\text{ km} = 5000\text{ m}$

Cable insulation resistance, $R = 0\cdot4\text{ M}\Omega = 0\cdot4\times10^6\Omega$

Conductor radius, $r_1 = 20/2 = 10\text{ mm}$

Internal sheath radius, $r_2 = 50/2 = 25\text{ mm}$

$\therefore$ Insulation resistance of the cables is

$$R = \frac{\rho}{2\pi l}\log_e\frac{r_2}{r_1}$$

or $$0\cdot4\times10^6 = \frac{\rho}{2\pi\times5000}\times\log_e\frac{25}{10}$$

$\therefore$ $$\rho = \mathbf{13.72\times10^9\ \Omega m}$$

TUTORIAL PROBLEMS

1. A single-core cable has a conductor diameter of 2.5 cm and insulation thickness of 1.2 cm. If the specific resistance of insulation is $4\cdot5\times10^{14}\ \Omega$ cm, calculate the insulation resistance per kilometre length of the cable. **[305·5 MΩ]**
2. A single core cable 3 km long has an insulation resistance of 1820 MΩ. If the conductor diameter is 1.5 cm and sheath diameter is 5 cm, calculate the resistivity of the dielectric in the cable. **[28.57 × 10^{12} Ωm]**
3. Determine the insulation resistance of a single-core cable of length 3 km and having conductor radius 12·5 mm, insulation thickness 10 mm and specific resistance of insulation of $5\times10^{12}\Omega$m. **[156 MΩ]**

11.8 Capacitance of a Single-Core Cable

A single-core cable can be considered to be equivalent to two long co-axial cylinders. The conductor (or core) of the cable is the inner cylinder while the outer cylinder is represented by lead sheath which is at earth potential. Consider a single core cable with conductor diameter d and inner sheath diameter D (Fig. 11.13). Let the charge per metre axial length of the cable be Q coulombs and ε be the permittivity of the insulation material between core and lead sheath. Obviously *$\varepsilon = \varepsilon_0\, \varepsilon_r$ where ε_r is the relative permittivity of the insulation.

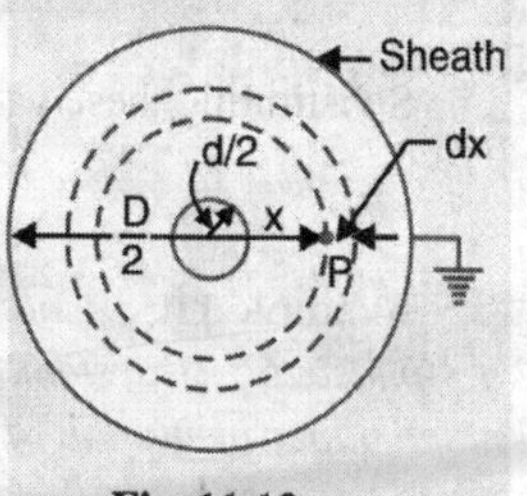

Fig. 11.13

Consider a cylinder of radius x metres and axial length 1 metre. The surface area of this cylinder is $= 2\,\pi\, x \times 1 = 2\,\pi\, x$ m^2

∴ Electric flux density at any point P on the considered cylinder is

$$D_x = \frac{Q}{2\pi x}\ \text{C/m}^2$$

Electric intensity at point P, $E_x = \dfrac{D_x}{\varepsilon} = \dfrac{Q}{2\pi x\,\varepsilon} = \dfrac{Q}{2\pi x \varepsilon_0 \varepsilon_r}$ volts/m

The work done in moving a unit positive charge from point P through a distance dx in the direction of electric field is $E_x\, dx$. Hence, the work done in moving a unit positive charge from conductor to sheath, which is the potential difference V between conductor and sheath, is given by :

$$V = \int_{d/2}^{D/2} E_x\, dx = \int_{d/2}^{D/2} \frac{Q}{2\pi x \varepsilon_0 \varepsilon_r}\, dx = \frac{Q}{2\pi\varepsilon_0\,\varepsilon_r} \log_e \frac{D}{d}$$

Capacitance of the cable is

$$C = \frac{Q}{V} = \frac{Q}{\dfrac{Q}{2\pi\,\varepsilon_0 \varepsilon_r} \log_e \dfrac{D}{d}}\ \text{F/m}$$

$$= \frac{2\pi\,\varepsilon_o \varepsilon_r}{\log_e (D/d)}\ \text{F/m}$$

$$= \frac{2\pi \times 8{\cdot}854 \times 10^{-12} \times \varepsilon_r}{2{\cdot}303\ \log_{10}(D/d)}\ \text{F/m}$$

$$= \frac{\varepsilon_r}{41{\cdot}4\ \log_{10}(D/d)} \times 10^{-9}\ \text{F/m}$$

If the cable has a length of l metres, then capacitance of the cable is

$$C = \frac{\varepsilon_r\, l}{41{\cdot}4\ \log_{10} \dfrac{D}{d}} \times 10^{-9}\ \text{F}$$

Example 11.4. *A single core cable has a conductor diameter of 1 cm and internal sheath diameter of 1·8 cm. If impregnated paper of relative permittivity 4 is used as the insulation, calculate the capacitance for 1 km length of the cable.*

Solution.

Capacitance of cable, $C = \dfrac{\varepsilon_r\, l}{41{\cdot}4\ \log_{10}(D/d)} \times 10^{-9}$ F

---

* It may be recalled $\varepsilon_r = \varepsilon/\varepsilon_0$ where ε_0 is the permittivity of free space. In the SI units, $\varepsilon_0 = 8.854 \times 10^{-12}$ F/m.

Here $\varepsilon_r = 4$; $l = 1000$ m

$D = 1{\cdot}8$ cm ; $d = 1$ cm

Substituting these values in the above expression, we get,

$$C = \frac{4 \times 1000}{41{\cdot}4 \log_{10}(1{\cdot}8/1)} \times 10^{-9} \text{ F} = 0{\cdot}378 \times 10^{-6} \text{ F} = \mathbf{0{\cdot}378\ \mu F}$$

Example 11.5. *Calculate the capacitance and charging current of a single core cable used on a 3-phase, 66 kV system. The cable is 1 km long having a core diameter of 10 cm and an impregnated paper insulation of thickness 7 cm. The relative permittivity of the insulation may be taken as 4 and the supply at 50 Hz.*

Solution.

Capacitance of cable, $$C = \frac{\varepsilon_r\, l}{41{\cdot}4 \log_{10}(D/d)} \times 10^{-9} \text{ F}$$

Here, $\varepsilon_r = 4$; $l = 1000$ m

$d = 10$ cm; $D = 10 + 2 \times 7 = 24$ cm

Substituting these values in the above expression,

$$C = \frac{4 \times 1000}{41{\cdot}4 \times \log_{10}(24/10)} \times 10^{-9} \text{ F} = 0{\cdot}254 \times 10^{-6} \text{ F} = \mathbf{0{\cdot}254\ \mu F}$$

Voltage between core and sheath is

$$V_{ph} = 66/\sqrt{3} = 38{\cdot}1 \text{ kV} = 38{\cdot}1 \times 10^3 \text{ V}$$

$$\text{Charging current} = V_{ph}/X_C = 2\pi f C V_{ph}$$

$$= 2\pi \times 50 \times 0{\cdot}254 \times 10^{-6} \times 38{\cdot}1 \times 10^3 \text{A} = \mathbf{3{\cdot}04\ A}$$

Example 11.6. *A 33 kV, 50 Hz, 3-phase underground cable, 4 km long uses three single core cables. Each of the conductor has a diameter of 2·5 cm and the radial thickness of insulation is 0·5 cm. Determine (i) capacitance of the cable/phase (ii) charging current/phase (iii) total charging kVAR. The relative permittivity of insulation is 3.*

Solution.

(*i*) Capacitance of cable/phase, $$C = \frac{\varepsilon_r\, l}{41{\cdot}4 \log_{10}(D/d)} \times 10^{-9} \text{ F}$$

Here $\varepsilon_r = 3$; $l = 4$ km $= 4000$ m

$d = 2{\cdot}5$ cm ; $D = 2.5 + 2 \times 0.5 = 3.5$ cm

Putting these values in the above expression, we get,

$$C = \frac{3 \times 4000 \times 10^{-9}}{41{\cdot}4 \times \log_{10}(3{\cdot}5/2{\cdot}5)} = \mathbf{1984 \times 10^{-9}\ F}$$

(*ii*) Voltage/phase, $$V_{ph} = \frac{33 \times 10^3}{\sqrt{3}} = 19{\cdot}05 \times 10^3 \text{ V}$$

Charging current/phase, $$I_C = \frac{V_{ph}}{X_C} = 2\pi f C V_{ph}$$

$$= 2\pi \times 50 \times 1984 \times 10^{-9} \times 19{\cdot}05 \times 10^3 = \mathbf{11{\cdot}87\ A}$$

(*iii*) Total charging kVAR $$= 3V_{ph} I_C = 3 \times 19{\cdot}05 \times 10^3 \times 11{\cdot}87 = \mathbf{678{\cdot}5 \times 10^3\ kVAR}$$

TUTORIAL PROBLEMS

1. A single core cable has a conductor diameter of 1 cm and internal sheath diameter of 1.8 cm. If the impregnated paper of relative permittivity 3 is used as insulation, calculate the capacitance for 1 km length of the cable. **[0.282 μF]**

2. Calculate the capacitance and charging current of a single core cable used on 3-phase, 66 kV system. The cable is 1 km long having a core diameter of 15 cm and impregnated paper insulation of thickness 22.5 cm. The relative permittivity of the insulation may be taken as 3.5 and supply at 50 Hz.

[0.144 μF ; 1.74 A]

3. An 11 kV, 50 Hz, single phase cable 2.5 km long, has a diameter of 20 mm and internal sheath radius of 15 mm. If the dielectric has a relative permittivity of 2.4, determine (*i*) capacitance (*ii*) charging current (*iii*) total charging kVAR. **[(*i*) 0.822 μF (*ii*) 2.84 A (*iii*) 31.24 kVAR]**

11.9 Dielectric Stress in a Single-Core Cable

Under operating conditions, the insulation of a cable is subjected to electrostatic forces. This is known as dielectric stress. The dielectric stress at any point in a cable is infact the potential gradient (or *electric intensity) at that point.

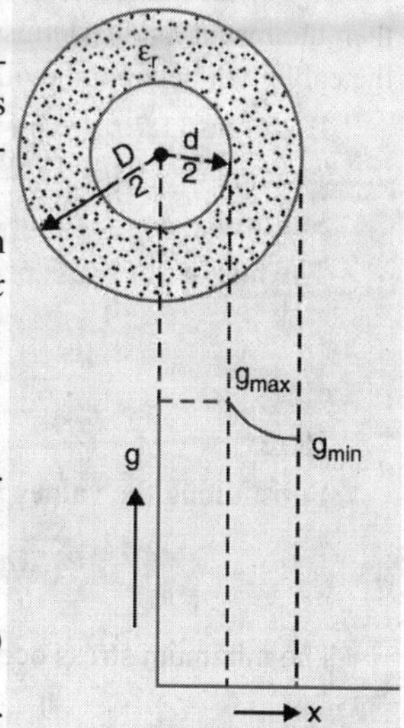

Fig. 11.14

Consider a single core cable with core diameter *d* and internal sheath diameter *D*. As proved in Art 11.8, the electric intensity at a point *x* metres from the centre of the cable is

$$E_x = \frac{Q}{2\pi\varepsilon_o\varepsilon_r x} \text{ volts/m}$$

By definition, electric intensity is equal to potential gradient. Therefore, potential gradient *g* at a point *x* metres from the centre of cable is

$$g = E_x$$

or
$$g = \frac{Q}{2\pi\varepsilon_o\varepsilon_r x} \text{ volts/m} \quad ...(i)$$

As proved in Art. 11.8, potential difference *V* between conductor and sheath is

$$V = \frac{Q}{2\pi\varepsilon_o\varepsilon_r}\log_e\frac{D}{d} \text{ volts}$$

or
$$Q = \frac{2\pi\varepsilon_o\varepsilon_r V}{\log_e\frac{D}{d}} \quad ...(ii)$$

Substituting the value of Q from exp. (*ii*) in exp. (*i*), we get,

$$g = \frac{\dfrac{2\pi\varepsilon_o\varepsilon_r V}{\log_e D/d}}{2\pi\varepsilon_o\varepsilon_r x} = \frac{V}{x\log_e\frac{D}{d}} \text{ volts/m} \quad ...(iii)$$

It is clear from exp. (*iii*) that potential gradient varies inversely as the distance *x*. Therefore, potential gradient will be maximum when *x* is minimum *i.e.,* when $x = d/2$ or at the surface of the conductor. On the other hand, potential gradient will be minimum at $x = D/2$ or at sheath surface.

∴ Maximum potential gradient is

$$g_{max} = \frac{2V}{d\log_e\frac{D}{d}} \text{ volts/m} \quad \text{[Putting } x = d/2 \text{ in exp. (iii)]}$$

Minimum potential gradient is

$$g_{min} = \frac{2V}{D\log_e\frac{D}{d}} \text{ volts/m} \quad \text{[Putting } x = D/2 \text{ in exp. (iii)]}$$

---

* It may be recalled that potential gradient at any point is equal to the electric intensity at that point.

$$\therefore \quad \frac{g_{max}}{g_{min}} = \frac{\dfrac{2V}{d \log_e D/d}}{\dfrac{2V}{D \log_e D/d}} = \frac{D}{d}$$

The variation of stress in the dielectric is shown in Fig. 11.14. It is clear that dielectric stress is maximum at the conductor surface and its value goes on decreasing as we move away from the conductor. It may be noted that maximum stress is an important consideration in the design of a cable. For instance, if a cable is to be operated at such a voltage that *maximum stress is 5 kV/mm, then the insulation used must have a dielectric strength of atleast 5 kV/mm, otherwise breakdown of the cable will become inevitable.

Example 11.7. *A 33 kV single core cable has a conductor diameter of 1 cm and a sheath of inside diameter 4 cm. Find the maximum and minimum stress in the insulation.*

Solution.

The maximum stress occurs at the conductor surface and its value is given by;

$$g_{max} = \frac{2V}{d \log_e \dfrac{D}{d}}$$

Here, V = 33 kV (r.m.s) ; d = 1 cm ; D = 4 cm

Substituting the values in the above expression, we get,

$$g_{max} = \frac{2 \times 33}{1 \times \log_e 4} \text{ kV } \dagger/\text{cm} = \mathbf{47{\cdot}61 \text{ kV/cm}} \textit{ r.m.s.}$$

The minimum stress occurs at the sheath and its value is give by ;

$$g_{min} = \frac{2V}{D \log_e \dfrac{D}{d}} = \frac{2 \times 33}{4 \times \log_e 4} \text{ kV/cm} = \mathbf{11{\cdot}9 \text{ kV/cm}} \textit{ r.m.s}$$

Alternatively ;

$$g_{min} = g_{max} \times \frac{d}{D} = 47{\cdot}61 \times 1/4 = \mathbf{11{\cdot}9 \text{ kV/cm}} \textit{ r.m.s.}$$

Example 11.8. *The maximum and minimum stresses in the dielectric of a single core cable are 40 kV/cm (r.m.s.) and 10 kV/cm (r.m.s.) respectively. If the conductor diameter is 2 cm, find :*

(i) thickness of insulation *(ii) operating voltage*

Solution.

Here, g_{max} = 40 kV/cm ; g_{min} = 10 kV/cm ; d = 2 cm ; D = ?

(*i*) As proved in Art. 11.9,

$$\frac{g_{max}}{g_{min}} = \frac{D}{d}$$

or

$$D = \frac{g_{max}}{g_{min}} \times d = \frac{40}{10} \times 2 = 8 \text{ cm}$$

$$\therefore \quad \text{Insulation thickness} = \frac{D-d}{2} = \frac{8-2}{2} = \mathbf{3 \text{ cm}}$$

(*ii*)

$$g_{max} = \frac{2V}{d \log_e \dfrac{D}{d}}$$

---

* Of course, it will occur at the conductor surface.

† Note that unit will be kV/cm. It is because V in the numerator is in kV and d in the denominator is in cm.

$$\therefore \quad V = \frac{g_{max}\, d \log_e \frac{D}{d}}{2} = \frac{40 \times 2 \log_e 4}{2} \text{ kV} = \mathbf{55.45\ kV}\ \textbf{\textit{r.m.s.}}$$

Example 11.9. *A single core cable for use on 11 kV, 50 Hz system has conductor area of 0·645 cm² and internal diameter of sheath is 2·18 cm. The permittivity of the dielectric used in the cable is 3·5. Find (i) the maximum electrostatic stress in the cable (ii) minimum electrostatic stress in the cable (iii) capacitance of the cable per km length (iv) charging current.*

Solution.

Area of cross-section of conductor, $a = 0.645 \text{ cm}^2$

Diameter of the conductor, $d = \sqrt{\frac{4a}{\pi}} = \sqrt{\frac{4 \times 0{\cdot}645}{\pi}} = 0{\cdot}906 \text{ cm}$

Internal diameter of sheath, $D = 2{\cdot}18 \text{ cm}$

(*i*) Maximum electrostatic stress in the cable is

$$g_{max} = \frac{2V}{d \log_e \frac{D}{d}} = \frac{2 \times 11}{0{\cdot}906 \log_e \frac{2{\cdot}18}{0{\cdot}906}} \text{ kV/cm} = \mathbf{27{\cdot}65\ kV/cm}\ \textbf{\textit{r.m.s.}}$$

(*ii*) Minimum electrostatic stress in the cable is

$$g_{min} = \frac{2\text{ V}}{D \log_e \frac{D}{d}} = \frac{2 \times 11}{2{\cdot}18 \log_e \frac{2{\cdot}18}{0{\cdot}906}} \text{ kV/cm} = \mathbf{11{\cdot}5\ kV/cm}\ \textbf{\textit{r.m.s.}}$$

(*iii*) Capacitance of cable, $C = \frac{\varepsilon_r\, l}{41{\cdot}4 \log_{10} \frac{D}{d}} \times 10^{-9} \text{ F}$

Here $\varepsilon_r = 3.5\,;\ l = 1 \text{ km} = 1000 \text{ m}$

$$\therefore \quad C = \frac{3.5 \times 1000}{41{\cdot}4 \log_{10} \frac{2{\cdot}18}{0{\cdot}906}} \times 10^{-9} = \mathbf{0{\cdot}22 \times 10^{-6}\ F}$$

(*iv*) Charging current, $I_C = \frac{V}{X_C} = 2\pi f C V = 2\pi \times 50 \times 0{\cdot}22 \times 10^{-6} \times 11000 = \mathbf{0{\cdot}76\ A}$

11.10 Most Economical Conductor Size in a Cable

It has already been shown that maximum stress in a cable occurs at the surface of the conductor. For safe working of the cable, dielectric strength of the insulation should be more than the maximum stress. Rewriting the expression for maximum stress, we get,

$$g_{max} = \frac{2V}{d \log_e \frac{D}{d}} \text{ volts/m} \quad \text{...(}i\text{)}$$

The values of working voltage V and internal sheath diameter D have to be kept fixed at certain values due to design considerations. This leaves conductor diameter d to be the only variable in exp. (*i*). For given values of V and D, the most economical conductor diameter will be one for which g_{max} has a minimum value. The value of g_{max} will be minimum when $d \log_e D/d$ is maximum *i.e.*

$$\frac{d}{dd}\left[d \log_e \frac{D}{d}\right] = 0$$

or $\log_e \frac{D}{d} + d.\frac{d}{D}.\frac{-D}{d^2} = 0$

or $\log_e (D/d) - 1 = 0$

or $\log_e (D/d) = 1$

or $(D/d) = e = 2{\cdot}718$

∴ Most economical conductor diameter is

$$d = \frac{D}{2.718}$$

and the value of g_{max} under this condition is

$$g_{max} = \frac{2V}{d} \text{ volts/m} \qquad [\text{Putting } \log_e D/d = 1 \text{ in exp. } (i)]$$

For low and medium voltage cables, the value of conductor diameter arrived at by this method (*i.e.*, $d = 2V/g_{max}$) is often too small from the point of view of current density. Therefore, the conductor diameter of such cables is determined from the consideration of safe current density. For high voltage cables, designs based on this theory give a very high value of *d,* much too large from the point of view of current carrying capacity and it is, therefore, advantageous to increase the conductor diameter to this value. There are three ways of doing this without using excessive copper :

(*i*) Using aluminium instead of copper because for the same current, diameter of aluminium will be more than that of copper.

(*ii*) Using copper wires stranded round a central core of hemp.

(*iii*) Using a central lead tube instead of hemp.

Example 11.10. *Find the most economical value of diameter of a single-core cable to be used on 50 kV, single-phase system. The maximum permissible stress in the dielectric is not to exceed 40 kV/cm.*

Solution.

Peak value of cable voltage, $V = 50 \times \sqrt{2} = 70.7$ kV

Maximum permissible stress, $g_{max} = 40$ kV/cm (assumed peak)

∴ Most economical conductor diameter is

$$d = \frac{2V}{g_{max}} = \frac{2 \times 70 \cdot 7}{40} = \mathbf{3 \cdot 53\ cm}$$

Example 11.11 *Find the most economical size of a single-core cable working on a 132 kV, 3-phase system, if a dielectric stress of 60 kV/cm can be allowed.*

Solution

Phase voltage of cable $= 132/\sqrt{3} = 76 \cdot 21$ kV

Peak value of phase voltage, $V = 76 \cdot 21 \times \sqrt{2} = 107 \cdot 78$ kV

Max. permissible stress, $g_{max} = 60$ kV/cm

∴ Most economical conductor diameter is

$$d = \frac{2V}{g_{max}} = \frac{2 \times 107 \cdot 78}{60} = \mathbf{3 \cdot 6\ cm}$$

Internal diameter of sheath, $D = 2 \cdot 718\, d = 2 \cdot 718 \times 3 \cdot 6 = \mathbf{9 \cdot 78\ cm}$

Therefore, the cable should have a conductor diameter of 3.6 cm and internal sheath diameter of 9·78 cm.

11.11 Grading of Cables

The process of achieving uniform electrostatic stress in the dielectric of cables is known as **grading of cables.**

It has already been shown that electrostatic stress in a single core cable has a maximum value (g_{max}) at the conductor surface and goes on decreasing as we move towards the sheath. The maximum voltage that can be safely applied to a cable depends upon g_{max} *i.e.,* electrostatic stress at the conductor surface. For safe working of a cable having homogeneous dielectric, the strength of dielectric must be more than g_{max}. If a dielectric of high strength is used for a cable, it is useful only near the conductor where stress is maximum. But as we move away from the conductor, the electrostatic stress decreases, so the dielectric will be unnecessarily overstrong.

The unequal stress distribution in a cable is undesirable for two reasons. Firstly, insulation of greater thickness is required which increases the cable size. Secondly, it may lead to the breakdown of insulation. In order to overcome above disadvantages, it is necessary to have a uniform stress distribution in cables. This can be achieved by distributing the stress in such a way that its value is increased in the outer layers of dielectric. This is known as grading of cables. The following are the two main methods of grading of cables :

(*i*) Capacitance grading (*ii*) Intersheath grading

11.12 Capacitance Grading

The process of achieving uniformity in the dielectric stress by using layers of different dielectrics is known as **capacitance grading.**

In capacitance grading, the homogeneous dielectric is replaced by a composite dielectric. The composite dielectric consists of various layers of different dielectrics in such a manner that relative permittivity ε_r of any layer is inversely proportional to its distance from the centre. Under such conditions, the value of potential gradient at any point in the dieletric is *constant and is independent of its distance from the centre. In other words, the dielectric stress in the cable is same everywhere and the grading is ideal one. However, ideal grading requires the use of an infinite number of dielectrics which is an impossible task. In practice, two or three dielectrics are used in the decreasing order of permittivity ; the dielectric of highest permittivity being used near the core.

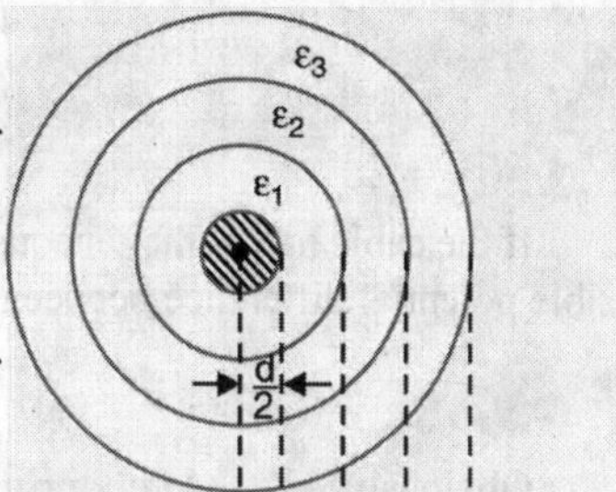

Fig. 11.15

The capacitance grading can be explained beautifully by referring to Fig. 11.15. There are three dielectrics of outer diameter d_1, d_2 and D and of relative permittivity ε_1, ε_2 and ε_3 respectively. If the permittivities are such that $\varepsilon_1 > \varepsilon_2 > \varepsilon_3$ and the three dielectrics are worked at the same maximum stress, then,

$$\frac{1}{\varepsilon_1 d} = \frac{1}{\varepsilon_2 d_1} = \frac{1}{\varepsilon_3 d_2}\,\dagger$$

or $$\varepsilon_1 d = \varepsilon_2 d_1 = \varepsilon_3 d_2$$

Potential difference across the inner layer is

---

* As $\varepsilon_r \propto \frac{1}{x}$ $\quad\therefore\quad$ $\varepsilon_r = k/x$ where k is a constant.

Potential gradient at a distance x from the centre

$$= \frac{Q}{2\pi\varepsilon_0\varepsilon_r x} = \frac{Q}{2\pi\varepsilon_0 (k/x)x} = \frac{Q}{2\pi\varepsilon_0 k} = \text{Constant}$$

This shows that if the condition $\varepsilon_r \propto 1/x$ is fulfilled, potential gradient will be constant throughout the dielectric of the cable.

† $g_{1max} = \frac{Q}{\pi\varepsilon_0\varepsilon_1 d}$; $\quad g_{2max} = \frac{Q}{\pi\varepsilon_0\varepsilon_2 d_1}$; $\quad g_{3max} = \frac{Q}{\pi\varepsilon_0\varepsilon_3 d_2}$

If $g_{1max} = g_{2max} = g_{3max} = g_{max}$ (say), then,

$$\frac{1}{\varepsilon_1 d} = \frac{1}{\varepsilon_2 d_1} = \frac{1}{\varepsilon_3 d_2}$$

$$V_1 = \int_{d/2}^{d_1/2} g\,dx = \int_{d/2}^{d_1/2} \frac{Q}{2\pi\varepsilon_0\varepsilon_1 x}dx$$

$$= \frac{Q}{2\pi\varepsilon_0\varepsilon_1}\log_e\frac{d_1}{d} = \frac{g_{max}}{2}d\log_e\frac{d_1}{d} \quad \left[\because \frac{Q}{2\pi\varepsilon_0\varepsilon_1} = \frac{*g_{max}}{2}d\right]$$

Similarly, potential across second layer (V_2) and third layer (V_3) is given by ;

$$V_2 = \frac{g_{max}}{2}d_1\log_e\frac{d_2}{d_1}$$

$$V_3 = \frac{g_{max}}{2}d_2\log_e\frac{D}{d_2}$$

Total p.d. between core and earthed sheath is

$$V = V_1 + V_2 + V_3$$

$$= \frac{g_{max}}{2}\left[d\log_e\frac{d_1}{d} + d_1\log_e\frac{d_2}{d_1} + d_2\log_e\frac{D}{d_2}\right]$$

If the cable had homogeneous dielectric, then, for the same values of d, D and g_{max}, the permissible potential difference between core and earthed sheath would have been

$$V' = \frac{g_{max}}{2}d\log_e\frac{D}{d}$$

Obviously, $V > V'$ *i.e.*, for given dimensions of the cable, a graded cable can be worked at a greater potential than non-graded cable. Alternatively, for the same safe potential, the size of graded cable will be less than that of non-graded cable. The following points may be noted :

(*i*) As the permissible values of g_{max} are peak values, therefore, all the voltages in above expressions should be taken as peak values and not the r.m.s. values.

(*ii*) If the maximum stress in the three dielectrics is not the same, then,

$$V = \frac{g_{1max}}{2}d\log_e\frac{d_1}{d} + \frac{g_{2max}}{2}d_1\log_e\frac{d_2}{d_1} + \frac{g_{3max}}{2}d_2\log_e\frac{D}{d_2}$$

The principal disadvantage of this method is that there are a few high grade dielectrics of reasonable cost whose permittivities vary over the required range.

Example 11.12. *A single-core lead sheathed cable is graded by using three dielectrics of relative permittivity 5, 4 and 3 respectively. The conductor diameter is 2 cm and overall diameter is 8 cm. If the three dielectrics are worked at the same maximum stress of 40 kV/cm, find the safe working voltage of the cable.*

What will be the value of safe working voltage for an ungraded cable, assuming the same conductor and overall diameter and the maximum dielectric stress ?

Solution.

Here, $d = 2$ cm ; $d_1 = ?$; $d_2 = ?$; $D = 8$ cm

$\varepsilon_1 = 5$; $\varepsilon_2 = 4$; $\varepsilon_3 = 3$; $g_{max} = 40$ kV/cm

Graded cable. As the maximum stress in the three dielectrics is the same,

$$\therefore \quad \varepsilon_1 d = \varepsilon_2 d_1 = \varepsilon_3 d_2$$

or $$5 \times 2 = 4 \times d_1 = 3 \times d_2$$

$$\therefore \quad d_1 = 2{\cdot}5 \text{ cm and } d_2 = 3{\cdot}34 \text{ cm}$$

---

* $g_{max} = \dfrac{Q}{\pi\varepsilon_0\varepsilon_1 d}$ $\quad\therefore\quad g_{max}\,d = \dfrac{Q}{\pi\varepsilon_0\varepsilon_1}$ or $\dfrac{g_{max}}{2}d = \dfrac{Q}{2\pi\varepsilon_0\varepsilon_1}$

Permissible peak voltage for the cable

$$= \frac{g_{max}}{2}\left[d \log_e \frac{d_1}{d} + d_1 \log_e \frac{d_2}{d_1} + d_2 \log_e \frac{D}{d_2}\right]$$

$$= \frac{40}{2}\left[2 \log_e \frac{2\cdot 5}{2} + 2\cdot 5 \log_e \frac{3\cdot 34}{2\cdot 5} + 3\cdot 34 \log_e \frac{8}{3\cdot 34}\right]$$

$$= 20\,[0{\cdot}4462 + 0{\cdot}7242 + 2{\cdot}92] \text{ kV}$$

$$= 20 \times 4{\cdot}0904 = 81{\cdot}808 \text{ kV}$$

∴ Safe working voltage (r.m.s.) for cable

$$= \frac{81\cdot 808}{\sqrt{2}} = \mathbf{57{\cdot}84\ kV}$$

Ungraded cable. Permissible peak voltage for the cable

$$= \frac{g_{max}}{2}\, d \log_e \frac{D}{d} = \frac{40}{2} \times 2 \log_e \frac{8}{2} \text{ kV} = 55{\cdot}44 \text{ kV}$$

∴ Safe working voltage (r.m.s.) for the cable

$$= \frac{55\cdot 44}{\sqrt{2}} = \mathbf{39{\cdot}2\ kV}$$

This example shows the utility of grading the cable. Thus for the same conductor diameter (d) and the same overall dimension (D), the graded cable can be operated at a voltage (57·84 – 39·20) = 18·64 kV (r.m.s.) higher than the homogeneous cable — an increase of about 47%.

Example 11.13. *A single core lead sheathed cable has a conductor diameter of 3 cm; the diameter of the cable being 9 cm. The cable is graded by using two dielectrics of relative permittivity 5 and 4 respectively with corresponding safe working stresses of 30 kV/cm and 20 kV/cm. Calculate the radial thickness of each insulation and the safe working voltage of the cable.*

Solution.

Here, $d = 3$ cm ; $d_1 = ?$; $D = 9$cm

$\varepsilon_1 = 5$; $\varepsilon_2 = 4$

$g_{1max} = 30$ kV/cm ; $g_{2max} = 20$ kV/cm

$$g_{1max} \propto \frac{1}{\varepsilon_1 d} \;;\quad g_{2max} \propto \frac{1}{\varepsilon_2 d_1}$$

$$\therefore \quad \frac{g_{1max}}{g_{2max}} = \frac{\varepsilon_2 d_1}{\varepsilon_1 d}$$

or

$$d_1 = \frac{g_{1max}}{g_{2max}} \times \frac{\varepsilon_1 d}{\varepsilon_2} = \frac{30}{20} \times \frac{5\times 3}{4} = 5{\cdot}625 \text{ cm}$$

∴ Radial thickness of inner dielectric

$$= \frac{d_1 - d}{2} = \frac{5\cdot 625 - 3}{2} = \mathbf{1{\cdot}312\ cm}$$

Radial thickness of outer dielectric

$$= \frac{D - d_1}{2} = \frac{9 - 5\cdot 625}{2} = \mathbf{1{\cdot}68\ cm}$$

Permissible peak voltage for the cable

$$= \frac{g_{1max}}{2} d \log_e \frac{d_1}{d} + \frac{g_{2max}}{2} d_1 \log_e \frac{D}{d_1}$$

$$= \frac{30}{2} \times 3 \log_e \frac{5\cdot 625}{3} + \frac{20}{2} \times 5\cdot 625 \log_e \frac{9}{5\cdot 625}$$

$$= 28{\cdot}28 + 26{\cdot}43 = 54{\cdot}71 \text{ kV}$$

∴ Safe working voltage (r.m.s.) for the cable

$$= 54.71/\sqrt{2} = \mathbf{38{\cdot}68\ kV}$$

Example 11.14. *A 66-kV single-core lead sheathed cable is graded by using two dielectrics of relative permittivity 5 and 3 respectively; thickness of each being 1 cm. The core diameter is 2 cm. Determine the maximum stress in the two dielectrics.*

Solution. Fig. 11.16 shows the composite dielectric of a capacitance graded cable. The potential difference V between conductor and earthed sheath is given by ;

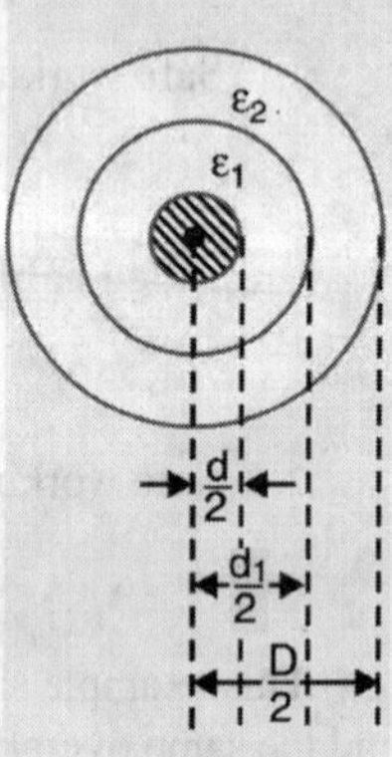

Fig. 11.16

$$V = \int_{d/2}^{d_1/2} g_1\,dx + \int_{d_1/2}^{D/2} g_2\,dx$$

$$= \int_{d/2}^{d_1/2} \frac{Q}{2\pi\varepsilon_0\varepsilon_1 x}\,dx + \int_{d_1/2}^{D/2} \frac{Q}{2\pi\varepsilon_0\varepsilon_2 x}\,dx$$

$$= \frac{Q}{2\pi\varepsilon_0}\left[\frac{1}{\varepsilon_1}\log_e\frac{d_1}{d} + \frac{1}{\varepsilon_2}\log_e\frac{D}{d_1}\right] \quad ...(i)$$

Now, $$g_{1max} = \frac{Q}{\pi\varepsilon_0\varepsilon_1 d} \quad ...(ii)$$

Putting the value of $Q = g_{1max}\,\pi\,\varepsilon_0\,\varepsilon_1\,d$ from exp. (*ii*) in exp. (*i*), we get,

$$V = \frac{g_{1max}\,\varepsilon_1 d}{2}\left[\frac{1}{\varepsilon_1}\log_e\frac{d_1}{d} + \frac{1}{\varepsilon_2}\log_e\frac{D}{d_1}\right]$$

or $$g_{1max} = \frac{2V}{d\left[\log_e\frac{d_1}{d} + \frac{\varepsilon_1}{\varepsilon_2}\log_e\frac{D}{d_1}\right]}$$

Here, $d = 2$ cm, $d_1 = 4$ cm, $D = 6$ cm ; $V = \frac{66}{\sqrt{3}} \times \sqrt{2} = 53.9$ kV, $\varepsilon_1 = 5$, $\varepsilon_2 = 3$

Substituting the values, we get,

$$g_{1max} = \frac{2 \times 53{\cdot}9}{2[\log_e 4/2 + 5/3\log_e 6/4]}\ \text{kV/cm}$$

$$= \frac{2 \times 53.9}{2[0{\cdot}6931 + 0{\cdot}6757]} = \mathbf{39{\cdot}38\ kV/cm}$$

Similarly, it can be *proved that :

$$g_{2max} = \frac{2V}{d_1\left[\frac{\varepsilon_2}{\varepsilon_1}\log_e\frac{d_1}{d} + \log_e\frac{D}{d_1}\right]} \quad ...(iii)$$

$$= \frac{2 \times 53.9}{4[3/5\log_e 4/2 + \log_e 6/4]}\ \text{kV/cm}$$

$$= \frac{2 \times 53{\cdot}9}{4[0{\cdot}4158 + 0{\cdot}4054]} = \mathbf{32{\cdot}81\ kV/cm}$$

11.13 Intersheath Grading

In this method of cable grading, a homogeneous dielectric is used, but it is divided into various layers by placing metallic intersheaths between the core and lead sheath. The intersheaths are held at suitable potentials which are inbetween the core potential and earth potential. This arrangement im-

* $g_{2max} = \frac{Q}{\pi\varepsilon_0\,\varepsilon_2 d_1}$. Putting the value of $Q = g_{2max}\,\pi\,\varepsilon_0\,\varepsilon_2 d_1$ in exp. (*i*), we get the exp. (*iii*).

proves voltage distribution in the dielectric of the cable and consequently more uniform potential gradient is obtained.

Consider a cable of core diameter d and outer lead sheath of diameter D. Suppose that two intersheaths of diameters d_1 and d_2 are inserted into the homogeneous dielectric and maintained at some fixed potentials. Let V_1, V_2 and V_3 respectively be the voltage between core and intersheath 1, between intersheath 1 and 2 and between intersheath 2 and outer lead sheath. As there is a definite potential difference between the inner and outer layers of each intersheath, therefore, each sheath can be treated like a homogeneous single core cable. As proved in Art. 11.9,

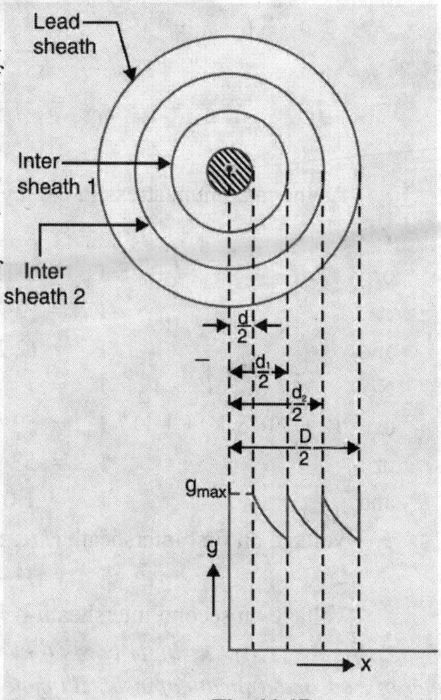

Fig. 11.17

Maximum stress between core and intersheath 1 is

$$g_{1max} = \frac{V_1}{\frac{d}{2}\log_e \frac{d_1}{d}}$$

Similarly,

$$g_{2max} = \frac{V_2}{\frac{d_1}{2}\log_e \frac{d_2}{d_1}}$$

$$g_{3max} = \frac{V_3}{\frac{d_2}{2}\log_e \frac{D}{d_2}}$$

Since the dielectric is homogeneous, the maximum stress in each layer is the same *i.e.*,

$$g_{1max} = g_{2max} = g_{3max} = g_{max} \text{ (say)}$$

$$\therefore \quad \frac{V_1}{\frac{d}{2}\log_e \frac{d_1}{d}} = \frac{V_2}{\frac{d_1}{2}\log_e \frac{d_2}{d_1}} = \frac{V_3}{\frac{d_2}{2}\log_e \frac{D}{d_2}}$$

As the cable behaves like three capacitors in series, therefore, all the potentials are in phase *i.e.* Voltage between conductor and earthed lead sheath is

$$V = V_1 + V_2 + V_3$$

Intersheath grading has three principal disadvantages. Firstly, there are complications in fixing the sheath potentials. Secondly, the intersheaths are likely to be damaged during transportation and installation which might result in local concentrations of potential gradient. Thirdly, there are considerable losses in the intersheaths due to charging currents. For these reasons, intersheath grading is rarely used.

Example 11.15. *A single core cable of conductor diameter 2 cm and lead sheath of diameter 5.3 cm is to be used on a 66 kV, 3-phase system. Two intersheaths of diameter 3·1 cm and 4·2 cm are introduced between the core and lead sheath. If the maximum stress in the layers is the same, find the voltages on the intersheaths.*

Solution.

Here, $d = 2 \text{ cm}; \quad d_1 = 3{\cdot}1 \text{ cm}; d_2 = 4{\cdot}2 \text{ cm}$

$$D = 5.3 \text{ cm}; \; V = \frac{66 \times \sqrt{2}}{\sqrt{3}} = 53{\cdot}9 \text{ kV}$$

$$g_{1max} = \frac{V_1}{\frac{d}{2}\log_e \frac{d_1}{d}} = \frac{V_1}{1 \times \log_e \frac{3{\cdot}1}{2}} = 2{\cdot}28\, V_1$$

$$g_{2max} = \frac{V_2}{\frac{d_1}{2}\log_e \frac{d_2}{d_1}} = \frac{V_2}{1.55 \log_e \frac{4.2}{3.1}} = 2{\cdot}12\ V_2$$

$$g_{3max} = \frac{V_3}{\frac{d_2}{2}\log_e \frac{D}{d_2}} = \frac{V_3}{2{\cdot}1\log_e \frac{5{\cdot}3}{4{\cdot}2}} = 2{\cdot}04\ V_3$$

As the maximum stress in the layers is the same,

$\therefore$ $g_{1max} = g_{2max} = g_{3max}$

or $2{\cdot}28\ V_1 = 2{\cdot}12\ V_2 = 2{\cdot}04\ V_3$

$\therefore$ $V_2 = (2{\cdot}28/2{\cdot}12)\ V_1 = 1{\cdot}075\ V_1$

and $V_3 = (2{\cdot}28/2{\cdot}04)\ V_1 = 1{\cdot}117\ V_1$

Now $V_1 + V_2 + V_3 = V$

or $V_1 + 1{\cdot}075\ V_1 + 1{\cdot}117\ V_1 = 53{\cdot}9$

or $V_1 = 53{\cdot}9/3{\cdot}192 = 16{\cdot}88$ kV

and $V_2 = 1{\cdot}075\ V_1 = 1{\cdot}075 \times 16{\cdot}88 = 18{\cdot}14$ kV

$\therefore$ Voltage on first intersheath (*i.e.,* near to the core)

$= V - V_1 = 53{\cdot}9 - 16{\cdot}88 =$ **37·02 kV**

Voltage on second intersheath $= V - V_1 - V_2 = 53{\cdot}9 - 16{\cdot}88 - 18{\cdot}14 =$ **18·88 kV**

Example 11.16. *A single-core 66 kV cable working on 3-phase system has a conductor diameter of 2 cm and a sheath of inside diameter 5.3 cm. If two intersheaths are introduced in such a way that the stress varies between the same maximum and minimum in the three layers, find :*

(*i*) *positions of intersheaths*

(*ii*) *voltage on the intersheaths*

(*iii*) *maximum and minimum stress*

Solution.

Here, $d = 2$ cm ; $D = 5.3$ cm ; $V = \dfrac{66 \times \sqrt{2}}{\sqrt{3}} = 53.9$ kV

(*i*) **Positions of intersheaths.** Suppose that diameters of intersheaths are d_1 and d_2 cm respectively. Let V_1, V_2 and V_3 respectively be the voltage between conductor and intersheath 1, between intersheath 1 and 2 and between intersheath 2 and outer lead sheath.

$$g_{1max} = \frac{V_1}{\frac{d}{2}\log_e \frac{d_1}{d}} \quad ; \quad g_{2max} = \frac{V_2}{\frac{d_1}{2}\log_e \frac{d_2}{d_1}} \quad ; \quad g_{3max} = \frac{V_3}{\frac{d_2}{2}\log_e \frac{D}{d_2}}$$

As the maximum stress in the three layers is the same,

$$\therefore \quad \frac{V_1}{\frac{d}{2}\log_e \frac{d_1}{d}} = \frac{V_2}{\frac{d_1}{2}\log_e \frac{d_2}{d_1}} = \frac{V_3}{\frac{d_2}{2}\log_e \frac{D}{d_2}} \quad \text{...}(i)$$

In order that stress may vary between the same maximum and minimum in the three layers, we have,

$$d_1/d = d_2/d_1 = D/d_2 \quad \text{...}(ii)$$

$$\therefore \quad \frac{V_1}{d} = \frac{V_2}{d_1} = \frac{V_3}{d_2} \quad \text{...*}(iii)$$

* This equation is obtained if we put the values of eq. (*ii*) in eq. (*i*).

From exp. (*ii*), we get,

$$d_1^2 = d \times d_2 = 2d_2 \qquad [\because d = 2 \text{ cm}]$$

or $$d_2 = d_1^2/2$$

and $$d_1 d_2 = D \times d = 5{\cdot}3 \times 2 = 10.6 \text{ cm}$$

or $$d_1 \times d_1^2/2 = 10{\cdot}6$$

or $$d_1 = (21{\cdot}2)^{1/3} = \mathbf{2{\cdot}76 \text{ cm}}$$

$\therefore$ $$d_2 = d_1^2/2 = (2{\cdot}76)^2/2 = \mathbf{3{\cdot}8 \text{ cm}}$$

Hence intersheaths of diameters 2.76 cm and 3.8 cm are required to be used.

(*ii*) Voltage on intersheaths

$$V = V_1 + V_2 + V_3$$

or $$53{\cdot}9 = V_1 + \frac{d_1}{d} V_1 + \frac{d_2}{d} V_1 \qquad \text{[From eq. (}iii\text{)]}$$

$$= V_1\left(1 + \frac{2{\cdot}76}{2} + \frac{3{\cdot}8}{2}\right) = 4{\cdot}28\, V_1$$

$\therefore$ $$V_1 = 53{\cdot}9/4{\cdot}28 = 12{\cdot}6 \text{ kV}$$

and $$V_2 = \frac{d_1}{d} \times V_1 = \frac{2.76}{2} \times 12{\cdot}6 = 17{\cdot}39 \text{ kV}$$

Voltage on first intersheath $= V - V_1 = 53{\cdot}9 - 12.6 = \mathbf{41{\cdot}3 \text{ kV } \mathit{max}}$

Voltage on second intersheath $= V - V_1 - V_2 = 53{\cdot}9 - 12{\cdot}6 - 17{\cdot}39 = \mathbf{23{\cdot}91 \text{ kV } \mathit{max}}$

(*iii*) Stresses in dielectrics

$$\text{Maximum stress} = \frac{V_1}{\frac{d}{2}\log_e \frac{d_1}{d}} = \frac{12{\cdot}6}{1 \times \log_e \frac{2{\cdot}76}{2}} \text{ kV/cm} = \mathbf{39 \text{ kV/cm}}$$

$$\text{Minimum stress} = \frac{V_1}{\frac{d_1}{2}\log_e \frac{d_1}{d}} = \frac{12{\cdot}6}{1{\cdot}38 \log_e \frac{2{\cdot}76}{2}} \text{ kV/cm} = \mathbf{28{\cdot}35 \text{ kV/cm}}$$

TUTORIAL PROBLEMS

1. A 33 kV, single-core cable has a conductor diameter of 1 cm and insulation of 1.5 cm. Find the maximum and minimum stress in the insulation. **[47·62 kV/cm (*r.m.s.*), 11·9 kV/cm (*r.m.s.*)]**
2. Find the economic size of a single-core cable working on 220 kV, 3-phase system. The maximum permissible stress in the dielectric is not to exceed 250 kV/cm. **[d = 1·43 cm, D = 3·88 cm]**
3. The inner conductor of a concentric cable has a diameter of 3 cm with insulation of diameter 8·5 cm. The cable is insulated with two materials having relative permittivities of 5 and 3 with corresponding safe working stresses of 38 kV/cm and 26 kV/cm. Calculate the radial thickness of insulating layers and the safe working voltage of the cable. **[2·15 cm, 0.6 cm, 46·1 kV (*r.m.s.*)]**
4. A single-core lead covered cable is to be designed for 66 kV to earth. Its conductor diameter is 2 cm and its three insulating materials have permittivities of 5, 4 and 3 respectively with the corresponding maximum safe working stress of 38 kV/cm (*r.m.s.*), 26 kV/cm. (*r.m.s.*) and 20 kV/cm (*r.m.s.*) respectively. Find the minimum diameter of lead sheath. **[8·3 cm]**
5. A single-core 66 kV cable has a conductor diameter of 2 cm and a sheath of inside diameter 5·3 cm. The cable has an inner layer of 1 cm thick of rubber of dielectric constant 4·5 and the rest impregnated paper of dielectric constant 3·6. Find the maximum stress in the rubber and in the paper. **[63 kV/cm, 39·5 kV/cm]**
6. A single-core cable working on 66 kV on 3-phase system has a conductor diameter of 2 cm and a sheath of inside diameter 5·3 cm. If two intersheaths are used, find the best positions, maximum stress and the voltage on the intersheaths. **[d_1 = 2·77 cm; d_2 = 3·84 cm ; 38·7 kV/cm ; V_1 = 41·1 kV, V_2 = 23·9 kV]**

11.14 Capacitance of 3-Core Cables

The capacitance of a cable system is much more important than that of overhead line because in cables (*i*) conductors are nearer to each other and to the earthed sheath (*ii*) they are separated by a dielectric of permittivity much greater than that of air. Fig. 11.18 shows a system of capacitances in a 3-core belted cable used for 3-phase system. Since potential difference exists between pairs of conductors and between each conductor and the sheath, electrostatic fields are set up in the cable as shown in Fig. 11.18 (*i*). These electrostatic fields give rise to core-core capacitances C_c and conductor-earth capacitances C_e as shown in Fig. 11.18 (*ii*). The three C_c are delta connected whereas the three C_e are star connected, the sheath forming the star point [See Fig. 11.18 (*iii*)].

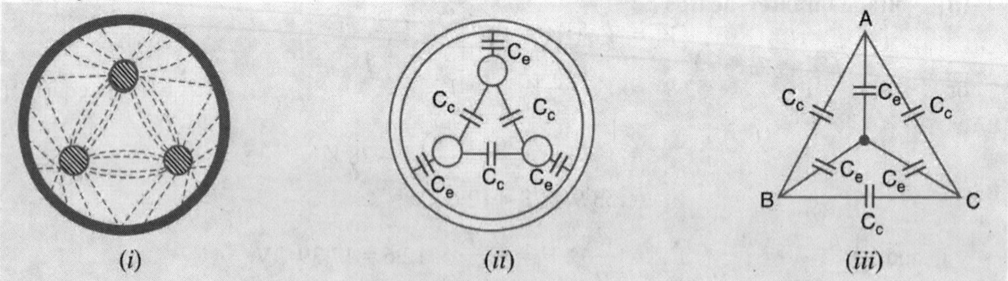

Fig. 11.18

They lay of a belted cable makes it reasonable to assume equality of each C_c and each C_e. The three delta connected capacitances C_c [See Fig. 11.19 (*i*)] can be converted into equivalent star connected capacitances as shown in Fig. 11.19 (*ii*). It can be easily *shown that equivalent star-capacitance C_{eq} is equal to three times the delta-capacitance C_c *i.e.* $C_{eq} = 3C_c$.

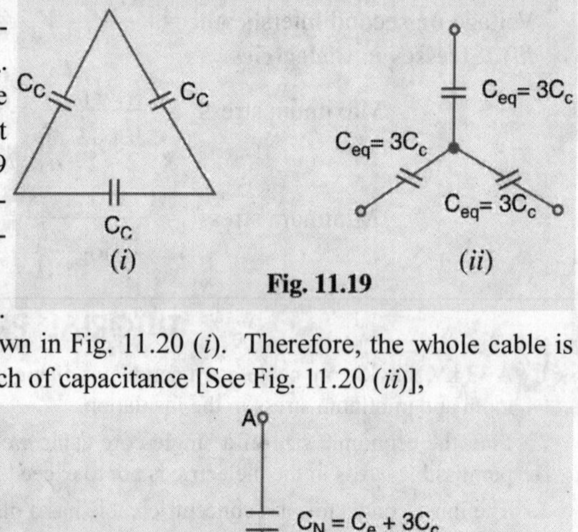

Fig. 11.19

The system of capacitances shown in Fig. 11.18 (*iii*) reduces to the equivalent circuit shown in Fig. 11.20 (*i*). Therefore, the whole cable is equivalent to three star-connected capacitors each of capacitance [See Fig. 11.20 (*ii*)],

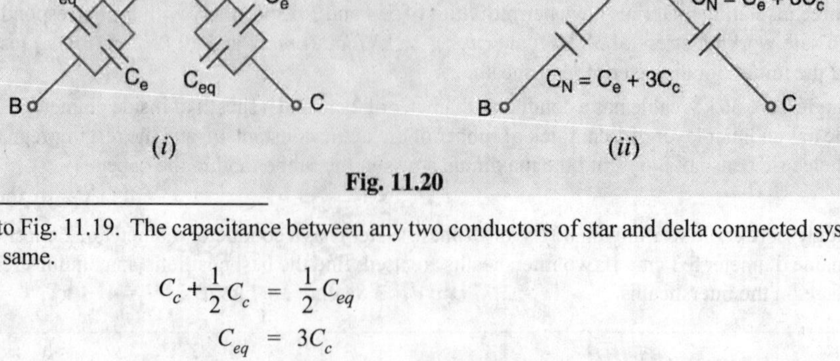

Fig. 11.20

* Refer to Fig. 11.19. The capacitance between any two conductors of star and delta connected system must be the same.

$$\therefore \quad C_c + \frac{1}{2}C_c = \frac{1}{2}C_{eq}$$

$$\text{or} \quad C_{eq} = 3C_c$$

$$C_N = C_e + C_{eq}$$
$$= C_e + 3C_c$$

If V_{ph} is the phase voltage, then charging current I_C is given by ;

$$I_C = \frac{V_{ph}}{\text{Capacitive reactance per phase}}$$
$$= 2\,\pi f V_{ph}\, C_N$$
$$= 2\,\pi f V_{ph}\,(C_e + 3C_c)$$

11.15 Measurements of C_e and C_c

Although core-core capacitance C_c and core-earth capacitance C_e can be obtained from the empirical formulas for belted cables, their values can also be determined by measurements. For this purpose, the following two measurements are required :

(*i*) In the first measurement, the three cores are bunched together (*i.e.* commoned) and the capacitance is measured between the bunched cores and the sheath. The bunching eliminates all the three capacitors C_c, leaving the three capacitors C_e in parallel. Therefore, if C_1 is the measured capacitance, this test yields :

$$C_1 = 3\,C_e$$

or
$$C_e = \frac{C_1}{3}$$

Knowing the value of C_1, the value of C_e can be determined.

(*ii*) In the second measurement, two cores are bunched with the sheath and capacitance is measured between them and the third core. This test yields $2C_c + C_e$. If C_2 is the measured capacitance, then,

$$C_2 = 2C_c + C_e$$

As the value of C_e is known from first test and C_2 is found experminentally, therefore, value of C_c can be determined.

It may be noted here that if value of $C_N (= C_e + 3C_c)$ is desired, it can be found directly by another test. In this test, the capacitance between two cores or lines is measured with the third core free or connected to the sheath. This eliminates one of the capacitors C_e so that if C_3 is the measured capacitance, then,

$$C_3 = C_c + \frac{C_c}{2} + \frac{C_e}{2}$$
$$= \frac{1}{2}\,(C_e + 3C_c)$$
$$= \frac{1}{2}\,C_N$$

Example 11.17. *The capacitance per kilometre of a 3-phase belted cable is 0·3 μF between the two cores with the third core connected to the lead sheath. Calculate the charging current taken by five kilometres of this cable when connected to a 3-phase, 50 Hz, 11 kV supply.*

Solution. The capacitance between a pair of cores with third core earthed for a length of 5 km is

$$C_3 = 0{\cdot}3 \times 5 = 1{\cdot}5\ \mu F$$

$$V_{ph} = \frac{11 \times 10^3}{\sqrt{3}} = 6351\ V;\ f = 50\text{ Hz}$$

As proved in Art 11.15, core to neutral capacitance C_N of this cable is given by :

$$C_N = 2C_3 = 2 \times 1{\cdot}5 = 3\ \mu F$$

$\therefore$ Charging current, $I_C = 2\pi f V_{ph} C_N$

$= 2\pi \times 50 \times 6351 \times 3 \times 10^{-6}$ A = **5·98 A**

Example 11.18. *The capacitances of a 3-phase belted cable are 12·6 μF between the three cores bunched together and the lead sheath and 7·4 μF between one core and the other two connected to sheath. Find the charging current drawn by the cable when connected to 66 kV, 50 Hz supply.*

Solution.

Here, $V_{ph} = \dfrac{66 \times 10^3}{\sqrt{3}} = 38105\ V$; $f = 50$ Hz ; $C_1 = 12·6\ \mu F$; $C_2 = 7·4\ \mu F$

Let core-core and core-earth capacitances of the cable be C_c and C_e respectively. As proved in Art. 11.15,

$$C_1 = 3\,C_e$$

$\therefore$ $C_e = C_1/3 = 12·6/3 = 4·2\ \mu F$

and $C_2 = 2C_c + C_e$

$\therefore$ $C_c = \dfrac{C_2 - C_e}{2} = \dfrac{7·4 - 4·2}{2} = 1·6\ \mu F$

$\therefore$ Core to neutral capacitance is

$$C_N = C_e + 3C_c = 4·2 + 3 \times 1·6 = 9\ \mu F$$

Charging current, $I_C = 2\pi f V_{ph} C_N$

$= 2\pi \times 50 \times 38105 \times 9 \times 10^{-6}\ A$ = **107·74 A**

Example 11.19. *The capacitance per kilometre of a 3-phase belted cable is 0·18 μF between two cores with the third core connected to sheath. Calculate the kVA taken by 20 km long cable when connected to 3-phase, 50 Hz, 3300 V supply.*

Solution. The capacitance between a pair of cores with third core earthed for a length of 20 km is

$$C_3 = 0·18 \times 20 = 3·6\ \mu F,\ V_{ph} = 3300/\sqrt{3} = 1905\ V;\ f = 50\ \text{Hz}$$

Core to neutral capacitance, $C_N = 2C_3 = 2 \times 3·6 = 7·2\ \mu F$

Charging current, $I_C = 2\pi f V_{ph} C_N$

$= 2\pi \times 50 \times 1905 \times 7·2 \times 10^{-6}$ A = 4·3 A

kVA taken by the cable $= 3\,V_{ph} I_C = 3 \times 1905 \times 4·3 \times 10^{-3}$ kVA = **24·57 kVA**

TUTORIAL PROBLEMS

1. The capacitances per kilometre of a 3-phase cable are 0·63 μF between the three cores bunched together and the sheath and 0·37 μF between one core and the other two connected to the sheath. Calculate the charging current taken by eight kilometres of this cable when connected to a 3-phase, 50 Hz, 6600 V supply. **[4·31 A]**
2. The capacitances of a 3-core belted type cable are measured as detailed below :
   (*i*) Between three cores bunched together and sheath is 8 μF
   (*ii*) Between a conductor and the other two connected to the sheath together is 6 μF.
   Calculate the capacitance per phase. **[23/3 μF]**
3. A 3-core, 3-phase belted cable tested for capacitance between a pair of cores on single phase, with the third core earthed, gave a capacitance of 0·4 μF per km. Calculate the charging current for 15 km length of this cable when connected to 22 kV, 3-phase, 50 Hz supply. **[48 A]**

11.16 Current-Carrying Capacity of Underground Cables

The safe current-carrying capacity of an underground cable is determined by the maximum permissible temperature rise. The cause of temperature rise is the losses that occur in a cable which appear as heat. These losses are :

(*i*) Copper losses in the conductors

(*ii*) Hysteresis losses in the dielectric

(*iii*) Eddy current losses in the sheath

The safe working conductor temperature is 65°C for armoured cables and 50°C for lead-sheathed cables laid in ducts. The maximum steady temperature conditions prevail when the heat generated in the cable is equal to the heat dissipated. The heat dissipation of the conductor losses is by conduction through the insulation to the sheath from which the total losses (including dielectric and sheath losses) may be conducted to the earth. Therefore, in order to find permissible current loading, the thermal resistivities of the insulation, the protective covering and the soil must be known.

11.17 Thermal Resistance

The thermal resistance between two points in a medium (*e.g.* insulation) is equal to temperature difference between these points divided by the heat flowing between them in a unit time *i.e.*

$$\text{Thermal resistance, } S = \frac{\text{Temperature difference}}{\text{Heat flowing in a unit time}}$$

In SI units, heat flowing in a unit time is measured in watts.

$$\therefore \quad \text{Thermal resistance, } S = \frac{\text{Temperature rise } (t)}{\text{Watts dissipated } (P)}$$

or

$$S = \frac{t}{P}$$

Clearly, the SI unit of thermal resistance is °C per watt. This is also called *thermal ohm.*

Like electrical resistance, thermal resistance is directly proportional to length l in the direction of transmission of heat and inversely proportional to the cross-section area a at right angles to that direction.

$$\therefore \quad S \propto \frac{l}{a}$$

or

$$S = k\frac{l}{a}$$

where k is the constant of proportionality and is known as *thermal resistivity.*

$$k = \frac{S\,a}{l}$$

$$\therefore \quad \text{Unit of } k = \frac{\text{Thermal ohm} \times m^2}{m} = \text{thermal ohm-metre}$$

11.18 Thermal Resistance of Dielectric of a Single-Core Cable

Let us now find the thermal resistance of the dielectric of a single-core cable.

Let r = radius of the core in metre

r_1 = inside radius of the sheath in metre

k = thermal resistivity of the insulation (*i.e.* dielectric)

Consider 1m length of the cable. The thermal resistance of small element of thickness dx at radius x is (See Fig. 11.21)

$$dS = k \times \frac{dx}{2\,\pi\,x}$$

$\therefore$ Thermal resistance of the dielectric is

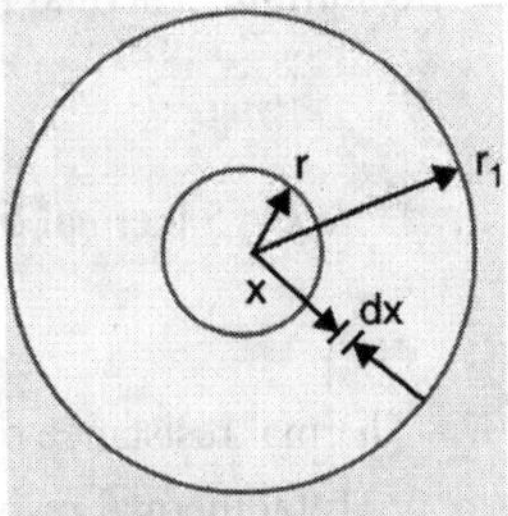

Fig. 11.21

$$S = \int_r^{r_1} k \times \frac{dx}{2\pi x}$$

$$= \frac{k}{2\pi}\int_r^{r_1} \frac{1}{x}\, dx$$

$$\therefore \quad S = \frac{k}{2\pi} \log_e \frac{r_1}{r} \text{ thermal ohms per metre length of the cable}$$

The thermal resistance of lead sheath is small and is generally neglected in calculations.

11.19 Permissible Current Loading

When considering heat dissipation in underground cables, the various thermal resistances providing a heat dissipation path are in series. Therefore, they add up like electrical resistances in series. Consider a cable laid in soil.

Let I = permissible current per conductor

n = number of conductors

R = electrical resistance per metre length of the conductor at the working temperature

S = total thermal resistance (*i.e.* sum of thermal resistances of dielectric and soil) per metre length

t = temperature difference (rise) between the conductor and the soil

Neglecting the dielectric and sheath losses, we have,

$$\text{Power dissipated} = n I^2 R$$

Now $$\text{Power dissipated} = \frac{\text{Temperature rise}}{\text{Thermal resistance}}$$

or $$n I^2 R = \frac{t}{S}$$

∴ Permissible current per conductor in given by;

$$I = \sqrt{\frac{t}{n R S}}$$

It should be noted that when cables are laid in proximity to each other, the permissible current is reduced further on account of mutual heating.

Example 11.20. *A single-core cable is laid in the ground, the core diameter being 30 mm and the dielectric thickness 40 mm. The thermal resistivity of the dielectric is 5 thermal ohm-metres and the thermal resistance between the sheath and the ground surface is 0·45 thermal ohm per metre length of the cable. Neglecting dielectric and sheath losses, estimate the maximum permissible current loading if the temperature difference between the conductor and the ground surface is not to exceed 55°C. The electrical resistance of the cable is 110 μΩ per metre length.*

Solution. Thermal resistance of the dielectric of the cable is

$$S_1 = \frac{k}{2\pi} \log_e \frac{r_1}{r} \text{ thermal-ohms per metre length of cable}$$

Here $k = 5$ thermal ohm-metres; $r = 30/2 = 15$ mm; $r_1 = 15 + 40 = 55$ mm

$$\therefore \quad S_1 = \frac{5}{2\pi} \log_e \frac{55}{15} = 1{\cdot}03 \text{ thermal ohms per metre length}$$

Thermal resistance of soil, S_2 = 0·45 thermal ohm per metre length (given)

∴ Total thermal resistance, $S = S_1 + S_2 = 1{\cdot}03 + 0{\cdot}45 = 1{\cdot}48$ thermal ohm per metre length

Now $$I = \sqrt{\frac{t}{n R S}}$$

Here $t = 55$ºC ; $n = 1$, $R = 110 \times 10^{-6}\ \Omega$; $S = 1{\cdot}48$

$$\therefore \quad I = \sqrt{\frac{55}{1 \times 110 \times 10^{-6} \times 1{\cdot}48}} = \mathbf{581\ A}$$

11.20 Types of Cable Faults

Cables are generally laid directly in the ground or in ducts in the underground distribution system. For this reason, there are little chances of faults in underground cables. However, if a fault does occur, it is difficult to locate and repair the fault because conductors are not visible. Nevertheless, the following are the faults most likely to occur in underground cables :

(*i*) Open-circuit fault
(*ii*) Short-circuit fault
(*iii*) Earth fault.

(*i*) *Open-circuit fault.* When there is a break in the conductor of a cable, it is called open-circuit fault. The open-circuit fault can be checked by a megger. For this purpose, the three conductors of the 3-core cable at the far end are shorted and earthed. Then resistance between each conductor and earth is measured by a megger. The megger will indicate zero resistance in the circuit of the conductor that is not broken. However, if the conductor is broken, the megger will indicate infinite resistance in its circuit.

(*ii*) *Short-circuit fault.* When two conductors of a multi-core cable come in electrical contact with each other due to insulation failure, it is called a short-circuit fault. Again, we can seek the help of a megger to check this fault. For this purpose, the two terminals of the megger are connected to any two conductors. If the megger gives zero reading, it indicates short-circuit fault between these conductors. The same step is repeated for other conductors taking two at a time.

(*iii*) *Earth fault.* When the conductor of a cable comes in contact with earth, it is called earth fault or ground fault. To identify this fault, one terminal of the megger is connected to the conductor and the other terminal connected to earth. If the megger indicates zero reading, it means the conductor is earthed. The same procedure is repeated for other conductors of the cable.

11.21 Loop Tests For Location of Faults in Underground Cables

There are several methods for locating the faults in underground cables. However, two popular methods known as loop tests are :

(*i*) Murray loop test
(*ii*) Varley loop test

These simple tests can be used to locate the earth fault or short-circuit fault in underground cables provided that a sound cable runs along the faulty cable. Both these tests employ the principle of Wheatstone bridge for fault location.

11.22 Murray Loop Test

The Murray loop test is the most common and accurate method of locating earth fault or short-circuit fault in underground cables.

(*i*) **Earth fault :** Fig. 11.22 shows the circuit diagram for locating the earth fault by Murray loop test. Here *AB* is the sound cable and *CD* is the faulty cable; the earth fault occuring at point *F*. The far end *D* of the faulty cable is joined to the far end *B* of the sound cable through a low resistance link. Two variable resistances P and Q are joined to ends A and C (See Fig. 11.22) respectively and serve as the ratio arms of the Wheatstone bridge.

Let R = resistance of the conductor loop upto the fault from the test end

X = resistance of the other length of the loop

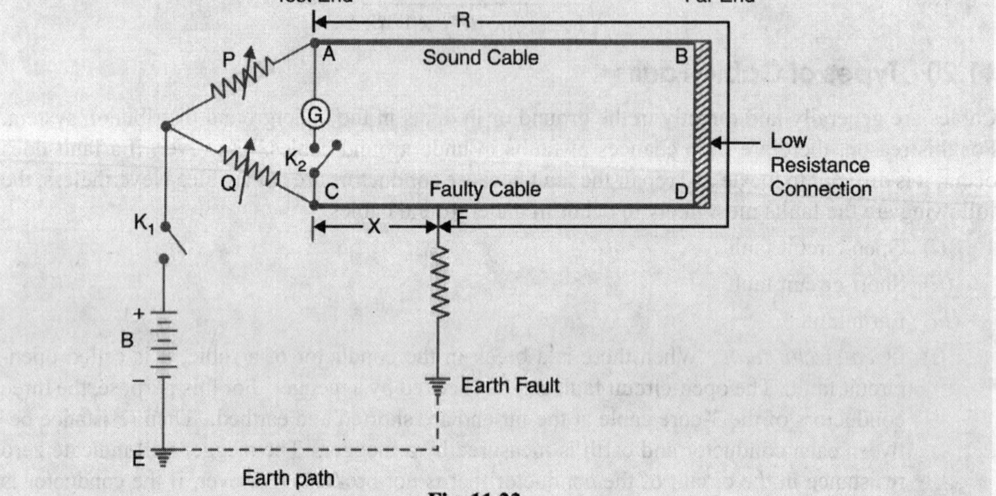

Fig. 11.22

Note that P, Q, R and X are the four arms of the Wheatstone bridge. The resistances P and Q are varied till the galvanometer indicates zero deflection.

In the balanced position of the bridge, we have,

$$\frac{P}{Q} = \frac{R}{X}$$

or $$\frac{P}{Q}+1 = \frac{R}{X}+1$$

or $$\frac{P+Q}{Q} = \frac{R+X}{X}$$

If r is the resistance of each cable, then $R + X = 2r$.

$$\therefore \quad \frac{P+Q}{Q} = \frac{2r}{X}$$

or $$X = \frac{Q}{P+Q}\times 2r$$

If l is the length of each cable in metres, then resistance per metre length of cable $= \frac{r}{l}$.

$\therefore$ Distance of fault point from test end is

$$d = \frac{X}{r/l} = \frac{Q}{P+Q}\times 2r \times \frac{l}{r} = \frac{Q}{P+Q}\times 2l$$

or $$d = \frac{Q}{P+Q}\times \text{(loop length)} \text{ *metres}$$

Thus the position of the fault is located. Note that resistance of the fault is in the battery circuit and not in the bridge circuit. Therefore, fault resistance does not affect the balancing of the bridge. However, if the fault resistance is high, the sensitivity of the bridge is reduced.

* Note that the term $Q/P+Q$ is dimensionless ; being the ratio of resistances.

(ii) **Short-circuit fault :** Fig. 11.23 shows the circuit diagram for locating the short-circuit fault by Murray loop test. Again P, Q, R and X are the four arms of the bridge. Note that fault resistance is in the battery circuit and not in the bridge circuit. The bridge in balanced by adjusting the resistances P and Q. In the balanced position of the bridge :

$$\frac{P}{Q} = \frac{R}{X}$$

or $$\frac{P+Q}{Q} = \frac{R+X}{X} = \frac{2r}{X}$$

$\therefore$ $$X = \frac{Q}{P+Q} \times 2r$$

or $$X = \frac{Q}{P+Q} \times \text{(loop length) metres}$$

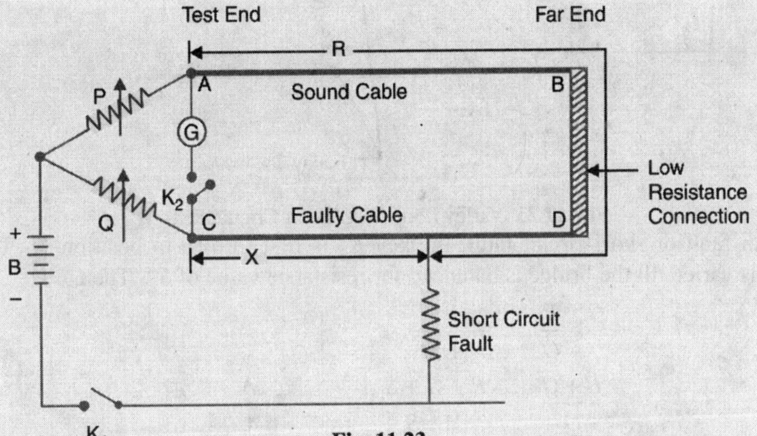

Fig. 11.23

Thus the position of the fault is located.

11.23 Varley Loop Test

The Varley loop test is also used to locate earth fault or short-circuit fault in underground cables. This test also employs Wheatstone bridge principle. It differs from Murray loop test in that here the ratio arms P and Q are fixed resistances. Balance is obtained by adjusting the variable resistance S

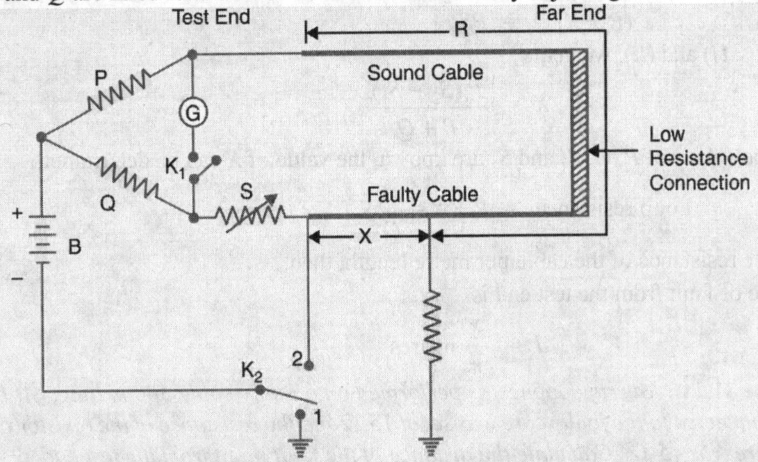

Fig. 11.24 Varley Loop Test (Earth Fault)

connected to the test end of the faulty cable. The connection diagrams for locating the earth fault and short-circuit fault by Varley loop test are shown in Figs. 11.24 and 11.25 respectively.

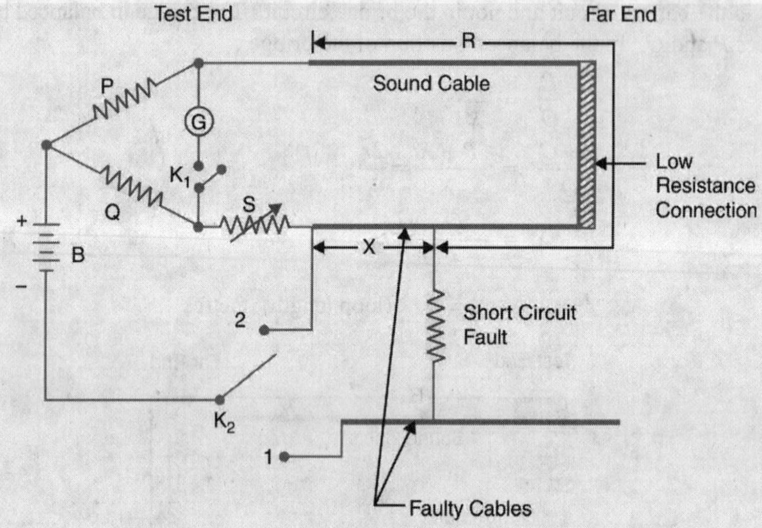

Fig. 11.25 Varley Loop Test (Short Circuit Test)

For earth fault or short-circuit fault, the key K_2 is first thrown to position 1. The variable resistance S is varied till the bridge is balanced for resistance value of S_1. Then,

$$\frac{P}{Q} = \frac{R}{X+S_1}$$

or $$\frac{P+Q}{Q} = \frac{R+X+S_1}{X+S_1}$$

or $$X = \frac{Q(R+X)-PS_1}{P+Q} \qquad ...(i)$$

Now key K_2 is thrown to position 2 (for earth fault or short-circuit fault) and bridge is balanced with new value of resistance S_2. Then,

$$\frac{P}{Q} = \frac{R+X}{S_2}$$

or $$(R+X)\,Q = PS_2 \qquad ...(ii)$$

From eqs. (*i*) and (*ii*), we get,

$$X = \frac{P(S_2-S_1)}{P+Q}$$

Since the values of P, Q, S_1 and S_2 are known, the value of X can be determined.

$$\text{Loop resistance} = R+X = \frac{P}{Q}S_2$$

If r is the resistance of the cable per metre length, then,

Distance of fault from the test end is

$$d = \frac{X}{r} \text{ metres}$$

Example 11.21. *Murray loop test is performed on a faulty cable 300 m long. At balance, the resistance connected to the faulty core was set at 15 Ω and the resistance of the resistor connected to the sound core was 45 Ω. Calculate the distance of the fault point from the test end.*

Solution.

Distance of the fault point from test end is

$$d = \frac{Q}{P+Q} \times \text{loop length}$$

Here $Q = 15\ \Omega$; $P = 45\ \Omega$; loop length $= 2 \times 300 = 600$ m

$$\therefore \qquad d = \frac{15}{45+15} \times 600 = \mathbf{150\ m}$$

Example 11.22. *In a test by Murray loop for ground fault on 500 m of cable having a resistance of 1·6 Ω/km, the faulty cable is looped with a sound cable of the same length and area of cross-section. If the ratio of the other two arms of the testing network at balance is 3 : 1, find the distance of the fault from the testing end of cables.*

Solution.

$$\frac{P}{Q} = 3 \quad \text{or} \quad \frac{P+Q}{Q} = 4$$

Distance of fault from test end is

$$d = \frac{Q}{P+Q} \times \text{loop length} = \frac{1}{4} \times (2 \times 500) = \mathbf{250\ m}$$

Example 11.23. *In a test for a fault to earth on a 500 m length of cable having a resistance of 1 Ω per 1000 m, the faulty cable is looped with a sound cable of the same length but having a resistance of 2·25 Ω per 1000 m. The resistance of the other two arms of the testing network at balance are in the ratio 2·75 : 1. Calculate the distance of the fault from the testing end of the cable.*

Solution. $$\frac{P}{Q} = 2{\cdot}75 \quad \text{or} \quad \frac{P+Q}{Q} = 2{\cdot}75 + 1 = 3{\cdot}75$$

$$\text{Resistance of loop} = \frac{1}{1000} \times 500 + \frac{2{\cdot}25}{1000} \times 500 = 1{\cdot}625\ \Omega$$

Resistance of faulty cable from test end upto fault point is

$$X = \frac{Q}{P+Q} \times (\text{loop resistance}) = \frac{1}{3{\cdot}75} \times 1{\cdot}625 = 0{\cdot}433\ \Omega$$

Distance of fault point from the testing end is

$$d = \frac{X}{1/1000} = 0{\cdot}433 \times 1000 = \mathbf{433\ m}$$

Example 11.24. *Varley loop test is performed to locate an earth fault on a 20 km long cable. The resistance per km of the single conductor is 20 Ω. The loop is completed with a similar healthy conductor. At balance, the variable resistance connected to the faulty conductor is 200 Ω. The fixed resistors have equal values. Calcualte the distance of the fault from the test end.*

Solution.

Resistance of faulty cable from test end to fault point is

$$X = \frac{Q(R+X) - PS}{P+Q}$$

Here $P = Q$; $S = 200\ \Omega$; $R + X = 20\ (20 + 20) = 800\ \Omega$

$$\therefore \qquad X = \frac{Q\,(800) - Q \times 200}{Q+Q} = 300\ \Omega$$

The resistance per km $= 20\ \Omega$

∴ Distance of fault from test end is

$$d = \frac{X}{20} = \frac{300}{20} = \mathbf{15\ km}$$

TUTORIAL PROBLEMS

1. The Murray loop test is used to locate an earth fault on one core of a two-core cable. The other core is used to complete the loop. When the network is balanced, the resistance connected to the faulty core has a value of 3·2 Ω. The other resistance arm has a value of 11·8 Ω. The fault is 42·7 m from the test end. Find the length of the cable. **[100 m]**
2. Murray loop test is performed to locate an earth fault on one core of a 2-core cable 100 m long. The other core is healthy and used to form the loop. At balance, the resistance connected to the faulty core was 4 Ω. The other resistance arm has a value of 16 Ω. Calculate the distance of the fault from the test end. **[40 m]**
3. The Varley loop test is used to find the position of an earth fault on a line of length 40 km. The resistance/km of a single line is 28 Ω. The fixed resistors have resistances of 250 Ω each. The fault is calculated to be 7 km from the test end. To what value of resistance was the variable resistor set? **[1848 Ω]**

SELF - TEST

1. **Fill in the blanks by inserting appropriate words/figures.**
   (*i*) The underground system is .... costly than the equivalent overhead line system.
   (*ii*) Voltage drop in cable system is less than that of equivalent overhead line because of .... of conductors in a cable.
   (*iii*) A metallic sheath is provided over the insulation to protect the cable from ....
   (*iv*) In single-core cables, armouring is not done in order to avoid ....
   (*v*) The most commonly used insulation in high-voltage cables is ....
   (*vi*) Belted cables are generally used upto .... kV.
   (*vii*) The working voltage level of belted cable is limited to 22 kV because of the .... set up in the dielectric.
   (*viii*) For voltages beyond 66 kV, solid type cables are unreliable because there is a danger of breakdown of insulation due to the ....
   (*ix*) If the length of a cable increases, its insulation resistance ....
   (*x*) Under operating conditions, the maximum stress in a cable is at ....
2. **Fill in the blanks by picking up correct words/figures from brackets.**
   (*i*) For voltages less than 66 kV, a 3-phase cable usually consists of .... [*3-core cable, 3 single-core cables*]
   (*ii*) If the length of a cable is doubled, its capacitance is .... [*doubled, halved, quadrupled*]
   (*iii*) A certain cable has an insulation of relative permittivity 2. If the insulation is replaced by one of relative permittivity 4, then capacitance of cable is .... [*doubled, halved*]
   (*iv*) The minimum dielectric stress in a cable is at .... [*conductor surface, lead sheath*]
   (*v*) If a cable of homogeneous insulation has maximum stress of 5 kV/mm, then the dielectric strength of insulation should be................. [*5-kV/mm, 2·5 kV/mm, 3 kV/mm*]
   (*vi*) In capacitance grading of cables, we use a .... dielectric. [*homogeneous, composite*]
   (*vii*) For the same safe potential, the size of a graded cable will be .... [*less, more*]
   (*viii*) For operating voltages beyond 66 kV, ... cables are used. [*Belted, S.L. type, oil-filled*]
   (*ix*) Voids in the layers of impregnated paper insualtion .... the breakdown voltage of the cable. [*increase, decrease*]
   (*x*) For voltages beyond 66 kV, 3-phase system usually employs................. [*3-core cable, 3 single core cables*]

ANSWERS TO SELF-TEST

1. (*i*) more (*ii*) closer spacing (*iii*) moisture (*iv*) excessive sheath losses (*v*) impregnated paper (*vi*) 11 (*vii*) tangential stresses (*viii*) presence of voids (*ix*) decreases (*x*) conductor surface
2. (*i*) 3-core cable (*ii*) doubled (*iii*) doubled (*iv*) lead sheath (*v*) 5 kV/mm (*vi*) composite (*vii*) less (*viii*) oil-filled (*ix*) decrease (*x*) 3 single-core cables

CHAPTER REVIEW TOPICS

1. Compare the merits and demerits of underground system versus overhead system.
2. With a neat diagram, show the various parts of a high voltage single-core cable.
3. What should be the desirable characteristics of insulating materials used in cables ?
4. Describe briefly some commonly used insulating materials for cables.
5. What is the most general criterion for the classification of cables ? Draw the sketch of a single-core low tension cable and label the various parts.
6. Draw a neat sketch of the cross-section of the following :
   (*i*) 3-core belted cable
   (*ii*) H-type cable
   (*iii*) S.L. type cable
7. What are the limitations of solid type cables ? How are these overcome in pressure cables ?
8. Write a brief note on oil-filled cables.
9. Describe the various methods of laying underground cables. What are the relative advantages and disadvantages of each method ?
10. Derive an expression for the insulation resistance of a single-core cable.
11. Deduce an expression for the capacitance of a single-core cable.
12. Show that maximum stress in a single-core cable is

$$\frac{2V}{d \log_e D/d}$$

where V is the operating voltage and d and D are the conductor and sheath diameter.
13. Prove that g_{max}/g_{min} in a single-core cable is equal to D/d.
14. Find an expression for the most economical conductor size of a single core cable.
15. Explain the following methods of cable grading :
   (*i*) Capacitance grading
   (*ii*) Intersheath grading
16. Write short notes on the following :
   (*i*) Laying of 11 kV underground power cable
   (*ii*) Capacitance grading in cables
   (*iii*) Capacitance of 3-core belted cables
17. Derive an expression for the thermal resistance of dielectric of a single-core cable.
18. What do you mean by permissible current loading of an underground cable ?
19. With a neat diagram, describe Murray loop test for the location of (*i*) earth fault (*ii*) short-circuit fault in an underground cable.
20. Describe Varley loop test for the location of earth fault and short-circuit fault in an underground cable.

DISCUSSION QUESTIONS

1. Overhead system can be operated at 400 kV or above but underground system offers problems at such voltages. Why ?
2. Why are VIR cables preferred to paper insulated cables for smaller installations ?
3. Why do we use 3 single-core cables and not 3-core cables for voltages beyond 66 kV ?
4. What is the mechanism of breakdown of an underground cable ?
5. How do voids in the insulation cause breakdown of the cable ?

CHAPTER

12

Distribution Systems – General

Introduction

The electrical energy produced at the gen erating station is conveyed to the consum ers through a network of transmission and distribution systems. It is often difficult to draw a line between the transmission and distribution systems of a large power system. It is impossible to distinguish the two merely by their voltage because what was considered as a high voltage a few years ago is now considered as a low voltage. In general, distribution system is that part of power system which distributes power to the consumers for utilisation.

The transmission and distribution systems are similar to man's circulatory system. The transmission system may be compared with arteries in the human body and distribution system with capillaries. They serve the same purpose of supplying the ultimate consumer in the city with the life-giving blood of civilisation–electricity. In this chapter, we shall confine our attention to the general introduction to distribution system.

12.1 Distribution System

That part of power system which distributes electric power for local use is known as **distribution system.**

In general, the distribution system is the electrical system between the sub-station fed by the transmission system and the consumers meters. It generally consists of *feeders, distributors* and the *service mains.* Fig. 12.1 shows the single line diagram of a typical low tension distribution system.

(*i*) *Feeders.* A feeder is a conductor which connects the sub-station (or localised generating station) to the area where power is to be distributed. Generally, no tappings are taken from the feeder so that current in it remains the same throughout. The main consideration in the design of a feeder is the current carrying capacity.

(*ii*) *Distributor.* A distributor is a conductor from which tappings are taken for supply to the consumers. In Fig. 12.1, *AB, BC, CD* and *DA* are the distributors. The current through a distributor is not constant because tappings are taken at various places along its length. While designing a distributor, voltage drop along its length is the main consideration since the statutory limit of voltage variations is ± 6% of rated value at the consumers' terminals.

(*iii*) *Service mains.* A service mains is generally a small cable which connects the distributor to the consumers' terminals.

12.2 Classification of Distribution Systems

A distribution system may be classified according to ;

(*i*) *Nature of current.* According to nature of current, distribution system may be classified as (*a*) d.c. distribution system (*b*) a.c. distribution system. Now-a-days, a.c. system is universally adopted for distribution of electric power as it is simpler and more economical than direct current method.

(*ii*) *Type of construction.* According to type of construction, distribution system may be classified as (*a*) overhead system (*b*) underground system. The overhead system is generally employed for distribution as it is 5 to 10 times cheaper than the equivalent underground system. In general, the underground system is used at places where overhead construction is impracticable or prohibited by the local laws.

(*iii*) *Scheme of connection.* According to scheme of connection, the distribution system may be classified as (*a*) radial system (*b*) ring main system (*c*) inter-connected system. Each scheme has its own advantages and disadvantages and those are discussed in Art.12.7.

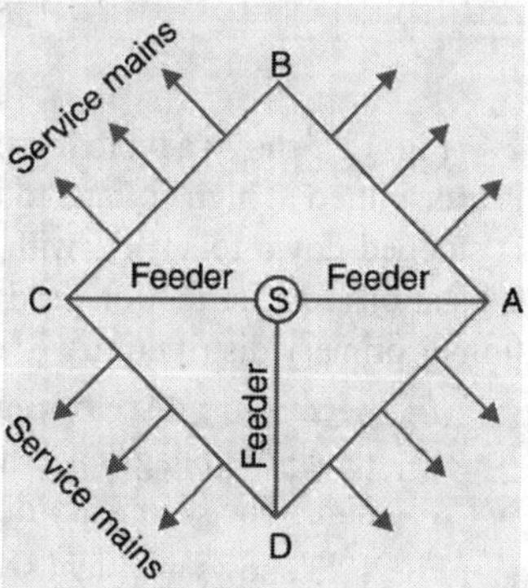

Fig. 12.1

12.3 A.C. Distribution

Now-a-days electrical energy is generated, transmitted and distributed in the form of alternating current. One important reason for the widespread use of alternating current in preference to direct current is the fact that alternating voltage can be conveniently changed in magnitude by means of a transformer. Transformer has made it possible to transmit a.c. power at high voltage and utilise it at a safe potential. High transmission and distribution voltages have greatly reduced the current in the conductors and the resulting line losses.

There is no definite line between transmission and distribution according to voltage or bulk capacity. However, in general, the a.c. distribution system is the electrical system between the step-down substation fed by the transmission system and the consumers' meters. The a.c. distribution system is classified into (*i*) primary distribution system and (*ii*) secondary distribution system.

(*i*) **Primary distribution system.** It is that part of a.c. distribution system which operates at voltages somewhat higher than general utilisation and handles large blocks of electrical energy than the average low-voltage consumer uses. The voltage used for primary distribution depends upon the amount of power to be conveyed and the distance of the substation required to be fed. The most commonly used primary distribution voltages are 11 kV, 6·6 kV and 3·3 kV. Due to economic considerations, primary distribution is carried out by 3-phase, 3-wire system.

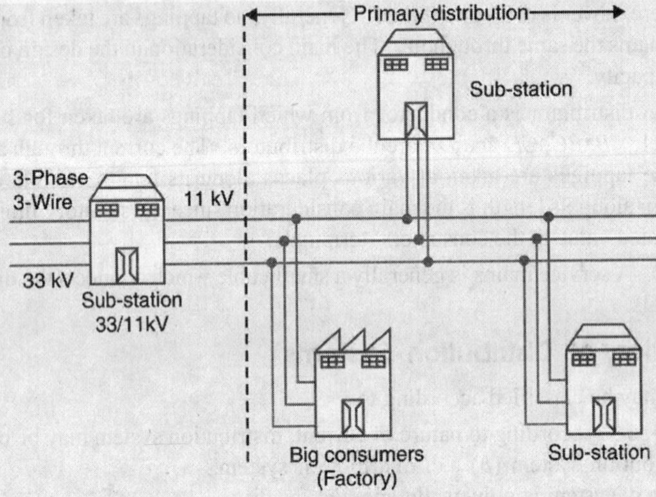

Fig. 12.2

Fig. 12.2 shows a typical primary distribution system. Electric power from the generating station is transmitted at high voltage to the substation located in or near the city. At this substation, voltage is stepped down to 11 kV with the help of step-down transformer. Power is supplied to various substations for distribution or to big consumers at this voltage. This forms the high voltage distribution or primary distribution.

(*ii*) **Secondary distribution system.** It is that part of a.c. distribution system which includes the range of voltages at which the ultimate consumer utilises the electrical energy delivered to him. The secondary distribution employs 400/230 V, 3-phase, 4-wire system.

Fig. 12.3 shows a typical secondary distribution system. The primary distribution circuit delivers power to various substations, called distribution substations. The substations are situated near the consumers' localities and contain step-down transformers. At each distribution substation, the voltage is stepped down to 400 V and power is delivered by 3-phase,4-wire a.c. system. The voltage between any two phases is 400 V and between any phase and neutral is 230 V. The single phase domestic loads are connected between any one phase and the neutral, whereas 3-phase 400 V motor loads are connected across 3-phase lines directly.

Power transformer

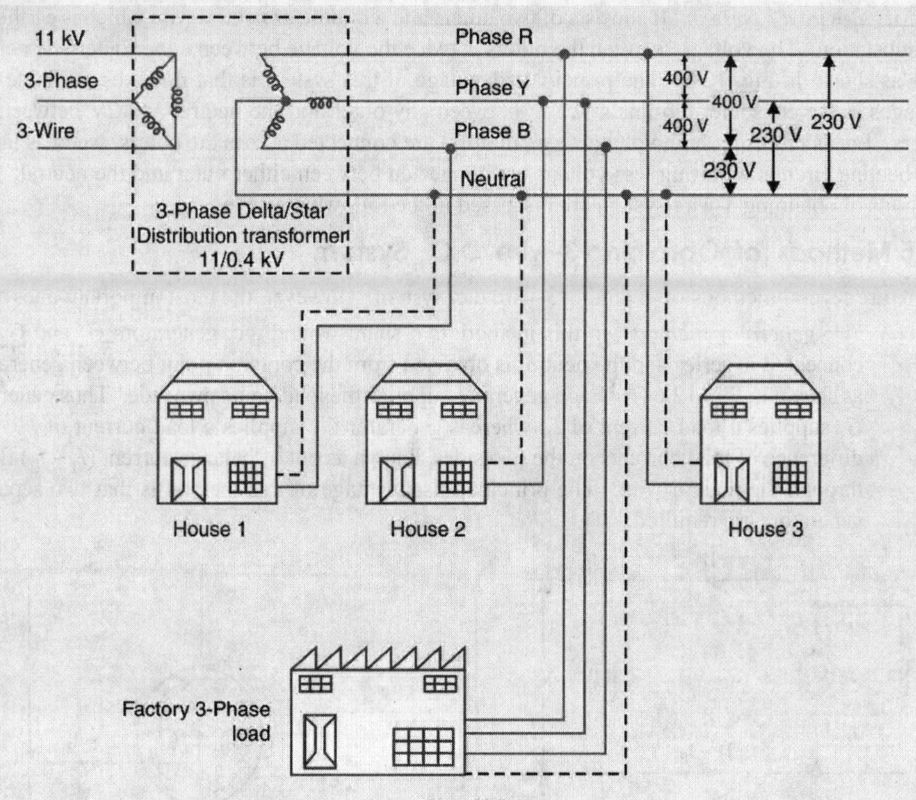

Fig. 12.3

12.4 D.C. Distribution

It is a common knowledge that electric power is almost exclusively generated, transmitted and distributed as a.c. However, for certain applications, d.c. supply is absolutely necessary. For instance, d.c. supply is required for the operation of variable speed machinery (*i.e.*, d.c. motors), for electrochemical work and for congested areas where storage battery reserves are necessary. For this purpose, a.c. power is converted into d.c. power at the substation by using converting machinery *e.g.*, mercury arc rectifiers, rotary converters and motor-generator sets. The d.c. supply from the substation may be obtained in the form of (*i*) 2-wire or (*ii*) 3-wire for distribution.

(*i*) *2-wire d.c. system.* As the name implies, this system of distribution consists of two wires. One is the outgoing or positive wire and the other is the return or negative wire. The loads such as lamps, motors etc. are connected in parallel between the two wires as shown in Fig. 12.4. This system is never used for transmission purposes due to low efficiency but may be employed for distribution of d.c. power.

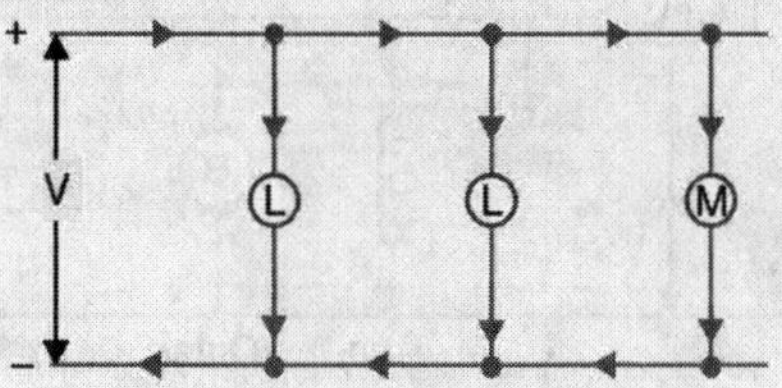

Fig. 12.4

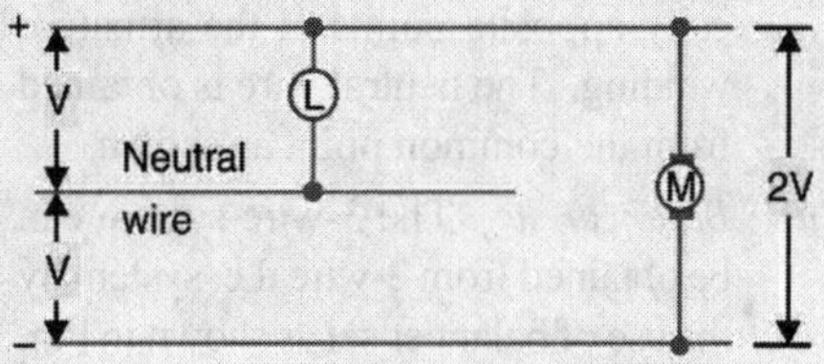

Fig. 12.5

(*ii*) *3-wire d.c. system.* It consists of two outers and a middle or neutral wire which is earthed at the substation. The voltage between the outers is twice the voltage between either outer and neutral wire as shown in Fig. 12.5. The principal advantage of this system is that it makes available two voltages at the consumer terminals *viz.*, *V* between any outer and the neutral and 2V between the outers. Loads requiring high voltage (*e.g.*, motors) are connected across the outers, whereas lamps and heating circuits requiring less voltage are connected between either outer and the neutral. The methods of obtaining 3-wire system are discussed in the following article.

12.5 Methods of Obtaining 3-wire D.C. System

There are several methods of obtaining 3-wire d.c. system. However, the most important ones are:

(*i*) *Two generator method.* In this method, two shunt wound d.c. generators G_1 and G_2 are connected in series and the neutral is obtained from the common point between generators as shown in Fig. 12.6 (*i*). Each generator supplies the load on its own side. Thus generator G_1 supplies a load current of I_1, whereas generator G_2 supplies a load current of I_2. The difference of load currents on the two sides, known as out of balance current $(I_1 - I_2)$ flows through the neutral wire. The principal disadvantage of this method is that two separate generators are required.

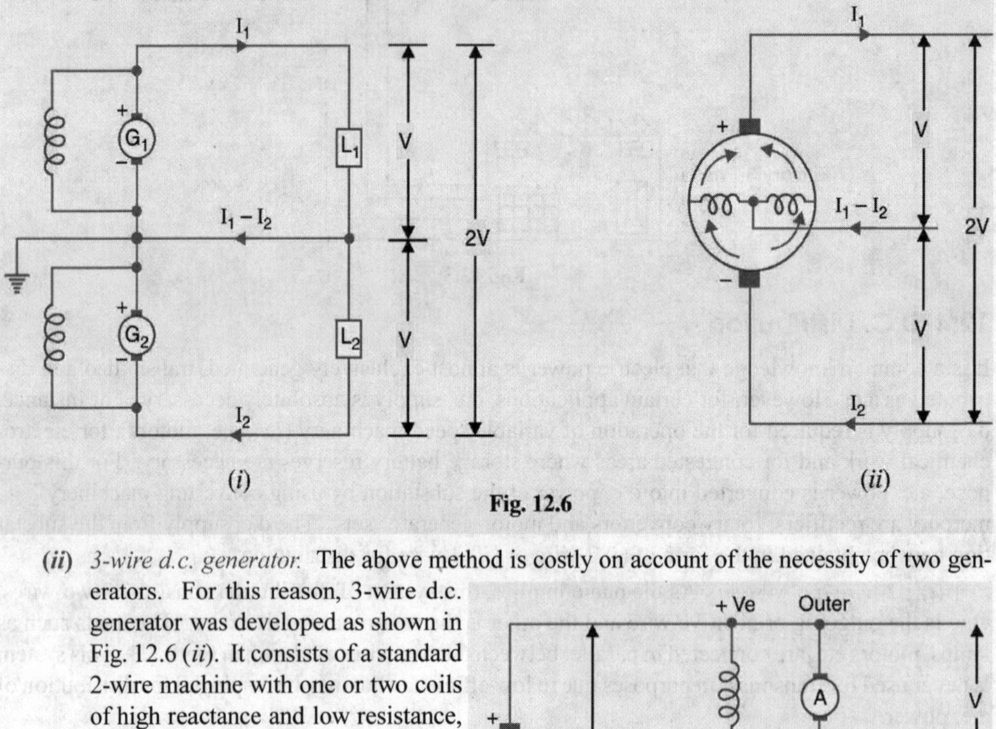

Fig. 12.6

(*ii*) *3-wire d.c. generator.* The above method is costly on account of the necessity of two generators. For this reason, 3-wire d.c. generator was developed as shown in Fig. 12.6 (*ii*). It consists of a standard 2-wire machine with one or two coils of high reactance and low resistance, connected permanently to diametrically opposite points of the armature winding. The neutral wire is obtained from the common point as shown.

(*iii*) *Balancer set.* The 3-wire system can be obtained from 2-wire d.c. system by the use of balancer set as shown in Fig. 12.7. *G* is the main 2-wire d.c. gen-

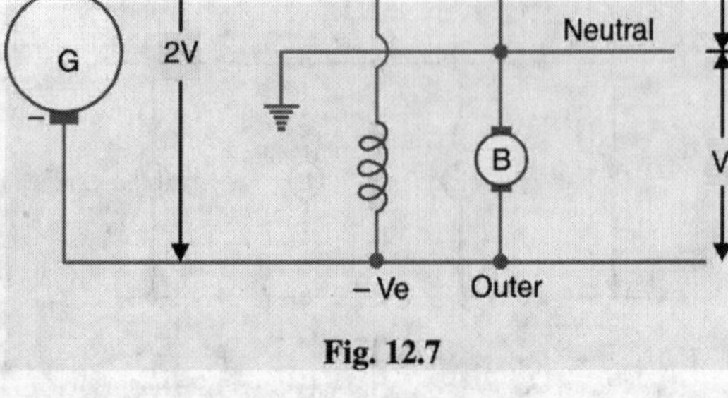

Fig. 12.7

erator and supplies power to the whole system. The balancer set consists of two identical d.c shunt machines A and B coupled mechanically with their armatures and field windings joined in series across the outers. The junction of their armatures is earthed and neutral wire is taken out from here. The balancer set has the additional advantage that it maintains the potential difference on two sides of neutral equal to each other. This method is discussed in detail in the next chapter.

12.6 Overhead Versus Underground System

The distribution system can be overhead or underground. Overhead lines are generally mounted on wooden, concrete or steel poles which are arranged to carry distribution transformers in addition to the conductors. The underground system uses conduits, cables and manholes under the surface of streets and sidewalks. The choice between overhead and underground system depends upon a number of widely differing factors. Therefore, it is desirable to make a comparison between the two.

(*i*) *Public safety.* The underground system is more safe than overhead system because all distribution wiring is placed underground and there are little chances of any hazard.

(*ii*) *Initial cost.* The underground system is more expensive due to the high cost of trenching, conduits, cables, manholes and other special equipment. The initial cost of an underground system may be five to ten times than that of an overhead system.

(*iii*) *Flexibility.* The overhead system is much more flexible than the underground system. In the latter case, manholes, duct lines etc., are permanently placed once installed and the load expansion can only be met by laying new lines. However, on an overhead system, poles, wires, transformers etc., can be easily shifted to meet the changes in load conditions.

(*iv*) *Faults.* The chances of faults in underground system are very rare as the cables are laid underground and are generally provided with better insulation.

(*v*) *Appearance.* The general appearance of an underground system is better as all the distribution lines are invisible. This factor is exerting considerable public pressure on electric supply companies to switch over to underground system.

(*vi*) *Fault location and repairs.* In general, there are little chances of faults in an underground system. However, if a fault does occur, it is difficult to locate and repair on this system. On an overhead system, the conductors are visible and easily accessible so that fault locations and repairs can be easily made.

(*vii*) *Current carrying capacity and voltage drop.* An overhead distribution conductor has a considerably higher current carrying capacity than an underground cable conductor of the same material and cross-section. On the other hand, underground cable conductor has much lower inductive reactance than that of an overhead conductor because of closer spacing of conductors.

(*viii*) *Useful life.* The useful life of underground system is much longer than that of an overhead system. An overhead system may have a useful life of 25 years, whereas an underground system may have a useful life of more than 50 years.

(*ix*) *Maintenance cost.* The maintenance cost of underground system is very low as compared with that of overhead system because of less chances of faults and service interruptions from wind, ice, lightning as well as from traffic hazards.

(*x*) *Interference with communication circuits.* An overhead system causes electromagnetic interference with the telephone lines. The power line currents are superimposed on speech currents, resulting in the potential of the communication channel being raised to an undesirable level. However, there is no such interference with the underground system.

It is clear from the above comparison that each system has its own advantages and disadvantages. However, comparative economics (*i.e.*, annual cost of operation) is the most powerful factor influencing the choice between underground and overhead system. The greater capital cost of underground system prohibits its use for distribution. But sometimes non-economic factors (*e.g.*, general appearance, public safety etc.) exert considerable influence on choosing underground system. In general, overhead system is adopted for distribution and the use of underground system is made only where overhead construction is impracticable or prohibited by local laws.

12.7 Connection Schemes of Distribution System

All distribution of electrical energy is done by constant voltage system. In practice, the following distribution circuits are generally used :

(*i*) **Radial System.** In this system, separate feeders radiate from a single substation and feed the distributors at one end only. Fig. 12.8 (*i*) shows a single line diagram of a radial system for d.c. distribution where a feeder *OC* supplies a distributor *AB* at point *A*. Obviously, the distributor is fed at one end only *i.e.*, point *A* is this case. Fig. 12.8 (*ii*) shows a single line diagram of radial system for a.c. distribution. The radial system is employed only when power is generated at low voltage and the substation is located at the centre of the load.

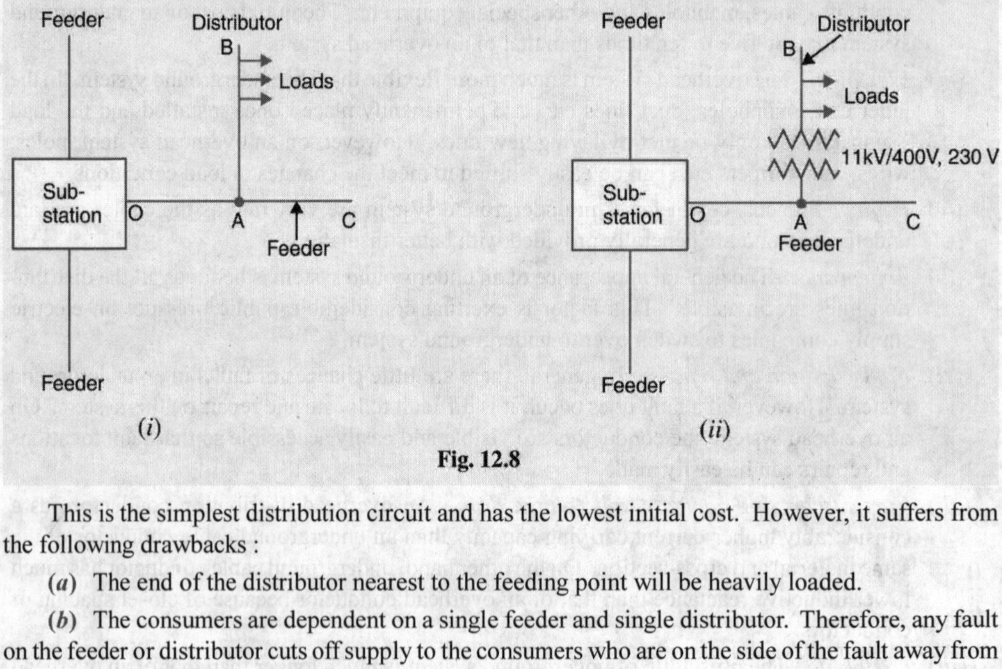

Fig. 12.8

This is the simplest distribution circuit and has the lowest initial cost. However, it suffers from the following drawbacks :

(*a*) The end of the distributor nearest to the feeding point will be heavily loaded.

(*b*) The consumers are dependent on a single feeder and single distributor. Therefore, any fault on the feeder or distributor cuts off supply to the consumers who are on the side of the fault away from the substation.

(*c*) The consumers at the distant end of the distributor would be subjected to serious voltage fluctuations when the load on the distributor changes.

Due to these limitations, this system is used for short distances only.

(*ii*) **Ring main system.** In this system, the primaries of distribution transformers form a loop. The loop circuit starts from the substation bus-bars, makes a loop through the area to be served, and returns to the substation. Fig. 12.9 shows the single line diagram of ring main system for a.c. distribution where substation supplies to the closed feeder LMNOPQRS. The distributors are tapped from different points *M*, *O* and *Q* of the feeder through distribution transformers. The ring main system has the following advantages :

(*a*) There are less voltage fluctuations at consumer's terminals.

(*b*) The system is very reliable as each distributor is fed *via* *two feeders. In the event of fault on any section of the feeder, the continuity of supply is maintained. For example, suppose that fault occurs at any point *F* of section SLM of the feeder. Then section SLM of the feeder can be isolated for repairs and at the same time continuity of supply is maintained to all the consumers *via* the feeder *SRQPONM*.

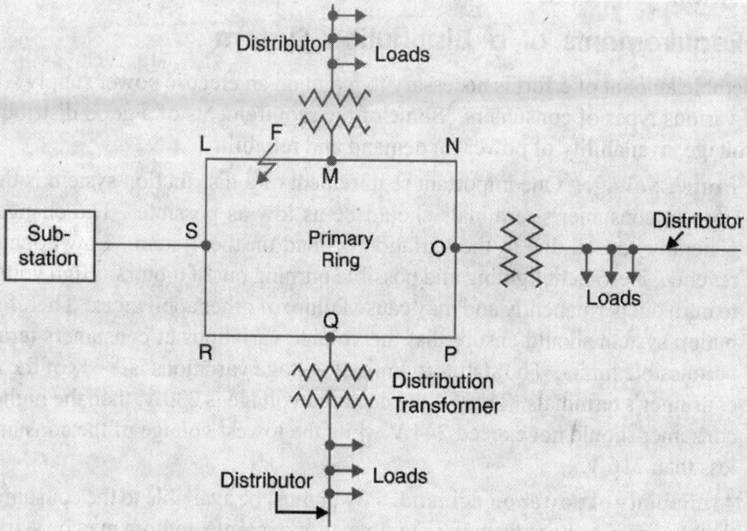

Fig. 12.9

(*iii*) **Interconnected system.** When the feeder ring is energised by two or more than two generating stations or substations, it is called inter-connected system. Fig. 12.10 shows the single line diagram of interconnected system where the closed feeder ring *ABCD* is supplied by two substations S_1 and S_2 at points *D* and *C* respectively. Distributors are connected to

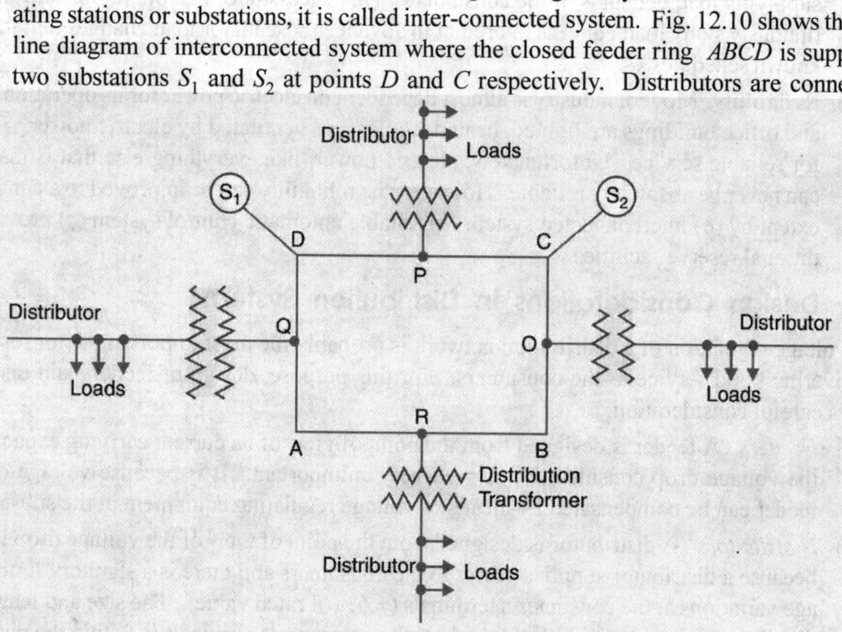

Fig. 12.10

* Thus the distributor from point *M* is supplied by the feeders SLM and SRQPONM.

points O, P, Q and R of the feeder ring through distribution transformers. The interconnected system has the following advantages :

(*a*) It increases the service reliability.

(*b*) Any area fed from one generating station during peak load hours can be fed from the other generating station. This reduces reserve power capacity and increases efficiency of the system.

12.8 Requirements of a Distribution System

A considerable amount of effort is necessary to maintain an electric power supply within the requirements of various types of consumers. Some of the requirements of a good distribution system are : proper voltage, availability of power on demand and reliability.

(*i*) **Proper voltage.** One important requirement of a distribution system is that voltage variations at consumer's terminals should be as low as possible. The changes in voltage are generally caused due to the variation of load on the system. Low voltage causes loss of revenue, inefficient lighting and possible burning out of motors. High voltage causes lamps to burn out permanently and may cause failure of other appliances. Therefore, a good distribution system should ensure that the voltage variations at consumers terminals are within permissible limits. The statutory limit of voltage variations is ± 6% of the rated value at the consumer's terminals. Thus, if the declared voltage is 230 V, then the highest voltage of the consumer should not exceed 244 V while the lowest voltage of the consumer should not be less than 216 V.

(*ii*) **Availability of power on demand.** Power must be available to the consumers in any amount that they may require from time to time. For example, motors may be started or shut down, lights may be turned on or off, without advance warning to the electric supply company. As electrical energy cannot be stored, therefore, the distribution system must be capable of supplying load demands of the consumers. This necessitates that operating staff must continuously study load patterns to predict in advance those major load changes that follow the known schedules.

(*iii*) **Reliability.** Modern industry is almost dependent on electric power for its operation. Homes and office buildings are lighted, heated, cooled and ventilated by electric power. This calls for reliable service. Unfortunately, electric power, like everything else that is man-made, can never be absolutely reliable. However, the reliability can be improved to a considerable extent by (*a*) interconnected system (*b*) reliable automatic control system (*c*) providing additional reserve facilities.

12.9 Design Considerations in Distribution System

Good voltage regulation of a distribution network is probably the most important factor responsible for delivering good service to the consumers. For this purpose, design of feeders and distributors requires careful consideration.

(*i*) *Feeders.* A feeder is designed from the point of view of its current carrying capacity while the voltage drop consideration is relatively unimportant. It is because voltage drop in a feeder can be compensated by means of voltage regulating equipment at the substation.

(*ii*) *Distributors.* A distributor is designed from the point of view of the voltage drop in it. It is because a distributor supplies power to the consumers and there is a statutory limit of voltage variations at the consumer's terminals (± 6% of rated value). The size and length of the distributor should be such that voltage at the consumer's terminals is within the permissible limits.

SELF - TEST

1. **Fill in the blanks by inserting appropriate words/figures.**
   (*i*) The underground system has .... initial cost than the overhead system.
   (*ii*) A ring main system of distribution is .... reliable than the radial system.
   (*iii*) The distribution transformer links the primary and .... distribution systems
   (*iv*) The most common system for secondary distribution is .... 3-phase, ... wire system.
   (*v*) The statutory limit for voltage variations at the consumer's terminals is .... % of rated value.
   (*vi*) The service mains connect the .... and the ....
   (*vii*) The overhead system is .... flexible than underground system.
2. **Fill in the blanks by picking up correct words/figures from brackets.**
   (*i*) The main consideration in the design of a feeder is the ....
   (*current carrying capacity, voltage drop*)
   (*ii*) A 3-wire d.c. distribution makes available .... voltages. (*one, two, three*)
   (*iii*) Now-a-days .... system is used for distribution. (*a, c, d.c.*)
   (*iv*) The interconnected system .... the reserve capacity of the systems. (*increases, decreases*)
   (*v*) The major part of investment on secondary distribution is made on ....
   (*Distribution transformers, conductors, pole fittings*)
   (*vi*) The chances of faults in underground system are .... as compared to overhead system.
   (*less, more*)

ANSWERS TO SELF-TEST

1. (*i*) more (*ii*) more (*iii*) secondary (*iv*) 400/230 V, 4 (*v*) = 6 (*vi*) distributor, consumer terminals (*vii*) more
2. (*i*) current carrying capacity (*ii*) two (*iii*) a.c. (*iv*) increases (*v*) distribution transformers (*vi*) less

CHAPTER REVIEW TOPICS

1. What do you understand by distribution system ?
2. Draw a single line diagram showing a typical distribution system.
3. Define and explain the terms : feeder, distributor and service mains.
4. Discuss the relative merits and demerits of underground and overhead systems.
5. Explain the following systems of distribution :
   (*i*) Radial system
   (*ii*) Ring main system
   (*iii*) Interconnected system
6. Discuss briefly the design considerations in distribution system.
7. With a neat diagram, explain the complete a.c. system for distribution of electrical energy.
8. Write short notes on the following :
   (*i*) Distribution transformers
   (*ii*) 3-wire d.c. distribution
   (*iii*) Primary distribution

DISCUSSION QUESTIONS

1. Can transmission and distribution systems be distinguished merely by their voltages ? Explain your answer.
2. It is suggested that since distribution transformer links the primary and utilisation voltage, secondary system is not essential. Is it a feasible proposition ?
3. What are the situations where the cost of underground system becomes comparable to overhead system ?
4. What are the effects of high primary voltage on the distribution system ?

CHAPTER

13

D.C. Distribution

Introduction

In the beginning of the electrical age, electricity was generated as a direct current and voltages were low. The resistance losses in the lines made it impracticable to transmit and distribute power for more than a few localities of the city. With the development of the transformer, a.c. has taken over the load formerly supplied by d.c. Now-a-days, electrical energy is generated, transmitted and distributed in the form of a.c. as an economical proposition. The transformer permits the transmission and distribution of a.c. power at high voltages. This has greatly reduced the current in the conductors (and hence their sizes) and the resulting line losses.

However, for certain applications, d.c. supply is absolutely necessary. For example, d.c. supply is required for the operation of variable speed machinery (*e.g.* d.c. motors), electro-chemical work and electric traction. For this purpose, a.c. power is converted into d.c. power at the sub-station by using converting machinery *e.g.* mercury are rectifiers, rotary converters and motor-generator sets. The d.c. supply from the sub-station is conveyed to the required places for distribution. In this chapter, we shall confine our attention to the various aspects of d.c. distribution.

13.1 Types of D.C. Distributors

The most general method of classifying d.c. distributors is the way they are fed by the feeders. On this basis, d.c. distributors are classified as:

(*i*) Distributor fed at one end

(*ii*) Distributor fed at both ends

(*iii*) Distributor fed at the centre

(*iv*) Ring distributor.

(*i*) **Distributor fed at one end.** In this type of feeding, the distributor is connected to the supply at one end and loads are taken at different points along the length of the distributor. Fig. 13.1 shows the single line diagram of a d.c. distributor *AB* fed at the end *A* (also known as *singly fed distributor*) and loads I_1, I_2 and I_3 tapped off at points *C*, *D* and *E* respectively.

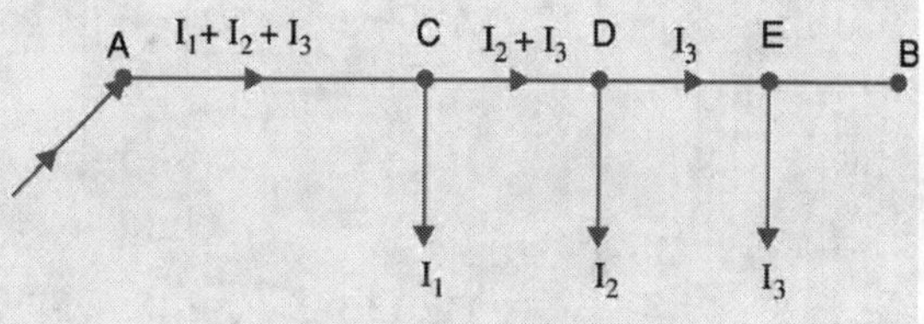

Fig. 13.1

The following points are worth noting in a singly fed distributor :

(*a*) The current in the various sections of the distributor away from feeding point goes on decreasing. Thus current in section *AC* is more than the current in section *CD* and current in section *CD* is more than the current in section *DE*.

(*b*) The voltage across the loads away from the feeding point goes on decreasing. Thus in Fig. 13.1, the minimum voltage occurs at the load point *E*.

(*c*) In case a fault occurs on any section of the distributor, the whole distributor will have to be disconnected from the supply mains. Therefore, continuity of supply is interrupted.

(*ii*) **Distributor fed at both ends.** In this type of feeding, the distributor is connected to the supply mains at both ends and loads are tapped off at different points along the length of the distributor. The voltage at the feeding points may or may not be equal. Fig. 13.2 shows a distributor *AB* fed at the ends *A* and *B* and loads of I_1, I_2 and I_3 tapped off at points *C*, *D* and *E* respectively. Here, the load voltage goes on decreasing as we move away from one feeding point *say A*, reaches minimum value and then again starts rising and reaches maximum value when we reach the other feeding point *B*. The minimum voltage occurs at some load point and is never fixed. It is shifted with the variation of load on different sections of the distributor.

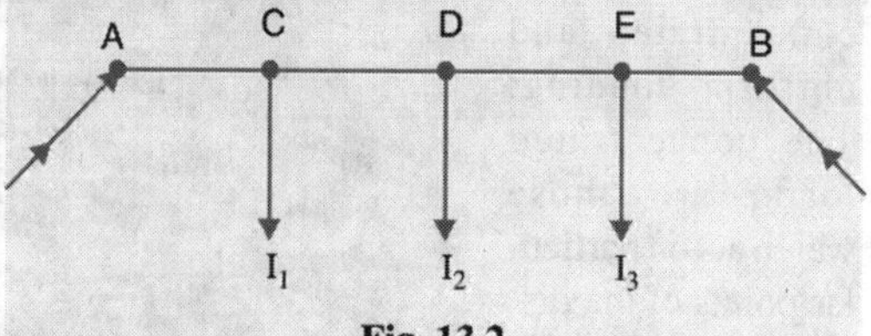

Fig. 13.2

Advantages

(*a*) If a fault occurs on any feeding point of the distributor, the continuity of supply is maintained from the other feeding point.

(*b*) In case of fault on any section of the distributor, the continuity of supply is maintained from the other feeding point.

(c) The area of X-section required for a doubly fed distributor is much less than that of a singly fed distributor.

(iii) **Distributor fed at the centre.** In this type of feeding, the centre of the distributor is connected to the supply mains as shown in Fig. 13.3. It is equivalent to two singly fed distributors, each distributor having a common feeding point and length equal to half of the total length.

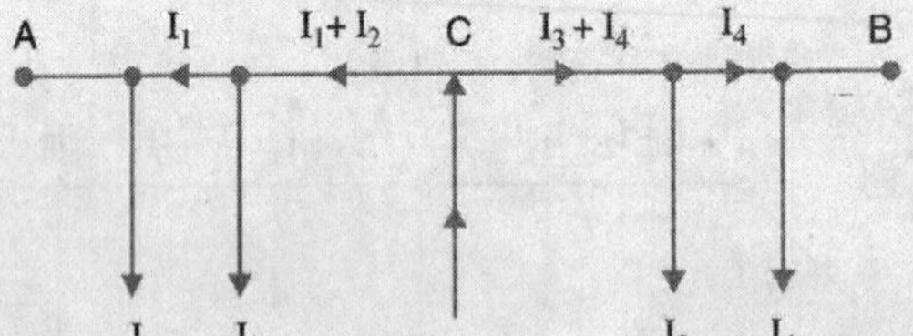

Fig. 13.3

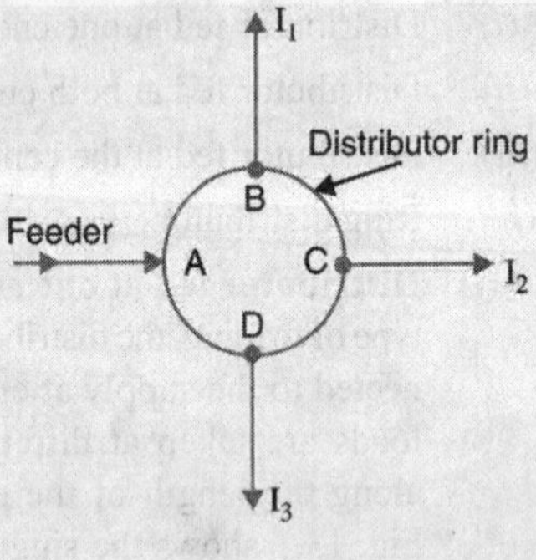

Fig. 13.4

(iv) **Ring mains.** In this type, the distributor is in the form of a closed ring as shown in Fig.13.4. It is equivalent to a straight distributor fed at both ends with equal voltages, the two ends being brought together to form a closed ring. The distributor ring may be fed at one or more than one point.

13.2 D.C. Distribution Calculations

In addition to the methods of feeding discussed above, a distributor may have (i) concentrated loading (ii) uniform loading (iii) both concentrated and uniform loading. The concentrated loads are those which act on particular points of the distributor. A common example of such loads is that tapped off for domestic use. On the other hand, distributed loads are those which act uniformly on all points of the distributor.

D.C. Load

Ideally, there are no distributed loads. However, a nearest example of distributed load is a large number of loads of same wattage connected to the distributor at equal distances.

In d.c. distribution calculations, one important point of interest is the determination of point of minimum potential on the distributor. The point where it occurs depends upon the loading conditions and the method of feeding the distributor. The distributor is so designed that the minimum potential on it is not less than 6% of rated voltage at the consumer's terminals. In the next sections, we shall discuss some important cases of d.c. distributors separately.

13.3 D.C. Distributor Fed at one End—Concentrated Loading

Fig. 13.5 shows the single line diagram of a 2-wire d.c. distributor *AB* fed at one end *A* and having concentrated loads I_1, I_2, I_3 and I_4 tapped off at points *C*, *D*, *E* and *F* respectively.

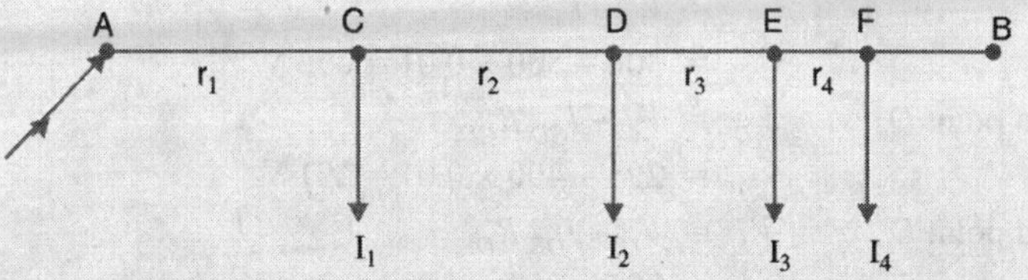

Fig. 13.5

Let r_1, r_2, r_3 and r_4 be the resistances of both wires (go and return) of the sections *AC*, *CD*, *DE* and *EF* of the distributor respectively.

Current fed from point *A* $= I_1 + I_2 + I_3 + I_4$

Current in section *AC* $= I_1 + I_2 + I_3 + I_4$

Current in section *CD* $= I_2 + I_3 + I_4$

Current in section *DE* $= I_3 + I_4$

Current in section *EF* $= I_4$

Voltage drop in section *AC* $= r_1 (I_1 + I_2 + I_3 + I_4)$

Voltage drop in section *CD* $= r_2 (I_2 + I_3 + I_4)$

Voltage drop in section *DE* $= r_3 (I_3 + I_4)$

Voltage drop in section *EF* $= r_4 I_4$

$\therefore$ Total voltage drop in the distributor

$$= r_1 (I_1 + I_2 + I_3 + I_4) + r_2 (I_2 + I_3 + I_4) + r_3 (I_3 + I_4) + r_4 I_4$$

It is easy to see that the minimum potential will occur at point *F* which is farthest from the feeding point *A*.

Example 13.1. *A 2-wire d.c. distributor cable AB is 2 km long and supplies loads of 100A, 150A, 200A and 50A situated 500 m, 1000 m, 1600 m and 2000 m from the feeding point A. Each conductor has a resistance of 0·01 Ω per 1000 m. Calculate the p.d. at each load point if a p.d. of 300 V is maintained at point A.*

Solution. Fig. 13.6 shows the single line diagram of the distributor with its tapped currents.

Resistance per 1000 m of distributor $= 2 \times 0{\cdot}01 = 0{\cdot}02\ \Omega$

Resistance of section *AC*, $R_{AC} = 0{\cdot}02 \times 500/1000 = 0{\cdot}01\ \Omega$

Resistance of section *CD*, $R_{CD} = 0{\cdot}02 \times 500/1000 = 0{\cdot}01\ \Omega$

Resistance of section *DE*, $R_{DE} = 0{\cdot}02 \times 600/1000 = 0{\cdot}012\ \Omega$

Resistance of section *EB*, $R_{EB} = 0{\cdot}02 \times 400/1000 = 0{\cdot}008\ \Omega$

Referring to Fig. 13.6, the currents in the various sections of the distributor are :

$I_{EB} = 50$ A ; $I_{DE} = 50 + 200 = 250$ A

$I_{CD} = 250 + 150 = 400$ A ; $I_{AC} = 400 + 100 = 500$ A

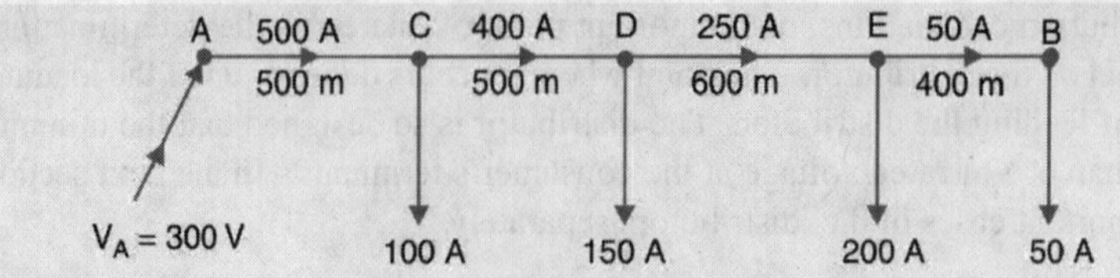

Fig. 13.6

P.D. at load point C, V_C = Voltage at A – Voltage drop in AC

$$= V_A - I_{AC} R_{AC}$$

$$= 300 - 500 \times 0{\cdot}01 = \mathbf{295\ V}$$

P.D. at load point D, $V_D = V_C - I_{CD} R_{CD}$

$$= 295 - 400 \times 0{\cdot}01 = \mathbf{291\ V}$$

P.D. at load point E, $V_E = V_D - I_{DE} R_{DE}$

$$= 291 - 250 \times 0{\cdot}012 = \mathbf{288\ V}$$

P.D. at load point B, $V_B = V_E - I_{EB} R_{EB}$

$$= 288 - 50 \times 0{\cdot}008 = \mathbf{287{\cdot}6\ V}$$

Example 13.2. *A 2-wire d.c. distributor AB is 300 metres long. It is fed at point A. The various loads and their positions are given below :*

At point	*distance from A in metres*	*concentrated load in amperes*
C	*40*	*30*
D	*100*	*40*
E	*150*	*100*
F	*250*	*50*

If the maximum permissible voltage drop is not to exceed 10 V, find the cross-sectional area of the distributor. Take $\rho = 1{\cdot}78 \times 10^{-8}\ \Omega m$.

Solution. The single line diagram of the distributor along with its tapped currents is shown in Fig. 13.7. Suppose that resistance of 100 m length of the distributor is r ohms. Then resistance of various sections of the distributor is :

$$R_{AC} = 0{\cdot}4\, r\, \Omega \ ; \ R_{CD} = 0{\cdot}6\, r\, \Omega \ ; \ R_{DE} = 0{\cdot}5\, r\, \Omega \ ; \ R_{EF} = r\, \Omega$$

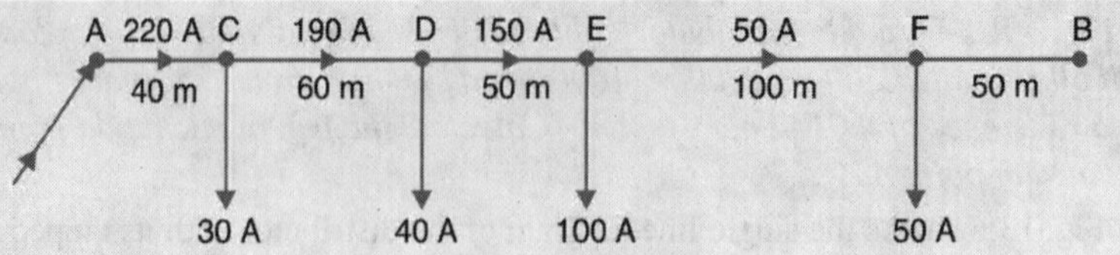

Fig. 13.7

Referring to Fig. 13.7, the currents in the various sections of the distributor are :

$$I_{AC} = 220\ \text{A} \ ; \ I_{CD} = 190\ \text{A} \ ; \ I_{DE} = 150\ \text{A} \ ; \ I_{EF} = 50\ \text{A}$$

Total voltage drop over the distributor

$$= I_{AC} R_{AC} + I_{CD} R_{CD} + I_{DE} R_{DE} + I_{EF} R_{EF}$$

$$= 220 \times 0{\cdot}4r + 190 \times 0{\cdot}6r + 150 \times 0{\cdot}5r + 50 \times r$$

$$= 327\, r$$

As the maximum permissible drop in the distributor is 10 V,

$$\therefore \qquad 10 = 327\, r$$

or $r = 10/327 = 0{\cdot}03058\ \Omega$

$$\text{X-sectional area of conductor} = \frac{*\rho\, l}{r/2} = \frac{1{\cdot}78\times10^{-8}\times100}{\dfrac{0{\cdot}03058}{2}} = 116{\cdot}4\times10^{-6}\ \text{m}^2 = \mathbf{1{\cdot}164\ cm^2}$$

Example 13.3. *Two tram cars (A & B) 2 km and 6 km away from a sub-station return 40 A and 20 A respectively to the rails. The sub-station voltage is 600 V d.c. The resistance of trolley wire is 0·25 Ω/km and that of track is 0·03 Ω/km. Calculate the voltage across each tram car.*

Solution. The tram car operates on d.c. supply. The positive wire is placed overhead while the rail track acts as the negative wire. Fig. 13.8 shows the single line diagram of the arrangement.

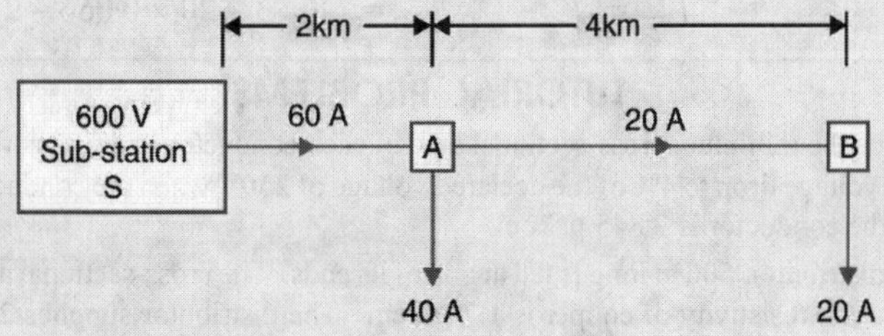

Fig. 13.8

Resistance of trolley wire and track/km

$= 0{\cdot}25 + 0{\cdot}03 = 0{\cdot}28\ \Omega$

Current in section SA $= 40 + 20 = 60$ A

Current in section AB $= 20$ A

Voltage drop in section SA $= 60 \times 0{\cdot}28 \times 2 = 33{\cdot}6$ V

Voltage drop in section AB $= 20 \times 0{\cdot}28 \times 4 = 22{\cdot}4$ V

∴ Voltage across tram A $= 600 - 33{\cdot}6 =$ **566·4 V**

Voltage across tram B $= 566{\cdot}4 - 22{\cdot}4 =$ **544 V**

Example 13.4. *The load distribution on a two-wire d.c. distributor is shown in Fig. 13.9. The cross-sectional area of each conductor is 0·27 cm². The end A is supplied at 250 V. Resistivity of the wire is ρ = 1·78 μ Ω cm. Calculate (i) the current in each section of the conductor (ii) the two-core resistance of each section (iii) the voltage at each tapping point.*

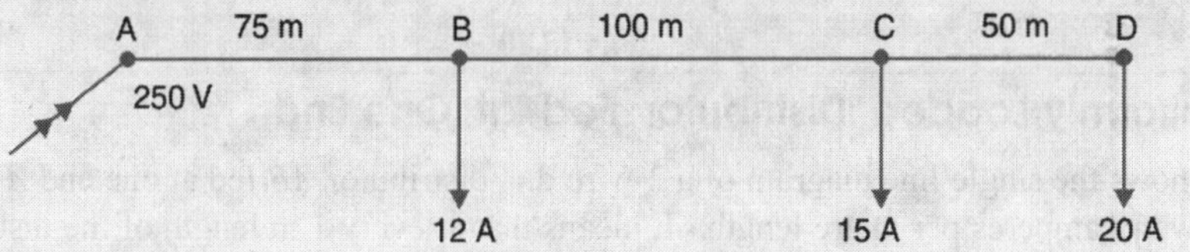

Fig. 13.9

Solution.

(*i*) Currents in the various sections are :

Section CD, I_{CD} = **20 A** ; section BC, $I_{BC} = 20 + 15 =$ **35 A**

Section AB, $I_{AB} = 20 + 15 + 12 =$ **47 A**

(*ii*) Single-core resistance of the section of 100 m length

$$= \rho\frac{l}{a} = 1{\cdot}78\times10^{-6}\times\frac{100\times100}{0{\cdot}27} = 0{\cdot}066\ \Omega$$

---

* Note that resistance of each conductor of $l = 100$ m is $r/2$.

The resistances of the various sections are :

$$R_{AB} = 0{\cdot}066 \times 0{\cdot}75 \times 2 = \mathbf{0{\cdot}099\ \Omega}\ ;\ R_{BC} = 0{\cdot}066 \times 2 = \mathbf{0{\cdot}132\ \Omega}$$

$$R_{CD} = 0{\cdot}066 \times 0{\cdot}5 \times 2 = \mathbf{0{\cdot}066\ \Omega}$$

(*iii*) Voltage at tapping point *B* is

$$V_B = V_A - I_{AB}\,R_{AB} = 250 - 47 \times 0{\cdot}099 = \mathbf{245{\cdot}35\ V}$$

Voltage at tapping point *C* is

$$V_C = V_B - I_{BC}\,R_{BC} = 245{\cdot}35 - 35 \times 0{\cdot}132 = \mathbf{240{\cdot}73\ V}$$

Voltage at tapping point *D* is

$$V_D = V_C - I_{CD}\,R_{CD} = 240{\cdot}73 - 20 \times 0{\cdot}066 = \mathbf{239{\cdot}41\ V}$$

TUTORIAL PROBLEMS

1. What should be the minimum cross-sectional area of each conductor in a two-core cable 100 m long to limit the total voltage drop to 4% of the declared voltage of 250V when the conductors carry 60A ? The resistivity of the conductor is 2·845 μΩ cm. **[0·34 cm²]**
2. A 2-wire d.c. distributor, 500 m long is fed at one of its ends. The cross-sectional area of each conductor is 3·4 cm² and the resistivity of copper is 1·7 μΩcm. The distributor supplies 200 A at a distance of 300m from the feeding point and 100 A at the terminus. Calculate the voltage at the feeding end if the voltage at the terminus is to be 230 V. **[241 V]**
3. A 2-wire d.c. distributor *AB* 500 metres long is fed from point *A* and is loaded as under :

Distance from feeding point *A* (in metres)	100	300	400	500
Load (amperes)	20	40	40	50

If the specific resistance of copper is $1{\cdot}7 \times 10^{-8}$ Ωm, what must be the cross-section of each wire in order that the voltage drop in the distributor shall not exceed 10 volts ? **[1·734 cm²]**

4. A 2-wire d.c. distributor is 250 m long. It is to be loaded as shown in Fig 13.10 at 50 m intervals. If the maximum voltage drop is not to exceed 10V and the resistivity of core material is 0·7 × 2·54 μΩ cm, determine the maximum cross-sectional area of each conductor. **[1·602 cm²]**

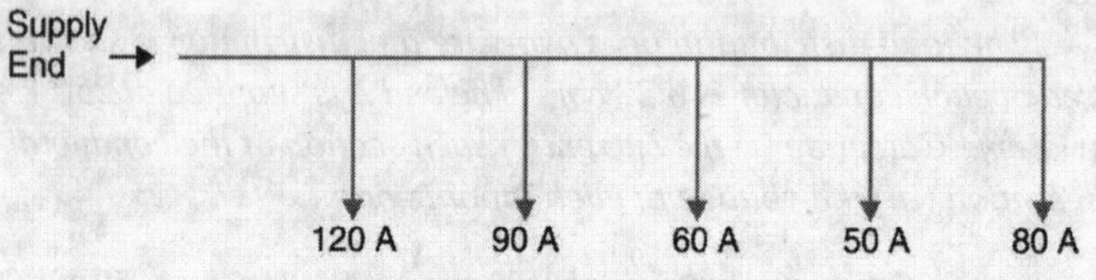

Fig. 13.10

13.4 Uniformly Loaded Distributor Fed at One End

Fig 13.11 shows the single line diagram of a 2-wire d.c. distributor *AB* fed at one end *A* and loaded uniformly with *i* amperes per metre length. It means that at every 1 m length of the distributor, the load tapped is *i* amperes. Let *l* metres be the length of the distributor and *r* ohm be the resistance per metre run.

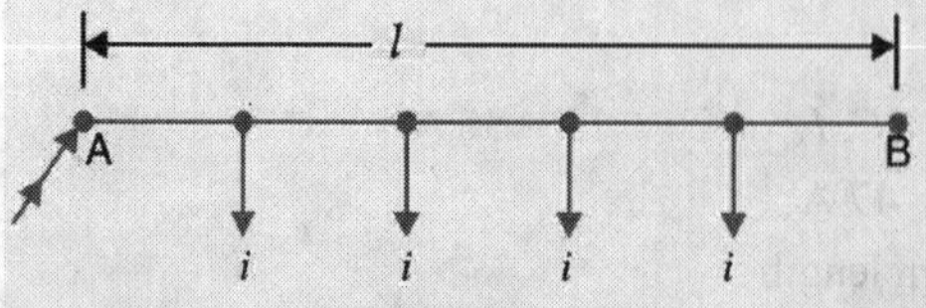

Fig. 13.11

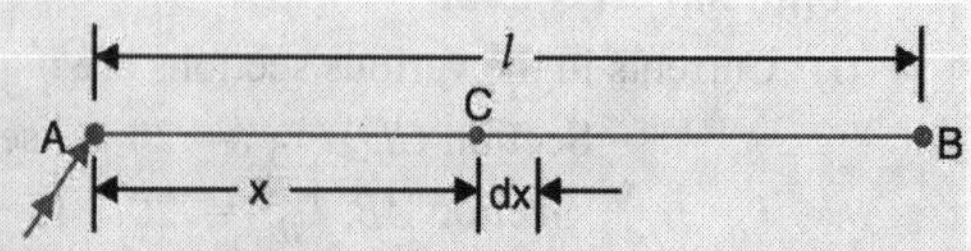

Fig. 13.12

Consider a point *C* on the distributor at a distance *x* metres from the feeding point *A* as shown in Fig. 13.12. Then current at point *C* is

$$= i\,l - i\,x \text{ amperes} = i\,(l - x) \text{ amperes}$$

Now, consider a small length dx near point C. Its resistance is $r\,dx$ and the voltage drop over length dx is

$$dv = i(l-x)\,r\,dx = i\,r\,(l-x)\,dx$$

Total voltage drop in the distributor upto point C is

$$v = \int_0^x i\,r\,(l-x)\,dx = i\,r\left(l\,x - \frac{x^2}{2}\right)$$

The voltage drop upto point B (*i.e.* over the whole distributor) can be obtained by putting $x = l$ in the above expression.

∴ Voltage drop over the distributor AB

$$= i\,r\left(l \times l - \frac{l^2}{2}\right)$$

$$= \tfrac{1}{2}\,i\,r\,l^2 = \tfrac{1}{2}\,(i\,l)\,(r\,l)$$

$$= \tfrac{1}{2}\,I\,R$$

where $i\,l = I$, the total current entering at point A

$r\,l = R$, the total resistance of the distributor

Thus, in a uniformly loaded distributor fed at one end, the total voltage drop is equal to that produced by the whole of the load assumed to be concentrated at the middle point.

Example 13.5. *A 2-wire d.c. distributor 200 metres long is uniformly loaded with 2A/metre. Resistance of single wire is 0·3 Ω/km. If the distributor is fed at one end, calculate :*

(i) the voltage drop upto a distance of 150 m from the feeding point

(ii) the maximum voltage drop

Solution.

Current loading, $i = 2$ A/m

Resistance of distributor per metre run,

$$r = 2 \times 0{\cdot}3/1000 = 0{\cdot}0006\ \Omega$$

Length of distributor, $l = 200$ m

(*i*) Voltage drop upto a distance x metres from feeding point

$$= i\,r\left(l\,x - \frac{x^2}{2}\right) \qquad \text{[See Art. 13·4]}$$

Here, $x = 150$ m

∴ Desired voltage drop $= 2 \times 0{\cdot}0006\left(200 \times 150 - \dfrac{150 \times 150}{2}\right) = \mathbf{22{\cdot}5\ V}$

(*ii*) Total current entering the distributor,

$$I = i \times l = 2 \times 200 = 400\ \text{A}$$

Total resistance of the distributor,

$$R = r \times l = 0{\cdot}0006 \times 200 = 0{\cdot}12\ \Omega$$

∴ Total drop over the distributor

$$= \frac{1}{2}\,I\,R = \frac{1}{2} \times 400 \times 0{\cdot}12 = \mathbf{24\ V}$$

Example 13.6. *A uniform 2-wire d.c. distributor 500 metres long is loaded with 0.4 ampere/metre and is fed at one end. If the maximum permissible voltage drop is not to exceed 10 V, find the cross-sectional area of the distributor conductor. Take* $\rho = 1{\cdot}7 \times 10^{-6}\ \Omega$ *cm.*

Solution.

Current entering the distributor, $I = i \times l = 0{\cdot}4 \times 500 = 200$ A

Max. permissible voltage drop $= 10$ V

Let r ohm be the resistance per metre length of the distributor (both wires).

Max. voltage drop $= \frac{1}{2} I R$

or $10 = \frac{1}{2} I r l$ $[\because R = r\,l]$

or $$r = \frac{2 \times 10}{I \times l} = \frac{2 \times 10}{200 \times 500} = 0{\cdot}2 \times 10^{-3}\ \Omega$$

$\therefore$ Area of cross-section of the distributor conductor is

$$a = \frac{\rho\, l}{r/2} = \frac{1{\cdot}7 \times 10^{-6} \times 100^{*} \times 2}{0{\cdot}2 \times 10^{-3}} = \mathbf{1{\cdot}7\ cm^2}$$

Example 13.7. *A 250 m, 2-wire d.c. distributor fed from one end is loaded uniformly at the rate of 1·6 A/metre. The resistance of each conductor is 0·0002 Ω per metre. Find the voltage necessary at feed point to maintain 250 V (i) at the far end (ii) at the mid-point of the distributor.*

Solution.

Current loading, $i = 1{\cdot}6$ A/m

Current entering the distributor, $I = i \times l = 1{\cdot}6 \times 250 = 400$ A

Resistance of the distributor per metre run

$$r = 2 \times 0{\cdot}0002 = 0{\cdot}0004\ \Omega$$

Total resistance of distributor, $R = r \times l = 0{\cdot}0004 \times 250 = 0{\cdot}1\ \Omega$

(*i*) Voltage drop over the entire distributor

$$= \frac{1}{2} I R = \frac{1}{2} \times 400 \times 0{\cdot}1 = 20 \text{ V}$$

$\therefore$ Voltage at feeding point $= 250 + 20 = \mathbf{270\ V}$

(*ii*) Voltage drop upto a distance of x metres from feeding point

$$= i\,r\left(l\,x - \frac{x^2}{2}\right)$$

Here $x = l/2 = 250/2 = 125$ m

$\therefore$ Voltage drop $$= 1{\cdot}6 \times 0{\cdot}0004\left(250 \times 125 - \frac{(125)^2}{2}\right) = 15 \text{ V}$$

$\therefore$ Voltage at feeding point $= 250 + 15 = \mathbf{265\ V}$

Example 13.8. *Derive an expression for the power loss in a uniformly loaded distributor fed at one end.*

Solution. Fig. 13.13 shows the single line diagram of a 2-wire d.c. distributor AB fed at end A and loaded uniformly with i amperes per metre length.

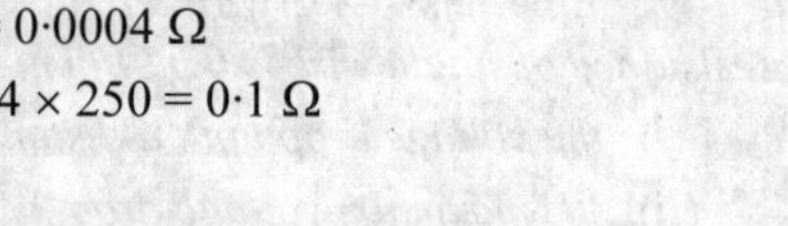

Fig. 13.13

Let l = length of the distributor in metres

r = resistance of distributor (both conductors) per metre run

Consider a small length dx of the distributor at point C at a distance x from the feeding end A. The small length dx will carry current which is tapped in the length CB.

$\therefore$ Current in $dx = i\,l - i\,x = i\,(l - x)$

* Because we have assumed that r ohm is the resistance of 1m (= 100 cm) length of the distributor.

Power loss in length dx = (current in length $dx)^2$ × Resistance of length dx

$$= [i(l-x)]^2 \times r\,dx$$

Total power loss P in the whole distributor is

$$P = \int_0^l [i(l-x)]^2\, r\,dx = \int_0^l i^2\left(l^2 + x^2 - 2\,lx\right) r\,dx$$

$$= i^2 r \int_0^l \left(l^2 + x^2 - 2\,lx\right) dx = i^2 r\left[l^2 x + \frac{x^3}{3} - \frac{2lx^2}{2}\right]_0^l$$

$$= i^2 r\left[l^3 + \frac{l^3}{3} - l^3\right] = i^2 \times \frac{r\,l^3}{3}$$

$$\therefore \quad P = \frac{i^2\, r\, l^3}{3}$$

Example 13.9. *Calculate the voltage at a distance of 200 m of a 300 m long distributor uniformly loaded at the rate of 0.75 A per metre. The distributor is fed at one end at 250 V. The resistance of the distributor (go and return) per metre is 0·00018 Ω. Also find the power loss in the distributor.*

Solution.

Voltage drop at a distance x from supply end

$$= i\,r\left(l\,x - \frac{x^2}{2}\right)$$

Here $i = 0{\cdot}75$ A/m; $l = 300$ m ; $x = 200$ m ; $r = 0{\cdot}00018\ \Omega$/m

$$\therefore \quad \text{Voltage drop} = 0{\cdot}75 \times 0{\cdot}00018\left[300 \times 200 - \frac{(200)^2}{2}\right] = 5{\cdot}4 \text{ V}$$

Voltage at a distance of 200 m from supply end

$$= 250 - 5{\cdot}4 = \mathbf{244{\cdot}6\ V}$$

Power loss in the distributor is

$$P = \frac{i^2\, r\, l^3}{3} = \frac{(0{\cdot}75)^2 \times 0{\cdot}00018 \times (300)^3}{3} = \mathbf{911{\cdot}25\ W}$$

TUTORIAL PROBLEMS

1. A 2-wire d.c. distributor 500 m long is loaded uniformly at the rate of 0·4A/m. If the voltage drop in the distributor is not to exceed 5V, calculate the area of X-section of each conductor required when the distributor is fed at one end. Take resistivity of conductor material as $1{\cdot}7 \times 10^{-8}$ Ωm. **[3·4 cm^2]**
2. A uniformly distributed load on a distributor of length 500 m is rated at 1 A per metre length. The distributor is fed from one end at 220V. Determine the voltage drop at a distance of 400 m from the feeding point. Assume a loop resistance of 2×10^{-5} Ω per metre. **[2·4 V]**
3. A 250 m, 2-wire d.c. distributor fed from one end is loaded uniformly at the rate of 0·8 A per metre. The resistance of each conductor is 0·0002 Ω per metre. Find the necessary voltage at the feeding point to maintain 250 V at the far end of the distributor. **[260 V]**

13.5 Distributor Fed at Both Ends — Concentrated Loading

Whenever possible, it is desirable that a long distributor should be fed at both ends instead of at one end only, since total voltage drop can be considerably reduced without increasing the cross-section of the conductor. The two ends of the distributor may be supplied with (*i*) equal voltages (*ii*) unequal voltages.

(*i*) **Two ends fed with equal voltages.** Consider a distributor *AB* fed at both ends with equal voltages *V* volts and having concentrated loads I_1, I_2, I_3, I_4 and I_5 at points *C, D, E, F* and *G* respectively as shown in Fig. 13.14. As we move away from one of the feeding points, *say A*, p.d. goes on decreasing till it reaches the minimum value at some load point, *say E*, and then again starts rising and becomes *V* volts as we reach the other feeding point *B*.

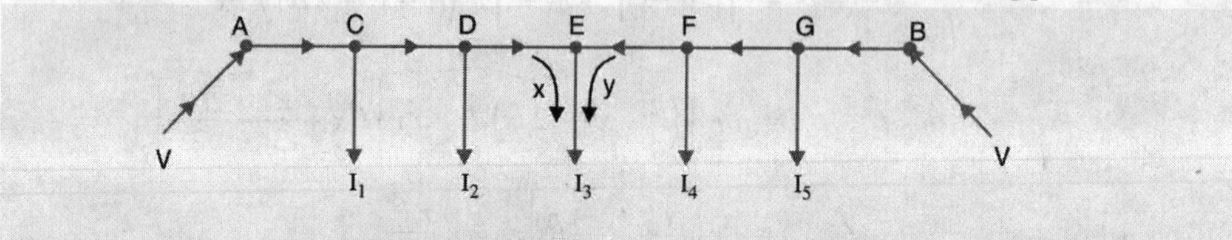

Fig. 13.14

All the currents tapped off between points *A* and *E* (minimum p.d. point) will be supplied from the feeding point *A* while those tapped off between *B* and *E* will be supplied from the feeding point *B*. The current tapped off at point *E* itself will be partly supplied from *A* and partly from *B*. If these currents are *x* and *y* respectively, then,

$$I_3 = x + y$$

Therefore, we arrive at a very important conclusion that at the point of minimum potential, current comes from both ends of the distributor.

Point of minimum potential. It is generally desired to locate the point of minimum potential. There is a simple method for it. Consider a distributor *AB* having three concentrated loads I_1, I_2 and I_3 at points *C, D* and *E* respectively. Suppose that current supplied by feeding end *A* is I_A. Then current distribution in the various sections of the distributor can be worked out as shown in Fig. 13.15 (*i*). Thus

$$I_{AC} = I_A; \qquad I_{CD} = I_A - I_1$$
$$I_{DE} = I_A - I_1 - I_2; \qquad I_{EB} = I_A - I_1 - I_2 - I_3$$

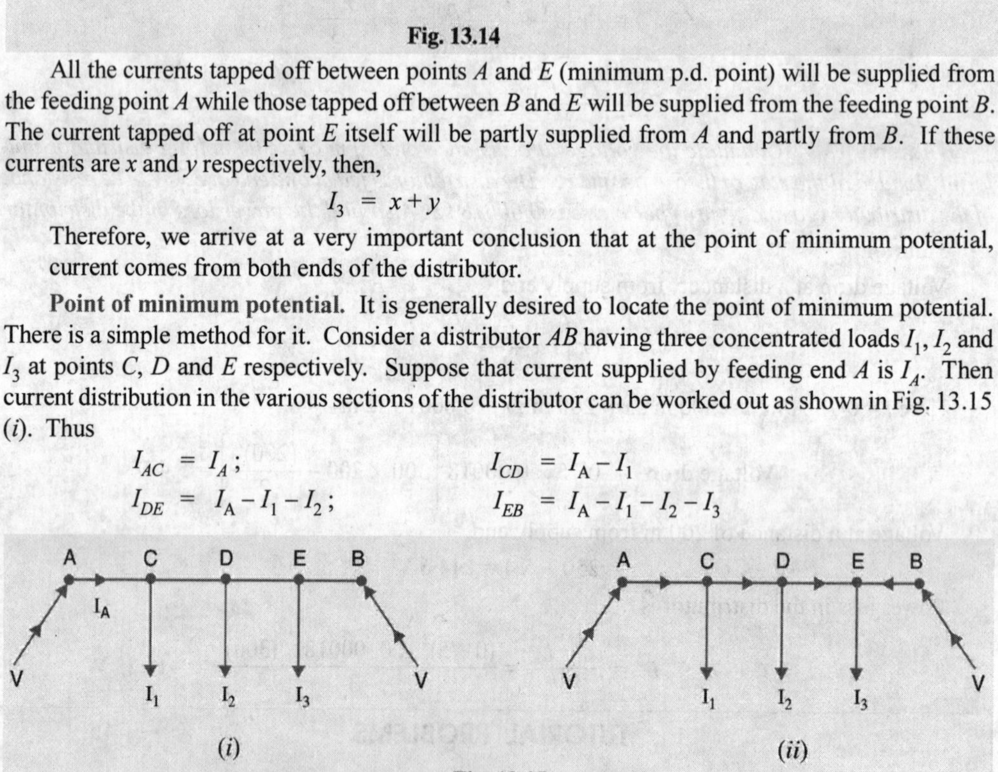

Fig. 13.15

Voltage drop between *A* and *B* = Voltage drop over *AB*

or $$V - V = I_A R_{AC} + (I_A - I_1) R_{CD} + (I_A - I_1 - I_2) R_{DE} + (I_A - I_1 - I_2 - I_3) R_{EB}$$

From this equation, the unknown I_A can be calculated as the values of other quantities are generally given. Suppose *actual* directions of currents in the various sections of the distributor are indicated as shown in Fig. 13.15 (*ii*). The load point where the currents are coming from both sides of the distributor is the point of minimum potential *i.e.* point *E* in this case

(*ii*) **Two ends fed with unequal voltages.** Fig. 13.16 shows the distributor *AB* fed with unequal voltages ; end *A* being fed at V_1 volts and end *B* at V_2 volts. The point of minimum potential can be found by following the same procedure as discussed above. Thus in this case,

Voltage drop between *A* and *B* = Voltage drop over *AB*

or $$V_1 - V_2 = \text{Voltage drop over } AB$$

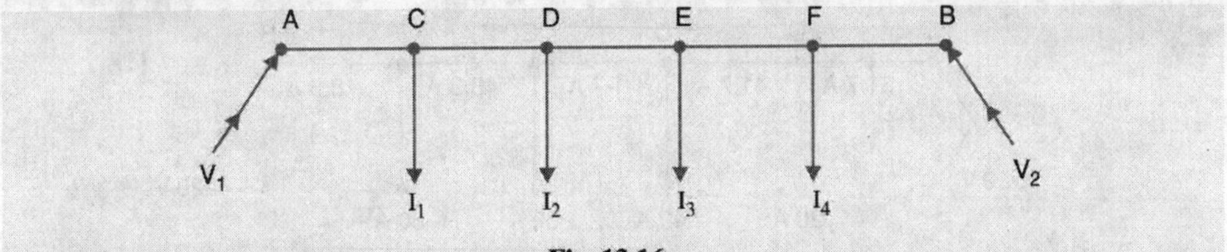

Fig. 13.16

Example 13.10. *A 2-wire d.c. street mains AB, 600 m long is fed from both ends at 220 V. Loads of 20 A, 40 A, 50 A and 30 A are tapped at distances of 100m, 250m, 400m and 500 m from the end A respectively. If the area of X-section of distributor conductor is 1cm², find the minimum consumer voltage. Take* $\rho = 1{\cdot}7 \times 10^{-6}\ \Omega$ *cm.*

Solution. Fig. 13.17 shows the distributor with its tapped currents. Let I_A amperes be the current supplied from the feeding end A. Then currents in the various sections of the distributor are as shown in Fig. 13.17.

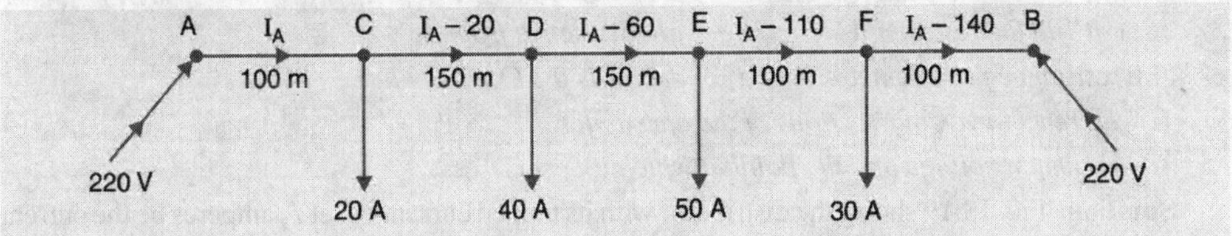

Fig. 13.17

Resistance of 1 m length of distributor

$$= 2 \times \frac{1{\cdot}7 \times 10^{-6} \times 100}{1} = 3{\cdot}4 \times 10^{-4}\ \Omega$$

$$\text{Resistance of section } AC,\ R_{AC} = (3{\cdot}4 \times 10^{-4}) \times 100 = 0{\cdot}034\ \Omega$$

$$\text{Resistance of section } CD,\ R_{CD} = (3{\cdot}4 \times 10^{-4}) \times 150 = 0{\cdot}051\ \Omega$$

$$\text{Resistance of section } DE,\ R_{DE} = (3{\cdot}4 \times 10^{-4}) \times 150 = 0{\cdot}051\ \Omega$$

$$\text{Resistance of section } EF,\ R_{EF} = (3{\cdot}4 \times 10^{-4}) \times 100 = 0{\cdot}034\ \Omega$$

$$\text{Resistance of section } FB,\ R_{FB} = (3{\cdot}4 \times 10^{-4}) \times 100 = 0{\cdot}034\ \Omega$$

$$\text{Voltage at } B = \text{Voltage at } A - \text{Drop over length } AB$$

or
$$V_B = V_A - [I_A R_{AC} + (I_A - 20)\, R_{CD} + (I_A - 60)\, R_{DE} + (I_A - 110)\, R_{EF} + (I_A - 140)\, R_{FB}]$$

or
$$220 = 220 - [0{\cdot}034\, I_A + 0{\cdot}051\,(I_A - 20) + 0{\cdot}051\,(I_A - 60) + 0{\cdot}034\,(I_A - 110) + 0{\cdot}034\,(I_A - 140)]$$

$$= 220 - [0{\cdot}204\, I_A - 12{\cdot}58]$$

or
$$0{\cdot}204\, I_A = 12{\cdot}58$$

$\therefore$
$$I_A = 12{\cdot}58/0{\cdot}204 = 61{\cdot}7\ \text{A}$$

The *actual distribution of currents in the various sections of the distributor is shown in Fig. 13.18. It is clear that currents are coming to load point E from both sides *i.e.* from point D and point F. Hence, E is the point of minimum potential.

$\therefore$ Minimum consumer voltage,

$$V_E = V_A - [I_{AC}\, R_{AC} + I_{CD}\, R_{CD} + I_{DE}\, R_{DE}]$$

---

* Knowing the value of I_A, current in any section can be determined. Thus,

$$\text{Current in section } CD,\ I_{CD} = I_A - 20 = 61{\cdot}7 - 20 = 41{\cdot}7\ A \text{ from } C \text{ to } D$$

$$\text{Current in section } EF,\ I_{EF} = I_A - 110 = 61{\cdot}7 - 110 = -48{\cdot}3\ \text{A from } E \text{ to } F$$

$$= 48{\cdot}3\ \text{A from } F \text{ to } E$$

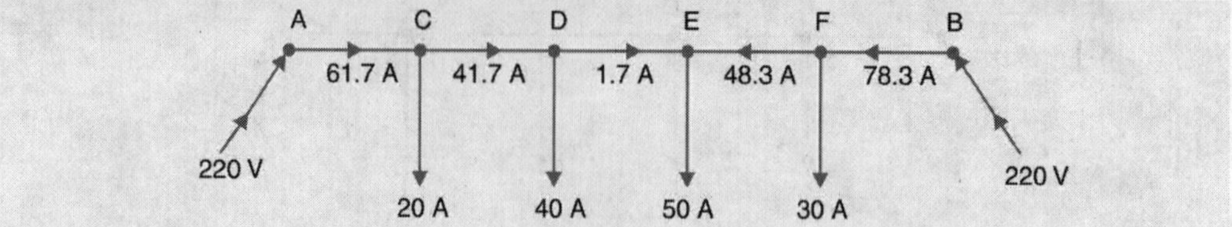

Fig. 13.18

$$= 220 - [61{\cdot}7 \times 0{\cdot}034 + 41{\cdot}7 \times 0{\cdot}051 + 1{\cdot}7 \times 0{\cdot}051]$$

$$= 220 - 4{\cdot}31 = \mathbf{215{\cdot}69\ V}$$

Example 13.11. *A 2-wire d.c. distributor AB is fed from both ends. At feeding point A, the voltage is maintained as at 230 V and at B 235 V. The total length of the distributor is 200 metres and loads are tapped off as under :*

25 A at 50 metres from A ; 50 A at 75 metres from A

30 A at 100 metres from A ; 40 A at 150 metres from A

The resistance per kilometre of one conductor is 0·3 Ω. Calculate :

(i) currents in various sections of the distributor

(ii) minimum voltage and the point at which it occurs

Solution. Fig. 13.19 shows the distributor with its tapped currents. Let I_A amperes be the current supplied from the feeding point A. Then currents in the various sections of the distributor are as shown in Fig 13.19.

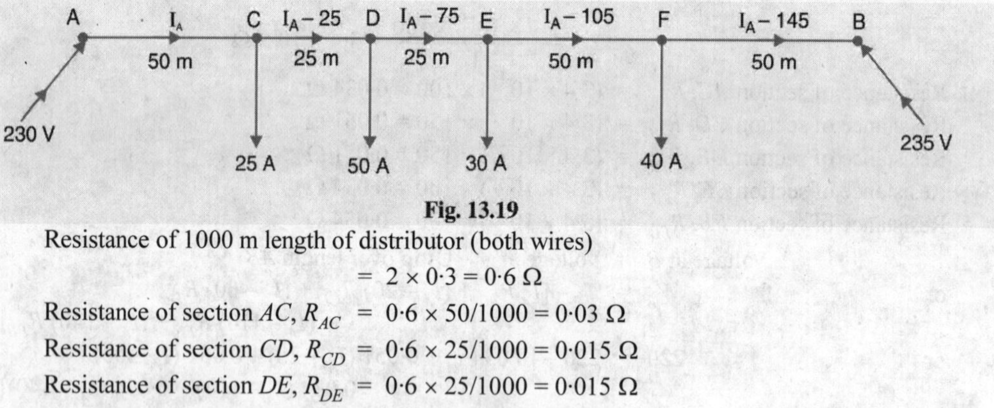

Fig. 13.19

Resistance of 1000 m length of distributor (both wires)

$$= 2 \times 0{\cdot}3 = 0{\cdot}6\ \Omega$$

Resistance of section AC, $R_{AC} = 0{\cdot}6 \times 50/1000 = 0{\cdot}03\ \Omega$

Resistance of section CD, $R_{CD} = 0{\cdot}6 \times 25/1000 = 0{\cdot}015\ \Omega$

Resistance of section DE, $R_{DE} = 0{\cdot}6 \times 25/1000 = 0{\cdot}015\ \Omega$

Resistance of section EF, $R_{EF} = 0{\cdot}6 \times 50/1000 = 0{\cdot}03\ \Omega$

Resistance of section FB, $R_{FB} = 0{\cdot}6 \times 50/1000 = 0{\cdot}03\ \Omega$

Voltage at B = Voltage at A – Drop over AB

or $$V_B = V_A - [I_A R_{AC} + (I_A - 25) R_{CD} + (I_A - 75) R_{DE} + (I_A - 105) R_{EF} + (I_A - 145) R_{FB}]$$

or $$235 = 230 - [0{\cdot}03\ I_A + 0{\cdot}015\ (I_A - 25) + 0{\cdot}015\ (I_A - 75) + 0{\cdot}03\ (I_A - 105) + 0{\cdot}03\ (I_A - 145)]$$

or $$235 = 230 - [0{\cdot}12\ I_A - 9]$$

$\therefore$ $$I_A = \frac{239 - 235}{0{\cdot}12} = 33{\cdot}34\ \text{A}$$

(*i*) $\therefore$ Current in section AC, $I_{AC} = I_A = \mathbf{33{\cdot}34\ A}$

Current in section CD, $I_{CD} = I_A - 25 = 33{\cdot}34 - 25 = \mathbf{8{\cdot}34\ A}$

Current in section DE, I_{DE} $= I_A - 75 = 33{\cdot}34 - 75 = -41{\cdot}66$ A *from D to E*

$= $ **41·66 A** *from E to D*

Current in section EF, I_{EF} $= I_A - 105 = 33{\cdot}34 - 105 = -71{\cdot}66$ A *from E to F*

$= $ **71·66 A** *from F to E*

Current in section FB, I_{FB} $= I_A - 145 = 33{\cdot}34 - 145 = -111{\cdot}66$ A *from F to B*

$= $ **111·66 A** *from B to F*

(*ii*) The actual distribution of currents in the various sections of the distributor is shown in Fig. 13.20. The currents are coming to load point D from both sides of the distributor. Therefore, load point D is the point of minimum potential.

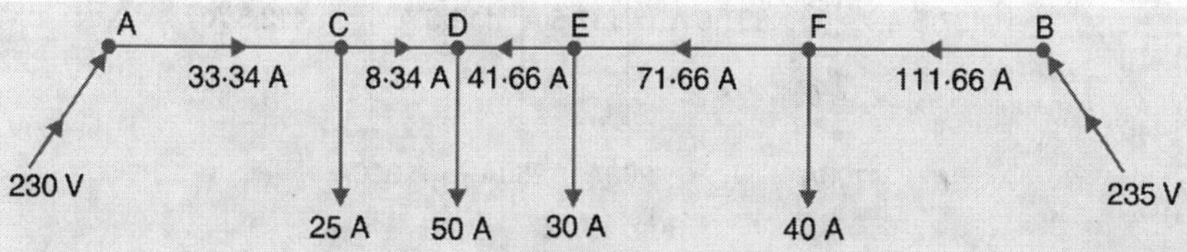

Fig. 13.20

Voltage at D, $V_D = V_A - [I_{AC} R_{AC} + I_{CD} R_{CD}]$

$= 230 - [33{\cdot}34 \times 0{\cdot}03 + 8{\cdot}34 \times 0{\cdot}015]$

$= 230 - 1{\cdot}125 = $ **228·875 V**

Example 13.12. *A two-wire d.c. distributor AB, 600 metres long is loaded as under :*

Distance from A (metres) :	*150*	*300*	*350*	*450*
Loads in Amperes :	*100*	*200*	*250*	*300*

The feeding point A is maintained at 440 V and that of B at 430 V. If each conductor has a resistance of 0·01 Ω per 100 metres, calculate :

(*i*) *the currents supplied from A to B,* (*ii*) *the power dissipated in the distributor.*

Solution. Fig. 13.21 shows the distributor with its tapped currents. Let I_A amperes be the current supplied from the feeding point A. Then currents in the various sections of the distributor are as shown in Fig.13.21.

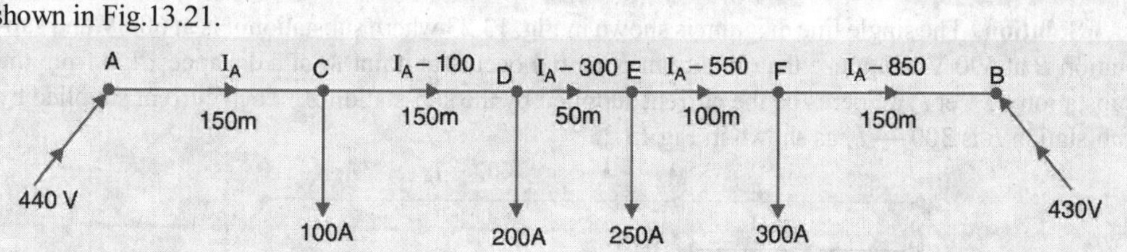

Fig. 13.21

Resistance of 100 m length of distributor (both wires)

$= 2 \times 0{\cdot}01 = 0{\cdot}02\ \Omega$

Resistance of section AC, $R_{AC} = 0{\cdot}02 \times 150/100 = 0{\cdot}03\ \Omega$

Resistance of section CD, $R_{CD} = 0{\cdot}02 \times 150/100 = 0{\cdot}03\ \Omega$

Resistance of section DE, $R_{DE} = 0{\cdot}02 \times 50/100 = 0{\cdot}01\ \Omega$

Resistance of section EF, $R_{EF} = 0{\cdot}02 \times 100/100 = 0{\cdot}02\ \Omega$

Resistance of section FB, $R_{FB} = 0{\cdot}02 \times 150/100 = 0{\cdot}03\ \Omega$

Voltage at B = Voltage at A — Drop over AB

or $V_B = V_A - [I_A R_{AC} + (I_A - 100) R_{CD} + (I_A - 300) R_{DE}$

$+ (I_A - 550) R_{EF} + (I_A - 850) R_{FB}]$

or $430 = 440 - [0{\cdot}03\, I_A + 0{\cdot}03\,(I_A - 100) + 0{\cdot}01\,(I_A - 300) + 0{\cdot}02\,(I_A - 550) + 0{\cdot}03\,(I_A - 850)]$

or $430 = 440 - [0{\cdot}12\, I_A - 42{\cdot}5]$

$\therefore \quad I_A = \dfrac{482{\cdot}5 - 430}{0{\cdot}12} = 437{\cdot}5 \text{ A}$

The actual distribution of currents in the various sections of the distributor is shown in Fig. 13.22. Incidentally, E is the point of minimum potential.

(*i*) Referring to Fig. 13.22, it is clear that

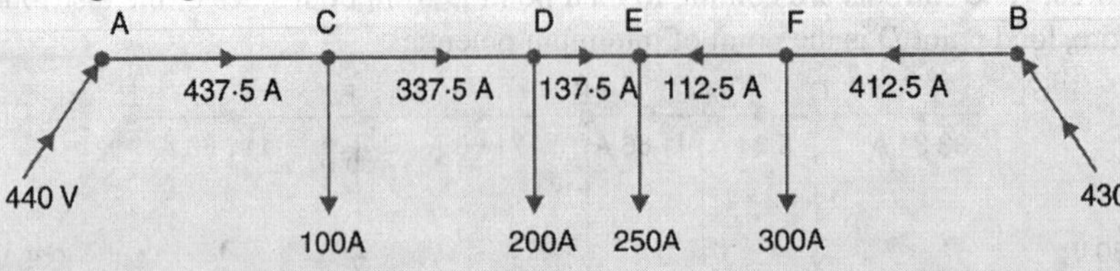

Fig. 13.22

Current supplied from end A, I_A = **437·5 A**

Current supplied from end B, I_B = **412·5 A**

(*ii*) Power loss in the distributor

$= I_{AC}^2 R_{AC} + I_{CD}^2 R_{CD} + I_{DE}^2 R_{DE} + I_{EF}^2 R_{EF} + I_{FB}^2 R_{FB}$

$= (437{\cdot}5)^2 \times 0{\cdot}03 + (337{\cdot}5)^2 \times 0{\cdot}03 + (137{\cdot}5)^2 \times 0{\cdot}01 + (112{\cdot}5)^2 \times 0{\cdot}02 + (412{\cdot}5)^2 \times 0{\cdot}03$

$= 5742 + 3417 + 189 + 253 + 5104 = 14{,}705$ watts = **14·705 kW**

Example 13.13. *An electric train runs between two sub-stations 6 km apart maintained at voltages 600 V and 590 V respectively and draws a constant current of 300 A while in motion. The track resistance of go and return path is 0·04 Ω/km. Calculate :*

(*i*) *the point along the track where minimum potential occurs*

(*ii*) *the current supplied by the two sub-stations when the train is at the point of minimum potential*

Solution. The single line diagram is shown in Fig. 13.23 where substation A is at 600 V and sub-station B at 590 V. Suppose that minimum potential occurs at point M at a distance x km from the substation A. Let I_A amperes be the current supplied by the sub-station A. Then current supplied by sub-station B is $300 - I_A$ as shown in Fig 13.23.

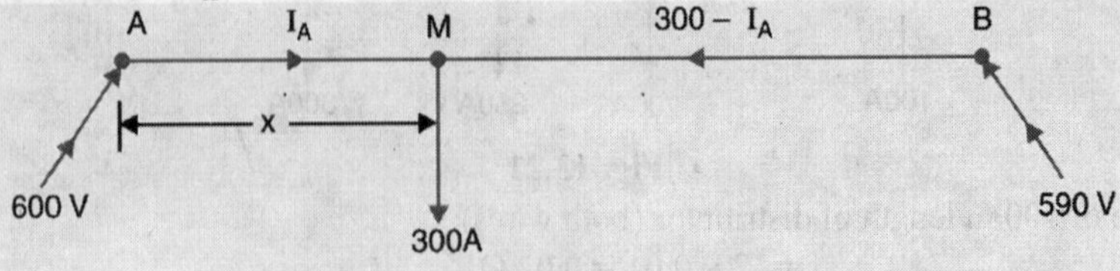

Fig. 13.23

Resistance of track (go and return path) per km

$= 0{\cdot}04\ \Omega$

Track resistance for section AM, $R_{AM} = 0{\cdot}04\, x\ \Omega$

Track resistance for section MB, $R_{MB} = 0{\cdot}04\,(6 - x)\Omega$

Potential at M, $V_M = V_A - I_A R_{AM}$ ... (*i*)

Also, Potential at M, $V_M = V_B - (300 - I_A)\, R_{MB}$ ... (*ii*)

From equations (*i*) and (*ii*), we get,

$$V_A - I_A R_{AM} = V_B - (300 - I_A) R_{MB}$$

or $$600 - 0{\cdot}04\, x\, I_A = 590 - (300 - I_A) \times 0{\cdot}04\,(6 - x)$$

or $$600 - 0{\cdot}04\, x\, I_A = 590 - 0{\cdot}04\,(1800 - 300\,x - 6\,I_A + I_A \times x)$$

or $$600 - 0{\cdot}04\, x\, I_A = 590 - 72 + 12\,x + 0{\cdot}24\,I_A - 0{\cdot}04\,xI_A$$

or $$0{\cdot}24\,I_A = 82 - 12\,x$$

or $$I_A = 341{\cdot}7 - 50\,x$$

Substituting the value of I_A in eq. (*i*), we get,

$$V_M = V_A - (341{\cdot}7 - 50\,x) \times 0{\cdot}04\,x$$

$\therefore$ $$V_M = 600 - 13{\cdot}7\,x + 2x^2 \qquad ...(iii)$$

(*i*) For V_M to be minimum, its differential coefficient *w.r.t.* x must be zero *i.e.*

$$\frac{d}{dx}(600 - 13{\cdot}7\,x + 2x^2) = 0$$

or $$0 - 13{\cdot}7 + 4x = 0$$

$\therefore$ $$x = 13{\cdot}7/4 = \mathbf{3{\cdot}425\ km}$$

i.e. minimum potential occurs at a distance of 3·425 km from the sub-station *A*.

(*ii*) $\therefore$ Current supplied by sub-station *A*

$$= 341{\cdot}7 - 50 \times 3{\cdot}425 = 341{\cdot}7 - 171{\cdot}25 = \mathbf{170{\cdot}45\ A}$$

Current supplied by sub-station *B* $= 300 - I_A = 300 - 170{\cdot}45 = \mathbf{129{\cdot}55\ A}$

TUTORIAL PROBLEMS

1. A 2-wire d.c. distributor *AB* is fed at both ends at the same voltage of 230 V. The length of the distributor is 500 metres and the loads are tapped off from the end A as shown below :

Load :	100 A	60 A	40 A	100 A
Distance :	50 m	150 m	250 m	400 m

If the maximum voltage drop of 5·5 V is to be allowed, find the X-sectional area of each conductor and point of minimum potential. Specific resistance of conductor material may be taken as $1{\cdot}73 \times 10^{-8}\ \Omega$ m. **[1·06 cm^2 ; 250 m from A]**

2. A d.c. distributor *AB* is fed at both ends. At feeding point *A*, the voltage is maintained at 235 V and at *B* at 236 V. The total length of the distributor is 200 metres and loads are tapped off as under :

   20 A at 50 m from *A*
   40 A at 75 m from *A*
   25 A at 100 m from *A*
   30 A at 150 m from *A*

   The resistance per kilometre of one conductor is 0·4 Ω. Calculate the minimum voltage and the point at which it occurs. **[232·175 V ; 75 m from point A]**

3. A two conductor main *AB*, 500 m in length is fed from both ends at 250 volts. Loads of 50 A, 60 A, 40 A and 30 A are tapped at distance of 100 m, 250 m, 350 m and 400 m from end *A* respectively. If the X-section of conductor be 1 cm^2 and specific resistance of the material of the conductor is 1·7 μ Ω cm, determine the minimum consumer voltage. **[245·07 V]**

13.6 Uniformly Loaded Distributor Fed at Both Ends

We shall now determine the voltage drop in a uniformly loaded distributor fed at both ends. There can be two cases *viz.* the distributor fed at both ends with (*i*) equal voltages (*ii*) unequal voltages. The two cases shall be discussed separately.

(*i*) **Distributor fed at both ends with equal voltages.** Consider a distributor *AB* of length *l* metres, having resistance *r* ohms per metre run and with uniform loading of *i* amperes per metre run as shown in Fig. 13.24. Let the distributor be fed at the feeding points *A* and *B* at equal voltages, say *V* volts. The total current supplied to the distributor is $i\,l$. As the two end voltages are equal, therefore, current supplied from each feeding point is $i\,l/2$ *i.e.*

Current supplied from each feeding point

$$= \frac{i\,l}{2}$$

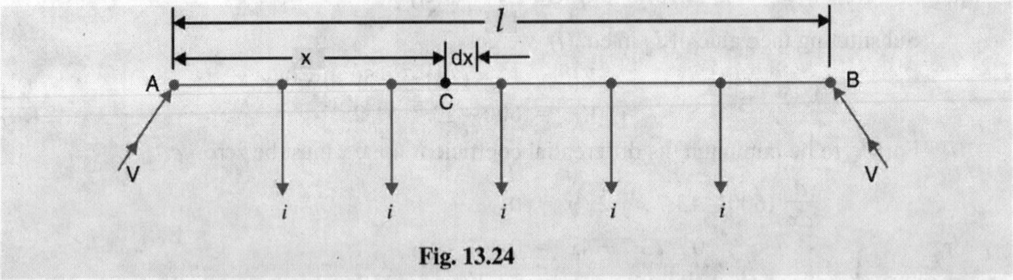

Fig. 13.24

Consider a point *C* at a distance *x* metres from the feeding point *A*. Then current at point *C* is

$$= \frac{i\,l}{2} - i\,x = i\left(\frac{l}{2} - x\right)$$

Now, consider a small length *dx* near point *C*. Its resistance is $r\,dx$ and the voltage drop over length *dx* is

$$dv = i\left(\frac{l}{2} - x\right) r\,dx = i\,r\left(\frac{l}{2} - x\right) dx$$

$\therefore$ Voltage drop upto point $C = \int_0^x i\,r\left(\frac{l}{2} - x\right) dx = i\,r\left(\frac{l\,x}{2} - \frac{x^2}{2}\right)$

$$= \frac{i\,r}{2}\left(l\,x - x^2\right)$$

Obviously, the point of minimum potential will be the mid-point. Therefore, maximum voltage drop will occur at mid-point *i.e.* where $x = l/2$.

$\therefore$ Max. voltage drop $= \frac{i\,r}{2}\left(l\,x - x^2\right)$

$$= \frac{i\,r}{2}\left(l \times \frac{l}{2} - \frac{l^2}{4}\right) \qquad \text{[Putting } x = l/2\text{]}$$

$$= \frac{1}{8}\,i\,r\,l^2 = \frac{1}{8}(i\,l)(r\,l) = \frac{1}{8}\,I\,R$$

where $i\,l = I$, the total current fed to the distributor from both ends

$r\,l = R$, the total resistance of the distributor

Minimum voltage $= V - \frac{I\,R}{8}$ volts

(ii) **Distributor fed at both ends with unequal voltages.** Consider a distributor *AB* of length *l* metres having resistance *r* ohms per metre run and with a uniform loading of *i* amperes per metre run as shown in Fig. 13.25. Let the distributor be fed from feeding points *A* and *B* at voltages V_A and V_B respectively.

Suppose that the point of minimum potential *C* is situated at a distance *x* metres from the feeding point *A*. Then current supplied by the feeding point *A* will be $*i\,x$.

* As *C* is at minimum potential, therefore, there is no current at this point. Consequently, current in section *AC* (*i.e.* $i\,x$) will be the current supplied by feeding point *A*.

∴ Voltage drop in section $AC = \frac{i\,r\,x^2}{2}$ volts

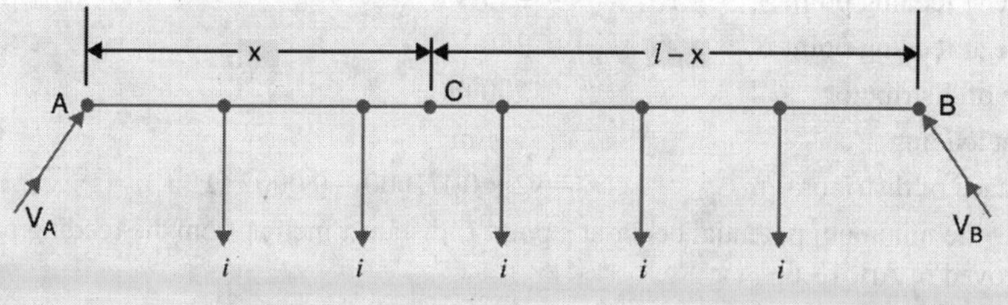

Fig. 13.25

As the distance of C from feeding point B is $(l - x)$, therefore, current fed from B is $i\,(l - x)$.

$$\therefore \quad \text{Voltage drop in section } BC = \frac{i\,r\,(l-x)^2}{2} \text{ volts}$$

$$\text{Voltage at point } C,\ V_C = V_A - \text{Drop over } AC$$

$$= V_A - \frac{i\,r\,x^2}{2} \qquad \text{...}(i)$$

$$\text{Also, voltage at point } C,\ V_C = V_B - \text{Drop over } BC$$

$$= V_B - \frac{i\,r\,(l-x)^2}{2} \qquad \text{...}(ii)$$

From equations (*i*) and (*ii*), we get,

$$V_A - \frac{i\,r\,x^2}{2} = V_B - \frac{i\,r\,(l-x)^2}{2}$$

Solving the equation for x, we get,

$$x = \frac{V_A - V_B}{i\,r\,l} + \frac{l}{2}$$

As all the quantities on the right hand side of the equation are known, therefore, the point on the distributor where minimum potential occurs can be calculated.

Example 13.14. *A two-wire d.c. distributor cable 1000 metres long is loaded with 0·5 A/metre. Resistance of each conductor is 0·05 Ω/km. Calculate the maximum voltage drop if the distributor is fed from both ends with equal voltages of 220 V. What is the minimum voltage and where it occurs ?*

Solution.

Current loading, $i = 0{\cdot}5$ A/m

Resistance of distributor/m, $r = 2 \times 0{\cdot}05/1000 = 0{\cdot}1 \times 10^{-3}\ \Omega$

Length of distributor, $l = 1000$ m

Total current supplied by distributor, $I = i\,l = 0{\cdot}5 \times 1000 = 500$ A

Total resistance of the distributor, $R = r\,l = 0{\cdot}1 \times 10^{-3} \times 1000 = 0{\cdot}1\ \Omega$

$$\therefore \quad \text{Max. voltage drop} = \frac{I\,R}{8} = \frac{500 \times 0{\cdot}1}{8} = \mathbf{6{\cdot}25\ V}$$

Minimum voltage will occur at the mid-point of the distributor and its value is

$$= 220 - 6{\cdot}25 = \mathbf{213{\cdot}75\ V}$$

Example 13.15. *A 2-wire d.c. distributor AB 500 metres long is fed from both ends and is loaded uniformly at the rate of 1·0 A/metre. At feeding point A, the voltage is maintained at 255 V and at B at 250 V. If the resistance of each conductor is 0·1 Ω per kilometre, determine :*

(*i*) *the minimum voltage and the point where it occurs*

(*ii*) *the currents supplied from feeding points A and B*

Solution. Fig. 13.26 shows the single line diagram of the distributor.

Voltage at feeding point A, $V_A = 255$ V

Voltage at feeding point B, $V_B = 250$ V

Length of distributor, $l = 500$ m

Current loading, $i = 1$ A/m

Resistance of distributor/m, $r = 2 \times 0{\cdot}1/1000 = 0{\cdot}0002\ \Omega$

(*i*) Let the minimum potential occur at a point C distant x metres from the feeding point A. As proved in Art. 13.6,

$$x = \frac{V_A - V_B}{i\,r\,l} + \frac{l}{2} = \frac{255 - 250}{1 \times 0{\cdot}0002 \times 500} + 500/2$$

$$= 50 + 250 = \mathbf{300\ m}$$

i.e. minimum potential occurs at 300 m from point A.

Minimum voltage, $$V_C = V_A - \frac{ir\,x^2}{2} = 255 - \frac{1 \times 0{\cdot}0002 \times (300)^2}{2}$$

$$= 255 - 9 = \mathbf{246\ V}$$

(*ii*) Current supplied from A = $i\,x = 1 \times 300 =$ **300 A**

Current supplied from B = $i\,(l - x) = 1\ (500 - 300) =$ **200 A**

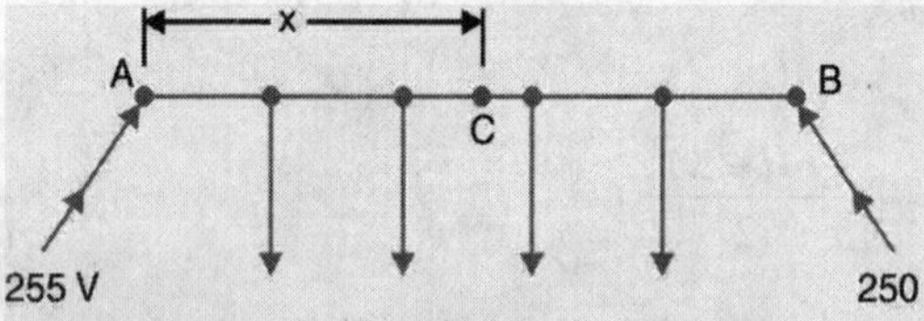

Fig. 13.26

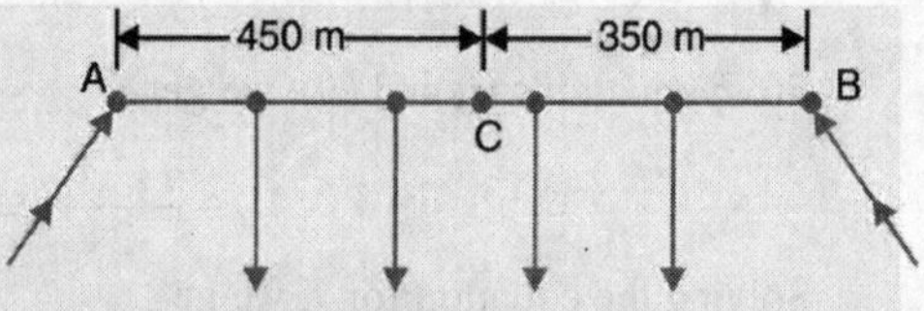

Fig. 13.27

Example 13.16. *A 800 metres 2-wire d.c. distributor AB fed from both ends is uniformly loaded at the rate of 1·25 A/metre run. Calculate the voltage at the feeding points A and B if the minimum potential of 220 V occurs at point C at a distance of 450 metres from the end A. Resistance of each conductor is 0·05 Ω/km.*

Solution. Fig. 13.27 shows the single line diagram of the distributor.

Current loading, $i = 1{\cdot}25$ A/m

Resistance of distributor/m, $r = 2 \times 0{\cdot}05/1000 = 0{\cdot}0001\ \Omega$

Voltage at C, $V_C = 220$ V

Length of distributor, $l = 800$ m

Distance of point C from A, $x = 450$ m

Voltage drop in section AC $$= \frac{i\,r\,x^2}{2} = \frac{1{\cdot}25 \times 0{\cdot}0001 \times (450)^2}{2} = 12{\cdot}65\ \text{V}$$

∴ Voltage at feeding point A, $V_A = 220 + 12{\cdot}65 =$ **232·65 V**

Voltage drop in section BC $$= \frac{i\,r\,(l - x)^2}{2} = \frac{1{\cdot}25 \times 0{\cdot}0001 \times (800 - 450)^2}{2}$$

$$= 7{\cdot}65\ \text{V}$$

∴ Voltage at feeding point B, $V_B = 220 + 7{\cdot}65 =$ **227·65 V**

Example 13.17.

(*i*) *A uniformly loaded distributor is fed at the centre. Show that maximum voltage drop = I R/8 where I is the total current fed to the distributor and R is the total resistance of the distributor.*

(*ii*) *A 2-wire d.c. distributor 1000 metres long is fed at the centre and is loaded uniformly at the rate of 1·25 A/metre. If the resistance of each conductor is 0·05 Ω/km, find the maximum voltage drop in the distributor.*

Solution. (*i*) Fig. 13.28 shows distributor *AB* fed at centre *C* and uniformly loaded with *i* amperes/metre. Let *l* metres be the length of the distributor and *r* ohms be the resistance per metre run. Obviously, maximum voltage drop will occur at either end.

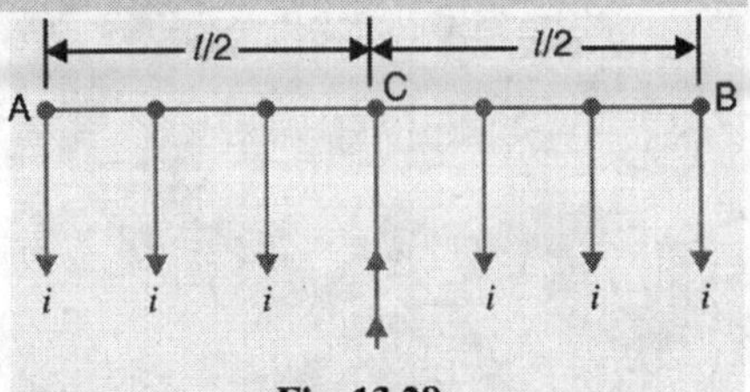

Fig. 13.28

∴ Max. voltage drop = Voltage drop in half distributor

$$= \frac{1}{2}\left(\frac{i\,l}{2}\right)\left(\frac{r\,l}{2}\right) = \frac{1}{8}(i\,l)(r\,l)$$

$$= \frac{1}{8}\,I R$$

where $i\,l$ = *I*, the total current fed to the distributor

$r\,l$ = *R*, the total resistance of the distributor

(*ii*) Total current fed to the distributor is

$$I = i\,l = 1{\cdot}25 \times 1000 = 1250 \text{ A}$$

Total resistance of the distributor is

$$R = r\,l = 2 \times 0{\cdot}05 \times 1 = 0{\cdot}1\ \Omega$$

$$\text{Max. voltage drop} = \frac{1}{8}\,I R = \frac{1}{8} \times 1250 \times 0{\cdot}1 = \mathbf{15.62\ V}$$

Example 13.18. *Derive an expression for the power loss in a uniformly loaded distributor fed at both ends with equal voltages.*

Solution. Consider a distributor *AB* of length *l* metres, having resistance *r* ohms per metre run with uniform loading of *i* amperes per metre run as shown in Fig.13.29. Let the distributor be fed at the feeding points *A* and *B* at equal voltages, say V volts. The total current supplied by the distributor is *i l*. As the two end voltages are equal, therefore, current supplied from each feeding point is *i l*/2.

$$\text{Current supplied from each feeding point} = \frac{i\,l}{2}$$

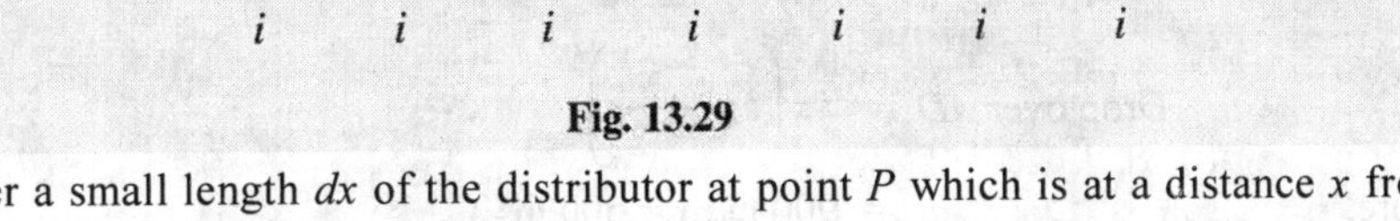

Fig. 13.29

Consider a small length *dx* of the distributor at point *P* which is at a distance *x* from the feeding end *A*.

$$\text{Resistance of length } dx = r\,dx$$

$$\text{Current in length } dx = \frac{i\,l}{2} - i\,x = i\left(\frac{l}{2} - x\right)$$

$$\text{Power loss in length } dx = (\text{current in } dx)^2 \times \text{Resistance of } dx$$

$$= \left[i\left(\frac{l}{2} - x\right)\right]^2 \times r\,dx$$

Total power loss in the distributor is

$$P = \int_0^l \left[i\left(\frac{l}{2} - x\right)\right]^2 r\,dx = i^2 r \int_0^l \left(\frac{l^2}{4} - l\,x + x^2\right) dx$$

$$= i^2 r\left[\frac{l^2 x}{4} - \frac{l\,x^2}{2} + \frac{x^3}{3}\right]_0^l = i^2 r\left[\frac{l^3}{4} - \frac{l^3}{2} + \frac{l^3}{3}\right]$$

$$\therefore \quad P = \frac{i^2\,r\,l^3}{12}$$

13.7 Distributor with Both Concentrated and Uniform Loading

There are several problems where a distributor has both concentrated and uniform loadings. In such situations, the total drop over any section of the distributor is equal to the sum of drops due to concentrated and uniform loading in that section. We shall solve a few problems by way of illustration.

Example 13.19. *A 2-wire d.c. distributor AB, 900 metres long is fed at A at 400 V and loads of 50 A, 100 A and 150 A are tapped off from C, D and E which are at a distance of 200 m, 500 m and 800 m from point A respectively. The distributor is also loaded uniformly at the rate of 0.5 A/m. If the resistance of distributor per metre (go and return) is 0.0001 Ω, calculate voltage (i) at point B and (ii) at point D.*

Solution. This problem can be solved in two stages. First, the drop at any point due to concentrated loading is found. To this is added the voltage drop due to uniform loading.

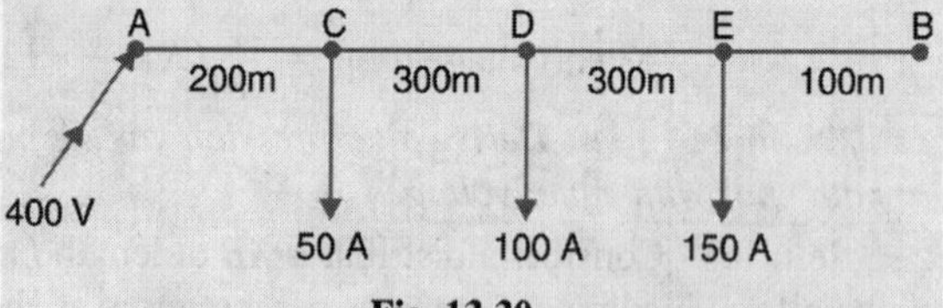

Fig. 13.30

Drops due to concentrated loads. Fig. 13.30 shows only the concentrated loads tapped off from the various points. The currents in the various sections are :

$$I_{AC} = 300 \text{ A} ; I_{CD} = 250 \text{ A} ; I_{DE} = 150 \text{ A}$$

$$\text{Drop in section } AC = I_{AC} R_{AC} = 300 \times (200 \times 0{\cdot}0001) = 6 \text{ V}$$

$$\text{Drop in section } CD = 250 \times (300 \times 0{\cdot}0001) = 7{\cdot}5 \text{ V}$$

$$\text{Drop in section } DE = 150 \times (300 \times 0{\cdot}0001) = 4{\cdot}5 \text{ V}$$

$$\text{Total drop over } AB = 6 + 7{\cdot}5 + 4{\cdot}5 = 18 \text{ V}$$

Drops due to uniform loading

$$\text{Drop over } AB = \frac{i\,r\,l^2}{2} = \frac{0{\cdot}5 \times 0{\cdot}0001 \times (900)^2}{2} = 20{\cdot}25 \text{ V}$$

$$\text{Drop over } AD = i\,r\left(l\,x - \frac{x^2}{2}\right)$$

Here, $l = 900 \text{ m} ; x = 500 \text{ m}$

$$\therefore \quad \text{Drop over } AD = 0{\cdot}5 \times 0{\cdot}0001\left(900 \times 500 - \frac{500^2}{2}\right) = 16{\cdot}25 \text{ V}$$

(*i*) Voltage at point B = V_A – Drop over AB due to conc. and uniform loadings

= 400 – (18 + 20·25) = **361·75 V**

(*ii*) Voltage at point D = V_A – Drop over AD due to conc. and uniform loadings

= 400 – (6 + 7·5 + 16·25) = **370·25 V**

Example 13.20. *Two conductors of a d.c. distributor cable AB 1000 m long have a total resistance of 0·1 Ω. The ends A and B are fed at 240 V. The cable is uniformly loaded at 0·5 A per metre length and has concentrated loads of 120 A, 60 A, 100 A and 40 A at points distant 200 m, 400 m, 700 m and 900 m respectively from the end A. Calculate (i) the point of minimum potential (ii) currents supplied from ends A and B (iii) the value of minimum potential.*

Solution.

Distributor resistance per metre length, $r = 0{\cdot}1/1000 = 10^{-4}\ \Omega$

Uniform current loading, $i = 0{\cdot}5$ A/m

(*i*) **Point of minimum potential.** The point of minimum potential is not affected by the uniform loading of the distributor. Therefore, let us consider the concentrated loads first as shown in Fig. 13.31. Suppose the current supplied by end A is I. Then currents in the various sections will be as shown in Fig. 13.31.

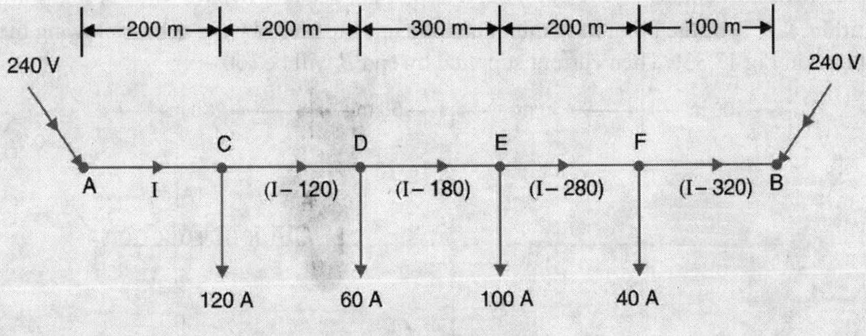

Fig. 13.31

$V_A - V_B$ = Drop over the distributor AB

$240 - 240 = I_{AC}\,R_{AC} + I_{CD}\,R_{CD} + I_{DE}\,R_{DE} + I_{EF}\,R_{EF} + I_{FB}\,R_{FB}$

or $0 = 10^{-4}\,[I \times 200 + (I - 120)\,200 + (I - 180)\,300 + (I - 280)\,200 + (I - 320) \times 100]$

or $0 = 1000\,I - 166000 \quad \therefore \quad I = 166000/1000 = 166$ A

The actual distribution of currents in the various sections of the distributor due to concentrated loading is shown in Fig. 13.32. It is clear from this figure that *D is the point of minimum potential.*

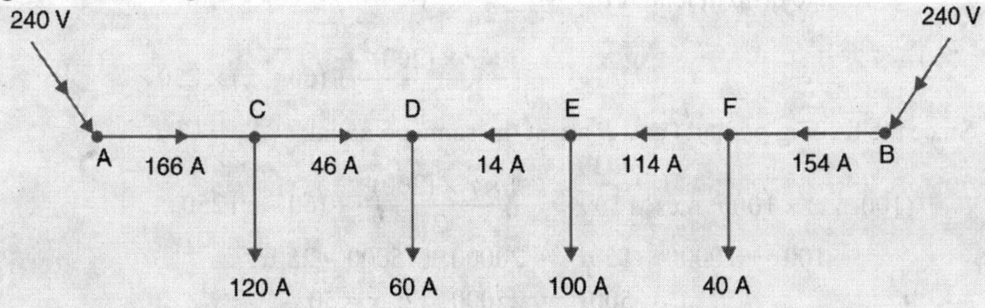

Fig. 13.32

(*ii*) The feeding point A will supply 166 A due to concentrated loading *plus* 0·5 × 400 = 200 A due to uniform loading.

∴ Current supplied by A, I_A = 166 + 200 = **366 A**

The feeding point B will supply a current of 154 A due to concentrated loading *plus* 0·5 × 600 = 300 A due to uniform loading.

∴ Current supplied by B, I_B = 154 + 300 = **454 A**

(*iii*) As stated above, D is the point of minimum potential.

∴ Minimum potential, V_D = V_A – Drop in AD due to conc. loading – Drop in AD due to uniform loading

Now, Drop in AD due to conc. loading $= I_{AC}R_{AC} + I_{CD}R_{CD}$

$= 166 \times 10^{-4} \times 200 + 46 \times 10^{-4} \times 200$

$= 3{\cdot}32 + 0{\cdot}92 = 4{\cdot}24$ V

Drop in AD due to uniform loading $= \dfrac{i\,r^*\,l^2}{2} = \dfrac{0{\cdot}5 \times 10^{-4} \times (400)^2}{2} = 4\ V$

∴ $V_D = 240 - 4{\cdot}24 - 4$ = **231·76 V**

Example 13.21. *A d.c. 2-wire distributor AB is 500m long and is fed at both ends at 240 V. The distributor is loaded as shown in Fig 13.33. The resistance of the distributor (go and return) is 0·001Ω per metre. Calculate (i) the point of minimum voltage and (ii) the value of this voltage.*

Solution. Let D be the point of **minimum potential and let x be the current flowing in section CD as shown in Fig 13.33. Then current supplied by end B will be $(60 - x)$.

Fig. 13.33

(*i*) If r is the resistance of the distributor (go and return) per metre length, then,

Voltage drop in length AD $= I_{AC}R_{AC} + I_{CD}R_{CD}$

$= (100 + x) \times 100\,r + x \times 150\,r$

Voltage drop in length BD $= \dfrac{i\,r\,l^2}{2} + (60 - x) \times 250\,r$

$= \dfrac{1 \times r \times (200)^2}{2} + (60 - x) \times 250\,r$

Since the feeding points A and B are at the same potential,

∴ $(100 + x) \times 100\,r + x \times 150\,r = \dfrac{1 \times r \times (200)^2}{2} + (60 - x)\,250\,r$

or $100x + 10000 + 150x = 20000 + 15000 - 250x$

or $500x = 25000$ ∴ $x = 50$ A

* Drop due to uniform loading can be determined by imagining that the distributor is cut into two at point D so that AD can be thought as a distributor fed at one end and loaded uniformly.

** You may carry out the calculation by assuming C to be point of minimum potential. The answer will be unaffected.

The actual directions of currents in the various sections of the distributor are shown in Fig. 13.34. Note that currents supplied by *A* and *B* meet at *D*. *Hence point D is the point of minimum potential.*

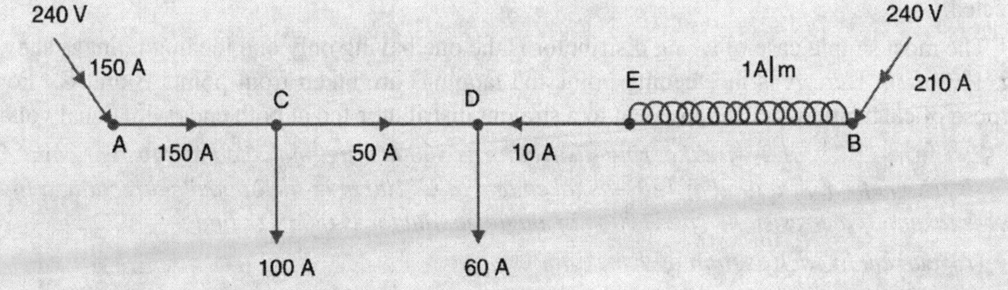

Fig. 13.34

(*ii*) Total current $= 160 + 1 \times 200 = 360$ A

Current supplied by $A, I_A = 100 + x = 100 + 50 = 150$ A

Current supplied by $B, I_B = 360 - 150 = 210$ A

Minimum potential, $V_D = V_A - I_{AC} R_{AC} - I_{CD} R_{CD}$

$= 240 - 150 \times (100 \times 0{\cdot}001) - 50 \times (150 \times 0{\cdot}001)$

$= 240 - 15 - 7{\cdot}5 = \mathbf{217{\cdot}5\ V}$

TUTORIAL PROBLEMS

1. A 2-wire d.c. distributor *AB*, 1000 m long has a total resistance of 0·1 Ω. The ends *A* and *B* are fed at 240 V. The distributor is uniformly loaded at 0.5 A/metre length and has concentrated loads of 120 A, 60 A, 100 A and 40 A at points distant 200, 400, 700 and 900 m respectively from end A. Calculate:
   (*i*) the point of minimum potential
   (*ii*) value of minimum potential
   (*iii*) current fed at both ends **[(*i*) 400 m from A (*ii*) 231·76 V (*iii*) $I_A = 366$ A ; $I_B = 454$ A]**
2. A 2-wire d.c. distributor *AB* is 300 metres long. The end *A* is fed at 205 V and end B at 200 V. The distributor is uniformly loaded at 0·15 A/metre length and has concentrated loads of 50 A, 60 A and 40 A at points distant 75, 175, 225 m respectively from the end *A*. The resistance of each conductor is 0·15 Ω per kilometre. Calculate :
   (*i*) the point of minimum potential
   (*ii*) currents fed at ends *A* and *B* **[(*i*) 175 m from A (*ii*) $I_A = 150$ A ; $I_B = 45$ A]**
3. A d.c. 2-wire distributor *AB* is 450 m long and is fed at both ends at 250 V. The distributor is loaded as shown in Fig. 13.35. The resistance of each conductor is 0·05 Ω per km. Find the point of minimum potential and its potential. **[261·74 m from A ; 247·35 V]**

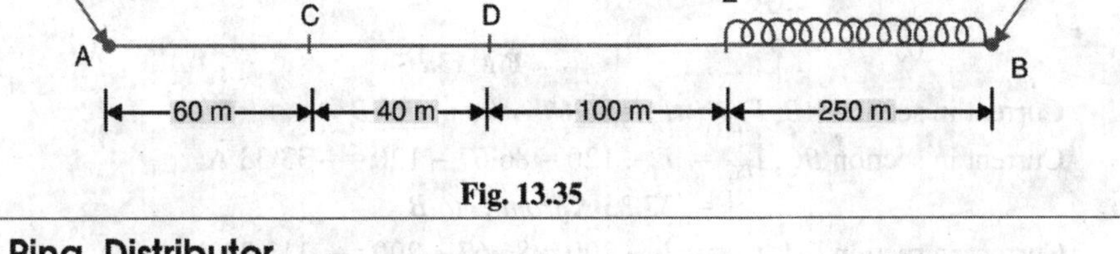

Fig. 13.35

13.8 Ring Distributor

A distributor arranged to form a closed loop and fed at one or more points is called a *ring distributor*. Such a distributor starts from one point, makes a loop through the area to be served, and returns to the original point. For the purpose of calculating voltage distribution, the distributor can be considered as consisting of a series of open distributors fed at both ends. The principal advantage of ring distributor is that by proper choice in the number of feeding points, great economy in copper can be affected.

The most simple case of a ring distributor is the one having only one feeding point as shown in Fig. 13.36(*ii*). Here A is the feeding point and tappings are taken from points B and C. For the purpose of calculations, it is equivalent to a straight distributor fed at both ends with equal voltages.

Example 13.22. *A 2-wire d.c. ring distributor is 300 m long and is fed at 240 V at point A. At point B, 150 m from A, a load of 120 A is taken and at C, 100 m in the opposite direction, a load of 80 A is taken. If the resistance per 100 m of single conductor is 0·03 Ω, find :*

(*i*) *current in each section of distributor*

(*ii*) *voltage at points B and C*

Solution.

Resistance per 100 m of distributor

$$= 2 \times 0{\cdot}03 = 0{\cdot}06\ \Omega$$

Resistance of section *AB*, $R_{AB} = 0{\cdot}06 \times 150/100 = 0{\cdot}09\ \Omega$

Resistance of section *BC*, $R_{BC} = 0{\cdot}06 \times 50/100 = 0{\cdot}03\ \Omega$

Resistance of section *CA*, $R_{CA} = 0{\cdot}06 \times 100/100 = 0{\cdot}06\ \Omega$

(*i*) Let us suppose that a current I_A flows in section *AB* of the distributor. Then currents in sections *BC* and *CA* will be $(I_A - 120)$ and $(I_A - 200)$ respectively as shown in Fig. 13.36 (*i*).

According to Kirchhoff's voltage law, the voltage drop in the closed loop *ABCA* is zero *i.e.*

$$I_{AB}\, R_{AB} + I_{BC}\, R_{BC} + I_{CA}\, R_{CA} = 0$$

or $$0{\cdot}09\, I_A + 0{\cdot}03\,(I_A - 120) + 0{\cdot}06\,(I_A - 200) = 0$$

or $$0{\cdot}18\, I_A = 15{\cdot}6$$

∴ $$I_A = 15{\cdot}6/0{\cdot}18 = 86{\cdot}67\ \text{A}$$

The actual distribution of currents is as shown in Fig. 13.36 (*ii*) from where it is seen that *B* is the point of minimum potential.

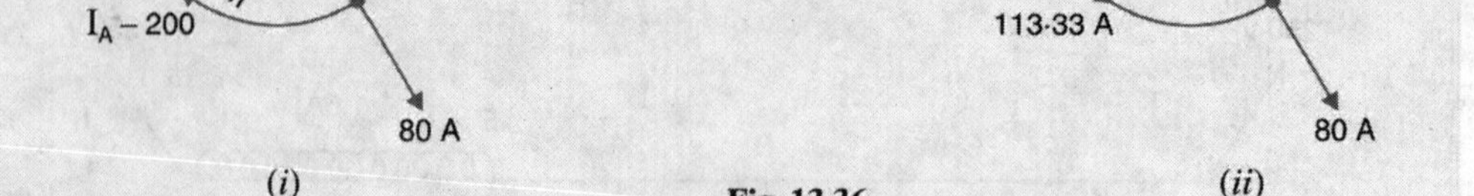

Fig. 13.36

Current in section *AB*, $I_{AB} = I_A =$ **86·67 A** *from A to B*

Current in section *BC*, $I_{BC} = I_A - 120 = 86{\cdot}67 - 120 = -\,33{\cdot}33$ A

$=$ **33.33 A** *from C to B*

Current in section *CA*, $I_{CA} = I_A - 200 = 86{\cdot}67 - 200 = -\,113{\cdot}33$ A

$=$ **113 ·33 A** *from A to C*

(*ii*) Voltage at point *B*, $V_B = V_A - I_{AB}\, R_{AB} = 240 - 86{\cdot}67 \times 0{\cdot}09 =$ **232·2 V**

Voltage at point C, $V_C = V_B + I_{BC} R_{BC}$

$= 232{\cdot}2 + 33{\cdot}33 \times 0{\cdot}03 = \mathbf{233{\cdot}2\ V}$

Example 13.23. *A 2-wire d.c. distributor ABCDEA in the form of a ring main is fed at point A at 220 V and is loaded as under :*

10A at B ; 20A at C ; 30A at D and 10 A at E.

The resistances of various sections (go and return) are : AB = 0·1 Ω ; BC = 0·05 Ω ; CD = 0·01 Ω ; DE = 0·025 Ω and EA = 0·075 Ω. Determine :

(i) the point of minimum potential

(ii) current in each section of distributor

Solution. Fig. 13.37 (*i*) shows the ring main distributor. Let us suppose that current I flows in section AB of the distributor. Then currents in the various sections of the distributor are as shown in Fig. 13.37 (*i*).

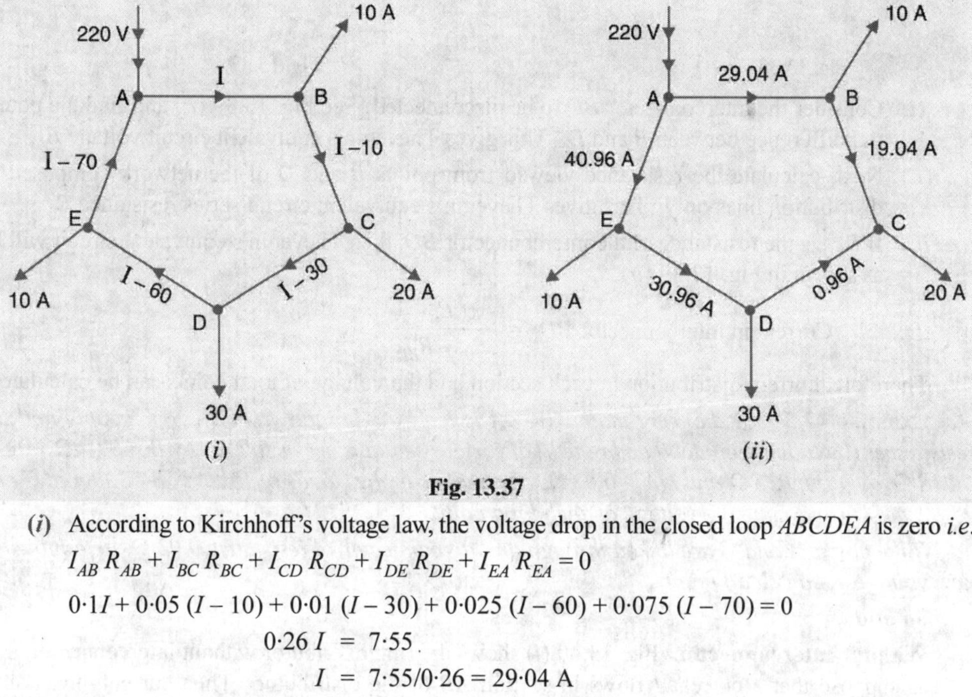

Fig. 13.37

(*i*) According to Kirchhoff's voltage law, the voltage drop in the closed loop $ABCDEA$ is zero *i.e.*

$$I_{AB} R_{AB} + I_{BC} R_{BC} + I_{CD} R_{CD} + I_{DE} R_{DE} + I_{EA} R_{EA} = 0$$

or $\quad 0{\cdot}1I + 0{\cdot}05\,(I - 10) + 0{\cdot}01\,(I - 30) + 0{\cdot}025\,(I - 60) + 0{\cdot}075\,(I - 70) = 0$

or $\quad 0{\cdot}26\,I = 7{\cdot}55$

$\therefore \quad I = 7{\cdot}55/0{\cdot}26 = 29{\cdot}04$ A

The actual distribution of currents is as shown in Fig. 13.37 (*ii*) from where it is clear that C is the point of minimum potential.

$\therefore$ C is the point of minimum potential.

(*ii*) Current in section AB = I = **29·04 A** *from A to B*

Current in section BC = $I - 10 = 29{\cdot}04 - 10$ = **19·04 A** *from B to C*

Current in section CD = $I - 30 = 29{\cdot}04 - 30 = -0{\cdot}96$ A = **0·96 A** *from D to C*

Current in section DE = $I - 60 = 29{\cdot}04 - 60 = -30{\cdot}96$ A = **30·96 A** *from E to D*

Current in section EA = $I - 70 = 29{\cdot}04 - 70 = -40{\cdot}96$ A = **40·96 A** *from A to E*

13.9 Ring Main Distributor with Interconnector

Sometimes a ring distributor has to serve a large area. In such a case, voltage drops in the various sections of the distributor may become excessive. In order to reduce voltage drops in various sections, distant points of the distributor are joined through a conductor called *interconnector*. Fig.

13.38 shows the ring distributor *ABCDEA*. The points *B* and *D* of the ring distributor are joined through an interconnector *BD*. There are several methods for solving such a network. However, the solution of such a network can be readily obtained by applying Thevenin's theorem. The steps of procedure are :

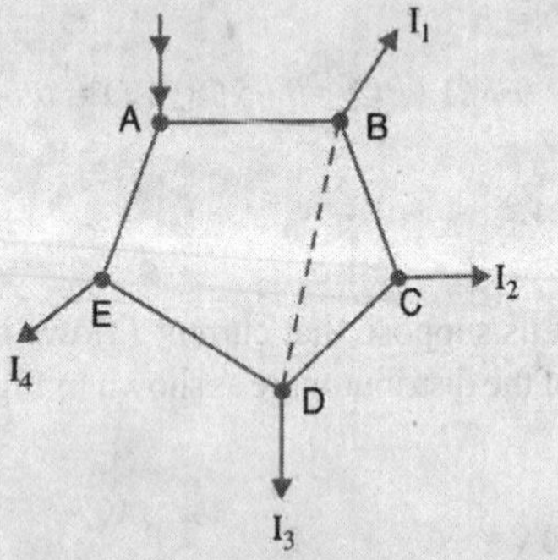

Fig. 13.38

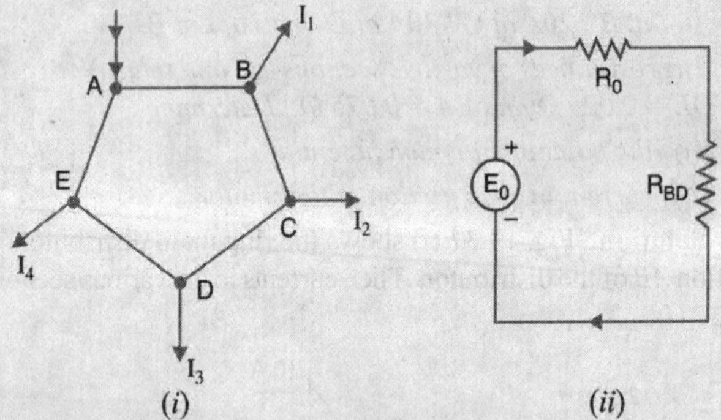

Fig. 13.39

(*i*) Consider the interconnector *BD* to be disconnected [See Fig. 13.39 (*i*)] and find the potential difference between *B* and *D*. This gives Thevenin's equivalent circuit voltage E_0.

(*ii*) Next, calculate the resistance viewed from points *B* and *D* of the network composed of distribution lines only. This gives Thevenin's equivalent circuit series resistance R_0.

(*iii*) If R_{BD} is the resistance of the interconnector *BD*, then Thevenin's equivalent circuit will be as shown in Fig. 13.39 (*ii*).

$$\therefore \quad \text{Current in interconnector } BD = \frac{E_0}{R_0 + R_{BD}}$$

Therefore, current distribution in each section and the voltage of load points can be calculated.

Example 13.24. *A d.c. ring main ABCDA is fed from point A from a 250 V supply and the resistances (including both lead and return) of various sections are as follows : AB = 0·02 Ω ; BC = 0·018 Ω ; CD = 0·025 Ω and DA = 0·02 Ω. The main supplies loads of 150 A at B ; 300 A at C and 250 A at D. Determine the voltage at each load point.*

If the points A and C are linked through an interconnector of resistance 0·02 Ω, determine the new voltage at each load point.

Solution.

Without Interconnector. Fig. 13.40 (*i*) shows the ring distributor without interconnector. Let us suppose that a current *I* flows in section *AB* of the distributor. Then currents in various sections of the distributor will be as shown in Fig. 13.40 (*i*).

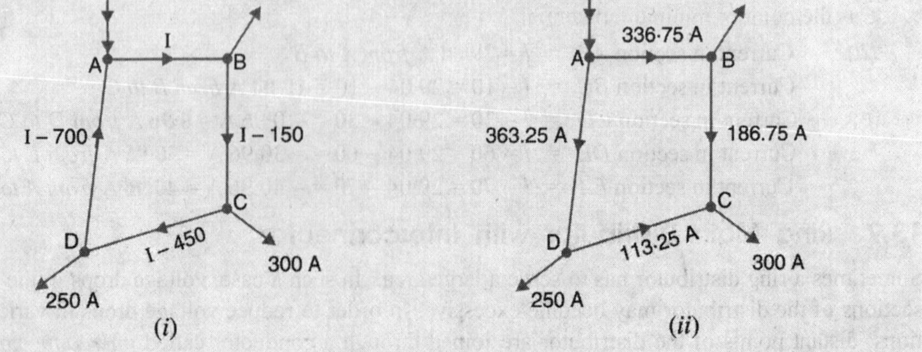

Fig. 13.40

According to Kirchhoff's voltage law, the voltage drop in the closed loop *ABCDA* is zero *i.e.*

$$I_{AB}R_{AB} + I_{BC}R_{BC} + I_{CD}R_{CD} + I_{DA}R_{DA} = 0$$

or $0{\cdot}02I + 0{\cdot}018\,(I - 150) + 0{\cdot}025\,(I - 450) + 0{\cdot}02\,(I - 700) = 0$

or $0{\cdot}083\,I = 27{\cdot}95$

$\therefore$ $I = 27{\cdot}95/0{\cdot}083 = 336{\cdot}75$ A

The actual distribution of currents is as shown in Fig. 13.40 (*ii*).

Voltage drop in *AB* = $336{\cdot}75 \times 0{\cdot}02 = 6{\cdot}735$ V

Voltage drop in *BC* = $186{\cdot}75 \times 0{\cdot}018 = 3{\cdot}361$ V

Voltage drop in *CD* = $113{\cdot}25 \times 0{\cdot}025 = 2{\cdot}831$ V

Voltage drop in *DA* = $363{\cdot}25 \times 0{\cdot}02 = 7{\cdot}265$ V

$\therefore$ Voltage at point *B* = $250 - 6{\cdot}735 =$ **243·265 V**

Voltage at point *C* = $243{\cdot}265 - 3{\cdot}361 =$ **239·904 V**

Voltage at point *D* = $239{\cdot}904 + 2{\cdot}831 =$ **242·735 V**

With Interconnector. Fig. 13.41 (*i*) shows the ring distributor with interconnector *AC*. The current in the interconnector can be found by applying Thevenin's theorem.

E_0 = Voltage between points *A* and *C*

= $250 - 239{\cdot}904 = 10{\cdot}096$ V

R_0 = Resistance viewed from points *A* and *C*

$$= \frac{(0\cdot02 + 0\cdot018)\,(0\cdot02 + 0\cdot025)}{(0\cdot02 + 0\cdot018) + (0\cdot02 + 0\cdot025)} = 0\cdot02\ \Omega$$

R_{AC} = Resistance of interconnector = $0{\cdot}02\ \Omega$

Thevenin's equivalent circuit is shown in Fig. 13.41 (*ii*). Current in interconnector *AC*

$$= \frac{E_0}{R_0 + R_{AC}} = \frac{10\cdot096}{0\cdot02 + 0\cdot02} = 252\cdot4\ A\ \textit{from A to C}$$

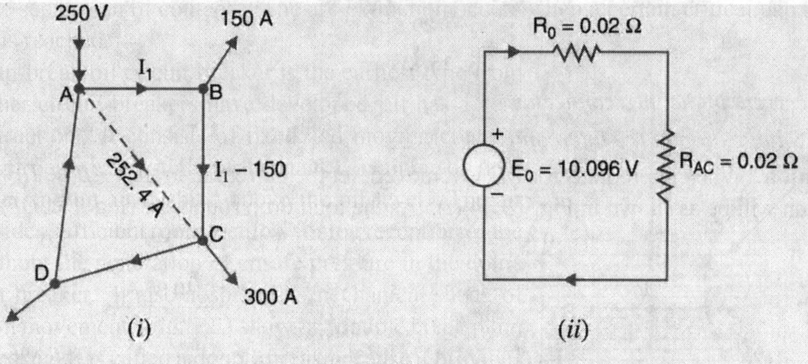

Fig. 13.41

Let us suppose that current in section *AB* is I_1. Then current in section *BC* will be $I_1 - 150$. As the voltage drop round the closed mesh *ABCA* is zero,

$\therefore$ $0{\cdot}02\,I_1 + 0{\cdot}018\,(I_1 - 150) - 0{\cdot}02 \times 252{\cdot}4 = 0$

or $0{\cdot}038\,I_1 = 7{\cdot}748$

$\therefore$ $I_1 = 7{\cdot}748/0{\cdot}038 = 203{\cdot}15$ A

The actual distribution of currents in the ring distributor with interconnector will be as shown in Fig. 13.42.

Drop in *AB* = $203{\cdot}15 \times 0{\cdot}02 = 4{\cdot}063$ V

Drop in BC = $53{\cdot}15 \times 0{\cdot}018$

= $0{\cdot}960$ V

Drop in AD = $244{\cdot}45 \times 0{\cdot}02 = 4{\cdot}9$ V

∴ Potential of B = $250 - 4{\cdot}063$

= **245·93 V**

Potential of C = $245{\cdot}93 - 0{\cdot}96$

= **244·97 V**

Potential of D = $250 - 4{\cdot}9 =$ **245·1 V**

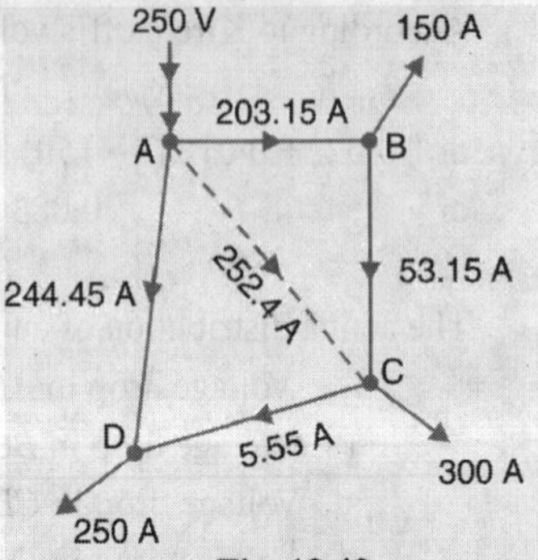

Fig. 13.42

It may be seen that with the use of interconnector, the voltage drops in the various sections of the distributor are reduced.

Example 13.25. *Fig. 13.43 shows a ring distributor with interconnector BD. The supply is given at point A. The resistances of go and return conductors of various sections are indicated in the figure. Calculate :*

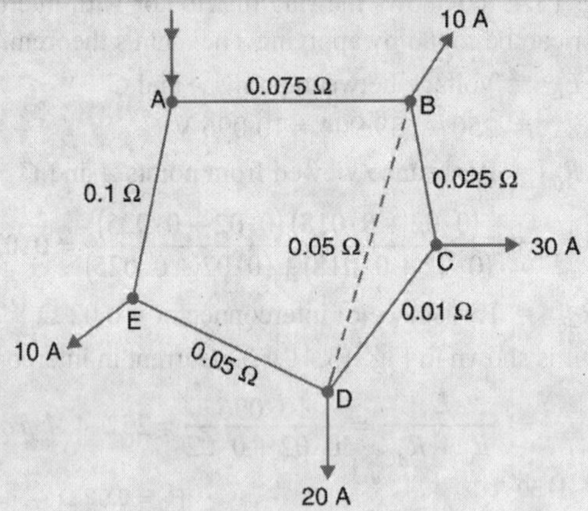

Fig. 13.43

(*i*) *current in the interconnector*

(*ii*) *voltage drop in the interconnector*

Solution. When interconnector BD is removed, let the current in branch AB be I. Then current distribution will be as shown in Fig. 13.44 (*i*). As the total drop round the ring $ABCDEA$ is zero,

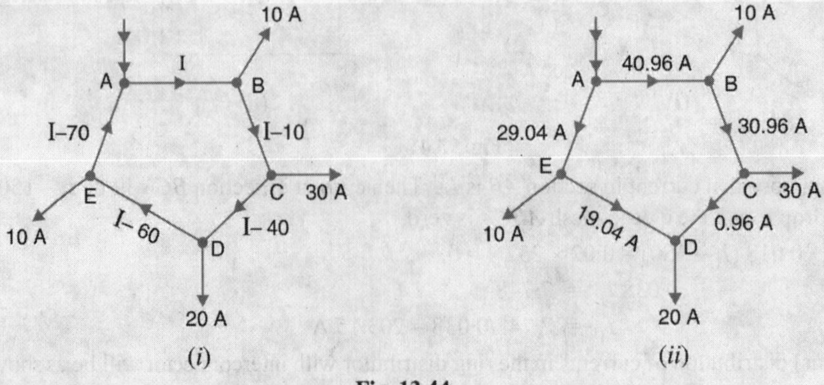

Fig. 13.44

∴ $0{\cdot}075\, I + 0{\cdot}025\,(I - 10) + 0{\cdot}01\,(I - 40) + 0{\cdot}05\,(I - 60) + 0{\cdot}1\,(I - 70) = 0$

or $\quad 0{\cdot}26\, I = 10{\cdot}65$

$$\therefore \quad I = \frac{10\cdot 65}{0\cdot 26} = 40\cdot 96 \text{ A}$$

The actual distribution of currents will be as shown in Fig. 13.44 (*ii*).

$$\text{Voltage drop along } BCD = 30{\cdot}96 \times 0{\cdot}025 + 0{\cdot}96 \times 0{\cdot}01$$
$$= 0{\cdot}774 + 0{\cdot}0096 = 0{\cdot}7836 \text{ V}$$

This is equal to Thevenin's open circuited voltage E_0 *i.e.*

$$E_0 = 0{\cdot}7836 \text{ V}$$

$$R_0 = \text{Resistance viewed from } B \text{ and } D$$
$$= \frac{(0\cdot 075 + 0\cdot 1 + 0\cdot 05)\,(0\cdot 025 + 0\cdot 01)}{(0\cdot 075 + 0\cdot 1 + 0\cdot 05) + (0\cdot 025 + 0\cdot 01)}$$
$$= \frac{(0\cdot 225)\,(0\cdot 035)}{0\cdot 225 + 0.035} = 0\cdot 03\ \Omega$$

(*i*) Current in interconnector *BD* is

$$I_{BD} = \frac{E_0}{R_0 + R_{BD}} = \frac{0\cdot 7836}{0\cdot 03 + 0.05} = \mathbf{9.\ 8\ A}$$

(*ii*) Voltage drop along interconnector *BD*

$$= I_{BD}\, R_{BD} = 9{\cdot}8 \times 0{\cdot}05 = \mathbf{0{\cdot}49\ V}$$

TUTORIAL PROBLEMS

1. A 300 m ring distributor has loads as shown in Fig. 13.45 where distances are in metres. The resistance of each conductor is 0·2 W per 1000 metres and the loads are tapped off at points *B*, *C* and *D* as shown. If the distributor is fed at A at 240 V, find voltages at *B*, *C* and *D*.
[V_B = 236·9 V ; V_C = 235·97 V ; V_D = 237·45 V]

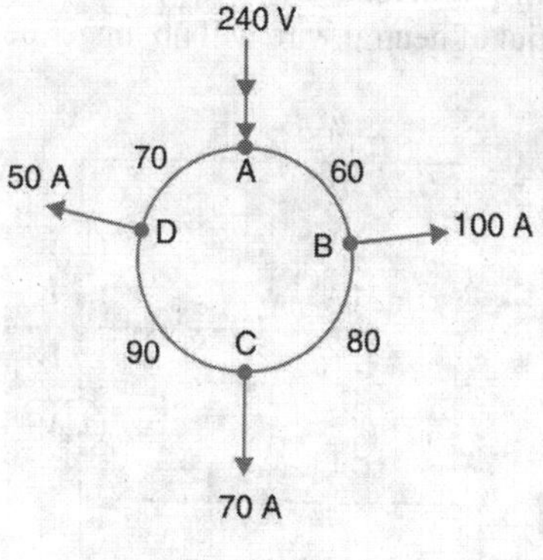

Fig. 13.45

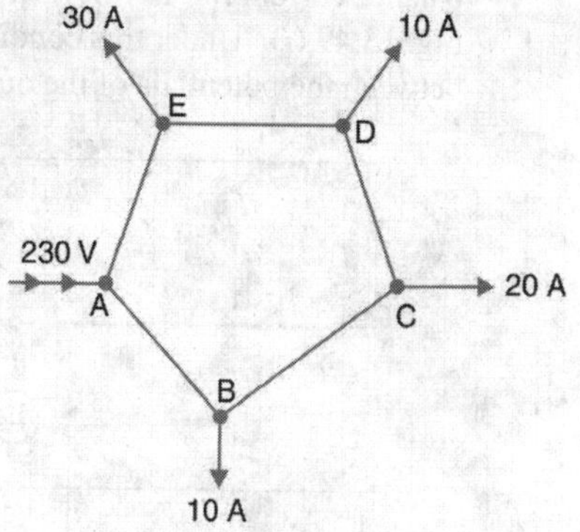

Fig. 13.46

2. A d.c. 2-wire ring main *ABCDEA* is fed from 230 V supply as shown in Fig. 13.46. The resistance of each section (go and return) *AB*, *BC*, *CD*, *DE* and *EA* is 0·1 W. The loads are tapped off as shown. Find the voltage at each load point.
[V_B = 227 V ; V_C = 225 V ; V_D = 225 V ; V_E = 226 V]

3. In the d.c. network shown in Fig.13.47, A is the feeding point and is maintained at 250 V. The resistances

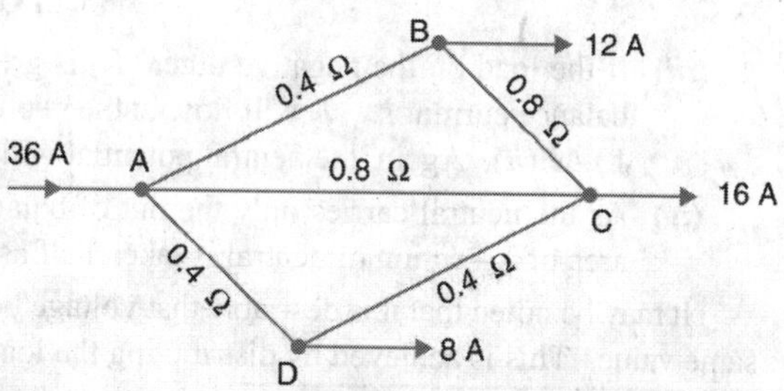

Fig. 13.47

of the various branches (go and return) are indicated in the figure. Determine the current in each branch. [$AB = 144A$; $BC = 2A$; $DC = 5A$; $AD = 13A$]

13.10 3-Wire D.C. System

The great disadvantage of direct current for general power purposes lies in the fact that its voltage cannot readily be changed, except by the use of rotating machinery, which in most cases is too expensive. The problem can be solved to a limited extent by the use of 3-wire d.c. system which makes available two voltages *viz.* V volts between any outer and neutral and 2V volts between the outers. Motor loads requiring high voltage are connected between the outers whereas lighting and heating loads requiring less voltage are connected between any one outer and the neutral. Due to the availability of two voltages, 3-wire system is preferred over 2-wire system for d.c. distribution.

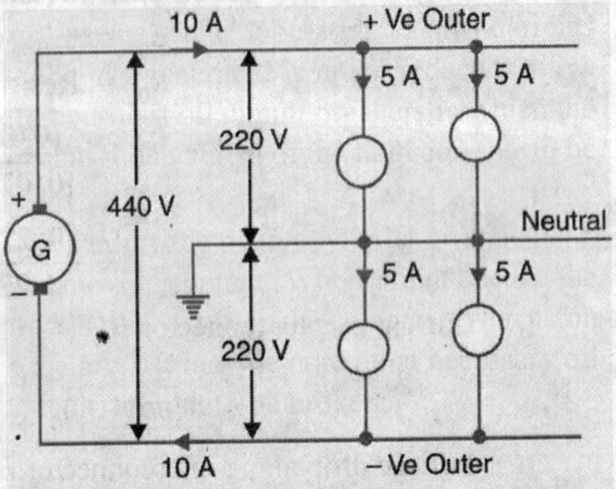

Fig. 13.48

Fig. 13.48 shows the general principles of a 3-wire d.c. system. It consists of two outers and a middle or neutral wire which is earthed at the generator end. The potential of the neutral wire is *half-way between the potentials of the outers. Thus, if p.d. between the outers is 440 V, then positive outer is at 220 V *above* the neutral and negative outer is 220 V *below* the neutral. The current in the neutral wire will depend upon the loads applied to the two sides.

(*i*) If the loads applied on both sides of the neutral are equal (*i.e.* balanced) as shown in Fig 13.48, the current in the neutral wire will be zero. Under these conditions, the potential of the neutral will be exactly half-way between the potential difference of the outers.

(*ii*) If the load on the positive outer (I_1) is greater than on the negative outer (I_2), then out of balance current $I_1 - I_2$ will flow in the neutral wire from load end to supply end as shown in Fig. 13.49 (*i*). Under this condition, the potential of neutral wire will no longer be midway between the potentials of the outers.

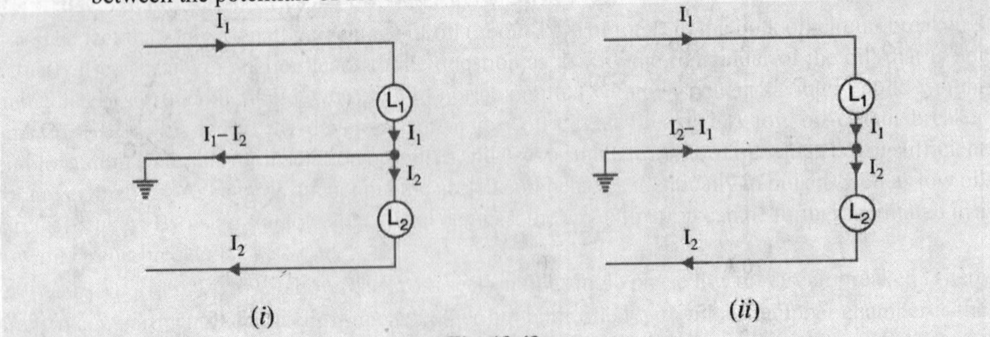

Fig. 13.49

(*iii*) If the load on the negative outer (I_2) is greater than on the positive outer (I_1), then out of balance current $I_2 - I_1$ will flow in the neutral from supply end to load end as shown in Fig. 13.49 (*ii*). Again, the neutral potential will not remain half-way between that of the outers.

(*iv*) As the neutral carries only the out of balance current which is generally small, therefore, area of X-section of neutral is taken half as compared to either of the outers.

It may be noted that it is desirable that voltage between any outer and the neutral should have the same value. This is achieved by distributing the loads equally on both sides of the neutral.

* For balanced loads *i.e.* equal loads on both sides of the neutral wire.

13.11 Current Distribution in 3-Wire D.C. System

Fig. 13.50 shows a 3-wire 500/250 V d.c. distributor. Typical values of loads have been assumed to make the treatment more illustrative. The motor requiring 500 V is connected across the outers and takes a current of 75 A. Other loads requiring lower voltage of 250 V are connected on both sides of the neutral.

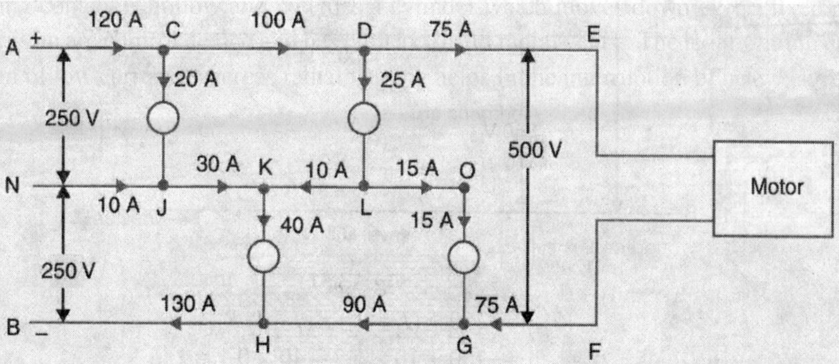

Fig. 13.50

Applying Kirchhoff's current law, it is clear that a current of 120 A enters the positive outer while 130 A comes out of the negative outer. Therefore, 130 – 120 = 10 A must flow *in* the neutral at point *N*. Once the magnitude and direction of current in the section *NJ* is known, the directions and magnitudes of currents in the other sections of the neutral can be easily determined. For instance, the currents meeting at point *K* must add up to 40 A to supply the load *KH*. As seen in Fig. 13.50, 20A of *CJ* and 10A of *NJ* flow towards *K*, the remaining 10A coming from point *L*. The current of 25A of load *DL* is divided into two parts ; 10A flowing along section *LK* and the remaining 15 A along the section *LO* to supply the load *OG*.

Load-point voltages. Knowing the currents in the various sections of the outers and neutral, the voltage at any load point can be determined provided resistances are known. As an illustration, let us calculate the voltage across load *CJ* of Fig.13.50. Applying Kirchhoff's voltage law to the loop *ACJNA*, we have,

[Algebraic sum of voltage drops] + [Algebraic sum of e.m.f.s.] = 0

or *[– drop in *AC* – voltage across *CJ* + drop in *NJ*] + [250] = 0

or Voltage across *CJ* = 250 – drop in *AC* + drop in *NJ*

Example 13.26. *A load supplied on 3-wire d.c. system takes a current of 50 A on the +ve side and 40 A on the negative side. The resistance of each outer wire is 0·1 Ω and the cross-section of middle wire is one-half of that of outer. If the system is supplied at 500/250 V, find the voltage at the load end between each outer and middle wire.*

Solution. Fig. 13.51 shows the current loading. Obviously, current in the neutral wire is 50 – 40 = 10A. As the X-sectional area of neutral is half that of outer, therefore, its resistance = 2 × 0·1 = 0·2 Ω.

Voltage at the load end on the +ve side,

$$V_{EL} = 250 - I_1 R_{AE} - (I_1 - I_2) R_{NL}$$
$$= 250 - 50 \times 0{\cdot}1 - (10) \times 0{\cdot}2 = \mathbf{243\ V}$$

Voltage at the load end on the –ve side,

* Remember, rise in potential should be considered positive while fall in potential should be considered negative. In section *AC*, current flows from *A* to *C* and hence there is fall in potential. In section *JN*, obviously, there is rise in potential.

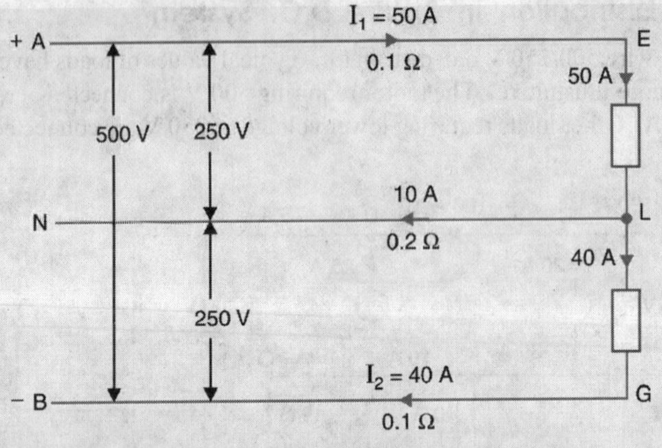

Fig. 13.51

$$V_{LG} = 250 + (I_1 - I_2) R_{NL} - I_2 R_{BG}$$
$$= 250 + 10 \times 0{\cdot}2 - 40 \times 0{\cdot}1 = \mathbf{248\ V}$$

Example 13.27. *A 3-wire d.c. distribution system supplies a load of 5 Ω resistance across the +ve outer and neutral and a load of 6 Ω resistance across –ve outer and neutral at the far end of the distributor. The resistance of each conductor is 0·1 Ω. If the voltage between any outer and neutral at the load end is to be kept at 240 V, find the feeding end voltages.*

Solution. Fig. 13.52 shows the 3-wire distribution system.

Current on +ve outer, $I_1 = 240/5 = 48\text{A}$

Current on –ve outer, $I_2 = 240/6 = 40\text{A}$

Current in neutral $= I_1 - I_2 = 48 - 40 = 8\text{A}$

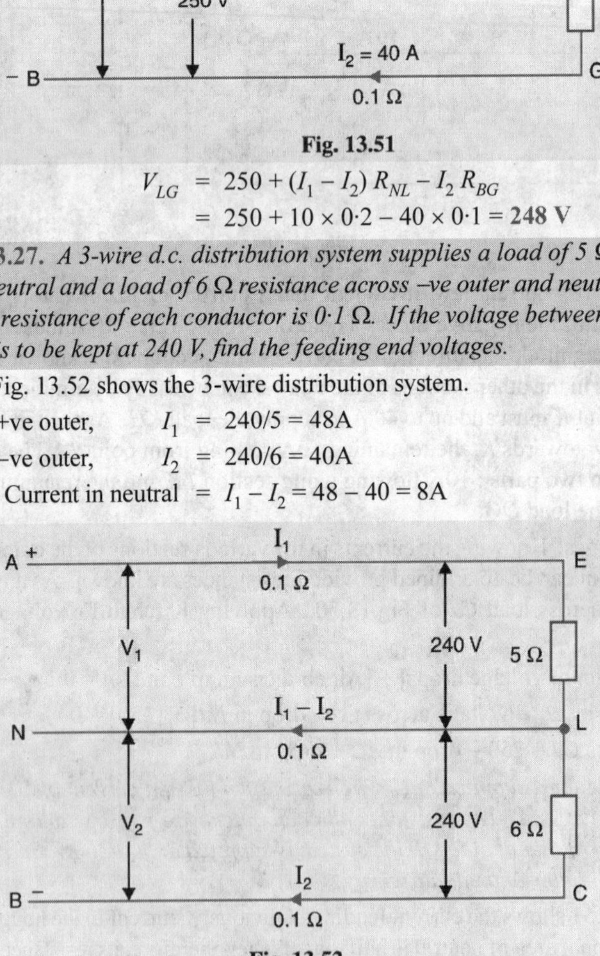

Fig. 13.52

Voltage between +ve outer and neutral at feeding end is

$$V_1 = V_{EL} + I_1 R_{AE} + (I_1 - I_2) R_{NL}$$
$$= 240 + 48 \times 0{\cdot}1 + 8 \times 0{\cdot}1 = \mathbf{245{\cdot}6\ V}$$

Voltage between –ve outer and neutral at feeding end is

$$V_2 = V_{LC} - (I_1 - I_2) R_{NL} + I_2 R_{BC}$$
$$= 240 - 8 \times 0{\cdot}1 + 40 \times 0{\cdot}1 = \mathbf{243{\cdot}2\ V}$$

Example 13.28. *A 3-wire 500/250 V d.c. system has a load of 35 kW between the positive lead and the middle wire and a load of 20 kW between the negative lead and the middle wire. If there is a break in the middle wire, calculate the voltage between the outers and the middle wire.*

Solution. Fig. 13.53 (*i*) shows the arrangement before the disconnection of the middle wire. The voltages on the two sides of the middle wire are equal *i.e.* 250 V. Let R_1 be the resistance of load on the +ve side and R_2 be the resistance of load on the negative side.

$$R_1 = \frac{(250)^2}{35 \times 10^3} = 1 \cdot 785\ \Omega\ ;\ R_2 = \frac{(250)^2}{20 \times 10^3} = 3 \cdot 125\ \Omega \qquad [\because R = V^2/P]$$

When there is a break in the middle wire, two resistances R_1 and R_2 are put in series across 500 V as shown in Fig. 13.53 (*ii*).

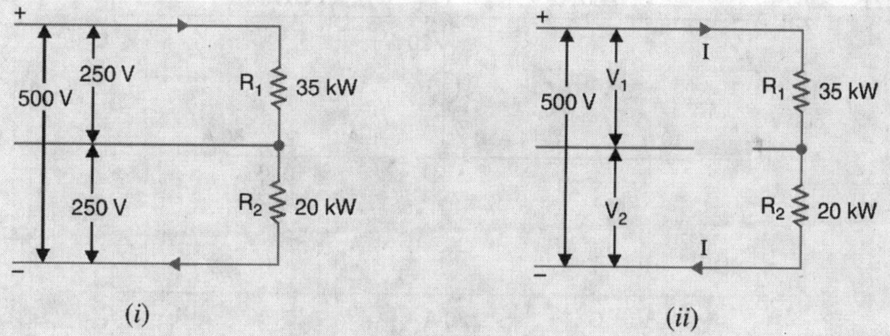

Fig. 13.53

$\therefore$ Circuit current, $$I = \frac{500}{R_1 + R_2} = \frac{500}{1 \cdot 785 + 3 \cdot 125} = 101 \cdot 83\ \text{A}$$

$\therefore$ Voltage across +ve outer and middle wire,

$$V_1 = I R_1 = 101 \cdot 83 \times 1 \cdot 785 = \mathbf{181 \cdot 8\ V}$$

Voltage across –ve outer and middle wire,

$$V_2 = I R_2 = 101 \cdot 83 \times 3 \cdot 125 = \mathbf{318 \cdot 2\ V}$$

Example 13.29. *A 3-wire, 500/250 V distributor is loaded as shown in Fig. 13.54. The resistance of each section is given in ohm. Find the voltage across each load point.*

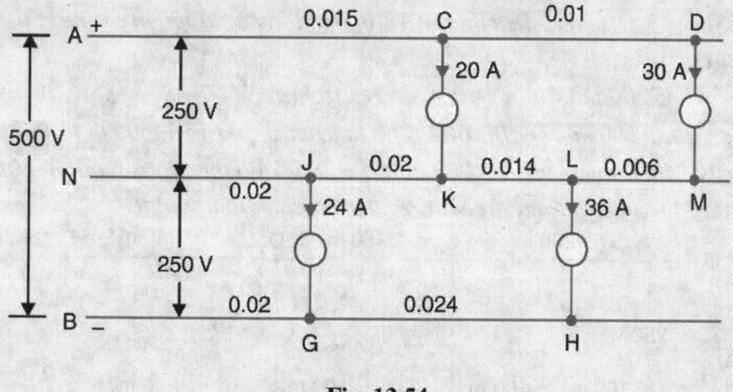

Fig. 13.54

Solution. From the current loading given in Fig. 13.54, we can find the magnitudes and directions of currents in the various sections by applying Kirchhoff's current law. Fig. 13.55 shows the magnitudes and directions of currents in the various sections.

The voltage drops in the various sections are worked out below.

Section	*Resistance* (Ω)	*Current* (*A*)	*Drop* (*V*)
AC	0·015	50	0·75
CD	0·01	30	0·3
ML	0·006	30	0·18
KL	0·014	6	0·084
KJ	0·02	14	0·28
NJ	0·02	10	0·2
HG	0·024	36	0·864
GB	0·02	60	1·2

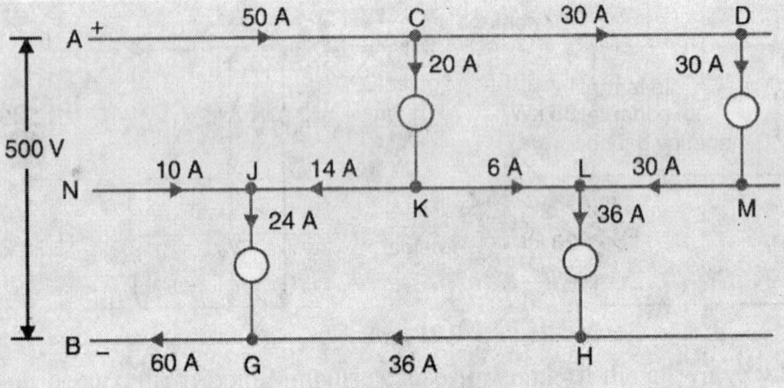

Fig. 13.55

Voltage across load CK = 250 – Drop in AC – Drop in KJ + Drop in NJ

= 250 – 0·75 – 0·28 + 0·2 = **249·17 V**

Voltage across load DM = 249·17 – Drop in CD – Drop in ML + Drop in KL

= 249·17 – 0·3 – 0·18 + 0·084 = **248·774 V**

Voltage across load JG = 250 – Drop in NJ – Drop in GB

= 250 – 0·2 – 1·2 = **248·6 V**

Voltage across load LH = 248·6 + Drop in KJ – Drop in KL – Drop in HG

= 248·6 + 0·28 – 0·084 – 0·864 = **247·932 V**

Example 13.30. *A 3-wire d.c. distributor AE 600 m long is supplied at end A at 500/250 V and is loaded as under :*

Positive side : 60A, 200 m from A ; 40 A, 360 m from A

Negative side : 20A, 100 m from B ; 60A, 260 m from B and 15A, 600 m from B

The resistance of each outer is 0·02 Ω per 100 metres and the cross-section of the neutral wire is the same as that of the outer. Find the voltage across each load point.

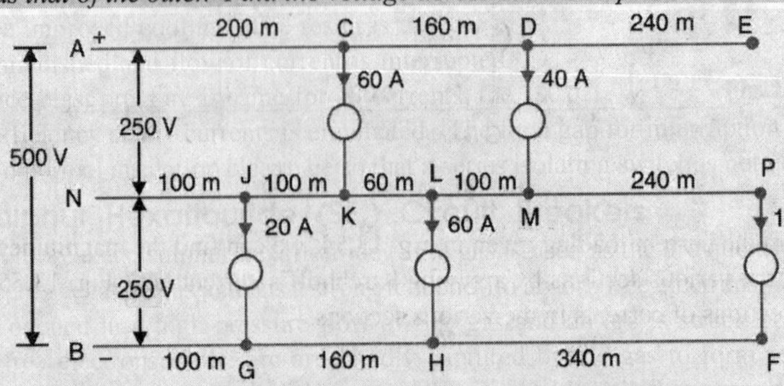

Fig. 13.56

Solution. From the current loading given in Fig. 13.56, we can find the magnitudes and directions of currents in the various sections by Kirchhoff's current law as shown in Fig. 13.57. As the neutral is of same cross-section, its resistance is 0·02 Ω per 100 metres. The voltage drops in the various sections are worked out below :

Section	*Resistance* (Ω)	*Current* (*A*)	*Drop* (*V*)
AC	0·02 × 200/100 = 0·04	100	4
CD	0·02 × 160/100 = 0·032	40	1·28
MP	0·02 × 240/100 = 0·048	15	0·72
ML	0·02 × 100/100 = 0·02	25	0·5
KL	0·02 × 60/100 = 0·012	35	0·42
KJ	0·02 × 100/100 = 0·02	25	0·5
JN	0·02 × 100/100 = 0·02	5	0·1
FH	0·02 × 340/100 = 0·068	15	1·02
HG	0·02 × 160/100 = 0·032	75	2·4
GB	0·02 × 100/100 = 0·02	95	1·9

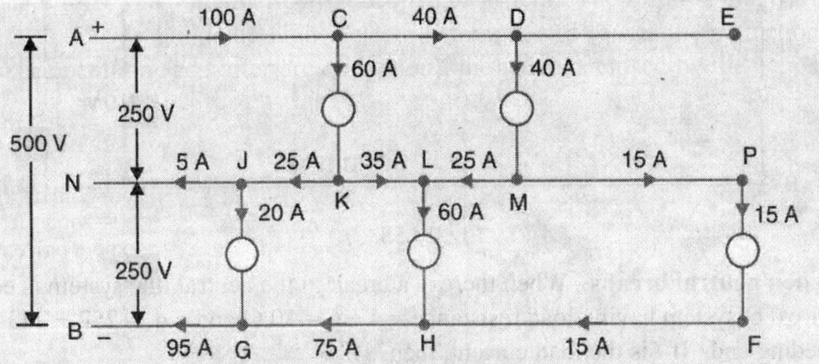

Fig. 13.57

Voltage across *CK* = 250 – Drop in *AC* – Drop in *KJ* – Drop in *JN*

= 250 – 4 – 0·5 – 0·1 = **245·4 V**

Voltage across *DM* = 245·4 – Drop in *CD* – Drop in *ML* + Drop in *KL*

= 245·4 – 1·28 – 0·5 + 0·42 = **244·04 V**

Voltage across *JG* = 250 + Drop in *JN* – Drop in *GB*

= 250 + 0·1 – 1·9 = **248·2 V**

Voltage across *LH* = 248·2 + Drop in *KJ* – Drop in *KL* – Drop in *HG*

= 248·2 + 0·5 – 0·42 – 2·4 = **245·88 V**

Voltage across *PF* = 245·88 + Drop in *ML* – Drop in *MP* – Drop in *FH*

= 245·88 + 0·5 – 0·72 – 1·02 = **244·64 V**

Example 13.31. *The 3-wire d.c. system supplies a load of 4 Ω resistance across +ve wire and the neutral wire and a load of 6 Ω resistance across –ve outer and the neutral at the far end of the distributor. The resistance of each conductor is 0.15 Ω and voltage across each outer and neutral is 240 V at the load end. Determine the load current and load voltages when there is a break in the (i) neutral wire (ii) positive outer (iii) negative outer. Assume that the load resistances and the feeding end voltages remain the same.*

Solution. Fig. 13.58 shows the conditions of the problem when the system is healthy. Let us find the feeding end voltages.

Current in the positive outer, $I_1 = 240/4 = 60$ A

Current in the negative outer, $I_2 = 240/6 = 40$ A

Current in the neutral wire $= I_1 - I_2 = 60 - 40 = 20$ A

Voltage between +ve outer and neutral at feeding point is

$$V_1 = V_{EL} + I_1 R_{AE} + (I_1 - I_2) R_{NL}$$
$$= 240 + 60 \times 0{\cdot}15 + 20 \times 0{\cdot}15 = 252 \text{ V}$$

Voltage between −ve outer and neutral at feeding point is

$$V_2 = V_{LC} - (I_1 - I_2) R_{NL} + I_2 R_{BC}$$
$$= 240 - 20 \times 0{\cdot}15 + 40 \times 0{\cdot}15 = 243 \text{ V}$$

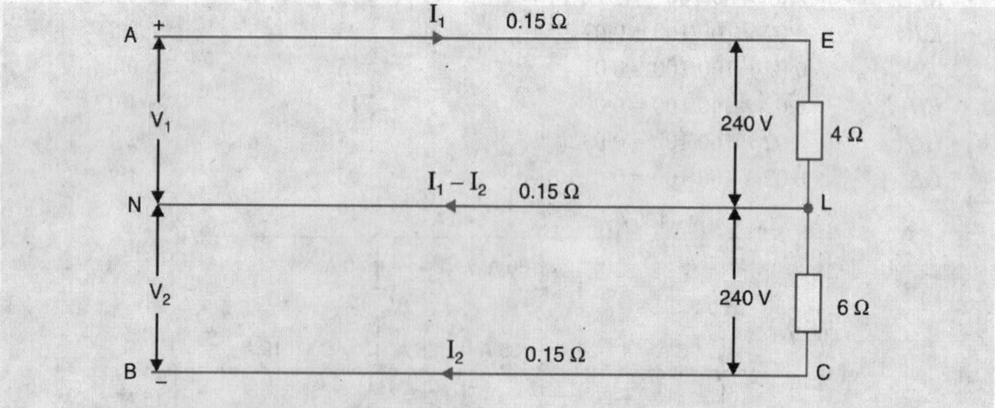

Fig. 13.58

(*i*) **When neutral breaks.** When there is a break in the neutral, the system is equivalent to 2-wire d.c. system having load resistance = 4 + 6 = 10 Ω and p.d. = 252 + 243 = 495 V at the feeding end. If I is the load current, then,

Total circuit resistance $= 10 + 0{\cdot}15 + 0{\cdot}15 = 10{\cdot}3\ \Omega$

∴ Load current, $I = 495/10{\cdot}3 =$ **48·06 A**

Voltage across 4 Ω resistance $= I \times 4 = 48{\cdot}06 \times 4 =$ **192·24 V**

Voltage across 6 Ω resistance $= I \times 6 = 48{\cdot}06 \times 6 =$ **288·36 V**

(*ii*) **When +ve outer breaks.** When there is a break in the +ve outer, there will be no current in 4 Ω load. The circuit is again 2-wire d.c. system but now load is 6 Ω and p.d. at the feeding point is 243 V.

Total circuit resistance $= 6 + 0{\cdot}15 + 0{\cdot}15 = 6{\cdot}3\ \Omega$

If I' is the load current, then,

$I' = 243/6{\cdot}3 =$ **38·57 A**

Voltage across 6 Ω $= I' \times 6 = 38{\cdot}57 \times 6 =$ **231·42 V**

(*iii*) **When −ve outer breaks.** When there is a break in the negative outer, there will be no current in 6 Ω load. The circuit is again 2-wire d.c. system but now load is 4 Ω and p.d. at the feeding point is 252 V.

Total circuit resistance $= 4 + 0{\cdot}15 + 0{\cdot}15 = 4{\cdot}3\ \Omega$

If I'' is the load current, then,

$I'' = 252/4{\cdot}3 =$ **58·6 A**

Voltage across 4 Ω $= I'' \times 4 = 58{\cdot}6 \times 4 =$ **234·42 V**

TUTORIAL PROBLEMS

1. A load supplied by a 3-wire d.c. distribution system takes a current of 600 A on the +ve side and 480 A on the –ve side. The resistance of the outer conductor is 0·015 Ω and that of the middle wire is 0·03 Ω. Determine the voltage at the load end between each outer and middle wire. Supply voltage is 500/250 V. **[+ve side : 237·4 V ; –ve side : 246·4 V]**
2. A 3-wire d.c. distributor, 250 m long, is supplied at end *P* at 500/250 V and is loaded as under :
   Positive side : 20 A, 150 m from *P* ; 30 A, 250 m from *P*
   Negative side : 24 A, 100 m from *P* ; 36 A, 220 m from *P*
   The resistance of each outer wire is 0·02 Ω per 100 m and the cross-section of the middle wire is one half that of the outer. Find the voltage across each load point.
   [Positive side : 248·62 V ; 247·83 V ; Negative side : 248·4 V ; 247·65 V]
3. The 3-wire d.c. system supplies a load of 4 Ω resistance across the +ve wire and the neutral and a load of 6 Ω resistance across –ve outer and the neutral at the far end of the distributor. The resistance of each conductor is 0·15 Ω and voltage across each outer and neutral is 240 V at the load end. Calculate the feeding end voltage. **[+ve side : 252 V ; –ve side : 243 V]**

13.12 Balancers in 3-Wire D.C. System

Although in a 3-wire d.c. system every effort is made to distribute the various loads equally on both sides of the neutral, yet it is difficult to achieve the exact balance. The result is that some current does flow in the neutral wire and consequently the voltages on the two sides of the neutral do not remain equal to each other. In order to maintain voltages on the two sides of the neutral equal to each other, a *balancer set* is used.

Circuit details. Fig. 13.59 shows the use of a balancer set in a 3-wire d.c. system. The balancer consists of two identical shunt wound machines *A* and *B* coupled mechanically and having their armature and field circuits connected in series across the outers. The neutral wire is connected to the junction of the armatures as shown. The circuit arrangement has two obvious advantages. Firstly, only one generator (*G*) is required which results in a great saving in cost. Secondly, the balancer set tends to equalise the voltages on the two sides of the neutral.

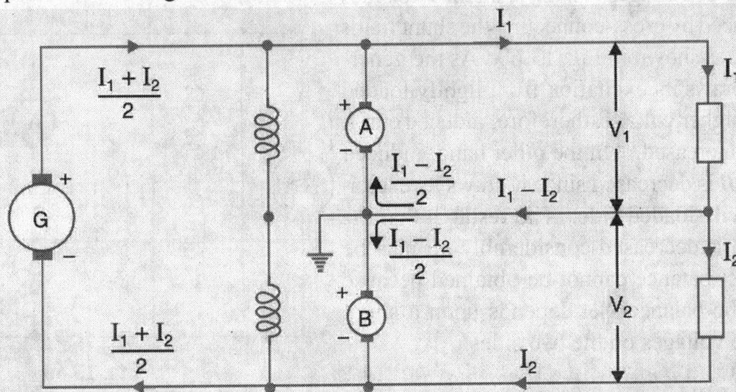

Fig. 13.59

Theory. Since the speeds and field currents of the two machines are equal, their back e.m.f.s have the same value. When the system is unloaded or when the loads on the two sides are the same (*i.e.* balanced), no current flows in the neutral wire. Hence, the two machines run as unloaded motors

When the load is unbalanced, the current supplied by the +ve outer will be different from that supplied by the negative outer. Suppose that load I_1 on the +ve outer is greater that the load I_2 on the –ve outer. Since the +ve side is more heavily loaded, p.d. on this side tends to fall below the e.m.f. of the balancer set. Therefore, machine A runs as a generator. On the other hand, p.d. on the lightly loaded –ve side rises above the e.m.f. of the balancer so that machine *B* runs as a motor. The result is that energy is transferred from lightly loaded side to the heavily loaded side, preventing the voltage across heavily loaded side form dropping very much below the normal value.

For the condition shown in Fig. 13.59, the machine *B* acts as a motor and machine *A* as a generator. The out of balance current $I_1 - I_2$ flows through the middle wire towards the balancers. Assuming the internal losses of the two machines to be negligible, then armature currents will be equal. Hence, one-half of the current in neutral *i.e.* $(I_1 - I_2)/2$ will flow through each machine as shown.

Let R_a = armature resistance of each machine

*E = induced e.m.f. of each machine

V_1 = terminal p.d. of machine *A* running as a generator

V_2 = terminal p.d. of machine *B* running as a motor

$$\therefore \quad V_1 = E - \left(\frac{I_1 - I_2}{2}\right) R_a$$

and

$$V_2 = E + \left(\frac{I_1 - I_2}{2}\right) R_a$$

Difference of voltages between two sides

$$= V_2 - V_1$$

$$= \left[E + \left(\frac{I_1 - I_2}{2}\right) R_a\right] - \left[E - \left(\frac{I_1 - I_2}{2}\right) R_a\right]$$

$$= (I_1 - I_2)\, R_a$$

It is clear that difference of voltages between the two sides of the system is proportional to

(*i*) the out-of-balance current $I_1 - I_2$

(*ii*) the armature resistance of balancer

Therefore, in order to keep the voltages on the two sides equal, R_a is kept small and loads are arranged on the two sides in such a way that out of balance current is as small as possible.

The difference of voltages $(V_2 - V_1)$ on the two sides can be further reduced by cross-connecting the shunt fields of the balancer set as shown in Fig. 13.60. As the generating machine *A* draws its excitation from lightly loaded side which is at a higher voltage, therefore, induced e.m.f. of the machine is increased. On the other hand, induced e.m.f. of machine *B* is decreased since it draws its excitation from the heavily loaded side. The result is that the difference $V_2 - V_1$ is decreased considerably. It may be noted that a perfect balance cannot be obtained because the operation of the balancer set depends upon a slight unbalancing of the voltages on the two sides.

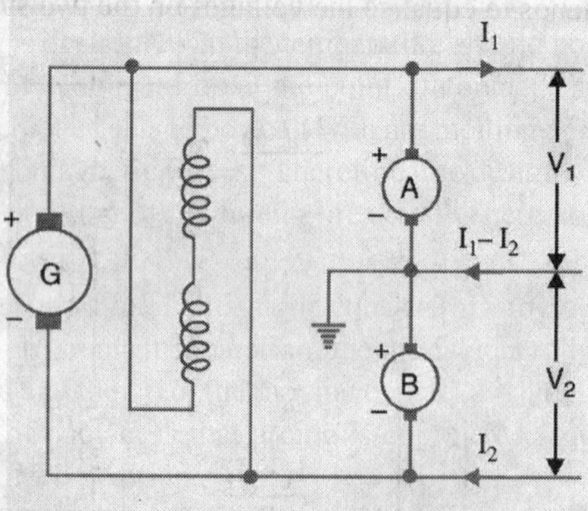

Fig. 13.60

Example 13.32. *A d.c. 3-wire system with 500 V between the outers supplies 1500 kW on the +ve outer and 2000 kW on the negative outer. If the losses in the machines are negligible, calculate:*

(*i*) *current in the neutral wire*

(*ii*) *total current supplied by main generator*

(*iii*) *current in each balancer armature*

Solution. The connections are shown in Fig. 13.61. As the negative side is more heavily loaded, therefore, machine *B* acts as a generator and machine *A* as a motor.

* Since both machines have the same excitation and run at the same speed, their induced e.m.f.s will be same.

Load current on +ve outer, $I_1 = 1500 \times 10^3/250 = 6000$ A

Load current on −ve outer, $I_2 = 2000 \times 10^3/250 = 8000$ A

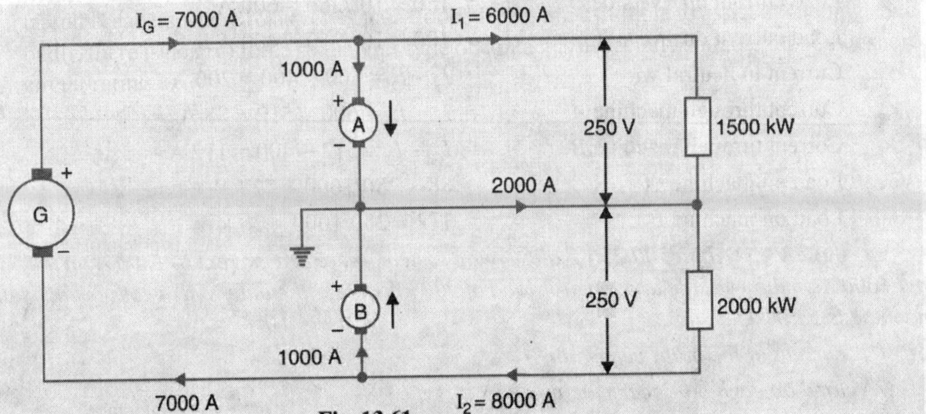

Fig. 13.61

(*i*) Current in the neutral $= I_2 - I_1 = 8000 - 6000 =$ **2000 A**

(*ii*) Total load on main generator $= 1500 + 2000 = 3500$ kW

∴ Current supplied by main generator, $I_G = 3500 \times 10^3/500 =$ **7000 A**

(*iii*) Current in machine $A = I_G - I_1 = 7000 - 6000 =$ **1000 A**

Current in machine $B = I_2 - I_G = 8000 - 7000 =$ **1000 A**

Example 13.33. *A d.c. 3-wire system with 500 V between outers has lighting loads of 150 kW on the positive side and 100 kW on the negative side. The loss in each balancer machine is 3 kW. Calculate :*

(*i*) *total load on the main generator*

(*ii*) *kW loading of each balancer machine*

Solution. The connections are shown in Fig. 13.62. As the positive side is more heavily loaded, therefore, machine *A* acts as a generator and machine *B* as a motor.

(*i*) Total load on the main generator

= load on +ve side + load on −ve side + losses

= 150 + 100 + 2 × 3 = **256 kW**

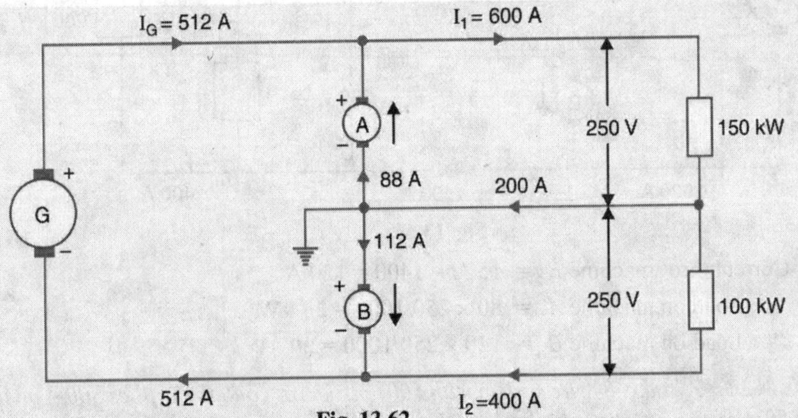

Fig. 13.62

(*ii*) Current supplied by the main generator,

$I_G = 256 \times 10^3/500 = 512$ A

Load current on +ve side, $I_1 = 150 \times 10^3/250 = 600$ A

Load current on –ve side, $I_2 = 100 \times 10^3/250 = 400$ A

Current in neutral wire $= I_1 - I_2 = 600 - 400 = 200$ A

Current through machine *A* $= I_1 - I_G = 600 - 512 = 88$ A

Current through machine *B* $= I_G - I_2 = 512 - 400 = 112$ A

∴ Load on machine *A* $= 88 \times 250/1000 =$ **22 kW**

Load on machine *B* $= 112 \times 250/1000 =$ **28 kW**

Example 13.34. *In a 500/250 V d.c. 3-wire system, there is a current of 1200 A on the +ve side and 1000 A on the –ve side and a motor load of 200 kW across the outers. The loss in each balancer machine is 5 kW. Calculate :*

(*i*) *current of the main generator*

(*ii*) *load on each balancer machine*

Solution. The connections are shown in Fig. 13.63. As the positive side is more heavily loaded, therefore, machine *A* acts as a generator and machine *B* as a motor.

Load on +ve side, $P_1 = 250 \times 1200/1000 = 300$ kW

Load on –ve side, $P_2 = 250 \times 1000/1000 = 250$ kW

Load on outers, $P_3 = 200$ kW

(*i*) Total load on the main generator $= P_1 + P_2 + P_3 +$ loss in balancer machines

$= 300 + 250 + 200 + 10 = 760$ kW

Current of main generator, $I_G = 760 \times 10^3/500 =$ **1520 A**

(*ii*) Current in neutral $= 1200 - 1000 = 200$ A

Current through machine *A* $= 1600 - 1520 = 80$ A

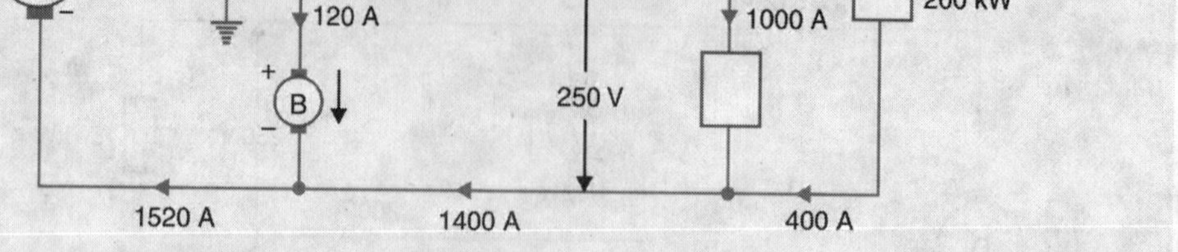

Fig. 13.63

Current thro' machine *B* $= 1520 - 1400 = 120$ A

Load on machine *A* $= 80 \times 250/1000 =$ **20 kW**

Load on machine *B* $= 120 \times 250/1000 =$ **30 kW**

Example 13.35. *A d.c. 3-wire system with 500 volts across outers supplies 800 A on the positive side and 550 A on the negative side and 1500 A across the outers. The rotary balancer has each an* *armature resistance of 0·2 Ω and takes 5 A on no load. Find :*

(*i*) *current loading of each balancer machine*

(*ii*) *the voltage across each balancer machine*

(*iii*) *total load on the main generator*

Solution. The connections are shown in Fig. 13.64. As the positive side is more heavily loaded, therefore, machine *A* acts as a generator and machine *B* as a motor.

Total current on +ve side = 800 + 1500 = 2300 A

Total current on −ve side = 550 + 1500 = 2050 A

Current in neutral wire = 800 − 550 = 250 A

Let the current through machines *A* and *B* be I_A and I_B respectively. Then $I_A + I_B$ must be equal to current in the neutral wire *i.e.*

$$I_A + I_B = 250$$

or $$I_A = 250 - I_B$$

(*i*) Let V_A and V_B be the p.d.s. of machines *A* and *B* respectively. Since machine *B* is driving the machine *A*, output of *B* supplies the losses in the set plus the output of machine *A* *i.e.*

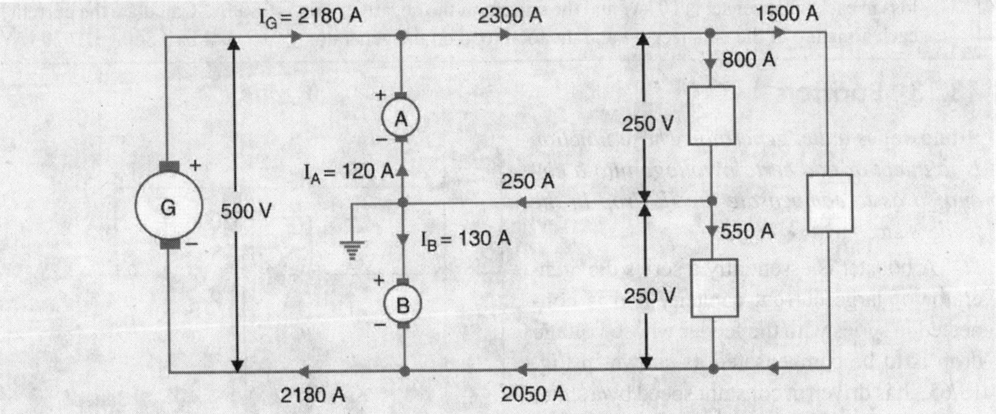

Fig. 13.64

Output of machine *B* = Output of machine A + *Losses in the set

or $$V_B I_B = V_A I_A + 500 \times 5 + I_A^2 R_A + I_B^2 R_B$$

∴ $$V_B I_B = V_A(250 - I_B) + 2500 + (250 - I_B)^2 \times 0{\cdot}2 + 0{\cdot}2\, I_B^2 \quad ...(i)$$

Each machine has same value of back e.m.f. *E* since their field currents and speeds are the same.

Back e.m.f., $E = 250 - 0{\cdot}2 \times 5 = 249$ V

Terminal p.d. across *A*, $V_A = E - I_A R_A = 249 - 0{\cdot}2\,(250 - I_B)$

∴ $$V_A = 199 + 0{\cdot}2\, I_B \quad ...(ii)$$

Terminal p.d. across *B*, $V_B = E + I_B R_B$

$$= 249 + 0{\cdot}2\, I_B \quad ...(iii)$$

Substituting the values of V_A and V_B in exp. (*i*), we get,

$$(249 + 0{\cdot}2\, I_B)\, I_B = (199 + 0{\cdot}2\, I_B)(250 - I_B) + 2500 + (250 - I_B)^2 \times 0{\cdot}2 + 0{\cdot}2\, I_B^2$$

or $$249\, I_B + 0{\cdot}2\, I_B^2 = 49{,}750 - 199\, I_B + 50\, I_B - 0{\cdot}2\, I_B^2 + 2500 + 12{,}500 + 0{\cdot}2\, I_B^2 - 100\, I_B + 0{\cdot}2\, I_B^2$$

* Losses in the set = No load losses + Copper losses

$$= 500 \times 5 + I_A^2 R_A + I_B^2 R_B$$

or $\quad 498\, I_B = 64{,}750$

$\therefore \quad I_B = 64750/498 = \mathbf{130\ A}$

and $\quad I_A = 250 - 130 = \mathbf{120\ A}$

(*ii*) Voltage across machine *A*, $\quad V_A = 199 + 0{\cdot}2\, I_B = 199 + 0{\cdot}2 \times 130 = \mathbf{225\ V}$

Voltage across machine *B*, $\quad V_B = 249 + 0{\cdot}2\, I_B = 249 + 0{\cdot}2 \times 130 = \mathbf{275\ V}$

(*iii*) Load on main generator $\quad = 2300 - I_A = 2300 - 120 = \mathbf{2180\ A}$

TUTORIAL PROBLEMS

1. A load on 3-wire d.c. system with 500 V between the outers consists of 1500 A on the positive side and 1300A on the negative side while motors connected across outers absorb 500 kW. Assuming each balancer machine has a loss of 3·75 kW, calculate the load on the main generator and each balancer machine. **[1207·5 kW ; 21·25 kW ; 28·75 kW]**
2. In a 500/250 V d.c. 3-wire system, there is a current of 2000 A on the +ve side, 1600A on the negative side and a load of 300 kW across the outers. The loss in each balancer set is 8 kW. Calculate the current in each armature of the balancer set and total load on the main generator. **[168A ; 232A ; 1216 kW]**
3. In a 500/250 volt 3-wire d.c. system, there is an out of balance load of 200 kW on the positive side. The loss in each balancer set is 10 kW and the current in the negative main is 2800A. Calculate the current in each armature of the balancer set and the total load on the generator. **[440A ; 360A ; 1620 kW]**

13.13 Boosters

A **booster** *is a d.c. generator whose function is to inject or add certain voltage into a circuit so as to compensate the IR drop in the feeders etc.*

A booster is essentially a series d.c. generator of large current capacity and is connected in series with the feeder whose voltage drop is to be compensated as shown in Fig. 13.65. It is driven at constant speed by a shunt motor working from the bus-bars. As the booster is a series generator, therefore, voltage generated by it is directly proportional to the field current which is here the feeder current. When the feeder current increases, the voltage drop in the feeder also increases. But increased feeder current results in greater field excitation of booster which injects higher voltage into the feeder to compensate the voltage drop. For exact compensation of voltage drop, the booster must be marked on the straight or linear portion of its voltage-current characteristics.

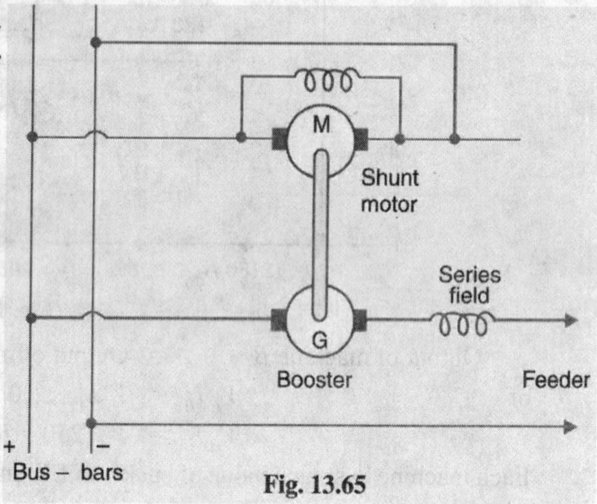

Fig. 13.65

It might be suggested to compensate the voltage drop in the feeder by overcompounding the generators instead of using a booster. Such a method is not practicable for feeders of different lengths because it will disturb the voltage of other feeders. The advantage of using a booster is that each feeder can be regulated independently — a great advantage if the feeders are of different lengths.

Example 13.36. *A 2-wire system has the voltage at the supply end maintained at 500 V. The line is 3 km long. If the full-load current is 120 A, what must be the booster voltage and output in order that the far end voltage may also be 500 V ? Take the resistance of the cable at the working temperature as 0·5 Ω/km.*

Solution.

Total resistance of line $= 0{\cdot}5 \times 3 = 1{\cdot}5\ \Omega$

F.L. voltage drop in the line $= 1{\cdot}5 \times 120 = 180$ V

$\therefore$ Terminal voltage of booster = **180 V**

$$\text{Output of booster} = \frac{120 \times 180}{1000}\ \text{kW} = \mathbf{21{\cdot}6\ kW}$$

13.14 Comparison of 3-Wire and 2-Wire D.C. Distribution

It is worthwhile to make a comparison between 3-wire and 2-wire systems for d.c. distribution. It will be shown that there is a great saving of conductor material if we use 3-wire system instead of 2-wire system for d.c. distribution. For comparison, it will be assumed that :

(*i*) the amount of power P transmitted is the same

(*ii*) the *voltage V at the consumer's terminals is the same

(*iii*) the distance of transmission is the same

(*iv*) the efficiency of transmission (and hence losses) is the same

(*v*) the 3-wire system is balanced *i.e.* no current in the neutral wire

(*vi*) the area of X-section of neutral wire is half the cross-section of outers in 3-wire system

Let R_2 = resistance of each conductor in 2-wire system

R_3 = resistance of each outer in 3-wire system

Current through outers in case of 3-wire system is

$$I_3 = P/2V$$

Total loss in two outers $= 2 I_3^2 R_3 = 2\,(P/2V)^2\, R_3$

Current in 2-wire system, $I_2 = P/V$

Total loss $= 2 I_2^2 R_2 = 2\,(P/V)^2\, R_2$

Since efficiency of transmission is the same, it means losses are the same *i.e.*

$$2\,(P/2V)^2\, R_3 = 2\,(P/V)^2\, R_2$$

$$\therefore \quad R_3 = 4\, R_2$$

Therefore, the area of X-section of outers in 3-wire case will be one-fourth of each conductor in 2-wire case.

Let a = area of X-section of each conductor is 2-wire case

Then $a/4$ = area of X-section of each outer in 3-wire case

and $a/8$ = area of X-section of neutral wire [assumption (*vi*) above]

If l is the length of the line, then,

$$\text{Volume of Cu for 3-wire system} = l\left(\frac{a}{4} + \frac{a}{4} + \frac{a}{8}\right) = \frac{5}{8}\,a\,l$$

$$\text{Volume of Cu for 2-wire system} = l\,(a + a) = 2\,a\,l$$

$$\therefore \quad \frac{\text{Volume of Cu for 3-wire system}}{\text{Volume of Cu for 2-wire system}} = \frac{5}{8}\,a\,l \times \frac{1}{2\,a\,l} = \frac{5}{16}$$

Hence a 3-wire system requires only 5/16 th (or 31·25%) as much copper as a 2-wire system.

Note. If the neutral has the same X-section as the outer, then,

$$\text{Volume of Cu for 3-wire system} = l\left(\frac{a}{4} + \frac{a}{4} + \frac{a}{4}\right) = \frac{3}{4}\,a\,l$$

$$\text{Volume of Cu for 2-wire system} = l\,(a + a) = 2\,a\,l$$

* Note that in case of 3-wire system, the voltage between the outers will be 2V.

$$\therefore \quad \frac{\text{Volume of Cu for 3 - wire system}}{\text{Volume of Cu for 2 - wire system}} = \frac{3}{4}\,a\,l \times \frac{1}{2\,a\,l} = \frac{3}{8} \text{ or } 37{\cdot}5\%$$

13.15 Ground Detectors

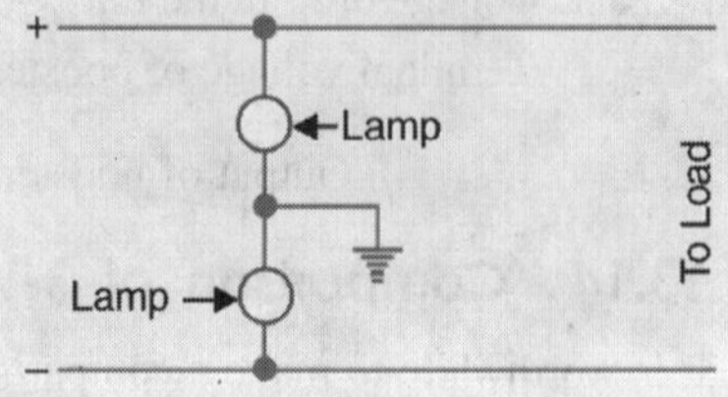

Fig. 13.66

Ground detectors are the devices that are used to detect/indicate the ground fault for ungrounded d.c. systems. When a ground fault occurs on such a system, immediate steps should be taken to clear it. If this is not done and a second ground fault happens, a short circuit occurs. Lamps are generally used for the detection of ground faults. They are connected for ungrounded 2-wire system as shown in Fig. 13.66. Each lamp should have a voltage rating equal to the line voltage. The two lamps in series, being subjected to half their rated voltage, will glow dimly. If a ground fault occurs on either wires, the lamp connected to the grounded wire will not glow while the other lamp will glow brightly.

SELF - TEST

1. **Fill in the blanks by inserting appropriate words/figures.**
   (*i*) In a singly fed distributor, if fault occurs on any section, the supply to all consumers has to be ....
   (*ii*) A ring main distributor fed at one end is equivalent to .... fed at both ends with equal voltages.
   (*iii*) A distributor is designed from .... considerations.
   (*iv*) The point of minimum potential of a uniformly loaded distributor fed at both ends with equal voltages will occur at ....
   (*v*) The d.c. interconnector is used .... the voltage drops in the various sections of the distributor.
   (*vi*) In a 3 wire d.c. system, the load on +ve side is 400A and on negative side it is 300A. Then current in neutral wire is ....
   (*vii*) In a balanced 3-wire d.c. system, the potential of neutral is .... between that of outers.
   (*viii*) A booster is used to .... voltage drop in feeders *etc.*
   (*ix*) Balancer set is used to maintain voltage on the two sides of the neutral ....
   (*x*) In a balanced 3-wire d.c. system, if voltage across the outers is 500 V, then voltage between any outer and neutral is ....
2. **Pick up the correct words/figures from brackets and fill in the blanks.**
   (*i*) The voltage drop in a doubly fed distributor is .... than the equivalent singly fed distributor. (*less, more*)
   (*ii*) In a 3-wire system, the area of X-section of neutral is generally .... of either outer. (*half, double*)
   (*iii*) If in a 3-wire d.c. system, the current in the neutral wire is zero, then voltage between any outer and neutral is .... (*the same, different*)
   (*iv*) A booster is connected in .... with the feeder. (*series, parallel*)
   (*v*) For exact compensation of voltage drop in the feeder, the booster must work on .... portion of its V—I characteristic. (*linear, non-linear*)
   (*vi*) The balancer machine connected to the heavily loaded side works as a .... (*generator, motor*)

ANSWERS TO SELF-TEST

1. (*i*) shut off (*ii*) straight distributor (*iii*) voltage drop (*iv*) mid-point (*v*) to reduce (*vi*) 100 A (*vii*) midway (*viii*) compensate (*ix*) equal to each other (*x*) 250 V
2. (*i*) less (*ii*) half (*iii*) the same (*iv*) series (*v*) linear (*vi*) generator

CHAPTER REVIEW TOPICS

1. Describe briefly the different types of d.c. distributors.
2. What are the advantages of a doubly fed distributor over singly fed distributor ?
3. Derive an expression for the voltage drop for a uniformly loaded distributor fed at one end.
4. What is the purpose of interconnector in a d.c. ring main distributor ?
5. Explain 3-wire d.c. system of distribution of electrical power.
6. What are the advantages of 3-wire distribution over 2-wire distribution ?
7. Show with a neat diagram how unbalanced loads in a 3-wire d.c. system cause unequal voltages on the two sides of the neutral.
8. Explain the use of rotary balancer in a 3-wire d.c. distribution system.
9. What is a booster ? With a neat diagram, explain how it can be used on a feeder.
10. Write short notes on the following :
    (*i*) Ring main distributor
    (*ii*) Current distribution in a 3-wire d.c. system
    (*iii*) Balancers

DISCUSSION QUESTIONS

1. What is the importance of minimum potential on the distributor ?
2. Why is 3-wire d.c. distribution preferred to 2-wire d.c. distribution ?
3. Which points of d.c. ring main should be connected through interconnector ?
4. Can we use compound generator as a booster ?
5. Why do we use a balancer set ?
6. Can exact balance of voltages to obtained with a balancer set ?

CHAPTER

14

A.C. Distribution

Introduction

In the beginning of electrical age, electricity was generated, transmitted and distributed as direct current. The principal disadvantage of d.c. system was that voltage level could not readily be changed, except by the use of rotating machinery, which in most cases was too expensive. With the development of transformer by George Westinghouse, a.c. system has become so predominant as to make d.c. system practically extinct in most parts of the world. The present day large power system has been possible only due to the adoption of a.c. system.

Now-a-days, electrical energy is generated, transmitted and distributed in the form of alternating current as an economical proposition. The electrical energy produced at the power station is transmitted at very high voltages by 3-phase, 3-wire system to step-down sub-stations for distribution. The distribution system consists of two parts *viz*. primary distribution and secondary distribution. The primary distribution circuit is 3-phase, 3-wire and operates at voltages (3·3 or 6·6 or 11kV) somewhat higher than general utilisation levels. It delivers power to the secondary distribution circuit through distribution transformers situated near consumers' localities. Each distribution transformer steps down the voltage to 400 V and power is distributed to ultimate consumers' by 400/230 V, 3-phase, 4-wire system. In this chapter, we shall focus our attention on the various aspects of a.c. distribution.

14.1 A.C. Distribution Calculations

A.C. distribution calculations differ from those of d.c. distribution in the following respects :

(*i*) In case of d.c. system, the voltage drop is due to resistance alone. However, in a.c. system, the voltage drops are due to the combined effects of resistance, inductance and capacitance.

(*ii*) In a d.c. system, additions and subtractions of currents or voltages are done arithmetically but in case of a.c. system, these operations are done vectorially.

(*iii*) In an a.c. system, power factor (p.f.) has to be taken into account. Loads tapped off form the distributor are generally at different power factors. There are two ways of referring power factor *viz*

(*a*) It may be referred to supply or receiving end voltage which is regarded as the reference vector.

(*b*) It may be referred to the voltage at the load point itself.

There are several ways of solving a.c. distribution problems. However, symbolic notation method has been found to be most convenient for this purpose. In this method, voltages, currents and impedances are expressed in complex notation and the calculations are made exactly as in d.c. distribution.

14.2 Methods of Solving A.C. Distribution Problems

In a.c. distribution calculations, power factors of various load currents have to be considered since currents in different sections of the distributor will be the vector sum of load currents and not the arithmetic sum. The power factors of load currents may be given (*i*) *w.r.t.* receiving or sending end voltage or (*ii*) *w.r.t.* to load voltage itself. Each case shall be discussed separately.

(*i*) **Power factors referred to receiving end voltage.** Consider an a.c. distributor *AB* with concentrated loads of I_1 and I_2 tapped off at points *C* and *B* as shown in Fig. 14.1. Taking the receiving end voltage V_B as the reference vector, let lagging power factors at *C* and *B* be $\cos\phi_1$ and $\cos\phi_2$ *w.r.t.* V_B. Let R_1, X_1 and R_2, X_2 be the resistance and reactance of sections *AC* and *CB* of the distributor.

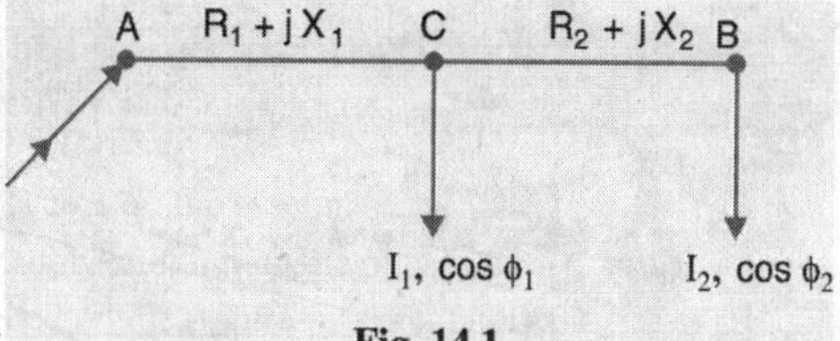

Fig. 14.1

Impedance of section *AC*, $\overrightarrow{Z_{AC}} = R_1 + jX_1$

Impedance of section *CB*, $\overrightarrow{Z_{CB}} = R_2 + jX_2$

Load current at point *C*, $\vec{I_1} = I_1(\cos\phi_1 - j\sin\phi_1)$

Load current at point *B*, $\vec{I_2} = I_2(\cos\phi_2 - j\sin\phi_2)$

Current in section *CB*, $\overrightarrow{I_{CB}} = \vec{I_2} = I_2(\cos\phi_2 - j\sin\phi_2)$

Current in section *AC*, $\overrightarrow{I_{AC}} = \vec{I_1} + \vec{I_2}$

$$= I_1(\cos\phi_1 - j\sin\phi_1) + I_2(\cos\phi_2 - j\sin\phi_2)$$

Voltage drop in section *CB*, $\overrightarrow{V_{CB}} = \overrightarrow{I_{CB}}\,\overrightarrow{Z_{CB}} = I_2(\cos\phi_2 - j\sin\phi_2)(R_2 + jX_2)$

Voltage drop in section *AC*, $\overrightarrow{V_{AC}} = \overrightarrow{I_{AC}}\,\overrightarrow{Z_{AC}} = \left(\vec{I_1} + \vec{I_2}\right)Z_{AC}$

$$= [I_1(\cos\phi_1 - j\sin\phi_1) + I_2(\cos\phi_2 - j\sin\phi_2)]\,[R_1 + jX_1]$$

Sending end voltage, $\vec{V_A} = \vec{V_B} + \vec{V_{CB}} + \vec{V_{AC}}$

Sending end current, $\vec{I_A} = \vec{I_1} + \vec{I_2}$

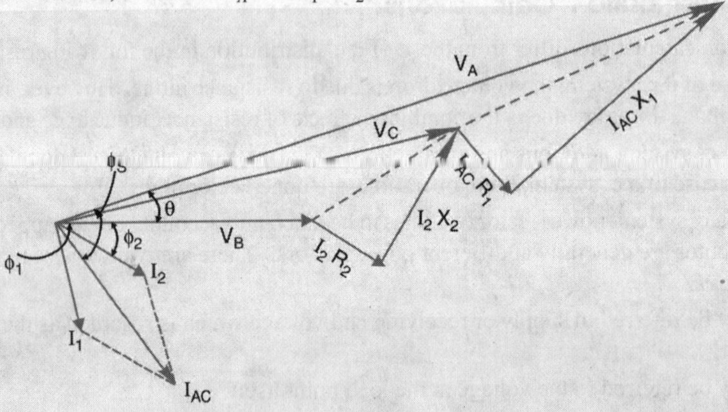

Fig. 14.2

The vector diagram of the a.c. distributor under these conditions is shown in Fig. 14.2. Here, the receiving end voltage V_B is taken as the reference vector. As power factors of loads are given *w.r.t.* V_B, therefore, I_1 and I_2 lag behind V_B by ϕ_1 and ϕ_2 respectively.

(*ii*) **Power factors referred to respective load voltages.** Suppose the power factors of loads in the previous Fig. 14.1 are referred to their respective load voltages. Then ϕ_1 is the phase angle between V_C and I_1 and ϕ_2 is the phase angle between V_B and I_2. The vector diagram under these conditions is shown in Fig. 14.3.

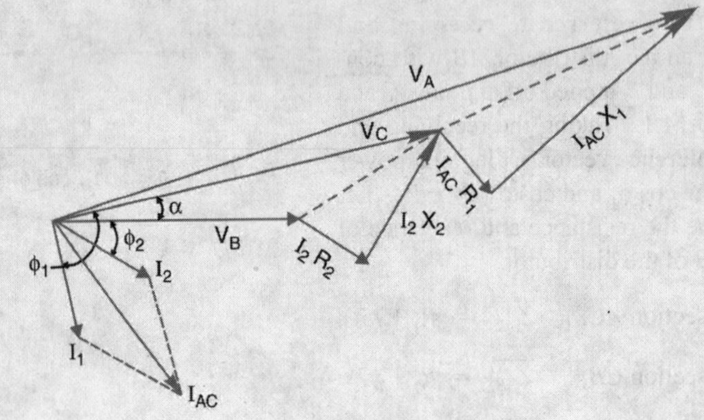

Fig. 14.3

Voltage drop in section $CB = \vec{I_2}\,\vec{Z_{CB}} = I_2(\cos\phi_2 - j\sin\phi_2)(R_2 + jX_2)$

Voltage at point $C = \vec{V_B}$ + Drop in section $CB = V_C\angle\alpha$ (say)

Now $\vec{I_1} = I_1\angle -\phi_1$ *w.r.t.* voltage V_C

$\therefore$ $\vec{I_1} = I_1\angle -(\phi_1 - \alpha)$ *w.r.t.* voltage V_B

i.e. $\vec{I_1} = I_1[\cos(\phi_1 - \alpha) - j\sin(\phi_1 - \alpha)]$

Now $\vec{I_{AC}} = \vec{I_1} + \vec{I_2}$

$$= I_1 [\cos(\phi_1 - \alpha) - j\sin(\phi_1 - \alpha)] + I_2 (\cos\phi_2 - j\sin\phi_2)$$

Voltage drop in section $AC = \overrightarrow{I_{AC}}\ \overrightarrow{Z_{AC}}$

$\therefore$ Voltage at point $A = V_B$ + Drop in CB + Drop in AC

Example 14.1. *A single phase a.c. distributor AB 300 metres long is fed from end A and is loaded as under :*

(i) 100 A at 0·707 p.f. lagging 200 m from point A

(ii) 200 A at 0·8 p.f. lagging 300 m from point A

The load resistance and reactance of the distributor is 0·2 Ω and 0·1 Ω per kilometre. Calculate the total voltage drop in the distributor. The load power factors refer to the voltage at the far end.

Solution. Fig. 14.4 shows the single line diagram of the distributor.

Impedance of distributor/km $= (0{\cdot}2 + j\,0{\cdot}1)\ \Omega$

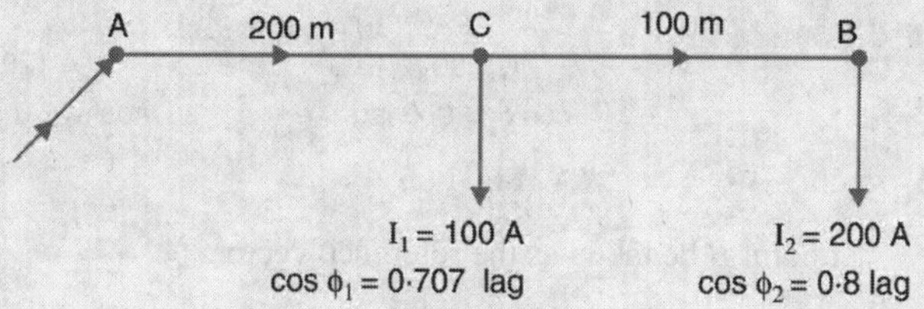

Fig. 14.4

Impedance of section AC, $\overrightarrow{Z_{AC}} = (0{\cdot}2 + j\,0{\cdot}1) \times 200/1000 = (0{\cdot}04 + j\,0{\cdot}02)\ \Omega$

Impedance of section CB, $\overrightarrow{Z_{CB}} = (0{\cdot}2 + j\,0{\cdot}1) \times 100/1000 = (0{\cdot}02 + j\,0{\cdot}01)\ \Omega$

Taking voltage at the far end B as the reference vector, we have,

Load current at point B, $\overrightarrow{I_2} = I_2(\cos\phi_2 - j\sin\phi_2) = 200\,(0{\cdot}8 - j\,0{\cdot}6)$
$= (160 - j\,120)$ A

Load current at point C, $\overrightarrow{I_1} = I_1(\cos\phi_1 - j\sin\phi_1) = 100\,(0{\cdot}707 - j\,0{\cdot}707)$
$= (70{\cdot}7 - j\,70{\cdot}7)$ A

Current in section CB, $\overrightarrow{I_{CB}} = \overrightarrow{I_2} = (160 - j\,120)$ A

Current in section AC, $\overrightarrow{I_{AC}} = \overrightarrow{I_1} + \overrightarrow{I_2} = (70{\cdot}7 - j\,70{\cdot}7) + (160 - j\,120)$
$= (230{\cdot}7 - j\,190{\cdot}7)$ A

Voltage drop in section CB, $\overrightarrow{V_{CB}} = \overrightarrow{I_{CB}}\ \overrightarrow{Z_{CB}} = (160 - j\,120)\,(0{\cdot}02 + j\,0{\cdot}01)$
$= (4{\cdot}4 - j\,0{\cdot}8)$ volts

Voltage drop in section AC, $\overrightarrow{V_{AC}} = \overrightarrow{I_{AC}}\ \overrightarrow{Z_{AC}} = (230{\cdot}7 - j\,190{\cdot}7)\,(0{\cdot}04 + j\,0{\cdot}02)$
$= (13{\cdot}04 - j\,3{\cdot}01)$ volts

Voltage drop in the distributor $= \overrightarrow{V_{AC}} + \overrightarrow{V_{CB}} = (13{\cdot}04 - j\,3{\cdot}01) + (4{\cdot}4 - j\,0{\cdot}8)$
$= (17{\cdot}44 - j\,3{\cdot}81)$ volts

Magnitude of drop $= \sqrt{(17{\cdot}44)^2 + (3{\cdot}81)^2}$ = **17·85 V**

Example 14.2. *A single phase distributor 2 kilometres long supplies a load of 120 A at 0·8 p.f. lagging at its far end and a load of 80 A at 0·9 p.f. lagging at its mid-point. Both power factors are*

referred to the voltage at the far end. The resistance and reactance per km (go and return) are 0·05 Ω and 0·1 Ω respectively. If the voltage at the far end is maintained at 230 V, calculate :

(i) voltage at the sending end

(ii) phase angle between voltages at the two ends.

Solution. Fig. 14.5 shows the distributor AB with C as the mid-point

Impedance of distributor/km $= (0·05 + j\,0·1)\ \Omega$

Impedance of section AC, $\vec{Z}_{AC} = (0·05 + j\,0·1) \times 1000/1000 = (0·05 + j\,0·1)\ \Omega$

Impedance of section CB, $\vec{Z}_{CB} = (0·05 + j\,0·1) \times 1000/1000 = (0·05 + j\,0·1)\ \Omega$

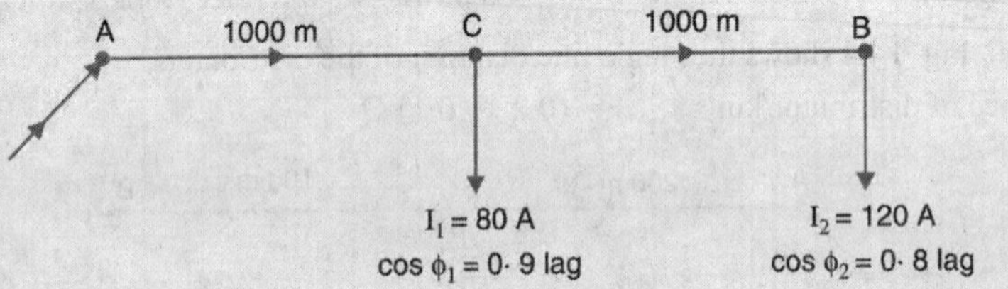

Fig. 14.5

Let the voltage V_B at point B be taken as the reference vector.

Then, $\vec{V_B} = 230 + j\,0$

(*i*) Load current at point B, $\vec{I_2} = 120\,(0·8 - j\,0·6) = 96 - j\,72$

Load current at point C, $\vec{I_1} = 80\,(0·9 - j\,0·436) = 72 - j\,34·88$

Current in section CB, $\vec{I_{CB}} = \vec{I_2} = 96 - j\,72$

Current in section AC, $\vec{I_{AC}} = \vec{I_1} + \vec{I_2} = (72 - j\,34·88) + (96 - j\,72)$

$= 168 - j\,106·88$

Drop in section CB, $\vec{V_{CB}} = \vec{I_{CB}}\,\vec{Z_{CB}} = (96 - j\,72)\,(0·05 + j\,0·1)$

$= 12 + j\,6$

Drop in section AC, $\vec{V_{AC}} = \vec{I_{AC}}\,\vec{Z_{AC}} = (168 - j\,106·88)\,(0·05 + j\,0·1)$

$= 19·08 + j\,11·45$

∴ Sending end voltage, $\vec{V_A} = \vec{V_B} + \vec{V_{CB}} + \vec{V_{AC}}$

$= (230 + j\,0) + (12 + j\,6) + (19.08 + j\,11.45)$

$= 261.08 + j\,17.45$

Its magnitude is $= \sqrt{(261·08)^2 + (17·45)^2} = \mathbf{261·67\ V}$

(*ii*) The phase difference θ between V_A and V_B is given by :

$$\tan\theta = \frac{17·45}{261·08} = 0·0668$$

∴ $\theta = \tan^{-1} 0·0668 = \mathbf{3·82°}$

Example 14.3. *A single phase distributor one km long has resistance and reactance per conductor of 0·1 Ω and 0·15 Ω respectively. At the far end, the voltage V_B = 200 V and the current is 100 A at a p.f. of 0·8 lagging. At the mid-point M of the distributor, a current of 100 A is tapped at a p.f.*

of 0·6 lagging with reference to the voltage V_M at the mid-point. Calculate :

(i) voltage at mid-point

(ii) sending end voltage V_A

(iii) phase angle between V_A and V_B

Solution. Fig. 14.6 shows the single line diagram of the distributor *AB* with *M* as the mid-point.

Total impedance of distributor $= 2(0{\cdot}1 + j\,0{\cdot}15) = (0{\cdot}2 + j\,0{\cdot}3)\ \Omega$

Impedance of section *AM*, $\overrightarrow{Z_{AM}} = (0{\cdot}1 + j\,0{\cdot}15)\ \Omega$

Impedance of section *MB*, $\overrightarrow{Z_{MB}} = (0{\cdot}1 + j\,0{\cdot}15)\ \Omega$

Let the voltage V_B at point *B* be taken as the reference vector.

Then, $\overrightarrow{V_B} = 200 + j\,0$

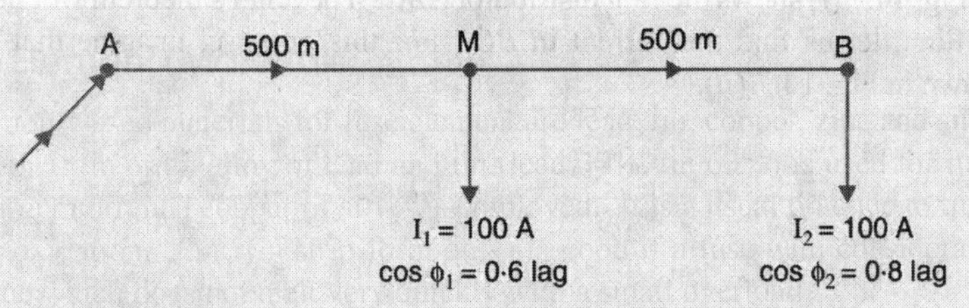

Fig. 14.6

(*i*) Load current at point *B*, $\overrightarrow{I_2} = 100\,(0{\cdot}8 - j\,0{\cdot}6) = 80 - j\,60$

Current in section *MB*, $\overrightarrow{I_{MB}} = \overrightarrow{I_2} = 80 - j\,60$

Drop in section *MB*, $\overrightarrow{V_{MB}} = \overrightarrow{I_{MB}}\ \overrightarrow{Z_{MB}}$

$= (80 - j\,60)\,(0{\cdot}1 + j\,0{\cdot}15) = 17 + j\,6$

∴ Voltage at point *M*, $\overrightarrow{V_M} = \overrightarrow{V_B} + \overrightarrow{V_{MB}} = (200 + j\,0) + (17 + j\,6)$

$= 217 + j\,6$

Its magnitude is $= \sqrt{(217)^2 + (6)^2} = \mathbf{217{\cdot}1\ V}$

Phase angle between V_M and V_B, $\alpha = \tan^{-1} 6/217 = \tan^{-1} 0{\cdot}0276 = 1{\cdot}58^o$

(*ii*) The load current I_1 has a lagging p.f. of 0·6 *w.r.t.* V_M. It lags behind V_M by an angle $\phi_1 = \cos^{-1} 0{\cdot}6 = 53{\cdot}13^o$

∴ Phase angle between I_1 and V_B, $\phi'_1 = \phi_1 - \alpha = 53{\cdot}13^o - 1{\cdot}58 = 51{\cdot}55^o$

Load current at *M*, $\overrightarrow{I_1} = I_1\,(\cos \phi_1' - j \sin \phi_1') = 100\,(\cos 51{\cdot}55^o - j \sin 51{\cdot}55^o)$

$= 62{\cdot}2 - j\,78{\cdot}3$

Current in section *AM*, $\overrightarrow{I_{AM}} = \overrightarrow{I_1} + \overrightarrow{I_2} = (62{\cdot}2 - j\,78{\cdot}3) + (80 - j\,60)$

$= 142{\cdot}2 - j\,138{\cdot}3$

Drop in section *AM*, $\overrightarrow{V_{AM}} = \overrightarrow{I_{AM}}\ \overrightarrow{Z_{AM}} = (142{\cdot}2 - j\,138{\cdot}3)\,(0{\cdot}1 + j\,0{\cdot}15)$

$= 34{\cdot}96 + j\,7{\cdot}5$

Sending end voltage, $\overrightarrow{V_A} = \overrightarrow{V_M} + \overrightarrow{V_{AM}} = (217 + j\,6) + (34{\cdot}96 + j\,7{\cdot}5)$

$$= 251{\cdot}96 + j\,13{\cdot}5$$

Its magnitude is $= \sqrt{(251{\cdot}96)^2 + (13{\cdot}5)^2} = \mathbf{252{\cdot}32\ V}$

(*iii*) The phase difference θ between V_A and V_B is given by :

$$\tan\theta = 13{\cdot}5/251{\cdot}96 = 0{\cdot}05358$$

$$\therefore \quad \theta = \tan^{-1} 0{\cdot}05358 = \mathbf{3{\cdot}07^o}$$

Hence supply voltage is 252·32 V and leads V_B by 3·07º.

Example 14.4. *A single phase ring distributor ABC is fed at A. The loads at B and C are 20 A at 0.8 p.f. lagging and 15 A at 0.6 p.f. lagging respectively ; both expressed with reference to the voltage at A. The total impedance of the three sections AB, BC and CA are (1 + j 1), (1+ j2) and (1 + j3) ohms respectively. Find the total current fed at A and the current in each section. Use Thevenin's theorem to obtain the results.*

Solution. Fig. 14.7 (*i*) shows the ring distributor *ABC*. Thevenin's theorem will be used to solve this problem. First, let us find the current in *BC*. For this purpose, imagine that section *BC* is removed as shown in Fig. 14.7 (*ii*).

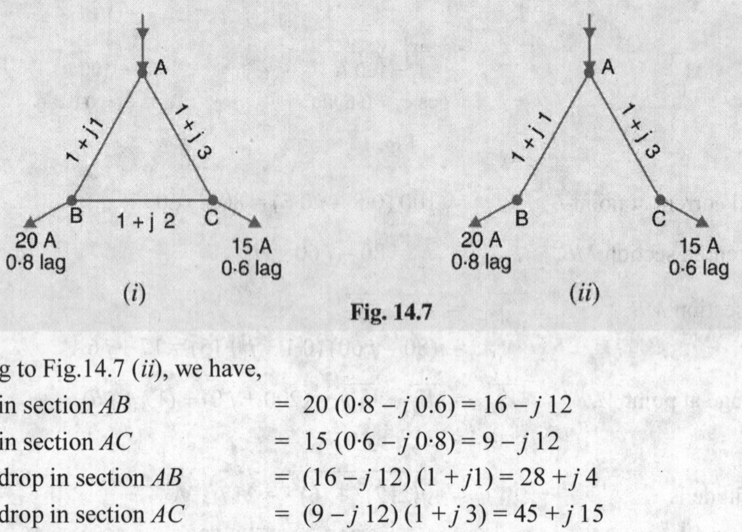

Fig. 14.7

Referring to Fig.14.7 (*ii*), we have,

Current in section *AB* $= 20\,(0{\cdot}8 - j\,0.6) = 16 - j\,12$

Current in section *AC* $= 15\,(0{\cdot}6 - j\,0{\cdot}8) = 9 - j\,12$

Voltage drop in section *AB* $= (16 - j\,12)\,(1 + j1) = 28 + j\,4$

Voltage drop in section *AC* $= (9 - j\,12)\,(1 + j\,3) = 45 + j\,15$

Obviously, point *B* is at higher potential than point *C*. The p.d. between *B* and *C* is Thevenin's equivalent circuit e.m.f. E_0 *i.e.*

Thevenin's equivalent circuit e.m.f., E_0 = p.d. between *B* and *C*

$$= (45 + j\,15) - (28 + j\,4) = 17 + j\,11$$

Thevenin's equivalent impedance Z_o can be found by looking into the network from points *B* and *C*.

Obviously, $Z_0 = (1 + j1) + (1 + j\,3) = 2 + j4$

$$\therefore \quad \text{Current in } BC = \frac{E_0}{Z_0 + \text{Impedance of } BC}$$

$$= \frac{17 + j11}{(2 + j4) + (1 + j2)} = \frac{17 + j11}{3 + j6}$$

$$= 2{\cdot}6 - j\,1{\cdot}53 = \mathbf{3\angle{-30{\cdot}48^o}\ A}$$

$$\text{Current in } AB = (16 - j\,12) + (2{\cdot}6 - j\,1{\cdot}53)$$

$= 18{\cdot}6 - j\,13{\cdot}53 = \mathbf{23\angle{-36{\cdot}03^\circ}\ A}$

Current in AC $= (9 - j\,12) - (2{\cdot}6 - j\,1{\cdot}53)$

$= 6{\cdot}4 - j\,10{\cdot}47 = \mathbf{12{\cdot}27\angle{-58{\cdot}56^\circ}\ A}$

Current fed at A $= (16 - j\,12) + (9 - j\,12)$

$= 25 - j\,24 = \mathbf{34{\cdot}65\angle{-43{\cdot}83^\circ}\ A}$

Example 14.5. *A 3-phase, 400V distributor AB is loaded as shown in Fig.14.8. The 3-phase load at point C takes 5A per phase at a p.f. of 0·8 lagging. At point B, a 3-phase, 400 V induction motor is connected which has an output of 10 H.P. with an efficiency of 90% and p.f. 0·85 lagging.*

If voltage at point B is to be maintained at 400 V, what should be the voltage at point A ? The resistance and reactance of the line are 1Ω and 0·5Ω per phase per kilometre respectively.

Solution. It is convenient to consider one phase only. Fig.14.8 shows the single line diagram of the distributor. Impedance of the distributor per phase per kilometre = $(1 + j\,0{\cdot}5)\ \Omega$.

Impedance of section AC, $\overrightarrow{Z_{AC}} = (1 + j\,0{\cdot}5) \times 600/1000 = (0{\cdot}6 + j\,0{\cdot}3)\ \Omega$

Impedance of section CB, $\overrightarrow{Z_{CB}} = (1 + j\,0{\cdot}5) \times 400/1000 = (0{\cdot}4 + j\,0{\cdot}2)\ \Omega$

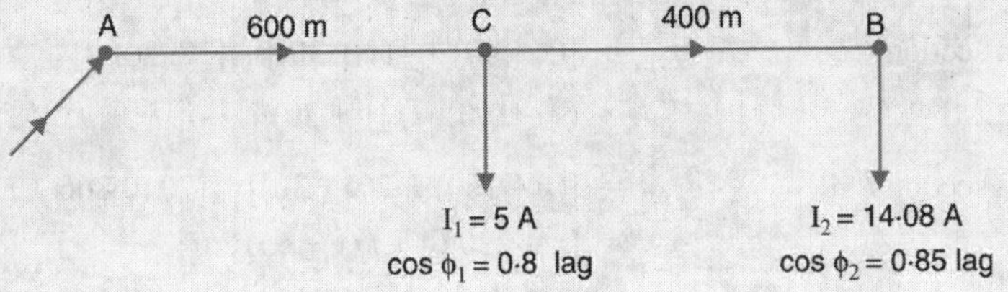

Fig. 14.8

Phase voltage at point B, $V_B = 400/\sqrt{3} = 231$ V

Let the voltage V_B at point B be taken as the reference vector.

Then, $\overrightarrow{V_B} = 231 + j\,0$

Line current at B $= \dfrac{\text{H.P.} \times 746}{\sqrt{3} \times \text{line voltage} \times \text{p.f.} \times \text{efficiency}}$

$= \dfrac{10 \times 746}{\sqrt{3} \times 400 \times 0{\cdot}85 \times 0{\cdot}9} = 14{\cdot}08$ A

$\therefore$ *Current/phase at B, $I_2 = 14{\cdot}08$ A

Load current at B, $\vec{I_2} = 14{\cdot}08\,(0{\cdot}85 - j\,0{\cdot}527) = 12 - j\,7{\cdot}4$

Load current at C, $\vec{I_1} = 5\,(0{\cdot}8 - j\,0{\cdot}6) = 4 - j\,3$

Current in section AC, $\overrightarrow{I_{AC}} = \vec{I_1} + \vec{I_2} = (4 - j\,3) + (12 - j\,7{\cdot}4)$

$= 16 - j\,10{\cdot}4$

Current in section CB, $\overrightarrow{I_{CB}} = \vec{I_2} = 12 - j\,7{\cdot}4$

Voltage drop in CB, $\vec{V}_{CB} = \vec{I}_{CB}\vec{Z}_{CB} = (12 - j\,7{\cdot}4)\,(0{\cdot}4 + j\,0{\cdot}2)$

$= 6{\cdot}28 - j\,0{\cdot}56$

Voltage drop in AC, $\vec{V}_{AC} = \vec{I}_{AC}\vec{Z}_{AC} = (16 - j\,10{\cdot}4)\,(0{\cdot}6 + j\,0{\cdot}3)$

$= 12{\cdot}72 - j\,1{\cdot}44$

* In a 3-phase system, if the type of connection is not mentioned, then star connection is understood.

Voltage at A per phase, $\vec{V}_A = \vec{V}_B + \vec{V}_{CB} + \vec{V}_{AC}$

$= (231 + j\,0) + (6{\cdot}28 - j\,0{\cdot}56) + (12{\cdot}72 - j\,1{\cdot}44)$

$= 250 - j\,2$

Magnitude of V_A/phase $= \sqrt{(250)^2 + (2)^2} = 250$ V

$\therefore$ Line voltage at A $= \sqrt{3} \times 250 = \mathbf{433\ V}$

Example 14.6. *A 3-phase ring main ABCD fed at A at 11 kV supplies balanced loads of 50 A at 0.8 p.f. lagging at B, 120 A at unity p.f. at C and 70 A at 0·866 lagging at D, the load currents being referred to the supply voltage at A. The impedances of the various sections are :*

Section AB $= (1 + j\,0{\cdot}6)\ \Omega$; *Section BC* $= (1{\cdot}2 + j\,0{\cdot}9)\ \Omega$

Section CD $= (0{\cdot}8 + j\,0{\cdot}5)\ \Omega$; *Section DA* $= (3 + j\,2)\ \Omega$

Calculate the currents in various sections and station bus-bar voltages at B, C and D.

Solution. Fig.14.9 shows one phase of the ring main. The problem will be solved by Kirchhoff's laws. Let current in section AB be $(x + j\,y)$.

$\therefore$ Current in section BC, $\overrightarrow{I_{BC}} = (x + j\,y) - 50\,(0{\cdot}8 - j\,0{\cdot}6) = (x - 40) + j\,(y + 30)$

Current in section CD, $\overrightarrow{I_{CD}} = [(x - 40) + j\,(y + 30)] - [120 + j\,0]$

$= (x - 160) + j\,(y + 30)$

Current in section DA, $\overrightarrow{I_{DA}} = [(x - 160) + j\,(y + 30)] - [70\,(0{\cdot}866 - j\,0{\cdot}5)]$

$= (x - 220{\cdot}6) + j\,(y + 65)$

Drop in section AB $= \overrightarrow{I_{AB}}\ \overrightarrow{Z_{AB}} = (x + jy)\,(1 + j0{\cdot}6)$

$= (x - 0{\cdot}6y) + j\,(0{\cdot}6x + y)$

Drop in section BC $= \overrightarrow{I_{BC}}\ \overrightarrow{Z_{BC}}$

$= [(x - 40) + j\,(y + 30)]\,[(1{\cdot}2 + j\,0{\cdot}9)]$

$= (1{\cdot}2x - 0{\cdot}9\,y - 75) + j\,(0{\cdot}9x + 1{\cdot}2\,y)$

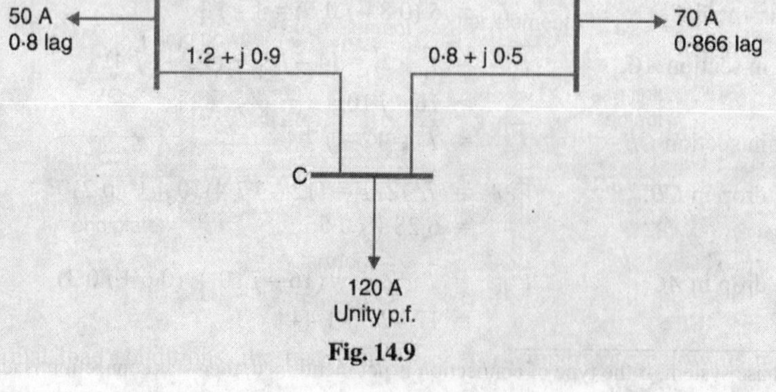

Fig. 14.9

$$\text{Drop in section } CD = \overrightarrow{I_{CD}}\ \overrightarrow{Z_{CD}}$$
$$= [(x - 160) + j\,(y + 30)]\ [(0{\cdot}8 + j\,0{\cdot}5)]$$
$$= (0{\cdot}8x - 0{\cdot}5y - 143) + j\,(0{\cdot}5x + 0{\cdot}8y - 56)$$

$$\text{Drop in section } DA = \overrightarrow{I_{DA}}\ \overrightarrow{Z_{DA}}$$
$$= [(x - 220{\cdot}6) + j\,(y + 65)]\ [(3 + j\,2)]$$
$$= (3x - 2y - 791{\cdot}8) + j\,(2x + 3y - 246{\cdot}2)$$

Applying Kirchhoff's voltage law to mesh *ABCDA*, we have,

Drop in *AB* + Drop in *BC* + Drop in *CD* + Drop in *DA* = 0

or
$$[(x - 0{\cdot}6y) + j\,(0{\cdot}6x + y)] + [(1{\cdot}2x - 0{\cdot}9y - 75) + j\,(0{\cdot}9x + 1{\cdot}2y)]$$
$$+ [(0{\cdot}8x - 0{\cdot}5y - 143) + j\,(0{\cdot}5x + 0{\cdot}8y - 56)]$$
$$+ [(3x - 2y - 791{\cdot}8) + j\,(2x + 3y - 246{\cdot}2)] = 0$$

or
$$(6x - 4y - 1009{\cdot}8) + j\,(4x + 6y - 302{\cdot}2) = 0$$

As the real (or active) and imaginary (or reactive) parts have to be separately zero,

$$\therefore \quad 6x - 4y - 1009{\cdot}8 = 0$$

and
$$4x + 6y - 302{\cdot}2 = 0$$

Solving for x and y, we have,

$$x = 139{\cdot}7 \text{ A} \ ; \ y = -42{\cdot}8 \text{ A}$$

Current in section *AB* $= \mathbf{(139{\cdot}7 - j\,42{\cdot}8)\ A}$

Current in section *BC* $= (x - 40) + j\,(y + 30)$

$$= (139{\cdot}7 - 40) + j\,(-42{\cdot}8 + 30) = \mathbf{(99{\cdot}7 - j\,12{\cdot}8)\ A}$$

Current in section *CD* $= (x - 160) + j\,(y + 30)$

$$= (139{\cdot}7 - 160) + j\,(-42{\cdot}8 + 30)$$
$$= \mathbf{(-20{\cdot}3 - j\,12{\cdot}8)\ A}$$

Current in section *DA* $= (x - 220{\cdot}6) + j\,(y + 65)$

$$= (139{\cdot}7 - 220{\cdot}6) + j\,(-42{\cdot}8 + 65)$$
$$= \mathbf{(-80{\cdot}9 + j\,22{\cdot}2)\ A}$$

Voltage at supply end *A*, $V_A = 11000/\sqrt{3} = 6351$ V/phase

$\therefore$ Voltage at station *B*, $\overrightarrow{V_B} = \overrightarrow{V_A} - \overrightarrow{I_{AB}}\ \overrightarrow{Z_{AB}}$

$$= (6351 + j\,0) - (139{\cdot}7 - j\,42{\cdot}8)\,(1 + j\,0{\cdot}6)$$
$$= \mathbf{(6185{\cdot}62 - j\,41{\cdot}02)\ volts/phase}$$

Voltage at station *C*, $\overrightarrow{V_C} = \overrightarrow{V_B} - \overrightarrow{I_{BC}}\ \overrightarrow{Z_{BC}}$

$$= (6185{\cdot}62 - j\,41{\cdot}02) - (99{\cdot}7 - j\,12{\cdot}8)\,(1{\cdot}2 + j\,0{\cdot}9)$$
$$= \mathbf{(6054{\cdot}46 - j\,115{\cdot}39)\ volts/phase}$$

Voltage at station *D*, $\overrightarrow{V_D} = \overrightarrow{V_C} - \overrightarrow{I_{CD}}\ \overrightarrow{Z_{CD}}$

$$= (6054{\cdot}46 - j\,115{\cdot}39) - (-20{\cdot}3 - j\,12{\cdot}8) \times (0{\cdot}8 + j\,0{\cdot}5)$$
$$= \mathbf{(6064{\cdot}3 - j\,95)\ volts/phase}$$

TUTORIAL PROBLEMS

1. A single phase distributor *AB* has a total impedance of $(0{\cdot}1 + j\,0{\cdot}2)$ ohm. At the far end *B*, a current of 80 A at 0·8 p.f. lagging and at mid-point *C* a current of 100 A at 0·6 p.f. lagging are tapped. If the voltage of the far end is maintained at 200 V, determine :
   (*i*) Supply end voltage V_A
   (*ii*) Phase angle between V_A and V_B
   The load power factors are w.r.t. the voltage at the far end. **[(*i*) 227·22 V (*ii*) 2°31′]**
2. A single-phase a.c. distributor *AB* is fed from end *A* and has a total impedance of $(0{\cdot}2 + j\,03)$ ohm. At the far end, the voltage $V_B = 240$ V and the current is 100 A at a p.f. of 0·8 lagging. At the mid-point *M*, a current of 100 A is tapped at a p.f. of 0·8 lagging with reference to the voltage V_M at the mid-point. Calculate the supply voltage V_A and phase angle between V_A and V_B. **[292 V, 2·6°]**
3. A single phase ring distributor *ABC* is fed at A. The loads at *B* and *C* are 40 A at 0·8 p.f. lagging and 60 A at 0·6 p.f. lagging respectively. Both power factors expressed are referred to the voltage at point A. The total impedance of sections *AB*, *BC* and *CA* are $2 + j1$, $2 + j3$ and $1 + j2$ ohms respectively. Determine the current in each section.
   [Current in $AB = (39{\cdot}54 - j\,25{\cdot}05)$ amp ; $BC = (7{\cdot}54 - j\,1{\cdot}05)$ amp ; $CA = (28{\cdot}46 - j\,46{\cdot}95)$ amp.]
4. A 3-phase ring distributor *ABCD* fed at A at 11 kV supplies balanced loads of 40 A at 0·8 p.f. lagging at *B*, 50 A at 0·707 p.f. lagging at C and 30 A at 0·8 p.f. lagging at D, the load currents being referred to the supply voltage at A.
   The impedances per phase of the various sections are :
   Section $AB = (1 + j\,2)\ \Omega$; Section $BC = (2 + j\,3)\ \Omega$
   Section $CD = (1 + j\,1)\ \Omega$; Section $DA = (3 + j\,4)\ \Omega$
   Calculate the currents in various sections and station bus-bar voltages at *B*, *C* and *D*.
   **[Current in $AB = (53{\cdot}8 - j\,46)$ amp ; $BC = (21{\cdot}8 - j\,22)$ amp.
   $CD = (-13{\cdot}55 + j\,13{\cdot}35)$ amp ; $DA = (-40{\cdot}55 - j\,26{\cdot}45)$ amp.
   $V_B = (6212{\cdot}5 - j\,61{\cdot}6)$ volts/phase ; $V_C = (6103 - j\,83)$ volts/phase
   $V_D = (6129{\cdot}8 - j\,82{\cdot}8)$ volts/phase]**

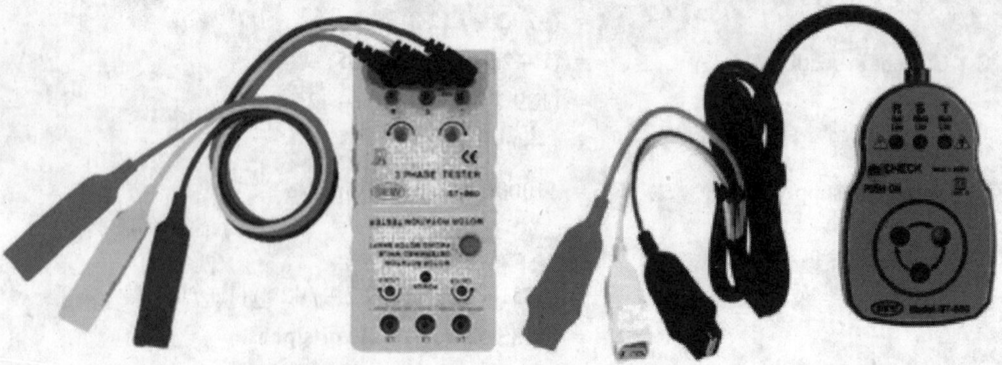

Phase Sequence Indicator

14.3 3-Phase Unbalanced Loads

The 3-phase loads that have the same impedance and power factor in each phase are called balanced loads. The problems on balanced loads can be solved by considering one phase only ; the conditions in the other two phases being similar. However, we may come across a situation when loads are unbalanced *i.e.* each load phase has different impedance and/or power factor. In that case, current and power in each phase will be different. In practice, we may come across the following unbalanced loads :

(*i*) Four-wire star-connected unbalanced load

(*ii*) Unbalanced Δ-connected load

(*iii*) Unbalanced 3-wire, *Y*-connected load

The 3-phase, 4-wire system is widely used for distribution of electric power in commercial and industrial buildings. The single phase load is connected between any line and neutral wire while a 3-phase load is connected across the three lines. The 3-phase, 4-wire system invariably carries *unbalanced loads. In this chapter, we shall only discuss this type of unbalanced load.

14.4 Four-Wire Star-Connected Unbalanced Loads

We can obtain this type of load in two ways. First, we may connect a 3-phase, 4-wire unbalanced load to a 3-phase, 4-wire supply as shown in Fig. 14.10. Note that star point *N* of the supply is connected to the load star point *N*′. Secondly, we may connect single phase loads between any line and the neutral wire as shown in Fig.14.11. This will also result in a 3-phase, 4-wire **unbalanced load because it is rarely possible that single phase loads on all the three phases have the same magnitude and power factor. Since the load is unbalanced, the line currents will be different in magnitude and displaced from one another by unequal angles. The current in the neutral wire will be the phasor sum of the three line currents *i.e.*

Current in neutral wire, $I_N = I_R + I_Y + I_B$ ...*phasor sum*

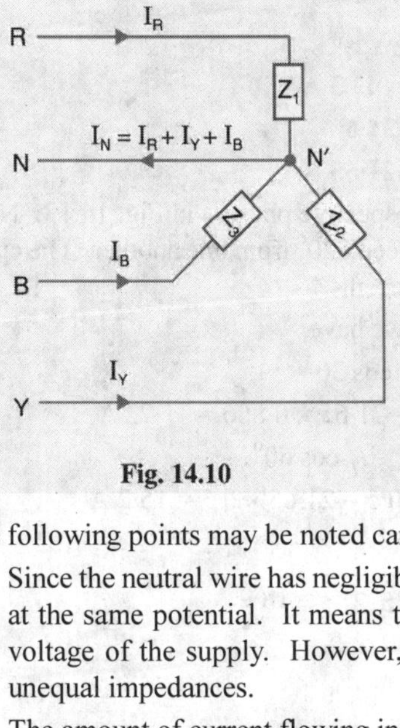

Fig. 14.10

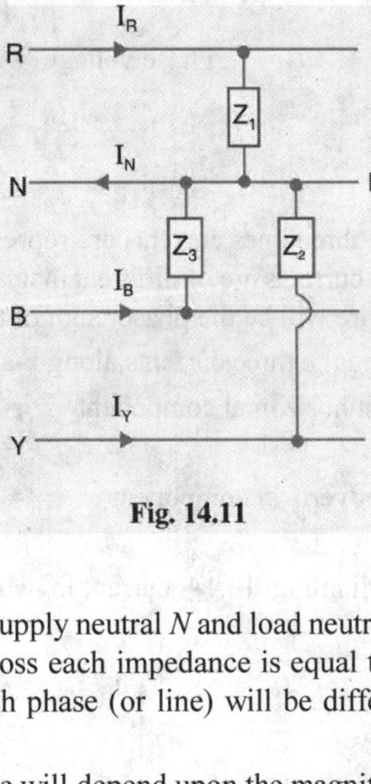

Fig. 14.11

The following points may be noted carefully :

(*i*) Since the neutral wire has negligible resistance, supply neutral *N* and load neutral *N*′ will be at the same potential. It means that voltage across each impedance is equal to the phase voltage of the supply. However, current in each phase (or line) will be different due to unequal impedances.

(*ii*) The amount of current flowing in the neutral wire will depend upon the magnitudes of line currents and their phasor relations. In most circuits encountered in practice, the neutral current is equal to or smaller than one of the line currents. The exceptions are those circuits having severe unbalance.

---

* No doubt 3-phase loads (*e.g.* 3-phase motors) connected to this supply are balanced but when we add single phase loads (*e.g.* lights, fans etc.), the balance is lost. It is because it is rarely possible that single phase loads on all the three phases have the same magnitude and power factor.

** In actual practice, we never have an unbalanced 3-phase, 4-wire load. Most of the 3-phase loads (*e.g.* 3-phase motors) are 3-phase, 3-wire and are balanced loads. In fact, these are the single phase loads on the 3-phase, 4-wire supply which constitute unbalanced, 4-wire *Y*-connected load.

Example 14.7. *Non-reactive loads of 10 kW, 8 kW and 5 kW are connected between the neutral and the red, yellow and blue phases respectively of a 3-phase, 4-wire system. The line voltage is 400V. Calculate (i) the current in each line and (ii) the current in the neutral wire.*

Solution. This is a case of unbalanced load so that the line currents (and hence the phase currents) in the three lines will be different. The current in the *neutral wire will be equal to the phasor sum of three line currents as shown in Fig. 14.12.

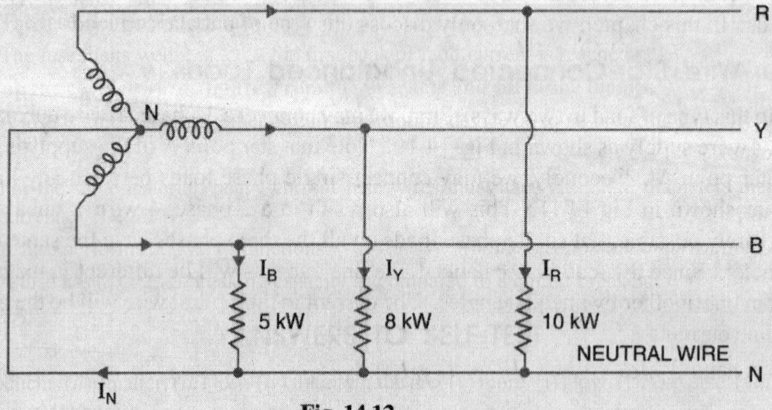

Fig. 14.12

(*i*) Phase voltage $= 400/\sqrt{3} = 231$ V

$$I_R = 10 \times 10^3/231 = \mathbf{43{\cdot}3\ A}$$

$$I_Y = 8 \times 10^3/231 = \mathbf{34{\cdot}6\ A}$$

$$I_B = 5 \times 10^3/231 = \mathbf{21{\cdot}65\ A}$$

(*ii*) The three lines currents are represented by the respective phasors in Fig. 14.13. Note that the three line currents are of different magnitude but displaced 120° from one another. The current in the neutral wire will be the phasor sum of the three line currents.

Resolving the three currents along *x*-axis and *y*-axis, we have,

Resultant horizontal component $= I_Y \cos 30^\circ - I_B \cos 30^\circ$

$= 34{\cdot}6 \times 0{\cdot}866 - 21{\cdot}65 \times 0{\cdot}866 = 11{\cdot}22$ A

Resultant vertical component $= I_R - I_Y \cos 60^\circ - I_B \cos 60^\circ$

$= 43{\cdot}3 - 34{\cdot}6 \times 0{\cdot}5 - 21{\cdot}65 \times 0{\cdot}5 = 15{\cdot}2$ A

As shown in Fig. 14.14, current in neutral wire is

$$I_N = \sqrt{(11{\cdot}22)^2 + (15{\cdot}2)^2} = \mathbf{18{\cdot}9\ A}$$

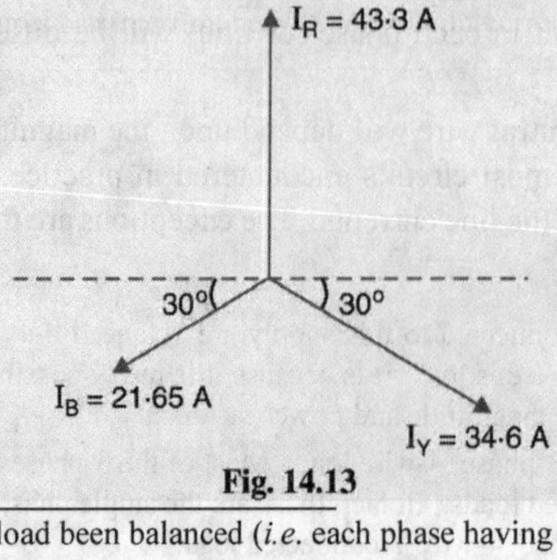

Fig. 14.13

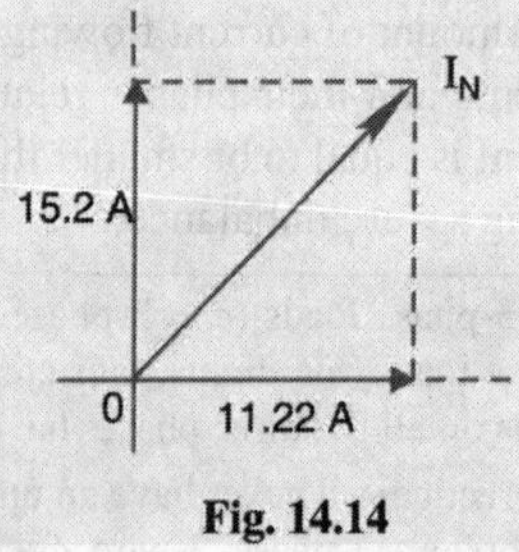

Fig. 14.14

* Had the load been balanced (*i.e.* each phase having identical load), the current in the neutral wire would have been zero.

Example 14.8. *A 3-phase, 4-wire system supplies power at 400 V and lighting at 230 V. If the lamps is use require 70, 84 and 33 amperes in each of the three lines, what should be the current in the neutral wire ? If a 3-phase motor is now started, taking 200 A from the lines at a p.f. of 0·2 lagging, what should be the total current in each line and the neutral wire ? Find also the total power supplied to the lamps and the motor.*

Solution. Fig. 14.15 shows the lamp load and motor load on 400 V/230 V, 3-phase, 4-wire sypply.

Lamp load alone. If there is lamp load alone, the line currents in phases *R, Y* and *B* are 70 *A*, 84 *A* and 33 *A* respectively. These currents will be 120° apart (assuming phase sequence *RYB*) as shown in Fig.14.16.

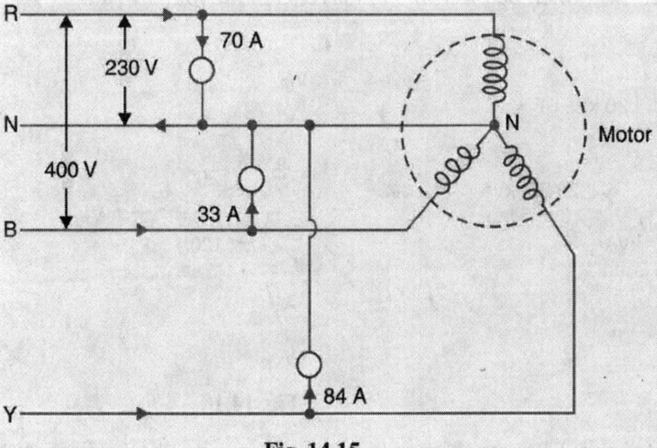

Fig. 14.15

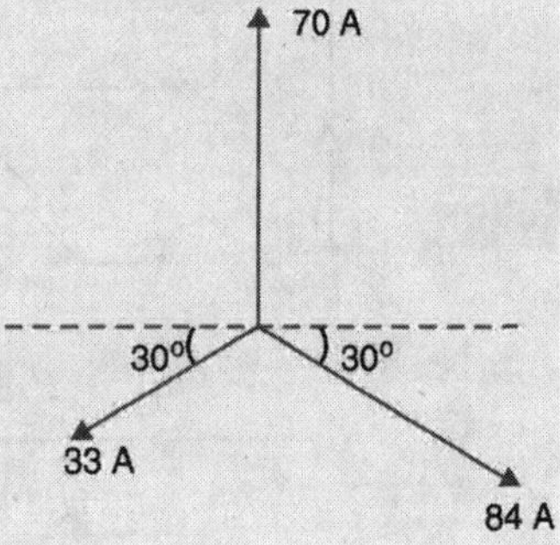

Fig. 14.16

Resultant *H*-component $= 84 \cos 30^\circ - 33 \cos 30^\circ = 44{\cdot}17$ A

Resultant V-component $= 70 - 33 \cos 60^\circ - 84 \cos 60^\circ = 11{\cdot}5$ A

∴ Neutral current, $I_N = \sqrt{(44{\cdot}17)^2 + (11{\cdot}5)^2} = \mathbf{45{\cdot}64\ A}$

Both lamp load and motor load

When motor load is also connected along with lighting load, there will be no change in current in the neutral wire. It is because the motor load is balanced and hence no current will flow in the neutral wire due to this load.

∴ Neutral current, $I_N = \mathbf{45{\cdot}64\ A}$ *...same as before*

The current in each line is the phasor sum of the line currents due to lamp load and motor load.

Active component of motor current $= 200 \times \cos \phi_m = 200 \times 0{\cdot}2 = 40$ A

Reactive component of motor current $= 200 \times \sin \phi_m = 200 \times 0{\cdot}98 = 196$ A

$$\therefore \quad I_R = \sqrt{(\text{sum of active comp.})^2 + (\text{reactive comp.})^2}$$

$$= \sqrt{(40 + 70)^2 + (196)^2} = \mathbf{224{\cdot}8\ A}$$

$$I_Y = \sqrt{(40 + 84)^2 + (196)^2} = \mathbf{232\ A}$$

$$I_B = \sqrt{(40 + 33)^2 + (196)^2} = \mathbf{209{\cdot}15\ A}$$

Power supplied

Power supplied to lamps $= 230\,(70 + 84 + 33) \times 1 = \mathbf{43010\ W}$ $(\because \cos \phi_L = 1)$

Power supplied to motor $= \sqrt{3}\ V_L I_L \cos \phi_m$

$= \sqrt{3} \times 400 \times 200 \times 0{\cdot}2 = \mathbf{27712\ W}$

Example 14.9. *The three line leads of a 400/230 V, 3-phase, 4-wire supply are designated as R, Y and B respectively. The fourth wire or neutral wire is designated as N. The phase sequence is RYB. Compute the currents in the four wires when the following loads are connected to this supply :*

From R to N : 20 kW, unity power factor

From Y to N : 28·75 kVA, 0·866 lag

From B to N : 28·75 kVA, 0·866 lead

If the load from B to N is removed, what will be the value of currents in the four wires ?

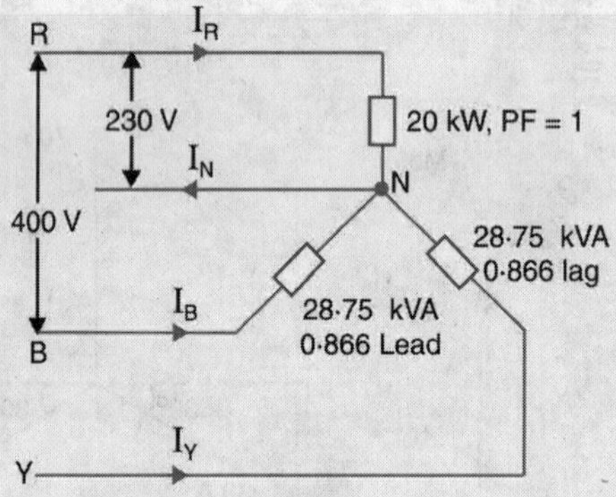

Fig. 14.17

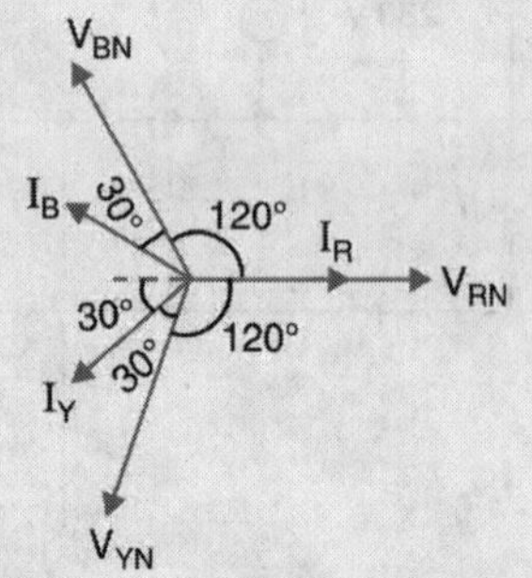

Fig. 14.18

Solution. Fig. 14.17 shows the circuit diagram whereas Fig.14.18 shows its phasor diagram. The current I_R is in phase with V_{RN}, current I_Y lags behind its phase voltage V_{YN} by $\cos^{-1} 0{\cdot}866 = 30^\circ$ and the current I_B leads its phase voltage V_{BN} by $\cos^{-1} 0{\cdot}866 = 30^\circ$.

$$I_R = 20 \times 10^3/230 = \mathbf{89{\cdot}96\ A}$$

$$I_Y = 28{\cdot}75 \times 10^3/230 = \mathbf{125\ A}$$

$$I_B = 28{\cdot}75 \times 10^3/230 = \mathbf{125\ A}$$

The current in the neutral wire will be equal to the phasor sum of the three line currents I_R, I_Y and I_B. Referring to the phasor diagram in Fig.14.18 and resolving these currents along *x*-axis and *y*-axis, we have,

Resultant *X*-component $= 86{\cdot}96 - 125 \cos 30^\circ - 125 \cos 30^\circ$

$= 86{\cdot}96 - 108{\cdot}25 - 108{\cdot}25 = -129{\cdot}54$ A

Resultant *Y*-component $= 0 + 125 \sin 30^\circ - 125 \sin 30^\circ = 0$

∴ Neutral current, $I_N = \sqrt{(-129 \cdot 54)^2 + (0)^2} = \mathbf{129{\cdot}54A}$

When load from *B* to *N* removed. When the load from *B* to *N* is removed, the various line currents are :

$I_R = \mathbf{86{\cdot}96A}$ in phase with V_{RN} ; $I_Y = \mathbf{125A}$ lagging V_{YN} by 30° ; $I_B = \mathbf{0\ A}$

The current in the neutral wire is equal to the phasor sum of these three line currents. Resolving the currents along *x*-axis and *y*-axis, we have,

Resultant *X*-component $= 86{\cdot}96 - 125 \cos 30^\circ = 86{\cdot}96 - 108{\cdot}25 = -21{\cdot}29$ A

Resultant *Y*-component $= 0 - 125 \sin 30^\circ = 0 - 125 \times 0{\cdot}5 = -62{\cdot}5$ A

∴ Neutral current, $I_N = \sqrt{(-21 \cdot 29)^2 + (-62 \cdot 5)^2} = \mathbf{66{\cdot}03\ A}$

Example 14.10. *A 3-phase, 4-wire distributor supplies a balanced voltage of 400/230 V to a load consisting of 30 A at p.f. 0·866 lagging for R-phase, 30 A at p.f. 0·866 leading for Y phase and 30 A at unity p.f. for B phase. The resistance of each line conductor is 0·2 Ω. The area of X-section of neutral is half of any line conductor. Calculate the supply end voltage for R phase. The phase sequence is RYB.*

Solution. The circuit diagram is shown in Fig. 14.19. Since neutral is half the cross-section, its resistance is 0·4 Ω. Considering the load end and taking V_R as the reference vector, the phase voltages can be written as :

$$\vec{V_R} = 230 \angle 0^\circ \text{ volts} \;;\; \vec{V_Y} = 230 \angle -120^\circ \text{ volts} \;;\; \vec{V_B} = 230 \angle 120^\circ \text{ volts}$$

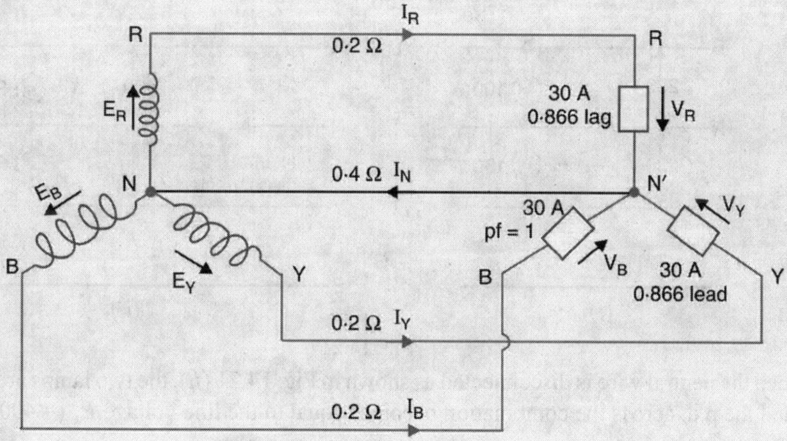

Fig. 14.19

The vector diagram of the circuit is shown in Fig. 14.20. The line current I_R lags behind V_R by an angle $\cos^{-1} 0{\cdot}866 = 30^\circ$. The current I_Y leads V_Y by 30° and the current I_B is in phase with V_B. Referring to the vector diagram of Fig.14.20, the line currents can be expressed as :

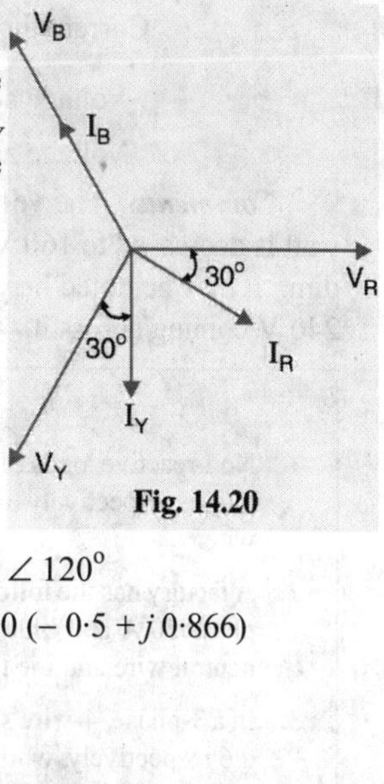

Fig. 14.20

$$\vec{I_R} = 30 \angle -30^\circ \text{ amperes}$$

$$\vec{I_Y} = 30 \angle -90^\circ \text{ amperes}$$

$$\vec{I_B} = 30 \angle 120^\circ \text{ amperes}$$

Current in neutral wire, $\vec{I_N} = \vec{I_R} + \vec{I_Y} + \vec{I_B}$

$$= 30 \angle -30^\circ + 30 \angle -90^\circ + 30 \angle 120^\circ$$

$$= 30\,(0{\cdot}866 - j\,0{\cdot}5) - 30\,(j) + 30\,(-0{\cdot}5 + j\,0{\cdot}866)$$

$$= 10{\cdot}98 - j\,19{\cdot}02$$

Let the supply voltage of phase R to neutral be $\vec{E_R}$. Then,

$$\vec{E_R} = \vec{V_R} + \text{Drop in } R \text{ phase} + \text{Drop in neutral}$$

$$= (230 + j\,0) + 0{\cdot}2 \times 30 \angle -30^\circ + (10{\cdot}98 - j\,19{\cdot}02) \times 0{\cdot}4$$

$$= 230 + 6\,(0{\cdot}866 - j\,0{\cdot}5) + 0{\cdot}4\,(10{\cdot}98 - j\,19{\cdot}02)$$

$$= 239{\cdot}588 - j\,10{\cdot}608$$

$$= \mathbf{239{\cdot}8 \angle -2{\cdot}54^\circ \text{ volts}}$$

Example 14.11. *In a 3-phase, 4-wire, 400/230 V system, a lamp of 100 watts is connected to one phase and neutral and a lamp of 150 watts is connected to the second phase and neutral. If the neutral wire is disconnected accidentally, what will be the voltage across each lamp ?*

Solution. Fig. 14.21 (*i*) shows the lamp connections. The lamp L_1 of 100 watts is connected between phase *R* and neutral whereas lamp L_2 of 150 watts is connected between phase *Y* and the neutral.

$$\text{Resistance of lamp } L_1, \quad R_1 = \frac{(230)^2}{100} = 529\ \Omega$$

$$\text{Resistance of lamp } L_2, \quad R_2 = \frac{(230)^2}{150} = 352{\cdot}67\ \Omega$$

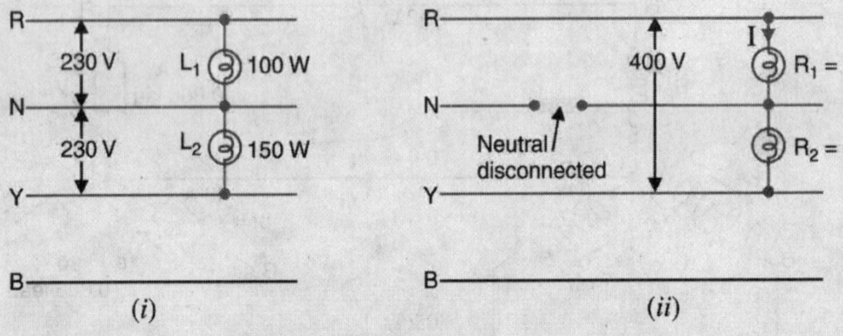

Fig. 14.21

When the neutral wire is disconnected as shown in Fig. 14.21 (*ii*), the two lamps are connected in series and the p.d. across the combination becomes equal to the line voltage E_L (= 400 V).

$$\text{Current through lamps, } I = \frac{E_L}{R_1 + R_2} = \frac{400}{529 + 352 \cdot 67} = 0{\cdot}454\ \text{A}$$

$$\text{Voltage across lamp } L_1 = I R_1 = 0{\cdot}454 \times 529 = \mathbf{240\ V}$$

$$\text{Voltage across lamp } L_2 = I R_2 = 0{\cdot}454 \times 352{\cdot}67 = \mathbf{160\ V}$$

Comments. The voltage across 100-watt lamp is increased to 240 V whereas that across 150-watt is decreased to 160 V. Therefore, 100-watt lamp becomes brighter and 150-watt lamp becomes dim. It may be noted here that if 100-watt lamp happens to be rated at 230 V, it may burn out due to 240 V coming across it.

TUTORIAL PROBLEMS

1. Non-reactive loads of 10 kW, 6kW and 4 kW are connected between the neutral and red, yellow and blue phases respectively of a 3-phase, 4-wire 400/230V supply. Find the current in each line and in the neutral wire. **[I_R = 43·3 A; I_Y = 26 A; I_B = 17·3 A; I_N = 22·9 A]**
2. A factory has the following loads with a power factor of 0·9 lagging in each case. Red phase 40 A, yellow phase 50 A and blue phase 60 A. If the supply is 400V, 3-phase, 4-wire, calculate the current in the neutral wire and the total power. **[17·3A, 31·2 kW]**
3. In a 3-phase, 4-wire system, two phases have currents of 10A and 6A at lagging power factors of 0·8 and 0·6 respectively, while the third phase is open-circuited. Calculate the current in the neutral wire. **[7A]**
4. A 3-phase, 4-wire system supplies a lighting load of 40A, 30A and 20A respectively in the three phases. If the line voltage is 400 V, determine the current in the neutral wire. **[17·32A]**

14.5. Ground Detectors

Ground detectors are the devices that are used to detect the ground fault for ungrounded a.c. systems.

When a ground fault occurs on such a system, immediate steps should be taken to clear it. If this is not done and a second ground fault happens, a short circuit occurs.

Fig.14.22 shows how lamps are connected to an ungrounded 3-phase system for the detection of ground fault. If ground fault occurs on any wire, the lamp connected to that wire will be dim and the lamps connected to healthy (ungrounded) wire will become brighter.

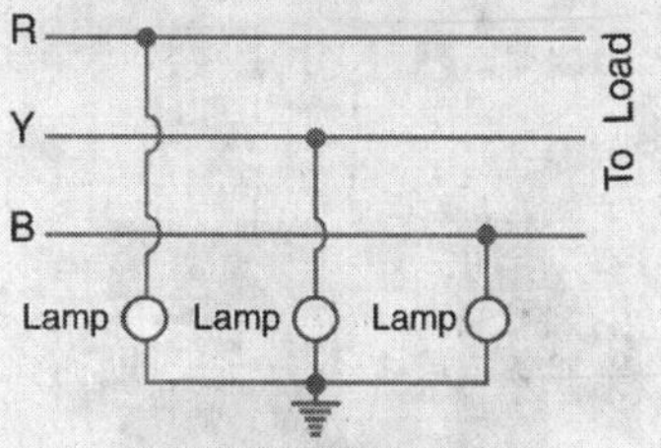

Fig. 14.22

SELF - TEST

1. **Fill in the blanks by inserting appropriate words/figures.**
   (i) The most common system for secondary distribution is 400/..... V, 3-phase, ... wire system.
   (ii) In a 3-phase, 4-wire a.c. system, if the loads are balanced, then current in the neutral wire is ....
   (iii) Distribution transformer links the .... and .... systems.
   (iv) The 3-phase, 3-wire a.c. system of distribution is used for .... loads.
   (v) For combined power and lighting load, ... system is used.
2. **Pick up the correct words/figures from brackets and fill in the blanks.**
   (i) 3-phase, 4-wire a.c. system of distribution is used for .... load. *(balanced, unbalanced)*
   (ii) In a balanced 3-phase, 4-wire a.c. system, the phase sequence is *RYB*. If the voltage of *R* phase = 230 $\angle$ 0° volts, then for *B* phase it will be .... *(230 $\angle$ – 120° volts, 230 $\angle$ 120° volts)*
   (iii) In a.c. system, additions and subtractions of currents are done .... *(vectorially, arithmetically)*
   (iv) The area of X-section of neutral is generally .... that of any line conductor. *(the same, half)*
   (v) For purely domestic loads, ... a.c. system is employed for distribution. *(single phase 2-wire, 3-phase 3-wire)*

ANSWERS TO SELF-TEST

1. (i) 230, 4 (ii) zero (iii) primary, secondary (iv) balanced (v) 3-phase 4-wire.
2. (i) unbalanced (ii) 230 $\angle$120° (iii) vectorially (iv) half (v) single phase 2-wire.

CHAPTER REVIEW TOPICS

1. How does a.c. distribution differ from d.c. distribution ?
2. What is the importance of load power factors in a.c. distribution ?
3. Describe briefly how will you solve a.c. distribution problems ?
4. Write short notes on the following :
   (i) Difference between d.c. and a.c. distribution
   (ii) Systems of a.c. distribution

DISCUSSION QUESTIONS

1. What are the undesirable effects of too much voltage variation on a distribution circuit ?
2. What are the effects of diversity factor on the maximum load of a distribution transformer ?
3. Where does the greatest current density occur in a distribution feeder ?
4. What is the controlling factor in determining the size of a distributor ?
5. In which situation is secondary distribution eliminated ?

CHAPTER

15

Voltage Control

Introduction

In a modern power system, electrical energy from the generating station is delivered to the ultimate consumers through a network of transmission and distribution. For satisfactory operation of motors, lamps and other loads, it is desirable that consumers are supplied with substantially constant voltage. Too wide variations of voltage may cause erratic operation or even malfunctioning of consumers' appliances. To safeguard the interest of the consumers, the government has enacted a law in this regard. The statutory limit of voltage variation is ± 6% of declared voltage at consumers' terminals.

The principal cause of voltage variation at consumer's premises is the change in load on the supply system. When the load on the system increases, the voltage at the consumer's terminals falls due to the increased voltage drop in (*i*) alternator synchronous impedance (*ii*) transmission line (*iii*) transformer impedance (*iv*) feeders and (*v*) distributors. The reverse would happen should the load on the system decrease. These voltage variations are undesirable and must be kept within the prescribed limits (*i.e.* ± 6% of the declared voltage). This is achieved by installing voltage regulating equipment at suitable places in the power system. The purpose of this chapter is to deal with important voltage control equipment and its increasing utility in this fast developing power system.

15.1 Importance of Voltage Control

When the load on the supply system changes, the voltage at the consumer's terminals also changes. The variations of voltage at the consumer's terminals are undesirable and must be kept within prescribed limits for the following reasons :

(*i*) In case of lighting load, the lamp characteristics are very sensitive to changes of voltage. For instance, if the supply voltage to an incandescent lamp decreases by 6% of rated value, then illuminating power may decrease by 20%. On the other hand, if the supply voltage is 6% above the rated value, the life of the lamp may be reduced by 50% due to rapid deterioration of the filament.

(*ii*) In case of power load consisting of induction motors, the voltage variations may cause erratic operation. If the supply voltage is above the normal, the motor may operate with a saturated magnetic circuit, with consequent large magnetising current, heating and low power factor. On the other hand, if the voltage is too low, it will reduce the starting torque of the motor considerably.

(*iii*) Too wide variations of voltage cause excessive heating of distribution transformers. This may reduce their ratings to a considerable extent.

It is clear from the above discussion that voltage variations in a power system must be kept to minimum level in order to deliver good service to the consumers. With the trend towards larger and larger interconnected system, it has become necessary to employ appropriate methods of voltage control.

15.2 Location of Voltage Control Equipment

In a modern power system, there are several elements between the generating station and the consumers. The voltage control equipment is used at more than one point in the system for two reasons. Firstly, the power network is very extensive and there is a considerable voltage drop in transmission and distribution systems. Secondly, the various circuits of the power system have dissimilar load characteristics. For these reasons , it is necessary to provide individual means of voltage control for each circuit or group of circuits. In practice, voltage control equipment is used at :

(*i*) generating stations

(*ii*) transformer stations

(*iii*) the feeders if the drop exceeds the permissible limits

15.3 Methods of Voltage Control

There are several methods of voltage control. In each method, the system voltage is changed in accordance with the load to obtain a fairly constant voltage at the consumer's end of the system. The following are the methods of voltage control in an *a.c. power system:

(*i*) By excitation control

(*ii*) By using tap changing transformers

(*iii*) Auto-transformer tap changing

(*iv*) Booster transformers

(*v*) Induction regulators

(*vi*) By synchronous condenser

Method (*i*) is used at the generating station only whereas methods (*ii*) to (*v*) can be used for

* Since the modern power system is a.c., voltage control for this system will be discussed. However, for a d.c. system, voltage control can be effected by (*i*) overcompounded generators and (*ii*) boosters.

transmission as well as primary distribution systems. However, methods (*vi*) is reserved for the voltage control of a transmission line. We shall discuss each method separately in the next sections.

15.4 Excitation Control

When the load on the supply system changes, the terminal voltage of the alternator also varies due to the changed voltage drop in the synchronous reactance of the armature. The voltage of the alternator can be kept constant by changing the *field current of the alternator in accordance with the load. This is known as *excitation control* method. The excitation of alternator can be controlled by the use of automatic or hand operated regulator acting in the field circuit of the alternator. The first method is preferred in modern practice. There are two main types of automatic voltage regulators *viz*.

(*i*) Tirril Regulator

(*ii*) Brown-Boveri Regulator

These regulators are based on the "overshooting the mark †principle" to enable them to respond quickly to the rapid fluctuations of load. When the load on the alternator increases, the regulator produces an increase in excitation more than is ultimately necessary. Before the voltage has the time to increase to the value corresponding to the increased excitation, the regulator reduces the excitation to the proper value.

15.5 Tirril Regulator

In this type of regulator, a fixed resistance is cut in and cut out of the exciter field circuit of the alternator. This is achieved by rapidly opening and closing a shunt circuit across the exciter rheostat. For this reason, it is also known as vibrating type voltage regulator.

Construction. Fig. 15.1 shows the essential parts of a Tirril voltage regulator. A rheostat R is provided in the exciter circuit and its value is set to give the required excitation. This rheostat is put in and out of the exciter circuit by the regulator, thus varying the exciter voltage to maintain the desired voltage of the alternator.

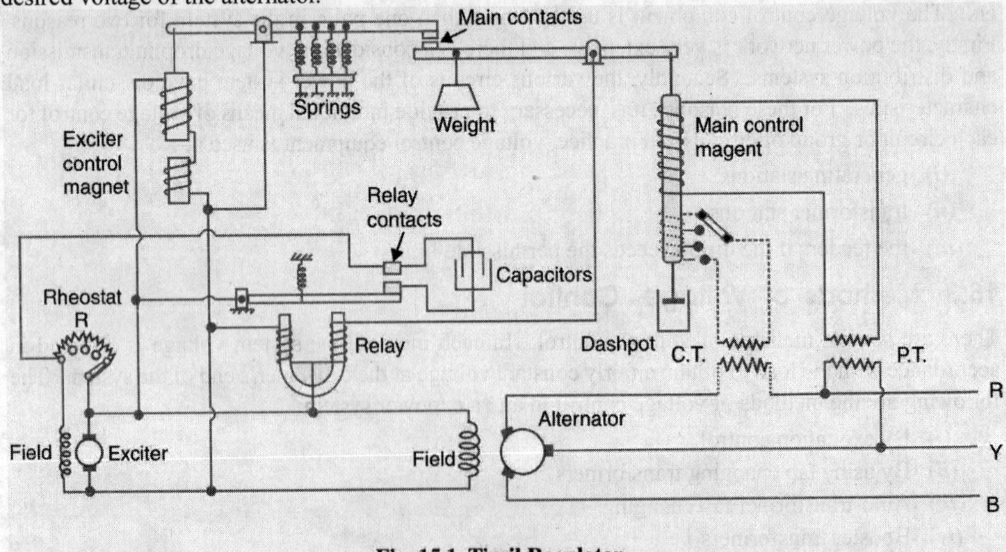

Fig. 15.1 Tirril Regulator

* As alternator has to be run at constant speed to obtain fixed frequency, therefore, induced e.m.f. of the alternator cannot be controlled by the adjustment of speed.

† The alternator has large inductance. If the exciter voltage is increased, the field current will take some time to reach the steady value. Therefore, response will not be quick. However, quick response is necessary to meet the rapid fluctuations of load. For this reason, this principle is used.

(i) *Main contact.* There are two levers at the top which carry the main contacts at the facing ends. The left-hand lever is controlled by the exciter magnet whereas the right hand lever is controlled by an a.c. magnet known as main control magnet.

(ii) *Exciter magnet.* This magnet is of the ordinary solenoid type and is connected across the exciter mains. Its exciting current is, therefore, proportional to the exciter voltage. The counter balancing force for the exciter magnet is provided by four coil springs.

(iii) *A. C. magnet.* It is also of solenoid type and is energised from a.c. bus-bars. It carries series as well as shunt excitation. This magnet is so adjusted that with normal load and voltage at the alternator, the pulls of the two coils are equal and opposite, thus keeping the right-hand lever in the horizontal position.

(iv) *Differential relay.* It essentially consists of a U-shaped relay magnet which operates the relay contacts. The relay magnet has two identical windings wound differentially on both the limbs. These windings are connected across the exciter mains–the left hand one permanently while the right hand one has its circuit completed only when the main contacts are closed. The relay contacts are arranged to shunt the exciter-field rheostat R. A capacitor is provided across the relay contacts to reduce the sparking at the time the relay contacts are opened.

Operation. The two control magnets (*i.e.* exciter magnet and a.c. magnet) are so adjusted that with normal load and voltage at the alternator, their pulls are equal, thus keeping the main contacts open. In this position of main contacts, the relay magnet remains energised and pulls down the armature carrying one relay contact. Consequently, relay contacts remain open and the exciter field rheostat is in the field circuit.

When the load on the alternator increases, its terminal voltage tends to fall. This causes the series excitation to predominate and the a.c. magnet pulls down the right-hand lever to close the main contacts. Consequently, the relay magnet is *de-energised and releases the armature carrying the relay contact. The relay contacts are closed and the rheostat R in the field circuit is short circuited. This increases the exciter-voltage and hence the excitation of the alternator. The increased excitation causes the alternator voltage to rise quickly. At the same time, the excitation of the exciter magnet is increased due to the increase in exciter voltage. Therefore, the left-hand lever is pulled down, opening the main contacts, energising the relay magnet and putting the rheostat R again in the field circuit before the alternator voltage has time to increase too far. The reverse would happen should the load on the alternator decrease.

It is worthwhile to mention here that exciter voltage is controlled by the rapid opening and closing of the relay contacts. As the regulator is worked on the overshooting the mark principle, therefore, the terminal voltage does not remain absolutely constant but oscillates between the maximum and minimum values. In fact, the regulator is so quick acting that voltage variations never exceed ± 1%.

15.6 Brown-Boveri Regulator

In this type of regulator, exciter field rheostat is varied continuously or in small steps instead of being first completely cut in and then completely cut out as in Tirril regulator. For this purpose, a regulating resistance is connected in series with the field circuit of the exciter. Fluctuations in the alternator voltage are detected by a control device which actuates a motor. The motor drives the regulating rheostat and cuts out or cuts in some resistance from the rheostat, thus changing the exciter and hence the alternator voltage.

Construction. Fig. 15.2 shows the schematic diagram of a Brown-Boveri voltage regulator. It

* Because the windings are wound differentially on the two limbs.

also works on the "overshooting the mark principle" and has the following four important parts :

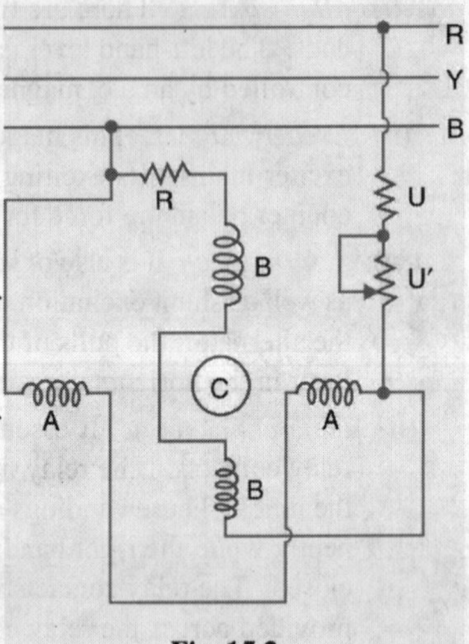

Fig. 15.2

(*i*) *Control system.* The control system is built on the principle of induction motor. It consists of two windings *A* and *B* on an annular core of laminated sheet steel. The winding *A* is excited from two of the generator terminals through resistances *U* and *U'* while a resistance *R* is inserted in the circuit of winding *B*. The ratio of resistance to reactance of the two windings are suitably adjusted so as to create a phase difference of currents in the two windings. Due to the phase difference of currents in the two windings, rotating magnetic field is set up. This produces electromagnetic torque on the thin aluminium drum *C* carried by steel spindle ; the latter being supported at both ends by jewel bearings. The torque on drum *C* varies with the terminal voltage of the alternator. The variable resistance *U'* can also vary the torque on the drum. If the resistance is increased, the torque is decreased and vice-versa. Therefore, the variable resistance *U'* provides a means by which the regulator may be set to operate at the desired voltage.

(*ii*) *Mechanical control torque.* The electric torque produced by the current in the split phase winding is opposed by a combination of two springs (main spring and auxiliary spring) which produce a constant mechanical torque irrespective of the position of the drum. Under steady deflected state, mechanical torque is equal and opposite to the electric torque.

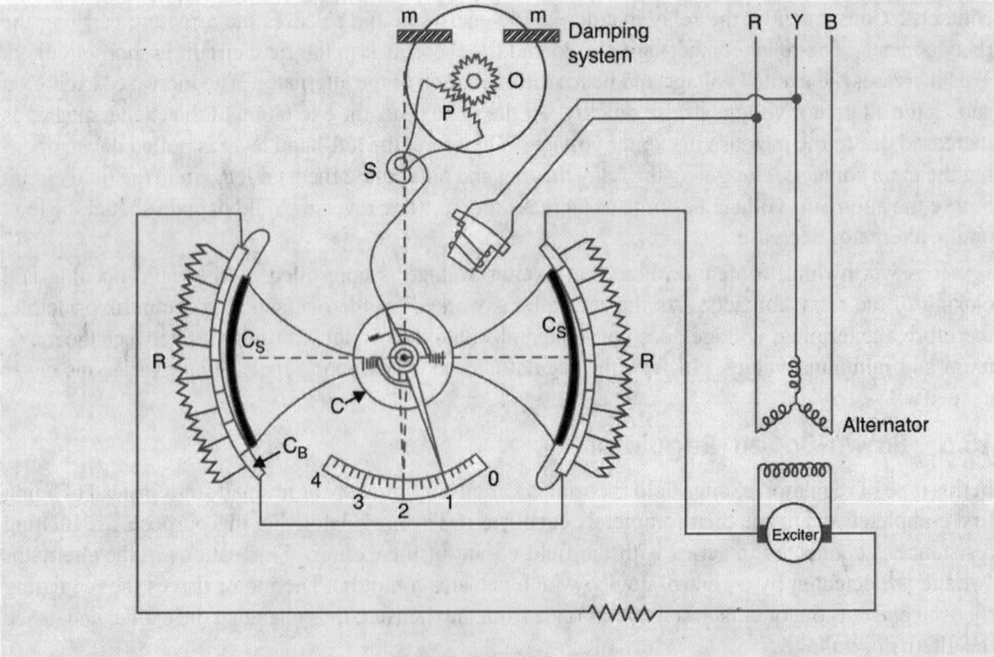

Fig. 15.3

(*iii*) *Operating system.* It consists of a field rheostat with contact device. The rheostat consists of a pair of resistance elements connected to the stationary contact blocks C_B. These two resistance sectors *R* are connected in series with each other and then in series with the field circuit of the exciter. On the inside surface of the contact blocks roll the contact sectors C_S. When the terminal voltage of the alternator changes, the electric torque acts on the drum. This causes the contact sectors to roll over the contact blocks, cutting in or cutting out rheostat resistance in the exciter field circuit.

(*iv*) *Damping torque.* The regulator is made stable by damping mechanism which consists of an aluminium disc *O* rotating between two permanent magnets *m*. The disc is geared to the rack of an aluminium sector *P* and is fastened to the aluminium drum *C* by means of a flexible spring *S* acting as the recall spring. If there is a change in the alternator voltage, the eddy currents induced in the disc *O* produce the necessary damping torque to resist quick response of the moving system.

Operation. Suppose that resistances *U* and *U'* are so adjusted that terminal voltage of the alternator is normal at position 1. In this position, the electrical torque is counterbalanced by the mechanical torque and the moving system is in equilibrium. It is assumed that electrical torque rotates the shaft in a clockwise direction.

Now imagine that the terminal voltage of the alternator rises due to decrease in load on the supply system. The increase in the alternator voltage will cause an increase in electrical torque which becomes greater than the mechanical torque. This causes the drum to rotate in clockwise direction, say to position 3. As a result, more resistance is inserted in the exciter circuit, thereby decreasing the field current and hence the terminal voltage of the alternator. Meanwhile, the recall spring *S* is tightened and provides a counter torque forcing the contact roller back to position 2 which is the equilibrium position. The damping system prevents the oscillations of the system about the equilibrium position.

15.7 Tap-Changing Transformers

The excitation control method is satisfactory only for relatively short lines. However, it is *not suitable for long lines as the voltage at the alternator terminals will have to be varied too much in order that the voltage at the far end of the line may be constant. Under such situations, the problem of voltage control can be solved by employing other methods. One important method is to use tap-changing transformer and is commonly employed where main transformer is necessary. In this method, a number of tappings are provided on the secondary of the transformer. The voltage drop in the line is supplied by changing the secondary e.m.f. of the transformer through the adjustment of its number of turns.

(*i*) **Off load tap-changing transformer.** Fig. 15.4 shows the arrangement where a number of tappings have been provided on the secondary. As the position of the tap is varied, the effective number of secondary turns is varied and hence the output voltage of the secondary can be changed. Thus referring to Fig. 15.4, when the movable arm makes contact with stud 1, the secondary voltage is minimum and when with stud 5, it is maximum. During the period of light load, the voltage across the primary is not much below the alternator voltage and the movable arm is placed on stud 1. When the load increases, the voltage across the primary drops, but the secondary voltage can be kept at the previous value by placing the movable arm on to a higher stud. Whenever a tapping is to be changed in this type of transformer, the load is kept off and hence the name off load tap-changing transformer.

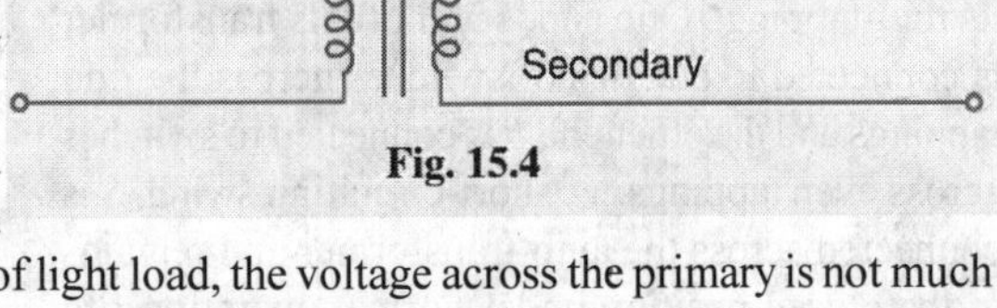

Fig. 15.4

---

* In a long line, difference in the receiving-end voltage between no load and full-load conditions is quite large.

The principal disadvantage of the circuit arrangement shown in Fig. 15.4 is that it cannot be used for tap-changing on load. Suppose for a moment that tapping is changed from position 1 to position 2 when the transformer is supplying load. If contact with stud 1 is broken before contact with stud 2 is made, there is break in the circuit and arcing results. On the other hand, if contact with stud 2 is made before contact with stud 1 is broken, the coils connected between these two tappings are short-circuited and carry damaging heavy currents. For this reason, the above circuit arrangement cannot be used for tap-changing on load.

***(ii)* On-load tap-changing transformer.** In supply system, tap-changing has normally to be performed on load so that there is no interruption to supply. Fig. 15.5 shows diagrammatically one type of on-load tap-changing transformer. The secondary consists of two equal parallel windings which have similar tappings $1a$ ...... $5a$ and $1b$ .... $5b$. In the normal working conditions, switches a, b and tappings with the same number remain closed and each secondary winding carries one-half of the total current. Referring to Fig. 15.5, the secondary voltage will be maximum when switches a, b and $5a$, $5b$ are closed. However, the secondary voltage will be minimum when switches a, b and $1a$, $1b$ are closed.

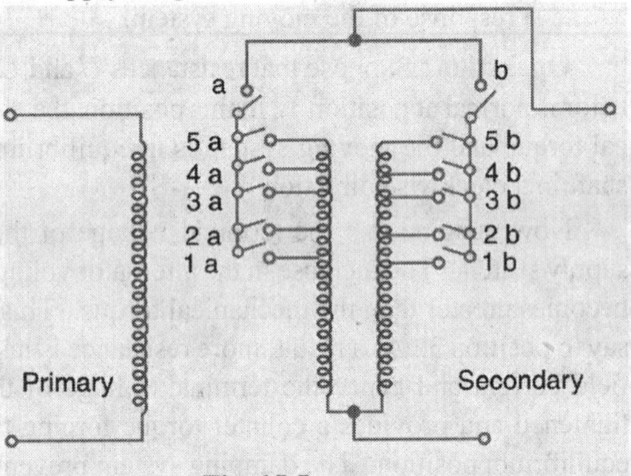

Fig. 15.5

Suppose that the transformer is working with tapping position at $4a$, $4b$ and it is desired to alter its position to $5a$, $5b$. For this purpose, one of the switches a and b, say a, is opened. This takes the secondary winding controlled by switch a out of the circuit. Now, the secondary winding controlled by switch b carries the total current which is twice its rated capacity. Then the tapping on the disconnected winding is changed to $5a$ and switch a is closed. After this, switch b is opened to disconnect its winding, tapping position on this winding is changed to $5b$ and then switch b is closed. In this way, tapping position is changed without interrupting the supply. This method has the following disadvantages :

(*i*) During switching, the impedance of transformer is increased and there will be a voltage surge.

(*ii*) There are twice as many tappings as the voltage steps.

15.8 Auto-Transformer Tap-changing

Fig. 15.6 shows diagrammatically auto-transformer tap changing. Here, a mid-tapped auto-transformer or reactor is used. One of the lines is connected to its mid-tapping. One end, say a of this transformer is connected to a series of switches across the odd tappings and the other end b is connected to switches across even tappings. A short-circuiting switch S is connected across the auto-transformer and remains in the closed position under normal operation. In the normal operation, there is *no inductive voltage drop across the auto-transformer. Referring to Fig. 15.6, it is clear that with switch 5 closed, minimum

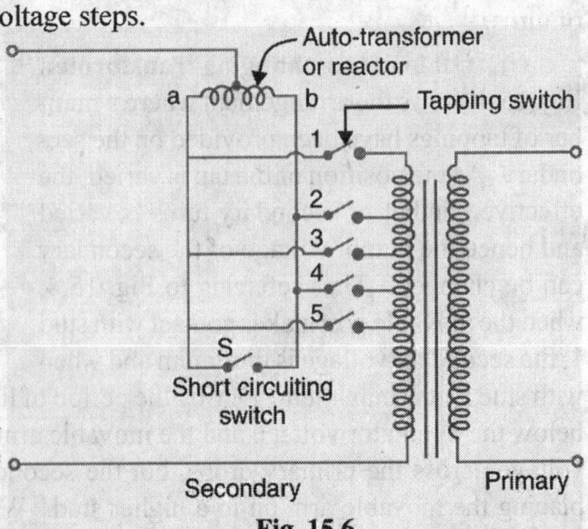

Fig. 15.6

* In the normal operation, switch S remains closed so that half the total current flows through each half of the reactor. Since the currents in each half of the reactor are in opposition, no resultant flux is set up and consequently there is no inductive voltage drop across it.

secondary turns are in the circuit and hence the output voltage will be the lowest. On the other hand, the output voltage will be maximum when switch 1 is closed.

Suppose now it is desired to alter the tapping point from position 5 to position 4 in order to raise the output voltage. For this purpose, short-circuiting switch *S* is opened, switch 4 is closed, then switch 5 is opened and finally short-circuiting switch is closed. In this way, tapping can be changed without interrupting the supply.

It is worthwhile to describe the electrical phenomenon occurring during the tap changing. When the short-circuiting switch is opened, the load current flows through one-half of the reactor coil so that there is a voltage drop across the reactor. When switch 4 is closed, the turns between points 4 and 5 are connected through the whole reactor winding. A circulating current flows through this local circuit but it is limited to a low value due to high reactance of the reactor.

15.9 Booster Transformer

Sometimes it is desired to control the voltage of a transmission line at a point far away from the main transformer. This can be conveniently achieved by the use of a booster transformer as shown in Fig. 15.7. The secondary of the booster transformer is connected in series with the line whose voltage is to be controlled. The primary of this transformer is supplied from a regulating transformer *fitted with on-load tap-changing gear. The booster transformer is connected in such a way that its secondary injects a voltage in phase with the line voltage.

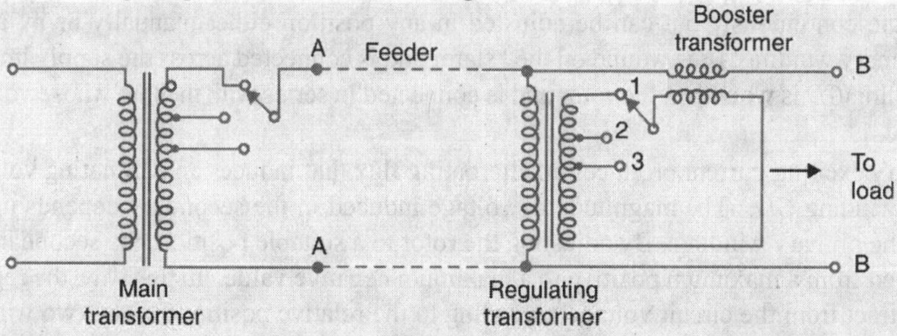

Fig. 15.7

The voltage at *AA* is maintained constant by tap-changing gear in the main transformer. However, there may be considerable voltage drop between *AA* and *BB* due to fairly long feeder and tapping of loads. The voltage at *BB* is controlled by the use of regulating transformer and booster transformer. By changing the tapping on the regulating transformer, the magnitude of the voltage injected into the line can be varied. This permits to keep the voltage at *BB* to the desired value. This method of voltage control has three disadvantages. Firstly, it is more expensive than the on-load tap-changing transformer. Secondly, it is less efficient owing to losses in the booster and thirdly more floor space is required. Fig. 15.8 shows a three-phase booster transformer.

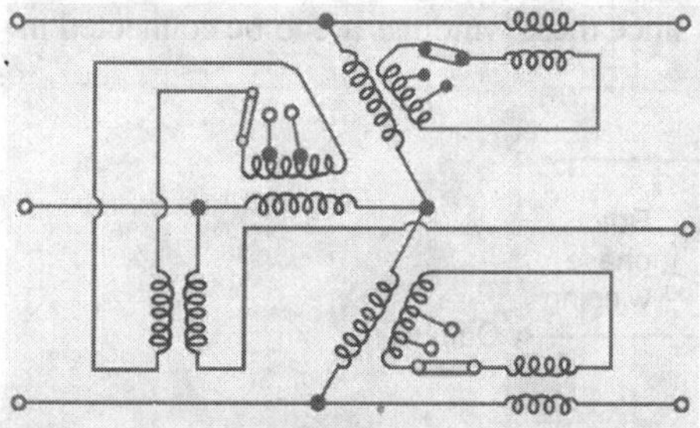

Fig. 15.8

Booster Transformer

* The on-load tap-changing gear is omitted from the diagram for the sake of simplicity.

15.10 Induction Regulators

An induction regulator is essentially a constant voltage transformer, one winding of which can be moved *w.r.t.* the other, thereby obtaining a variable secondary voltage. The primary winding is connected across the supply while the secondary winding is connected in series with the line whose voltage is to be controlled. When the position of one winding is changed w.r.t. the other, the secondary voltage injected into the line also changes. There are two types of induction regulators *viz.* single phase and 3-phase.

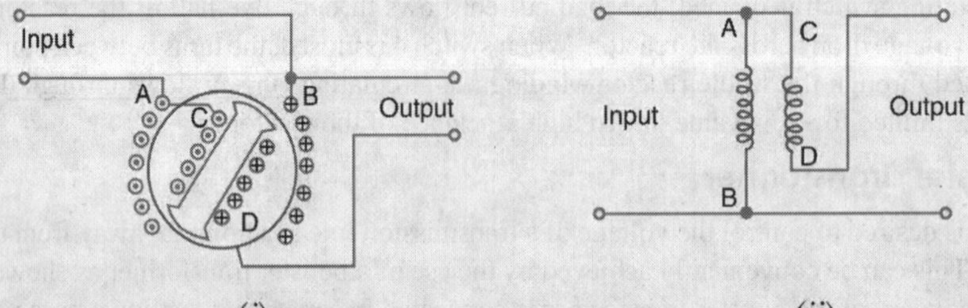

Fig. 15.9

(*i*) **Single-phase induction regulator.** A single phase induction regulator is illustrated in Fig. 15.9. In construction, it is similar to a single phase induction motor except that the rotor is not allowed to rotate continuously but can be adjusted in any position either manually or by a small motor. The primary winding *AB* is wound on the *stator and is connected across the supply line. The secondary winding *CD* is wound on the rotor and is connected in series with the line whose voltage is to be controlled.

The primary exciting current produces an alternating flux that induces an alternating voltage in the secondary winding *CD*. The magnitude of voltage induced in the secondary depends upon its position *w.r.t.* the primary winding. By adjusting the rotor to a suitable position, the secondary voltage can be varied from a maximum positive to a maximum negative value. In this way, the regulator can add or subtract from the circuit voltage according to the relative positions of the two windings. Owing to their greater flexibility, single phase regulators are frequently used for voltage control of distribution primary feeders.

(*ii*) **Three-phase induction regulator.** In construction, a 3-phase induction regulator is similar to a 3-phase induction motor with wound rotor except that the rotor is not allowed to rotate continuously but can be held in any position by means of a worm gear. The primary windings either in star or delta are wound on the stator and are connected across the supply. The secondary windings are wound on the rotor and the six terminals are brought out since these windings are to be connected in series with the line whose voltage is to be controlled.

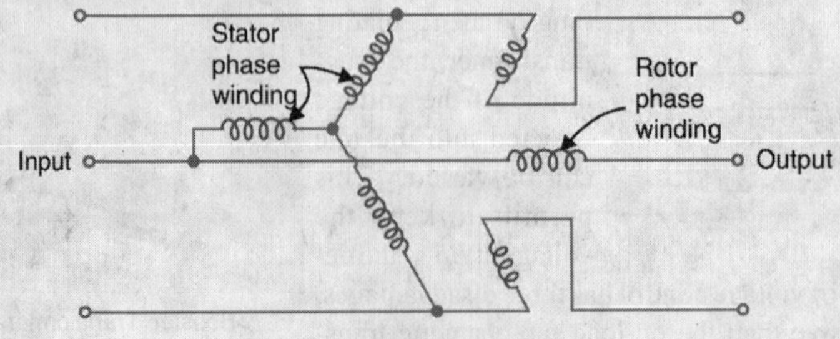

Fig. 15.10

* From electrical point of view, it is immaterial whether the rotor or stator carries the primary winding.

When polyphase currents flow through the primary windings, a rotating field is set up which induces an e.m.f. in each phase of rotor winding. As the rotor is turned, the magnitude of the rotating flux is not changed; hence the rotor e.m.f. per phase remains constant. However, the variation of the position of the rotor will affect the phase of the rotor e.m.f. *w.r.t.* the applied voltage as shown in Fig. 15.11. The input primary voltage per phase is V_p and the boost introduced by the regulator is V_r. The output voltage V is the vector sum of V_p and V_r. Three phase induction regulators are used to regulate the voltage of feeders and in connection with high voltage oil testing transformers.

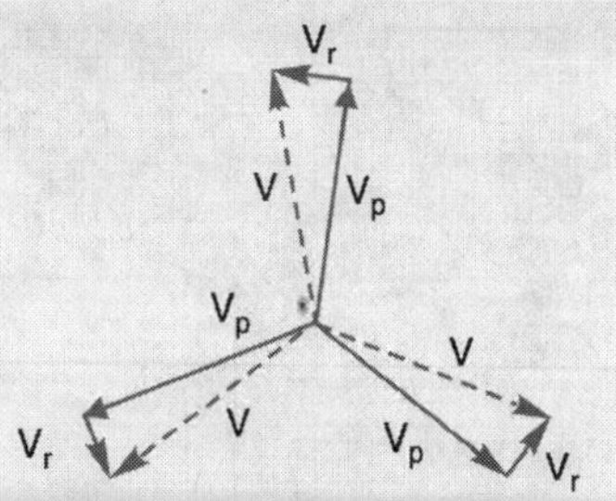

Fig. 15.11

15.11 Voltage Control by Synchronous Condenser

The voltage at the receiving end of a transmission line can be controlled by installing specially designed synchronous motors called *synchronous condensers at the receiving end of the line. The synchronous condenser supplies wattless leading kVA to the line depending upon the excitation of the motor. This wattless leading kVA partly or fully cancels the wattless lagging kVA of the line, thus controlling the voltage drop in the line. In this way, voltage at the receiving end of a transmission line can be kept constant as the load on the system changes.

For simplicity, consider a short transmission line where the effects of capacitance are neglected. Therefore, the line has only resistance and inductance. Let V_1 and V_2 be the per phase sending end and receiving end voltages respectively. Let I_2 be the load current at a lagging power factor of cos ϕ_2.

(*i*) **Without synchronous condenser.** Fig. 15.12 (*i*) shows the transmission line with resistance R and inductive reactance X per phase. The load current I_2 can be resolved into two rectangular components *viz* I_p in phase with V_2 and I_q at right angles to V_2 [See Fig. 15.12 (*ii*)]. Each component will produce resistive and reactive drops ; the resistive drops being in phase with and the reactive drops in quadrature leading with the corresponding currents. The vector addition of these voltage drops to V_2 gives the sending end voltage V_1.

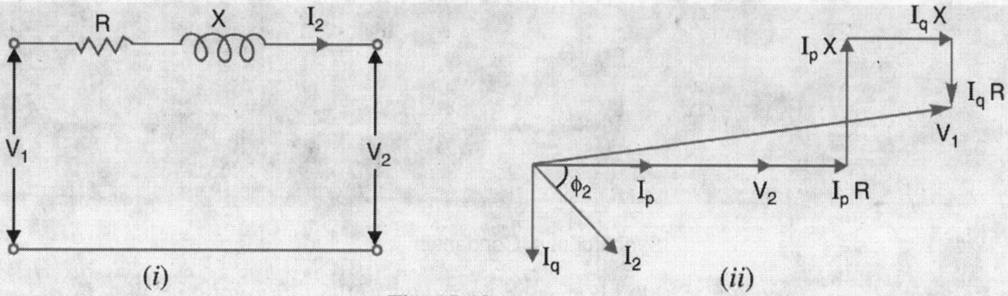

Fig. 15.12

(*ii*) **With synchronous condenser.** Now suppose that a synchronous condenser taking a leading current **I_m is connected at the receiving end of the line. The vector diagram of the circuit becomes as shown in Fig. 15.13. Note that since I_m and I_q are in direct opposition and that I_m must be greater than I_q, the four drops due to these two currents simplify to :

* By changing the excitation of a synchronous motor, it can be made to take a leading power factor. A synchronous motor at no load and taking a leading power factor is known as a *synchronous condenser.* It is so called because the characteristics of the motor then resemble with that of a condenser.

** Neglecting the losses of the synchronous condenser, I_m will lead V_2 by 90°.

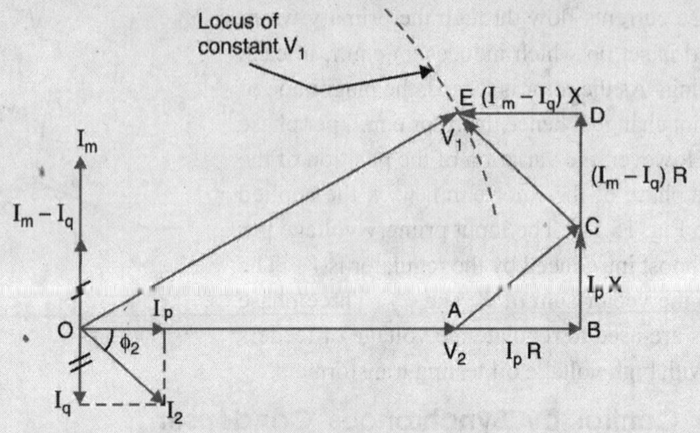

Fig. 15.13

$(I_m - I_q)\, R$ in phase with I_m

and $(I_m - I_q)\, X$ in quadrature leading with I_m

From the vector diagram, the relation between V_1 and V_2 is given by ;

$$OE^2 = (OA + AB - DE)^2 + (BC + CD)^2$$

or
$$V_1^2 = [V_2 + I_pR - (I_m - I_q)\, X]^2 + [I_pX + (I_m - I_q)\, R]^2$$

From this equation, the value of I_m can be calculated to obtain any desired ratio of V_1/V_2 for a given load current and power factor.

$$\text{kVAR capacity of condenser} = \frac{3\, V_2\, I_m}{1000}$$

Synchronous Condenser

Example 15.1. *A load of 10,000 kW at a power factor of 0·8 lagging is supplied by a 3-phase line whose voltage has to be maintained at 33kV at each end. If the line resistance and reactance per phase are 5 Ω and 10 Ω respectively, calculate the capacity of the synchronous condenser to be installed for the purpose. Comment on the result.*

Solution.

$$\text{Load current, } I_2 = \frac{10{,}000 \times 10^3}{\sqrt{3} \times 33 \times 10^3 \times 0{\cdot}8} = 218 \text{ A}$$

$$\therefore \quad I_p = I_2 \cos \phi_2 = 218 \times 0{\cdot}8 = 174{\cdot}4 \text{ A}$$

$$I_q = I_2 \sin \phi_2 = 218 \times 0{\cdot}6 = 130{\cdot}8 \text{ A}$$

$$R = 5\,\Omega\,;\; X = 10\,\Omega$$

Sending-end voltage/phase, V_1 = Receiving end voltage/phase (V_2)

$$= \frac{33 \times 10^3}{\sqrt{3}} = 19{,}053 \text{ V}$$

Let I_m be the current taken by the synchronous condenser. Referring to Fig. 15.13,

$$(19{,}053)^2 = [19{,}053 + 174{\cdot}4 \times 5 - 10\,(I_m - 130{\cdot}8)]^2 + [174{\cdot}4 \times 10 + (I_m - 130{\cdot}8)5]^2$$

Solving this equation, we get, I_m = 231 A

$$\text{Capacity of synchronous condenser} = \frac{3V_2\,I_m}{1000}\text{ kVAR} = \frac{3 \times 19{,}053 \times 231}{1000}\text{ kVAR}$$

$$= \textbf{13,203 kVAR}$$

Comments. This example shows that kVA capacity of the synchronous condenser is considerably greater than the kVA capacity of the load *viz* 13203 against 10000/0·8 = 12,500. Since the cost of synchronous condenser is usually very high, it would not be an economical proposition to have the same sending end and receiving end voltages. In practice, the synchronous condenser is operated in such a way so as to allow a small drop in the line.

Example 15.2. *A 3-phase overhead line has resistance and reactance per phase of 5 Ω and 20 Ω respectively. The load at the receiving end is 25 MW at 33 kV and a power factor of 0·8 lagging. Find the capacity of the synchronous condenser required for this load condition if it is connected at the receiving end and the line voltages at both ends are maintained at 33 kV.*

Solution.

$$\text{Load current, } I_2 = \frac{25 \times 10^6}{\sqrt{3} \times 33000 \times 0{\cdot}8} = 546{\cdot}8 \text{ A}$$

$$\therefore \quad I_p = I_2 \cos\phi_2 = 546{\cdot}8 \times 0{\cdot}8 = 437{\cdot}4 \text{ A}$$

$$I_q = I_2 \sin\phi_2 = 546{\cdot}8 \times 0{\cdot}6 = 328{\cdot}1 \text{ A}$$

$$R = 5\,\Omega\,;\, X = 20\,\Omega$$

Sending end voltage/phase, V_1 = Receiving end voltage/phase, V_2

$$= \frac{33 \times 10^3}{\sqrt{3}} = 19053 \text{ V}$$

Let I_m be the current taken by the synchronous condenser. Then,

$$V_1^2 = [V_2 + I_pR - (I_m - I_q) \times X]^2 + [I_pX + (I_m - I_q)\,R]^2$$

or

$$(19053)^2 = [19053 + 437{\cdot}4 \times 5 - (I_m - 328{\cdot}1) \times 20]^2 + [437{\cdot}4 \times 20 + (I_m - 328{\cdot}1) \times 5]^2$$

On solving this equation, we get, I_m = 579·5 A

$$\text{Capacity of synchronous condenser} = \frac{3\,V_2\,I_m}{10^6}\text{ MVAR} = \frac{3 \times 19{,}053 \times 579{\cdot}5}{10^6} = \textbf{33·13 MVAR}$$

TUTORIAL PROBLEMS

1. A 3-phase line having an impedance of (5 + *j* 20) ohms per phase delivers a load of 30 MW at a p.f. of 0·8 lagging and voltage 33 kV. Determine the capacity of the synchronous condenser required to be installed at the receiving end if voltage at the sending end is to be maintained at 33 kV. **[42·78 MVAR]**
2. A 12500 kVA load is supplied at a power factor of 0·8 lagging by a 3-phase transmission line whose voltage is to be maintained at 33 kV at both ends. Determine the capacity of the synchronous condenser to be installed at the receiving end. The impedance of the line is (4 + *j* 12) ohms per phase. **[11490 kVAR]**

SELF - TEST

1. **Fill in the blanks by inserting appropriate words/figures :**
   (*i*) The statutory limit of voltage variations is .... of the declared voltage at consumer's terminals.
   (*ii*) In the automatic voltage regulators used at the generating station, the .... principle is used.
   (*iii*) The voltage variations in a Brown-Boveri regulator never exceed ....
   (*iv*) In a Tirril regulator, a .... resistance is cut in and out of the exciter field circuit of the alternator.
   (*v*) In practice, tap-changing is performed on load so that there is .... to supply.
   (*vi*) Induction regulators are used for voltage control in .... system.
   (*vii*) A synchronous condenser is generally installed at the .... end of a transmission line.
   (*viii*) The principal cause of voltage variation is the change of .... on the system.
   (*ix*) In a Tirril regulator, capacitor is provided across the relay contacts to reduce .... at the time relay contacts are opened.
2. **Pick up the correct words/figures from brackets and fill in the blanks :**
   (*i*) The voltage control equipment is used at .... in the power system. (*one point, more than one point*)
   (*ii*) The mechanical control torque in a Brown-Boveri regulator is .... whatever may be the position of drum. (*constant, variable*)
   (*iii*) The excitation control method is suitable only for .... lines. (*short, long*)
   (*iv*) It is .... to maintain the same voltage at both ends of a transmission line by synchronous condenser method. (*economical, not economical*).

ANSWERS TO SELF-TEST

1. (*i*) ± 6% (*ii*) overshooting the mark (*iii*) ± 1% (*iv*) regulating (*v*) no interruption (*vi*) primary distribution (*vii*) receiving (*viii*) load (*ix*) arcing
2. (*i*) more than one point (*ii*) constant (*iii*) short (*iv*) not economical.

CHAPTER REVIEW TOPICS

1. Discuss the importance of voltage control in the modern power system.
2. What are the various methods of voltage control in a power system ?
3. Describe with the aid of neat sketch the construction and working of a Tirril regulator.
4. Explain the construction and working of Brown-Boveri regulator with a neat sketch.
5. Describe the off-load tap changing transformer method of voltage control. What are the limitations of the method ?
6. Explain with a neat sketch :
   (*i*) On-load tap-changing transformer
   (*ii*) Auto-transformer tap-changing
7. What do you understand by induction regulators ? Describe single phase and three phase induction regulators.
8. Describe the synchronous condenser method of voltage control for a transmission line. Illustrate your answer with a vector diagram.

DISCUSSION QUESTIONS

1. Voltage control equipment is generally located at more than one point. Why ?
2. Tap-changing is generally performed on load. Why ?
3. Why do we use overshooting the mark principle in automatic voltage regulators ?

CHAPTER

16

Introduction to Switchgear

16.1 Switchgear

16.2 Essential Features of Switchgear

16.3 Switchgear Equipment

16.4 Bus-Bar Arrangements

16.5 Switchgear Accommodation

16.6 Short-Circuit

16.7 Short-Circuit Currents

16.8 Faults in a Power System

General

A great demand for electrical energy is a notable feature of modern civilisation. Most of this energy is needed for lighting, heating, domestic appliances, industrial electrical machinery and electric traction. The importance of electric supply in everyday life has reached such a stage that it is desirable to protect the power system from harm during fault conditions and to ensure maximum continuity of supply. For this purpose, means must be provided to switch on or off generators, transmission lines, distributors and other equipment under both normal and abnormal conditions. This is achieved by an apparatus called *switchgear.* A switchgear essentially consists of switching and protecting devices such as switches, fuses, circuit breakers, relays *etc*.

During normal operation, switchgear permits to switch on or off generators, transmission lines, distributors and other electrical equipment. On the other hand, when a failure (*e.g.* short circuit) occurs on any part of power system, a heavy current flows through the equipment, threatening damage to the equipment and interruption of service to the customers. However, the switchgear detects the fault and disconnects the unhealthy section from the system. In this way, switchgear protects the system from the damage and ensures continuity of supply. In this chapter, we shall present the elementary introduction to switchgear.

16.1 Switchgear

The apparatus used for switching, controlling and protecting the electrical circuits and equipment is known as **switchgear**.

The switchgear equipment is essentially concerned with switching and interrupting currents either under normal or abnormal operating conditions. The tumbler switch with ordinary fuse is the simplest form of switchgear and is used to control and protect lights and other equipment in homes, offices *etc*. For circuits of higher rating, a high-rupturing capacity (H.R.C.) fuse in conjuction with a switch may serve the purpose of controlling and protecting the circuit. However, such a switchgear cannot be used profitably on high voltage system (3·3 kV) for two reasons. Firstly, when a fuse blows, it takes sometime to replace it and consequently there is interruption of service to the customers. Secondly, the fuse cannot successfully interrupt large fault currents that result from the faults on high voltage system.

With the advancement of power system, lines and other equipments operate at high voltages and carry large currents. When a short circuit occurs on the system, heavy current flowing through the equipment may cause considerable damage. In order to interrupt such heavy fault currents, *automatic circuit breakers* (or simply circuit breakers) are used. A circuit breaker is a switchgear which can open or close an electrical circuit under both normal and abnormal conditions. Even in instances where a fuse is adequate, as regards to breaking capacity, a circuit breaker may be preferable. It is because a circuit breaker can close circuits, as well as break them without replacement and thus has wider range of use altogether than a fuse.

16.2 Essential Features of Switchgear

The essential features of switchgear are :

(*i*) **Complete reliability.** With the continued trend of interconnection and the increasing capacity of generating stations, the need for a reliable switchgear has become of paramount importance. This is not surprising because switchgear is added to the power system to improve the reliability. When fault occurs on any part of the power system, the switchgear must operate to isolate the faulty section from the remainder circuit.

(*ii*) **Absolutely certain discrimination.** When fault occurs on any section of the power system, the switchgear must be able to discriminate between the faulty section and the healthy section. It should isolate the faulty section from the system without affecting the healthy section. This will ensure continuity of supply.

(*iii*) **Quick operation.** When fault occurs on any part of the power system, the switchgear must operate quickly so that no damage is done to generators, transformers and other equipment by the short-circuit currents. If fault is not cleared by switchgear quickly, it is likely to spread into healthy parts, thus endangering complete shut down of the system.

(*iv*) **Provision for manual control.** A switchgear must have provision for manual control. In case the electrical (or electronics) control fails, the necessary operation can be carried out through manual control.

(*v*) **Provision for instruments.** There must be provision for instruments which may be required. These may be in the form of ammeter or voltmeter on the unit itself or the necessary current and voltage transformers for connecting to the main switchboard or a separate instrument panel.

16.3 Switchgear Equipment

Switchgear covers a wide range of equipment concerned with switching and interrupting currents under both normal and abnormal conditions. It includes switches, fuses, circuit breakers, relays and other equipment. A brief account of these devices is given below. However, the reader may find the detailed discussion on them in the subsequent chapters.

1. Switches. A switch is a device which is used to open or close an electrical circuit in a convenient way. It can be used under full-load or no-load conditions *but* it cannot interrupt the fault currents. When the contacts of a switch are opened, an *arc is produced in the air between the contacts. This is particularly true for circuits of high voltage and large current capacity. The switches may be classified into (*i*) air switches (*ii*) oil switches. The contacts of the former are opened in air and that of the latter are opened in oil.

(*i*) *Air-break switch.* It is an air switch and is designed to open a circuit *under load*. In order to quench the arc that occurs on opening such a switch, special arcing horns are provided. Arcing horns are pieces of metals between which arc is formed during opening operation. As the switch opens, these horns are spread farther and farther apart. Consequently, the arc is lengthened, cooled and interrupted. Air-break switches are generally used outdoor for circuits of medium capacity such as lines supplying an industrial load from a main transmission line or feeder.

Air - break Switch

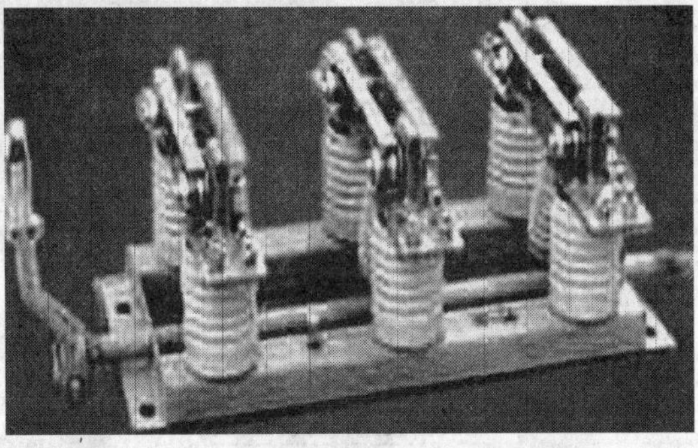

Isolator Switch

(*ii*) *Isolator or disconnecting switch.* It is essentially a knife switch and is designed to open a circuit under *no load*. Its main purpose is to isolate one portion of the circuit from the other and is not intended to be opened while current is flowing in the line. Such switches are generally used on both sides of circuit breakers in order that repairs and replacement of circuit breakers can be made without any danger. They should never be opened until the circuit breaker in the same circuit has been opened and should always be closed before the circuit breaker is closed.

(*iii*) *Oil switches.* As the name implies, the contacts of such switches are opened under oil, usually transformer oil. The effect of oil is to cool and quench the arc that tends to form when the circuit is opened. These switches are used for circuits of high voltage and large current carrying capacities.

2. Fuses. A fuse is a short piece of wire or thin strip which melts when excessive current flows through it for sufficient time. It is inserted in series with the circuit to be protected. Under normal operating conditions, the fuse element it at a temperature below its melting point. Therefore, it carries the normal load current without overheating. However, when a short circuit or overload occurs, the current through the fuse element increases beyond its rated capacity. This raises the temperature and the fuse element melts (or blows out), disconnecting the circuit protected by it. In

* Generally, the load contains reactive elements (inductance and capacitance). The sudden change of current in the circuit due to breaking results in the production of large induced e.m.f. (several thousand volts) which ruptures the insulation of air inbetween the contacts, causing a spark.

this way, a fuse protects the machines and equipment from damage due to excessive currents. It is worthwhile to note that a fuse performs both detection and interruption functions.

3. **Circuit breakers.** A circuit breaker is an equipment which can open or close a circuit under all conditions *viz*. no load, full load and fault conditions. It is so designed that it can be operated manually (or by remote control) under normal conditions and automatically under fault conditions. For the latter operation, a relay circuit is used with a circuit breaker. Fig. 16.1 (*i*) shows the parts of a typical oil circuit breaker whereas Fig. 16.1 (*ii*) shows its control by a relay circuit. The circuit breaker essentially consists of moving and fixed contacts enclosed in strong metal tank and immersed in oil, known as transformer oil.

Under normal operating conditions, the contacts remain closed and the circuit breaker carries the full-load current continuously. In this condition, the e.m.f. in the secondary winding of current transformer (C.T.) is insufficient to operate the trip coil of the breaker but the contacts can be opened (and hence the circuit can be opened) by manual or remote control. When a fault occurs, the resulting overcurrent in the C.T. primary winding increases the secondary e.m.f. This energises the trip coil of the breaker and moving contacts are pulled down, thus opening the contacts and hence the circuit. The arc produced during the opening operation is quenched by the oil. It is interesting to note that relay performs the function of detecting a fault whereas the circuit breaker does the actual circuit interruption.

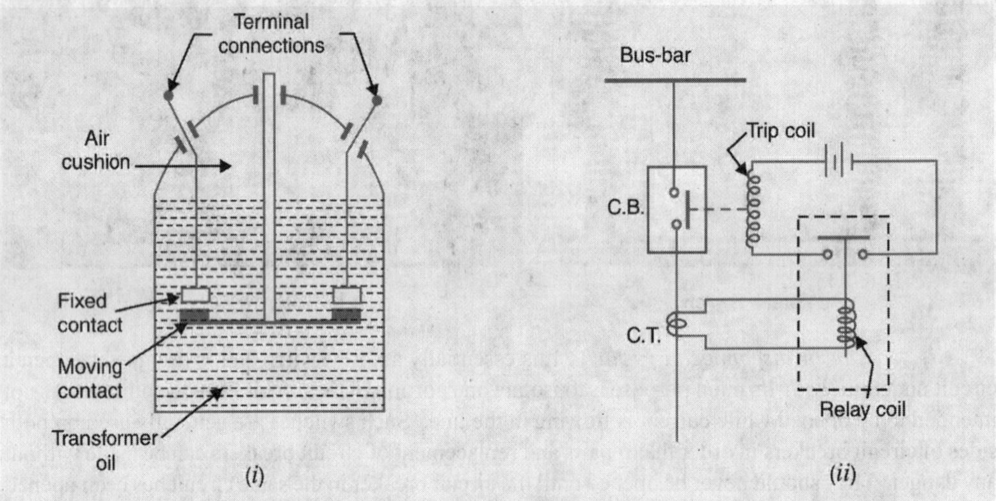

Fig. 16.1

4. **Relays.** A relay is a device which detects the fault and supplies information to the breaker for circuit interruption. Fig. 16.1 (*ii*) shows a typical relay circuit. It can be divided into three parts *viz*.

(*i*) The primary winding of a *current transformer (C.T.) which is connected in series with the circuit to be protected. The primary winding often consists of the main conductor itself.

(*ii*) The second circuit is the secondary winding of C.T. connected to the relay operating coil.

(*iii*) The third circuit is the tripping circuit which consists of a source of supply, trip coil of circuit breaker and the relay stationary contacts.

Under normal load conditions, the e.m.f. of the secondary winding of C.T. is small and the current flowing in the relay operating coil is insufficient to close the relay contacts. This keeps the trip coil of the circuit breaker unenergised. Consequently, the contacts of the circuit breaker remain closed and it carries the normal load current. When a fault occurs, a large current flows through the

* The circuit to be protected carries a large current (normal or fault current) and is reduced to a suitable value for relay operation with the help of a current transformer.

primary of C.T. This increases the secondary e.m.f. and hence the current through the relay operating coil. The relay contacts are closed and the trip coil of the circuit breaker is energised to open the contacts of the circuit breaker.

16.4 Bus-Bar Arrangements

When a number of generators or feeders operating at the same voltage have to be directly connected electrically, bus-bars are used as the common electrical component. *Bus-bars are copper rods or thin walled tubes and operate at constant voltage. We shall discuss some important bus-bars arrangements used for power stations and sub-stations. All the diagrams refer to 3-phase arrangement but are shown in single-phase for simplicity.

(1) Single Bus-bar System. The single bus-bar system has the simplest design and is used for power stations. It is also used in small outdoor stations having relatively few outgoing or incoming feeders and lines. Fig. 16.2 shows the single bus-bar system for a typical power station. The generators, outgoing lines and transformers are connected to the bus-bar. Each generator and feeder is controlled by a circuit breaker. The isolators permit to isolate generators, feeders and circuit breakers from the bus-bar for maintenance. The chief advantages of this type of arrangement are low initial cost, less maintenance and simple operation.

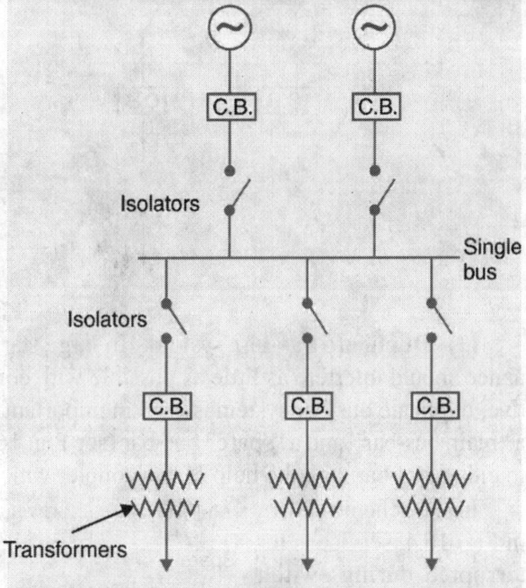

Fig. 16.2

Disadvantages. Single bus-bar system has the following three principal disadvantages :

(*i*) The bus-bar cannot be cleaned, repaired or tested without de-energising the whole system.

(*ii*) If a fault occurs on the bus-bar itself, there is complete interruption of supply.

(*iii*) Any fault on the system is fed by all the generating capacity, resulting in very large fault currents.

(2) Single bus-bar system with Sectionalisation. In large generating stations where several units are installed, it is a common practice to sectionalise the bus so that fault on any section of the bus-bar will not cause complete shut down. This is illustrated in Fig. 16.3 which shows the bus-bar divided into two sections connected by a circuit breaker and isolators. Three principal advantages are claimed for this arrangement. Firstly, if a fault occurs on any section of the bus-bar, that section can be isolated without affecting the supply to other sections. Secondly, if a fault occurs on any feeder, the fault current is much **lower than with unsectionalised bus-bar. This permits the use of circuit breakers of lower capacity in the feeders. Thirdly, repairs and maintenance of any section of the bus-bar can be carried out by de-energising that section only, eliminating the possibility of complete shut-down.

It is worthwhile to keep in mind that a circuit breaker should be used as the sectionalising switch so that uncoupling of the bus-bars may be carried out safely during load transfer. Moreover, the circuit breaker itself should be provided with isolators on both sides so that its maintenance can be done while the bus-bars are alive.

---

* The term bus is derived from the word omnibus, meaning collector of things. Thus, electrical bus-bar is the collector of electrical energy at one location.

** because a feeder fault is fed from one section only.

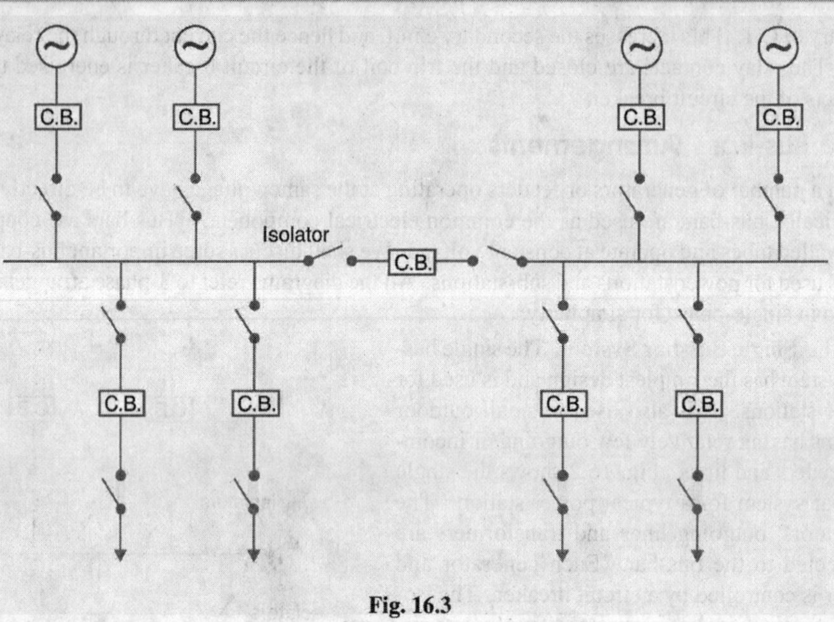

Fig. 16.3

(3) Duplicate bus-bar system. In large stations, it is important that breakdowns and maintenance should interfere as little as possible with continuity of supply. In order to achieve this objective, duplicate bus-bar system is used in important stations. Such a system consists of two bus-bars, a "main bus-bar" and a "spare" bus-bar (see Fig. 16.4). Each generator and feeder may be connected to either bus-bar with the help of bus coupler which consists of a circuit breaker and isolators.

In the scheme shown in Fig. 16.4, service is interrupted during switch over from one bus to another. However, if it were desired to switch a circuit from one to another without interruption of service, there would have to be two circuit breakers per circuit. Such an arrangement will be too expensive.

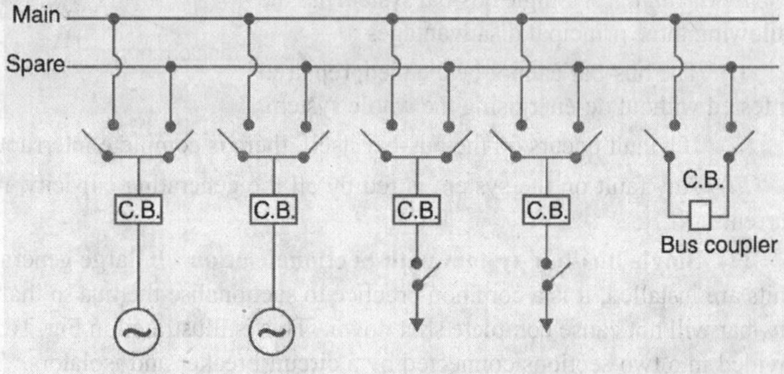

Fig. 16.4

Advantages

(*i*) If repair and maintenance it to be carried on the main bus, the supply need not be interrupted as the entire load can be transferred to the spare bus.

(*ii*) The testing of feeder circuit breakers can be done by putting them on spare bus-bar, thus keeping the main bus-bar undisturbed.

(*iii*) If a fault occurs on the bus-bar, the continuity of supply to the circuit can be maintained by transferring it to the other bus-bar.

16.5 Switchgear Accommodation

The main components of a switchgear are circuit breakers, switches, bus-bars, instruments and instrument transformers. It is necessary to house the switchgear in power stations and sub-stations in such a way so as to safeguard personnel during operation and maintenance and to ensure that the effects of fault on any section of the gear are confined to a limited region. Depending upon the voltage to be handled, switchgear may be broadly classified into (*i*) outdoor type (*ii*) indoor type.

(*i*) *Outdoor type.* For voltages beyond 66 kV, switchgear equipment is installed outdoor. It is because for such voltages, the clearances between conductors and the space required for switches, circuit breakers, transformers and others equipment become so great that it is not economical to install all such equipment indoor.

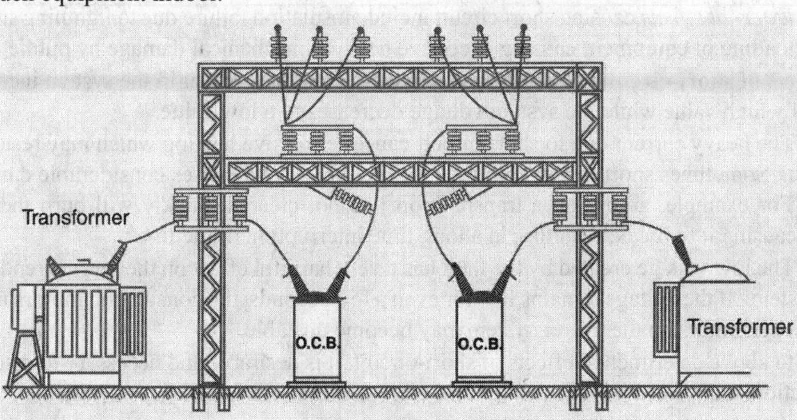

Fig. 16.5

Fig. 16.5 shows a typical outdoor sub-station with switchgear equipment. The circuit breakers, isolators, transformers and bus-bars occupy considerable space on account of large electrical clearance associated with high voltages.

(*ii*) *Indoor type.* For voltages below 66 kV, switchgear is generally installed indoor because of economic considerations. The indoor switchgear is generally of metal-clad type. In this type of construction, all live parts are completely enclosed in an earthed metal casing. The primary object of this practice is the definite localisation and restriction of any fault to its place of origin.

16.6 Short-Circuit

Whenever a fault occurs on a network such that a large current flows in one or more phases, a **short-circuit** *is said to have occurred.*

When a short circuit occurs, a heavy current called short circuit current flows through the circuit. This can be beautifully illustrated by referring to Fig. 16.6 where a single phase generator of voltage V and internal impedance Z_i is supplying to a load Z. Under normal conditions, the current in the circuit is limited by *load impedance Z. However, if the load terminals get shorted due to any reason, the circuit impedance is reduced to a very low value ; being Z_i in this case. As Z_i is very small, therefore, a large current flows through the circuit. This is called short-circuit current. It is worthwhile to make a distinction between a ***short-circuit* and an *overload*. When a short-circuit occurs, the voltage at fault point is reduced to zero and current of abnormally high magnitude flows through the network to the point of fault. On the other hand, an overload means that loads greater than the designed values have been imposed on the system. Under such conditions, the voltage at the overload point may be low, but not zero. The undervoltage conditions may extend for some distance beyond the overload point into the remainder of the system. The currents in the overloaded

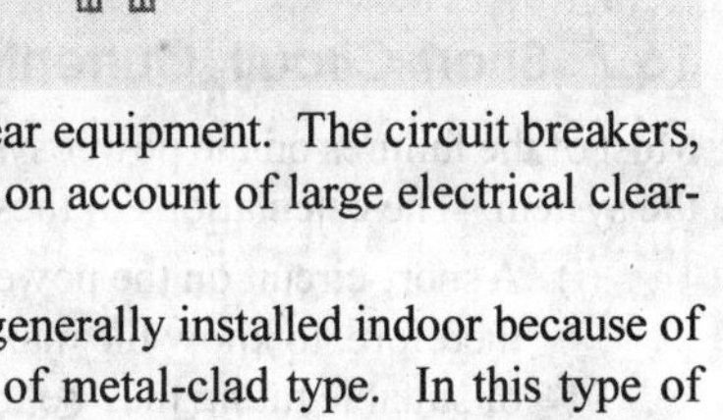

Fig. 16.6

* As internal impedance Z_i of the generator is generally small.

** Note that path of current is shortened and hence the name short circuit current.

equipment are high but are substantially lower than that in the case of a short-circuit.

Causes of short-circuit. A short circuit in the power system is the result of some kind of abnormal conditions in the system. It may be caused due to internal and/or external effects.

(*i*) *Internal effects* are caused by breakdown of equipment or transmission lines, from deterioration of insulation in a generator, transformer *etc*. Such troubles may be due to ageing of insulation, inadequate design or improper installation.

(*ii*) *External effects* causing short circuit include insulation failure due to lightning surges, overloading of equipment causing excessive heating; mechanical damage by public *etc*.

Effects of short-circuit. When a short-circuit occurs, the current in the system increases to an abnormally high value while the system voltage decreases to a low value.

(*i*) The heavy current due to short-circuit causes excessive heating which may result in fire or explosion. Sometimes short-circuit takes the form of an arc and causes considerable damage to the system. For example, an arc on a transmission line not cleared quickly will burn the conductor severely causing it to break, resulting in a long time interruption of the line.

(*ii*) The low voltage created by the fault has a very harmful effect on the service rendered by the power system. If the voltage remains low for even a few seconds, the consumers' motors may be shut down and generators on the power system may become unstable.

Due to above deterimental effects of short-circuit, it is desirable and necessary to disconnect the faulty section and restore normal voltage and current conditions as quickly as possible.

16.7 Short-Circuit Currents

Most of the failures on the power system lead to short-circuit fault and cause heavy current to flow in the system. The calculations of these short-circuit currents are important for the following reasons :

(*i*) A short-circuit on the power system is cleared by a circuit breaker or a fuse. It is necessary, therefore, to know the maximum possible values of short-circuit current so that switchgear of suitable rating may be installed to interrupt them.

(*ii*) The magnitude of short-circuit current determines the setting and sometimes the types and location of protective system.

(*iii*) The magnitude of short-circuit current determines the size of the protective reactors which must be inserted in the system so that the circuit breaker is able to withstand the fault current.

(*iv*) The calculation of short-circuit currents enables us to make proper selection of the associated apparatus (*e.g.* bus-bars, current transformers etc.) so that they can withstand the forces that arise due to the occurrence of short circuits.

16.8 Faults in a Power System

A fault occurs when two or more conductors that normally operate with a potential difference come in contact with each other. These faults may be caused by sudden failure of a piece of equipment, accidental damage or short-circuit to overhead lines or by insulation failure resulting from lightning surges. Irrespective of the causes, the faults in a 3-phase system can be classified into two main categories *viz*.

(*i*) Symmetrical faults (*ii*) Unsymmetrical faults

(*i*) **Symmetrical faults.** That fault which gives rise to symmetrical fault currents (*i.e.* equal faults currents with 120° displacement) is called a symmetrical fault. The most common example of symmetrical fault is when all the three conductors of a 3-phase line are brought together simultaneously into a short-circuit condition. The method of calculating fault currents for symmetrical faults is discussed in chapter 17.

(*ii*) **Unsymmetrical faults.** Those faults which give rise to unsymmetrical currents (*i.e.* unequal line currents with unequal displacement) are called unsymmetrical faults. The unsymmetrical faults may take one of the following forms :

(*a*) Single line-to-ground fault (*b*) Line-to-line fault (*c*) Double line-to-ground fault

The great majority of faults on the power system are of unsymmetrical nature; the most common type being a short-circuit from one line to ground. The calculations of such fault currents are made by "symmetrical components" method. This is fully discussed in chapter 18.

SELF - TEST

1. **Fill in the blanks by inserting appropriate words/figures.**
   (*i*) A fuse is a .... device.
   (*ii*) A circuit breaker is a .... device.
   (*iii*) An isolator is designed to open a circuit under ....
   (*iv*) When a switch is opened, ... is produced.
   (*v*) Under normal operating conditions, the contacts of the circuit breaker remain ....
   (*vi*) Under fault conditions, ... supplies information to the circuit breaker to open.
   (*vii*) If a fault occurs on the bus itself in a single bus-bar system, then there is complete ....
   (*viii*) The sectionalised bus-bar system gives .... fault current than that of unsectionalised bus-bar.
   (*ix*) For greater flexibility, ... bus-bar system is used.
   (*x*) The outdoor type switchgear is generally used for voltages beyond .... kV.
2. **Pick up the correct words/figures from brackets and fill in the blanks.**
   (*i*) A fuse performs .... functions. (*both detection and interruption, interruption*)
   (*ii*) The circuit breaker performs .... function. (*detection, circuit interruption*)
   (*iii*) For voltages beyond 66 kV, switchgear equipment is installed .... (*indoor, outdoor*)
   (*iv*) Bus-bars operate at .... voltage. (*constant, variable*)
   (*v*) Isolator is an .... switch. (*air, oil*)

ANSWERS TO SELF-TEST

1. (*i*) protective (*ii*) circuit interrupting (*iii*) no load (*iv*) arc (*v*) closed (*vi*) relay (*vii*) shut down (*viii*) lower (*ix*) duplicate (*x*) 66
2. (*i*) both detection and interruption (*ii*) circuit interruption (*iii*) outdoor (*iv*) constant (*v*) air

CHAPTER REVIEW TOPICS

1. What do you understand by switchgear ?
2. Discuss the various types of switches.
3. What is the difference between
   (*i*) a switch and circuit breaker
   (*ii*) a fuse and circuit breaker ?
4. Discuss the different types of bus-bar arrangements.
5. Explain the various methods of accommodating high-voltage switchgear.

DISCUSSION QUESTIONS

1. What are the limitations of a fuse ?
2. Why do we use C.T. in the relay circuit ?
3. What is the necessity of bus-bar ?
4. Why do we use isolators on both sides of the circuit breaker ?
5. Why are isolators not opened on load ?
6. Which faults ___ symmetrical or unsymmetrical ___ are more frequent in power system and why ?
7. Suddenly a circuit carries a current 20 times the normal current. Is there possibility of short-circuit or overload ?

CHAPTER

17

Symmetrical Fault Calculations

Introduction

Most of the faults on the power system lead to a short-circuit condition. When such a condition occurs, a heavy current (called short circuit current) flows through the equipment, causing considerable damage to the equipment and interruption of service to the consumers. There is probably no other subject of greater importance to an electrical engineer than the question of determination of short circuit currents under fault conditions. The choice of apparatus and the design and arrangement of practically every equipment in the power system depends upon short-circuit current considerations. In this chapter, we shall confine our discussion to fault currents due to symmetrical faults.

17.1 Symmetrical Faults on 3-Phase System

That fault on the power system which gives rise to symmetrical fault currents (i.e. equal fault currents in the lines with 120° displacement) is called a **symmetrical fault.**

The symmetrical fault occurs when all the three conductors of a 3-phase line are brought together simultaneously into a short-circuit condition as shown in Fig. 17.1. This type of fault gives rise to symmetrical currents *i.e.* equal fault currents with 120° displacement. Thus referring to Fig. 17.1, fault currents I_R, I_Y and I_B will be equal in magnitude with 120° displacement among them. Because of balanced nature of fault, only one* phase need be considered in calculations since condition in the other two phases will also be similar. The following points may be particularly noted :

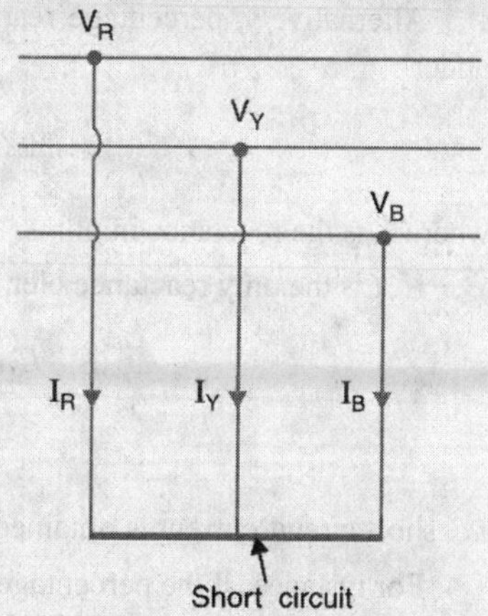

Fig. 17.1

(*i*) The symmetrical fault rarely occurs in practice as majority of the faults are of unsymmetrical nature. However, symmetrical fault calculations are being discussed in this chapter to enable the reader to understand the problems that short circuit conditions present to the power system.

(*ii*) The symmetrical fault is the most severe and imposes more heavy duty on the circuit breaker.

17.2 Limitation of Fault Current

When a short circuit occurs at any point in a system, the short-circuit current is limited by the impedance of the system upto the point of fault. Thus referring to Fig. 17.2, if a fault occurs on the feeder at point *F*, then the short circuit current from the generating station will have a value limited by the impedance of generator and transformer and the impedance of the line between the generator and the point of fault. This shows that the knowledge of the impedances of various equipment and circuits in the line of the system is very important for the determination of short-circuit currents.

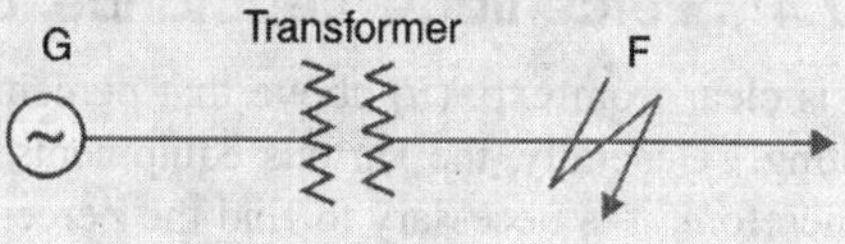

Fig. 17.2

In many situations, the impedances limiting the fault current are largely reactive, such as transformers, reactors and generators. Cables and lines are mostly resistive, but where the total reactance in calculations exceeds 3 times the resistance, the latter is usually neglected. The error introduced by this assumption will not exceed 5%.

17.3 Percentage Reactance

The reactance of generators, transformers, reactors *etc.* is usually expressed in *percentage reactance* to permit rapid short circuit calculations. The percentage reactance of a circuit is defined as under :

It is the percentage of the total phase-voltage dropped in the circuit when full-load current is flowing i.e.,

$$\%X = \frac{I\,X}{V} \times 100 \qquad ...(i)$$

where I = full-load current

V = phase voltage

X = reactance in ohms per phase

---

* Balanced three-phase faults, like balanced 3-φ loads, may be handled on a line-to-neutral basis.

Alternatively, percentage reactance (%X) can also be expressed in terms* of kVA and kV as under :

$$\%X = \frac{(\text{kVA})\,X}{10\,(\text{kV})^2} \qquad ...(ii)$$

where X is the reactance in ohms.

If X is the only reactance element in the circuit, then short-circuit current is given by ;

$$I_{SC} = \frac{V}{X}$$

$$= I \times \left(\frac{100}{\%X}\right) \qquad \text{[By putting the value of } X \text{ from exp. } (i)]$$

i.e. short circuit current is obtained by multiplying the full-load current by 100/% X.

For instance, if the percentage reactance of an element is 20% and the full-load current is 50 A, then short-circuit current will be 50 × 100/20 = 250 A when only that element is in the circuit.

It may be worthwhile to mention here the advantage of using percentage reactance instead of ohmic reactance in short-circuit calculations. Percentage reactance values remain unchanged as they are referred through transformers, unlike ohmic reactances which become multiplied or divided by the square of transformation ratio. This makes the procedure simple and permits quick calculations.

17.4 Percentage Reactance and Base kVA

It is clear from exp. (*ii*) above that percentage reactance of an equipment depends upon its kVA rating. Generally, the various equipments used in the power system have different kVA ratings. Therefore, it is necessary to find the percentage reactances of all the elements on a common kVA rating. This common kVA rating is known as *base kVA*. The value of this base kVA is quite unimportant and may be :

(*i*) equal to that of the largest plant

(*ii*) equal to the total plant capacity

(*iii*) any arbitrary value

The conversion can be effected by using the following relation :

$$\%\text{ age reactance at base kVA} = \frac{\text{Base kVA}}{\text{Rated kVA}} \times \%\text{ age reactance at rated kVA}$$

Thus, a 1000 kVA transformer with 5% reactance will have a reactance of 10% at 2000 kVA base.

Illustration. The fact that the value of base kVA does not affect the short circuit current needs illustration. Consider a 3-phase transmission line operating at 66 kV and connected through a 1000 kVA transformer with 5% reactance to a generating station bus-bar. The generator is of 2500 kVA with 10% reactance. The single line diagram of the system is shown in Fig. 17.3. Suppose a short-circuit fault between three phases occurs

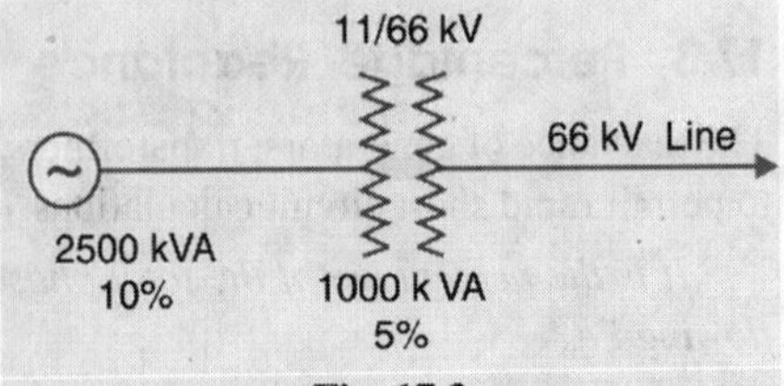

Fig. 17.3

---

* From exp. (*i*), $X = \dfrac{(\%X)\,V}{100I} = \dfrac{(\%X)\,V \times V}{100 \times VI} = \dfrac{(\%X)\left(\dfrac{V}{1000}\right)\left(\dfrac{V}{1000}\right) \times 1000}{100 \times \left(\dfrac{V}{1000}\right) \times I} = \dfrac{(\%X)\,(\text{kV})^2 \times 10}{\text{kVA}}$

$\therefore \quad \%X = \dfrac{(\text{kVA})\,X}{10\,(\text{kV})^2}$

at the high voltage terminals of transformer. It will be shown that whatever value of base kVA we may choose, the value of short-circuit current will be the same.

(*i*) Suppose we choose 2500 kVA as the common base kVA. On this base value, the reactances of the various elements in the system will be :

Reactance of transformer at 2500 kVA base

$$= 5 \times 2500^*/1000 = 12{\cdot}5\%$$

Reactance of generator at 2500 kVA base

$$= 10 \times 2500/2500 = 10\%$$

Total percentage reactance on the common base kVA

$$\%X = 12{\cdot}5 + 10 = 22{\cdot}5\%$$

The full† load current corresponding to 2500 kVA base at 66 kV is given by ;

$$I = \frac{2500 \times 1000}{\sqrt{3} \times 66 \times 1000} = 21{\cdot}87 \text{ A}$$

$\therefore$ Short-circuit current, $I_{SC} = I \times \dfrac{100}{\%X} = 21{\cdot}87 \times \dfrac{100}{22{\cdot}5} = \mathbf{97{\cdot}2\ A}$

(*ii*) Now, suppose we choose 5000 kVA as the common base value.

Reactance of transformer at 5000 kVA base

$$= 5 \times 5000/1000 = 25\%$$

Reactance of generator at 5000 kVA base

$$= 10 \times 5000/2500 = 20\%$$

Total percentage reactance on the common base kVA

$$\%X = 25 + 20 = 45\%$$

Full-load current corresponding to 5000 kVA at 66 kV is

$$I = \frac{5000 \times 1000}{\sqrt{3} \times 66 \times 1000} = 43{\cdot}74 \text{ A}$$

$\therefore$ Short-circuit current, $I_{SC} = I \times \dfrac{100}{\%X} = 43{\cdot}74 \times \dfrac{100}{45} = \mathbf{97{\cdot}2\ A}$

which is the same as in the previous case.

From the above illustration, it is clear that whatever may be the value of base kVA, short-circuit current is the same. However, in the interest of simplicity, numerically convenient value for the base kVA should be chosen.

17.5 Short-Circuit kVA

Although the potential at the point of fault is zero, it is a normal practice to express the short-circuit current in terms of short-circuit kVA based on the normal system voltage at the point of fault.

The product of normal system voltage and short-circuit current at the point of fault expressed in kVA is known as **short-circuit kVA.**

Let V = normal phase voltage in volts

I = full-load current in amperes at base kVA

$\%X$ = percentage reactance of the system on base kVA upto the fault point

---

* % reactance at base kVA = $\dfrac{\text{base kVA}}{\text{rated kVA}} \times$ % reactance at rated kVA

† Full-load current has to be found out for the base kVA selected.

As proved in Art 17.3,

Short-circuit current, $I_{SC} = I\left(\frac{100}{\%X}\right)$

$\therefore$ Short-circuit kVA for 3-phase circuit

$$= \frac{3\,V\,I_{SC}}{1000}$$

$$= \frac{3\,V\,I}{1000} \times \frac{100}{\%X}$$

$$= \text{Base kVA} \times \frac{100}{\%X}$$

i.e. short-circuit kVA is obtained by multiplying the base kVA by 100/% *X*.

17.6 Reactor Control of Short-Circuit Currents

With the fast expanding power system, the fault level (*i.e.* the power available to flow into a fault) is also rising. The circuit breakers connected in the power system must be capable of dealing with maximum possible short-circuit currents that can occur at their points of connection. Generally, the reactance of the system under fault conditions is low and fault currents may rise to a dangerously high value. If no steps are taken to limit the value of these short-circuit currents, not only will the duty required of circuit breakers be excessively heavy, but also damage to lines and other equipment will almost certainly occur.

In order to limit the short-circuit currents to a value which the circuit breakers can handle, additional reactances known as *reactors* are connected in series with the system at suitable points. A reactor is a coil of number of turns designed to have a large inductance as compared to its ohmic resistance. The forces on the turns of these reactors under short-circuit conditions are considerable and, therefore, the windings must be solidly braced. It may be added that due to very small resistance of reactors, there is very little change in the efficiency of the system.

Advantages

(*i*) Reactors limit the flow of short-circuit current and thus protect the equipment from over-heating as well as from failure due to destructive mechanical forces.

(*ii*) Troubles are localised or isolated at the point where they originate without communicating their disturbing effects to other parts of the power system. This increases the chances of continuity of supply.

(*iii*) They permit the installation of circuit breakers of lower rating.

17.7 Location of Reactors

Short circuit current limiting reactors may be connected (*i*) in series with each generator (*ii*) in series with each feeder and (*iii*) in bus-bars. No definite statement can be given as to which one of the above locations is preferable; each installation has its own particular demands which must be carefully considered before a choice of reactor location can be made.

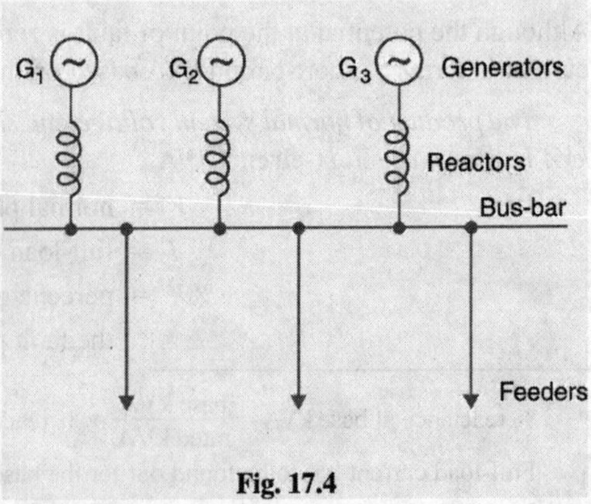

Fig. 17.4

(1) Generator reactors. When the reactors are connected in series with each generator, they are known as *generator reactors* (see Fig. 17.4). In this case, the reactor may be considered as a part of leakage reactance of the generator ; hence its effect is to protect the generator in the case of any short-circuit beyond the reactors.

Disadvantages

(*i*) There is a constant voltage drop and power loss in the reactors even during normal operation.

(*ii*) If a bus-bar or feeder fault occurs close to the bus-bar, the voltage at the bus-bar will be reduced to a low value, thereby causing the generators to fall out of step.

(*iii*) If a fault occurs on any feeder, the continuity of supply to other is likely to be affected.

Due to these disadvantages and also since modern power station generators have sufficiently large leakage reactance to protect them against short-circuit, it is not a common practice to use separate reactors for the generators.

(2) **Feeder reactors.** When the reactors are connected in series with each feeder, they are known as *feeder reactors* (see Fig. 17.5). Since most of the short-circuits occur on feeders, a large number of reactors are used for such circuits. Two principal advantages are claimed for feeder reactors. Firstly, if a fault occurs on any feeder, the voltage drop in its reactor will not affect the bus-bars voltage so that there is a little tendency for the generator to lose synchronism. Secondly, the fault on a feeder will not affect other feeders and consequently the effects of fault are localised.

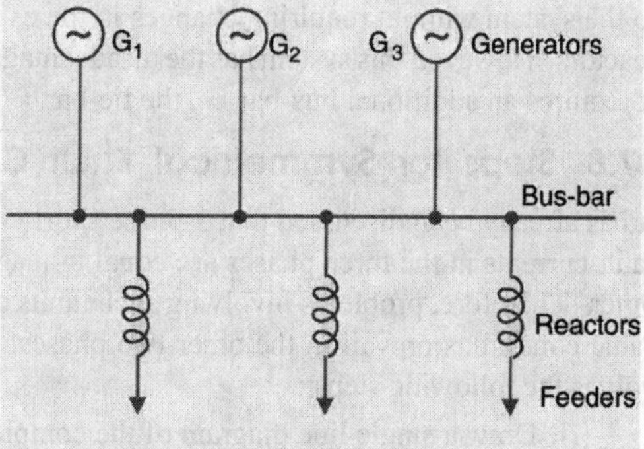

Fig. 17.5

Disadvantages

(*i*) There is a constant power loss and voltage drop in the reactors even during normal operation.

(*ii*) If a short-circuit occurs at the bus-bars, no protection is provided to the generators. However, this is of little importance because such faults are rare and modern generators have considerable leakage reactance to enable them to withstand short-circuit across their terminals.

(*iii*) If the number of generators is increased, the size of feeder reactors will have to be increased to keep the short-circuit currents within the ratings of the feeder circuit breakers.

(3) **Bus-bar reactors.** The above two methods of locating reactors suffer from the disadvantage that there is considerable voltage drop and power loss in the reactors even during normal operation. This disadvantage can be overcome by locating the reactors in the bus-bars. There are two methods for this purpose, namely ; Ring system and Tie-Bar system.

(*i*) *Ring system.* In this system, bus-bar is divided into sections and these sections are connected through reactors as shown in Fig. 17.6. Generally, one feeder is fed from one generator only. Under normal operating conditions, each generator will supply its own section of the load and very little power will be fed by other generators. This results in low power loss and voltage drop in the reactors. However, the principal advantage of the system is that if a fault occurs on any feeder, only one generator (to which the particular feeder is connected) mainly feeds the fault current while the current fed

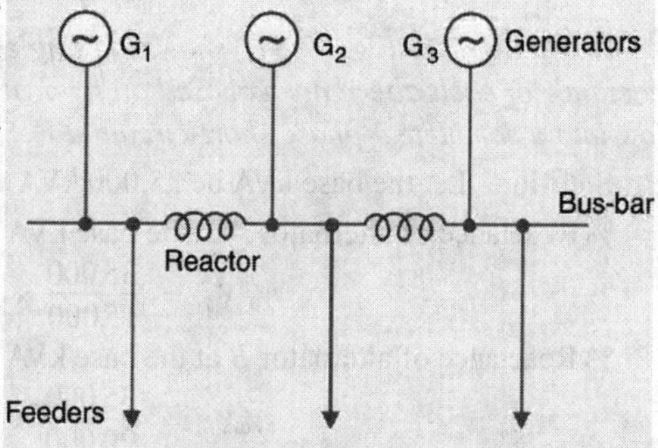

Fig. 17.6

from other generators is small due to the presence of reactors. Therefore, only that section of bus-bar is affected to which the feeder is connected, the other sections being able to continue in normal operation.

(*ii*) *Tie-Bar system.* Fig. 17.7 shows the tie-bar system. Comparing the ring system with tie-bar system, it is clear that in the tie-bar system, there are effectively two reactors in series between sections so that reactors must have approximately half the reactance of those used in a comparable ring system. Another advantage of tie-bar system is that additional generators may be connected to the system without requiring changes in the existing reactors. However, this system has the disadvantage that it requires an additional bus-bar *i.e.* the tie-bar.

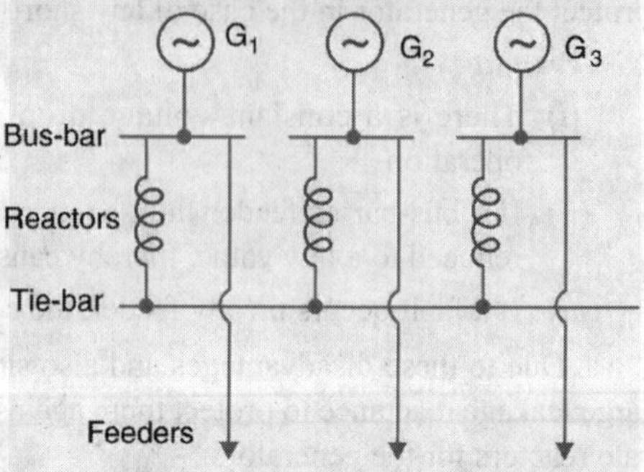

Fig. 17.7

17.8 Steps for Symmetrical Fault Calculations

It has already been discussed that 3-phase short-circuit faults result in symmetrical fault currents *i.e.* fault currents in the three phases are equal in magnitude but displaced 120° electrical from one another. Therefore, problems involving such faults can be solved by considering one phase only as the same conditions prevail in the other two phases. The procedure for the solution of such faults involves the following steps :

(*i*) Draw a single line diagram of the complete network indicating the rating, voltage and percentage reactance of each element of the network.

(*ii*) Choose a numerically convenient value of base kVA and convert all percentage reactances to this base value.

(*iii*) Corresponding to the single line diagram of the network, draw the reactance diagram showing one phase of the system and the neutral. Indicate the % reactances on the base kVA in the reactance diagram. The transformer in the system should be represented by a reactance in series.

(*iv*) Find the total % reactance of the network upto the point of fault. Let it be *X*%.

(*v*) Find the full-load current corresponding to the selected base kVA and the normal system voltage at the fault point. Let it be *I*.

(*vi*) Then various short-circuit calculations are :

$$\text{Short-circuit current, } I_{SC} = I \times \frac{100}{\%X}$$

$$\text{Short-circuit kVA} = \text{Base kVA} \times \frac{100}{\%X}$$

Example 17.1. *Fig. 17.8 (i) shows the single line diagram of a 3-phase system. The percentage reactance of each alternator is based on its own capacity. Find the short-circuit current that will flow into a complete 3-phase short-circuit at F.*

Solution. Let the base kVA be 35,000 kVA.

% Reactance of alternator *A* at the base kVA is

$$\% X_A = \frac{35{,}000}{15{,}000} \times 30 = 70\%$$

% Reactance of alternator *B* at the base kVA is

$$\% X_B = \frac{35{,}000}{20{,}000} \times 50 = 87{\cdot}5\ \%$$

Line current corresponding to 35000 kVA at 12 kV is

$$I = \frac{35{,}000 \times 10^3}{\sqrt{3} \times 12 \times 10^3} = 1684 \text{ A}$$

Fig. 17.8 (*ii*) shows the reactance* diagram of the network at the selected base kVA.

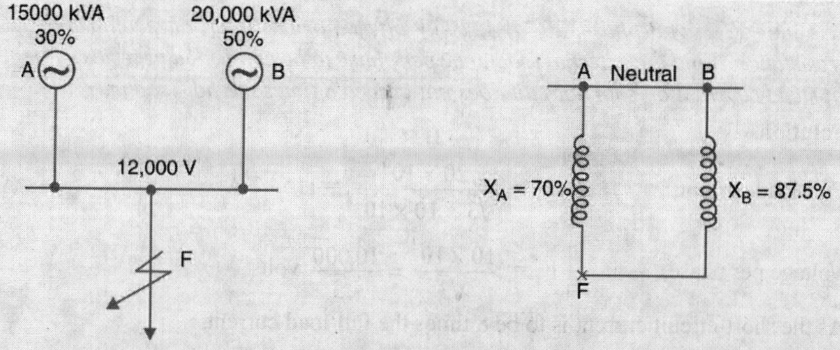

Single-line diagram
(*i*)

Reactance diagram at 35,000 kVA
(*ii*)

Fig. 17.8

Total % reactance from generator neutral up to fault point is

$$\%X = X_A \parallel X_B$$

$$= \frac{X_A X_B}{X_A + X_B} = \frac{70 \times 87 \cdot 5}{70 + 87 \cdot 5} = 38 \cdot 89 \ \%$$

∴ Short-circuit current, $I_{SC} = I \times \frac{100}{\%X} = 1684 \times \frac{100}{38 \cdot 89} = \mathbf{4330\ A}$

Alternate method. The problem can also be solved by component short-circuit current method. Each alternator supplies short circuit current to the fault. The total current fed to the fault is the sum of the two.

Full-load current delivered by alternator *A*,

$$I_A = \frac{\text{Rated kVA of alternator } A}{\sqrt{3} \times \text{Bus-bar voltage}}$$

$$= \frac{15000 \times 10^3}{\sqrt{3} \times 12 \times 10^3} = 721 \cdot 7 \text{ A}$$

∴ Short-circuit current fed to fault by alternator *A*,

$$I_{SA} = I_A \times \frac{100}{\% \text{ Reactance}^{**} \text{of A}}$$

$$= 721 \cdot 7 \times (100/30) = 2405 \cdot 5 \text{ A}$$

Full-load line current delivered by alternator *B*,

$$I_B = \frac{20000 \times 10^3}{\sqrt{3} \times 12 \times 10^3} = 962 \cdot 28 \text{ A}$$

∴ Short-circuit current fed to fault by alternator *B*,

$$I_{SB} = 962 \cdot 28 \times 100/50 = 1924 \cdot 5 \text{ A}$$

∴ Total short-circuit current fed to fault,

$$I_{SC} = I_{SA} + I_{SB} = 2405 \cdot 5 + 1924 \cdot 5 = \mathbf{4330\ A}$$

* Note that the diagram shows one phase and neutral. The percentage reactances of the alternators are in parallel.

** At its own rated capacity.

Comments. In simple problems, either of the two methods can be used with about the same degree of ease. However, in complicated networks, it will be found that the first method (known as equivalent % reactance method) has greater advantage owing to the ease by which calculations can be carried out.

Example 17.2. *A 3-phase, 20 MVA, 10 kV alternator has internal reactance of 5% and negligible resistance. Find the external reactance per phase to be connected in series with the alternator so that steady current on short-circuit does not exceed 8 times the full load current.*

Solution.

Full-load current, $I = \dfrac{20 \times 10^6}{\sqrt{3} \times 10 \times 10^3} = 1154{\cdot}7\text{ A}$

Voltage per phase, $V = \dfrac{10 \times 10^3}{\sqrt{3}} = \dfrac{10{,}000}{\sqrt{3}}$ volts

As the short-circuit current is to be 8 times the full-load current,

∴ Total percentage reactance required

$$= \frac{\text{Full - load current}}{\text{Short - circuit current}} \times 100$$

$$= \left(\frac{1}{8}\right) \times 100 = 12{\cdot}5\ \%$$

∴ External percentage reactance required

$$= 12{\cdot}5 - 5 = 7{\cdot}5\%$$

Let $X\ \Omega$ be the per phase external reactance required.

Now, percentage reactance $= \dfrac{I\,X}{V} \times 100$

or $$7{\cdot}5 = \frac{1154{\cdot}7X}{\dfrac{10{,}000}{\sqrt{3}}} \times 100$$

∴ $$X = \frac{7{\cdot}5 \times 10000}{\sqrt{3} \times 100 \times 1154{\cdot}7} = 0{\cdot}375\ \Omega$$

Example 17.3. *A 3-phase transmission line operating at 10 kV and having a resistance of 1Ω and reactance of 4 Ω is connected to the generating station bus-bars through 5 MVA step-up transformer having a reactance of 5%. The bus-bars are supplied by a 10 MVA alternator having 10% reactance. Calculate the short-circuit kVA fed to symmetrical fault between phases if it occurs*

(*i*) *at the load end of transmission line*

(*ii*) *at the high voltage terminals of the transformer*

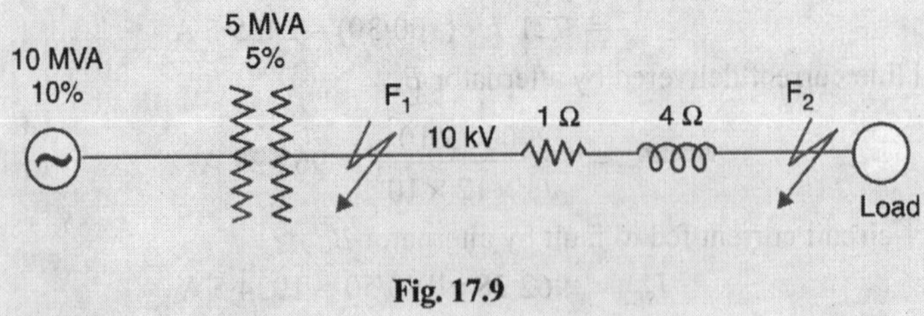

Fig. 17.9

Solution. Fig. 17.9 shows the single line diagram of the network. Let 10,000 kVA be the base kVA.

% reactance of alternator on base kVA,

$$\% X_A = \frac{10{,}000}{10 \times 10^3} \times 10 = 10\%$$

% reactance of transformer on base kVA,

$$\% X_T = \frac{10{,}000}{5 \times 10^3} \times 5 = 10\%$$

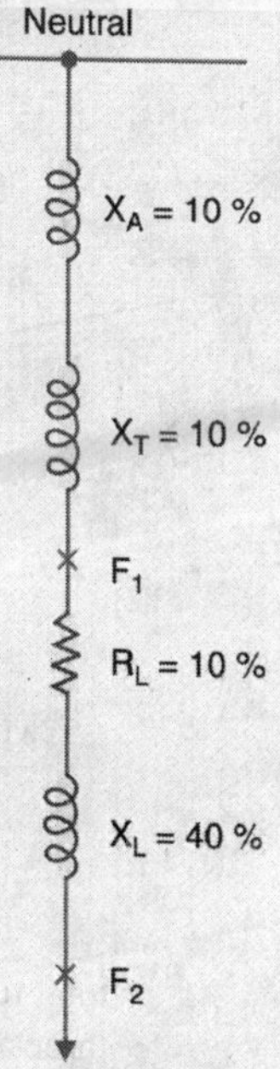

Fig. 17.10

The line impedance is given in ohms. It can be converted into percentage impedance by using exp. (*ii*) of Art. 17.3.

% reactance of transmission line is

$$\% X_L = \frac{(\text{kVA}) \times \text{reactance in } \Omega}{10\,(\text{kV})^2}$$

$$= \frac{10{,}000 \times 4}{10 \times (10)^2} = 40\%$$

% age resistance of transmission line,

$$\% R_L = \frac{10{,}000 \times 1}{10 \times (10)^2} = 10\%$$

(*i*) The reactance diagram of the network on the selected base kVA is shown in Fig. 17.10. For a fault at the end of a transmission line (point F_2),

$$\text{Total \% reactance} = \% X_A + \% X_T + \% X_L$$

$$= 10 + 10 + 40 = 60\%$$

$$\text{\% resistance} = 10\%$$

∴ % impedance from generator neutral upto fault point F_2

$$= \sqrt{(60)^2 + (10)^2} = 60{\cdot}83\%$$

∴ Short-circuit kVA = 10,000 × 100/60·83 = **16,440 kVA**

(*ii*) For a fault at the high voltage terminals of the transformer (point F_1),

Total % reactance from generator neutral upto fault point F_1

$$= \% X_A + \% X_T = 10 + 10 = 20\%$$

∴ Short-circuit kVA = 10,000 × 100/20 = **50,000 kVA**

Example 17.4. *The plant capacity of a 3-phase generating station consists of two 10,000 kVA generators of reactance 12% each and one 5000 kVA generator of reactance 18%. The generators are connected to the station bus-bars from which load is taken through three 5000 kVA step-up transformers each having a reactance of 5%. Determine the maximum fault MVA which the circuit breakers on (i) low voltage side and (ii) high voltage side may have to deal with.*

Solution. Fig. 17.11 shows the single line diagram of the network. Let 10,000 kVA be the base kVA.

The percentage reactance of generators *A*, *B* and *C* and that of each transformer on the selected base kVA is

$$\% X_A = 12 \times 10{,}000/10{,}000 = 12\%$$

$$\% X_B = 12 \times 10{,}000/10{,}000 = 12\%$$

$$\% X_C = 18 \times 10{,}000/5{,}000 = 36\%$$

$$\% X_T = 5 \times 10{,}000/5{,}000 = 10\%$$

(*i*) When the fault occurs on the low voltage side of the transformer (point F_1 in Fig. 17.11), the reactance diagram at the selected base kVA will be as shown in Fig. 17.12. Obviously, the

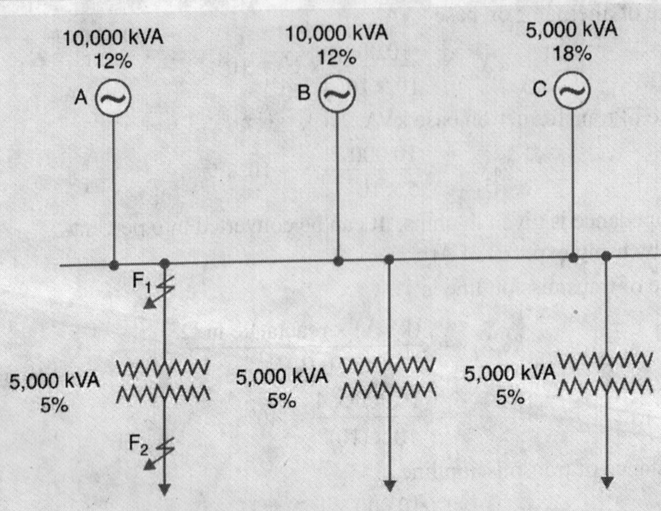

Fig. 17.11

total reactance upto the point of fault F_1 is the parallel combination of the reactances of the three alternators *i.e.*

Total % reactance form generator neutral upto fault point F_1

$$= \% X_A \parallel \% X_B \parallel \% X_C$$

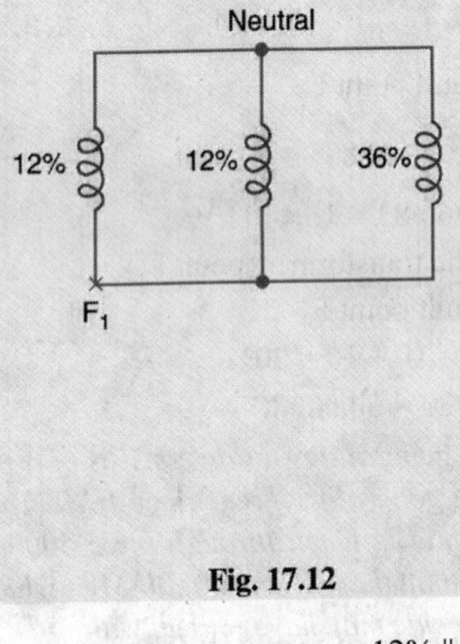

Fig. 17.12

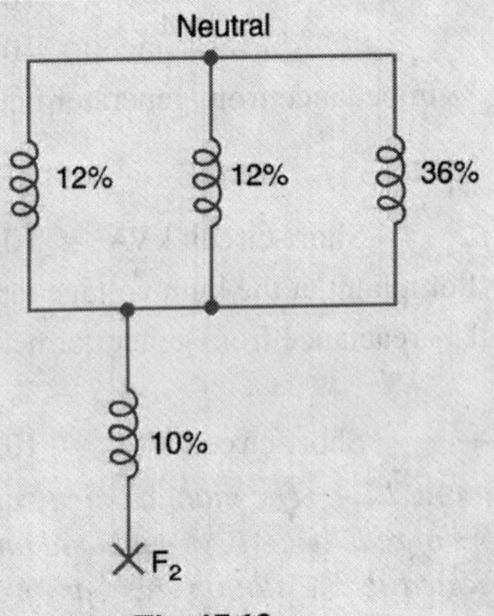

Fig. 17.13

$$= 12\% \parallel 12\% \parallel 36\% = \frac{6 \times 36}{6 + 36} = 5{\cdot}14\%$$

$$\therefore \quad \text{Fault MVA} = 10{,}000 \times \frac{100}{5{\cdot}14} \times \frac{1}{1000} = \mathbf{194{\cdot}5}$$

(*ii*) When the fault occurs on the high voltage side of the transformer (point F_2 in Fig. 17.11), the reactance diagram will be as shown in Fig. 17.13.

Total % reactance from generator neutral upto fault point F_2

$$= 5{\cdot}14 + 10 = 15{\cdot}14\%$$

$$\therefore \quad \text{Fault MVA} = 10{,}000 \times \frac{100}{15{\cdot}14} \times \frac{1}{1000} = \mathbf{66}$$

It may be noted that circuit breakers of lower ratings will be required on the high voltage side of the transformers.

Example 17.5. *The section bus-bars A and B are linked by a bus-bar reactor rated at 5000 kVA with 10% reactance. On bus-bar A, there are two generators each of 10,000 kVA with 10% reactance and on B two generators each of 8000 kVA with 12% reactance. Find the steady MVA fed into a dead short circuit between all phases on B with bus-bar reactor in the circuit.*

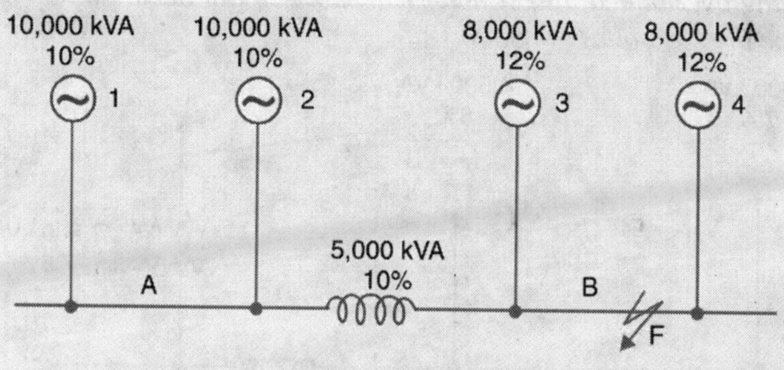

Fig. 17.14

Solution. Fig. 17.14 shows the single line diagram of the network.

Let 10,000 kVA be the base kVA.

% Reactance of generator 1 or 2 on the base kVA

$$= 10 \times 10{,}000/10{,}000 = 10\%$$

% Reactance of generator 3 or 4 on the base kVA

$$= 12 \times 10{,}000/8000 = 15\%$$

% Reactance of bus-bar reactor on the base kVA

$$= 10 \times 10{,}000/5000 = 20\%$$

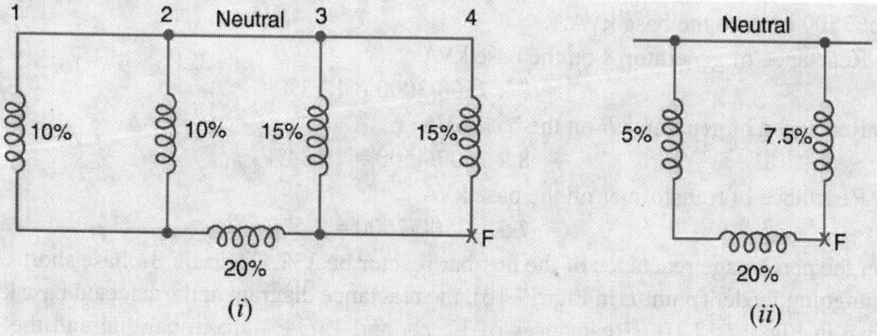

Fig. 17.15

When fault occurs on section *B* (point *F* in Fig. 17.14), the reactance diagram at the selected base kVA will be as shown in Fig. 17.15 (*i*). This series parallel circuit is further reduced to Fig. 17.15 (*ii*). Referring to Fig. 17.15 (*ii*), it is clear that reactance from generator neutral upto the fault point *F* is (5% + 20%) in parallel with 7·5% *i.e.*

Total % reactance from generator neutral upto fault point *F*

$$= (5\,\% + 20\%) \parallel 7{\cdot}5\%$$

$$= \frac{25 \times 7 \cdot 5}{25 + 7 \cdot 5} = 5{\cdot}77\%$$

$\therefore$ Fault kVA $= 10{,}000 \times 100/5{\cdot}77 = 1{,}73{,}310$

or Fault MVA $= \mathbf{173{\cdot}31}$

Example 17.6. *A small generating station has two alternators of 3000 kVA and 4500 kVA and percentage reactances of 7% and 8% respectively. The circuit breakers have a rupturing capacity of 150 MVA. It is desired to extend the system by a supply from the grid via a transformer of 7500 kVA and 7·5% reactance. Find the reactance of the reactor connected in the bus-bar section to prevent the circuit breakers being overloaded, if a symmetrical short-circuit occurs on an outgoing feeder connected to it. Assume the bus voltage = 3300 V.*

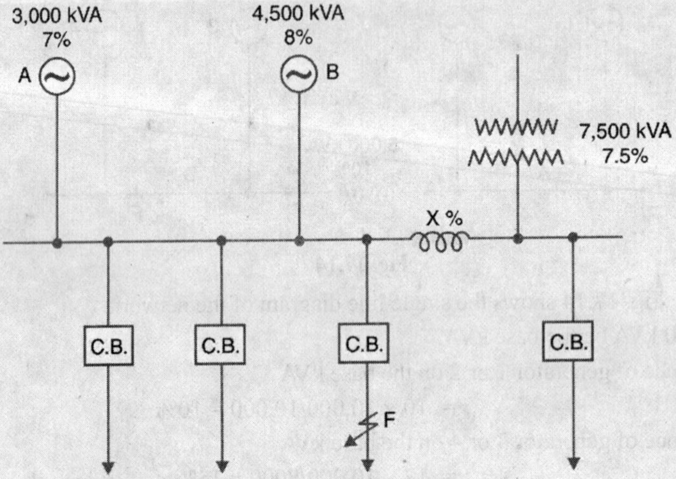

Fig. 17.16

Solution. Fig. 17.16 shows the single line diagram of the system.

Let 7500 kVA be the base kVA.

% Reactance of generator *A* on the base kVA

$= 7 \times 7500/3000 = 17{\cdot}5\%$

% Reactance of generator *B* on the base kVA

$= 8 \times 7500/4500 = 13{\cdot}34\%$

% Reactance of transformer on the base kVA

$= 7{\cdot}5 \times 7500/7500 = 7{\cdot}5\%$

Let the percentage reactance of the bus-bar reactor be *X*%. Then for 3-phase short-circuit fault on an outgoing feeder (point *F* in Fig. 17.16), the reactance diagram at the selected base kVA will be as shown in Fig. 17.17 (*i*). Reactances of 17·5% and 13·34% are in parallel and the equivalent reactance* = 7·57%. The circuit then reduces to Fig. 17.17 (*ii*).

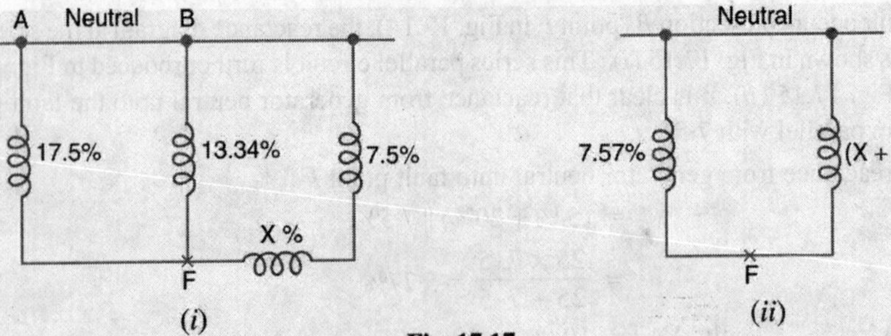

Fig. 17.17

* Equivalent reactance $= \dfrac{17{\cdot}5 \times 13{\cdot}34}{17{\cdot}5 + 13{\cdot}34} = 7{\cdot}57\%$

Total % reactance from generator neutral to fault point F

$$= 7{\cdot}57\% \parallel (X + 7{\cdot}5)\%$$

$$= \frac{7{\cdot}57\,(X+7{\cdot}5)}{X+15{\cdot}07}\,\%$$

$$\text{Short-circuit kVA} = 7500 \times 100 \times \frac{X+15{\cdot}07}{7{\cdot}57\,(X+7{\cdot}5)}$$

But the short-circuit kVA should not exceed 150×10^3 kVA, the rupturing capacity of the breaker.

$$\therefore \quad 150 \times 10^3 = \frac{7500 \times 100 \times (X+15{\cdot}07)}{7{\cdot}57\,(X+7{\cdot}5)}$$

or $$7{\cdot}57\,(X+7{\cdot}5) = 5\,(X+15{\cdot}07)$$

or $$7{\cdot}57\,X + 56{\cdot}77 = 5\,X + 75{\cdot}35$$

or $$X = \frac{75{\cdot}35 - 56{\cdot}77}{7{\cdot}57 - 5} = 7{\cdot}23\%$$

The %age reactance can be converted into reactance in ohms by the following expression :

$$\%\,X = \frac{(\text{kVA})\,X}{10\,(\text{kV})^2}$$

or $$7{\cdot}23 = \frac{7500\,X}{10\,(3{\cdot}3)^2}$$

or $$X = \frac{7{\cdot}23 \times 10 \times (3{\cdot}3)^2}{7500} = \mathbf{0{\cdot}105\ \Omega}$$

i.e. reactance of the reactor per phase should be $0{\cdot}105\ \Omega$.

Example 17.7. *The estimated short-circuit MVA at the bus-bars of a generating station A is 1500 MVA and of another station B is 1200 MVA. The generated voltage at each station is 33 kV. If these stations are interconnected through a line having a reactance of 1Ω and negligible resistance, calculate the possible short-circuit MVA at both stations.*

Solution. Fig. 17.18 (*i*) shows the single line diagram of the system.

Let the base MVA be 100.

% Reactance of station A on the base MVA

$$\%\,X_A = \frac{\text{Base MVA}}{\text{Short-circuit MVA}} \times 100$$

$$= (100/1500) \times 100 = 6{\cdot}67\%$$

% Reactance of station B on the base MVA

$$\%\,X_B = (100/1200) \times 100 = 8{\cdot}33\%$$

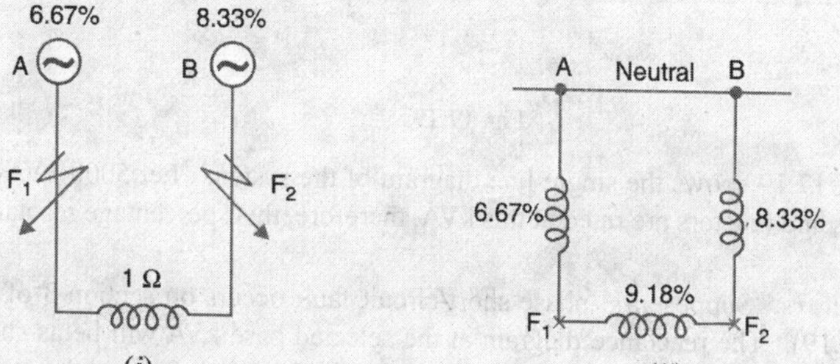

Fig. 17.18

% Reactance of interconnector on the base MVA

$$\% X_I = \frac{(\text{kVA})\, X}{10\, (\text{kV})^2} = \frac{(100 \times 10^3) \times 1}{10 \times (33)^2} = 9{\cdot}18\%$$

The reactance diagram is shown in Fig. 17.18 (*ii*).

Fault on station A. When a fault occurs on power station *A* (point F_1), the total percentage reactance from generator neutral upto fault point F_1 is the parallel combination of (8·33% + 9·18%) and 6·67% *i.e.*

Total % reactance upto fault point F_1

$$= \frac{17{\cdot}51 \times 6{\cdot}67}{17{\cdot}51 + 6{\cdot}67} = 4{\cdot}83\%$$

$$\therefore \quad \text{Short-circuit MVA} = \text{Base MVA} \times \frac{100}{\%\ \text{Reactance}}$$

$$= 100 \times 100/4{\cdot}83 = \mathbf{2070}$$

Fault on station B. When a fault occurs on power station *B* (point F_2), the total percentage reactance from generator neutral upto fault point F_2 is the parallel combination of (6·67% + 9·18%) and 8·33% *i.e.*

Total % reactance upto fault point F_2

$$= \frac{15{\cdot}85 \times 8{\cdot}33}{15{\cdot}85 + 8{\cdot}33} = 5{\cdot}46\%$$

$$\therefore \quad \text{Short-circuit MVA} = 100 \times 100/5{\cdot}46 = \mathbf{1831}$$

Example 17.8. *A generating station has three section bus-bars connected with a tie-bar through 6% reactors rated at 5000 kVA. Each generator is of 5000 kVA with 12% reactance and is connected to one section of bus-bars. Find the total steady input to a dead short-circuit between the lines on one of the sections of bus-bar (i) with reactors and (ii) without reactors.*

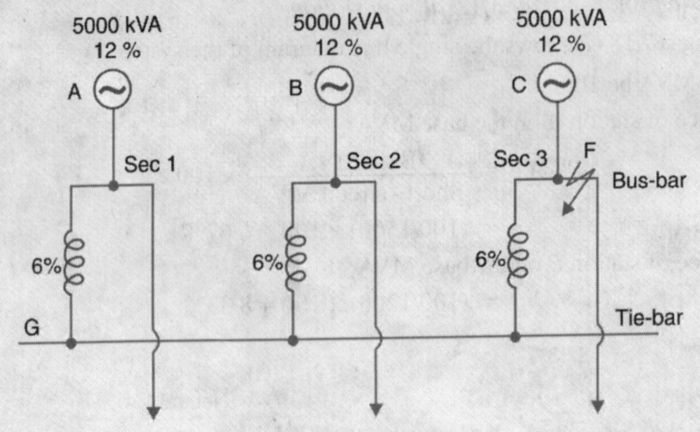

Fig. 17.19

Solution. Fig. 17.19 shows the single line diagram of the system. Let 5000 kVA be the base kVA. As generators and reactors are rated at this kVA, therefore, their percentage reactances remain the same.

(*i*) **With reactors.** Suppose a 3-phase short-circuit fault occurs on section 3 of the bus-bar (point *F* in Fig. 17.19). The reactance diagram at the selected base kVA will be as shown in Fig. 17.20 (*i*). The equivalent reactance of the parallel branches is (18/2)% = 9% and the circuit reduces to Fig. 17.20 (*ii*).

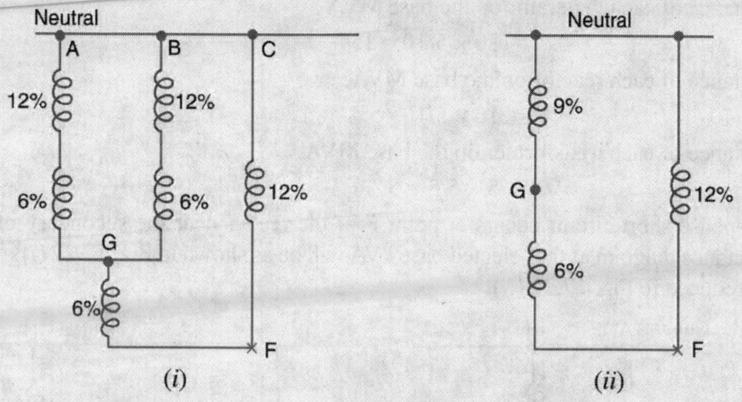

Fig. 17.20

Total % reactance from generator neutral upto fault point F

$$= (9\% + 6\%) \parallel 12\% = \frac{15 \times 12}{15 + 12} = 6{\cdot}67\%$$

∴ Short-circuit input $= 5000 \times 100/6{\cdot}67$

$= 74{,}962 \text{ kVA} = \mathbf{74{\cdot}962 \text{ MVA}}$

(*ii*) **Without reactors.** Suppose no reactors are used. Then for a fault on section 3, the total reactance upto the fault point will be a parallel combination of the reactances of the three generators *i.e.*

Total % reactance upto fault point F

$= 12\% \parallel 12\% \parallel 12\% = 12/3 = 4\%$

∴ Short-circuit input $= 5000 \times 100/4 = 1{,}25{,}000 \text{ kVA} = \mathbf{125 \text{ MVA}}$

Example 17.9. *A generating station is laid out as shown in Fig. 17.21. The ratings and percentage reactances of different elements are as indicated. If a 3-phase short-circuit occurs on any feeder near transformer secondary (e.g. point F), find the short-circuit MVA fed to the fault.*

Solution. Fig. 17.21 shows the single line diagram of the system.

Let us choose 5 MVA as the base value.

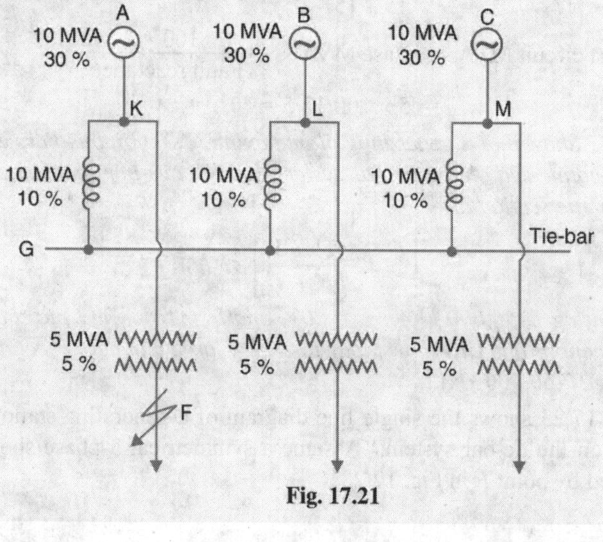

Fig. 17.21

%age reactance of each generator on the base MVA

$$= 30 \times 5/10 = 15\%$$

%age reactance of each reactor on the base MVA

$$= 10 \times 5/10 = 5\%$$

%age reactance of each transformer on the base MVA

$$= 5 \times 5/5 = 5\%$$

When a 3-phase short-circuit occurs at point F of the feeder near the secondary of the transformer, the reactance diagram at the selected base kVA will be as shown in Fig. 17.22 (*i*). This circuit can be further reduced to Fig. 17.22 (*ii*).

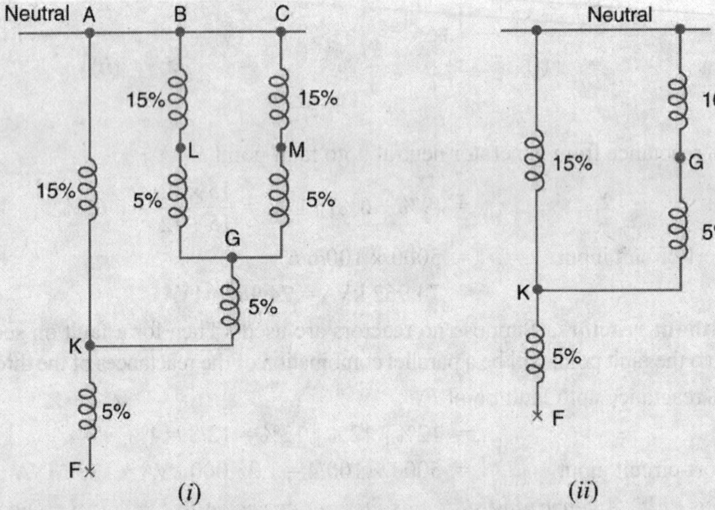

Fig. 17.22

Total %age reactance from generator neutral upto fault point *F*

$$= (10\% + 5\%) \parallel 15\% + 5\%$$

$$= \frac{15 \times 15}{15 + 15} + 5 = 7\cdot5 + 5 = 12\cdot5\%$$

$$\text{Short circuit MVA} = \text{Base MVA} \times \frac{100}{\%\ \text{Fault reactance}}$$

$$= 5 \times 100/12\cdot5 = 40$$

Example 17.10. *Show that a generating plant having N section bus-bars each rated at Q kVA with x% reactance, connected on the tie-bar system through bus-bar reactances of b% has a total short-circuit kVA on one section of*

$$\left[\frac{Q}{x} + \frac{Q(N-1)}{bN + x}\right] \times 100$$

If the section rating is 50,000 kVA ; x = 20% and b = 10%, find the short-circuit kVA with (i) 3 sections (ii) 9 sections and (iii) show that with a very large number of sections, the short-circuit kVA would not exceed 7,50,000 kVA.

Solution. Fig. 17.23 shows the single line diagram of a generating station having *N* bus-bar sections connected on the tie-bar system. Assume a symmetrical 3-phase short-circuit occurs on section 1 as indicated by point *F* in Fig. 17.23.

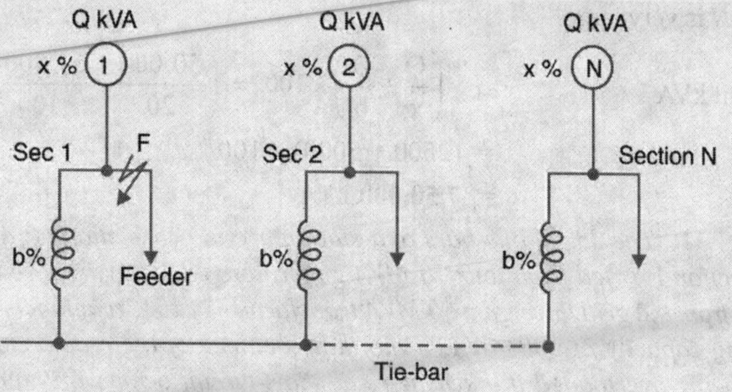

Fig. 17.23

The reactance diagram of the system is shown in Fig. 17.24. For a fault on section 1 (point F), the remaining $(N-1)$ sections are in parallel and their equivalent percentage reactance is $(x+b/N-1)\%$. The circuit of Fig. 17.24 is, therefore, reduced to the one shown in Fig. 17.25. Referring to Fig. 17.25, the total percentage reactance between the generator neutral and fault point F is given by :

% age reactance from generator neutral upto fault point F

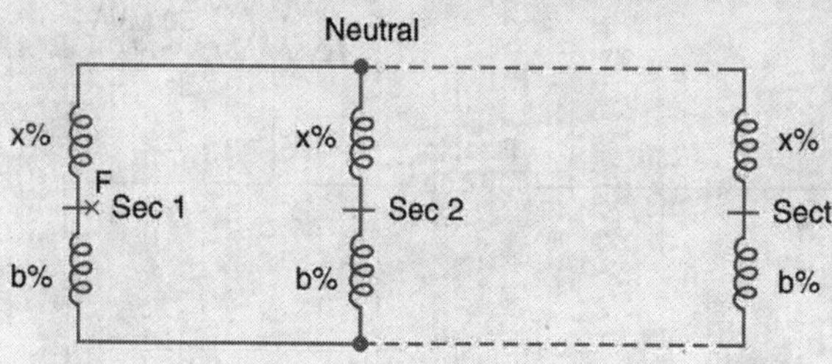

Fig. 17.24

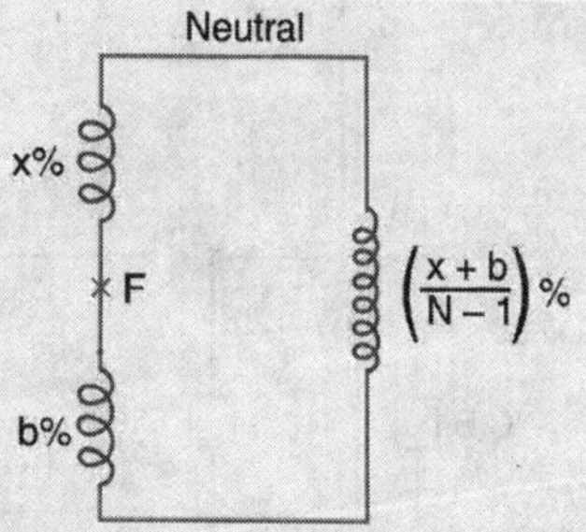

Fig. 17.25

$$= \left\{b + \frac{x+b}{N-1}\right\}\% \,||\, x\% = \frac{\left[b + \frac{x+b}{N-1}\right]x}{b + \frac{x+b}{N-1} + x}\%$$

$$= \frac{x\,(x+bN)}{N(x+b)}\%$$

Suppose the base kVA is Q.

$$\therefore \quad \text{Short-circuit kVA} = Q \times 100 \times \frac{N\,(x+b)}{x\,(x+bN)}$$

$$= \left[\frac{Q}{x} + \frac{Q\,(N-1)}{bN+x}\right] \times 100 \qquad \text{... (i) Q.E.D.}$$

(*i*) With 3 sections, $N = 3$

$$\therefore \quad \text{Short-circuit kVA} = \left[\frac{50{,}000}{20} + \frac{50{,}000\,(3-1)}{10 \times 3 + 20}\right] \times 100 = \mathbf{4{,}50{,}000\ kVA}$$

(*ii*) With 9 sections, $N = 9$

$$\therefore \quad \text{Short-circuit kVA} = \left[\frac{50{,}000}{20} + \frac{50{,}000\,(9-1)}{10 \times 9 + 20}\right] \times 100 = \mathbf{6{,}13{,}636\ kVA}$$

(*iii*) When *N* is very large,

$$\text{Short-circuit kVA} = {}^{*}\left(\frac{Q}{x}+\frac{Q}{b}\right)\times 100 = \left(\frac{50,000}{20}+\frac{50,000}{10}\right)\times 100$$

$$= (2500 + 5000) \times 100$$

$$= \mathbf{7,50,000\ kVA}$$ *Q.E.D.*

Example 17.11. *The 33 kV bus-bars of a station are in two sections P and Q separated by a reactor. The section P is fed from four 10 MVA generators each having a reactance of 20%. The section Q is fed from the grid through a 50 MVA transformer of 10% reactance. The circuit breakers have a rupturing capacity of 500 MVA. Find the reactance of the reactor to prevent the circuit breakers from being overloaded if a symmetrical short-circuit occurs on an outgoing feeder connected to A. Take base MVA as 50 MVA.*

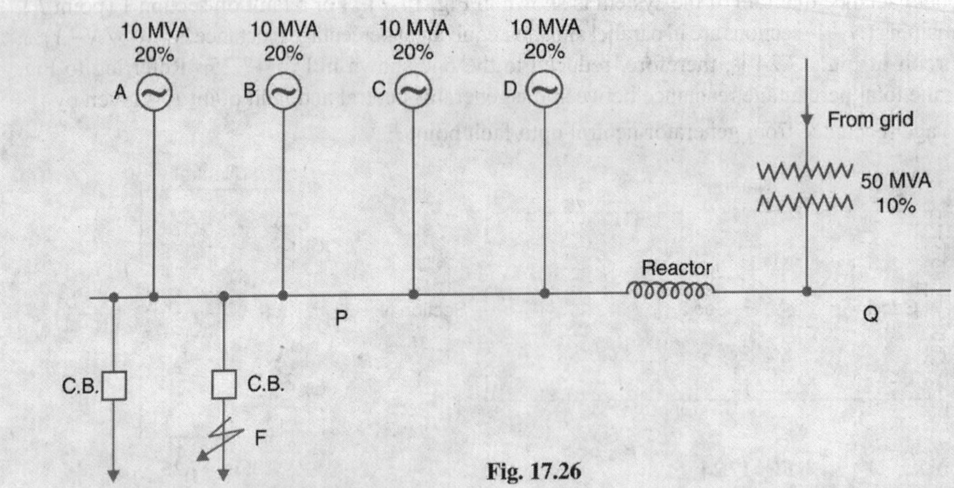

Fig. 17.26

Solution. Fig. 17.26 shows the single line diagram of the network. Suppose the fault occurs at point *F* on an outgoing feeder connected to section *P*. As per the given condition, the short-circuit MVA at *F* should not exceed 500 MVA.

% reactance of each of the generator on base MVA

$$= \frac{50}{10}\times 20 = 100\ \%$$

$\therefore$ $\% X_A = 100\ \%$; $\% X_B = 100\ \%$; $\% X_C = 100\ \%$; $\% X_D = 100\ \%$

% reactance of the transformer on base MVA is

$$\% X_T = \frac{50}{50}\times 10 = 10\%$$

Suppose the required reactance of the reactor is *X*% on 50 MVA base. When the fault occurs at point *F*, the reactance diagram at the selected base MVA will be as shown in Fig. 17.27 (*i*). Clearly, the reactances of the four generators are in parallel and their equivalent reactance = 100/4 = 25%. The circuit then reduces to that shown in Fig. 17.27 (*ii*).

* Referring to exp. (*i*) above, if *N* is very large, then *x* can be neglected as compared to *bN*. Also *N* – 1 may be taken as *N*.

$$\therefore \quad \text{Short-circuit kVA} = \left(\frac{Q}{x}+\frac{QN}{bN}\right)\times 100 = \left(\frac{Q}{x}+\frac{Q}{b}\right)\times 100$$

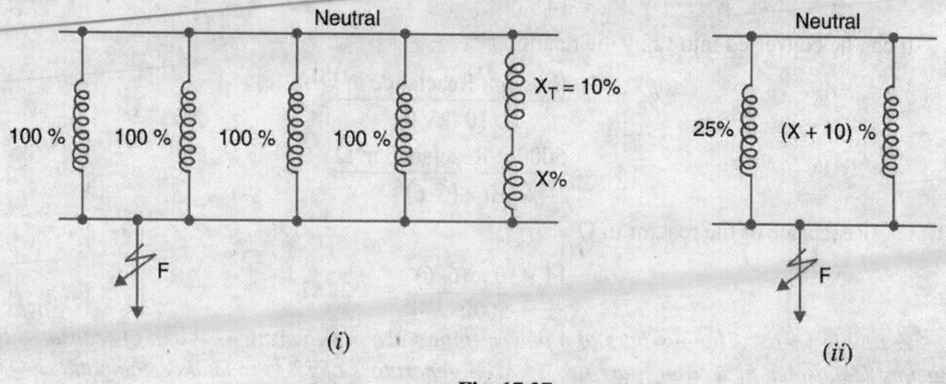

Fig. 17.27

Total percentage reactance from generator neutral to fault point F

$$= \frac{25\,(X+10)}{25+(X+10)} \quad ...(i)$$

Now fault MVA at F is not to exceed 500 MVA.

$$\text{Fault MVA} = \text{Base MVA} \times \frac{100}{\text{Required \% reactance}}$$

or
$$500 = 50 \times \frac{100}{\text{Required \% reactance}}$$

$$\therefore \quad \text{Required \% reactance} = \frac{50 \times 100}{500} = 10\,\%$$

This means that total % reactance from generator neutral to fault point F should be 10 % *i.e.*

$$10 = \frac{25\,(X+10)}{25+(X+10)}$$

$$\therefore \quad X = \frac{100}{15}\,\%$$

$$\text{\% reactance} = \frac{(\text{Base kVA}) \times \text{Reactance in } \Omega}{10\,(\text{kV})^2}$$

or
$$\frac{100}{15} = \frac{(50 \times 10^3) \times \text{Reactance in } \Omega}{10 \times (33)^2}$$

$\therefore$ Reactance of the reactor in Ω

$$= 10 \times (33)^2 \times \frac{100}{15} \times \frac{1}{50 \times 10^3} = \mathbf{1{\cdot}452\ \Omega}$$

Example 17.12. *A 3-phase alternator can supply a maximum load of 5000 kVA at 6600 V. The machine has internal reactance of 6%. Find the reactance per phase of the limiting reactor if the steady apparent power (kVA) on short-circuit is not be exceed 5 times the full-load value.*

Solution. Let the base kVA be 5000 kVA. Let the % reactance of the reactor be X% on the base kVA. When short-circuit fault occurs, the total % reactance from generator neutral to fault point = (X + 6) %. The short-circuit kVA is not to exceed 5 × 5000 kVA.

$$\text{Fault kVA} = \text{Base kVA} \times \frac{100}{\text{\% Reactance}}$$

or
$$5 \times 5000 = 5000 \times \frac{100}{(X+6)}$$

$$\therefore \quad X = 14\ \%$$

It can be converted into Ω by the relation :

$$\%\,X = \frac{(\text{kVA}) \times \text{Reactance in } \Omega}{10\,(\text{kV})^2}$$

$$14 = \frac{5000 \times \text{Reactance in } \Omega}{10 \times (6\cdot 6)^2}$$

∴ Reactance of the reactor in Ω

$$= \frac{14 \times 10 \times (6\cdot 6)^2}{5000} = \mathbf{1{\cdot}22\ \Omega}$$

Example 17.13. *The bus-bars of a power station are in two sections P and Q separated by a reactor. Connected in section P are two 15 MVA generators of 12 % reactance each and to Q one 8 MVA generator of 10% reactance. The reactor is rated at 10 MVA and 15% reactance. Feeders are connected to bus-bar P through transformers, each rated at 5 MVA and 4% reactance. Determine the maximum short-circuit MVA with which circuit breakers on the outgoing side of the transformers have to deal.*

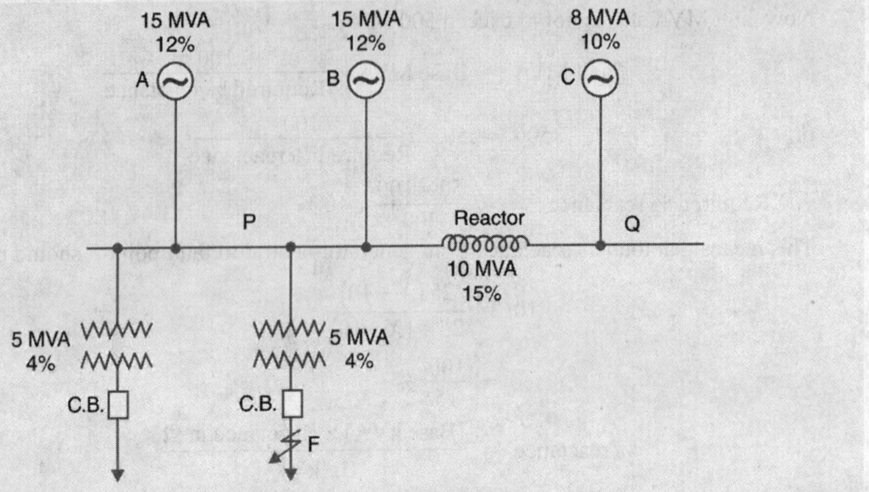

Fig. 17.28

Solution. Fig. 17.28 shows the single line diagram of the network. There may be several transformers but only two are shown for clarity of the figure. Suppose short-circuit fault occurs at point *F* on the outgoing side of one transformer as shown in Fig. 17.28. The circuit breaker will have to handle short-circuit MVA at point *F*.

Let 10 MVA be the base MVA. The percentage reactance of various elements on the selected base MVA will be :

$$\%\,X_A = \frac{10}{15} \times 12 = 8\ \%$$

$$\%\,X_B = \frac{10}{15} \times 12 = 8\ \%$$

$$\%\,X_C = \frac{10}{8} \times 10 = 12{\cdot}5\ \%$$

$$\%\,X_T = \frac{10}{5} \times 4 = 8\ \%$$

When fault occurs at point F, the reactance diagram at the selected base MVA will be as shown in Fig. 17.29 (*i*). The reactances of generators A and B are in parallel and their equivalent reactance = 8%/2 = 4%. The circuit then reduces to the one shown in Fig. 17.29 (*ii*). This further reduces to the one shown in Fig. 17.29 (*iii*).

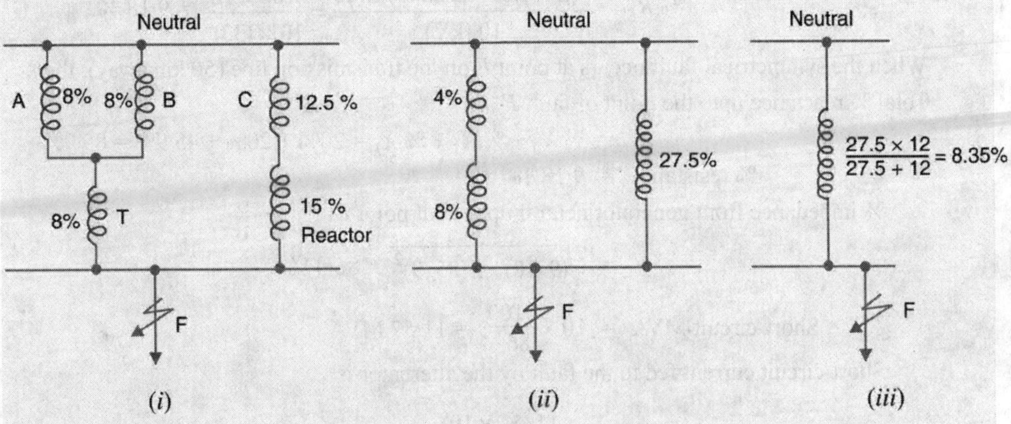

Fig. 17.29

∴ Total % reactance from generator neutral upto fault point F

$$= 12\ \%\ \|\ 27{\cdot}5\ \% = \frac{12 \times 27{\cdot}5}{12 + 27{\cdot}5} = 8{\cdot}35\ \%$$

$$\text{Fault MVA} = \text{Base MVA} \times \frac{100}{\%\ \text{Reactance}} = 10 \times \frac{100}{8{\cdot}35} = \mathbf{119{\cdot}8\ MVA}$$

Example 17.14. *A 10 MVA, 6·6 kV, 3-phase star-connected alternator having a reactance of 20% is connected through a 5 MVA, 6·6 kV/33 kV transformer of 10% reactance to a transmission line having a resistance and reactance per conductor per kilometre of 0·2 Ω and 1 Ω respectively. Fifty kilometres along the line, a short-circuit occurs between the three conductors. Find the current fed to the fault by the alternator.*

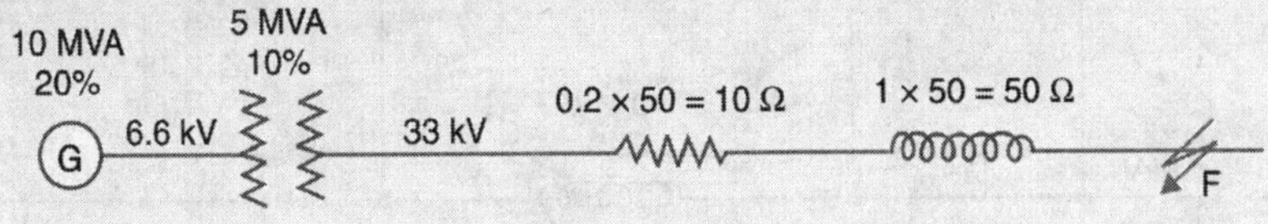

Fig. 17.30

Solution. Fig. 17.30 shows the single line diagram of the network. Let 10 MVA (= 10^4 kVA) be the base MVA.

% reactance of the alternator on base MVA is

$$\%\ X_A = \frac{10}{10} \times 20 = 20\ \%$$

% reactance of the transformer on base MVA is

$$\%\ X_T = \frac{10}{5} \times 10 = 20\ \%$$

% reactance of the transmission line is

$$\% X_L = \frac{(\text{kVA}) \times \text{reactance in } \Omega}{10\,(\text{kV})^2} = \frac{10^4 \times 50}{10 \times (33)^2} = 45{\cdot}9\%$$

% resistance of the transmission line is

$$\% R_L = \frac{(\text{kVA}) \times \text{resistance in } \Omega}{10\,(\text{kV})^2} = \frac{10^4 \times 10}{10 \times (33)^2} = 9{\cdot}18\ \%$$

When the symmetrical fault occurs at point F on the transmission line (50 km away), then,

Total % reactance upto the point of fault F

$$= \% X_A + \% X_T + \% X_L = 20\% + 20\% + 45{\cdot}9\% = 85{\cdot}9\%$$

$$\%\ \text{resistance} = 9{\cdot}18\ \%$$

∴ % impedance from generator neutral upto fault point F

$$= \sqrt{(9{\cdot}18)^2 + (85{\cdot}9)^2} = 86{\cdot}4\ \%$$

$$\text{Short-circuit MVA} = 10 \times \frac{100}{86{\cdot}4} = 11{\cdot}57\ \text{MVA}$$

∴ Short-circuit current fed to the fault by the alternator is

$$I_{SC} = \frac{11{\cdot}57 \times 10^6}{\sqrt{3} \times 6{\cdot}6 \times 1000} = \mathbf{1012\ A}$$

Example 17.15. *An 11 kV generating station has four identical 3-phase alternators A, B, C and D, each of 10 MVA capacity and 12% reactance. There are two sections of bus-bar, P and Q linked by a reactor rated at 10 MVA with 24% reactance. Alternators A and B are connected to bus-bar section P and alternators C and D to bus-bar section Q. From each section, load is taken through a number of 6 MVA, 11 kV/66 kV step-up transformers, each having a reactance of 3%. Calculate the current fed into fault if a short-circuit occurs on all phases near the high-voltage terminals of one of the transformers at the bus-bar section Q.*

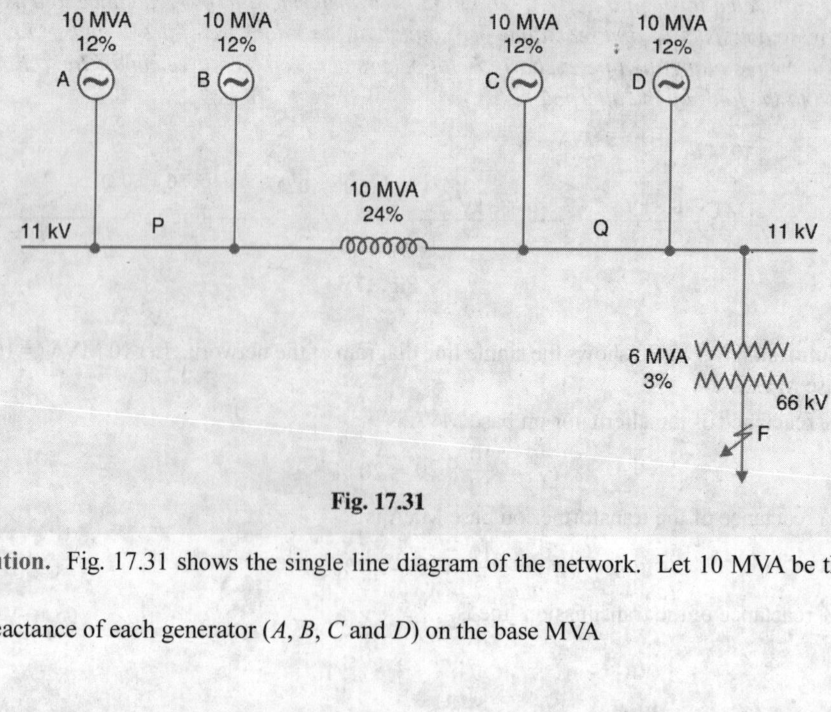

Fig. 17.31

Solution. Fig. 17.31 shows the single line diagram of the network. Let 10 MVA be the base MVA.

% reactance of each generator (A, B, C and D) on the base MVA

$$= \frac{10}{10} \times 12 = 12\%$$

% reactance of the reactor on the base MVA

$$= \frac{10}{10} \times 24 = 24\%$$

% reactance of the transformer on the base MVA

$$= \frac{10}{6} \times 3 = 5\%$$

When fault occurs at point F, the reactance diagram on the selected base MVA will be as shown in Fig. 17.32 (*i*). This further reduces to the circuit shown in Fig. 17.32 (*ii*).

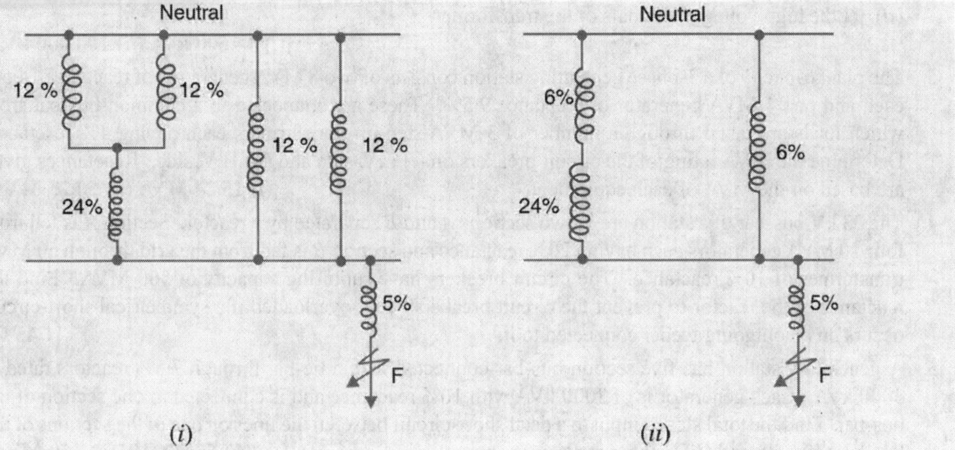

Fig. 17.32

$\therefore$ % reactance from generator neutral upto fault point F

$$= \frac{30 \times 6}{30 + 6} + 5 = 5 + 5 = 10\%$$

$$\text{Fault MVA} = 10 \times \frac{100}{10} = 100 \text{ MVA}$$

Short-circuit current, $I_{SC} = \dfrac{100 \times 10^6}{\sqrt{3} \times 66000} = \mathbf{875\ A}$

Oil circuit breaker

TUTORIAL PROBLEMS

1. A 3-phase, 30 MVA, 33 kV alternator has internal reactance of 4% and negligible resistance. Find the external reactance per phase to be connected in series with the alternator so that steady current on short-circuit does not exceed 10 times the full load current. **[2·178 Ω]**
2. A 3-phase transmission line operating at 33 kV and having a resistance of 5 Ω and reactance of 20 Ω is connected to the generating station through 15,000 kVA step-up transformer. Connected to the bus-bar are two alternators, one of 10,000 kVA with 10% reactance and another of 5000 kVA with 7·5% reactance. Calculate the short-circuit kVA fed to the symmetrical fault between phases if it occurs
   (*i*) at the load end of transmission line
   (*ii*) at the high voltage terminals of the transformer
   [(*i*) 44,500 kVA (*ii*) 100,000 kVA]
3. The plant capacity of a 3-phase generating station consists of two 8 MVA generators of reactance 14·5% each and one 4 MVA generator of reactance 9·5%. These are connected to a common bus-bar from which loads are taken through a number of 3 MVA step-up transformers each having 4% reactance. Determine the MVA rating of the circuit breakers on (*i*) L.V. side and (*ii*) H.V. side. Reactances given are based on the MVA of each equipment. **[(*i*) 15·24 MVA (*ii*) 50·25 MVA]**
4. The 33 kV bus-bar of a station are in two sections *A* and *B* separated by a reactor. Section *A* is fed from four 10 MVA generators each having 20% reactance and section *B* is fed from the grid through 50 MVA transformer of 10% reactance. The circuit breakers have rupturing capacity of 500 MVA. Find the reactance of the reactor to prevent the circuit breakers being overloaded if a symmetrical short-circuit occurs on an outgoing feeder connected to it. **[1·45 Ω]**
5. A generating station has five section bus-bar connected with a tie-bar through 7·5% reactors rated at 3000 kVA. Each generator is of 3000 kVA with 10% reactance and is connected to one section of the bus-bar. Find the total steady input to a dead short-circuit between the lines on one of the sections of the bus-bars (*i*) with and (*ii*) without reactors. **[(*i*) 55·3 MVA (*ii*) 150 MVA]**
6. A generating station has four bus-bar sections. Each section is connected to tie-bar though 20% reactors rated at 200 MVA. Generators of total capacity 100 MVA and 20% reactance are connected to each bus-bar section. Calculate the MVA fed to a fault under short-circuit condition one one of the bus-bars. **[1000 MVA]**

SELF - TEST

1. **Fill in the blanks by inserting appropriate words/figures.**
   (*i*) When a short-circuit occurs, a .... current flows through the system.
   (*ii*) The most serious result of a major uncleared short-circuit fault is the ....
   (*iii*) When all the three phases are short-circuited, it gives rise to .... currents.
   (*iv*) The rating of a circuit breaker is generally determined on the basis of .... short circuit currents.
   (*v*) The most common type of fault in overhead lines is ....
   (*vi*) The .... short-circuit fault gives very heavy duty on the circuit breaker.
   (*vii*) If the % age reactance upto the fault point is 20%, then short-circuit current will be .... times the full-load current.
   (*viii*) A 1000 kVA transformer with 5% reactance will have a reactance of .... at 2000 kVA base.
   (*ix*) Short-circuit kVA is obtained by multiplying the base kVA by ....
   (*x*) Reactors are used at various points in the power system to ....
2. **Pick up the correct words/figures from the brackets and fill in the blanks.**
   (*i*) .... fault gives rise to symmetrical fault currents.
   (*single phase to ground, phase to phase, 3-phase short-circuit*)
   (*ii*) Percentage reactances .... as they are referred through transformers.
   (*remain unchanged, are changed*)

(*iii*) If the % age reactance of the system upto the fault point is 20% and the base kVA is 10,000, then short-circuit kVA is .... (*50,000, 10,000, 2,000*)

(*iv*) The use of reactors permits installation of circuit breakers of .... ratings. (*lower, higher*)

(*v*) A 20,000 kVA transformer with 10% reactance will have a reactance of .... at 10,000 kVA base. (*5%, 20%, 10%*)

ANSWERS TO SELF-TEST

1. (*i*) heavy (*ii*) fire (*iii*) symmetrical fault (*iv*) symmetrical (*v*) phase-to-ground fault (*vi*) 3-phase (*vii*) 5 (*viii*) 10% (*ix*) 100/% X (*x*) limit short-circuit current
2. (*i*) 3-phase short-circuit (*ii*) remain unchanged (*iii*) 50,000 (*iv*) lower (*v*) 5%

CHAPTER REVIEW TOPICS

1. What do you understand by a short-circuit ? Discuss the possible causes of short-circuit in the power system.
2. Explain the harmful effects of short-circuit fault on the power system.
3. What is the importance of short-circuit calculations ?
4. Discuss the possible faults on overhead lines.
5. What do you understand by percentage reactance ? Why do we prefer to express the reactances of various elements in percentage values for short-circuit calculations ?
6. What is the importance of base kVA in short-circuit calculations ?
7. Why do we use reactors in the power system ? Discuss their advantages.
8. Explain the various methods of connecting short-circuit current limiting reactors in the power system.

DISCUSSION QUESTIONS

1. Why do we choose a base kVA in short-circuit calculations ?
2. What is the advantage of expressing reactances in percentage values ?
3. Why do we decide the rating of a circuit breaker on the basis of symmetrical short-circuit currents ?
4. Will the value of short-circuit current change if we take different base kVAs' ? Explain your answer.
5. Can feeder reactors permit the use of circuit breakers of lower ratings ?

CHAPTER

18

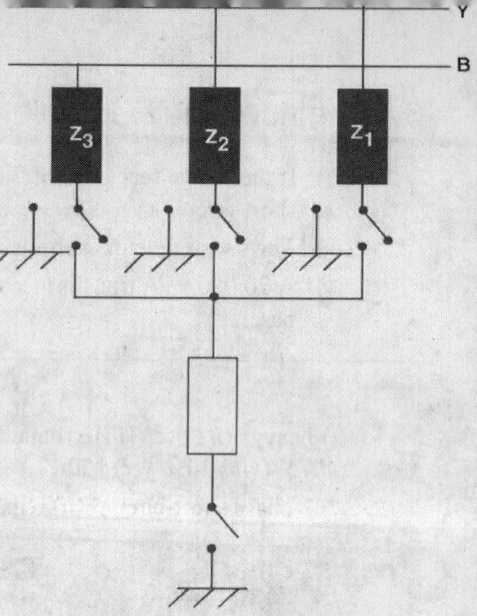

Unsymmetrical Fault Calculations

Introduction

In the previous chapter, attention was confined to the analysis of symmetrical faults *e.g.* all three lines short-circuited ($L — L — L$) *or* all three lines short-circuited with an earth connection at the fault ($L — L — L — G$). When such a fault occurs, it gives rise to symmetrical fault currents *i.e.* fault currents in the three lines are equal in magnitude and displaced 120° electrical from one another. Although symmetrical faults are the most severe and impose heavy duty on the circuit breakers, yet the analysis of such faults can be made with a fair degree of ease. It is because the balanced nature of fault permits to consider only one phase in calculations ; the conditions in the other two phases being similar.

The great majority of faults on the power system are of unsymmetrical nature; the most common type being a short-circuit from one line to ground. When such a fault occurs, it gives rise to unsymmetrical currents *i.e.* the magnitude of fault currents in the three lines are different having unequal phase displacement. The calculation procedure known as *method of symmetrical components* is used to determine the currents and voltages on the occurrence of an unsymmetrical fault. In this chapter, we shall focus our attention on the analysis of unsymmetrical faults.

18.1 Unsymmetrical Faults on 3-Phase System

Those faults on the power system which give rise to unsymmetrical fault currents (i.e. unequal fault currents in the lines with unequal phase displacement) are known as **unsymmetrical faults.**

On the occurrence of an unsymmetrical fault, the currents in the three lines become unequal and so is the phase displacement among them. It may be noted that the term 'unsymmetry' applies only to the fault itself and the resulting line currents. However, the system impedances and the source voltages are always symmetrical* through its main elements *viz.* generators, transmission lines, synchoronous reactors *etc.* There are three ways in which unsymmetrical faults may occur in a power system (see Fig. 18.1).

(*i*) Single line-to-ground fault ($L — G$)
(*ii*) Line-to-line fault ($L — L$)
(*iii*) Doube line-to-ground fault ($L — L — G$)

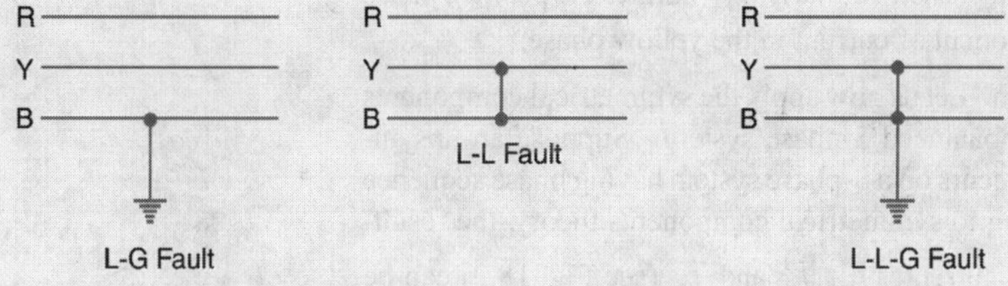

(*i*) (*ii*) (*iii*)

Fig. 18.1

The solution of unsymmetrical fault problems can be obtained by either (*a*) Kirchhoff's laws or (*b*) Symmetrical components method. The latter method is preferred because of the following reasons :

(*i*) It is a simple method and gives more generality to be given to fault performance studies.
(*ii*) It provides a useful tool for the protection engineers, particularly in connection with tracing out of fault currents.

Electronic earth fault indicator

18.2 Symmetrical Components Method

In 1918, Dr. C.L. Fortescue, an American scientist, showed that any unbalanced system of 3-phase currents (or voltages) may be regarded as being composed** of three separate sets of balanced vectors *viz.*

---

* In other words, no piece of equipment ever has a red phase impedance which differs from a yellow phase impedance.

** This has come to be known as *symmetrical component theory.* This is a general theory and is applicable to any three vector system whose resultant is zero.

(*i*) a balanced *system of 3-phase currents having positive† (or normal) phase sequence. These are called *positive phase sequence components.*

(*ii*) a balanced system of 3-phase currents having the opposite or negative phase sequence. These are called *negative phase sequence components.*

(*iii*) a system of three currents equal in magnitude and having zero phase displacement. These are called *zero phase sequence components.*

The positive, negative and zero phase sequence components are called the *symmetrical components* of the original unbalanced system. The term 'symmetrical' is appropriate because the unbalanced 3-phase system has been resolved into three sets of balanced (or symmetrical) components. The subscripts 1, 2 and 0 are generally used to indicate positive, negative and zero phase sequence components respectively. For instance, $\overrightarrow{I_{R0}}$ indicates the zero phase sequence component of the current in the red phase. Similarly, $\overrightarrow{I_{Y1}}$ implies the positive phase sequence component of current in the yellow phase.

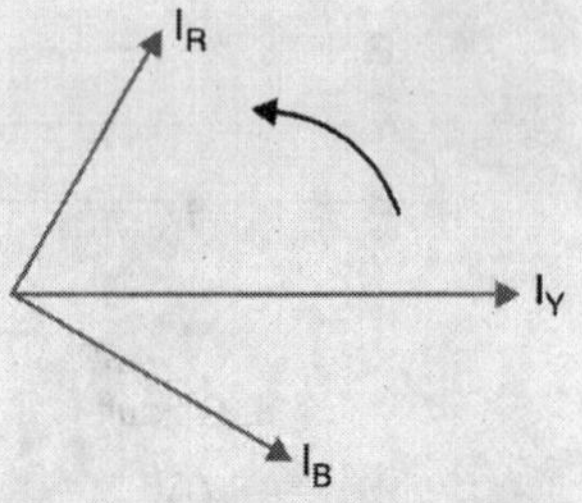

Fig. 18.2

Illustration. Let us now apply the symmetrical components theory to an unbalanced 3-phase system. Suppose an unsymmetrical fault occurs on a 3-phase system having phase sequence *RYB*. According to symmetrical components theory, the resulting unbalanced currentes $\overrightarrow{I_R}$, $\overrightarrow{I_Y}$ and $\overrightarrow{I_B}$ (see Fig. 18.2) can be resolved into :

(*i*) a balanced system of 3-phase currents, $\overrightarrow{I_{R1}}$, $\overrightarrow{I_{Y1}}$ and $\overrightarrow{I_{B1}}$ having positive phase sequence (*i.e. RYB*) as shown in Fig. 18.3 (*i*). These are the positive phase sequence components.

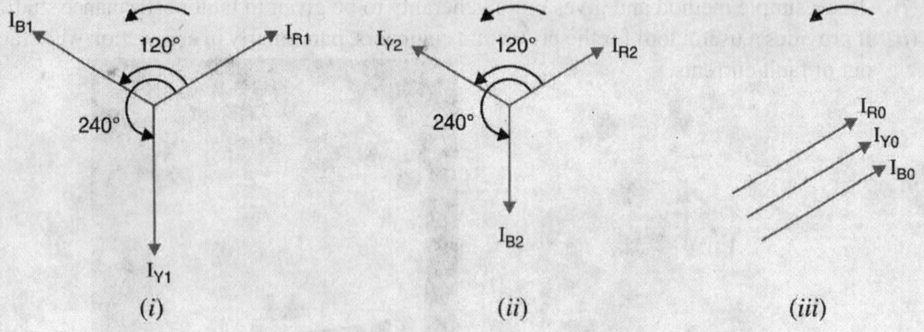

Fig. 18.3

(*ii*) a balanced system of 3-phase currents $\overrightarrow{I_{R2}}$, $\overrightarrow{I_{Y2}}$ and $\overrightarrow{I_{B2}}$ having negative phase sequence (*i.e. RBY*) as shown in Fig. 18.3 (*ii*). These are the negative phase sequence components.

(*iii*) a system of three currents $\overrightarrow{I_{R0}}$, $\overrightarrow{I_{Y0}}$ and $\overrightarrow{I_{B0}}$ equal in magnitude with zero phase displacement from each other as shown in Fig. 18.3 (*iii*). These are the zero phase sequence components.

* A balanced system of 3-phase currents implies that three currents are equal in magnitude having 120° displacement from each other.

† Positive phase sequence means that phase sequence is the same as that of the original 3-phase system.

The current in any phase is equal to the vector sum of positive, negative and zero phase sequence currents in that *phase as shown in Fig. 18.4.

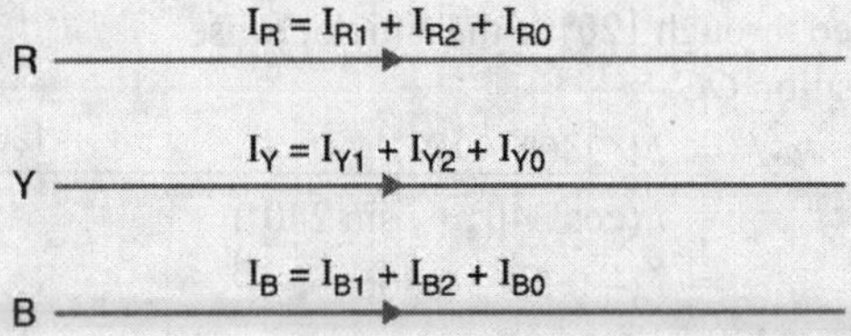

Fig. 18.4

$$\vec{I_R} = \vec{I_{R1}} + \vec{I_{R2}} + \vec{I_{R0}}$$

$$\vec{I_Y} = \vec{I_{Y1}} + \vec{I_{Y2}} + \vec{I_{Y0}}$$

$$\vec{I_B} = \vec{I_{B1}} + \vec{I_{B2}} + \vec{I_{B0}}$$

The following points may be noted :

(*i*) The positive phase sequence currents ($\vec{I_{R1}}$, $\vec{I_{Y1}}$ and $\vec{I_{B1}}$), negative phase sequence currents ($\vec{I_{R2}}$, $\vec{I_{Y2}}$ and $\vec{I_{B2}}$) and zero phase sequence currents ($\vec{I_{R0}}$, $\vec{I_{Y0}}$ and $\vec{I_{B0}}$) separately form balanced system of currents. Hence, they are called symmetrical components of the unbalanced system.

(*ii*) The symmetrical component theory applies equally to 3-phase currents and voltages both phase and line values.

(*iii*) The symmetrical components do not have separate existence. They are only mathematical components of unbalanced currents (or voltages) which actually flow in the system.

(*iv*) In a balanced 3-phase system, negative and zero phase sequence currents are zero. This is demonstrated in example 18.7.

18.3 Operator 'a'

As the symmetrical component theory involves the concept of 120° displacement in the positive sequence set and negative sequence set, therefore, it is desirable to evolve some operator which should cause 120° rotation. For this purpose, operator '*a*' (symbols *h* or λ are sometimes used instead of '*a*') is used. It is defined as under :

The ***operator '*a*'** *is one, which when multiplied to a vector rotates the vector through 120° in the anticlockwise direction.*

Consider a vector *I* represented by *OA* as shown in Fig. 18.5. If this vector is multiplied by operator '*a*', the vector is rotated through 120° in the anticlockwise direction and assumes the position *OB*.

$$\therefore \quad a\,I = I\angle 120^\circ$$

$$= I\,(\cos 120^\circ + j \sin 120^\circ)$$

* Star connected system being considered in Fig. 18.4.

** Just as the operator *j* rotates a vector through 90° in the anticlockwise direction.

$$= \quad I(-0{\cdot}5 + j\,0{\cdot}866)$$

$$\therefore \quad a = -0{\cdot}5 + j\,0{\cdot}866 \qquad \ldots (i)$$

If the vector assuming position *OB* is multiplied by operator '*a*', the vector is further rotated through 120º in the anticlockwise direction and assumes the position *OC*.

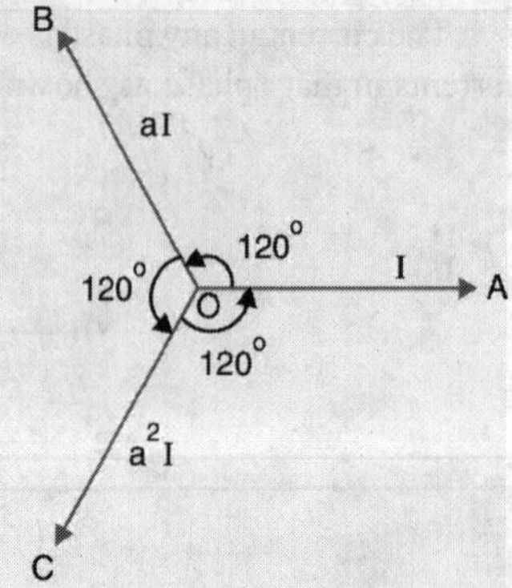

Fig. 18.5

$$\therefore \quad a^2 I = I\angle 240º$$

$$= I(\cos 240º + j\sin 240º)$$

$$= I(-0{\cdot}5 - j\,0{\cdot}866)$$

$$\therefore \quad a^2 = -0{\cdot}5 - j\,0{\cdot}866 \qquad \ldots (ii)$$

Thus the operator 'a^2' will turn the vector through 240º in the anticlockwise direction. This is the same as turning the vector through 120º in clockwise direction.

$$\therefore \quad a^2 I = I\angle -120º$$

$$\text{Similarly,} \quad a^3 I = I\angle 360º$$

$$= I(\cos 360º + j\sin 360º)$$

$$\therefore \quad a^3 = 1 \qquad \ldots (iii)$$

Properties of Operator '*a*'

(*i*) Adding exps. (*i*) and (*ii*), we get,

$$a + a^2 = (-0{\cdot}5 + j\,0{\cdot}866) + (-0{\cdot}5 - j\,0{\cdot}866) = -1$$

$$\therefore \quad 1 + a + a^2 = 0$$

(*ii*) Subtracting exp. (*ii*) from exp. (*i*), we get,

$$a - a^2 = (-0{\cdot}5 + j\,0{\cdot}866) - (-0{\cdot}5 - j\,0{\cdot}866) = j\,1{\cdot}732$$

$$\therefore \quad a - a^2 = j\sqrt{3}$$

18.4 Symmetrical Components in Terms of Phase Currents

The unbalanced phase currents in a 3-phase system can be expressed in terms of symmetrical components as under :

$$\overrightarrow{I_R} = \overrightarrow{I_{R1}} + \overrightarrow{I_{R2}} + \overrightarrow{I_{R0}}$$

$$\overrightarrow{I_Y} = \overrightarrow{I_{Y1}} + \overrightarrow{I_{Y2}} + \overrightarrow{I_{Y0}}$$

$$\overrightarrow{I_B} = \overrightarrow{I_{B1}} + \overrightarrow{I_{B2}} + \overrightarrow{I_{B0}}$$

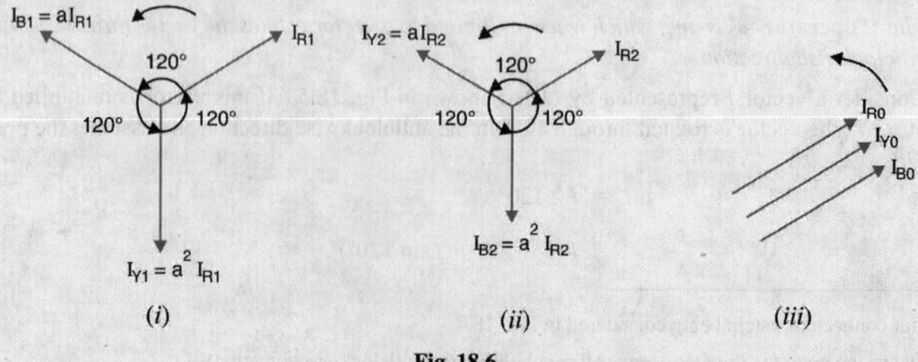

Fig. 18.6

Fig. 18.6 shows the vector representation of symmetrical components. It is usually profitable in calculations to express the symmetrical components in terms of unbalanced phase currents. Let us express the symmetrical components of *R*-phase in terms of phase currents $\vec{I_R}$, $\vec{I_Y}$ and $\vec{I_B}$. For this purpose, express all symmetrical components of *Y* and *B* phases in terms of the symmetrical components of R-phase by means of operator '*a*' as shown in Fig. 18.6.

Note that the positive sequence set shown in Fig. 18.6 (*i*) can be expressed in terms of $\vec{I_{R1}}$ by means of operator *a*. Thus positive sequence current $\vec{I_{B1}}$ in phase *B* leads $\vec{I_{R1}}$ by 120º and, therefore, $\vec{I_{B1}} = a\,\vec{I_{R1}}$. Similarly, positive sequence current in phase *Y* is 240º ahead of $\vec{I_{R1}}$ so that $\vec{I_{Y1}} = a^2\,\vec{I_{R1}}$. In an exactly similar manner, the negative sequence set can be expressed in terms of $\vec{I_{R2}}$ by means of operator '*a*' as shown in Fig. 18.6(*ii*). It is clear from Fig. 18.6 that :

$$\vec{I_R} = \vec{I_{R1}} + \vec{I_{R2}} + \vec{I_{R0}} \qquad ...(i)$$

$$\vec{I_Y} = \vec{I_{Y1}} + \vec{I_{Y2}} + \vec{I_{Y0}}$$

$$= a^2\,\vec{I_{R1}} + a\,\vec{I_{R2}} + \vec{I_{R0}} \qquad ...(ii)$$

$$\vec{I_B} = \vec{I_{B1}} + \vec{I_{B2}} + \vec{I_{B0}}$$

$$= a\,\vec{I_{R1}} + a^2\,\vec{I_{R2}} + \vec{I_{R0}} \qquad ...(iii)$$

(*i*) **Zero sequence current.** By adding exps. (*i*), (*ii*) and (*iii*), we get,

$$\vec{I_R} + \vec{I_Y} + \vec{I_B} = \vec{I_{R1}}\,(1 + a^2 + a) + \vec{I_{R2}}\,(1 + a + a^2) + 3\vec{I_{R0}}$$

$$= \vec{I_{R1}}\,(0) + \vec{I_{R2}}\,(0) + 3\,\vec{I_{R0}} = 3\,\vec{I_{R0}} \qquad (\because 1 + a + a^2 = 0)$$

$$\therefore \quad \vec{I_{R0}} = \frac{1}{3}\left(\vec{I_R} + \vec{I_Y} + \vec{I_B}\right)$$

As the red phase is always taken as the reference phase, therefore, subscript *R* is usually omitted.

$$\therefore \quad \vec{I_0} = \frac{1}{3}\left(\vec{I_R} + \vec{I_Y} + \vec{I_B}\right)$$

(*ii*) **Positive sequence current.** Multiply exp.(*ii*) by '*a*' and exp. (*iii*) by 'a^2' and then adding these exps. to exp. (*i*), we get,

$$\vec{I_R} + a\,\vec{I_Y} + a^2\,\vec{I_B} = \vec{I_{R1}}\,(1 + a^3 + a^3) + \vec{I_{R2}}\,(1 + a^2 + a^4) + \vec{I_{R0}}\,(1 + a + a^2)$$

$$= 3\,\vec{I_{R1}} + \vec{I_{R2}}\,(0^*) + \vec{I_{R0}}\,(0) = 3\,\vec{I_{R1}}$$

$$\therefore \quad \vec{I_{R1}} = \frac{1}{3}\left(\vec{I_R} + a\,\vec{I_Y} + a^2\vec{I_B}\right)$$

Omitting the subscript *R*, we have,

$$\vec{I_1} = \frac{1}{3}\left(\vec{I_R} + a\,\vec{I_Y} + a^2\vec{I_B}\right)$$

(*iii*) **Negative sequence current.** Multiply exp. (*ii*) by 'a^2' and exp. (*iii*) by '*a*' and then adding these exps. to (*i*), we get,

$$\vec{I_R} + a^2\vec{I_Y} + a\,\vec{I_B} = \vec{I_{R1}}\,(1 + a^4 + a^2) + \vec{I_{R2}}\,(1 + a^3 + a^3) + \vec{I_{R0}}\,(1 + a^2 + a)$$

$$= \vec{I_{R1}}\,(0) + \vec{I_{R2}}\,(3) + \vec{I_{R0}}\,(0) = 3\,\vec{I_{R2}}$$

$$\therefore \quad \vec{I_{R2}} = \frac{1}{3}\left(\vec{I_R} + a^2\,\vec{I_Y} + a\,\vec{I_B}\right)$$

$$\text{or} \quad \vec{I_2} = \frac{1}{3}\left(\vec{I_R} + a^2\,\vec{I_Y} + a\,\vec{I_B}\right)$$

---

* $a^4 = a^3 \times a = 1 \times a = a$

$\therefore \quad 1 + a^2 + a^4 = 1 + a^2 + a = 0$

The following points may be noted carefully :

(*i*) The currents $\vec{I_1}$, $\vec{I_2}$ and $\vec{I_0}$ are the symmetrical components of *R*-phase. Because of the symmetry of each set, the symmetrical components of yellow and blue phases can be easily known.

(*ii*) Although the treatment has been made considering currents, the method applies equally to voltages. Thus the symmetrical voltage components of *R*-phase in terms of phase voltages shall be :

$$\vec{E_0} = \frac{1}{3}\left(\vec{E_R} + \vec{E_Y} + \vec{E_B}\right)$$

$$\vec{E_1} = \frac{1}{3}\left(\vec{E_R} + a\,\vec{E_Y} + a^2\,\vec{E_B}\right)$$

$$\vec{E_2} = \frac{1}{3}\left(\vec{E_R} + a^2\,\vec{E_Y} + a\,\vec{E_B}\right)$$

18.5 Some Facts about Sequence Currents

It is now desirable to get the readers acquainted with the following facts about positive, negative and zero phase sequence currents :

(*i*) A balanced 3-phase system consists of positive sequence components only; the negative and zero sequence components being zero.

(*ii*) The presence of negative or zero sequence currents in a 3-phase system introduces unsymmetry and is indicative of an abnormal condition of the circuit in which these components are found.

(*iii*) The vector sum of the positive and negative sequence currents of an unbalanced 3-phase system is zero. The resultant solely consists of three zero sequence currents *i.e.*

Vector sum of all sequence currents in 3-phase unbalanced system

$$= \vec{I_{R0}} + \vec{I_{Y0}} + \vec{I_{B0}}$$

(*iv*) In a 3-phase, 4 wire unbalanced system, the magnitude of zero sequence components is one-third of the current in the neutral wire *i.e.*

$$\text{Zero sequence current} = \frac{1}{3}\,[\text{Current in neutral wire}]$$

In the absence of path through the neutral of a 3-phase system, the neutral current is zero and the line currents contain no zero -sequence components. *A delta-connected load provides no path to the neutral and the line currents flowing to delta-connected load can contain no zero-sequence components.*

(*v*) In a 3-phase unbalanced system, the magnitude of negative sequence components cannot exceed that of the positive sequence components. If the negative sequence components were the greater, the phase sequence of the resultant system would be reversed.

(*vi*) The current of a single phase load drawn from a 3-phase system comprises equal positive, negative and zero sequence components.

Example 18.1. *Prove that :*

(*i*) $\dfrac{1-a^2}{a-a^2} = -a$ (*ii*) $\dfrac{1-a}{1+a^2} = 1-a^2$

Solution.

(*i*) $$\frac{1-a^2}{a-a^2} = \frac{(1+a)(1-a)}{a(1-a)} = \frac{1+a}{a} = -\frac{a^2}{a} = -a \qquad (\because 1 + a + a^2 = 0)$$

(ii) $\dfrac{1-a}{1+a^2} = \dfrac{1-a}{-a} = \dfrac{(1-a)(-a^2)}{(-a)(-a^2)} = \dfrac{-a^2+a^3}{a^3} = 1-a^2$

Example 18.2. *In a 3-phase, 4-wire system, the currents in R, Y and B lines under abnormal conditions of loading are as under :*

$\vec{I_R} = 100 \angle 30^o A$; $\vec{I_Y} = 50 \angle 300^o A$; $\vec{I_B} = 30 \angle 180^o A$

Calculate the positive, negative and zero sequence currents in the R-line and return current in the neutral wire.

Solution. Let $\vec{I_0}, \vec{I_1}$ and $\vec{I_2}$ be the zero, positive and negative sequence currents respectively of the line current in red line.

$$
\begin{aligned}
\therefore \quad \vec{I_0} &= \tfrac{1}{3}\left[\vec{I_R} + \vec{I_Y} + \vec{I_B}\right] \\
&= \tfrac{1}{3}[100 \angle 30^o + 50 \angle 300^o + 30 \angle 180^o] \\
&= \tfrac{1}{3} \, {}^*[(86{\cdot}60 + j\,50) + (25 - j\,43{\cdot}3) + (-30 + j0)] \\
&= \tfrac{1}{3}\,[\,81{\cdot}6 + j\,6{\cdot}7] \\
&= (27{\cdot}2 + j\,2{\cdot}23) = \mathbf{27{\cdot}29 \angle 4{\cdot}68^o\ A}
\end{aligned}
$$

$$
\begin{aligned}
\vec{I_1} &= \tfrac{1}{3}\left[\vec{I_R} + a\,\vec{I_Y} + a^2\,\vec{I_B}\right] \\
&= \tfrac{1}{3}\,[\,100 \angle 30^o + 1 \angle 120^o \times 50 \angle 300^o + 1 \angle -120^o \times 30 \angle 180^o] \\
&= \tfrac{1}{3}\,[100 \angle 30^o + 50 \angle 60^o + 30 \angle 60^o] \\
&= \tfrac{1}{3}[(86{\cdot}6 + j\,50) + (25 + j\,43{\cdot}3) + (15 + j\,25{\cdot}98)] \\
&= \tfrac{1}{3}\,[126{\cdot}6 + j\,119{\cdot}28] \\
&= (42{\cdot}2 + j\,39{\cdot}76) = \mathbf{57{\cdot}98 \angle 43{\cdot}3^o\ A}
\end{aligned}
$$

$$
\begin{aligned}
\vec{I_2} &= \tfrac{1}{3}[\vec{I_R} + a^2\,\vec{I_Y} + a\,\vec{I_B}] \\
&= \tfrac{1}{3}[100 \angle 30^o + 1 \angle -120^o \times 50 \angle 300^o + 1 \angle 120^o \times 30 \angle 180^o] \\
&= \tfrac{1}{3}[100 \angle 30^o + 50 \angle 180^o + 30 \angle 300^o] \\
&= \tfrac{1}{3}[(86{\cdot}6 + j\,50) + (-50 + j\,0) + (15 - j\,25{\cdot}98)] \\
&= \tfrac{1}{3}[51{\cdot}6 + j\,24{\cdot}02] \\
&= (17{\cdot}2 + j\,8) = \mathbf{18{\cdot}96 \angle 24{\cdot}9^o\ A}
\end{aligned}
$$

Current in the neutral wire $= \vec{I_R} + \vec{I_Y} + \vec{I_B} = (81{\cdot}6 + j\,6{\cdot}7) = \mathbf{81{\cdot}87 \angle 4{\cdot}7^o\ A}$

Example 18.3. *The currents in a 3-phase unbalanced system are :*

$\vec{I_R} = (12 + j\,6)\ A$; $\vec{I_Y} = (12 - j\,12)\ A$; $\vec{I_B} = (-15 + j\,10)\ A$

The phase sequence in RYB. Calculate the zero, positive and negative sequence components of the currents.

Solution.

Red phase

Zero phase sequence component,

$$\vec{I_{R0}} = \frac{1}{3}\left[\vec{I_R} + \vec{I_Y} + \vec{I_B}\right]$$

---

* With the help of scientific calculator, *polar form* can be directly changed to *rectangular form* and *vice-versa.*

$$= \frac{1}{3}[(12 + j\,6) + (12 - j\,12) + (-15 + j\,10)]$$

$$= \frac{1}{3}[9 + j\,4] = \mathbf{(3 + j\,1{\cdot}33)}\ A$$

Positive phase sequence component is

$$\overrightarrow{I_{R1}} = \frac{1}{3}\left[\overrightarrow{I_R} + a\,\overrightarrow{I_Y} + a^2\,\overrightarrow{I_B}\right]$$

$$= \frac{1}{3}[(12 + j\,6) + (-0{\cdot}5 + j\,0{\cdot}866)(12 - j\,12) + {}^*(-0{\cdot}5 - j\,0{\cdot}866)(-15 + j\,10)]$$

$$= \frac{1}{3}[32{\cdot}55 + j\,30{\cdot}39] = \mathbf{(10{\cdot}85 + j\,10{\cdot}13)\ A}$$

Negative phase sequence component is

$$\overrightarrow{I_{R2}} = \frac{1}{3}\left[\overrightarrow{I_R} + a^2\,\overrightarrow{I_Y} + a\,\overrightarrow{I_B}\right]$$

$$= \frac{1}{3}[(12 + j\,6) + (-0{\cdot}5 - j\,0{\cdot}866)(12 - j\,12) + (-0{\cdot}5 + j\,0{\cdot}866)(-15 + j\,10)]$$

$$= \frac{1}{3}[-5{\cdot}55 - j\,16{\cdot}41] = \mathbf{(-1{\cdot}85 - j\,5{\cdot}47)}\ A$$

Yellow phase

Zero phase sequence component is

$$\overrightarrow{I_{Y0}} = \overrightarrow{I_{R0}} = \mathbf{(3 + j\,1{\cdot}33)}\ A$$

Positive phase sequence component is

$$\overrightarrow{I_{Y1}} = a^2\,\overrightarrow{I_{R1}} = (-0{\cdot}5 - j\,0{\cdot}866)(10{\cdot}85 + j\,10{\cdot}13)$$

$$= \mathbf{(3{\cdot}35 - j\,14{\cdot}4)}\ A$$

Negative phase sequence component is

$$\overrightarrow{I_{Y2}} = a\,\overrightarrow{I_{R2}} = (-0{\cdot}5 + j\,0{\cdot}866)(-1{\cdot}85 - j\,5{\cdot}47) = \mathbf{(5{\cdot}7 + j\,1{\cdot}13)}\ A$$

Blue phase

Zero phase sequence component is

$$\overrightarrow{I_{B0}} = \overrightarrow{I_{Y0}} = \overrightarrow{I_{R0}} = \mathbf{(3 + j\,1{\cdot}33)}\ A$$

Positive phase sequence component is

$$\overrightarrow{I_{B1}} = a^2\,\overrightarrow{I_{R1}} = (-0{\cdot}5 + j\,0{\cdot}866)(10{\cdot}85 + j\,10{\cdot}13) = \mathbf{(-14{\cdot}2 + j\,4{\cdot}31)}\ A$$

Negative phase sequence component is

$$\overrightarrow{I_{B2}} = a^2\,\overrightarrow{I_{R2}} = (-0{\cdot}5 - j\,0{\cdot}866)(-1{\cdot}85 - j\,5{\cdot}47) = \mathbf{(-3{\cdot}82 + j\,4{\cdot}34)}\ A$$

Example 18.4. *The sequence voltages in the red phase are as under :*

$\overrightarrow{E_{R0}} = 100$ V ; $\overrightarrow{E_{R1}} = (200 - j\,100)$ V ; $\overrightarrow{E_{R2}} = -100$ V

Find the phase voltages $\overrightarrow{E_R}$, $\overrightarrow{E_Y}$ and $\overrightarrow{E_B}$.

Solution. In the polar form, we have,

$\overrightarrow{E_{R0}} = 100\ \angle 0°$ V; $\overrightarrow{E_{R1}} = 223{\cdot}6\ \angle{-26{\cdot}56°}$ V; $\overrightarrow{E_{R2}} = 100\ \angle 180°$ V

$$\overrightarrow{E_R} = \overrightarrow{E_{R0}} + \overrightarrow{E_{R1}} + \overrightarrow{E_{R2}}$$

$$= 100 + (200 - j\,100) + (-100)$$

$$= 200 - j\,100 = \mathbf{223{\cdot}6\ \angle{-26{\cdot}56°}\ volts}$$

---

* $a = -0{\cdot}5 + j\,0{\cdot}866$ and $a^2 = -0{\cdot}5 - j\,0{\cdot}866$

$$\vec{E_Y} = \vec{E_{R0}} + a^2\,\vec{E_{R1}} + a\,\vec{E_{R2}}$$
$$= 100\angle 0° + 1\angle 240° \times 223{\cdot}6\ \angle -26{\cdot}56° + 1\ \angle 120° \times 100\angle 180°$$
$$= 100\ \angle 0° + 223{\cdot}6\ \angle 213{\cdot}44° + 100\ \angle 300°$$
$$= (100 + j\,0) + (-186{\cdot}58 - j\,123{\cdot}2) + (50 - j\,86{\cdot}6)$$
$$= -36{\cdot}58 - j\,209{\cdot}8 = \mathbf{213\ \angle -99{\cdot}89°\ volts}$$

$$\vec{E_B} = \vec{E_{R0}} + a\,\vec{E_{R1}} + a^2\,\vec{E_{R2}}$$
$$= 100\angle 0° + 1\angle 120° \times 223{\cdot}6\ \angle -26{\cdot}56° + 1\ \angle 240° \times 100\ \angle 180°$$
$$= 100\ \angle 0° + 223{\cdot}6\ \angle 93{\cdot}44° + 100\ \angle 420°$$
$$= (100 + j\,0) + (-13{\cdot}4 + j\,223{\cdot}2) + (50 + j\,86{\cdot}6)$$
$$= 136{\cdot}6 + j\,309{\cdot}8 = \mathbf{338{\cdot}57\ \angle 66{\cdot}2°\ volts}$$

Example 18.5. *The zero and positive sequence components of red phase are as under :*

$\vec{E_{R0}} = (0{\cdot}5 - j\,0{\cdot}866)\ V$; $\vec{E_{R1}} = 2\ \angle 0°\ V$

If the phase voltage $\vec{E_R} = 3\ \angle 0°$ V, *find the negative sequence component of red phase and the phase voltages* $\vec{E_Y}$ *and* $\vec{E_B}$.

Solution.

$$\vec{E_R} = \vec{E_{R0}} + \vec{E_{R1}} + \vec{E_{R2}}$$

or
$$3 = (0{\cdot}5 - j\,0{\cdot}866) + 2 + \vec{E_{R2}}$$

∴ Negative sequence component in *R*-phase is

$$\vec{E_{R2}} = 0{\cdot}5 + j\,0{\cdot}866 = \mathbf{1\ \angle 60°\ volts}$$

In polar form,
$$\vec{E_{R0}} = 0{\cdot}5 - j\,0{\cdot}866 = 1\ \angle -60°$$

Now
$$\vec{E_Y} = \vec{E_{R0}} + a^2\,\vec{E_{R1}} + a\,\vec{E_{R2}}$$
$$= [1\ \angle -60°] + [1\angle 240° \times 2\angle 0°] + [1\ \angle 120° \times 1\ \angle 60°]$$
$$= 1\ \angle -60° + 2\,\angle 240° + 1\ \angle 180°$$
$$= (0{\cdot}5 - j\,0{\cdot}866) + (-1 - j\,1{\cdot}732) + (-1 + j\,0)$$
$$= -1{\cdot}5 - j\,2{\cdot}598$$
$$= \mathbf{3\ \angle -120°\ volts}$$

$$\vec{E_B} = \vec{E_{R0}} + a\,\vec{E_{R1}} + a^2\,\vec{E_{R2}}$$
$$= [1\ \angle -60°] + [1\ \angle 120° \times 2\ \angle 0°] + [1\ \angle 240° \times 1\ \angle 60°]$$
$$= 1\ \angle -60° + 2\ \angle 120° + 1\ \angle 300°$$
$$= (0{\cdot}5 - j\,0{\cdot}866) + (-1 + j\,1{\cdot}732) + (0{\cdot}5 - j\,0{\cdot}866)$$
$$= \mathbf{0\ volt}$$

Example 18.6. *The current from neutral to ground connection is 12 A. Calculate the zero phase sequence components in phases.*

Solution. We know that zero sequence components in all phases have the same value and that each component is equal to one-third the current in the neutral wire.

∴ Zero sequence current in each phase

$$= \frac{1}{3} \times 12 = \mathbf{4\ A}$$

Example 18.7. *A balanced star connected load takes 90 A from a balanced 3-phase, 4-wire supply. If the fuses in the Y and B phases are removed, find the symmetrical components of the line currents*

(i) before the fuses are removed *(ii) after the fuses are removed*

Solution. Fig. 18.7. shows the star-connected system with fuses in phases *B* and *Y*.

(*i*) **Before removal of fuses.** Before fuses are removed from *Y* and *B* lines, the system is balanced and current in each line is 90 A.

$\therefore \quad \vec{I_R} = 90 \angle 0° \text{A}$; $\vec{I_Y} = 90 \angle 240° \text{ A}$; $\vec{I_B} = 90 \angle 120° \text{A}$

Since the system is balanced, it will have only positive sequence currents *i.e.*, negative sequence and zero sequence components will be zero in the three lines. This can be readily established.

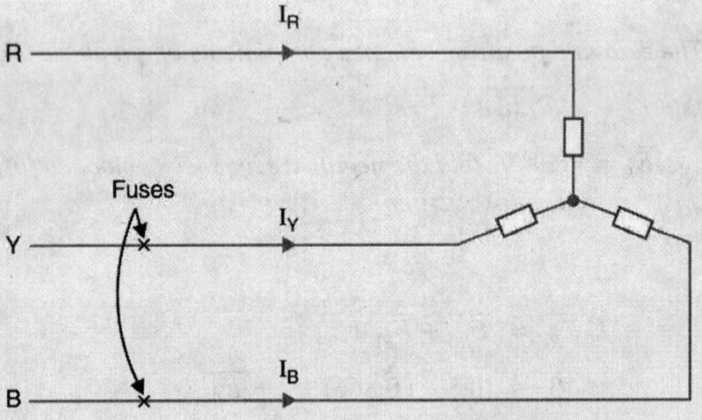

Fig. 18.7

$$\vec{I_{R0}} = \vec{I_{Y0}} = \vec{I_{B0}}$$

$$= \frac{1}{3}\left[\vec{I_R} + \vec{I_Y} + \vec{I_B}\right] = \frac{1}{3}[90\angle 0° + 90\angle 240° + 90\angle 120°]$$

$$= \frac{1}{3}[90\angle 0° + 90\angle -120° + 90 \angle 120°] = 0 \text{ A}$$

Hence zero sequence components in three lines are zero.

$$\vec{I_{R2}} = \frac{1}{3}\left[\vec{I_R} + a^2\,\vec{I_Y} + a\,\vec{I_B}\right]$$

$$= \frac{1}{3}[\,90\angle 0° + 1\,\angle -120° \times 90\angle 240° + 1\,\angle 120° \times 90\angle 120°]$$

$$= \frac{1}{3}[90\angle 0° + 90\angle 120° + 90\angle 240°]$$

$$= \frac{1}{3}[\,90\angle 0° + 90\angle 120° + 90\angle -120°] = 0 \text{ A}$$

Also $\quad \vec{I_{Y2}} = a\,\vec{I_{R2}} = 1\,\angle 120° \times 0 = 0 \text{ A}$

and $\quad \vec{I_{B2}} = a^2\,\vec{I_{R2}} = 1\,\angle 240° \times 0 = 0 \text{ A}$

Hence negative sequence components in the three lines are also zero. It can be easily shown that three positive sequence components will have the following values :

$\vec{I_{R1}} = \vec{I_R} = 90\angle 0° \text{ A}$; $\vec{I_{Y1}} = \vec{I_Y} = 90\angle 240° \text{ A}$; $\vec{I_{B1}} = \vec{I_B} = 90\,\angle 120° \text{ A}$

(ii) After removal of fuses. When the fuses are removed in Y and B phases, the system becomes unbalanced with line currents as under :

$$\vec{I_R} = 90\angle 0^\circ \text{A} \;;\; \vec{I_Y} = \vec{I_B} = 0 \text{ A}$$

The sequence currents in the three lines can be found out as under :

$$\vec{I_{R0}} = \vec{I_{Y0}} = \vec{I_{B0}}$$
$$= \frac{1}{3}\left[\vec{I_R} + \vec{I_Y} + \vec{I_B}\right]$$
$$= \frac{1}{3}[90\angle 0^\circ + 0 + 0] = \mathbf{30\ \angle 0^\circ\ A}$$

i.e. zero sequence current in each line is 30 $\angle 0^\circ$ A.

$$\vec{I_{R1}} = \frac{1}{3}\left[\vec{I_R} + a\,\vec{I_Y} + a^2\,\vec{I_B}\right]$$
$$= \frac{1}{3}[90\angle 0^\circ + 0 + 0] = \mathbf{30\ \angle 0^\circ A}$$
$$\vec{I_{Y1}} = a^2\,\vec{I_{R1}} = 1\ \angle 240^\circ \times 30\ \angle 0^\circ = \mathbf{30\ \angle 240^\circ A}$$
$$\vec{I_{B1}} = a\,\vec{I_{R1}} = 1\ \angle 120^\circ \times 30\ \angle 0^\circ = \mathbf{30\ \angle 120^\circ A}$$
$$\vec{I_{R2}} = \frac{1}{3}\left[\vec{I_R} + a^2\,\vec{I_Y} + a\,\vec{I_B}\right]$$
$$= \frac{1}{3}[90\ \angle 0^\circ + 0 + 0] = \mathbf{30\ \angle 0^\circ A}$$
$$\vec{I_{Y2}} = a\,\vec{I_{R2}} = 1\ \angle 120^\circ \times 30\angle 0^\circ = \mathbf{30\ \angle 120^\circ\ A}$$
$$\vec{I_{B2}} = a^2\,\vec{I_{R2}} = 1\ \angle 240^\circ \times 30\angle 0^\circ = \mathbf{30\ \angle 240^\circ\ A}$$

The reader may wonder how sequence currents can flow in the yellow and blue lines when fuses are removed in them. The answer is that these components do not have separate existence. They are only the mathematical components of the current which does exist. Thus the current in the yellow line is zero and this can be readily established from its sequence components :

$$\vec{I_Y} = \vec{I_{Y0}} + \vec{I_{Y1}} + \vec{I_{Y2}}$$
$$= 30\ \angle 0^\circ + 30\ \angle 240^\circ + 30\ \angle 120^\circ$$
$$= 30\ \angle 0^\circ + 30\ \angle -120^\circ + 30\ \angle 120^\circ = 0 \text{ A}$$

Similary, it can be proved that sum of sequence currents in the blue line is zero and that is what the circuit reveals.

Example 18.8. *A 3-ϕ, 4-wire-system supplies loads which are unequally distributed in the three phases. An analysis of the current flowing in R, Y and B lines shows that in R line, positive phase sequence component is 200 $\angle 0^\circ$ A and the negative phase sequence component is 100 $\angle 60^\circ$A. The total observed current flowing back to the supply in the neutral conductor is 300 $\angle 300^\circ$ A. Calculate the currents in the three lines.*

Solution.

Zero phase sequence current in R-line is

$$\vec{I_{R0}} = \frac{1}{3} \times \text{Current in neutral wire}$$
$$= \frac{1}{3} \times 300\ \angle 300^\circ = 100\ \angle 300^\circ \text{A}$$

Positive phase sequence current in R-line is

$$\vec{I_{R1}} = 200\ \angle 0^\circ \text{A}$$

Negative phase sequence current in R-line is

$$\overrightarrow{I_{R2}} = 100 \angle 60^\circ \text{A}$$

$\therefore$ Current in the R-line, $\overrightarrow{I_R} = \overrightarrow{I_{R0}} + \overrightarrow{I_{R1}} + \overrightarrow{I_{R2}} = 100 \angle 300^\circ + 200 \angle 0^\circ + 100 \angle 60^\circ$

$$= (50 - j\,86{\cdot}6) + (200 + j\,0) + (50 + j\,86{\cdot}6) = \mathbf{300 \angle 0^\circ\ A}$$

Current in the Y-line, $\overrightarrow{I_Y} = \overrightarrow{I_{R0}} + a^2\,\overrightarrow{I_{R1}} + a\,\overrightarrow{I_{R2}}$

$$= 100 \angle 300^\circ + 1 \angle 240^\circ \times 200 \angle 0^\circ + 1\angle 120^\circ \times 100\angle 60^\circ$$
$$= 100 \angle 300^\circ + 200 \angle 240^\circ + 100 \angle 180^\circ$$
$$= (50 - j\,86{\cdot}6) + (-100 - j\,173{\cdot}2) + (-100 + j\,0)$$
$$= -150 - j\,259{\cdot}8 = \mathbf{300 \angle{-120^\circ}\ A}$$

Current in B line, $\overrightarrow{I_B} = \overrightarrow{I_{R0}} + a\,\overrightarrow{I_{R1}} + a^2\,\overrightarrow{I_{R2}}$

$$= 100 \angle 300^\circ + 1 \angle 120^\circ \times 200 \angle 0^\circ + 1 \angle 240^\circ \times 100 \angle 60^\circ$$
$$= 100 \angle 300^\circ + 200 \angle 120^\circ + 100 \angle 300^\circ$$
$$= (50 - j\,86{\cdot}6) + (-100 + j\,173{\cdot}2) + (50 - j\,86{\cdot}6) = \mathbf{0\ A}$$

Example 18.9. *One conductor of a 3-phase line is open. The current flowing to the Δ-connected load through the line R is 10 A. With the current in line R [See Fig. 18.8] as reference and assuming that line B is open, find the symmetrical components of the line currents.*

Solution. The line currents are :

$\overrightarrow{I_R} = 10\angle 0^\circ$ A ; $\overrightarrow{I_Y} = 10\angle 180^\circ$ A ; $\overrightarrow{I_B} = 0$ A

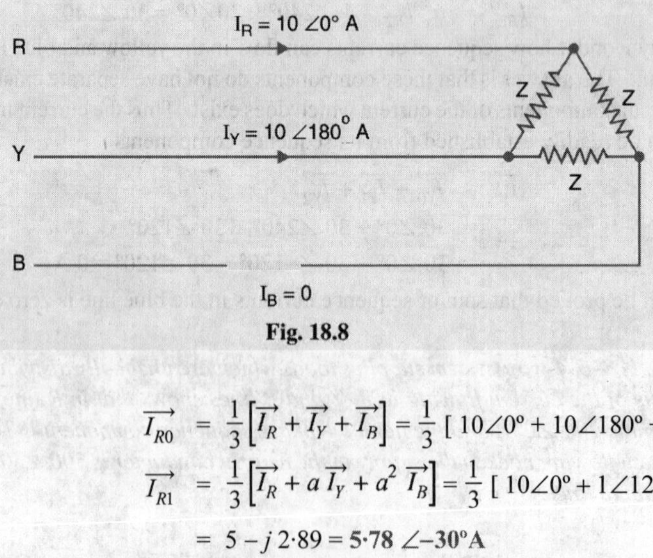

Fig. 18.8

***R*-line**

$$\overrightarrow{I_{R0}} = \frac{1}{3}\left[\overrightarrow{I_R} + \overrightarrow{I_Y} + \overrightarrow{I_B}\right] = \frac{1}{3}\,[\,10\angle 0^\circ + 10\angle 180^\circ + 0] = \mathbf{0\ A}$$

$$\overrightarrow{I_{R1}} = \frac{1}{3}\left[\overrightarrow{I_R} + a\,\overrightarrow{I_Y} + a^2\,\overrightarrow{I_B}\right] = \frac{1}{3}\,[\,10\angle 0^\circ + 1\angle 120^\circ \times 10\angle 180^\circ + 0]$$

$$= 5 - j\,2{\cdot}89 = \mathbf{5{\cdot}78 \angle{-30^\circ} A}$$

$$\overrightarrow{I_{R2}} = \frac{1}{3}\left[\overrightarrow{I_R} + a^2\,\overrightarrow{I_Y} + a\,\overrightarrow{I_B}\right] = \frac{1}{3}\,[\,10\angle 0^\circ + 1\angle 240^\circ \times 10\angle 180^\circ + 0]$$

$$= 5 + j\,2{\cdot}89 = \mathbf{5{\cdot}78 \angle 30^\circ A}$$

***Y*-line**

$$\overrightarrow{I_{Y0}} = \overrightarrow{I_{R0}} = \mathbf{0\ A}$$

$$\overrightarrow{I_{Y1}} = a^2\,\overrightarrow{I_{R1}} = 1 \angle 240^\circ \times 5{\cdot}78 \angle{-30^\circ} = \mathbf{5{\cdot}78 \angle{-150^\circ}\ A}$$

$$\overrightarrow{I_{Y2}} = a\,\overrightarrow{I_{R2}} = 1\angle 120^\circ \times 5{\cdot}78\angle 30^\circ = \mathbf{5{\cdot}78\angle 150^\circ A}$$

***B*-line**

$$\overrightarrow{I_{B0}} = \overrightarrow{I_{R0}} = \mathbf{0\ A}$$

$$\overrightarrow{I_{B1}} = a\,\overrightarrow{I_{R1}} = 1\angle 120^\circ \times 5{\cdot}78\angle -30^\circ = \mathbf{5{\cdot}78\angle 90^\circ\ A}$$

$$\overrightarrow{I_{B2}} = a^2\,\overrightarrow{I_{R2}} = 1\angle 240^\circ \times 5{\cdot}78\angle 30^\circ = \mathbf{5{\cdot}78\angle -90^\circ A}$$

Note that components I_{B1} and I_{B2} have finite values although the line B is open and can carry no net current. As expected, the sum of I_{B1} and I_{B2} is zero. However, the sum of components in line R is $10\angle 0^\circ$A and the sum of components in line Y is $10\angle 180^\circ$A.

Example 18.10. *Three resistors of 5Ω, 10Ω and 20Ω are connected in delta across the three phases of a balanced 100 volts supply. What are the sequence components in the resistors and in supply lines ?*

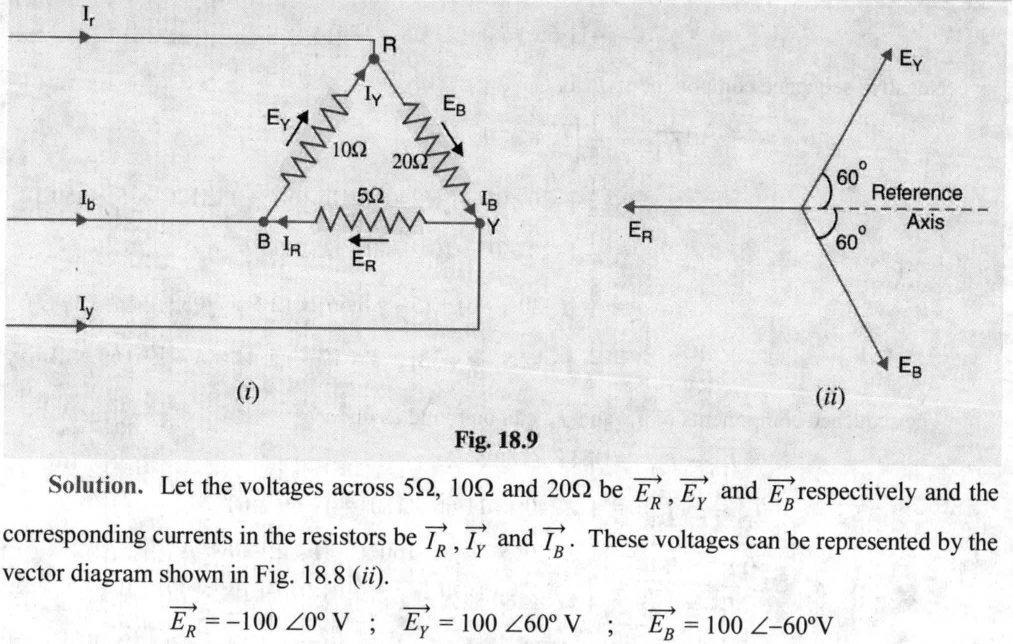

Fig. 18.9

Solution. Let the voltages across 5Ω, 10Ω and 20Ω be $\overrightarrow{E_R}$, $\overrightarrow{E_Y}$ and $\overrightarrow{E_B}$ respectively and the corresponding currents in the resistors be $\overrightarrow{I_R}$, $\overrightarrow{I_Y}$ and $\overrightarrow{I_B}$. These voltages can be represented by the vector diagram shown in Fig. 18.8 (*ii*).

$$\overrightarrow{E_R} = -100\angle 0^\circ\ \text{V}\ ; \quad \overrightarrow{E_Y} = 100\angle 60^\circ\ \text{V}\ ; \quad \overrightarrow{E_B} = 100\angle -60^\circ\text{V}$$

Current in 5Ω, $\overrightarrow{I_R} = \overrightarrow{E_R}/5 = \dfrac{-100\angle 0^\circ}{5} = -20\angle 0^\circ$ A

Current in 10Ω, $\overrightarrow{I_Y} = \overrightarrow{E_Y}/10 = \dfrac{100\angle 60^\circ}{10} = 10\angle 60^\circ$A

Current in 20Ω, $\overrightarrow{I_B} = \overrightarrow{E_B}/20 = \dfrac{100\angle -60^\circ}{20} = 5\angle -60^\circ$A

Sequence currents in resistors

Zero sequence component of $\overrightarrow{I_R}$ is

$$\overrightarrow{I_{R0}} = \frac{1}{3}\left[\overrightarrow{I_R} + \overrightarrow{I_Y} + \overrightarrow{I_B}\right]$$

$$= \frac{1}{3}[-20\angle 0^\circ + 10\angle 60^\circ + 5\angle -60^\circ]$$

$$= \frac{1}{3}[(-20 + j\,0) + (5 + j\,8{\cdot}66) + (2{\cdot}5 - j\,4{\cdot}33)]$$

$$= \frac{1}{3}[-12 \cdot 5 + j\,4 \cdot 33]$$
$$= -4 \cdot 17 + j\,1 \cdot 44 = \mathbf{4 \cdot 41\ \angle 160 \cdot 9^{\circ} A}$$

Positive sequence component of $\overrightarrow{I_R}$ is

$$\overrightarrow{I_{R1}} = \frac{1}{3}\left[\overrightarrow{I_R} + a\,\overrightarrow{I_Y} + a^2\,\overrightarrow{I_B}\right]$$
$$= \frac{1}{3}[-20\ \angle 0^{\circ} + 1\ \angle 120^{\circ} \times 10\ \angle 60^{\circ} + 1\ \angle 240^{\circ} \times 5\ \angle -60^{\circ}]$$
$$= \frac{1}{3}[-20\ \angle 0^{\circ} + 10\ \angle 180^{\circ} + 5\ \angle 180^{\circ}]$$
$$= \frac{1}{3}[(-20 + j\ 0) + (-10 + j\,0) + (-5 + j\,0)]$$
$$= \frac{1}{3}[-35 + j\,0]$$
$$= -11 \cdot 66 + j\,0 = \mathbf{11 \cdot 66\ \angle 180^{\circ} A}$$

Negative sequence component of $\overrightarrow{I_R}$ is

$$\overrightarrow{I_{R2}} = \frac{1}{3}\left[\overrightarrow{I_R} + a^2\,\overrightarrow{I_Y} + a\,\overrightarrow{I_B}\right]$$
$$= \frac{1}{3}[-20\angle 0^{\circ} + 1\ \angle 240^{\circ} \times 10\ \angle 60^{\circ} + 1\ \angle 120^{\circ} \times 5\ \angle -60^{\circ}]$$
$$= \frac{1}{3}[-20\ \angle 0^{\circ} + 10\ \angle 300^{\circ} + 5\ \angle 60^{\circ}]$$
$$= \frac{1}{3}[(-20 + j\,0) + (5 - j\,8 \cdot 66) + (2 \cdot 5 + j\,4 \cdot 33)]$$
$$= \frac{1}{3}[-12 \cdot 5 - j\,4 \cdot 33] = -4 \cdot 17 - j\,1 \cdot 44 = \mathbf{4 \cdot 4\ \angle -160 \cdot 9^{\circ}\ A}$$

The sequence components of $\overrightarrow{I_Y}$ and $\overrightarrow{I_B}$ can be found as under :

$$\overrightarrow{I_{Y0}} = \overrightarrow{I_{R0}} = \mathbf{4 \cdot 41\ \angle 160 \cdot 9^{\circ}\ A}$$
$$\overrightarrow{I_{Y1}} = a^2\,\overrightarrow{I_{R1}} = 1\ \angle 240^{\circ} \times 11 \cdot 66\ \angle 180^{\circ} = \mathbf{11 \cdot 66\ \angle 60^{\circ}\ A}$$
$$\overrightarrow{I_{Y2}} = a\,\overrightarrow{I_{R2}} = 1\ \angle 120^{\circ} \times 4 \cdot 4\ \angle -160 \cdot 9^{\circ} = \mathbf{4 \cdot 4\ \angle -40 \cdot 9^{\circ}\ A}$$
$$\overrightarrow{I_{B0}} = \overrightarrow{I_{R0}} = \mathbf{4 \cdot 41\ \angle 160 \cdot 9^{\circ}\ A}$$
$$\overrightarrow{I_{B1}} = a\,\overrightarrow{I_{R1}} = 1\ \angle 120^{\circ} \times 11 \cdot 66\ \angle 180^{\circ} = \mathbf{11 \cdot 66\ \angle 300^{\circ}\ A}$$
$$\overrightarrow{I_{B2}} = a^2\,\overrightarrow{I_{R2}} = 1\ \angle 240^{\circ} \times 4 \cdot 4\ \angle -160 \cdot 9^{\circ} = \mathbf{4 \cdot 4\ \angle 79 \cdot 1^{\circ}\ A}$$

Sequence currents in supply lines

Line current in R-line, $\overrightarrow{I_r} = \overrightarrow{I_B} - \overrightarrow{I_Y} = 5\ \angle -60^{\circ} - 10\ \angle 60^{\circ}$

$$= (2 \cdot 5 - j\,4 \cdot 33) - (5 + j\,8 \cdot 66)$$
$$= -2 \cdot 5 - j\,12 \cdot 99 = 13 \cdot 22\ \angle -100 \cdot 9^{\circ}\ A$$

Line current in Y-line, $\overrightarrow{I_y} = \overrightarrow{I_R} - \overrightarrow{I_B} = -20\angle 0^{\circ} - 5\ \angle -60^{\circ}$

$$= (-20 + j\,0) - (2 \cdot 5 - j\,4 \cdot 33)$$
$$= -22 \cdot 5 + j\,4 \cdot 33 = 22 \cdot 91\ \angle 169^{\circ}\ A$$

Line current in B-line, $\overrightarrow{I_b} = \overrightarrow{I_Y} - \overrightarrow{I_R} = 10\angle 60^{\circ} - (-20\angle 0^{\circ})$

$$= (5 + j\,8 \cdot 66) - (-20 + j\,0)$$
$$= 25 + j\,8 \cdot 66 = 26 \cdot 45\ \angle 19 \cdot 1^{\circ} A$$

Zero sequence component of $\vec{I_r}$ is

$$\overrightarrow{I_{r0}}\,* = \frac{1}{3}\left(\vec{I_r} + \vec{I_y} + \vec{I_b}\right) = \frac{1}{3}[0] = \mathbf{0A**}$$

Positive sequence component of $\vec{I_r}$ is

$$\overrightarrow{I_{r1}} = \frac{1}{3}\left(\vec{I_r} + a\,\vec{I_y} + a^2\,\vec{I_b}\right)$$

$$= \frac{1}{3}\left[\left(\vec{I_B} - \vec{I_Y}\right) + a\left(\vec{I_R} - \vec{I_B}\right) + a^2\left(\vec{I_Y} - \vec{I_R}\right)\right]$$

$$= \frac{1}{3}\left[a\left(\vec{I_R} + a\,\vec{I_Y} + a^2\,\vec{I_B}\right) - a^2\left(\vec{I_R} + a\,\vec{I_Y} + a^2\,\vec{I_B}\right)\right]$$

$$[\because a^3 = 1 \text{ and } a^4 = a]$$

$$= \frac{1}{3}\left[\left(a - a^2\right)\left(\vec{I_R} + a\,\vec{I_Y} + a^2\,\vec{I_B}\right)\right]$$

Now $a - a^2 = j\sqrt{3}$ and $\vec{I_R} + a\,\vec{I_Y} + a^2\,\vec{I_B} = 3\,\overrightarrow{I_{R1}}$

$$\therefore \quad \overrightarrow{I_{r1}} = \frac{1}{3}\left[\left(j\sqrt{3}\right)\left(3\,\overrightarrow{I_{R1}}\right)\right]$$

$$= j\sqrt{3}\,\overrightarrow{I_{R1}} = j\sqrt{3}\,(-11{\cdot}66 + j\,0)$$

$$= -j\,20{\cdot}2 = \mathbf{20{\cdot}2\ \angle -90^\circ\ A}$$

Negative sequence component of $\vec{I_r}$ is

$$\overrightarrow{I_{r2}} = \frac{1}{3}\left[\vec{I_r} + a^2\,\vec{I_y} + a\,\vec{I_b}\right]$$

$$= \frac{1}{3}\left[\left(\vec{I_B} - \vec{I_Y}\right) + a^2\left(\vec{I_R} - \vec{I_B}\right) + a\left(\vec{I_Y} - \vec{I_R}\right)\right]$$

$$= \frac{1}{3}\left[a^2\left(\vec{I_R} + a^2\,\vec{I_Y} + a\,\vec{I_B}\right) - a\left(\vec{I_R} + a^2\,\vec{I_Y} + a\,\vec{I_B}\right)\right]$$

$$= \frac{1}{3}\left[\left(a^2 - a\right)\left(\vec{I_R} + a^2\,\vec{I_Y} + a\,\vec{I_B}\right)\right]$$

Now $a^2 - a = -j\sqrt{3}$ and $\vec{I_R} + a^2\,\vec{I_Y} + a\,\vec{I_B} = 3\,\overrightarrow{I_{R2}}$

$$\therefore \quad \overrightarrow{I_{r2}} = \frac{1}{3}\left(-j\sqrt{3} \times 3\,\overrightarrow{I_{R2}}\right) = -j\sqrt{3} \times \overrightarrow{I_{R2}}$$

$$= -j\sqrt{3} \times (-4{\cdot}17 - j\,1{\cdot}44)$$

$$= -2{\cdot}5 + j\,7{\cdot}2 = \mathbf{7{\cdot}62\ \angle 109{\cdot}1^\circ A}$$

Note. Incidentally, we have the formulas for relation among sequence components in the phases and lines.

$$\overrightarrow{I_{r1}} = j\sqrt{3}\,\overrightarrow{I_{R1}}\ ; \quad \overrightarrow{I_{r2}} = -j\sqrt{3}\,\overrightarrow{I_{R2}}$$

Example 18.11. *A delta connected load is supplied from a 3-phase supply. The fuse in the B line is removed and current in the other two lines is 20 A. Find the symmetrical components of line currents.*

Solution. Let R, Y and B be the supply lines. When fuse in the line B is removed, the various line currents are :

---

* Since vector sum of $\vec{I_r} + \vec{I_y} + \vec{I_b} = 0$, $\overrightarrow{I_{r0}} = 0$

** This shows that in delta formation, the zero sequence currents are present in phases but they disappear in line currents. As line current is the difference of two phase currents, therefore, the zero sequence components cancel out.

$$\vec{I_r} = 20\angle 0^\circ \text{ A};\ \vec{I_y} = 20\angle 180^\circ \text{ A};\ \vec{I_b} = 0 \text{ A}$$

***R*-line**

$$\vec{I_{r0}} = \frac{1}{3}\left[\vec{I_r} + \vec{I_y} + \vec{I_b}\right] = \frac{1}{3}[20\angle 0^\circ + 20\angle 180^\circ + 0]$$

$$= \frac{1}{3}[(20 + j0) + (-20 + j0) + 0] = \frac{1}{3}[0] = \mathbf{0A}$$

$$\vec{I_{r1}} = \frac{1}{3}\left[\vec{I_r} + a\,\vec{I_y} + a^2\,\vec{I_b}\right]$$

$$= \frac{1}{3}[20\angle 0^\circ + 1\angle 120^\circ \times 20\angle 180^\circ + 0]$$

$$= \frac{1}{3}[20\angle 0^\circ + 20\angle 300^\circ] = \frac{1}{3}[(20 + j0) + (10 - j\,17{\cdot}32)]$$

$$= \frac{1}{3}[30 - j\,17{\cdot}32] = 10 - j\,5{\cdot}77 = \mathbf{11{\cdot}54\angle -30^\circ A}$$

$$\vec{I_{r2}} = \frac{1}{3}\left[\vec{I_r} + a^2\,\vec{I_y} + a\,\vec{I_b}\right]$$

$$= \frac{1}{3}[20\angle 0^\circ + 1\angle 240^\circ \times 20\angle 180^\circ + 0]$$

$$= \frac{1}{3}[20\angle 0^\circ + 20\angle 60^\circ] = \frac{1}{3}[(20 + j0) + (10 + j\,17{\cdot}32)]$$

$$= \frac{1}{3}[30 + j\,17{\cdot}32] = 10 + j\,5{\cdot}77 = \mathbf{11{\cdot}54\angle 30^\circ A}$$

***Y*-line**

$$\vec{I_{y0}} = \vec{I_{r0}} = \mathbf{0A}$$

$$\vec{I_{y1}} = a^2\,\vec{I_{r1}} = 1\angle 240^\circ \times 11{\cdot}54\angle -30^\circ = \mathbf{11{\cdot}54\angle 210^\circ\ A}$$

$$\vec{I_{y2}} = a\,\vec{I_{r2}} = 1\angle 120^\circ \times 11{\cdot}54\angle 30^\circ = \mathbf{11{\cdot}54\angle 150^\circ\ A}$$

***B*-line**

$$\vec{I_{bo}} = \vec{I_{ro}} = \mathbf{0A}$$

$$\vec{I_{b1}} = a\,\vec{I_{r1}} = 1\angle 120^\circ \times 11{\cdot}54\angle -30^\circ = \mathbf{11{\cdot}54\angle 90^\circ A}$$

$$\vec{I_{b2}} = a^2\,\vec{I_{r2}} = 1\angle 240^\circ \times 11{\cdot}54\angle 30^\circ = \mathbf{11{\cdot}54\angle 270^\circ\ A}$$

Example 18.12. *Three impedances of 5—j10, 6 + j 5 and 3 + j15 ohms are connected in star to red, yellow and blue lines of a 3300 V, 3-phase, 3-wire supply. The phase sequence is RYB. Calculate the line current I_R.*

Solution. This is a case of unbalanced 3-phase star connected load supplied from a balanced 3-phase supply. Since the phase sequence is *RY B*,

$$\therefore \quad \vec{V_{RY}} = 3300\angle 0^\circ \text{ V};\quad \vec{V_{YB}} = a^2\,\vec{V_{RY}} = 3300\angle 240^\circ \text{ V}$$

Let $\vec{V_R}$, $\vec{V_Y}$ and $\vec{V_B}$ be the voltages across impedances in *R*, *Y* and *B* phases respectively and $\vec{I_R}$, $\vec{I_Y}$ and $\vec{I_B}$ the resulting line currents.

$$\therefore \quad \vec{V_R} - \vec{V_Y} = \vec{V_{RY}} = 3300 + j\,0$$

$$\text{and} \quad \vec{V_Y} - \vec{V_B} = \vec{V_{YB}} = 3300(-0{\cdot}5 - j\,0{\cdot}866)$$

$$\text{Since} \quad \vec{I_R} + \vec{I_Y} + \vec{I_B} = 0\text{A}$$

$\therefore \quad \overrightarrow{I_{R0}} = \overrightarrow{I_{Y0}} = \overrightarrow{I_{B0}} = 0\text{A}$

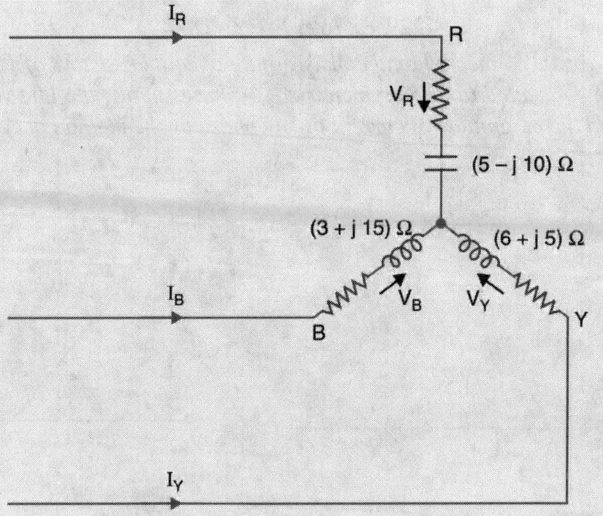

Fig. 18.10

$$\overrightarrow{V_R} = (5 - j10)\,\overrightarrow{I_R} = (5 - j10)\left(\overrightarrow{I_{R0}} + \overrightarrow{I_{R1}} + \overrightarrow{I_{R2}}\right)$$

$$= (5 - j10)\left(\overrightarrow{I_{R1}} + \overrightarrow{I_{R2}}\right) \qquad ...(i)$$

$$\overrightarrow{V_Y} = (6 + j5)\,\overrightarrow{I_Y} = (6 + j5)\left(\overrightarrow{I_{Y0}} + \overrightarrow{I_{Y1}} + \overrightarrow{I_{Y2}}\right)$$

$$= (6 + j5)\left(\overrightarrow{I_{Y1}} + \overrightarrow{I_{Y2}}\right)$$

$$= (6 + j5)\left(a^2\,\overrightarrow{I_{R1}} + a\,\overrightarrow{I_{R2}}\right) \qquad ...(ii)$$

$$\overrightarrow{V_B} = (3 + j15)\,\overrightarrow{I_B} = (3 + j15)\left(\overrightarrow{I_{B0}} + \overrightarrow{I_{B1}} + \overrightarrow{I_{B2}}\right)$$

$$= (3 + j15)\left(\overrightarrow{I_{B1}} + \overrightarrow{I_{B2}}\right)$$

$$= (3 + j15)\left(a\,\overrightarrow{I_{R1}} + a^2\,\overrightarrow{I_{R2}}\right) \qquad ...(iii)$$

Subtracting exp. (*ii*) from exp. (*i*), we get,

$$\overrightarrow{V_R} - \overrightarrow{V_Y} = \left(\overrightarrow{I_{R1}} + \overrightarrow{I_{R2}}\right)(5 - j10) - \left(a^2\,\overrightarrow{I_{R1}} + a\,\overrightarrow{I_{R2}}\right)(6 + j5)$$

or $$3300 = (3{\cdot}67 - j\,2{\cdot}3)\,\overrightarrow{I_{R1}} + (12{\cdot}33 - j\,12{\cdot}7)\,\overrightarrow{I_{R2}} \qquad ...(iv)$$

Subtracting exp. (*iii*) from exp. (*ii*), we get,

$$\overrightarrow{V_Y} - \overrightarrow{V_B} = \left(a^2\,\overrightarrow{I_{R1}} + a\,\overrightarrow{I_{R2}}\right)(6 + j5) - \left(a\,\overrightarrow{I_{R1}} + a^2\,\overrightarrow{I_{R2}}\right)(3 + j15)$$

or $$3300(-0{\cdot}5 - j\,0{\cdot}866) = (15{\cdot}8 - j\,2{\cdot}8)\,\overrightarrow{I_{R1}} - (18{\cdot}84 - j\,12{\cdot}8)\,\overrightarrow{I_{R2}}$$

or $$-1650 - j\,2858 = (15{\cdot}8 - j\,2{\cdot}8)\,\overrightarrow{I_{R1}} - (18{\cdot}84 - j\,12{\cdot}8)\,\overrightarrow{I_{R2}} \qquad ...(v)$$

Solving exps. (*iv*) and (*v*), we get,

$$\overrightarrow{I_{R1}} = 134 - j\,65$$

and $$\overrightarrow{I_{R2}} = 95 + j\,141$$

$$\therefore \quad \vec{I_R} = \vec{I_{R1}} + \vec{I_{R2}} = (134 - j\,65) + (95 + j\,141)$$
$$= 229 + j\,76 = \mathbf{241 \angle{-18{\cdot}4^\circ}\ A}$$

Example 18.13. *A star connected load consists of three equal resistors of 1 Ω resistance. The load is assumed to be connected to an unsymmetrical 3-phase supply, the line voltages are 200 V, 346 V and 400 V. Find the magnitude of current in any phase by the method of symmetrical components.*

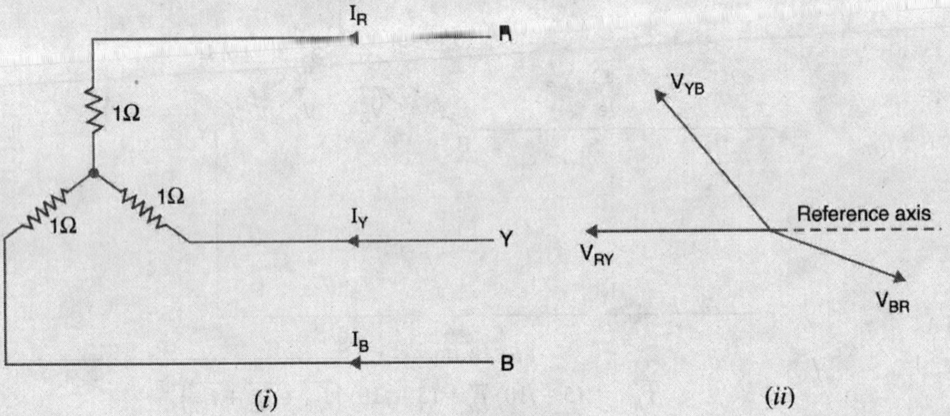

Fig. 18.11

Solution. This is a case of a balanced star-connected load supplied from an unbalanced 3-phase supply. Fig. 18.11 (*i*) shows the balanced star-connected load receiving unbalanced supply. Fig. 18.11 (*ii*) shows the vector diagram. Since the vector sum of three voltages is zero, these can be represented by the three sides of a triangle as shown in Fig. 18.12. Referring to Fig. 18.12, it is clear that :

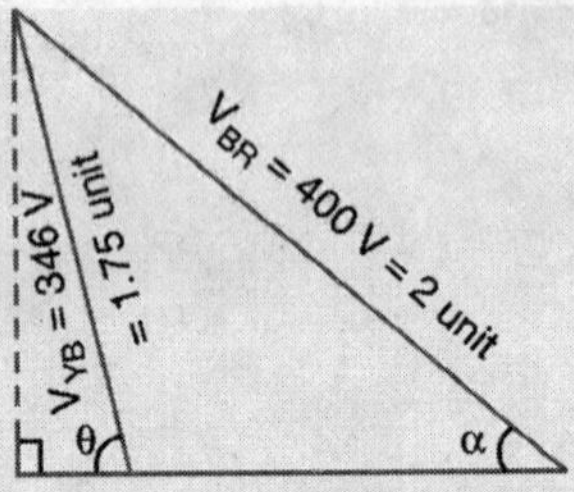

Fig. 18.12

$$(2)^2 = (1 + 1{\cdot}75 \cos\theta)^2 + (1{\cdot}75 \sin\theta)^2$$

or $$4 = 1 + (1{\cdot}75)^2 (\cos^2\theta + \sin^2\theta) + 2 \times 1{\cdot}75 \cos\theta$$

or $$4 = 1 + 3 \times 1 + 3{\cdot}5 \cos\theta$$

$$\therefore \quad \cos\theta = (4 - 4)/3{\cdot}5 = 0$$

$$\therefore \quad \theta = 90^\circ$$

and $$\cos\alpha = \frac{1 + 1{\cdot}75\cos\theta}{2} = \frac{1+0}{2} = 0{\cdot}5$$

$$\therefore \quad \alpha = 60^\circ$$

As the phase sequence is *RYB*, therefore, various line voltages are :

$$\vec{V_{RY}} = 200 \angle 180^\circ = (-200 + j0)\ \text{V}$$
$$\vec{V_{YB}} = 346 \angle 180^\circ - 90^\circ = 346 \angle 90^\circ = (0 + j\,346)\ \text{V}$$
$$\vec{V_{BR}} = 400 \angle{-60^\circ} = (200 - j\,346)\ \text{V}$$

The current in any phase (or line) is equal to phase voltage divided by resistance in that phase.

$$\therefore \quad \text{Line current, } \vec{I_R} = \frac{200 \angle 180^\circ}{1 \times \sqrt{3}} = 115{\cdot}47 \angle 180^\circ\ \text{A}$$

$$\text{Line current, } \vec{I_Y} = \frac{346 \angle 90^\circ}{1 \times \sqrt{3}} = 199{\cdot}77 \angle 90^\circ\ \text{A}$$

$$\text{Line current, } \vec{I_B} = \frac{400\angle -60^\circ}{1 \times \sqrt{3}} = 230{\cdot}94\angle -60^\circ\text{A}$$

Sequence components in red phase are :

$$\vec{I_{R0}} = \frac{1}{3}\left[\vec{I_R} + \vec{I_Y} + \vec{I_B}\right]$$
$$= \frac{1}{3}[115{\cdot}47\angle 180^\circ + 199{\cdot}77\angle 90^\circ + 230{\cdot}94\angle -60^\circ]$$
$$= \frac{1}{3}[(-115{\cdot}47 + j\,0) + (0 + j\,199{\cdot}77) + (115{\cdot}47 - j\,199{\cdot}99)]$$
$$= \frac{1}{3}[0] = \mathbf{0A}$$

$$\vec{I_{R1}} = \frac{1}{3}\left[\vec{I_R} + a\,\vec{I_Y} + a^2\,\vec{I_B}\right]$$
$$= \frac{1}{3}[115{\cdot}47\angle 180^\circ + 1\angle 120^\circ \times 199{\cdot}99\angle 90^\circ + 1\angle 240^\circ \times 230{\cdot}94\angle -60^\circ]$$
$$= \frac{1}{3}[115{\cdot}47\angle 180^\circ + 199{\cdot}99\angle 210^\circ + 230{\cdot}94\angle 180^\circ]$$
$$= \frac{1}{3}[(-115{\cdot}47 + j0) + (-173 - j\,99{\cdot}99) + (-230{\cdot}94 + j0)]$$
$$= \frac{1}{3}[-519{\cdot}4 - j\,99{\cdot}99] = -173{\cdot}13 - j\,33{\cdot}3 = \mathbf{176{\cdot}3\angle -169^\circ A}$$

$$\vec{I_{R2}} = \frac{1}{3}\left[\vec{I_R} + a^2\,\vec{I_Y} + a\,\vec{I_B}\right]$$
$$= \frac{1}{3}[115{\cdot}47\angle 180^\circ + 1\angle 240^\circ \times 199{\cdot}99\angle 90^\circ + 1\angle 120^\circ \times 230{\cdot}94\angle -60^\circ]$$
$$= \frac{1}{3}[115{\cdot}47\angle 180^\circ + 199{\cdot}99\angle 330^\circ + 230{\cdot}94\angle 60^\circ]$$
$$= \frac{1}{3}[(-115{\cdot}47 + j0) + (173 - j\,99{\cdot}99) + (115{\cdot}47 + j\,199{\cdot}99)]$$
$$= \frac{1}{3}[173 + j\,100] = 57{\cdot}66 + j\,33{\cdot}3 = \mathbf{66{\cdot}58\angle 30^\circ\ A}$$

TUTORIAL PROBLEMS

1. In a 3-phase, 4-wire system, currents in *R*, *Y* and *B* lines under abnormal conditions of loading are:
   $I_R = 150\angle 45^\circ$ A ; $I_Y = 250\angle 150^\circ$ A ; $I_B = 100\angle 300^\circ$ A
   Calculate the zero, positive and negative phase sequence currents in the *R*-line and return current in the neutral connection.
   [$I_{R0} = 52{\cdot}2\angle 112{\cdot}7^\circ$ A ; $I_{R1} = 48{\cdot}02\angle -87{\cdot}6^\circ$ A; $I_{R2} = 163{\cdot}21\angle 40{\cdot}45^\circ$ A; $I_N = 156{\cdot}6\angle 112{\cdot}7^\circ$ A]
2. In a 3-phase system, the phase voltages are as under :
   $E_R = 1\angle 0^\circ$ V ; $E_B = 1\angle -120^\circ$ V ; $E_Y = 0$ V
   Find the zero, positive and negative phase sequence components in the *R*-phase.
   [$E_{R0} = -0{\cdot}33\angle 120^\circ$V; $E_{R2} = -0{\cdot}33\angle 240^\circ$V; $E_{R1} = 0{\cdot}66\angle 0^\circ$ V]
3. The currents in a 3-phase unbalanced system are :
   $I_R = (80 + j\,0)$ A ; $I_Y = (-10 - j\,60)$ A ; $I_B = (70 + j\,60)$ A
   The phase sequence is *RYB*. Calculate the zero, positive and negative sequence components of the red line current and determine the current in the neutral wire.
   [$I_{R0} = 0$A; $I_{R1} = 76{\cdot}58\angle 13^\circ$A ; $I_{R2} = 18{\cdot}12\angle -72{\cdot}6^\circ$A; $I_N = 0$A]
4. A 3-phase, 4-wire system supplies loads which are unequally distributed in the three phases. An analysis of the circuit shows that positive and negative phase sequence components of the current in the red line are as under :

$I_{R1} = (7 \cdot 89 + j\,0 \cdot 732)$ A ; $I_{R2} = (2 \cdot 11 - j\,2 \cdot 732)$ A

The total observed current flowing back to supply in the neutral conductor is zero. Calculate the current in the three lines. **[$I_R = (10 - j\,2)$A; $I_Y = (-2 - j\,4)$ A; $I_B = (-8 + j\,6)$ A]**

18.6 Sequence Impedances

Each element of power system will offer impedance to different phase sequence components of current which may not be the same. For example, the impedance which any piece of equipment offers to positive sequence current will not necessarily be the same as offered to negative sequence current or zero sequence current. Therefore, in unsymmetrical fault calculations, each piece of equipment will have three values of impedance—one corresponding to each sequence current *viz.*

(*i*) Positive sequence impedance (Z_1)

(*ii*) Negative sequence impedance (Z_2)

(*iii*) Zero sequence impedance (Z_0)

The impedance offered by an equipment or circuit to positive sequence current is called positive sequence impedance and is represented by Z_1. Similarly, impedances offered by any circuit or equipment to negative and zero sequence currents are respectively called negative sequence impedance (Z_2) and zero sequence impedance (Z_0).

The following points may be noted :

(*a*) In a 3-phase balanced system, each piece of equipment or circuit offers only one impedance–the one offered to positive or normal sequence current. This is expected because of the absence of negative and zero sequence currents in the 3-phase balanced system.

(*b*) In a 3-phase unbalanced system, each piece of equipment or circuit will have three values of impedance *viz.* positive sequence impedance, negative sequence impedance and zero sequence impedance.

(*c*) The positive and negative sequence impedances of linear, symmetrical and static circuits (*e.g.* transmission lines, cables, transformers and static loads) are equal and are the same as those used in the analysis of balanced conditions. This is due to the fact that impedance of such circuits is independent of the phase order, provided the applied voltages are balanced. It may be noted that positive and negative sequence impedances of rotating machines (*e.g.* synchronous and induction motors) are normally different.

(*d*) The zero sequence impedance depends upon the path taken by the zero sequence current. As this path is generally different from the path taken by the positive and negative sequence currents, therefore, zero sequence impedance is usually different from positive or negative sequence impedance.

18.7 Sequence Impedances of Power System Elements

The concept of impedances of various elements of power system (*e.g.* generators, transformers, transmission lines *etc.*) to positive, negative and zero sequence currents is of considerable importance in determining the fault currents in a 3-phase unbalanced system. A complete consideration of this topic does not fall within the scope of this book, but a short preliminary explanation may be of interest here. The following three main pieces of equipment will be considered :

(*i*) Synchronous generators

(*ii*) Transformers

(*iii*) Transmission lines

(*i*) **Synchronous generators.** The positive, negative and zero sequence impedances of rotating machines are generally different. The positive sequence impedance of a synchronous generator is equal to the synchronous impedance of the machine. The negative sequence impedance is much less than the positive sequence impedance. The zero sequence impedance is a variable item and if its value is not given, it may be assumed to be equal to the positive sequence impedance. In short :

Negative sequence impedance < Positive sequence impedance

Zero sequence impedance = Variable item

= may be taken equal to +ve sequence impedance if its value is not given

It may be worthwhile to mention here that any impedance Z_e in the earth connection of a star-connected system has the effect to introduce an impedance of 3 Z_e per phase. It is because the three equal zero-sequence currents, being in phase, do not sum to zero at the star point, but they flow back along the neutral earth connection.

***(ii)* Transformers.** Since transformers have the same impedance with reversed phase rotation, their positive and negative sequence impedances are equal; this value being equal to the impedance of the transformer. However, the zero sequence impedance depends upon earth connection. If there is a through circuit for earth current, zero sequence impedance will be equal to positive sequence impedance otherwise it will be infinite. In short,

Positive sequence impedance = Negative sequence impedance

= Impedance of Transformer

Zero sequence impedance = Positive sequence impedance, if there is circuit for earth current

= Infinite, if there is no through circuit for earth current.

***(iii)* Transmission lines.** The positive sequence and negative sequence impedance of a line are the same; this value being equal to the normal impedance of the line. This is expected because the phase rotation of the currents does not make any difference in the constants of the line. However, the zero sequence impedance is usually much greater than the positive or negative sequence impedance. In short :

Positive sequence impedance = Negative sequence impedance

= Impedance of the line

Zero sequence impedance = Variable item

= may be taken as three times the +ve sequence impedance if its value is not given

18.8 Analysis of Unsymmetrical Faults

In the analysis of unsymmetrical faults, the following assumptions will be made :

(i) The generated e.m.f. system is of positive sequence only.

(ii) No current flows in the network other than due to fault *i.e.* load currents are neglected.

(iii) The impedance of the fault is zero.

(iv) Phase *R* shall be taken as the reference phase.

In each case of unsymmetrical fault, e.m.f.s' per phase are denoted by E_R, E_Y and E_B and the terminal p.d. per phase by V_R, V_Y and V_B.

18.9 Single Line-to-Ground Fault

Consider a 3-phase system with an earthed neutral. Let a single line-to-ground fault occur on the red phase as shown in Fig. 18.13. It is clear from this figure that :

$$^*\vec{V_R} = 0 \text{ and } \vec{I_B} = \vec{I_Y} = 0$$

* Note that V_R is the terminal potential of phase *R i.e.* p.d. between *N* and *R*. Under line-to-ground fault, it will obviously be zero.

The sequence currents in the red phase in terms of line currents shall be :

$$\vec{I_0} = \frac{1}{3}\left(\vec{I_R} + \vec{I_Y} + \vec{I_B}\right) = \frac{1}{3}\vec{I_R}$$

$$\vec{I_1} = \frac{1}{3}\left(\vec{I_R} + a\,\vec{I_Y} + a^2\,\vec{I_B}\right) = \frac{1}{3}\vec{I_R}$$

$$\vec{I_2} = \frac{1}{3}\left(\vec{I_R} + a^2\,\vec{I_Y} + a\,\vec{I_B}\right) = \frac{1}{3}\vec{I_R}$$

$$\therefore \quad \vec{I_0} = \vec{I_1} = \vec{I_2} = \frac{1}{3}\vec{I_R}$$

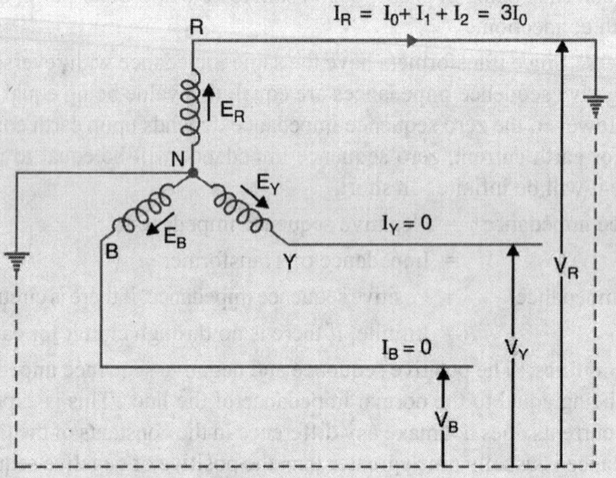

Fig. 18.13

Fault current. First of all expression for fault current $\vec{I_R}$ will be derived. Let $\vec{Z_1}$, $\vec{Z_2}$ and $\vec{Z_0}$ be the positive, negative and zero sequence impedances of the generator respectively. Consider the closed loop *NREN*. As the sequence currents produce voltage drops due only to their respective sequence impedances, therefore, we have,

$$\vec{E_R} = \vec{I_1}\vec{Z_1} + \vec{I_2}\vec{Z_2} + \vec{I_0}\vec{Z_0} + \vec{V_R}$$

As $\quad \vec{V_R} = 0$ and $\vec{I_1} = \vec{I_2} = \vec{I_0}$

$$\therefore \quad \vec{E_R} = \vec{I_0}\left(\vec{Z_1} + \vec{Z_2} + \vec{Z_0}\right)$$

or

$$\vec{I_0} = \frac{\vec{E_R}}{\vec{Z_1} + \vec{Z_2} + \vec{Z_0}}$$

$\therefore$ Fault current,

$$\vec{I_R} = 3\,\vec{I_0} = \frac{3\,\vec{E_R}}{\vec{Z_1} + \vec{Z_2} + \vec{Z_0}} \quad ...(i)$$

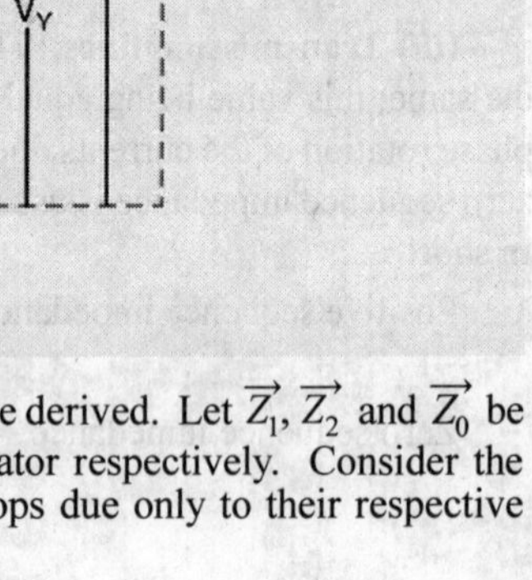

Fig. 18.14

Examination of exp. (*i*) shows that the equivalent circuit from which fault current may be calculated is as given in Fig. 18.14. It is clear that fault current is obtained by connecting the phase sequence impedances in series across an imaginary generator of voltage 3 E_R. This is a wonderful part of the method of symmetrical components and makes the analysis easy and interesting. In fact, this method permits to bring any unsymmetrical fault into a simple circuit of interconnection of sequence impedances appropriate to the fault condition prevailing.

The assumption made in arriving at exp. (*i*) is that the fault impedance is zero. However, if the fault impedance is Z_e, then expression for fault current becomes :

$$\vec{I_R} = \frac{3\,\vec{E_R}}{\vec{Z_1} + \vec{Z_2} + \vec{Z_0} + 3\,\vec{Z_e}}$$

It may be added here that if the neutral is not grounded, then zero sequence impedance will be infinite and the fault current is zero. This is expected because now no path exists for the flow of fault current.

Phase voltages at fault. Now let us calculate the phase voltages at fault (*i.e.* voltage between each line and fault). Since the generated e.m.f. system is of positive sequence only, the sequence components of e.m.f. in *R*-phase are :

$$\vec{E_0} = 0\ ;\ \vec{E_2} = 0 \text{ and } \vec{E_1} = \vec{E_R}$$

The sequence voltages at the fault for *R*-phase are :

$$\vec{V_1} = \vec{E_R} - \vec{I_1}\,\vec{Z_1} = \vec{E_R} - \frac{\vec{E_R}\,\vec{Z_1}}{\vec{Z_1} + \vec{Z_2} + \vec{Z_0}}$$

$$\therefore \quad \vec{V_1} = \frac{\vec{Z_2} + \vec{Z_0}}{\vec{Z_1} + \vec{Z_2} + \vec{Z_0}}\,\vec{E_R}$$

$$\vec{V_2} = 0 - \vec{Z_2}\,\vec{I_2} = \frac{-\vec{Z_2}}{\vec{Z_1} + \vec{Z_2} + \vec{Z_0}}\,\vec{E_R}$$

$$\vec{V_0} = 0 - \vec{I_0}\,\vec{Z_0} = \frac{-\vec{Z_0}}{\vec{Z_1} + \vec{Z_2} + \vec{Z_0}}\,\vec{E_R}$$

It can be readily seen that $\vec{V_1} + \vec{V_2} + \vec{V_0} = 0$. This is expected because *R*-phase is shorted to ground.

$\therefore$ The phase voltages at fault are :

$$\vec{V_R} = \vec{V_0} + \vec{V_1} + \vec{V_2} = 0$$

$$\vec{V_Y} = \vec{V_0} + a^2\,\vec{V_1} + a\,\vec{V_2}$$

$$\vec{V_B} = \vec{V_0} + a\,\vec{V_1} + a^2\,\vec{V_2}$$

Summary of Results. For line (*R*-phase)-to-ground fault :

(*i*) $\vec{I_R}$ = Fault current = $\dfrac{3\,\vec{E_R}}{\vec{Z_1} + \vec{Z_2} + \vec{Z_0}}$; $\vec{I_Y} = 0$; $\vec{I_B} = 0$

(*ii*)

$$\vec{V_R} = 0$$

$$\vec{V_Y} = \vec{V_0} + a^2\,\vec{V_1} + a\,\vec{V_2}$$

$$\vec{V_B} = \vec{V_0} + a\,\vec{V_1} + a^2\,\vec{V_2}$$

18.10 Line-to-Line Fault

Consider a line-to-line fault between the blue (*B*) and yellow (*Y*) lines as shown in Fig. 18.15. The conditions created by this fault lead to :

$$\vec{V_Y} = \vec{V_B}\ ;\ \vec{I_R} = 0 \text{ and } \vec{I_Y} + \vec{I_B} = 0$$

Again taking *R*-phase as the reference, we have,

$$\vec{I_0} = \frac{1}{3}\left(\vec{I_R} + \vec{I_Y} + \vec{I_B}\right) = 0$$

Now $\vec{V_Y} = \vec{V_B}$

Expressing in terms of sequence components of red line, we have,

$$\vec{V_0} + a^2\,\vec{V_1} + a\,\vec{V_2} = \vec{V_0} + a\,\vec{V_1} + a^2\,\vec{V_2}$$

or $\vec{V_1}\,(a^2 - a) = \vec{V_2}\,(a^2 - a)$

$\therefore$ $\vec{V_1} = \vec{V_2}$ ...(*i*)

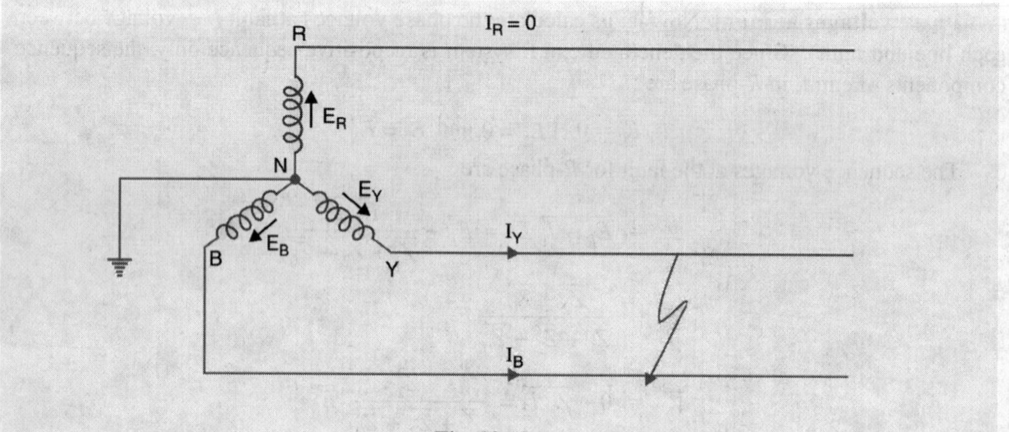

Fig. 18.15

Also $\vec{I_Y} + \vec{I_B} = 0$

or $\left(\vec{I_0} + a^2\,\vec{I_1} + a\,\vec{I_2}\right) + \left(\vec{I_0} + a\,\vec{I_1} + a^2\,\vec{I_2}\right) = 0$

or $(a^2 + a)\left(\vec{I_1} + \vec{I_2}\right) + 2\vec{I_0} = 0$

or $\vec{I_1} + \vec{I_2} = 0$ $[\because I_0 = 0]$ ...(*ii*)

Fault current. Examination of exp. (*i*) and exp (*ii*) reveals that sequence impedances should be connected as shown in Fig. 18.16. It is clear from the figure that :

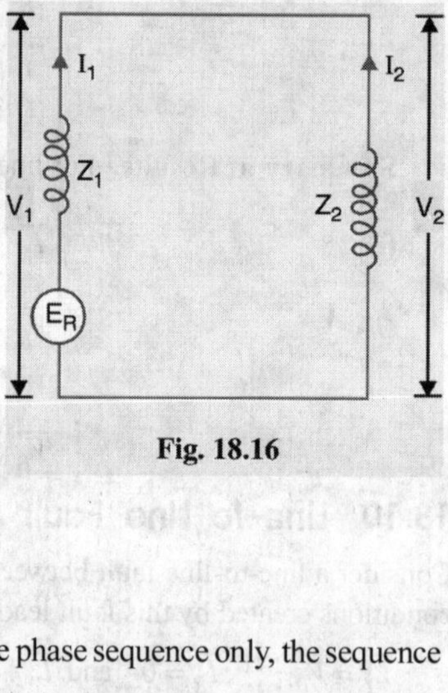

Fig. 18.16

$$\vec{I_1} = -\vec{I_2} = \frac{\vec{E_R}}{\vec{Z_1} + \vec{Z_2}}$$

Fault current, $\vec{I_Y} = \vec{I_0} + a^2\,\vec{I_1} + a\,\vec{I_2}$

$$= 0 + a^2\left(\frac{\vec{E_R}}{\vec{Z_1} + \vec{Z_2}}\right) + a\left(\frac{-\vec{E_R}}{\vec{Z_1} + \vec{Z_2}}\right)$$

$$= (a^2 - a)\frac{\vec{E_R}}{\vec{Z_1} + \vec{Z_2}}$$

$$= \frac{-j\sqrt{3}\,\vec{E_R}}{\vec{Z_1} + \vec{Z_2}} = -\vec{I_B}$$

Phase voltages. Since the generated e.m.f. system is of positive phase sequence only, the sequence components of e.m.f. in *R*-phase are :

$\vec{E_0} = 0$; $\vec{E_2} = 0$ and $\vec{E_1} = \vec{E_R}$

The sequence voltages at the fault for R-phase are :

$$\vec{V_1} = \vec{E_R} - \vec{I_1}\,\vec{Z_1} = \vec{E_R} - \left(\frac{\vec{E_R}}{\vec{Z_1}+\vec{Z_2}}\right)\vec{Z_1}$$

$$\therefore \quad \vec{V_1} = \frac{\vec{Z_2}}{\vec{Z_1}+\vec{Z_2}}\,\vec{E_R}$$

$$\vec{V_2} = 0 - \vec{I_2}\,\vec{Z_2} = \frac{\vec{Z_2}}{\vec{Z_1}+\vec{Z_2}}\,\vec{E_R}$$

$$\vec{V_0} = 0 - \vec{I_0}\,\vec{Z_0} = 0$$

The phase voltages at fault are :

$$\vec{V_R} = \vec{V_0} + \vec{V_1} + \vec{V_2}$$

$$= 0 + \frac{\vec{Z_2}}{\vec{Z_1}+\vec{Z_2}}\,\vec{E_R} + \frac{\vec{Z_2}}{\vec{Z_1}+\vec{Z_2}}\,\vec{E_R}$$

$$\therefore \quad \vec{V_R} = \frac{2\,\vec{Z_2}}{\vec{Z_1}+\vec{Z_2}}\,\vec{E_R}$$

$$\vec{V_Y} = \vec{V_0} + a^2\,\vec{V_1} + a\,\vec{V_2}$$

$$= 0 + a^2\left(\frac{\vec{Z_2}}{\vec{Z_1}+\vec{Z_2}}\,\vec{E_R}\right) + a\left(\frac{\vec{Z_2}}{\vec{Z_1}+\vec{Z_2}}\,\vec{E_R}\right)$$

$$= (a^2 + a)\left(\frac{\vec{Z_2}}{\vec{Z_1}+\vec{Z_2}}\,\vec{E_R}\right)$$

$$\therefore \quad \vec{V_Y} = -\frac{\vec{Z_2}}{\vec{Z_1}+\vec{Z_2}}\,\vec{E_R} \qquad (\because a^2 + a = -1)$$

$$\vec{V_B} = \vec{V_0} + a\,\vec{V_1} + a^2\,\vec{V_2}$$

$$= 0 + a\left(\frac{\vec{Z_2}}{\vec{Z_1}+\vec{Z_2}}\,\vec{E_R}\right) + a^2\left(\frac{\vec{Z_2}}{\vec{Z_1}+\vec{Z_2}}\,\vec{E_R}\right)$$

$$= (a^2 + a)\left(\frac{\vec{Z_2}}{\vec{Z_1}+\vec{Z_2}}\,\vec{E_R}\right)$$

$$\therefore \quad \vec{V_B} = -\frac{\vec{Z_2}}{\vec{Z_1}+\vec{Z_2}}\,\vec{E_R}$$

Summary of Results. For line-to-line fault (Blue and Yellow lines) :

(*i*) $\vec{I_R} = 0$; $\vec{I_Y} = -\vec{I_B} = \dfrac{-j\sqrt{3}\,\vec{E_R}}{\vec{Z_1}+\vec{Z_2}}$

(*ii*) $\vec{V_Y} = \vec{V_B} = -\dfrac{\vec{Z_2}}{\vec{Z_1}+\vec{Z_2}}\,\vec{E_R}$ and $\vec{V_R} = \dfrac{2\,\vec{Z_2}}{\vec{Z_1}+\vec{Z_2}}\,\vec{E_R}$

18.11 Double Line-to-Ground Fault

Consider the double line-to-ground fault involving Y–B lines and earth as shown in Fig. 18.17. The conditions created by this fault lead to :

$$\vec{I_R} = 0 \quad ; \quad \vec{V_Y} = \vec{V_B} = 0$$

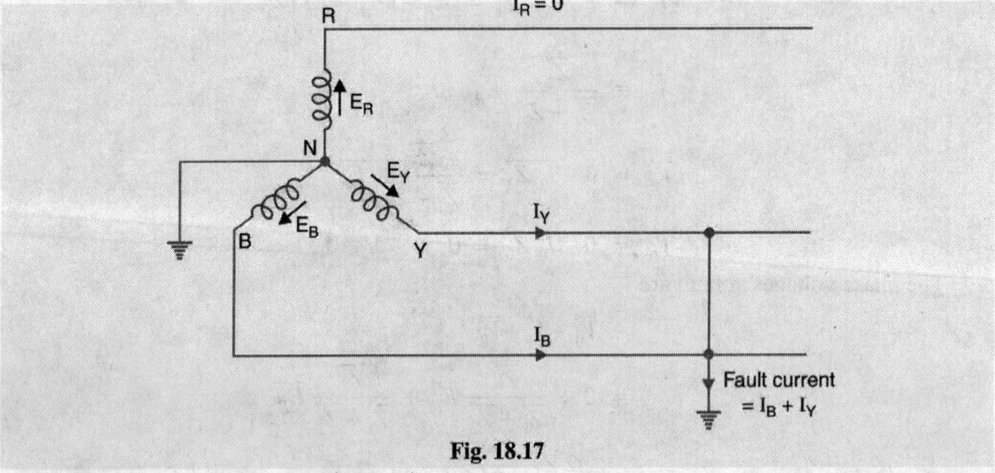

Fig. 18.17

Since $\vec{V_Y} = \vec{V_B} = 0$, it is implied that :

$$\vec{V_1} = \vec{V_2} = \vec{V_0} = \frac{1}{3}\vec{V_R} \qquad ...(i)$$

Also $\vec{I_R} = \vec{I_1} + \vec{I_2} + \vec{I_0} = 0 \quad (given) \qquad ...(ii)$

Fault current. Examination of exp. (*i*) and exp. (*ii*) reveals that sequence impedances should be *connected as shown in Fig. 18.18. It is clear that :

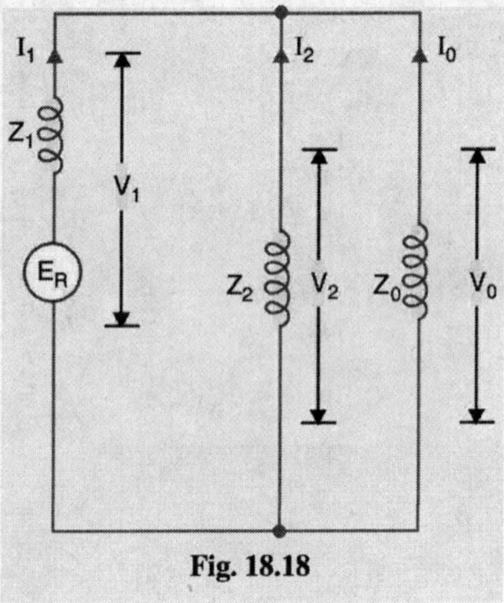

Fig. 18.18

$$\vec{I_1} = \frac{\vec{E_R}}{\vec{Z_1} + \dfrac{\vec{Z_2}\,\vec{Z_0}}{\vec{Z_2} + \vec{Z_0}}}$$

$$\vec{I_2} = -\vec{I_1}\,\frac{\vec{Z_0}}{\vec{Z_2} + \vec{Z_0}}$$

$$\vec{I_0} = -\vec{I_1}\,\frac{\vec{Z_2}}{\vec{Z_2} + \vec{Z_0}}$$

Fault current, $\vec{I_F} = \vec{I_Y} + \vec{I_B} = 3\,\vec{I_0}\ ^{**} = 3\left(-\vec{I_1}\,\dfrac{\vec{Z_2}}{\vec{Z_2} + \vec{Z_0}}\right)$

$$= -\frac{3\,\vec{Z_2}}{\vec{Z_2} + \vec{Z_0}} \times \frac{\vec{E_R}}{\vec{Z_1} + \dfrac{\vec{Z_2}\vec{Z_0}}{\vec{Z_2} + \vec{Z_0}}}$$

$$= -\frac{3\,\vec{Z_2}\,\vec{E_R}}{\vec{Z_0}\,\vec{Z_1} + \vec{Z_0}\,\vec{Z_2} + \vec{Z_1}\,\vec{Z_2}}$$

---

* Since $\vec{V_1} = \vec{V_2} = \vec{V_0} = \frac{1}{3}\vec{V_R}$, sequence impedances must be in parallel.

** $\vec{I_0} = \frac{1}{3}\left(\vec{I_R} + \vec{I_Y} + \vec{I_B}\right) = \frac{1}{3}$ (0 + Fault Current)

∴ Fault current $= 3\,\vec{I_0}$

Phase Voltages. The sequence voltages for phase R are :

$$\vec{V_1} = \vec{E_R} - \vec{I_1}\,\vec{Z_1}\,; \quad \vec{V_2} = 0 - \vec{I_2}\,\vec{Z_2}\,; \quad \vec{V_0} = 0 - \vec{I_0}\,\vec{Z_0}$$

Now $$\vec{V_1} = \vec{V_2} = \vec{V_0} = \frac{1}{3}\vec{V_R}$$

$\therefore$ $$\vec{V_R} = \vec{V_1} + \vec{V_2} + \vec{V_0} = 3\,\vec{V_2}$$

$$\vec{V_Y} = a^2\,\vec{V_1} + a\,\vec{V_2} + \vec{V_0} = (a^2 + a + 1)\,\vec{V_2} \qquad (\because \vec{V_1} = \vec{V_2} = \vec{V_0})$$

$$= 0 \times \vec{V_2} = 0 \qquad (\because a^2 + a + 1 = 0)$$

$$\vec{V_B} = a\,\vec{V_1} + a^2\,\vec{V_2} + \vec{V_0} = (a + a^2 + 1)\,\vec{V_2} = 0$$

Example 18.14. *A 3-phase, 10 MVA, 11 kV generator with a solidly earthed neutral point supplies a feeder. The relevant impedances of the generator and feeder in ohms are as under :*

	Generator	*feeder*
Positive sequence impedance	*j* 1·2	*j* 1·0
Negative sequence impedance	*j* 0·9	*j* 1·0
Zero sequence impedance	*j* 0·4	*j* 3·0

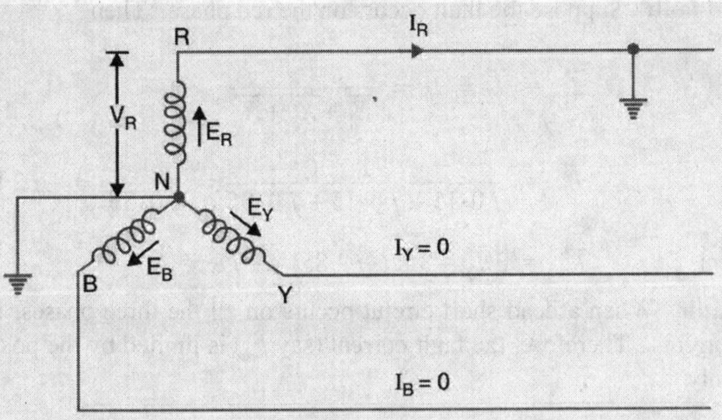

Fig. 18.19

If a fault from one phase to earth occurs on the far end of the feeder, calculate

(i) *the magnitude of fault current*

(ii) *line to neutral voltage at the generator terminal*

Solution. The circuit diagram is shown in Fig. 18.19. The fault is assumed to occur on the red phase. Taking red phase as the reference,

Phase e.m.f. of R-phase, $\vec{E_R} = 11 \times 10^3/\sqrt{3} = 6350$ V

(*i*) The total impedance to any sequence current is the sum of generator and feeder impedances to that sequence current.

$\therefore$ $$\text{Total } \vec{Z_1} = j\,1{\cdot}2 + j\,1{\cdot}0 = j\,2{\cdot}2\ \Omega$$

$$\text{Total } \vec{Z_2} = j\,0{\cdot}9 + j\,1{\cdot}0 = j\,1{\cdot}9\ \Omega$$

$$\text{Total } \vec{Z_0} = j\,0{\cdot}4 + j\,3{\cdot}0 = j\,3{\cdot}4\ \Omega$$

For a line-to-ground fault, we have,

$$\vec{I_1} = \vec{I_2} = \vec{I_0} = \frac{\vec{E_R}}{\vec{Z_1} + \vec{Z_2} + \vec{Z_0}} = \frac{6350}{j\,2{\cdot}2 + j\,1{\cdot}9 + j\,3{\cdot}4}$$

$$= \frac{6350}{j\,7 \cdot 5} = -j\,846 \text{ A}$$

∴ Fault current, $\vec{I_R} = 3\,\vec{I_0} = 3 \times (-j\,846) = \mathbf{-j\,2538\ A}$

(*ii*) Line-to-neutral voltage of *R*-phase,

$$\vec{V_R} = \vec{E_R} - \vec{I_1}\,\vec{Z_1} - \vec{I_2}\,\vec{Z_2} - \vec{I_0}\,\vec{Z_0}$$

where $\vec{Z_0}$, $\vec{Z_1}$ and $\vec{Z_2}$ are the sequence impedances of generator.

$$= \vec{E_R} - \vec{I_0}\left(\vec{Z_1} + \vec{Z_2} + \vec{Z_0}\right)$$

$$= 6350 - (-j\,846)\,(j\,1 \cdot 2 + j\,0 \cdot 9 + j\,0 \cdot 4)$$

$$= 6350 + j\,846\,(j\,2 \cdot 5)$$

$$= 6350 - 2115 = \mathbf{4235\ V}$$

Example 18.15. *A 3-phase, 11 kV, 10 MVA alternator has sequence reactances of $X_0 = 0 \cdot 05$ p.u., $X_1 = 0 \cdot 15$ p.u. and $X_2 = 0 \cdot 15$ p.u. If the generator is on no load, find the ratio of fault currents for L-G fault to that when all the 3-phases are dead short-circuited.*

Solution. Taking red phase as the reference, let its phase e.m.f. be $\vec{E_R} = 1$ p.u.

Line-to-ground fault. Suppose the fault occurs on the red phase. Then,

$$\vec{I_1} = \vec{I_2} = \vec{I_0} = \frac{\vec{E_R}}{\vec{X_1} + \vec{X_2} + \vec{X_0}}$$

$$\therefore \quad \vec{I_0} = \frac{1}{j\,0 \cdot 15 + j\,0 \cdot 15 + j\,0 \cdot 05} = \frac{1}{j\,0 \cdot 35} = -j\,2 \cdot 85$$

∴ Fault current, $\vec{I_R} = 3\,\vec{I_0} = 3 \times (-j\,2 \cdot 85) = -j\,8 \cdot 55$ A

Three phase fault. When a dead short circuit occurs on all the three phases, it gives rise to symmetrical fault currents. Therefore, the fault current (say I_{sh}) is limited by the positive sequence reactance (*i.e.* X_1) only.

∴ Fault current, $\vec{I_{sh}} = \frac{\vec{E_R}}{\vec{X_1}} = \frac{1}{j\,0 \cdot 15} = -j\,6 \cdot 66$

Ratio of two fault currents $= \frac{\vec{I_R}}{\vec{I_{sh}}} = \frac{-j\,8 \cdot 55}{-j\,6 \cdot 66} = \mathbf{1 \cdot 284}$

i.e. single line-to-ground fault current is 1·284 times that due to dead short circuit on the 3-phases.

Example 18.16. *A 3-phase, 11 kV, 25 MVA generator with $X_0 = 0 \cdot 05$ p.u., $X_1 = 0 \cdot 2$ p.u. and $X_2 = 0 \cdot 2$ p.u. is grounded through a reactance of 0·3 Ω. Calculate the fault current for a single line to ground fault.*

Solution. Fig. 18.20 shows the circuit diagram. The fault is assumed to occur on the red phase. Taking red phase as the reference, let its phase e.m.f. be $\vec{E_R} = 1$ p.u.

First of all, convert the reactance X_n into p.u. value from the following relation :

$$\text{*p.u. value of } X_n = X_n \text{ in ohms} \times \frac{\text{kVA rating}}{(\text{kV})^2 \times 1000}$$

---

* $\%X_n = X_n$ in ohms $\times \frac{\text{kVA rating}}{(\text{kV})^2 \times 10}$. If this value is divided by 100, we get p.u. value.

$$= 0{\cdot}3 \times \frac{25{,}000}{(11)^2 \times 1000} = 0{\cdot}062 \text{ p.u.}$$

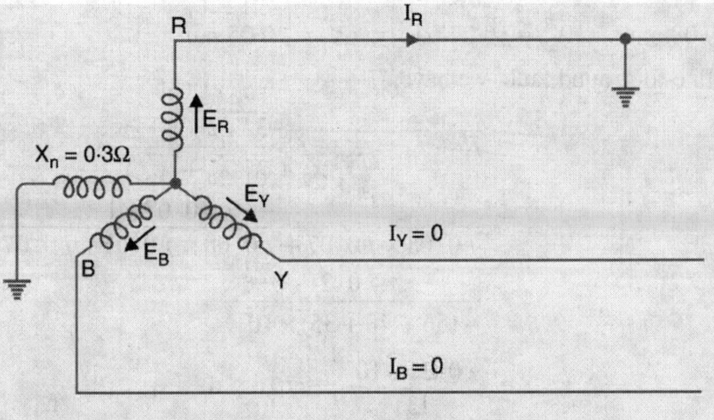

Fig. 18.20

For a line-to-ground fault, we have,

$$\vec{I_1} = \vec{I_2} = \vec{I_0} = \frac{\vec{E_R}}{\vec{X_1} + \vec{X_2} + (\vec{X_0} + 3\,\vec{X_n})}$$

$$= \frac{1}{j\,0{\cdot}2 + j\,0{\cdot}2 + j(0{\cdot}05 + 3 \times 0{\cdot}062)}$$

$$= \frac{1}{j\,0{\cdot}636} = -j\,1{\cdot}572 \text{ p.u.}$$

Fault current, $\vec{I_R} = 3\,\vec{I_0} = 3 \times (-j\,1{\cdot}572) = -j\,4{\cdot}716$ p.u.

∴ Fault current in amperes = Rated current × p.u. value

$$= \frac{25 \times 10^6}{\sqrt{3} \times 11 \times 10^3} \times 4{\cdot}716 = \mathbf{6188\ A}$$

Example 18.17. *A 3-phase, 3-wire system has a normal voltage of 10·4 kV between the lines. It is supplied by a generator having positive, negative and zero sequence reactances of 0·6, 0·5 and 0·2 Ω per phase respectively. Calculate the fault current which flows when a line-to-line fault occurs at the generator terminals.*

Solution. Suppose the short circuit fault occurs between yellow and blue phases. Taking red phase as the reference, its phase e.m.f. is :

Phase e.m.f. of R-phase, $\vec{E_R} = 10{\cdot}4 \times 10^3/\sqrt{3} = 6000$ V

Now $\vec{X_1} = j\,0{\cdot}6\ \Omega$; $\vec{X_2} = j\,0{\cdot}5\ \Omega$; $\vec{X_0} = j\,0{\cdot}2\ \Omega$

For line-to-line fault, we have,

Fault current, $I_F = \dfrac{\sqrt{3}\,E_R}{X_1 + X_2}$ (*in magnitude*)

$$= \frac{\sqrt{3} \times 6000}{(0{\cdot}6 + 0{\cdot}5)} = \mathbf{9447{\cdot}5\ A}$$

Example 18.18. *The per unit values of positive, negative and zero sequence reactances of a network at fault are 0·08, 0·07 and 0·05. Determine the fault current if the fault is double line-to-ground.*

Solution. Suppose the fault involves yellow and blue phases and the ground. Taking red phase as the reference, let its phase e.m.f. be $\vec{E_R} = 1$ p.u.

Now, $\vec{X_1} = j\,0{\cdot}08$ p.u. ; $\vec{X_2} = j\,0{\cdot}07$ p.u. ; $\vec{X_0} = j\,0{\cdot}05$ p.u.

For a double line-to-ground fault, we have,

Fault current,
$$\vec{I_F} = \vec{I_Y} + \vec{I_B} = \frac{-3\vec{X_2}\,\vec{E_R}}{\vec{X_1}\,\vec{X_2} + \vec{X_1}\,\vec{X_0} + \vec{X_2}\,\vec{X_0}}$$
$$= \frac{-3 \times j\,0\cdot 07 \times 1}{j\,0\cdot 08 \times j\,0\cdot 07 + j\,0\cdot 08 \times j\,0\cdot 05 + j\,0\cdot 07 \times j\,0\cdot 05}$$
$$= \frac{-j\,0\cdot 21}{-(56+40+35)\times 10^{-4}}$$
$$= \frac{j\,0\cdot 21 \times 10^4}{131} = \mathbf{j\,16\ p.u.}$$

Example 18.19. *A 20 MVA, 11 kV, 3-phase, 50 Hz generator has its neutral earthed through a 5% reactor. It is in parallel with another identical generator having isolated neutral. Each generator has a positive sequence reactance of 20%, negative sequence reactance of 10% and zero sequence reactance of 15%. If a line to ground short circuit occurs in the common bus-bar, determine the fault current.*

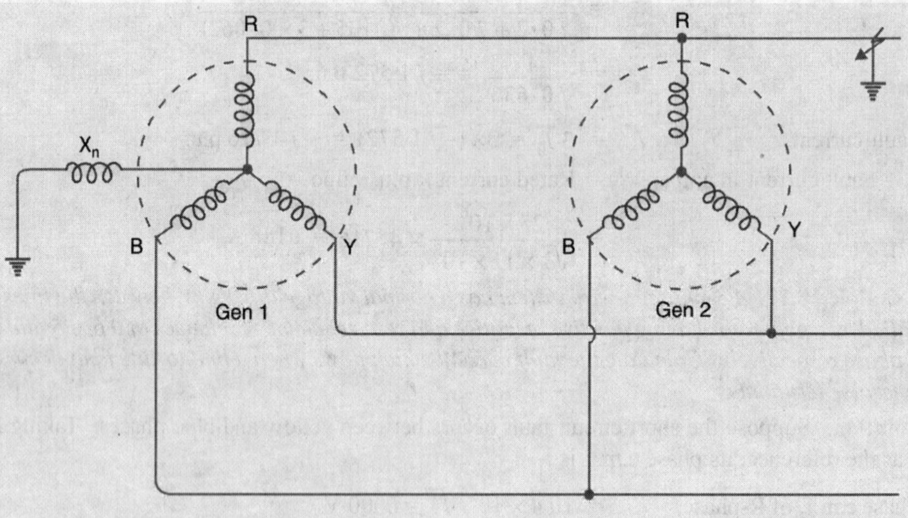

Fig. 18.21

Solution. Fig. 18.21 shows the two generators in parallel. The generator 1 has its neutral earthed through a reactance (= 5%) whereas generator 2 has ungrounded neutral. The earth fault is assumed to occur on the red phase. Taking red phase as the reference, its phase e.m.f. $E_R = 11 \times 10^3/\sqrt{3} = 6351$ V. For a line to ground fault, the *equivalent circuit will be as shown in Fig. 18.22 (*i*) which further reduces to the circuit shown in Fig. 18.22 (*ii*).

* Note the equivalent circuit diagram. The positive sequence reactances (20%) of two generators are in parallel and so are their negative sequence reactances (10%). The zero sequence reactance of generator 2 is zero because its neutral is ungrounded. However, the zero sequence reactance of generator 1 = 15% + 3 × 5% = 30%.

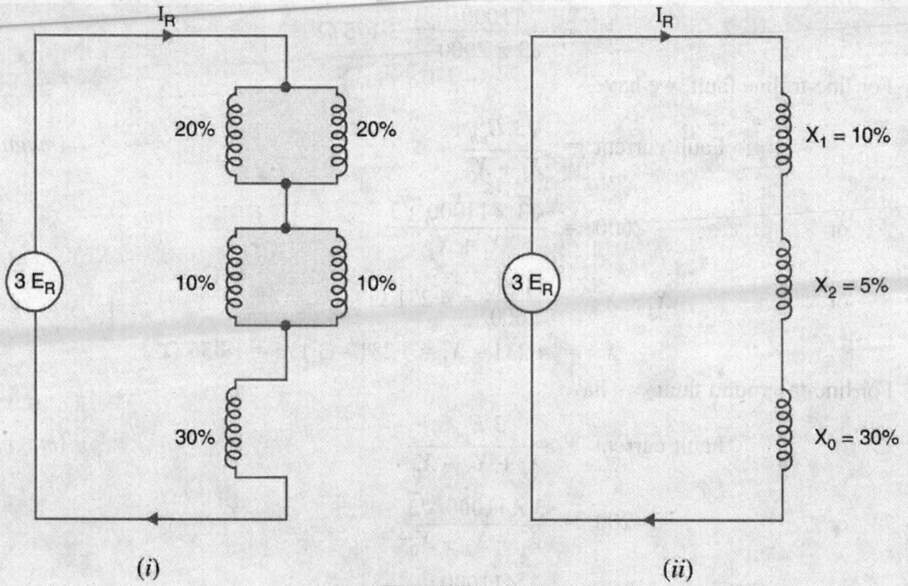

Fig. 18.22

The percentage reactances in Fig. 18.22 (*ii*) can be converted into ohmic values as under :

$$X_1 = \% \text{ reactance} \times \frac{(\text{Voltage in kV})^2 \times 10}{\text{Base kVA}}$$

$$= 10 \times \frac{(11)^2 \times 10}{10^3 \times 20} = 0{\cdot}605\ \Omega$$

$$X_2 = 5 \times \frac{(11)^2 \times 10}{10^3 \times 20} = 0{\cdot}3025\ \Omega$$

$$X_0 = 30 \times \frac{(11)^2 \times 10}{10^3 \times 20} = 1{\cdot}815\ \Omega$$

$$\therefore \quad \text{Fault current, } \vec{I_R} = \frac{3\,\vec{E_R}}{\vec{X_1} + \vec{X_2} + \vec{X_0}} = \frac{3 \times 6351}{j\,0{\cdot}605 + j\,0{\cdot}3025 + j\,1{\cdot}815}$$

$$= \frac{19053}{j\,2{\cdot}7225} = -j\,\mathbf{6998\ A}$$

Example 18.20. *A 50 MVA, 11 kV three-phase alternator was subjected to different types of faults. The fault currents are as under :*

3-phase fault = 2000 A ; Line-to-Line fault = 2600 A ; Line-to-ground fault = 4200 A

The generator neutral is solidly grounded. Find the values of the three sequence reactances of the alternator. Ignore resistances.

Solution. Let X_1, X_2 and X_0 be the positive, negative and zero sequence reactances respectively of the alternaor.

For 3-phase fault, Fault current $= \dfrac{E_{ph}}{X_1}$ (*magnitude*)

or $$2000 = \frac{11000/\sqrt{3}}{X_1}$$

$$\therefore \quad X_1 = \frac{11000}{\sqrt{3} \times 2000} = \mathbf{3{\cdot}175\ \Omega}$$

For line-to-line fault, we have,

$$\text{Fault current} = \frac{\sqrt{3}\, E_{ph}}{X_1 + X_2} \quad (\textit{magnitude})$$

$$\text{or} \quad 2600 = \frac{\sqrt{3} \times 11000/\sqrt{3}}{X_1 + X_2}$$

$$\text{or} \quad X_1 + X_2 = \frac{11000}{2600} = 4{\cdot}231\ \Omega$$

$$\therefore \quad X_2 = 4{\cdot}231 - X_1 = 4{\cdot}231 - 3{\cdot}175 = \mathbf{1{\cdot}056\ \Omega}$$

For line-to-ground fault, we have,

$$\text{Fault current} = \frac{3\, E_{ph}}{X_1 + X_2 + X_0} \quad (\textit{magnitude})$$

$$\text{or} \quad 4200 = \frac{3 \times 11000/\sqrt{3}}{X_1 + X_2 + X_0}$$

$$\text{or} \quad X_1 + X_2 + X_0 = \frac{3 \times 11000}{\sqrt{3} \times 4200} = 4{\cdot}536\ \Omega$$

$$\therefore \quad X_0 = 4{\cdot}536 - X_1 - X_2 = 4{\cdot}536 - 3{\cdot}175 - 1{\cdot}056 = \mathbf{0{\cdot}305\ \Omega}$$

TUTORIAL PROBLEMS

1. A 3-phase, 75 MVA, 0·8 p.f. (lagging), 11·8 kV star-connected alternator having its star point solidly earthed supplies a feeder. The relevant per-unit (p.u.) impedances, based on the rated phase voltage and phase current of the alternator are as follows :

	Generator	Feeder
Positive sequence impedance (p.u.)	$j\,1{\cdot}7$	$j\,0{\cdot}1$
Negative sequence impedance (p.u.)	$j\,0{\cdot}18$	$j\,0{\cdot}1$
Zero sequence impedance (p.u.)	$j\,0{\cdot}12$	$j\,0{\cdot}3$

Determine the fault current for a one line-to-earth fault occuring at the far end of the feeder. The generated e.m.f. per phase is of positive sequence only and is equal to the rated voltage per phase. **[4400 A]**

2. A 3-phase, 75 MVA, 11·8 kV star-connected alternator with a solidly earthed neutral point has the following p.u. impedances based on rated phase voltage and rated phase current :

Positive phase sequence impedance	=	$j\,2$ p.u.
Negative phase sequence impedance	=	$j\,0{\cdot}16$ p.u.
Zero phase sequence impedance	=	$j\,0{\cdot}08$ p.u.

Determine the steady-state fault current for the following : (*i*) 3-phase symmetrical short-circuit (*ii*) one line-to-earth fault (*iii*) two line-to-earth fault. The generated e.m.f. per phase is equal to the rated voltage. **[(*i*)1840 A (*ii*) 4920 A (*iii*) 3580 A]**

3. The per unit values of positive, negative and zero sequence reactances of a network at fault are 0·08, 0·07 and 0·05 respectively. Determine the fault current if fault is line-to-line-to-ground. **[*j* 16 p.u.]**

18.12 Sequence Networks

The analysis of an unsymmetrical fault by symmetrical components method can be conveniently done by drawing sequence networks. A sequence network of a particular sequence current in a given power system is the path for the flow of that sequence current in the system. It is composed of impedances offered to that sequence current in the system. Since there are three sequence currents (*viz.* positive sequence current, negative sequence current and zero sequence current), there will be three sequence networks for a given power system, namely ;

1. Positive sequence network
2. Negative sequence network
3. Zero sequence network

1. **Positive sequence network.** The positive sequence network for a given power system shows all the paths for the flow of positive sequence currents in the system. It is represented by one-line diagram and is composed of impedances offered to the positive sequence currents. While drawing the positive sequence network of a given power system, the following points may be kept in view:

(*i*) Each generator in the system is represented by the generated voltage in series with appropriate reactance and resistance.

(*ii*) Current limiting impedances between the generator's neutral and ground pass no positive sequence current and hence are not included in the positive sequence network.

(*iii*) All resistances and magnetising currents for each transformer are neglected as a matter of simplicity.

(*iv*) For transmission lines, the shunt capacitances and resistances are generally neglected.

(*v*) Motor loads are included in the network as generated e.m.f. in series with appropriate reactance.

2. **Negative sequence network.** The negative sequence network for a given power system shows all the paths for the flow of negative sequence currents in the system. It is also represented by one-line diagram and is composed of impedances offered to the negative sequence currents. The negative sequence network can be readily obtained from positive sequence network with the following modifications :

(*i*) Omit the e.m.fs. of 3-phase generators and motors in the positive sequence network. It is because these devices have only positive sequence-generated voltages.

(*ii*) Change, if necessary, the impedances that represent rotating machinery in the positive sequence network. It is because negative sequence impedance of rotating machinery is generally different from that of positive sequence impedance.

(*iii*) Current limiting impedances between generator's neutral and ground pass no negative sequence current and hence are not included in the negative sequence network.

(*iv*) For static devices such as transmission lines and transformers, the negative sequence impedances have the same value as the corresponding positive sequence impedances.

3. **Zero sequence network.** The zero sequence network for a given power system shows all the paths for the flow of zero sequence currents. The zero sequence network of a system depends upon the nature of connections of the 3-phase windings of the components in the system. The following points may be noted about zero sequence network :

(*i*) The zero sequence currents will flow only if there is a return path *i.e.* path from neutral to ground or to another neutral point in the circuit.

(*ii*) In the case of a system with no return path for zero sequence currents, these currents cannot exist.

18.13 Reference Bus for Sequence Networks

While drawing the sequence networks, it is necessary to specify the reference potential *w.r.t.* which all sequence voltage drops are to be taken. For this purpose, the reader may keep in mind the following points :

(*i*) For positive or negative sequence networks, the neutral of the generator is taken as the reference bus. This is logical because positive or negative sequence components represent balanced sets and hence all the neutral points must be at the same potential for either positive or negative sequence currents.

(*ii*) For zero sequence network, the reference bus is the ground at the generator.

Example 18.21. *An unloaded generator is grounded through a reactance Z_n as shown in Fig. 18.23. If a single line-to-ground fault occurs, draw (i) the positive sequence network (ii) negative sequence network and (iii) zero sequence network.*

Solution. Fig. 18.23 shows the unloaded generator with single line-to-ground fault. We shall now draw the sequence networks for this system.

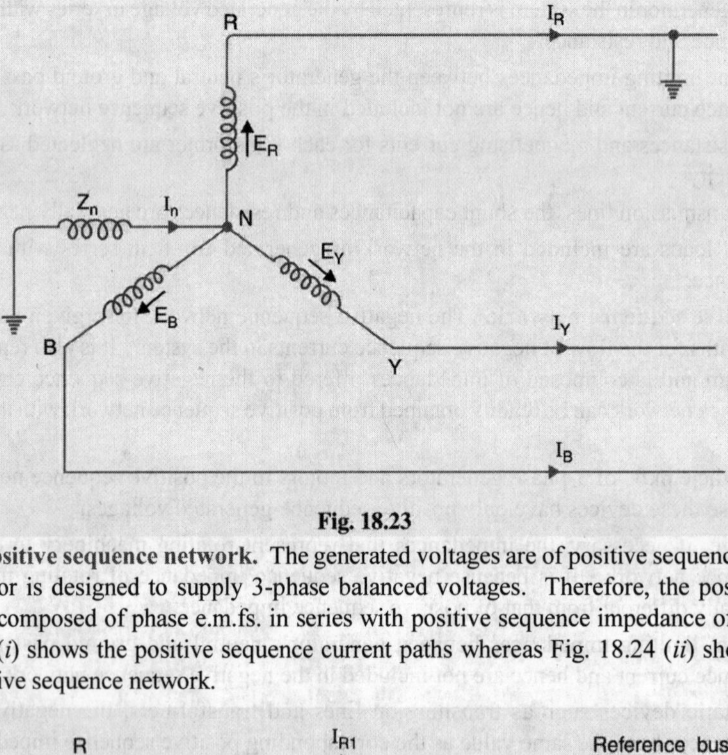

Fig. 18.23

(*i*) **Positive sequence network.** The generated voltages are of positive sequence only becasue the generator is designed to supply 3-phase balanced voltages. Therefore, the positive sequence network is composed of phase e.m.fs. in series with positive sequence impedance of the generator. Fig. 18.24 (*i*) shows the positive sequence current paths whereas Fig. 18.24 (*ii*) shows the single-phase positive sequence network.

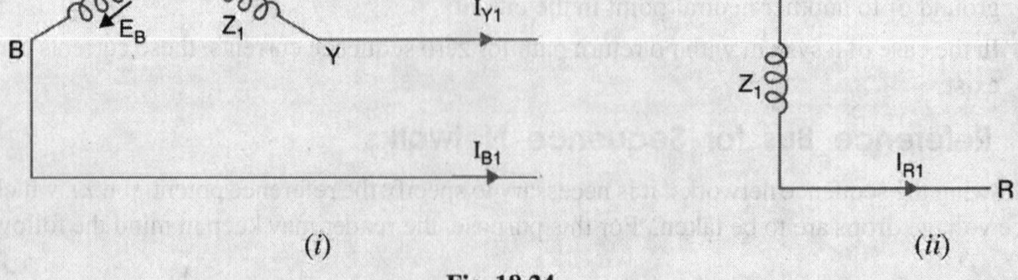

Fig. 18.24

(*ii*) **Negative sequence network.** A negative sequence network contains no e.m.f. but includes the impedances of the generator to negative sequence currents. Thus negative sequence network is readily obtained by omitting e.m.fs. in the positive sequence network. Fig. 18.25 (*i*) shows the negative sequence current paths whereas Fig. 18.25 (*ii*) shows the single-phase negative sequence network.

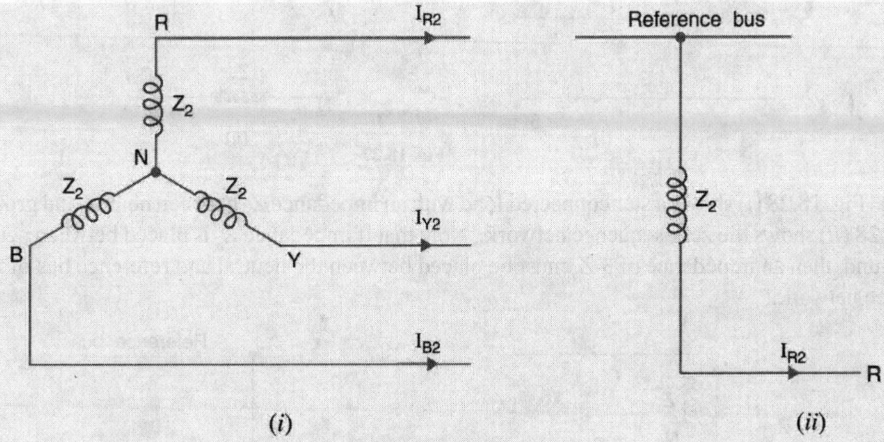

Fig. 18.25

(*iii*) **Zero sequence network.** The zero sequence currents flow through phases as well as through the reactance Z_n as shown in Fig. 18.26 (*i*). It is clear that current flowing in impedance Z_n is $3I_{R0}$. It is because $I_{R0} = I_{B0} = I_{Y0}$.

∴ Voltage drop of zero sequence current from R to ground

$$= -3I_{R0}Z_n - I_{R0}Z_0$$

$$= -I_0(3Z_n + Z_0)$$

Therefore, the per phase impedance to zero sequence current is $3Z_n + Z_0$. Fig. 18.26 (*ii*) shows the zero sequence network.

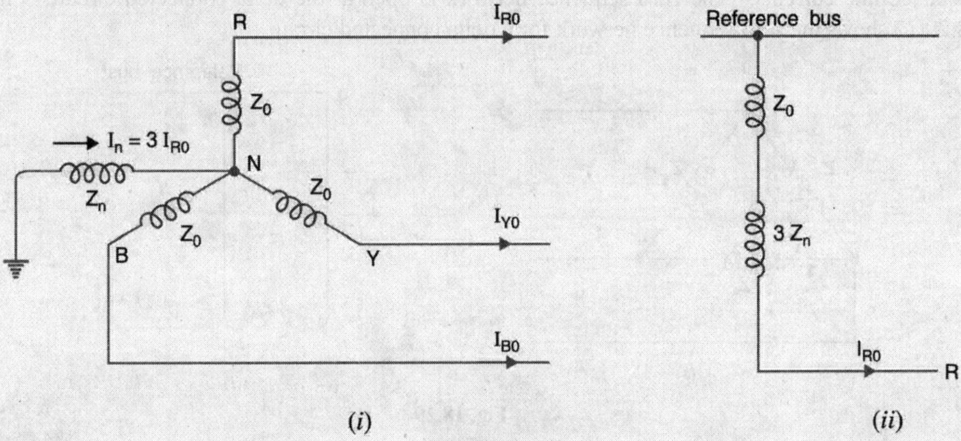

Fig. 18.26

Example 18.22. *Draw the zero sequence network for (i) star-connected load with no earth connection (ii) star-connected load with Z_n from neutral to ground (iii) delta-connected load.*

Solution. (*i*) Fig. 18.27 (*i*) shows the star connected load with no earth connection. In this case, neutral current is zero and no zero sequence current can exist. Fig. 18.27 (*ii*) shows the zero sequence network.

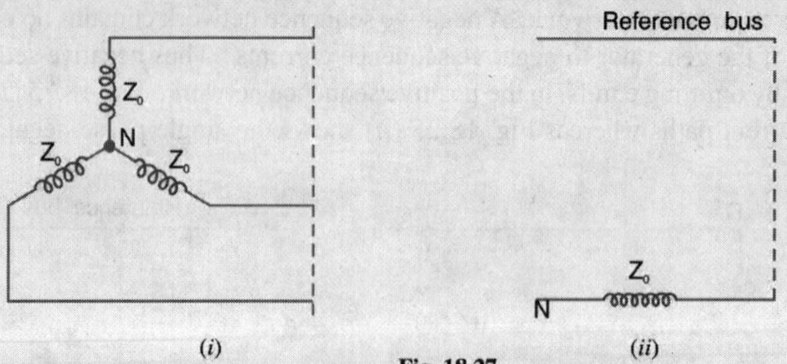

Fig. 18.27

(*ii*) Fig. 18.28 (*i*) shows a star connected load with an impedance Z_n between neutral and ground. Fig. 18.28 (*ii*) shows the zero sequence network. Note that if impedance Z_n is placed between neutral and ground, then an impedance of 3 Z_n must be placed between the neutral and reference bus of zero sequence network.

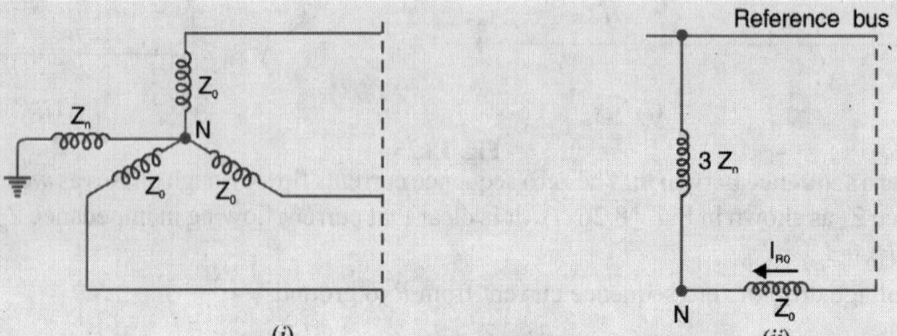

Fig. 18.28

(*iii*) Since a delta connected load provides no return path, zero sequence currents cannot exist in the phase windings. In other words, a delta connected circuit provides infinite impedance to zero sequence line currents. The zero sequence network is open at the delta connected circuit. Fig. 18.29(*ii*) shows the zero sequence network for a delta connected circuit.

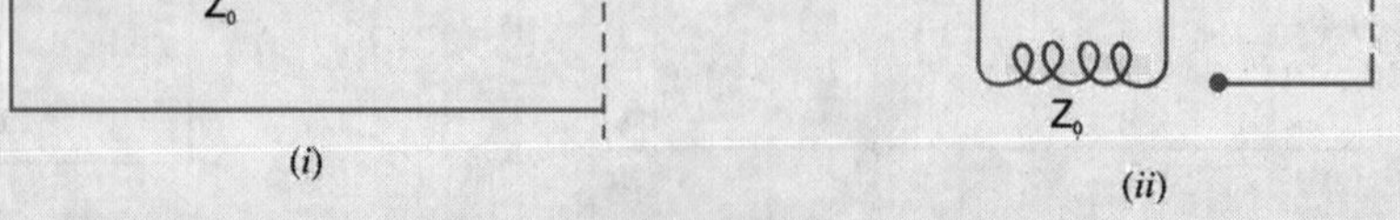

Fig. 18.29

SELF-TEST

1. Fill in the blanks by appropriate words/figures.

(*i*) The most common type of 3ϕ unsymmetrical fault is ....

(*ii*) In a balanced 3–ϕ system, negative and zero phase sequence currents are ....

(*iii*) In a 3-phase, 4-wire unbalanced system, the magnitude of zero sequence current is .... of the current in the neutral wire.

(*iv*) The positive sequence impedance of a transmission line is .... to the negative sequence impedance.

(*v*) The zero sequence impedance of different elements of power system is generally ....

2. Pick up the correct words/figures from the brackets and fill in the blanks.

(*i*) A symmetrical fault on a power system is .... severe than an unsymmetrical fault. (*more, less*)

(*ii*) The operator 'a' rotates the vector through .... in the anticlockwise direction. (*90°, 120°, 180°*)

(*iii*) $a - a^2 =$ .... $(j\sqrt{3}, -j\sqrt{3}, 1)$

(*iv*) On the occurrence of an unsymmetrical fault, the positive sequence component is always ...... than that of negative sequence component. (*more, less*)

(*v*) The zero sequence impedance of an element in a power system is generally .... the positive or negative sequence impedance. (*the same as, different from*)

ANSWERS TO SELF-TEST

1. (*i*) Single line-to-ground (*ii*) zero (*iii*) one-third (*iv*) equal (*v*) different
2. (*i*) more (*ii*) 120° (*iii*) $j\sqrt{3}$ (*iv*) more (*v*) different from

CHAPTER REVIEW TOPICS

1. What is a 3-φ unsymmetrical fault ? Discuss the different types of unsymmetrical faults that can occur on a 3-φ system.
2. Discuss the 'symmetrical components method' to analyse an unbalanced 3-φ system.
3. What is operator '*a*' ? Show that :
   (*i*) $a^2 = -0{\cdot}5 - j\,0{\cdot}866$ (*ii*) $a^3 = 1$
   (*iii*) $1 + a + a^2 = 0$ (*iv*) $a - a^2 = j\sqrt{3}$
4. Express unbalanced phase currents in a 3-φ system in terms of symmetrical components.
5. What do you understand by positive, negative and zero sequence impedances ? Discuss them with reference to synchronous generators, transformers and transmission lines.
6. Derive an expression for fault current for single line-to-ground fault by symmetrical components method.
7. Derive an expression for fault current for line-to-line fault by symmetrical components method.
8. Derive an expression for fault current for doube line-to-ground fault by symmetrical components method.
9. What do you understand by sequence networks ? What is their importance in unsymmetrical fault calculations ?
10. Write short notes on the following :
    (*i*) Positive sequence network
    (*ii*) Negative sequence network
    (*iii*) Zero sequence network

DISCUSSION QUESTIONS

1. Why is 3-φ symmetrical fault more severe than a 3-φ unsymmetrical fault ?
2. In a 3-φ system, it has been found that negative sequence components and zero sequence components are absent. What do you conclude from it ?
3. Do the sequence components physically exist in a 3-φ system ?
4. Why do we prefer to analyse unsymmetrical faults by symmetrical components method ?
5. The positive sequence network of a power system is similar to the negative sequence network. What do you infer from it ?

CHAPTER

19

Circuit Breakers

Introduction

During the operation of power system, it is often desirable and necessary to switch on or off the various circuits (*e.g.,* transmission lines, distributors, generating plants etc.) under both normal and abnormal conditions. In earlier days, this function used to be performed by a switch and a fuse placed in series with the circuit. However, such a means of control presents two disadvantages. Firstly, when a fuse blows out, it takes quite sometime to replace it and restore supply to the customers. Secondly, a fuse cannot successfully interrupt heavy fault currents that result from faults on modern high-voltage and large capacity circuits. Due to these disadvantages, the use of switches and fuses is limited to low-voltage and small capacity circuits where frequent operations are not expected *e.g.,* for switching and protection of distribution transformers, lighting circuits, branch circuits of distribution lines etc.

With the advancement of power system, the lines and other equipment operate at very high voltages and carry large currents. The arrangement of switches along with fuses cannot serve the desired function of swithgear in such high capacity circuits. This necessitates to employ a more dependable means of control such as is obtained by the use of *circuit breakers*. A circuit breaker can make or break a circuit either manually or automatically under all conditions *viz*., no-load, full-load and short-circuit conditions. This characteristic of the circuit breaker has made it a very useful equipment for switching and protection of various parts of the power system. In this chapter, we shall deal with the various types of circuit breakers and their increasing applications as control devices.

19.1 Circuit Breakers

A circuit breaker is a piece of equipment which can

(*i*) make or break a circuit either manually or by remote control under normal conditions

(*ii*) break a circuit *automatically* under fault conditions

(*iii*) make a circuit either manually or by remote control under fault conditions

Thus a circuit breaker incorporates manual (or remote control) as well as automatic control for switching functions. The latter control employs relays and operates only under fault conditions. The mechanism of opening of the circuit breaker under fault conditions has already been briefed in chapter 16.

Operating principle. A circuit breaker essentially consists of fixed and moving contacts, called electrodes. Under normal operating conditions, these contacts remain closed and will not open automatically until and unless the system becomes faulty. Of course, the contacts can be opened manually or by remote control whenever desired. When a fault occurs on any part of the system, the trip coils of the circuit breaker get energised and the moving contacts are pulled apart by some mechanism, thus opening the circuit.

When the contacts of a circuit breaker are separated under fault conditions, an arc is struck between them. The current is thus able to continue until the discharge ceases. The production of arc not only delays the current interruption process but it also generates enormous heat which may cause damage to the system or to the circuit breaker itself. Therefore, the main problem in a circuit breaker is to extinguish the arc within the shortest possible time so that heat generated by it may not reach a dangerous value.

19.2 Arc Phenomenon

When a short-circuit occurs, a heavy current flows through the contacts of the *circuit breaker before they are opened by the protective system. At the instant when the contacts begin to separate, the contact area decreases rapidly and large fault current causes increased current density and hence rise in temperature. The heat produced in the medium between contacts (usually the medium is oil or air) is sufficient to ionise the air or vapourise and ionise the oil. The ionised air or vapour acts as conductor and an arc is struck between the contacts. The p.d. between the contacts is quite small and is just sufficient to maintain the arc. The arc provides a low resistance path and consequently the current in the circuit remains uninterrupted so long as the arc persists.

During the arcing period, the current flowing between the contacts depends upon the arc resistance. The greater the arc resistance, the smaller the current that flows between the contacts. The arc resistance depends upon the following factors :

(*i*) *Degree of ionisation*— the arc resistance increases with the decrease in the number of ionised particles between the contacts.

(*ii*) *Length of the arc*— the arc resistance increases with the length of the arc *i.e.,* separation of contacts.

---

* **Important note.** In single phase circuits (*e.g.,* lighting circuits etc.), a switch is located in only one of the two conductors to lamps. However, in power circuits, a circuit interrupting device (*e.g.,* a circuit breaker) is put in each *phase* or conductor. These are sometimes called three-pole circuit breakers. In the discussion that follows, we shall confine ourselves to a single-pole device, it is being understood that three such units will be provided in three-phase installation.

(*iii*) *Cross-section of arc*— the arc resistance increases with the decrease in area of X-section of the arc.

19.3 Principles of Arc Extinction

Before discussing the methods of arc extinction, it is necessary to examine the factors responsible for the maintenance of arc between the contacts. These are :

(*i*) p.d. between the contacts

(*ii*) ionised particles between contacts

Taking these in turn,

(*i*) When the contacts have a small separation, the p.d. between them is sufficient to maintain the arc. One way to extinguish the arc is to separate the contacts to such a distance that p.d. becomes inadequate to maintain the arc. However, this method is impracticable in high voltage system where a separation of many metres may be required.

(*ii*) The ionised particles between the contacts tend to maintain the arc. If the arc path is deionised, the arc extinction will be facilitated. This may be achieved by cooling the arc or by bodily removing the ionised particles from the space between the contacts.

19.4 Methods of Arc Extinction

There are two methods of extinguishing the arc in circuit breakers *viz.*

1. High resistance method. 2. Low resistance or current zero method

1. **High resistance method.** In this method, arc resistance is made to increase with time so that current is reduced to a value insufficient to maintain the arc. Consequently, the current is interrupted or the arc is extinguished. The principal disadvantage of this method is that enormous energy is dissipated in the arc. Therefore, it is employed only in d.c. circuit breakers and low-capacity a.c. circuit breakers.

The resistance of the arc may be increased by :

(*i*) *Lengthening the arc.* The resistance of the arc is directly proportional to its length. The length of the arc can be increased by increasing the gap between contacts.

(*ii*) *Cooling the arc.* Cooling helps in the deionisation of the medium between the contacts. This increases the arc resistance. Efficient cooling may be obtained by a gas blast directed along the arc.

(*iii*) *Reducing X-section of the arc.* If the area of X-section of the arc is reduced, the voltage necessary to maintain the arc is increased. In other words, the resistance of the arc path is increased. The cross-section of the arc can be reduced by letting the arc pass through a narrow opening or by having smaller area of contacts.

(*iv*) *Splitting the arc.* The resistance of the arc can be increased by splitting the arc into a number of smaller arcs in series. Each one of these arcs experiences the effect of lengthening and cooling. The arc may be split by introducing some conducting plates between the contacts.

2. **Low resistance or Current zero method.** This method is employed for arc extinction in a.c. circuits only. In this method, arc resistance is kept low until current is zero where the arc extinguishes naturally and is prevented from restriking inspite of the rising voltage across the contacts. All modern high power a.c. circuit breakers employ this method for arc extinction.

In an a.c. system, current drops to zero after every half-cycle. At every current zero, the arc extinguishes for a brief moment. Now the medium between the contacts contains ions and electrons so that it has small dielectric strength and can be easily broken down by the rising contact voltage known as *restriking voltage.* If such a breakdown does occur, the arc will persist for another half-cycle. If immediately after current zero, the dielectric strength of the medium between contacts is built up more rapidly than the voltage across the contacts, the arc fails to restrike and the current will be interrupted. The rapid increase of dielectric strength of the medium near current zero can be achieved by :

(*a*) causing the ionised particles in the space between contacts to recombine into neutral molecules.

(*b*) sweeping the ionised particles away and replacing them by un-ionised particles

Therefore, the real problem in a.c. arc interruption is to rapidly deionise the medium between contacts as soon as the current becomes zero so that the rising contact voltage or restriking voltage cannot breakdown the space between contacts. The de-ionisation of the medium can be achieved by:

(*i*) *lengthening of the gap.* The dielectric strength of the medium is proportional to the length of the gap between contacts. Therefore, by opening the contacts rapidly, higher dielectric strength of the medium can be achieved.

(*ii*) *high pressure.* If the pressure in the vicinity of the arc is increased, the density of the particles constituting the discharge also increases. The increased density of particles causes higher rate of de-ionisation and consequently the dielectric strength of the medium between contacts is increased.

(*iii*) *cooling.* Natural combination of ionised particles takes place more rapidly if they are allowed to cool. Therefore, dielectric strength of the medium between the contacts can be increased by cooling the arc.

(*iv*) *blast effect.* If the ionised particles between the contacts are swept away and replaced by un-ionised particles, the dielectric strength of the medium can be increased considerably. This may be achieved by a gas blast directed along the discharge or by forcing oil into the contact space.

19.5 Important Terms

The following are the important terms much used in the circuit breaker analysis :

(*i*) **Arc Voltage.** *It is the voltage that appears across the contacts of the circuit breaker during the arcing period.*

As soon as the contacts of the circuit breaker separate, an arc is formed. The voltage that appears across the contacts during arcing period is called the arc voltage. Its value is low except for the *period the fault current is at or near zero current point. At current zero, the arc voltage rises rapidly to peak value and this peak voltage tends to maintain the current flow in the form of arc.

(*ii*) **Restriking voltage.** *It is the transient voltage that appears across the contacts at or near current zero during arcing period.*

At current zero, a high-frequency transient voltage appears across the contacts and is caused by the rapid distribution of energy between the magnetic and electric fields associated with the plant and transmission lines of the system. This transient voltage is known as restriking voltage (Fig. 19.1). The current interruption in the circuit depends upon this voltage. If the restriking voltage rises more rapidly than the dielectric strength of the medium between the contacts, the arc will persist for another half-cycle. On the other hand, if the dielectric strength of the medium builds up more rapidly than the restriking voltage, the arc fails to restrike and the current will be interrupted.

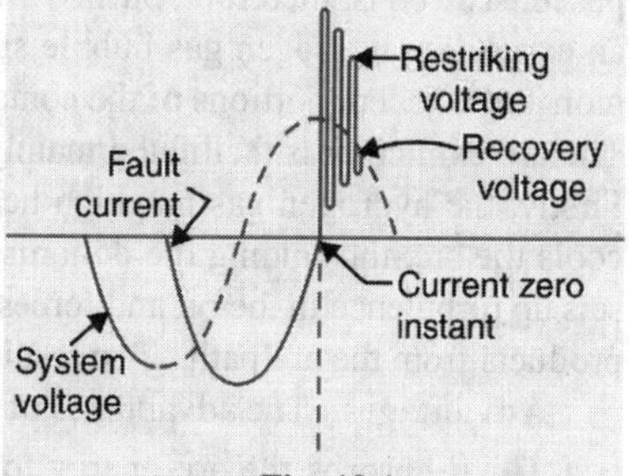

Fig. 19.1

(*iii*) **Recovery voltage.** *It is the normal frequency (50 Hz) r.m.s. voltage that appears across the contacts of the circuit breaker after final arc extinction. It is approximately equal to the system voltage.*

* The power system contains an appreciable amount of inductance. At the instant of current zero point, the fault currrent lags behind the arc voltage by 90º. Obviously, the arc voltge must have a peak value at this instant.

When contacts of circuit breaker are opened, current drops to zero after every half cycle. At some current zero, the contacts are separated sufficiently apart and dielectric strength of the medium between the contacts attains a high value due to the removal of ionised particles. At such an instant, the medium between the contacts is strong enough to prevent the breakdown by the restriking voltage. Consequently, the final arc extinction takes place and circuit current is interrupted. Immediately after final current interruption, the voltage that appears across the contacts has a transient part (See Fig. 19.1). However, these transient oscillations subside rapidly due to the damping effect of system resistance and normal circuit voltage appears across the contacts. The voltage across the contacts is of normal frequency and is known as recovery voltage.

19.6 Classification of Circuit Breakers

There are several ways of classifying the circuit breakers. However, the most general way of classification is on the basis of medium used for arc extinction. The medium used for arc extinction is usually oil, air, sulphur hexafluoride (SF_6) or vacuum. Accordingly, circuit breakers may be classified into :

(*i*) *Oil circuit breakers* which employ some insulating oil (*e.g.*, transformer oil) for arc extinction.

(*ii*) *Air-blast circuit breakers* in which high pressure air-blast is used for extinguishing the arc.

(*iii*) *Sulphur hexafluroide circuit breakers* in which sulphur hexafluoride (SF_6) gas is used for arc extinction.

(*iv*) *Vacuum circuit breakers* in which vacuum is used for arc extinction.

Each type of circuit breaker has its own advantages and disadvantages. In the following sections, we shall discuss the construction and working of these circuit breakers with special emphasis on the way the arc extinction is facilitated.

19.7 Oil Circuit Breakers

In such circuit breakers, some insulating oil (*e.g.*, transformer oil) is used as an arc quenching medium. The contacts are opened under oil and an arc is struck between them. The heat of the arc evaporates the surrounding oil and dissociates it into a substantial volume of gaseous *hydrogen at high pressure. The hydrogen gas occupies a volume about one thousand times that of the oil decomposed. The oil is, therefore, pushed away from the arc and an expanding hydrogen gas bubble surrounds the arc region and adjacent portions of the contacts (See Fig. 19.2). The arc extinction is facilitated mainly by two processes. Firstly, the hydrogen gas has high heat conductivity and cools the arc, thus aiding the de-ionisation of the medium between the contacts. Secondly, the gas sets up turbulence in the oil and forces it into the space between contacts, thus eliminating the arcing products from the arc path. The result is that arc is extinguished and circuit current †interrupted.

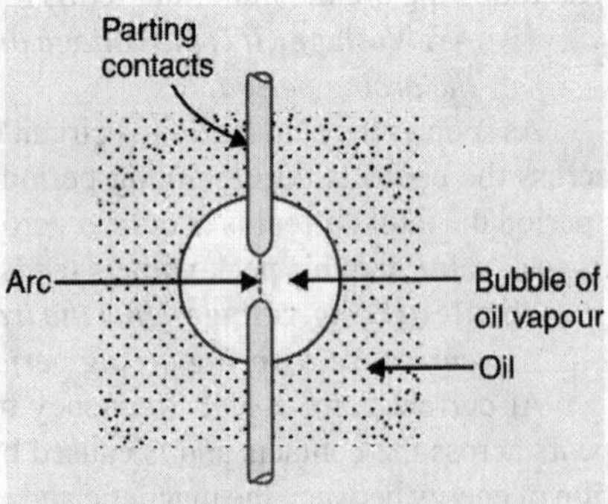

Fig. 19.2

Advantages. The advantages of oil as an arc quenching medium are :

(*i*) It absorbs the arc energy to decompose the oil into gases which have excellent cooling properties.

* Mainly hydrogen gas is produced as a result of oil decomposition. However, a small percentage of methane, ethylene and acetylene is also generated.

† **Important.** The reader may note that arc itself is employed for its extinction. Therefore, it should not be regarded as an unfortunate manifestation. It must also be realised that in the absence of the arc, the current flowing in the circuit would be interrupted instantaneously, and due to the rapid collapse of associated magnetic field, very high voltages would be induced which would severely stress the insulation on the system. On the other hand, the arc permits the circuit interruption to take place at some current zero and thus without inducing potentials of dangerous values.

(*ii*) It acts as an insulator and permits smaller clearance between live conductors and earthed components.

(*iii*) The surrounding oil presents cooling surface in close proximity to the arc.

Disadvantages. The disadvantages of oil as an arc quenching medium are :

(*i*) It is inflammable and there is a risk of a fire.

(*ii*) It may form an explosive mixture with air

(*iii*) The arcing products (*e.g.,* carbon) remain in the oil and its quality deteriorates with successive operations. This necessitates periodic checking and replacement of oil.

19.8 Types of Oil Circuit Breakers

The oil circuit breakers find extensive use in the power system. These can be classified into the following types :

(*i*) *Bulk oil circuit breakers* which use a large quantity of oil. The oil has to serve two purposes. Firstly, it extinguishes the arc during opening of contacts and secondly, it insulates the current conducting parts from one another and from the earthed tank. Such circuit breakers may be classified into :

(*a*) Plain break oil circuit breakers (*b*) Arc control oil circuit breakers.

In the former type, no special means is available for controlling the arc and the contacts are directly exposed to the whole of the oil in the tank. However, in the latter type, special arc control devices are employed to get the beneficial action of the arc as efficiently as possible.

(*ii*) *Low oil circuit breakers* which use minimum amount of oil. In such circuit breakers, oil is used *only* for arc extinction; the current conducting parts are insulated by air or porcelain or organic insulating material.

19.9 Plain Break Oil Circuit Breakers

A plain-break oil circuit breaker involves the simple process of separating the contacts under the whole of the oil in the tank. There is no special system for arc control other than the increase in length caused by the separation of contacts. The arc extinction occurs when a certain critical gap between the contacts is reached.

The plain-break oil circuit breaker is the earliest type from which all other circuit breakers have developed. It has a very simple construction. It consists of fixed and moving contacts enclosed in a strong weather-tight earthed tank containing oil upto a certain level and an air cushion above the oil level. The air cushion provides sufficient room to allow for the reception of the arc gases without the generation of unsafe pressure in the dome of the circuit breaker. It also absorbs the mechanical shock of the upward oil movement. Fig. 19.3 shows a *double break plain oil circuit breaker. It is called a double break because it provides two breaks in series.

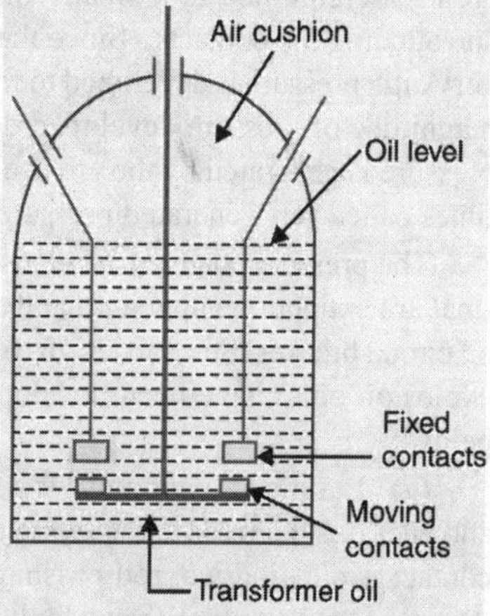

Fig. 19.3

Under normal operating conditions, the fixed and moving contacts remain closed and the breaker carries the normal circuit current. When a fault occurs, the moving contacts are pulled down by the protective system and an arc is struck which vapourises the oil mainly into hydrogen gas. The arc extinction is facilitated by the following processes :

* This type of construction increases the effective speed of arc lengthening and permits to divide the transient re-striking voltage over two breaks so that only half appears across each.

(*i*) The hydrogen gas bubble generated around the arc cools the arc column and aids the de-ionisation of the medium between the contacts.

(*ii*) The gas sets up turbulence in the oil and helps in eliminating the arcing products from the arc path.

(*iii*) As the arc lengthens due to the separating contacts, the dielectric strength of the medium is increased.

The result of these actions is that at some critical gap length, the arc is extinguished and the circuit current is interrupted.

Disadvantages

(*i*) There is no special control over the arc other than the increase in length by separating the moving contacts. Therefore, for successful interruption, long arc length is necessary.

(*ii*) These breakers have long and inconsistent arcing times.

(*iii*) These breakers do not permit high speed interruption.

Due to these disadvantages, plain-break oil circuit breakers are used only for low-voltage applications where high breaking-capacities are not important. It is a usual practice to use such breakers for low capacity installations for voltages not exceeding †11 kV.

19.10 Arc Control Oil Circuit Breakers

In case of plain-break oil circuit breaker discussed above, there is very little artificial control over the arc. Therefore, comparatively long arc length is essential in order that turbulence in the oil caused by the gas may assist in quenching it. However, it is necessary and desirable that final arc extinction should occur while the contact gap is still short. For this purpose, some arc control is incorporated and the breakers are then called arc control circuit breakers. There are two types of such breakers, namely :

(*i*) *Self-blast oil circuit breakers*— in which arc control is provided by internal means *i.e.* the arc itself is employed for its own extinction efficiently.

(*ii*) *Forced-blast oil circuit breakers*— in which arc control is provided by mechanical means external to the circuit breaker.

(*i*) **Self-blast oil circuit breakers.** In this type of circuit breaker, the gases produced during arcing are confined to a small volume by the use of an insulating rigid pressure chamber or pot surrounding the contacts. Since the space available for the arc gases is restricted by the chamber, a very high pressure is developed to force the oil and gas through or around the arc to extinguish it. The magnitude of pressure developed depends upon the value of fault current to be interrupted. As the pressure is generated by the arc itself, therefore, such breakers are sometimes called self-generated pressure oil circuit breakers.

The pressure chamber is relatively cheap to make and gives reduced final arc extinction gap length and arcing time as against the plain-break oil circuit breaker. Several designs of pressure chambers (sometimes called explosion pots) have been developed and a few of them are described below :

(*a*) **Plain explosion pot.** It is a rigid cylinder of insulating material and encloses the fixed and moving contacts (See Fig. 19.4). The moving contact is a cylindrical rod passing through a restricted opening (called throat) at the bottom. When a fault occurs, the contacts get separated and an arc is struck between them. The heat of the arc decomposes oil into a gas at very high pressure in the pot. This high pressure forces the oil and

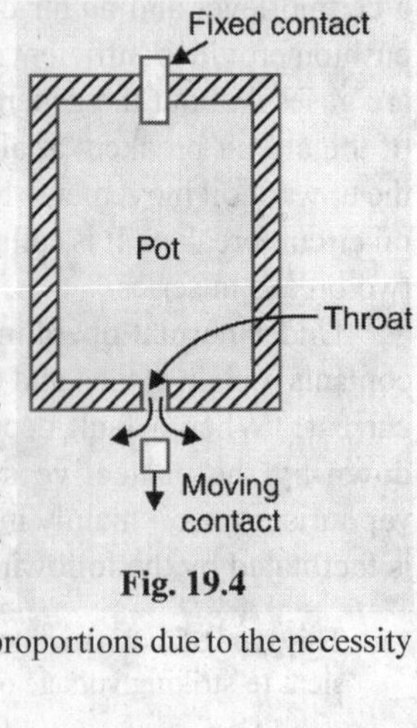

Fig. 19.4

† At relatively high voltages, the size of such a breaker assumes unduly large proportions due to the necessity of very long gap between the contacts for successful arc extinction.

gas through and round the arc to extinguish it. If the final arc extinction does not take place while the moving contact is still within the pot, it occurs immediately after the moving contact leaves the pot. It is because emergence of the moving contact from the pot is followed by a violent rush of gas and oil through the throat producing rapid extinction.

The principal limitation of this type of pot is that it cannot be used for very low or for very high fault currents. With low fault currents, the pressure developed is small, thereby increasing the arcing time. On the other hand, with high fault currents, the gas is produced so rapidly that explosion pot is liable to burst due to high pressure. For this reason, plain explosion pot operates well on moderate short-circuit currents only where the rate of gas evolution is moderate.

(b) Cross jet explosion pot. This type of pot is just a modification of plain explosion pot and is illustrated in Fig. 19.5. It is made of insulating material and has channels on one side which act as arc splitters. The arc splitters help in increasing the arc length, thus facilitating arc extinction. When a fault occurs, the moving contact of the circuit breaker begins to separate. As the moving contact is withdrawn, the arc is initially struck in the top of the pot. The gas generated by the arc exerts pressure on the oil in the back passage. When the moving contact uncovers the arc splitter ducts, fresh oil is forced **across* the arc path. The arc is, therefore, driven sideways into the "arc splitters" which increase the arc length, causing arc extinction.

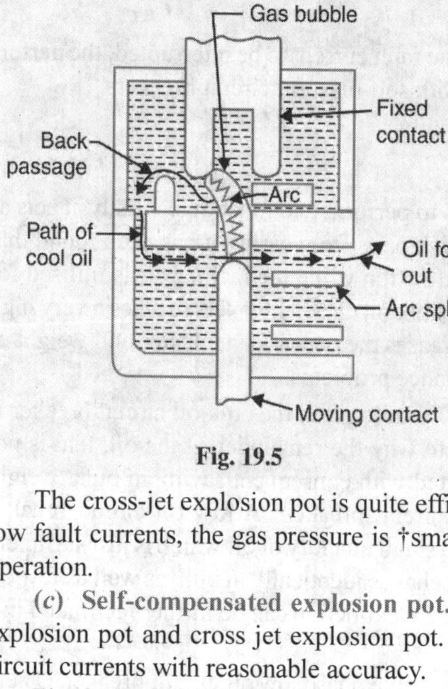

Fig. 19.5

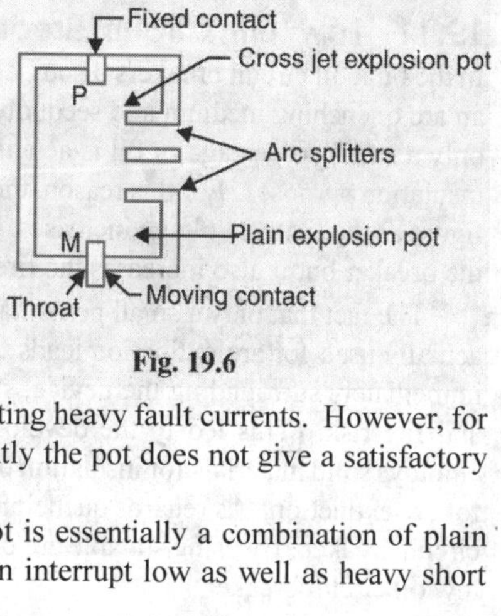

Fig. 19.6

The cross-jet explosion pot is quite efficient for interrupting heavy fault currents. However, for low fault currents, the gas pressure is †small and consequently the pot does not give a satisfactory operation.

(c) Self-compensated explosion pot. This type of pot is essentially a combination of plain explosion pot and cross jet explosion pot. Therefore, it can interrupt low as well as heavy short circuit currents with reasonable accuracy.

Fig. 19.6 shows the schematic diagram of self-compensated explosion pot. It consists of two chambers, the upper chamber is the cross-jet explosion pot with two arc splitter ducts while the lower one is the plain explosion pot. When the short-circuit current is heavy, the rate of generation of gas is very high and the device behaves as a cross-jet explosion pot. The arc extinction takes place when the moving contact uncovers the first or second arc splitter duct. However, on low short-circuit currents, the rate of gas generation is small and the tip of the moving contact has the time to reach the lower chamber. During this time, the gas builds up sufficient pressure as there is very little leakage through

* Since the jet of oil is forced at right angles to the arc path, this type of pot is referred to as cross-jet explosion pot.

† The rate at which oil moves into the path of arc is a function of gas pressure. The gas pressure depends upon the value of fault current. Lower the fault current, lesser the gas pressure generated and vice-versa.

arc splitter ducts due to the obstruction offered by the arc path and right angle bends. When the moving contact comes out of the throat, the arc is extinguished by plain pot action.

It may be noted that as the severity of the short-circuit current increases, the device operates less and less as a plain explosion pot and more and more as a cross-jet explosion pot. Thus the tendency is to make the control self-compensating over the full range of fault currents to be interrupted.

(ii) **Forced-blast oil circuit breakers.** In the self-blast oil circuit breakers discussed above, the arc itself generates the necessary pressure to force the oil across the arc path. The major limitation of such breakers is that arcing times tend to be long and inconsistent when operating against currents considerably less than the rated currents. It is becasue the gas generated is much reduced at low values of fault currents. This difficulty is overcome in forced-blast oil circuit breakers in which the necessary pressure is generated by external mechanical means independent of the fault currents to be broken.

In a forced -blast oil circuit breaker, oil pressure is created by the piston-cylinder arrangement. The movement of the piston is mechanically coupled to the moving contact. When a fault occurs, the contacts get separated by the protective system and an arc is struck between the contacts. The piston forces a jet of oil towards the contact gap to extinguish the arc. It may be noted that necessary oil pressure produced does not in any way depend upon the fault current to be broken.

Advantages

(*a*) Since oil pressure developed is independent of the fault current to be interrupted, the performance at low currents is more consistent than with self-blast oil circuit breakers.

(*b*) The quantity of oil required is reduced considerably.

19.11 Low OIl Circuit Breakers

In the bulk oil circuit breakers discussed so far, the oil has to perform two functions. Firstly, it acts as an arc quenching medium and secondly, it insulates the live parts from earth. It has been found that only a small percentage of oil is actually used for arc extinction while the major part is utilised for insulation purposes. For this reason, the quantity of oil in bulk oil circuit breakers reaches a very high figure as the system voltage increases. This not only increases the expenses, tank size and weight of the breaker but it also increases the fire risk and maintenance problems.

The fact that only a small percentage of oil (about 10% of total) in the bulk oil circuit breaker is actually used for arc extinction leads to the question as to why the remainder of the oil, that is not immediately surrounding the device, should not be omitted with consequent saving in bulk, weight and fire risk. This led to the development of low-oil circuit breaker. A low oil circuit breaker employs solid materials for insulation purposes and uses a small quantity of oil which is just sufficient for arc extinction. As regards quenching the arc, the oil behaves identically in bulk as well as low oil circuit breaker. By using suitable arc control devices, the arc extinction can be further facilitated in a low oil circuit breaker.

Construction. Fig 19.7 shows the cross section of a single phase low oil circuit breaker. There are two compartments separated from each other but both filled with oil. The upper chamber is the circuit breaking chamber while the lower one is the supporting chamber. The two chambers are separated by a partition and oil from one chamber is prevented from mixing with the other chamber. This arrangement permits two advantages. Firstly, the circuit breaking chamber requires a small volume of oil which is just enough for arc extinction. Secondly, the amount of oil to be replaced is reduced as the oil in the supporting chamber does not get contaminated by the arc.

(*i*) *Supporting chamber.* It is a porcelain chamber mounted on a metal chamber. It is filled with oil which is physically separated from the oil in the circuit breaking compartment. The oil inside the supporting chamber and the annular space formed between the porcelain insulation and bakelised paper is employed for insulation purposes only.

(*ii*) *Circuit-breaking chamber.* It is a porcelain enclosure mounted on the top of the supporting compartment. It is filled with oil and has the following parts :

(*a*) upper and lower fixed contacts

(*b*) moving contact

(*c*) turbulator

The moving contact is hollow and includes a cylinder which moves down over a fixed piston. The turbulator is an arc control device and has both axial and radial vents. The axial venting ensures the interruption of low currents whereas radial venting helps in the interruption of heavy currents.

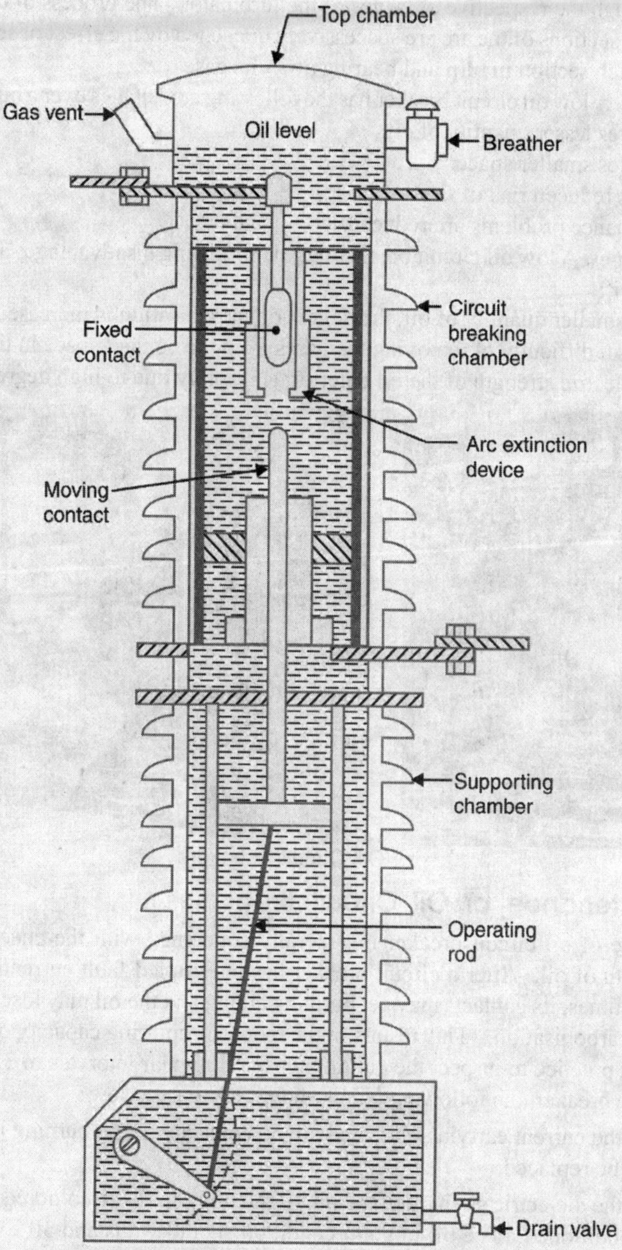

Fig. 19.7 Low-oil Circuit Breaker

(*iii*) *Top chamber.* It is a metal chamber and is mouted on the circuit-breaking chamber. It provides expansion space for the oil in the circuit breaking compartment. The top chamber is also provided with a separator which prevents any loss of oil by centrifugal action caused by circuit breaker operation during fault conditions.

Operation. Under normal operating conditions, the moving contact remains engaged with the upper fixed contact. When a fault occurs, the moving contact is pulled down by the tripping springs and an arc is struck. The arc energy vaporises the oil and produces gases under high pressure. This action constrains the oil to pass through a central hole in the moving contact and results in forcing series of oil through the respective passages of the turbulator. The process of turbulation is orderly one, in which the sections of the arc are successively quenched by the effect of separate streams of oil moving across each section in turn and bearing away its gases.

Advantages. A low oil circuit breaker has the following advantages over a bulk oil circuit breaker:

(*i*) It requires lesser quantity of oil.

(*ii*) It requires smaller space.

(*iii*) There is reduced risk of fire.

(*iv*) Maintenance problems are reduced.

Disadvantages. A low oil circuit breaker has the following disadvantages as compared to a bulk oil circuit breaker :

(*i*) Due to smaller quantity of oil, the degree of carbonisation is increased.

(*ii*) There is a difficulty of removing the gases from the contact space in time.

(*iii*) The dielectric strength of the oil deteriorates rapidly due to high degree of carbonisation.

Oil Circuit Breakers

19.12 Maintenance of Oil Circuit Breakers

The maintenance of oil circuit breaker is generally concerned with the checking of contacts and dielectric strength of oil. After a circuit breaker has interrupted fault currents a few times or load currents several times, its contacts may get burnt by arcing and the oil may lose some of its dielectric strength due to carbonisation. This results in the reduced rupturing capacity of the breaker. Therefore, it is a good practice to inspect the circuit breaker at regular intervals of 3 or 6 months. During inspection of the breaker, the following points should be kept in view :

(*i*) Check the current carrying parts and arcing contacts. If the burning is severe, the contacts should be replaced.

(*ii*) Check the dielectric strength of the oil. If the oil is badly discoloured, it should be changed or reconditioned. The oil in good condition should withstand 30 kV for one minute in a standard oil testing cup with 4 mm gap between electrodes.

(iii) Check the insulation for possible damage. Clean the surface and remove carbon deposits with a strong and dry fabric.
(iv) Check the oil level.
(v) Check closing and tripping mechanism.

19.13 Air-Blast Circuit Breakers

These breakers employ a high pressure *air-blast as an arc quenching medium. The contacts are opened in a flow of air-blast established by the opening of blast valve. The air-blast cools the arc and sweeps away the arcing products to the atomsphere. This rapidly increases the dielectric strength of the medium between contacts and prevents from re-establishing the arc. Consequently, the arc is extinguished and flow of current is interrupted.

Advantages. An air-blast circuit breaker has the following advantages over an oil circuit breaker:

(i) The risk of fire is eliminated.
(ii) The arcing products are completely removed by the blast whereas the oil deteriorates with successive operations; the expense of regular oil replacement is avoided.
(iii) The growth of dielectric strength is so rapid that final contact gap needed for arc extinction is very small. This reduces the size of the device.
(iv) The arcing time is very small due to the rapid build up of dielectric strength between contacts. Therefore, the arc energy is only a fraction of that in oil circuit breakers, thus resulting in less burning of contacts.
(v) Due to lesser arc energy, air-blast circuit breakers are very suitable for conditions where frequent operation is required.
(vi) The energy supplied for arc extinction is obtained from high pressure air and is independent of the current to be interrupted.

Disadvantages. The use of air as the arc quenching medium offers the following disadvantges :

(i) The air has relatively inferior arc extinguishing properties.
(ii) The air-blast circuit breakers are very sensitive to the variations in the rate of rise of restriking voltage.
(iii) Considerable maintenance is required for the compressor plant which supplies the air-blast.

The air blast circuit breakers are finding wide applications in high voltage installations. Majority of the circuit breakers for voltages beyond 110 kV are of this type.

19.14 Types of Air-Blast Circuit Breakers

Depending upon the direction of air-blast in relation to the arc, air-blast circuit breakers are classified into :

(i) *Axial-blast type* in which the air-blast is directed along the arc path as shown in Fig. 19.8(*i*).

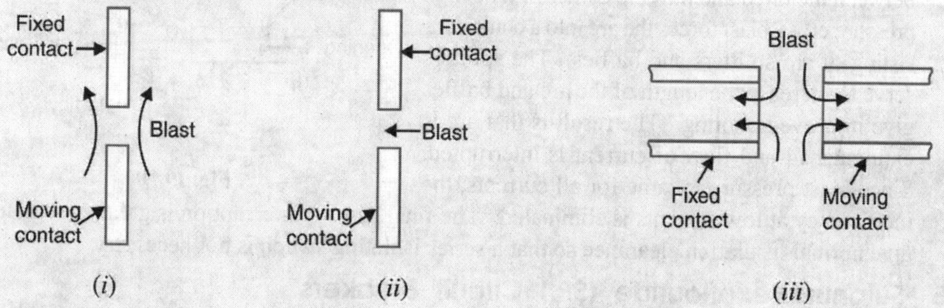

Fig. 19.8

* Other gases such as nitrogen, carbon dioxide and hyrdogen can also be used. The circuit breaking properties of nitrogen are about similar to air and there is no added advantage of using it. Carbon dioxide tends to freeze and hydrogen gas is very costly. Therefore, air is used as the circuit breaking medium.

(*ii*) *Cross-blast type* in which the air-blast is directed at right angles to the arc path as shown in Fig. 19.8 (*ii*).

(*iii*) *Radial-blast type* in which the air-blast is directed radially as shown in Fig. 19.8 (*iii*).

(*i*) **Axial-blast air circuit breaker.** Fig 19.9 shows the essential components of a typical axial-blast air circuit breaker. The fixed and moving contacts are held in the closed position by spring pressure under normal conditions. The air reservoir is connected to the arcing chamber through an air valve. This valve remains closed under normal conditions but opens automatically by the tripping impulse when a fault occurs on the system.

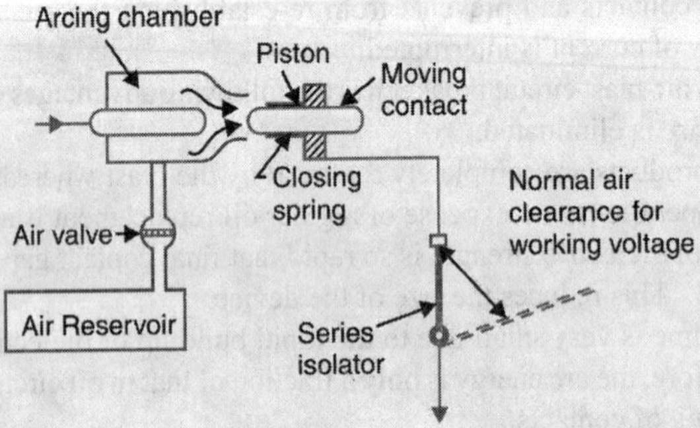

Fig. 19.9

When a fault occurs, the tripping impulse causes opening of the air valve which connects the circuit breaker reservoir to the arcing chamber. The high pressure air entering the arcing chamber pushes away the moving contact against spring pressure. The moving contact is separated and an arc is struck. At the same time, high pressure air blast flows along the arc and takes away the ionised gases along with it. Consequently, the arc is extinguished and current flow is interrupted.

It may be noted that in such circuit breakers, the contact separation required for interruption is generally small (1·75 cm or so). Such a small gap may constitute inadequate clearance for the normal service voltage. Therefore, an isolating switch is incorporated as a part of this type of circuit breaker. This switch opens immediately after fault interruption to provide the necessary clearance for insulation.

(*ii*) **Cross-blast air breaker.** In this type of circuit breaker, an air-blast is directed at right angles to the arc. The cross-blast lengthens and forces the arc into a suitable chute for arc extinction. Fig. 19.10 shows the essential parts of a typical cross-blast air circuit breaker. When the moving contact is withdrawn, an arc is struck between the fixed and moving contacts. The high pressure cross-blast forces the arc into a chute consisting of arc splitters and baffles. The splitters serve to increase the length of the arc and baffles give improved cooling. The result is that arc is extinguished and flow of current is interrupted. Since blast pressure is same for all currents, the inefficiency at low currents is eliminated. The final gap for interruption is great enough to give normal insulation clearance so that a series isolating switch is not necessary.

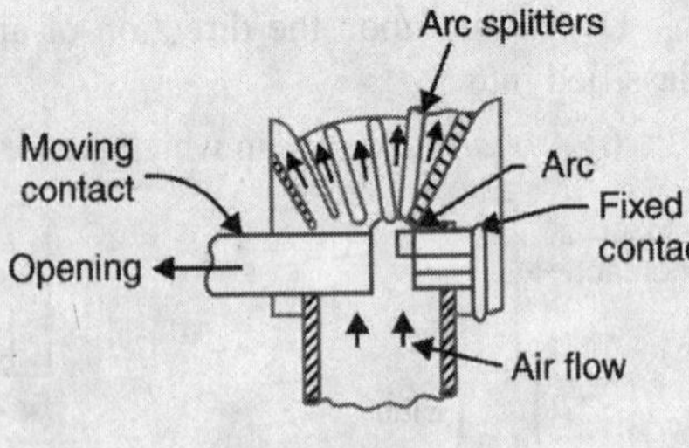

Fig. 19.10

19.15 Sulphur Hexaflouride (SF_6) Circuit Breakers

In such circuit breakers, sulphur hexaflouride (SF_6) gas is used as the arc quenching medium. The SF_6 is an electro-negative gas and has a strong tendency to absorb free electrons. The contacts of the breaker are opened in a high pressure flow of SF_6 gas and an arc is struck between them. The conducting free electrons in the arc are rapidly captured by the gas to form relatively immobile negative ions. This loss of conducting electrons in the arc quickly builds up enough insulation strength to extinguish the arc. The SF_6 circuit breakers have been found to be very effective for high power and high voltage service.

Construction. Fig. 19.11 shows the parts of a typical SF_6 circuit breaker. It consists of fixed and moving contacts enclosed in a chamber (called arc interruption chamber) containing SF_6 gas. This chamber is connected to SF_6 gas reservior. When the contacts of breaker are opened, the valve mechanism permits a high pressure SF_6 gas from the reservoir to flow towards the arc interruption chamber. The fixed contact is a hollow cylindrical current carrying contact fitted with an arc horn. The moving contact is also a hollow cylinder with rectangular holes in the sides to permit the SF_6 gas to let out through these holes after flowing along and across the arc. The tips of fixed contact, moving contact and arcing horn are coated with copper-tungsten arc resistant material. Since SF_6 gas is costly, it is reconditioned and reclaimed by suitable auxiliary sytem after each operation of the breaker.

Working. In the closed position of the breaker, the contacts remain surrounded by SF_6 gas at a pressure of about 2·8 kg/cm^2. When the breaker operates, the moving contact is pulled apart and an arc is struck between the contacts. The movement of the moving contact is synchronised with the opening of a valve which permits SF_6 gas at 14 kg/cm^2 pressure from the reservoir to the arc interruption chamber. The high pressure flow of SF_6 rapidly absorbs the free electrons in the arc path to form immobile negative ions which are ineffective as charge carriers. The result is that the medium between the contacts quickly builds up high dielectric strength and causes the extinction of the arc. After the breaker operation (*i.e.*, after arc extinction), the valve is closed by the action of a set of springs.

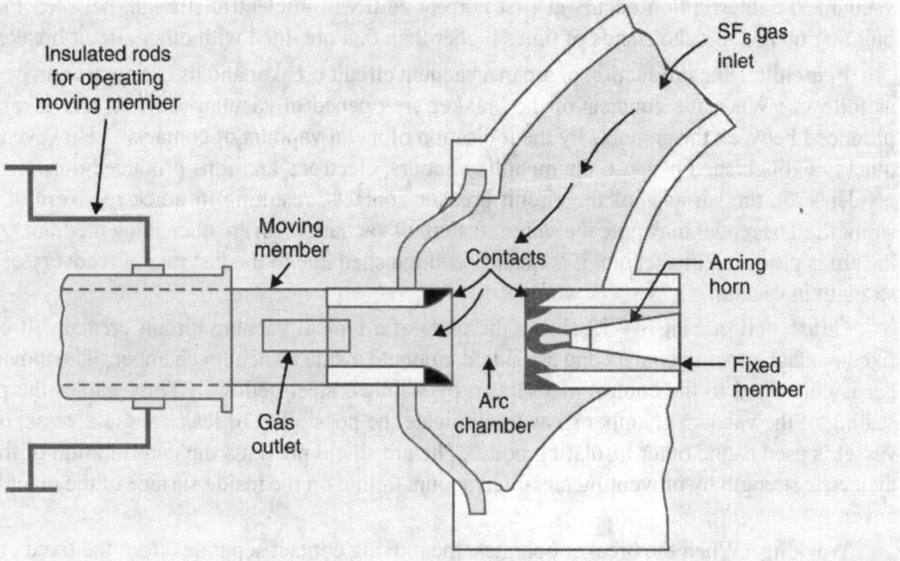

Fig. 19.11

Advantages. Due to the superior arc quenching properties of SF_6 gas, the SF_6 circuit breakers have many advantages over oil or air circuit breakers. Some of them are listed below :

(*i*) Due to the superior arc quenching property of SF_6, such circuit breakers have very short arcing time.

(*ii*) Since the dielectric strength of SF_6 gas is 2 to 3 times that of air, such breakers can interrupt much larger currents.

(*iii*) The SF_6 circuit breaker gives noiselss operation due to its closed gas circuit and no exhaust to atmosphere unlike the air blast circuit breaker.

(*iv*) The closed gas enclosure keeps the interior dry so that there is no moisture problem.
(*v*) There is no risk of fire in such breakers because SF_6 gas is non-inflammable.
(*vi*) There are no carbon deposits so that tracking and insulation problems are eliminated.
(*vii*) The SF_6 breakers have low maintenance cost, light foundation requirements and minimum auxiliary equipment.
(*viii*) Since SF_6 breakers are totally enclosed and sealed from atmosphere, they are particularly suitable where explosion hazard exists *e.g.,* coal mines.

Disadvantages

(*i*) SF_6 breakers are costly due to the high cost of SF_6.
(*ii*) Since SF_6 gas has to be reconditioned after every operation of the breaker, additional equipment is requried for this purpose.

Applications. A typical SF_6 circuit breaker consists of interrupter units each capable of dealing with currents upto 60 kA and voltages in the range of 50—80 kV. A number of units are connected in series according to the system voltage. SF_6 circuit breakers have been developed for voltages 115 kV to 230 kV, power ratings 10 MVA to 20 MVA and interrupting time less than 3 cycles.

19.16 Vacuum Circuit Breakers (VCB)

In such breakers, vacuum (degree of vacuum being in the range from 10^{-7} to 10^{-5} torr) is used as the arc quenching medium. Since vacuum offers the highest insulating strength, it has far superior arc quenching properties than any other medium. For example, when contacts of a breaker are opened in vacuum, the interruption occurs at first current zero with dielectric strength between the contacts building up at a rate thousands of times higher than that obtained with other circuit breakers.

Principle. The production of arc in a vacuum circuit breaker and its extinction can be explained as follows : When the contacts of the breaker are opened in vacuum (10^{-7} to 10^{-5} torr), an arc is produced between the contacts by the ionisation of metal vapours of contacts*. However, the arc is quickly extinguished because the metallic vapours, electrons and ions produced during arc rapidly condense on the surfaces of the circuit breaker contacts, resulting in quick recovery of dielectric strength. The reader may note the salient feature of vacuum as an arc quenching medium. As soon as the arc is produced in vacuum, it is quickly extinguished due to the fast rate of recovery of dielectric strength in vacuum.

Construction. Fig. 19.12 shows the parts of a typical vacuum circuit breaker. It consists of fixed contact, moving contact and arc shield mounted inside a vacuum chamber. The movable member is connected to the control mechanism by stainless steel bellows. This enables the permanent sealing of the vacuum chamber so as to eliminate the possibility of leak. A glass vessel or ceramic vessel is used as the outer insulating body. The arc shield prevents the deterioration of the internal dielectric strength by preventing metallic vapours falling on the inside surface of the outer insulating cover.

Working. When the breaker operates, the moving contact separates from the fixed contact and an arc is struck between the contacts. The production of arc is due to the ionisation of metal ions and depends very much upon the material of contacts. The arc is quickly extinguished because the metallic vapours, electrons and ions produced during arc are diffused in a short time and seized by the surfaces of moving and fixed members and shields. Since vacuum has very fast rate of recovery of dielectric strength, the arc extinction in a vacuum breaker occurs with a short contact separation (say 0·625 cm).

---

* This can be explained by *Field emission theory.* The microscopic sharp projections on the surface of contacts of the circuit breaker act as points of high intensity field, resulting in the emission of electrons. Since the sectional area of each projection is very small, the current density becomes very large and the heat produced causes melting of projections.

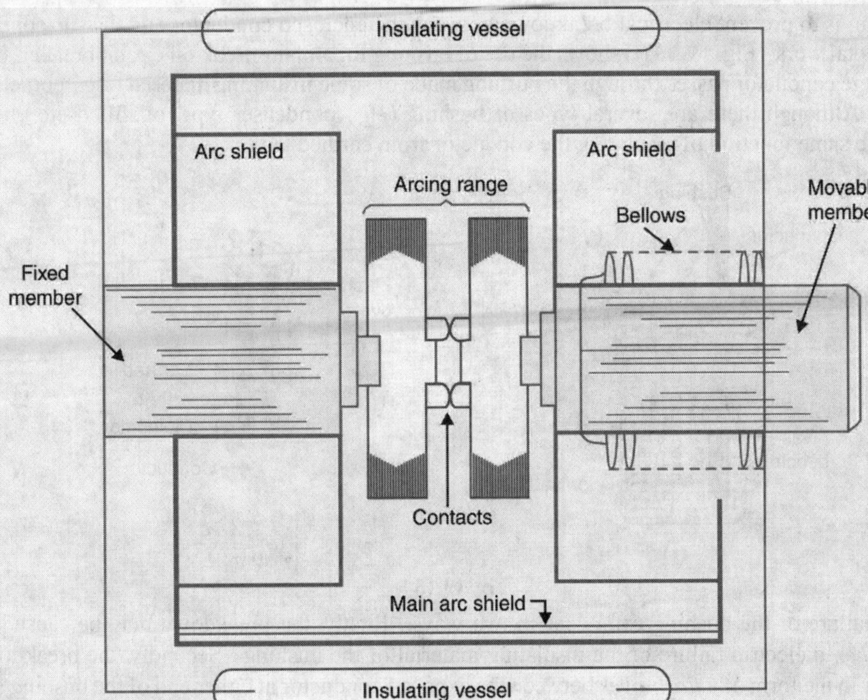

Fig. 19.12

Advantages. Vacuum circuit breakers have the following advantages :

(*i*) They are compact, reliable and have longer life.

(*ii*) There are no fire hazards.

(*iii*) There is no generation of gas during and after operation.

(*iv*) They can interrupt any fault current. The outstanding feature of a VCB is that it can break any heavy fault current perfectly just before the contacts reach the definite open position.

(*v*) They require little maintenance and are quiet in operation.

(*vi*) They can successfully withstand lightning surges.

(*vii*) They have low arc energy.

(*viii*) They have low inertia and hence require smaller power for control mechanism.

Applications. For a country like India, where distances are quite large and accessibility to remote areas difficult, the installation of such outdoor, maintenance free circuit breakers should prove a definite advantage. Vacuum circuit breakers are being employed for outdoor applications ranging from 22 kV to 66 kV. Even with limited rating of say 60 to 100 MVA, they are suitable for a majority of applications in rural areas.

19.17 Switchgear Components

The following are some important components common to most of the circuit breakers :

(*i*) Bushings

(*ii*) Circuit breaker contacts

(*iii*) Instrument transformers

(*iv*) Bus-bars and conductors

(*i*) **Bushings.** When a high voltage conductor passes through a metal sheet or frame which is at earth potential, the necessary insulation is provided in the form of bushing. The primary function of the bushing is to prevent electrical breakdown between the enclosed conductor and the surrounding earthed metal work. Fig. 19.13 (*i*) shows the use of bushing for a plain-break oil circuit breaker. The high voltage conductor passes through the bushing made of some insulating material (*e.g.*, porcelain, steatite). Although there are several types of bushing (*e.g.*, condenser type, oil filled etc.), they perfom the same function of insulating the conductor from earthed tank.

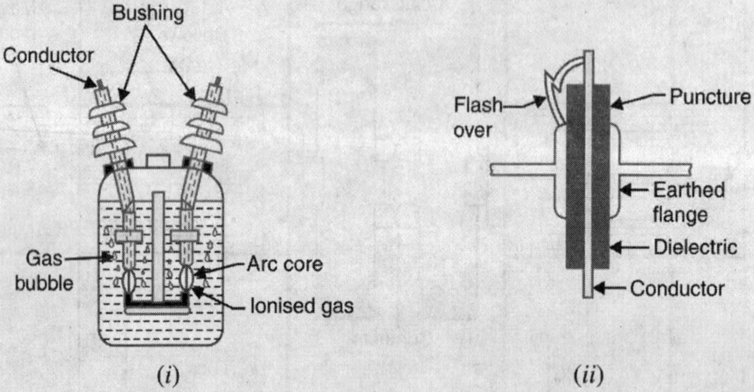

Fig. 19.13

The failure of the bushing can occur in two ways. Firstly, the breakdown may be caused by *puncture i.e.*, dielectric failure of the insulating material of the bushing. Secondly, the breakdown may occur in the form of a *flash-over* between the exposed conductor at either end of the bushing and the earthed metal. Fig. 19.13 (*ii*) illustrates these two possibilities. The bushings are so designed that flash-over takes place before they get punctured. It is because the puncture generally renders the bushing insulation unserviceable and incapable of withstanding the normal voltage. On the other hand, a flash-over may result in comparatively harmless burning of the surface of the bushing which can then continue to give adequate service pending replacement.

(*ii*) **Circuit breaker contacts.** The circuit breaker contacts are required to carry normal as well as short-circuit current. In carrying the normal current, it is desirable that the temperature should not rise above the specified limits and that there should be low voltage drop at the point of contact. In carrying breaking and making short-circuit currents, the chief effects to be dealt with are melting and vapourisation by the heat of the arc and those due to electromagnetic forces. Therefore, the design of contacts is of considerable importance for satisfactory operation of the circuit breakers. There are three types of circuit breaker contacts *viz.*

(*a*) *Tulip type contacts.* Fig. 19.14 (*i*) shows the Tulip type contact. It consists of moving contact which moves inside the fixed contacts. At contact separation, the arc is generally established between the tips of the fixed contacts and the tip of the moving contact as shown in Fig. 19.14 (*ii*). The advantage of this type of contact is that arcing is confined to the regions which are not in contact in the fully engaged position.

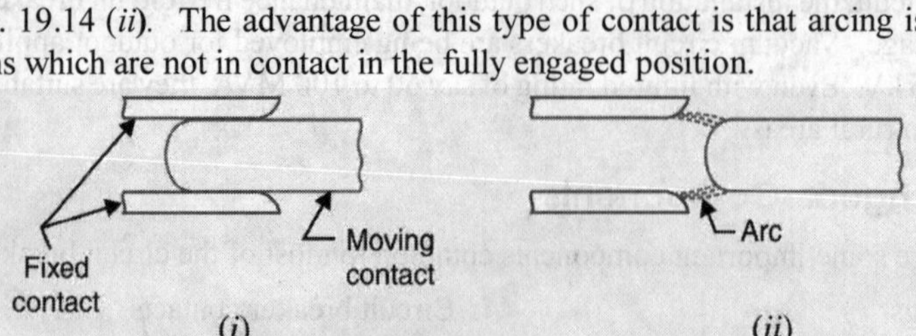

Fig. 19.14

(*b*) *Finger and wedge contacts.* Fig. 19.15 (*i*) shows the finger and wedge type contact. This type of contact is largely used for low-voltage oil circuit breakers owing to the general unsuitability for use with arc control devices.

(*c*) *Butt contacts.* Fig. 19.15 (*ii*) shows the butt type contact and is formed by the springs and the moving contact. It possesses two advantages. Firstly, spring pressure is available to assist contact separation. This is useful in single-break oil circuit breakers and air-blast circuit breakers where relatively small "loop" forces are available to assist in opening. Secondly, there is no grip force so that this type of contact is especially suitable for higher short-circuit rating.

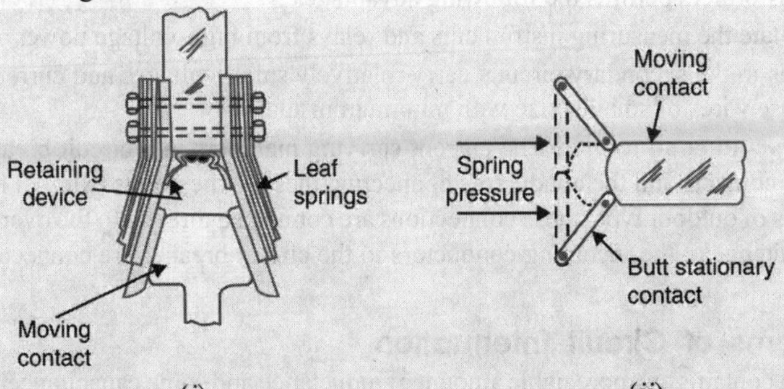

Fig. 19.15

(*iii*) **Instrument transformers.** In a modern power system, the circuits operate at very high voltages and carry current of thousands of amperes. The measuring instruments and protective devices cannot work satisfactorily if mounted directly on the power lines. This difficulty is overcome by installing instrument transformers on the power lines. The function of these instrument transformers is to transform voltages or currents in the power lines to values which are convenient for the operation of measuring instruments and relays. There are two types of instrument transformers *viz.*

(*a*) Current transformer (C.T.)

(*b*) Potential transformer (P.T.)

The primary of current transformer is connected in the power line. The secondary winding provides for the instruments and relays a current which is a constant fraction of the current in the line.

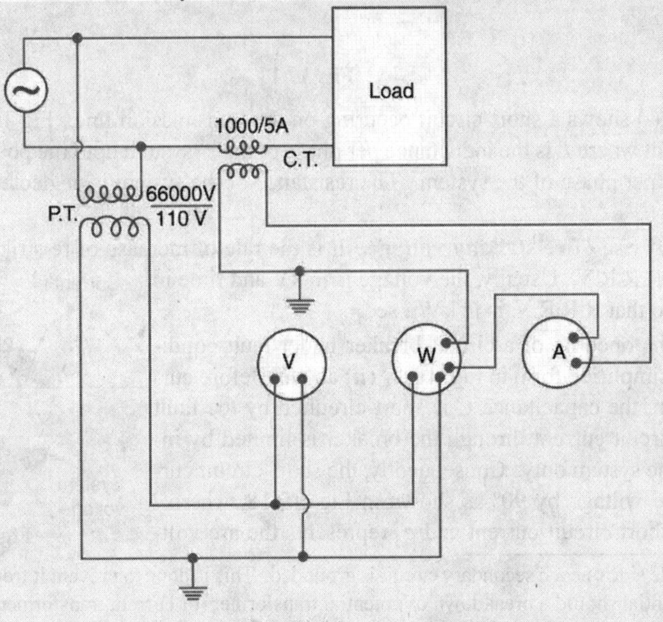

Fig. 19.16

Similarly, a potential transformer is connected with its primary in the power line. The secondary provides for the instruments and relays a voltage which is a known fraction of the line voltage.

Fig. 19.16 shows the use of instrument transformers. The *potential transformer rated 66,000/110V provides a voltage supply for the potential coils of voltmeter and wattmeter. The current transformer rated 1000/5 A supplies current to the current coils of wattmeter and ammeter. The use of instrument transformers permits the following advantages :

(*a*) They isolate the measuring instruments and relays from high-voltage power circuits.

(*b*) The leads in the secondary circuits carry relatively small voltages and currents. This permits to use wires of smaller size with minimum insulation.

(*iv*) Bus-bars and conductors. The current carrying members in a circuit breaker consist of fixed and moving contacts and the conductors connecting these to the points external to the breaker. If the switchgear is of outdoor type, these connections are connected directly to the overhead lines. In case of indoor switchgear, the incoming conductors to the circuit breaker are connected to the bus-bars.

19.18 Problems of Circuit Interruption

The power system contains an appreciable amount of inductance and some capacitance. When a fault occurs, the energy stored in the system can be considerable. Interruption of fault current by a circuit breaker will result in most of the stored energy dissipated within the circuit breaker, the remainder being dissipated during oscillatory surges in the system. The oscillatory surges are undesirable and, therefore, the circuit breaker must be designed to dissipate as much of the stored energy as possible.

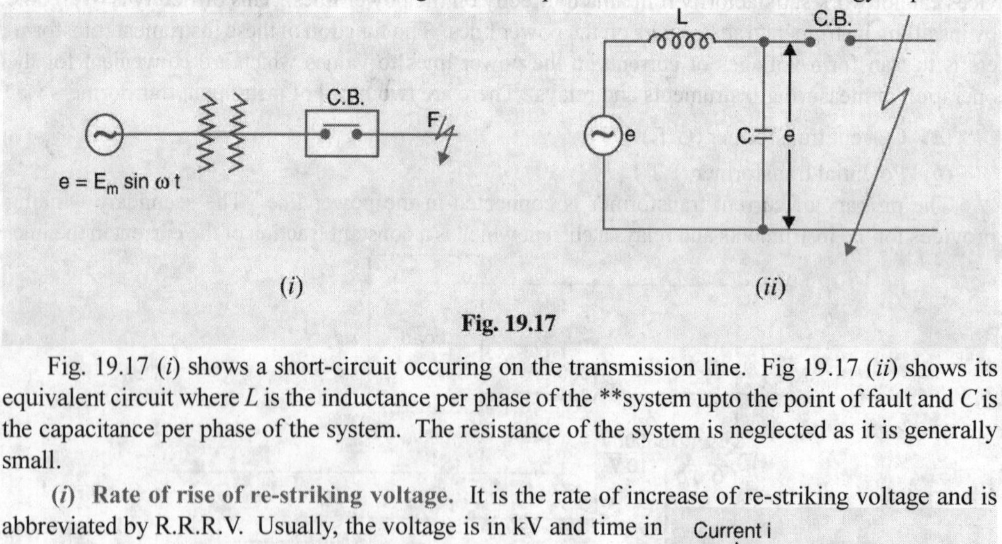

Fig. 19.17

Fig. 19.17 (*i*) shows a short-circuit occuring on the transmission line. Fig 19.17 (*ii*) shows its equivalent circuit where L is the inductance per phase of the **system upto the point of fault and C is the capacitance per phase of the system. The resistance of the system is neglected as it is generally small.

(*i*) Rate of rise of re-striking voltage. It is the rate of increase of re-striking voltage and is abbreviated by R.R.R.V. Usually, the voltage is in kV and time in microseconds so that R.R.R.V. is in kV/μ sec.

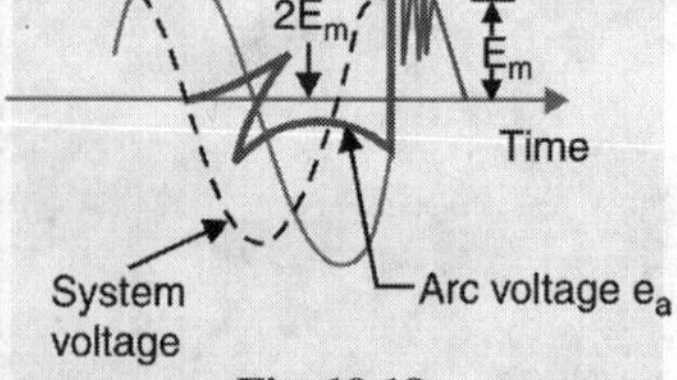

Fig. 19.18

Consider the opening of a circuit breaker under fault conditions shown in simplified form in Fig. 19.17 (*ii*) above. Before current interruption, the capacitance C is short-circuited by the fault and the short-circuit current through the breaker is limited by inductance L of the system only. Consequently, the short-circuit current will lag the voltage by 90° as shown in Fig. 19.18, where i represents the short-circuit current and e_a represents the arc volt-

* Note that one side of each secondary circuit is grounded. This is done to prevent it from becoming charged at high potential should a breakdown of potential transformer (or current transformer) insulation occur.

** Generator, transformer and transmission line in this case.

age. It may be seen that in this condition, the *entire generator voltage appears across inductance L.

When the contacts are opened and the arc finally extinguishes at some current zero, the generator voltage e is suddenly applied to the inductance and capacitance in series. This $L–C$ combination forms an oscillatory circuit and produces a transient of frequency :

$$f_n = \frac{1}{2\,\pi\,\sqrt{LC}}$$

which appears across the capacitor C and hence across the contacts of the circuit breaker. This transient voltage, as already noted, is known as re-striking voltage and may reach an instantaneous peak value twice the peak phase-neutral voltage *i.e.* $2\,E_m$. The system losses cause the oscillations to decay fairly rapidly but the initial overshoot increases the possibility of re-striking the arc.

It is the rate of rise of re-striking voltage (R.R.R.V.) which decides whether the arc will re-strike or not. If R.R.R.V. is greater than the rate of rise of dielectric strength between the contacts, the arc will re-strike. However, the arc will fail to re-strike if R.R.R.V. is less than the rate of increase of dielectric strength between the contacts of the breaker. The value of R.R.R.V. depends upon :

(*a*) recovery voltage

(*b*) natural frequency of oscillations

For a short-circuit occuring near the power station bus-bars, C being small, the natural frequency f_n $(= 1/2\,\pi\,\sqrt{LC})$ will be high. Consequently, R.R.R.V. will attain a large value. Thus the worst condition for a circuit breaker would be that when the fault takes place near the bus-bars.

(*ii*) **Current chopping.** *It is the phenomenon of current interruption before the natural current zero is reached.*

Current chopping **mainly occurs in air-blast circuit breakers because they retain the same extinguishing power irrespective of the magnitude of the current to be interrupted. When breaking low currents (*e.g.*, transformer magnetising current) with such breakers, the powerful de-ionising effect of air-blast causes the current to fall abruptly to zero well before the natural current zero is reached. This phenomenon is known as current chopping and results in the production of high voltage transient across the contacts of the circuit breaker as discussed below :

Consider again Fig. 19.17 (*ii*) repeated as Fig. 19.19 (*i*). Suppose the arc current is i when it is chopped down to zero value as shown by point a in Fig. 19.19 (*ii*). As the chop occurs at current i, therefore, the energy stored in inductance is $L\,i^2/2$. This energy will be transferred to the capacitance C, charging the latter to a prospective voltage e given by :

$$\frac{1}{2}\,L\,i^2 = \frac{C\,e^2}{2}$$

or $$e = i\sqrt{\frac{L}{C}} \text{ volts} \qquad ...(i)$$

The prospectve voltage e is very *** high as compared to the dielectric strength gained by the gap so that the breaker restrikes. As the de-ionising force is still in action, therefore, chop occurs again but the arc current this time is smaller than the previous case. This induces a lower prospective

---

* Because capacitance is short-circuited by the fault.

** Current chopping is not common in oil circuit breakers because in most of them, arc control is proportional to the fault current to be interrupted. In other words, the exinguishing power in such breakers is proportional to the current to be interrupted.

*** For example, if L and C are 4mH and 0·001 μF respectively, a current chop of magnitude 50 A would induce a voltage of

$$e = i\sqrt{\frac{L}{C}} = 50\sqrt{\frac{4\times10^{-3}}{0\cdot001\times10^{-6}}} = 100\times10^3 \text{ volts} = 100 \text{ kV}$$

voltage to re-ignite the arc. In fact, several chops may occur until a low enough current is interrupted which produces insufficient induced voltage to re-strike across the breaker gap. Consequently, the final interruption of current takes place.

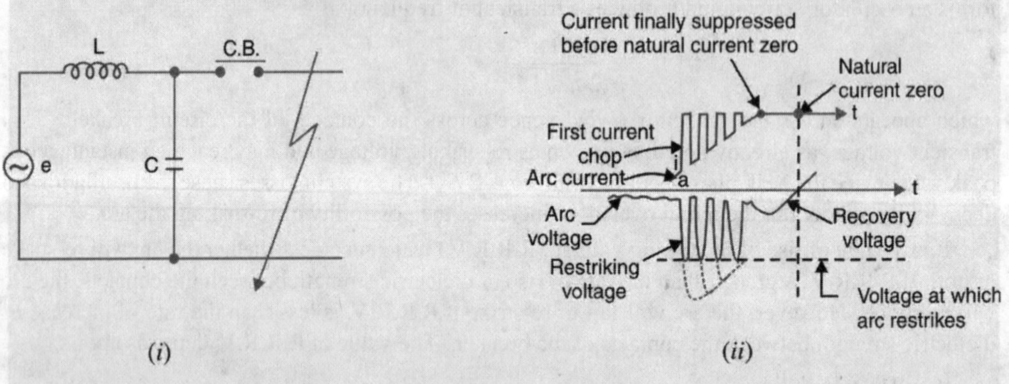

Fig. 19.19

Excessive voltage surges due to current chopping are prevented by shunting the contacts of the breaker with a resistor (*resistance switching*) such that reignition is unlikely to occur. This is explained in Art 19.19.

***(iii)* Capacitive current breaking.** Another cause of excessive voltage surges in the circuit breakers is the interruption of capacitive currents. Examples of such instances are opening of an unloaded long transmission line, disconnecting a capacitor bank used for power factor improvement etc. Consider the simple equivalent circuit of an unloaded transmission line shown in Fig. 19.20. Such a line, although unloaded in the normal sense, will actually carry a capacitive current I on account of appreciable amount of capacitance C between the line and the earth.

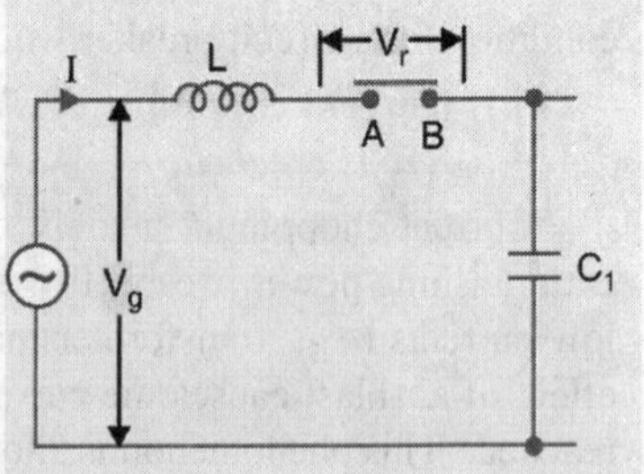

Fig. 19.20

Let us suppose that the line is opened by the circuit breaker at the instant when line capacitive current is zero [point 1 in Fig. 19.21]. At this instant, the generator voltage V_g will be maximum (*i.e.,* V_{gm}) lagging behind the current by 90°. The opening of the line leaves a standing charge on it (*i.e.,* end B of the line) and the capacitor C_1 is charged to V_{gm}. However, the generator end of the line (*i.e.,* end A of the line) continues its normal sinusoidal variations. The voltage V_r across the circuit breaker will be the difference between the voltages on the respective sides. Its initial value is zero (point 1) and increases slowly in the beginning. But half a cycle later [point R in Fig. 19.21], the potential of the circuit breaker contact 'A' becomes maximum negative which causes the voltage across the breaker (V_r) to become 2 V_{gm}. This voltage may be sufficient to restrike the arc. The two previously separated parts of the circuit will now be joined by an arc of very low resistance. The line capacitance discharges at once to reduce the voltage across the circuit breaker, thus setting up high frequency transient. The peak value of the initial transient will be twice the voltage at that instant *i.e.,* $-4\ V_{gm}$. This will cause the transmission voltage to swing to $-4\ V_{gm}$ to $+V_{gm}$ *i.e.,* $-3V_{gm}$.

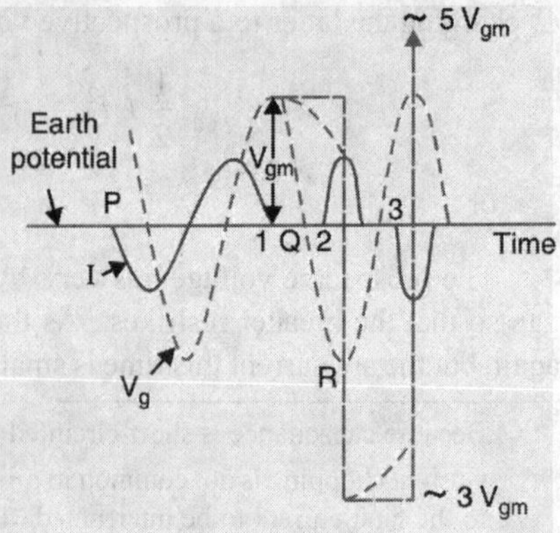

Fig. 19.21

The re-strike arc current quickly reaches its first zero as it varies at natural frequency. The voltage on the line is now $-3\,V_{gm}$ and once again the two halves of the circuit are separated and the line is isolated at this potential. After about half a cycle further, the aforesaid events are repeated even on more formidable scale and the line may be left with a potential of $5\,V_{gm}$ above earth potential. Theoretically, this phenomenon may proceed infinitely increasing the voltage by successive increment of 2 times V_{gm}.

While the above description relates to the worst possible conditions, it is obvious that if the gap breakdown strength does not increase rapidly enough, successive re-strikes can build up a dangerous voltage in the open circuit line. However, due to leakage and corona loss, the maximum voltage on the line in such cases is limited to 5 V_{gm}.

19.19 Resistance Switching

It has been discussed above that current chopping, capacitive current breaking etc. give rise to severe voltage oscillations. These excessive voltage surges during circuit interruption can be prevented by the use of shunt resistance R connected across the circuit breaker contacts as shown in the equivalent circuit in Fig. 19.22. This is known as resistance switching.

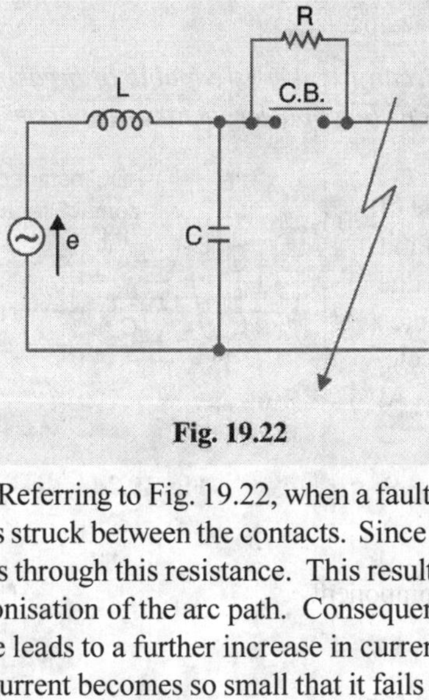

Fig. 19.22

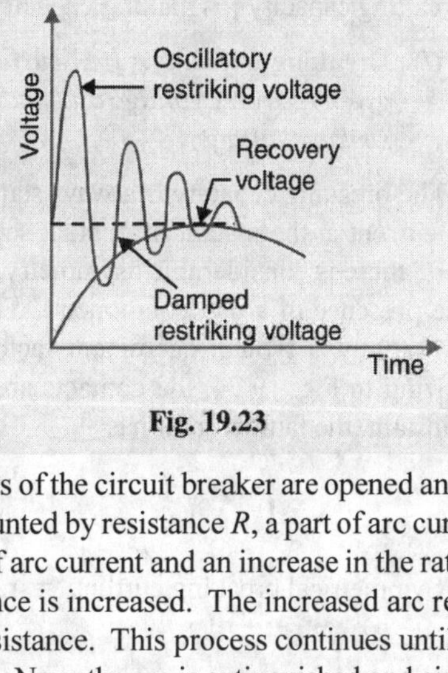

Fig. 19.23

Referring to Fig. 19.22, when a fault occurs, the contacts of the circuit breaker are opened and an arc is struck between the contacts. Since the contacts are shunted by resistance R, a part of arc current flows through this resistance. This results in the decrease of arc current and an increase in the rate of de-ionisation of the arc path. Consequently, the arc resistance is increased. The increased arc resistance leads to a further increase in current through shunt resistance. This process continues until the arc current becomes so small that it fails to maintain the arc. Now, the arc is extinguished and circuit current is interrupted.

The shunt resistor also helps in limiting the oscillatory growth of re-striking voltage. It can be proved mathematically that natural frequency of oscillations of the circuit shown in Fig. 19.22 is given by :

$$f_n = \frac{1}{2\pi}\sqrt{\frac{1}{LC} - \frac{1}{4\,R^2C^2}}$$

The effect of shunt resistance R is to prevent the oscillatory growth of re-striking voltage and cause it to grow exponentially upto recovery voltage. This is being most effective when the value of R is so chosen that the circuit is critically damped. The value of R required for critical damping is $0{\cdot}5\sqrt{L/C}$. Fig. 19.23 shows the oscillatory growth and exponential growth when the circuit is critically damped.

To sum up, resistors across breaker contacts may be used to perform one or more of the following functions :

(*i*) To reduce the rate of rise of re-striking voltage and the peak value of re-striking voltage.

(*ii*) To reduce the voltage surges due to current chopping and capacitive current breaking.

(*iii*) To ensure even sharing of re-striking voltage transient acorss the various breaks in multi-break circuit breakers.

It may be noted that value of resistance required to perform each function is usually different. However, it is often necessary to compromise and make one resistor do more than one of these functions.

19.20 Circuit Breaker Ratings

A circuit breaker may be called upon to operate under all conditions. However, major duties are imposed on the circuit breaker when there is a fault on the system in which it is connected. Under fault conditions, a circuit breaker is required to perform the following three duties :

(*i*) It must be capable of opening the faulty circuit and breaking the fault current.

(*ii*) It must be capable of being closed on to a fault.

(*iii*) It must be capable of carrying fault current for a short time while another circuit breaker (in series) is clearing the fault.

Corresponding to the above mentioned duties, the circuit breakers have three ratings *viz.* (*i*) breaking capacity (*ii*) making capacity and (*iii*) short-time capacity.

(*i*) **Breaking capacity.** *It is current (r.m.s.) that a circuit breaker is capable of breaking at given recovery voltage and under specified conditions (e.g., power factor, rate of rise of re-striking voltage).*

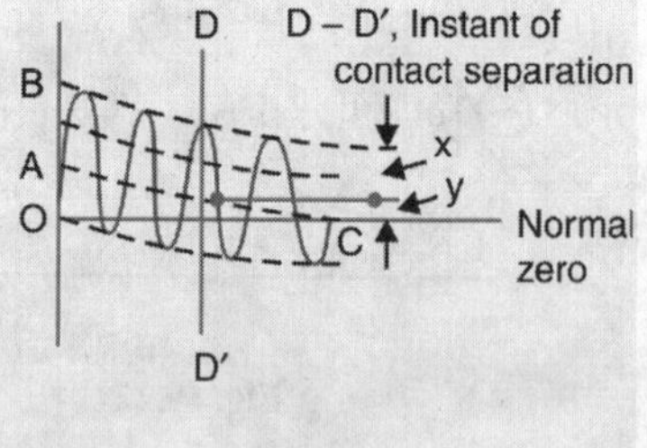

Fig. 19.24

The breaking capacity is always stated at the r.m.s. value of fault current at the instant of contact separation. When a fault occurs, there is considerable asymmetry in the fault current due to the presence of a d.c. component. The d.c. component dies away rapidly, a typical decrement factor being 0·8 per cycle. Referring to Fig. 19.24, the contacts are separated at DD′. At this instant, the fault current has

x = maximum value of a.c. component

y = d.c. component

∴ Symmetrical breaking current = r.m.s. value of a.c. component

$$= \frac{x}{\sqrt{2}}$$

Asymmetrical breaking current = r.m.s. value of total current

$$= \sqrt{\left(\frac{x}{\sqrt{2}}\right)^2 + y^2}$$

It is a common practice to express the breaking capacity in MVA by taking into account the rated breaking current and rated service voltage. Thus, if I is the rated breaking current in amperes and V is the rated service line voltage in volts, then for a 3-phase circuit,

$$\text{Breaking capacity} = \sqrt{3} \times V \times I \times 10^{-6} \text{ MVA}$$

In India (or Britain), it is a usual practice to take breaking current equal to the symmetrical breaking current. However, American practice is to take breaking current equal to asymmetrical breaking current. Thus the American rating given to a circuit breaker is higher than the Indian or British rating.

It seems to be illogical to give breaking capacity in MVA since it is obtained from the product of short-circuit current and rated service voltage. When the short-circuit current is flowing, there is only a small voltage across the breaker contacts, while the service voltage appears across the contacts only after the current has been interrupted. Thus MVA rating is the product of two quantities which do not exist simultaneously in the circuit.

Therefore, the *agreed international standard of specifying breaking capacity is defined as the rated symmetrical breaking current at a rated voltage.

(*ii*) **Making capacity.** There is always a possibility of closing or making the circuit under short-circuit conditions. The capacity of a breaker to "make" current depends upon its ability to withstand and close successfully against the effects of electromagnetic forces. These forces are proportional to the square of maximum instantaneous current on closing. Therefore, making capacity is stated in terms of a peak value of current instead of r.m.s. value.

The peak value of current (including d.c. component) during the first cycle of current wave after the closure of circuit breaker is known as **making capacity.**

It may be noted that the definition is concerned with the first cycle of current wave on closing the circuit breaker. This is because the maximum value of fault current possibly occurs in the first cycle only when maximum asymmetry occurs in any phase of the breaker. In other words, the making current is equal to the maximum value of asymmetrical current. To find this value, we must multiply symmetrical breaking current by $\sqrt{2}$ to convert this from r.m.s. to peak, and then by 1·8 to include the "doubling effect" of maximum asymmetry. The total multiplication factor becomes $\sqrt{2} \times 1{\cdot}8 = 2{\cdot}55$.

$\therefore$ Making capacity $= 2{\cdot}55 \times$ Symmetrical breaking capacity

(*iii*) **Short-time rating.** *It is the period for which the circuit breaker is able to carry fault current while remaining closed.*

Sometimes a fault on the system is of very temporary nature and persists for 1 or 2 seconds after which the fault is automatically cleared. In the interest of continuity of supply, the breaker should not trip in such situations. This means that circuit breakers should be able to carry high current safely for some specified period while remaining closed *i.e.,* they should have proven short-time rating. However, if the fault persists for a duration longer than the specified time limit, the circuit breaker will trip, disconnecting the faulty section.

The short-time rating of a circuit breaker depends upon its ability to withstand (*a*) the electromagnetic force effects and (*b*) the temperature rise. The oil circuit breakers have a specified limit of 3 seconds when the ratio of symmetrical breaking current to the rated normal current does not exceed 40. However, if this ratio is more than 40, then the specified limit is 1 second.

Normal current rating. It is the r.m.s. value of current which the circuit breaker is capable of carrying continuously at its rated frequency under specified conditions. The only limitation in this case is the temperature rise of current-carrying parts.

Example 19.1. *A circuit breaker is rated as 1500 A, 1000 MVA, 33 kV, 3-second, 3-phase oil circuit breaker. Find (i) rated normal current (ii) breaking capacity (iii) rated symmetrical breaking current (iv) rated making current (v) short-time rating (vi) rated service voltage.*

Solution.

(*i*) Rated normal current = **1500 A**

(*ii*) Breaking capacity = **1000 MVA**

* The proponents of MVA rating, however, advance another view which does not seem to be so illogical. Though this MVA rating does not appear at the breaker contacts, it does exist in the windings of the generating plant where both generated voltage and current exist simultaneously. The generated voltage is absorbed in the impedance of the circuit. Since the impedance is distributed, the voltage cannot be measured anywhere while the faulty current is flowing. Therefore, the only way to obtain its value is to measure the recovery voltage immediately after current interruption. This argument seems to have some justification.

(*iii*) Rated symmetrical breaking current = $\frac{1000 \times 10^6}{\sqrt{3} \times 33 \times 10^3}$ = **17496 A** (r.m.s.)

(*iv*) Rated making current = 2·55 × 17496 = **44614 A** (peak)

(*v*) Short-time rating = **17496A** for 3 seconds

(*vi*) Rated service voltage = **33 kV** (r.m.s.)

Example 19.2. *A 50 Hz, 11 kV, 3-phase alternator with earthed neutral has a reactance of 5 ohms per phase and is connected to a bus-bar through a circuit breaker. The distributed capacitance upto circuit breaker between phase and neutral in 0·01 μF. Determine*

(*i*) *peak re-striking voltage across the contacts of the breaker*

(*ii*) *frequency of oscillations*

(*iii*) *the average rate of rise of re-striking voltage upto the first peak*

Solution.

Inductance per phase, $$L = \frac{X_L}{2\pi f} = \frac{5}{2\pi \times 50} = 0{\cdot}0159 \text{ H}$$

Capacitance per phase, $$C = 0{\cdot}01 \ \mu\text{F} = 10^{-8} \text{ F}$$

(*i*) Maximum value of recovery voltage (phase to neutral)

$$E_{max} = \sqrt{2} \times \frac{11}{\sqrt{3}} = 8{\cdot}98 \text{ kV}$$

∴ Peak re-striking voltage $= 2\,E_{max} = 2 \times 8{\cdot}98 =$ **17·96 kV**

(*ii*) Frequency of oscillations is

$$f_n = \frac{1}{2\pi\sqrt{LC}} = \frac{1}{2\pi\sqrt{0{\cdot}0159 \times 10^{-8}}} = \mathbf{12{,}628 \text{ Hz}}$$

(*iii*) Peak re-striking voltage occurs at a time *t* given by ;

$$t = \frac{1}{2\,f_n} = \pi\sqrt{LC} = \pi\sqrt{0{\cdot}0159 \times 10^{-8}} = 39{\cdot}6 \times 10^{-6} \text{ sec} = 39{\cdot}6 \ \mu \text{ sec}$$

∴ Average rate of rise of re-striking voltage

$$= \frac{\text{Peak re-striking voltage}}{\text{Time upto first peak}} = \frac{17{\cdot}96 \text{ kV}}{39{\cdot}6 \ \mu \text{ sec}}$$

$$= 0{\cdot}453 \text{ kV}/\mu \text{ sec} = \mathbf{453 \times 10^3 \text{ kV/sec}}$$

Example 19.3. *In a short circuit test on a circuit breaker, the following readings were obtained on single frequency transient :*

(*i*) *time to reach the peak re-striking voltage, 50 μ sec*

(*ii*) *the peak re-striking voltage, 100 kV*

Determine the average RRRV and frequency of oscillations

Solution.

$$\text{Average RRRV} = \frac{\text{Peak re-striking voltage}}{\text{Time to reach peak value}}$$

$$= \frac{100 \text{ kV}}{50 \ \mu \text{ sec}} = 2 \text{ kV}/\mu \text{ Sec} = \mathbf{2 \times 10^6 \text{ kV/sec}}$$

Natural frequency of oscillations,

$$f_n = \frac{1}{2 \times \text{Time to reach peak value}}$$

$$= \frac{1}{2 \times 50\ \mu\ \text{sec}}$$

$$= \frac{1}{2 \times 50 \times 10^{-6}} = \mathbf{10{,}000\ Hz}$$

Example 19.4. *An air-blast circuit breaker is designed to interrupt a transformer magnetising current of 11 A (r.m.s.) chops the current at an instantaneous value of 7 A. If the values of L and C in the circuit are 35·2 H and 0·0023 μF, find the value of voltage that appears across the contacts of the breaker. Assume that all the inductive energy is transferred to the capacitance.*

Solution.

Voltage across breaker contacts at chopping is

$$e = i\sqrt{\frac{L}{C}} \qquad \text{(See. Art. 19.18)}$$

Here, $i = 7$ A ; $L = 35{\cdot}2$ H and $C = 0{\cdot}0023$ μF

$$\therefore \quad e = 7\sqrt{\frac{35\cdot 2}{0\cdot 0023 \times 10^{-6}}} \text{ volts} = 866 \times 10^3 \text{ V} = \mathbf{866\ kV}$$

TUTORIAL PROBLEMS

1. A circuit breaker is rated 2500 A, 1500 MVA, 33 kV, 3 sec, 3-phase oil circuit breaker. Determine (*i*) the rated normal current (*ii*) breaking current (*iii*) making current (*iv*) short time rating current.
[(*i*) 2500 A (*ii*) 26243 A (*iii*) 66920 A (*iv*) 26243 A for 3 sec]
2. A circuit breaker is rated at 1500 A, 2000 MVA, 33 kV, 3 sec, 3-phase oil circuit breaker. Determine (*i*)the rated normal current (*ii*) breaking current (*iii*) making current (*iv*) short time rating current.
[(*i*)1500A (*ii*) 34·29 kA (*iii*) 89·22 kA (*iv*) 34·29 kA for 3 sec]

SELF - TEST

1. **Fill in the blanks by inserting appropriate words/figures.**
   (*i*) A circuit breaker opens .... when a fault occurs on the system.
   (*ii*) A circuit breaker can .... the circuit immediately after automatic operation.
   (*iii*) When the contacts of a circuit breaker are opened on the occurrence of a fault, an .... is struck.
   (*iv*) An a.c. circuit is more easily interrupted than a d.c. circuit because alternating current provides ....
   (*v*) If the length of the arc increases, its resistance is ....
   (*vi*) If dielectric strength of the medium between contacts builds up more rapidly than the re-striking voltage, then arc will be ....
   (*vii*) In an oil circuit breaker, ... is used as the arc quenching medium.
   (*viii*) The quantity of oil needed for arc control oil circuit breakers is .... than that of plain break oil-circuit breaker.
   (*ix*) Current chopping mainly occurs in .... circuit breakers.
   (*x*) Capacitive current breaking results in ....
2. **Pick up the correct words/figures from the brackets and fill in the blanks.**
   (*i*) Cross jet explosion pot breaker can interrupt .... short circuit currents efficiently. (*heavy, low*)
   (*ii*) In forced blast oil circuit breakers, the extinguishing force is .... the fault current to be interrupted. (*dependent upon, independent of*)
   (*iii*) In low oil circuit breakers, ... is used for insulation purposes. (*solid material, oil*)
   (*iv*) Forced-blast circuit breakers have .... speed of circuit interruption. (*low, high*)

ANSWERS TO SELF-TEST

1. (*i*) automatically (*ii*) remake (*iii*) arc (*iv*) natural current zeros (*v*) increased (*vi*) extinguished (*vii*) some mineral oil (*viii*) less (*ix*) air-blast (*x*) voltage surges
2. (*i*) heavy (*ii*) independent of (*iii*) solid material (*iv*) high

CHAPTER REVIEW TOPICS

1. What is a circuit breaker ? Describe its operating principle.
2. Discuss the arc phenomenon in a circuit breaker.
3. Explain the various methods of arc extinction in a circuit breaker.
4. Define and explain the following terms as applied to circuit breakers :
   (*i*) Arc voltage
   (*ii*) Restriking voltage
   (*iii*) Recovery voltage
5. Describe briefly the action of an oil circuit breaker. How does oil help in arc extinction ?
6. Discuss the advantages and disadvantages of oil circuit breakers.
7. Explain with neat sketches the construction and working of the following circuit breakers :
   (*i*) Plain explosion pot
   (*ii*) Cross jet explosion pot
   (*iii*) Self-compensated explosion pot
8. Explain the difference between bulk oil circuit breakers and low-oil circuit breakers.
9. Discuss the constructional details and operation of a typical low-oil circuit breaker ? What are its relative merits and demerits ?
10. Discuss the principle of operation of an air-blast circuit breaker. What are the advantages and disadvantages of using air as the arc quenching medium ?
11. Explain briefly the following types of air-blast circuit breakers :
    (*i*) Axial-blast type
    (*ii*) Cross-blast type
12. What are the important components common to most of circuit breakers ? Discuss each component briefly.
13. Write a short note on the rate of re-striking voltage indicating its importance in the arc extinction.
14. Discuss the phenomenon of
    (*i*) Current chopping
    (*ii*) Capacitive current breaking
15. Write short notes on the following :
    (*i*) Resistance switching
    (*ii*) Circuit breaker ratings
    (*iii*) Circuit interruption problems

DISCUSSION QUESTIONS

1. Why do plain break oil circuit breakers have a low speed of circuit interruption ?
2. Is it logical to express the breaking capacity of a circuit breaker in MVA ? Discuss.
3. Why is current interruption easier in an a.c. circuit than in a d.c. circuit ?
4. Why are circuit breakers designed to have a short-time rating ?
5. Is arc production in a circuit breaker unfortunate mainfestation ?

CHAPTER

20

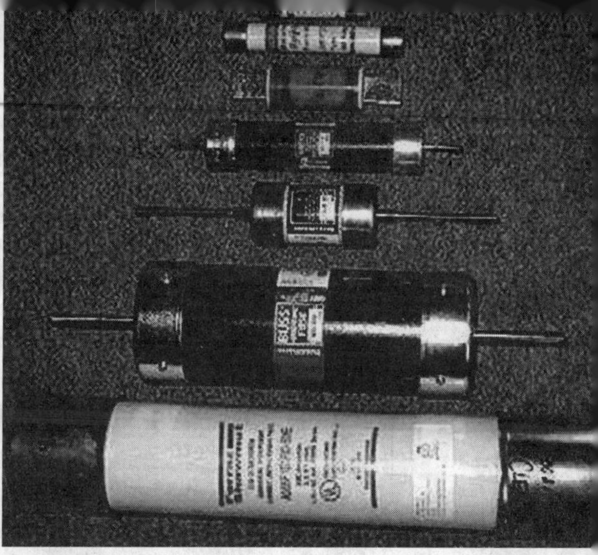

Fuses

Introduction

It was discussed in the previous chapter that a circuit breaker interrupts the circuit automatically on the occurrence of a short-circuit fault. The same function can also be performed by a fuse, though with lesser reliability and efficiency. Invented in 1890 by Edison, fuse is the cheapest form of protection against excessive currents. Many improvements have been made since the invention of first crude model. Now-a-days, several types of fuses are available which find extensive use in low to moderate voltage applications where frequent operations are not expected or where the use of circuit breaker is uneconomical. In this chapter, we shall confine our attention to the various types of fuses and their applications in the fast expanding power system.

20.1 Fuses

A **fuse** *is a short piece of metal, inserted in the circuit, which melts when excessive current flows through it and thus breaks the circuit.*

The fuse element is generally made of materials having low melting point, high conductivity and least deterioration due to oxidation *e.g.*, silver, copper etc. It is inserted in series with the circuit to be protected. Under normal operating conditions, the fuse element is at a temperature below its melting point. Therefore, it carries the normal current without overheating. However, when a short-circuit or overload occurs, the current through the fuse increases beyond its rated value. This raises the temperature and fuse element melts (or blows out), disconnecting the circuit protected by it. In this way, a fuse protects the machines and equipment from damage due to excessive currents.

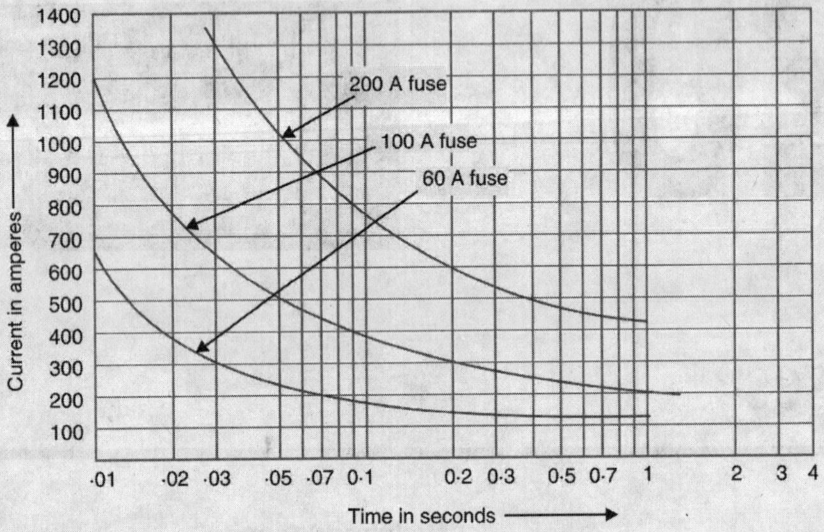

Fig. 20.1

The time required to blow out the fuse depends upon the magnitude of excessive current. The greater the current, the smaller is the time taken by the fuse to blow out. In other words, a fuse has inverse time-current characteristics as shown in Fig. 20.1. Such a characteristic permits its use for overcurrent protection.

Advantages

(*i*) It is the cheapest form of protection available.

(*ii*) It requires no maintenance.

(*iii*) Its operation is inherently completely automatic unlike a circuit breaker which requires an elaborate equipment for automatic action.

(*iv*) It can break heavy short-circuit currents without noise or smoke.

(*v*) The smaller sizes of fuse element impose a current limiting effect under short-circuit conditions.

(*vi*) The inverse time-current characteristic of a fuse makes it suitable for overcurrent protection.

(*vii*) The minimum time of operation can be made much shorter than with the circuit breakers.

Disadvantages

(*i*) Considerable time is lost in rewiring or replacing a fuse after operation.

(*ii*) On heavy short-circuits, *discrimination between fuses in series cannot be obtained unless there is sufficient difference in the sizes of the fuses concerned.

(*iii*) The current-time characteristic of a fuse cannot always be co-related with that of the protected apparatus.

---

* Discrimination between two fuses is said to occur if on the occurrence of a short-circuit or overcurrent fault, only the desired fuse operates.

20.2 Desirable Characteristics of Fuse Element

The function of a fuse is to carry the normal current without overheating but when the current exceeds its normal value, it rapidly heats up to melting point and disconnects the circuit protected by it. In order that it may perform this function satisfactorily, the fuse element should have the following desirable characteristics :

(*i*) low melting point *e.g.*, tin, lead.
(*ii*) high conductivity *e.g.*, silver, copper.
(*iii*) free from deterioration due to oxidation *e.g.*, silver.
(*iv*) low cost *e.g.*, lead, tin, copper.

The above discussion reveals that no material possesses all the characteristics. For instance, lead has low melting point but it has high specific resistance and is liable to oxidation. Similarly, copper has high conductivity and low cost but oxidises rapidly. Therefore, a compromise is made in the selection of material for a fuse.

20.3 Fuse Element Materials

The most commonly used materials for fuse element are lead, tin, copper, zinc and silver. For small currents upto 10 A, tin or an alloy of lead and tin (lead 37%, tin 63%) is used for making the fuse element. For larger currents, copper or silver is employed. It is a usual practice to tin the copper to protect it from oxidation. Zinc (in strip form only) is good if a fuse with considerable time-lag is required *i.e.*, one which does not melt very quickly with a small overload.

The present trend is to use silver despite its high cost due to the following reasons :

(*i*) It is comparatively free from oxidation.
(*ii*) It does not deteriorate when used in dry air.
(*iii*) The coefficient of expansion of silver is so small that no critical fatigue occurs. Therefore, the fuse element can carry the rated current continuously for a long time.

(*iv*) The conductivity of silver is very high. Therefore, for a given rating of fuse element, the mass of silver metal required is smaller than that of other materials. This minimises the problem of clearing the mass of vapourised material set free on fusion and thus permits fast operating speed.

(*v*) Due to comparatively low specific heat, silver fusible elements can be raised from normal temperature to vapourisation quicker than other fusible elements. Moreover, the resistance of silver increases abruptly as the melting temperature is reached, thus making the transition from melting to vapourisation almost instantaneous. Consequently, operation becomes very much faster at higher currents.

(*vi*) Silver vapourises at a temperature much lower than the one at which its vapour will readily ionise. Therefore, when an arc is formed through the vapourised portion of the element, the arc path has high resistance. As a result, short-circuit current is quickly interrupted.

20.4 Important Terms

The following terms are much used in the analysis of fuses :

(*i*) **Current rating of fuse element.** It is the current which the fuse element can normally carry without overheating or melting. It depends upon the temperature rise of the contacts of the fuse holder, fuse material and the surroundings of the fuse.

(*ii*) **Fusing current.** It is the minimum current at which the fuse element melts and thus disconnects the circuit protected by it. Obviously, its value will be more than the current rating of the fuse element.

For a round wire, the approximate relationship between fusing current I and diameter d of the wire is

$$I = k\,d^{3/2}$$

where k is a constant, called the *fuse constant*. Its value depends upon the metal of which the fuse element is made. Sir W.H. Preece found the value of k for different materials as given in the table below :

S. No.	Material	Value of k	
		d in cm	d in mm
1	Copper	2530	80
2	Aluminium	1873	59
3	Tin	405·5	12·8
4	Lead	340·6	10·8

The fusing current depends upon the various factors such as :

(*a*) material of fuse element

(*b*) length – the smaller the length, the greater the current because a short fuse can easily conduct away all the heat

(*c*) diameter

(*d*) size and location of terminals

(*e*) previous history

(*f*) type of enclosure used

(*iii*) **Fusing factor.** It is the ratio of minimum fusing current to the current rating of the fuse element *i.e.*

$$\text{Fusing factor} = \frac{\text{Minimum fusing current}}{\text{Current rating of fuse}}$$

Its value is always more than one. The smaller the fusing factor, the greater is the difficulty in avoiding deterioration due to overheating and oxidation at rated carrying current. For a semi-enclosed or rewirable fuse which employs copper wire as the fuse element, the fusing factor is usually 2. Lower values of fusing factor can be employed for enclosed type cartridge fuses using silver or bimetallic elements.

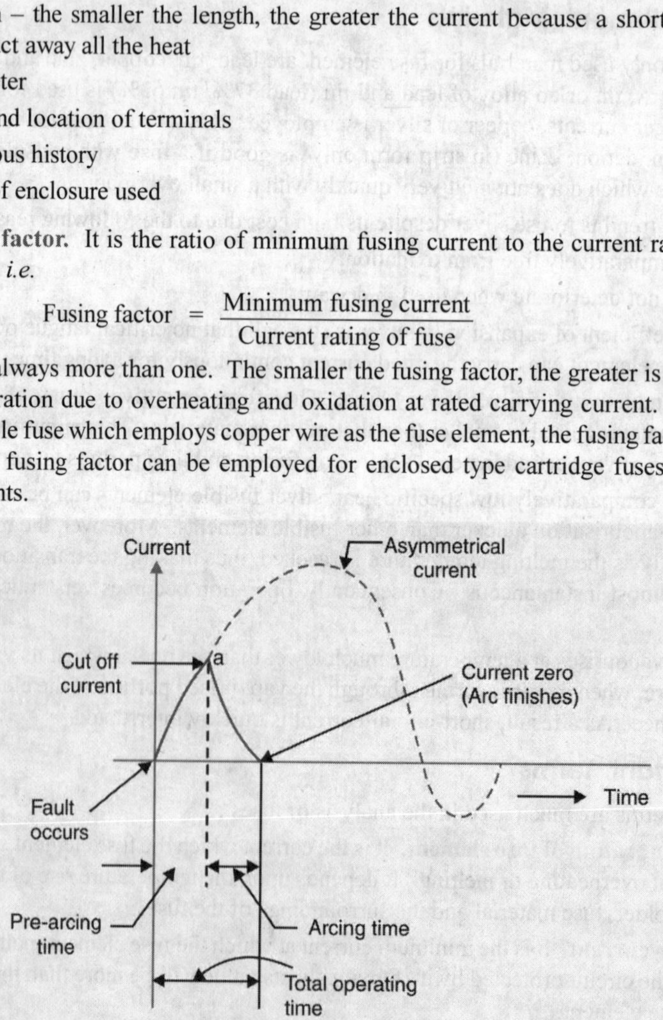

Fig. 20.2

(*iv*) **Prospective Current.** Fig. 20.2 shows how a.c. current is cut off by a fuse. The fault current would normally have a very large first loop, but it actually generates sufficient energy to melt the fuseable element well before the peak of this first loop is reached. The *r.m.s.* value of the first loop of fault current is known as prospective current. Therefore, prospective current can be defined as under:

It is the r.m.s. value of the first loop of the fault current obtained if the fuse is replaced by an ordinary conductor of negligible resistance.

(*v*) **Cut-off current.** It is the maximum value of fault current actually reached before the fuse melts.

On the occurrence of a fault, the fault current has a very large first loop due to a fair degree of asymmetry. The heat generated is sufficient to melt the fuse element well before the peak of first loop is reached (point '*a*' in Fig. 20.2). The current corresponding to point '*a*' is the cut off current. The cut off value depends upon :

(*a*) current rating of fuse
(*b*) value of prospective current
(*c*) asymmetry of short-circuit current

It may be mentioned here that outstanding feature of fuse action is the breaking of circuit *before* the fault current reaches its first peak. This gives the fuse a great advantage over a circuit breaker since the most severe thermal and electro-magnetic effects of short-circuit currents (which occur at the peak value of prospective current) are not experienced with fuses. Therefore, the circuits protected by fuses can be designed to withstand maximum current equal to the cut-off value. This consideration together with the relative cheapness of fuses allows much saving in cost.

(*vi*) **Pre-arcing time.** It is the time between the commencement of fault and the instant when cut off occurs.

When a fault occurs, the fault current rises rapidly and generates heat in the fuse element. As the fault current reaches the cut off value, the fuse element melts and an arc in initiated. The time from the start of the fault to the instant the arc is initiated is known as pre-arcing time. The pre-arcing time is generally small : a typical value being 0·001 second

(*vii*) **Arcing time.** This is the time between the end of pre-arcing time and the instant when the arc is extinguished.

(*viii*) **Total operating time.** It is the sum of pre-arcing and arcing times.

It may be noted that operating time of a fuse is generally quite low (say 0·002 sec.) as compared to a circuit breaker (say 0·2 sec or so). This is an added advantage of a fuse over a circuit breaker. A fuse in series with a circuit breaker of low-breaking capacity is a useful and economical arrangement to provide adequate short-circuit protection. It is because the fuse will blow under fault conditions before the circuit breaker has the time to operate.

(*ix*) **Breaking capacity.** It is the r.m.s. value of a.c. component of maximum prospective current that a fuse can deal with at rated service voltage.

20.5 Types of Fuses

Fuse is the simplest current interrupting device for protection against excessive currents. Since the invention of first fuse by Edison, several improvements have been made and now-a-days, a variety of fuses are available. Some fuses also incorporate means for extinguishing the arc that appears when the fuse element melts. In general, fuses may be classified into :

(*i*) Low voltages fuses (*ii*) High voltage fuses

It is a usual practice to provide isolating switches in series with fuses where it is necessary to permit fuses to be replaced or rewired with safety. If such means of isolation are not available, the fuses must be so shielded as to protect the user against accidental contact with the live metal when the fuse carrier is being inserted or removed.

20.6 Low Voltage Fuses

Low voltage fuses can be subdivided into two classes *viz.*, (*i*) semi-enclosed rewireable fuse (*ii*) high rupturing capacity (H.R.C.) cartridge fuse.

1. Semi-enclosed rewireable fuse. Rewireable fuse (also known as kit-kat type) is used where low values of fault current are to be interrupted. It consists of (*i*) a base and (*ii*) a fuse carrier. The base is of porcelain and carries the fixed contacts to which the incoming and outgoing phase wires are connected. The fuse carrier is also of porcelain and holds the fuse element (tinned copper wire) between its terminals. The fuse carrier can be inserted in or taken out of the base when desired.

When a fault occurs, the fuse element is blown out and the circuit is interrupted. The fuse carrier is taken out and the blown out fuse element is replaced by the new one. The fuse carrier is then re-inserted in the base to restore the supply. This type of fuse has two advantages. Firstly, the detachable fuse carrier permits the replacement of fuse element without any danger of coming in contact with live parts. Secondly, the cost of replacement is negligible.

Disadvantages

(*i*) There is a possibility of renewal by the fuse wire of wrong size or by improper material.

(*ii*) This type of fuse has a low-breaking capacity and hence cannot be used in circuits of high fault level.

(*iii*) The fuse element is subjected to deterioration due to oxidation through the continuous heating up of the element. Therefore, after some time, the current rating of the fuse is decreased *i.e.*, the fuse operates at a lower current than originally rated.

(*iv*) The protective capacity of such a fuse is uncertain as it is affected by the ambient conditions.

(*v*) Accurate calibration of the fuse wire is not possible because fusing current very much depends upon the length of the fuse element.

Semi-enclosed rewireable fuses are made upto 500 A rated current, but their breaking capacity is low *e.g.*, on 400 V service, the breaking capacity is about 4000 A. Therefore, the use of this type of fuses is limited to domestic and lighting loads.

2. High-Rupturing capacity (H.R.C.) cartridge fuse. The primary objection of low and uncertain breaking capacity of semi-enclosed rewireable fuses is overcome in H.R.C. cartridge fuse. Fig. 20.3 shows the essential parts of a typical H.R.C. cartridge fuse. It consists of a heat resisting ceramic body having metal end-caps to which is welded silver current-carrying element. The space within the body surrounding the element is completely packed with a filling powder. The filling material may be chalk, plaster of paris, quartz or marble dust and acts as an arc quenching and cooling medium.

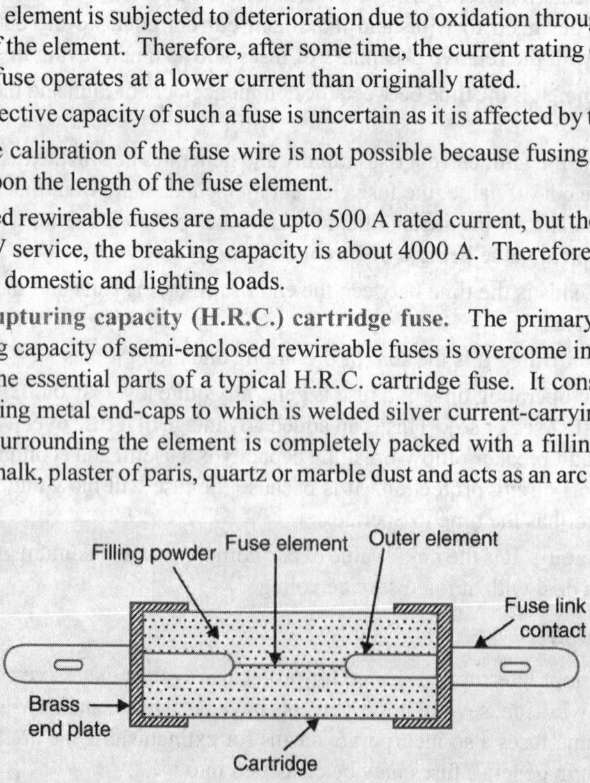

Fig. 20.3

Under normal load conditions, the fuse element is at a temperature below its melting point.

Therefore, it carries the normal current without overheating. When a fault occurs, the current increases and the fuse element melts before the fault current reaches its first peak. The heat produced in the process vapourises the melted silver element. The chemical reaction between the silver vapour and the filling powder results in the formation of a high resistance substance which helps in quenching the arc.

Advantages

(*i*) They are capable of clearing high as well as low fault currents.

(*ii*) They do not deteriorate with age.

(*iii*) They have high speed of operation.

(*iv*) They provide reliable discrimination.

(*v*) They require no maintenance.

(*vi*) They are cheaper than other circuit interrupting devices of equal breaking capacity.

(*vii*) They permit consistent performance.

Disadvantages

(*i*) They have to be replaced after each operation.

(*ii*) Heat produced by the arc may affect the associated switches.

3. **H.R.C. fuse with tripping device.** Sometime, H.R.C. cartridge fuse is provided with a tripping device. When the fuse blows out under fault conditions, the tripping device causes the circuit breaker to operate. Fig. 20.4 shows the essential parts of a H.R.C. fuse with a tripping device. The body of the fuse is of ceramic material with a metallic cap rigidly fixed at each end. These are connected by a number of silver fuse elements. At one end is a plunger which under fault conditions hits the tripping mechanism of the circuit breaker and causes it to operate. The plunger is electrically connected through a fusible link, chemical charge and a tungsten wire to the other end of the cap as shown.

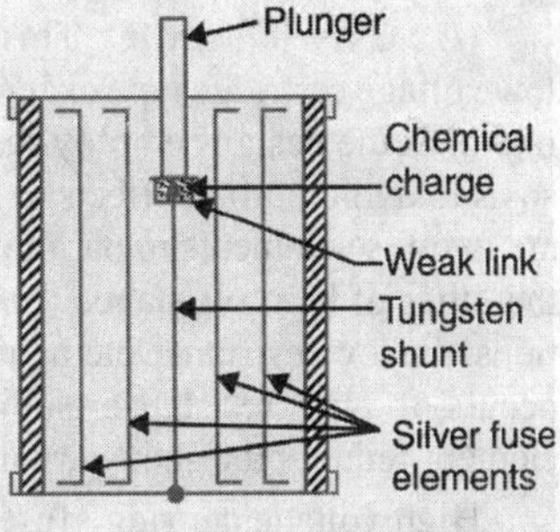

Fig. 20.4

When a fault occurs, the silver fuse elements are the first to be blown out and then current is transferred to the tungsten wire. The weak link in series with the tungsten wire gets fused and causes the chemical charge to be detonated. This forces the plunger outward to operate the circuit breaker. The travel of the plunger is so set that it is not ejected from the fuse body under fault conditions.

Advantages. H.R.C. fuse with a tripping device has the following advantages over a H.R.C. fuse without tripping device :

(*i*) In case of a single phase fault on a three-phase system, the plunger operates the tripping mechanism of circuit breaker to open all the three phases and thus prevents "single phasing".

(*ii*) The effects of full short circuit current need not be considered in the choice of circuit breaker. This permits the use of a relatively inexpensive circuit breaker.

(*iii*) The fuse-tripped circuit breaker is generally capable of dealing with fairly small fault currents itself. This avoids the necessity for replacing the fuse except after highest currents for which it is intended.

Low voltage H.R.C. fuses may be built with a breaking capacity of 16,000 A to 30,000 A at 440V. They are extensively used on low-voltage distribution system against over-load and short-circuit conditions.

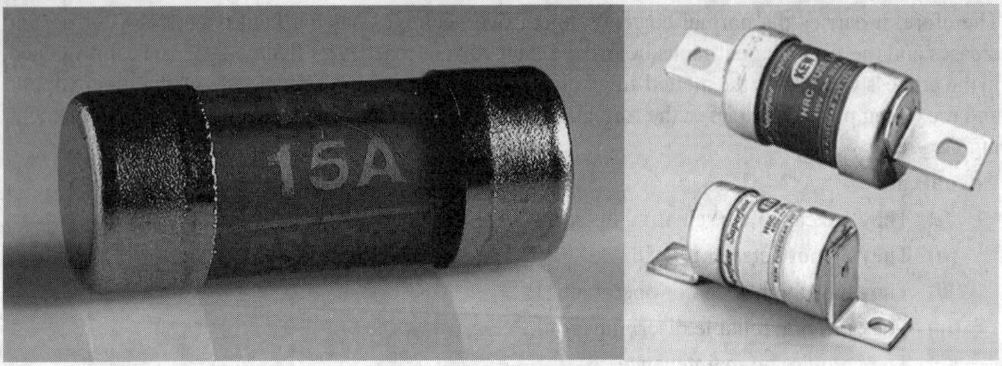

HRC Fuse

20.7 High Voltage Fuses

The low-voltage fuses discussed so far have low normal current rating and breaking capacity. Therefore, they cannot be successfully used on modern high voltage circuits. Intensive research by the manufacturers and supply engineers has led to the development of high voltage fuses. Some of the high voltage fuses are :

(*i*) **Cartridge type.** This is similar in general construction to the low voltage cartridge type except that special design features are incorporated. Some designs employ fuse elements wound in the form of a helix so as to avoid corona effects at higher voltages. On some designs, there are two fuse elements in parallel ; one of low resistance (silver wire) and the other of high resistance (tungsten wire). Under normal load conditions, the low resistance element carries the normal current. When a fault occurs, the low-resistance element is blown out and the high resistance element reduces the short-circuit current and finally breaks the circuit.

High voltage cartridge fuses are used upto 33 kV with breaking capacity of about 8700 A at that voltage. Rating of the order of 200 A at 6·6 kV and 11 kV and 50 A at 33 kV are also available.

(*ii*) **Liquid type.** These fuses are filled with carbon tetrachloride and have the widest range of application to h.v. systems. They may be used for circuits upto about 100 A rated current on systems upto 132 kV and may have breaking capacities of the order of 6100 A.

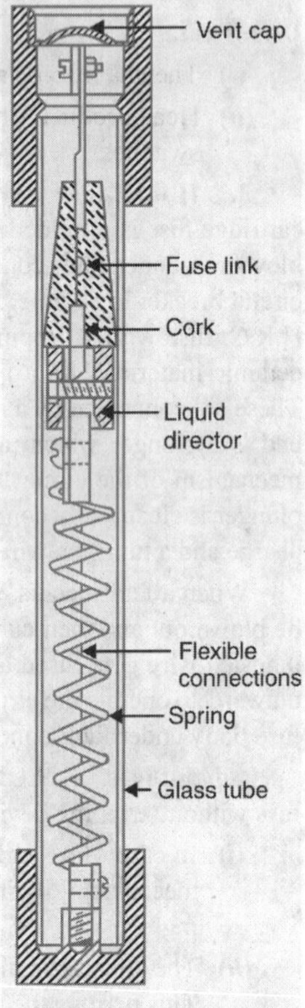

Fig. 20.5

Fig. 20.5 shows the essential parts of the liquid fuse. It consists of a glass tube filled with carbon tetrachloride solution and sealed at both ends with brass caps. The fuse wire is sealed at one end of the tube and the other end of the wire is held by a strong phosphor bronze spiral spring fixed at the other end of the glass tube. When the current exceeds the prescribed limit, the fuse wire is blown out. As the fuse melts, the spring retracts part of it through a baffle (or liquid director) and draws it well into the liquid. The small quantity of gas generated at the point of fusion forces some part of liquid into the passage through baffle and there it effectively extinguishes the arc.

(*iii*) **Metal clad fuses.** Metal clad oil-immersed fuses have been developed with the object of providing a substitute for the oil circuit breaker. Such fuses can be used for very high voltage circuits and operate most satisfactorily under short-circuit conditions approaching their rated capacity.

20.8 Current Carrying Capacity of Fuse Element

The current carrying capacity of a fuse element mainly depends on the metal used and the cross-sectional area but is affected also by the length, the state of surface and the surroundings of the fuse. When the fuse element attains steady temperature,

Heat produced per sec = Heat lost per second by convection, radiation and conduction

or $$I^2R = \text{Constant} \times \text{Effective surface area}$$

or $$I^2\left(\rho\frac{l}{a}\right) = \text{constant} \times d \times l$$

where d = diameter of fuse element

l = length of fuse element

$$\therefore \quad I^2\frac{\rho\, l}{(\pi/4)\, d^2} = \text{constant} \times d \times l$$

or $$I^2 = \text{constant} \times d^3$$

or $$I^2 \propto d^3 \qquad ...(i)$$

Expression (*i*) is known as ordinary *fuse law*.

Example 20.1. *A fuse wire of circular cross-section has a radius of 0·8 mm. The wire blows off at a current of 8A. Calculate the radius of the wire that will blow off at a current of 1A.*

Solution.

$$I^2 \propto r^3$$

$$\therefore \quad \left(\frac{I_2}{I_1}\right)^2 = \left(\frac{r_2}{r_1}\right)^3$$

or $$r_2 = r_1 \times \left(\frac{I_2}{I_1}\right)^{2/3} = 0\cdot 8 \times \left(\frac{1}{8}\right)^{2/3} = \mathbf{0\cdot 2\ mm}$$

20.9 Difference Between a Fuse and Circuit Breaker

It is worthwhile to indicate the salient differences between a fuse and a circuit breaker in the tabular form.

S. No.	Particular	Fuse	Circuit breaker
1.	*Function*	It performs both detection and interruption functions.	It performs interruption function only. The detection of fault is made by relay system.
2.	*Operation*	Inherently completely automatic.	Requires elaborate equipment (*i.e.* relays) for automatic action.
3.	*Breaking capacity*	Small	Very large
4.	*Operating time*	Very small (0·002 sec or so)	Comparatively large (0·1 to 0·2 sec)
5.	*Replacement*	Requires replacement after every operation.	No replacement after operation.

SELF - TEST

1. **Fill in the blanks by inserting appropriate words/figures**

(*i*) Fuses are generally used in circuits where .... operations are not expected.

(*ii*) The minimum time of operation of a fuse is .... than that of a circuit breaker.

(*iii*) A fuse element should have .... melting point.
(*iv*) The disadvantages of tin fuse element is that its vapour tends to .... when it blows out.
(*v*) The value of fusing factor is always .... than unity
(*vi*) Semi-enclosed rewireable fuses have .... breaking capacity.
(*vii*) A fuse has .... time current characteristics.
(*viii*) The action of a fuse is .... completely automatic.
(*ix*) The fuse element is generally made of ....
(*x*) The fuse melts well .... the first peak of fault current is reached.

2. **Pick up the correct words/figures from the brackets and fill in the blanks.**
(*i*) A fuse is .... than other circuit interrupting device of equal breaking capacity. (*cheaper, costlier*)
(*ii*) For the same material, heavy-current fuse wires must have .... diameters than for smaller currents. (*larger, smaller*)
(*iii*) A fuse performs .... functions. (*both detection and interruption, interruption*)
(*iv*) A fuse has .... breaking capacity as compared to a circuit breaker. (*low, high*)

ANSWERS TO SELF-TEST

1. (*i*) frequent (*ii*) smaller (*iii*) low (*iv*) maintain the arc (*v*) more (*vi*) low (*vii*) inverse (*viii*) inherently (*ix*) silver (*x*) before
2. (*i*) cheaper (*ii*) larger (*iii*) both detection and interruption (*iv*) low

CHAPTER REVIEW TOPICS

1. What is a fuse ? Discuss the advantages and disadvantages of a fuse.
2. Why do we prefer silver as a fuse element ?
3. Define and explain the following terms :
(*i*) fusing current (*ii*) cut off current (*iii*) operating time (*iv*) breaking capacity
4. Write short notes on the following :
(*i*) Semi-enclosed rewireable fuse
(*ii*) H.R.C. cartridge fuse
(*iii*) Difference between a fuse and circuit breaker

DISCUSSION QUESTIONS

1. Why are circuit breakers preferred to fuses ?
2. Why fuses cannot provide adequate discrimination on heavy short-circuit ?
3. Why fuses can interrupt heavy short-circuit currents successfully ?

CHAPTER

21

Protective Relays

Introduction

In a power system consisting of generators, transformers, transmission and distribution circuits, it is inevitable that sooner or later some failure will occur somewhere in the system. When a failure occurs on any part of the system, it must be quickly detected and disconnected from the system. There are two principal reasons for it. Firstly, if the fault is not cleared quickly, it may cause unnecessary interruption of service to the customers. Secondly, rapid disconnection of faulted apparatus limits the amount of damage to it and prevents the effects of fault from spreading into the system.

The detection of a fault and disconnection of a faulty section or apparatus can be achieved by using *fuses* or *relays in conjunction with circuit breakers*. A fuse performs both detection and interruption functions automatically but its use is limited for the protection of low-voltage circuits only. For high voltage circuits (say above 3·3 kV), relays and circuit breakers are employed to serve the desired function of automatic protective gear. The relays detect the fault and supply information to the circuit breaker which performs the function of circuit interruption. In this chapter, we shall focus our attention on the various types of relays and their increasing use for the protection of power system.

21.1 Protective Relays

***A protective relay** is a device that detects the fault and initiates the operation of the circuit breaker to ioslate the defective element from the rest of the system.*

The relays detect the abnormal conditions in the electrical circuits by constantly measuring the electrical quantities which are different under normal and fault conditions. The electrical quantities which may change under fault conditions are voltage, current, frequency and phase angle. Through the changes in one or more of these quantities, the faults signal their presence, type and location to the protective relays. Having detected the fault, the relay operates to close the trip circuit of the breaker. This results in the opening of the breaker and disconnection of the faulty circuit.

A typical relay circuit is shown in Fig. 21.1. This diagram shows one phase of 3-phase system for simplicity. The relay circuit connections can be divided into three parts *viz.*

(*i*) First part is the primary winding of a current transformer (C.T.) which is connected in series with the line to be protected.

(*ii*) Second part consists of secondary winding of C.T. and the relay operating coil.

(*iii*) Third part is the tripping circuit which may be either a.c. or d.c. It consists of a source of supply, the trip coil of the circuit breaker and the relay stationary contacts.

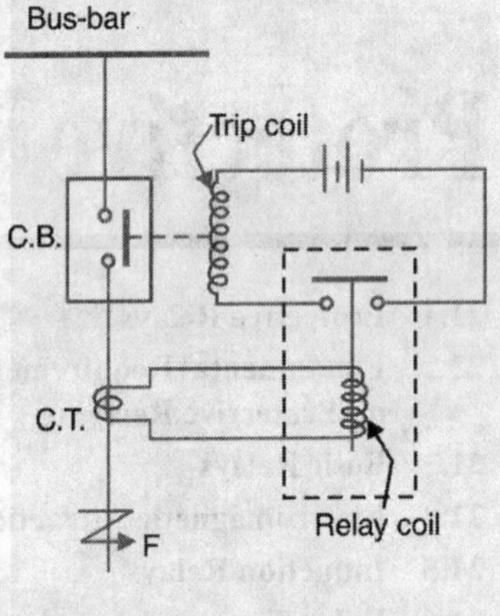

Fig. 21.1

When a short circuit occurs at point F on the transmission line, the current flowing in the line increases to an enormous value. This results in a heavy current flow through the relay coil, causing the relay to operate by closing its contacts. This in turn closes the trip circuit of the breaker, making the circuit breaker open and isolating the faulty section from the rest of the system. In this way, the relay ensures the safety of the circuit equipment from damage and normal working of the healthy portion of the system.

21.2 Fundamental Requirements of Protective Relaying

The principal function of protective relaying is to cause the prompt removal from service of any element of the power system when it starts to operate in an abnormal manner or interfere with the effective operation of the rest of the system. In order that protective relay system may perform this function satisfactorily, it should have the following qualities :

(*i*) selectivity (*ii*) speed (*iii*) sensitivity
(*iv*) reliability (*v*) simplicity (*vi*) economy

(*i*) **Selectivity.** *It is the ability of the protective system to select correctly that part of the system in trouble and disconnect the faulty part without disturbing the rest of the system.*

A well designed and efficient relay system should be selective *i.e.* it should be able to detect the point at which the fault occurs and cause the opening of the circuit breakers closest to the fault with minimum or no damage to the system. This can be illustrated by referring to the single line diagram of a portion of a typical power system shown in Fig. 21.2. It may be seen that circuit breakers are located in the connections to each power system element in order to make it possible to disconnect only the faulty section. Thus, if a fault occurs at bus-bars on the last zone, then only breakers nearest to the fault *viz.* 10, 11, 12 and 13 should open. In fact, opening of any other breaker to clear the fault will lead to a greater part of the system being disconnected.

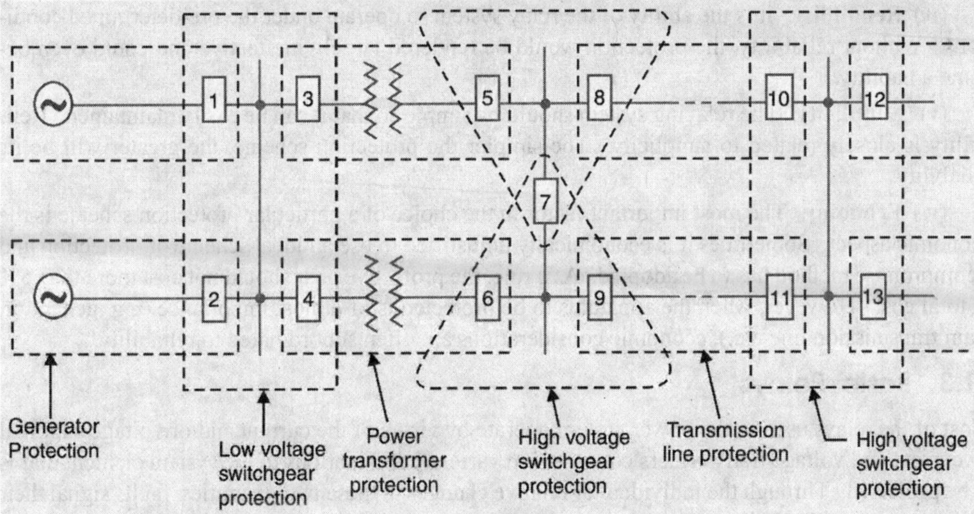

Fig. 21.2

In order to provide selectivity to the system, it is a usual practice to divide the entire system into several protection zones. When a fault occurs in a given zone, then only the circuit breakers within that zone will be opened. This will isolate only the faulty circuit or apparatus, leaving the healthy circuits intact. The system can be divided into the following protection zones :

(*a*) generators (*b*) low-tension switchgear (*c*) transformers

(*d*) high-tension switchgear (*e*) transmission lines

It may be seen in Fig. 21.2 that there is certain amount of overlap between the adjacent protection zones. For a failure within the region where two adjacent zones overlap, more breakers will be opened than the minimum necessary to disconnect the faulty section. But if there were no overlap, a failure in the region between zones would not lie in either region and, therefore, no breaker would be opened. For this reason, a certain amount of overlap* is provided between the adjacent zones.

***(ii)* Speed.** The relay system should disconnect the faulty section as fast as possible for the following reasons :

(*a*) Electrical apparatus may be damaged if they are made to carry the fault currents for a long time.

(*b*) A failure on the system leads to a great reduction in the system voltage. If the faulty section is not disconnected quickly, then the low voltage created by the fault may shut down consumers' motors and the generators on the system may become unstable.

(*c*) The high speed relay system decreases the possibility of development of one type of fault into the other more severe type.

***(iii)* Sensitivity.** *It is the ability of the relay system to operate with low value of actuating quantity.*

Sensitivity of a relay is a function of the volt-amperes input to the coil of the relay necessary to cause its operation. The smaller the volt-ampere input required to cause relay operation, the more sensitive is the relay. Thus, a 1 VA relay is more sensitive than a 3 VA relay. It is desirable that relay system should be sensitive so that it operates with low values of volt-ampere input.

* The overlap is usually small and the probability of failure in this region is low. Therefore, opening of too many breakers will be quite infrequent.

(*iv*) **Reliability.** It is the ability of the relay system to operate under the pre-determined conditions. Without reliability, the protection would be rendered largely ineffective and could even become a liability.

(*v*) **Simplicity.** The relaying system should be simple so that it can be easily maintained. Reliability is closely related to simplicity. The simpler the protection scheme, the greater will be its reliability.

(*vi*) **Economy.** The most important factor in the choice of a particular protection scheme is the economic aspect. Sometimes it is economically unjustified to use an ideal scheme of protection and a compromise method has to be adopted. As a rule, the protective gear should not cost more than 5% of total cost. However, when the apparatus to be protected is of utmost importance (*e.g.* generator, main transmission line etc.), economic considerations are often subordinated to reliability.

21.3 Basic Relays

Most of the relays used in the power system operate by virtue of the current and/or voltage supplied by current and voltage transformers connected in various combinations to the system element that is to be protected. Through the individual or relative changes in these two quantities, faults signal their presence, type and location to the protective relays. Having detected the fault, the relay operates the trip circuit which results in the opening of the circuit breaker and hence in the disconnection of the faulty circuit.

Most of the relays in service on electric power system today are of electro-mechanical type. They work on the following two main operating principles :

(*i*) Electromagnetic attraction

(*ii*) Electromagnetic induction

21.4 Electromagnetic Attraction Relays

Electromagnetic attraction relays operate by virtue of an armature being attracted to the poles of an electromagnet or a plunger being drawn into a solenoid. Such relays may be actuated by d.c. or a.c. quantities. The important types of electromagnetic attraction relays are :

(*i*) **Attracted armature type relay.** Fig. 21.3 shows the schematic arrangement of an attracted armature type relay. It consists of a laminated electromagnet *M* carrying a coil *C* and a pivoted laminated armature. The armature is balanced by a counterweight and carries a pair of spring contact fingers at its free end. Under normal operating conditions, the current through the relay coil *C* is such that counterweight holds the armature in the position shown. However, when a short-circuit occurs, the current through the relay coil increases sufficiently and the relay armature is attracted upwards. The contacts on the relay armature bridge a pair of stationary contacts attached to the relay frame. This completes the trip circuit which results in the opening of the circuit breaker and, therefore, in the disconnection of the faulty circuit.

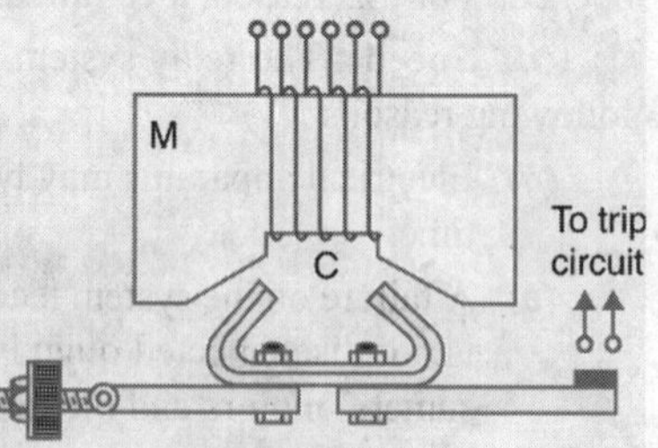

Fig. 21.3

The minimum current at which the relay armature is attracted to close the trip circuit is called *pickup* current. It is a usual practice to provide a number of tappings on the relay coil so that the number of turns in use and hence the setting value at which the relay operates can be varied.

(*ii*) **Solenoid type relay.** Fig. 21.4 shows the schematic arrangement of a solenoid type relay. It consists of a solenoid and movable iron plunger arranged as shown. Under normal operating conditions, the current through the relay coil *C* is such that it holds the plunger by gravity or spring in the position shown. However, on the occurrence of a fault, the current through the relay coil becomes

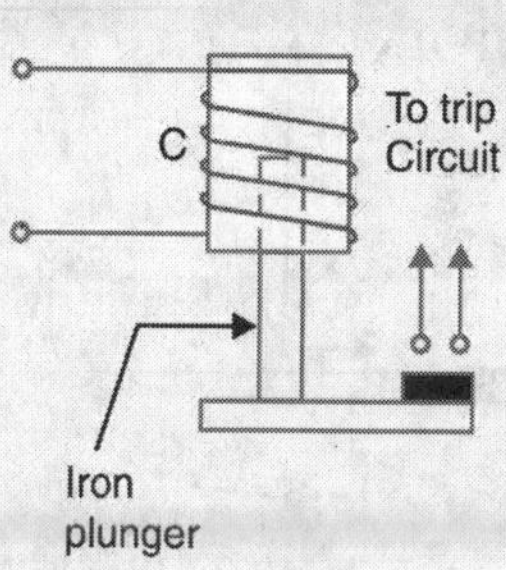

Fig. 21.4

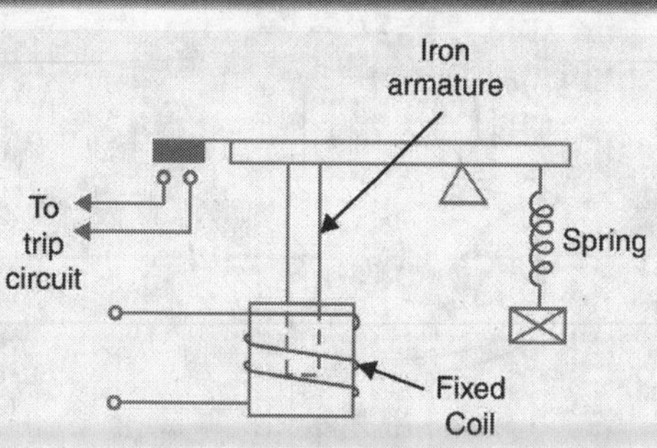

Fig. 21.5

more than the pickup value, causing the plunger to be attracted to the solenoid. The upward movement of the plunger closes the trip circuit, thus opening the circuit breaker and disconnecting the faulty circuit.

(*iii*) **Balanced beam type relay.** Fig. 21.5 shows the schematic arrangement of a balanced beam type relay. It consists of an iron armature fastened to a balance beam. Under normal operating conditions, the current through the relay coil is such that the beam is held in the horizontal position by the spring. However, when a fault occurs, the current through the relay coil becomes greater than the pickup value and the beam is attracted to close the trip circuit. This causes the opening of the circuit breaker to isolate the faulty circuit.

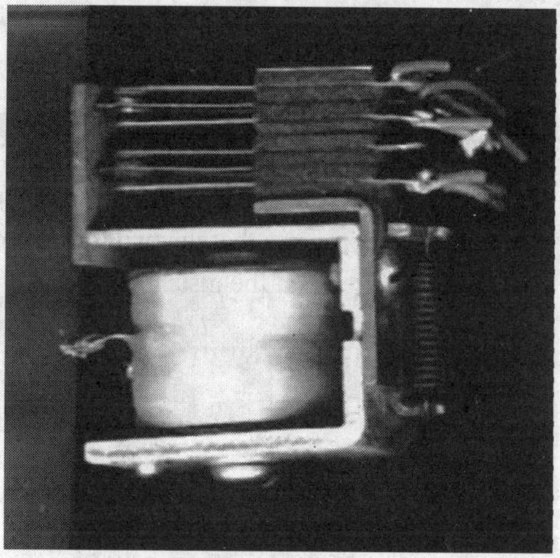

Electromagnetic relay

21.5 Induction Relays

Electromagnetic induction relays operate on the principle of induction motor and are widely used for protective relaying purposes involving a.c. quantities. They are not used with d.c. quantities owing to the principle of operation. An induction relay essentially consists of a pivoted aluminium disc placed in two alternating magnetic fields of the same frequency but displaced in time and space. The torque is produced in the disc by the interaction of one of the magnetic fields with the currents induced in the disc by the other.

To understand the production of torque in an induction relay, refer to the elementary arrangement shown in Fig. 21.6 (*i*). The two a.c. fluxes ϕ_2 and ϕ_1 differing in phase by an angle α induce e.m.f.s' in the disc and cause the circulation of eddy currents i_2 and i_1 respectively. These currents lag behind their respective fluxes by 90°.

Referring to Fig. 21.6 (*ii*) where the two a.c. fluxes and induced currents are shown separately for clarity, let

$$\phi_1 = \phi_{1max} \sin \omega t$$

$$\phi_2 = \phi_{2max} \sin (\omega t + \alpha)$$

where ϕ_1 and ϕ_2 are the instantaneous values of fluxes and ϕ_2 leads ϕ_1 by an angle α.

Assuming that the paths in which the rotor currents flow have negligible self-inductance, the rotor currents will be in phase with their voltages.

$$\therefore \quad i_1 \propto \frac{d\phi_1}{dt} \propto \frac{d}{dt}(\phi_{1max} \sin \omega t)$$

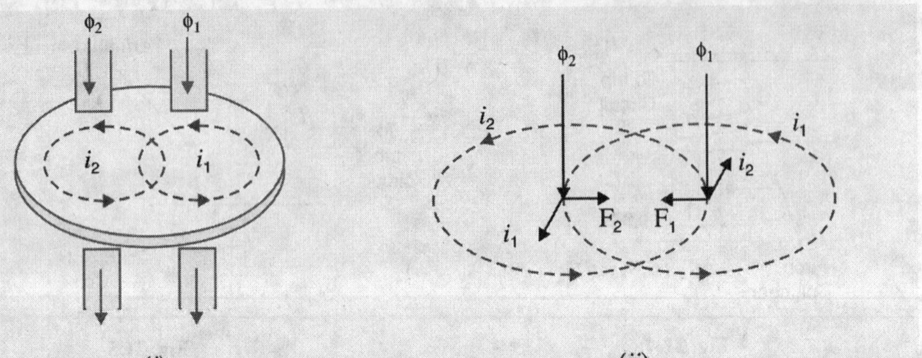

Fig. 21.6

$$\propto \phi_{1max} \cos \omega t$$

and $$i_2 \propto \frac{d\phi_2}{dt} \propto \phi_{2max} \cos(\omega t + \alpha)$$

Now, $$F_1 \propto \phi_1 i_2 \quad \text{and} \quad F_2 \propto \phi_2 i_1$$

Fig. 21.6 (*ii*) shows that the two forces are in opposition.

∴ Net force F at the instant considered is

$$F \propto F_2 - F_1$$
$$\propto \phi_2 i_1 - \phi_1 i_2$$
$$\propto \phi_{2max} \sin(\omega t + \alpha)\, \phi_{1max} \cos \omega t - \phi_{1max} \sin \omega t\, \phi_{2max} \cos(\omega t + \alpha)$$
$$\propto \phi_{1max}\, \phi_{2max} [\sin(\omega t + \alpha) \cos \omega t - \sin \omega t \cos(\omega t + \alpha)]$$
$$\propto \phi_{1max}\, \phi_{2max} \sin \alpha$$
$$\propto \phi_1 \phi_2 \sin \alpha \qquad ...(i)$$

where ϕ_1 and ϕ_2 are the r.m.s. values of the fluxes.

The following points may be noted from exp. (*i*) :

(*a*) The greater the phase angle α between the fluxes, the greater is the net force applied to the disc. Obviously, the maximum force will be produced when the two fluxes are 90° out of phase.

(*b*) The net force is the same at every instant. This fact does not depend upon the assumptions made in arriving at exp. (*i*).

(*c*) The direction of net force and hence the direction of motion of the disc depends upon which flux is leading.

The following three types of structures are commonly used for obtaining the phase difference in the fluxes and hence the operating torque in induction relays :

(*i*) shaded-pole structure

(*ii*) watthour-meter or double winding structure

(*iii*) induction cup structure

(*i*) **Shaded-pole structure.** The general arrangement of shaded-pole structure is shown in Fig. 21.7. It consists of a pivoted aluminium disc free to rotate in the air-gap of an electromagnet. One-half of each pole of the magnet is surrounded by a copper band known as *shading ring.* The alternating flux ϕ_s in the shaded protion of the poles will, owing to the reaction of the current induced in the ring, lag behind the flux ϕ_u in the unshaded portion by an angle α. These two a.c. fluxes differing in phase will produce the necessary torque to rotate the disc. As proved earlier, the driving torque T is given by;

$$T \propto \phi_s \, \phi_u \sin \alpha$$

Assuming the fluxes ϕ_s and ϕ_u to be proportional to the current I in the relay coil,

$$T \propto I^2 \sin \alpha$$

This shows that driving torque is proportional to the square of current in the relay coil.

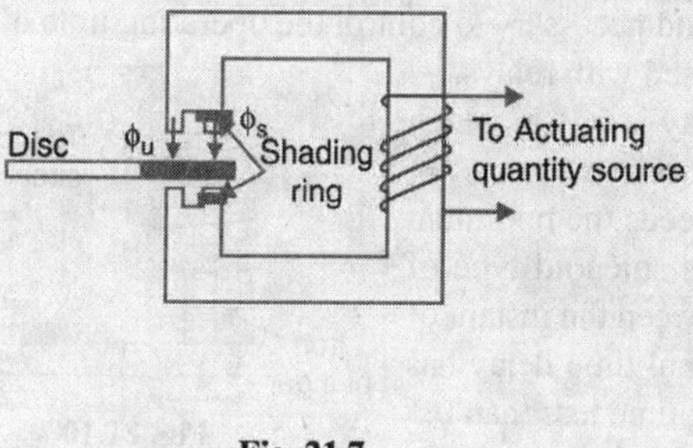

Fig. 21.7

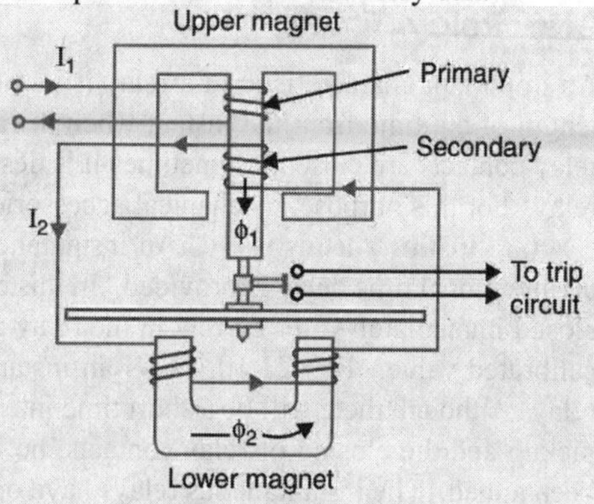

Fig. 21.8

(*ii*) **Watthour-meter structure.** This structure gets its name from the fact that it is used in watt-hour meters. The general arrangement of this type of relay is shown in Fig. 21.8. It consists of a pivoted aluminium disc arranged to rotate freely between the poles of two electromagnets. The upper electromagnet carries two windings ; the pirmary and the secondary. The primary winding carries the relay current I_1 while the secondary winding is connected to the winding of the lower magnet. The primary current induces e.m.f. in the secondary and so circulates a current I_2 in it. The flux ϕ_2 induced in the lower magnet by the current in the secondary winding of the upper magnet will lag behind ϕ_1 by an angle α. The two fluxes ϕ_1 and ϕ_2 differing in phase by α will produce a driving torque on the disc proportional to $\phi_1 \phi_2 \sin \alpha$.

An important feature of this type of relay is that its operation can be controlled by opening or closing the secondary winding circuit. If this circuit is opened, no flux can be set by the lower magnet however great the vaule of current in the pirmary winding may be and consequently no torque will be produced. Therefore, the relay can be made inoperative by opening its secondary winding circuit.

(*iii*) **Induction cup structure.** Fig. 21.9 shows the general arrangement of an induction cup structure. It most closely resembles an induction motor, except that the rotor iron is stationary, only the rotor conductor portion being free to rotate.

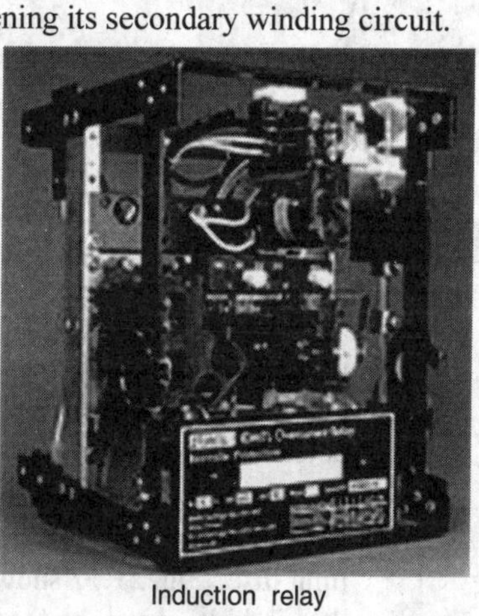

Induction relay

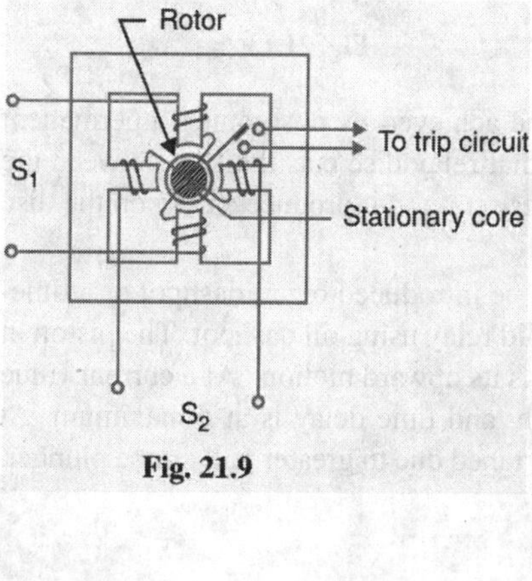

Fig. 21.9

The moving element is a hollow cylindrical rotor which turns on its axis. The rotating field is produced by two pairs of coils wound on four poles as shown. The rotating field induces currents in the cup to provide the necessary driving torque. If ϕ_1 and ϕ_2 represent the fluxes produced by the respective pairs of poles, then torque produced is proportional to $\phi_1 \phi_2 \sin \alpha$ where α is the phase difference between the two fluxes. A control spring and the back stop for closing of the contacts carried on an arm are attached to the spindle of the cup to prevent the continuous rotation.

Induction cup structures are more efficient torque producers than either the shaded-pole or the watthour meter structures. Therefore, this type of relay has very high speed and may have an operating time less then 0·1 second.

21.6 Relay Timing

An important characteristic of a relay is its time of operation. By 'the time of operation' is meant length of the time from the instant when the actuating element is energised to the instant when the relay contacts are closed. Sometimes it is desirable and necessary to control the operating time of a relay. For this purpose, mechanical accessories are used with relays.

(*i*) **Instantaneous relay.** An instantaneous relay is one in which no intentional time delay is provided. In this case, the relay contacts are closed immediately after current in the relay coil exceeds the minimum calibrated value. Fig. 21.10 shows an instantaneous solenoid type of relay. Although there will be a short time interval between the instant of pickup and the closing of relay contacts, no intentional time delay has been added. The instantaneous relays have operating time less than 0·1 second. The instantaneous relay is effective only where the impedance between the relay and source is small compared to the protected section impedance. The operating time of instantaneous relay is sometimes expressed in cycles based on the power-system frequency *e.g.* one-cycle would be 1/50 second in a 50-cycle system.

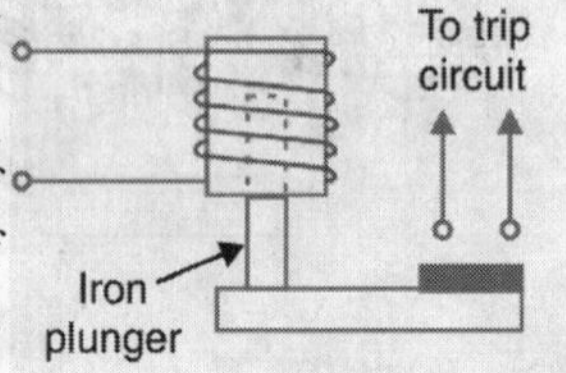

Fig. 21.10

(*ii*) **Inverse-time relay.** An inverse-time relay is one in which the operating time is approximately inversely proportional to the magnitude of the actuating quantity. Fig. 21.11 shows the time-current characteristics of an inverse current relay. At values of current less than pickup, the relay never operates. At higher values, the time of operation of the relay decreases steadily with the increase of current. The inverse-time delay can be achieved by associating mechanical accessories with relays.

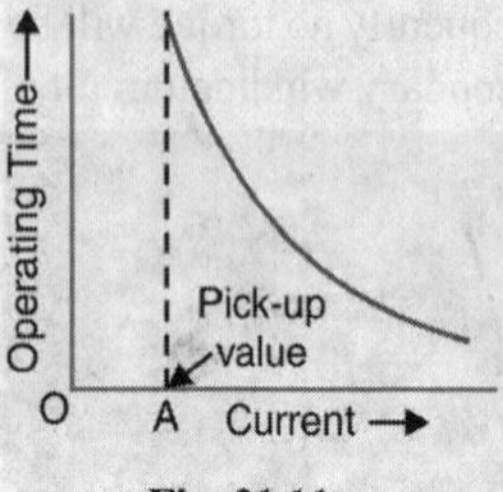

Fig. 21.11

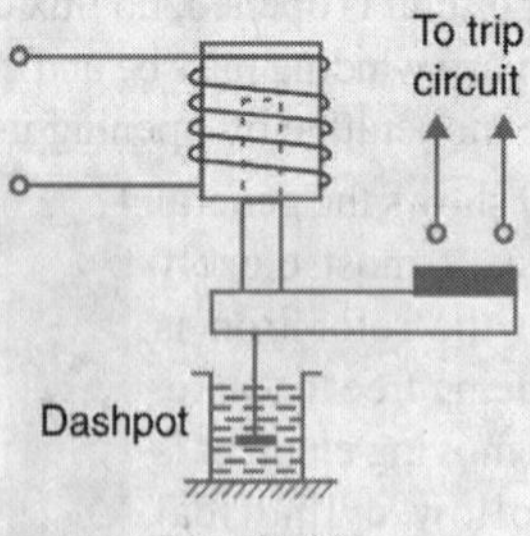

Fig. 21.12

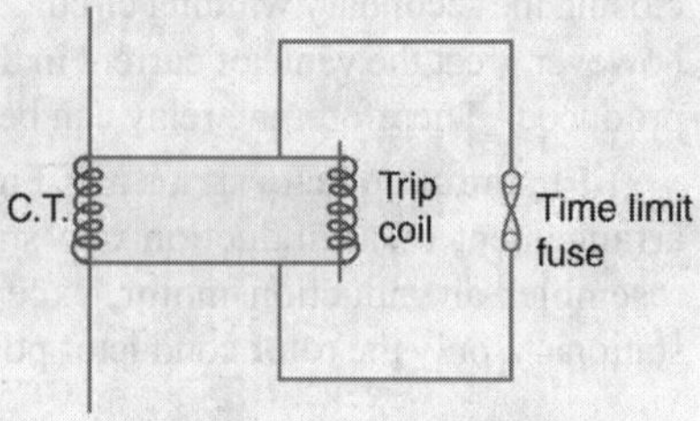

Fig. 21.13

(*a*) In an induction relay, the inverse-time delay can be achieved by positioning a permanent magnet (known as a *drag magnet*) in such a way that relay disc cuts the flux between the poles of the magnet. When the disc moves, currents set up in it produce a drag on the disc which slows its motion.

(*b*) In other types of relays, the inverse time delay can be introduced by oil dashpot or a time-limit fuse. Fig. 21.12 shows an inverse time solenoid relay using oil dashpot. The piston in the oil dashpot attached to the moving plunger slows its upward motion. At a current value just equal to the pickup, the plunger moves slowly and time delay is at a maximum. At higher values of relay current, the delay time is shortened due to greater pull on the plunger.

The inverse-time characteristic can also be obtained by connecting a time-limit fuse in parallel with the trip coil terminals as shown in Fig. 21.13. The shunt path formed by time-limit fuse is of negligible impedance as compared with the relatively high impedance of the trip coil. Therefore, so long as the fuse remains intact, it will divert practically the whole secondary current of CT from the trip oil. When the secondary current exceeds the current carrying capacity of the fuse, the fuse will blow and the whole current will pass through the trip coil, thus opening the circuit breaker. The time-lag between the incidence of excess current and the tripping of the breaker is governed by the characteristics of the fuse. Careful selection of the fuse can give the desired inverse-time characteristics, although necessity for replacement after operation is a disadvantage.

(*iii*) Definite time lag relay. In this type of relay, there is a definite time elapse between the instant of pickup and the closing of relay contacts. This particular time setting is independent of the amount of current through the relay coil ; being the same for all values of current in excess of the pickup value. It may be worthwhile to mention here that practically all inverse-time relays are also provided with definite minimum time feature in order that the relay may never become instantaneous in its action for very long overloads.

21.7 Important Terms

It is desirable to define and explain some important terms much used in connection with relays.

(*i*) Pick-up current. It is the minimum current in the relay coil at which the relay starts to operate. So long as the current in the relay is less than the pick-up value, the relay does not operate and the breaker controlled by it remains in the closed position. However, when the relay coil current is equal to or greater than the pickup value, the relay operates to energise the trip coil which opens the circuit breaker.

(*ii*) Current setting. It is often desirable to adjust the pick-up current to any required value. This is known as current setting and is usually achieved by the use of tappings on the relay operating coil. The taps are brought out to a plug bridge as shown in Fig. 21.14. The plug bridge permits to alter the number of turns on the relay coil. This changes the torque on the disc and hence the time of operation of the relay. The values assigned to each tap are expressed in terms of percentage full-load rating of C.T. with which the relay is associated and represents the value *above* which the disc commences to rotate and finally closes the trip circuit.

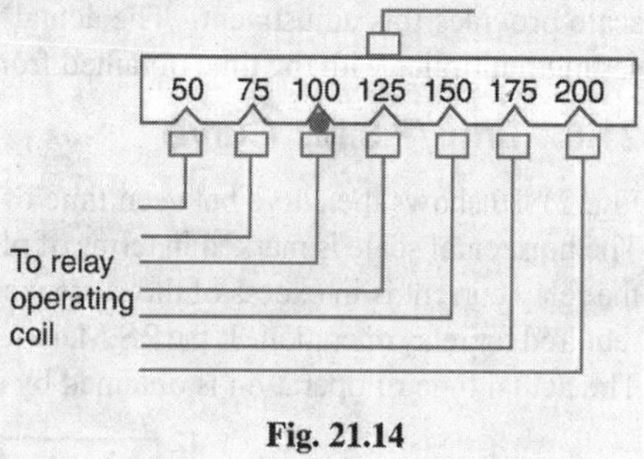

Fig. 21.14

$\therefore$ Pick-up current = Rated secondary current of C.T. × Current setting

For example, suppose that an overcurrent relay having current setting of 125% is connected to a supply circuit through a current transformer of 400/5. The rated secondary current of C.T. is 5 amperes. Therefore, the pick-up value will be 25% more than 5 A *i.e.* $5 \times 1{\cdot}25 = 6{\cdot}25$ A. It means that with above current setting, the relay will actually operate for a relay coil current equal to or greater than 6·25 A.

The current plug settings usually range from 50% to 200% in steps of 25% for overcurrent relays and 10% to 70% in steps of 10% for earth leakage relays. The desired current setting is obtained by inserting a plug between the jaws of a bridge type socket at the tap value required.

(*iii*) Plug-setting multiplier (P.S.M.). It is the ratio of fault current in relay coil to the pick-up current *i.e.*

$$\text{P.S.M.} = \frac{\text{Fault current in relay coil}}{\text{Pick-up current}}$$

$$= \frac{\text{Fault current in relay coil}}{\text{Rated secondary current of } CT \times \text{Current setting}}$$

For example, suppose that a relay is connected to a 400/5 current transformer and set at 150%. With a primary fault current of 2400 A, the plug-setting multiplier can be calculated as under :

$$\text{Pick-up value} = \text{Rated secondary current of } CT \times \text{Current setting}$$

$$= 5 \times 1{\cdot}5 = 7{\cdot}5 \text{ A}$$

$$\text{Fault current in relay coil} = 2400 \times \frac{5}{400} = 30 \text{ A}$$

$$\therefore \quad \text{P.S.M.} = 30/7{\cdot}5 = 4$$

(iv) **Time-setting multiplier.** A relay is generally provided with control to adjust the time of operation. This adjustment is known as time-setting multiplier. The time-setting dial is calibrated from 0 to 1 in steps of 0.05 sec (see Fig. 21.15). These figures are multipliers to be used to convert the time derived from time/P.S.M. curve into the actual operating time. Thus if the time setting is 0·1 and the time obtained from the time/P.S.M. curve is 3 seconds, then actual relay operating time = 3 × 0·1 = 0·3 second. For instance, in an induction relay, the time of operation is controlled by adjusting the amount of travel of the disc from its reset position to its pickup position. This is achieved by the adjustment of the position of a movable backstop which controls the travel of the disc and thereby varies the time in which the relay will close its contacts for given values of fault current. A so-called "time dial" with an evenly divided scale provides this adjustment. The acutal time of operation is calculated by multiplying the time setting multiplier with the time obtained from time/P.S.M. curve of the relay.

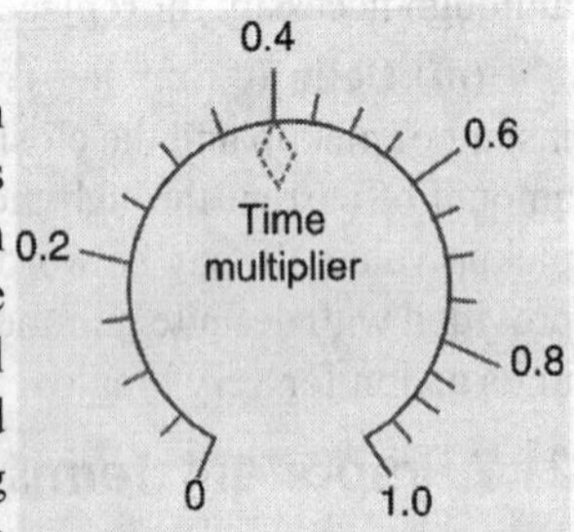

Fig. 21.15

21.8 Time/P.S.M. Curve

Fig. 21.16 shows the curve between time of operation and plug setting multiplier of a typical relay. The horizontal scale is marked in terms of plug-setting multiplier and represents the number of times the relay current is in excess of the current setting. The vertical scale is marked in terms of the time required for relay operation. If the P.S.M. is 10, then the time of operation (from the curve) is 3 seconds. The actual time of operation is obtained by multiplying this time by the time-setting multiplier.

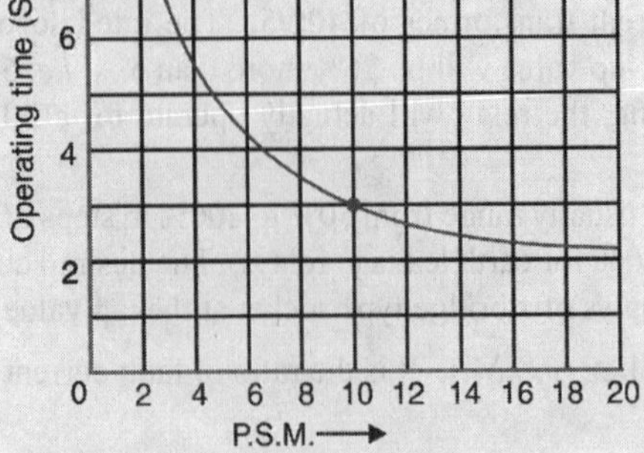

Fig. 21.16

It is evident from Fig. 21.16 that for lower values of overcurrent, time of operation varies inversely with the current but as the current approaches 20 times full-load value, the operating time of relay tends to become constant. This feature is necessary in order to ensure discrimination on very heavy fault currents flowing through sound feeders.

21.9 Calculation of Relay Operating Time

In order to calculate the actual relay operating time, the following things must be known :

(*a*) Time/P.S.M. curve
(*b*) Current setting
(*c*) Time setting
(*d*) Fault current
(*e*) Current transformer ratio

The procedure for calculating the actual relay operating time is as follows :

(*i*) Convert the fault current into the relay coil current by using the current transformer ratio.
(*ii*) Express the relay current as a multiple of current setting *i.e.* calculate the P.S.M.
(*iii*) From the Time/P.S.M. curve of the relay, read off the time of operation for the calculated P.S.M.
(*iv*) Determine the actual time of operation by multiplying the above time of the relay by time-setting multiplier in use.

Example 21.1. *Determine the time of operation of a 5-ampere, 3-second overcurrent relay having a current setting of 125% and a time setting multiplier of 0·6 connected to supply circuit through a 400/5 current transformer when the circuit carries a fault current of 4000 A. Use the curve shown in Fig. 21.16.*

Solution.

$$\text{Rated secondary current of C.T.} = 5 \text{ A}$$

$$\text{Pickup current} = 5 \times 1{\cdot}25 = 6{\cdot}25 \text{ A}$$

$$\text{Fault current in relay coil} = 4000 \times \frac{5}{400} = 50 \text{ A}$$

$$\therefore \quad \text{Plug-setting multiplier (P.S.M.)} = \frac{50}{6{\cdot}25} = 8$$

Corresponding to the plug-setting multiplier of 8 (See Fig. 21.16), the time of operation is 3.5 seconds.

$$\therefore \quad \text{Actual relay operating time} = 3{\cdot}5 \times \text{Time-setting} = 3{\cdot}5 \times 0{\cdot}6 = \mathbf{2{\cdot}1\ seconds}$$

21.10 Functional Relay Types

Most of the relays in service on power system today operate on the principle of electromagnetic attraction or electromagnetic induction. Regardless of the principle involved, relays are generally classified according to the function they are called upon to perform in the protection of elelctric power circuits. For example, a relay which recognises overcurrent in a circuit (*i.e.* current greater than that which can be tolerated) and initiates corrective measures would be termed as an *overcurrent relay* irrespective of the relay design. Similarly an *overvoltage relay* is one which recognises overvoltage in a circuit and initiates the corrective measures. Although there are several types of special-function relays, only the following important types will be discussed in this chapter :

(*i*) Induction type overcurrent relays
(*ii*) Induction type reverse power relays
(*iii*) Distance relays
(*iv*) Differential relays
(*v*) Translay scheme

21.11 Induction Type Overcurrent Relay (*non-directional*)

This type of relay works on the induction principle and initiates corrective measures when current in the circuit exceeds the predetermined value. The actuating source is a current in the circuit supplied to the relay from a current transformer. These relays are used on a.c. circuits only and can operate for fault current flow in either direction.

Constructional details. Fig. 21.17 shows the important constructional details of a typical non-directional induction type overcurrent relay. It consists of a metallic (aluminium) disc which is free to rotate inbetween the poles of two electromagnets. The upper electromagnet has a primary and a secondary winding. The primary is connected to the secondary of a C.T. in the line to be protected and is tapped at intervals. The tappings are connected to a plug-setting bridge by which the number of active turns on the relay operating coil can be varied, thereby giving the desired current setting. The secondary winding is energised by induction from primary and is connected in series with the winding on the lower magnet. The controlling torque is provided by a spiral spring.

The spindle of the disc carries a moving contact which bridges two fixed contacts (connected to trip circuit) when the disc rotates through a pre-set angle. This angle can be adjusted to any value between 0° and 360°. By adjusting this angle, the travel of the moving contact can be adjusted and hence the relay can be given any desired time setting.

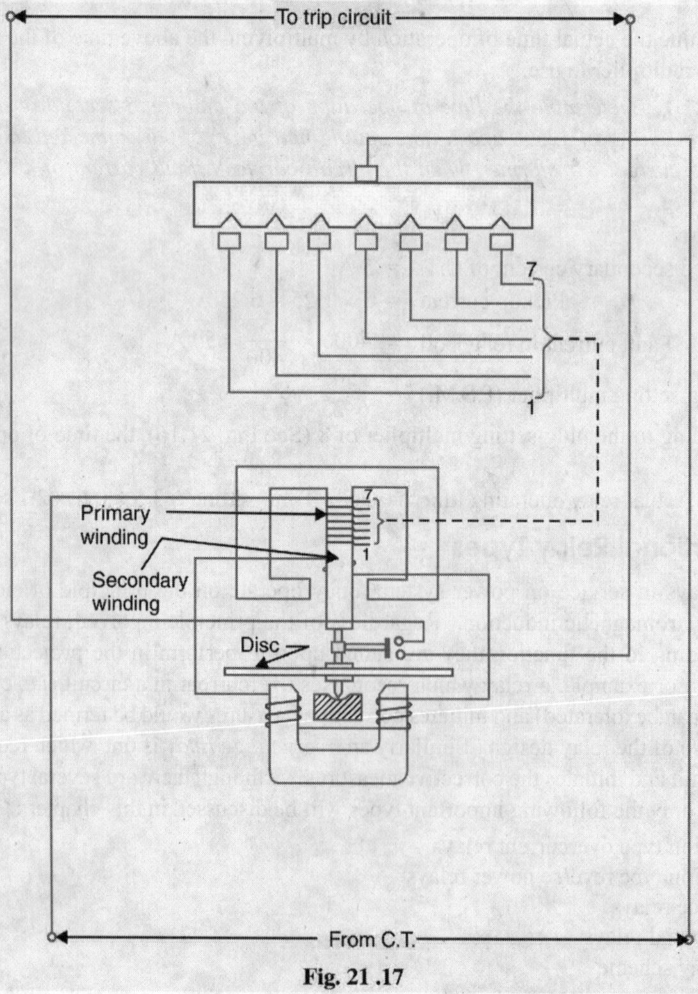

Fig. 21.17

Operation. The driving torque on the aluminium disc is set up due to the induction principle as discussed in Art. 21.5. This torque is opposed by the restraining torque provided by the spring. Under normal operating conditions, restraining torque is greater than the driving torque produced by the relay coil current. Therefore, the aluminium disc remains stationary. However, if the current in the protected circuit exceeds the pre-set value, the driving torque becomes greater than the restraining torque. Consequently, the disc rotates and the moving contact bridges the fixed contacts when the disc has rotated through a pre-set angle. The trip circuit operates the circuit breaker which isolates the faulty section.

21.12 Induction Type Directional Power Relay

This type of relay operates when power in the circuit flows in a specific direction. Unlike a *non-directional overcurrent relay, a directional power relay is so designed that it obtains its operating torque by the interaction of magnetic fields derived from both voltage and current source of the circuit it protects. Thus this type of relay is essentially a wattmeter and the direction of the torque set up in the relay depends upon the direction of the current relative to the voltage with which it is associated.

Constructional details. Fig. 21.18 shows the essential parts of a typical induction type directional power relay. It consists of an aluminum disc which is free to rotate inbetween the poles of two electromagnets. The upper electromagnet carries a winding (called *potential coil*) on the central limb which is connected through a potential transformer (P.T.) to the circuit voltage source. The lower electromagnet has a separate winding (called *current coil*) connected to the secondary of C.T. in the line to be protected. The current coil is provided with a number of tappings connected to the plug-setting bridge (not shown for clarity). This permits to have any desired current setting. The restraining torque is provided by a spiral spring.

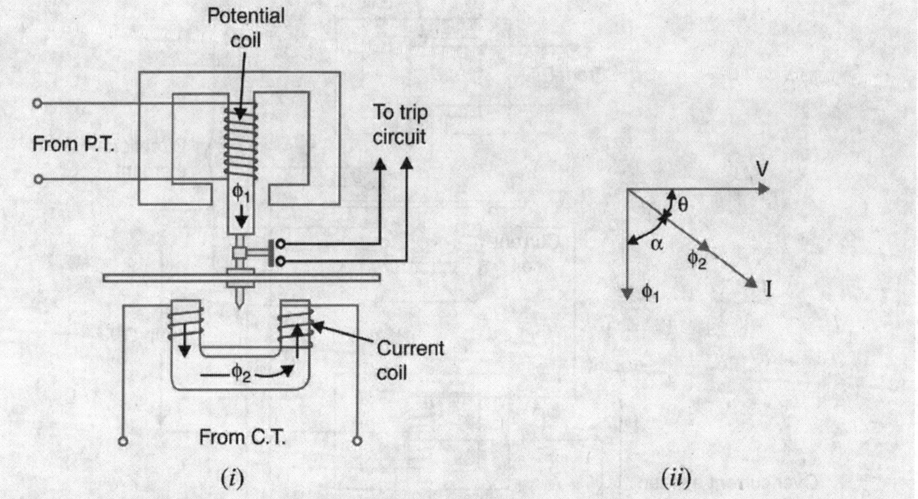

Fig. 21.18

The spindle of the disc carries a moving contact which bridges two fixed contacts when the disc has rotated through a pre-set angle. By adjusting this angle, the travel of the moving disc can be adjusted and hence any desired time-setting can be given to the relay.

Operation. The flux ϕ_1 due to current in the potential coil will be nearly 90° lagging behind the applied voltage V. The flux ϕ_2 due to current coil will be nearly in phase with the operating current I

* where the driving torque is due to the interaction of magnetic fields derived from current in the relay winding.

[See vector diagram in Fig. 21.18 (*ii*)]. The interaction of fluxes ϕ_1 and ϕ_2 with the eddy currents induced in the disc produces a driving torque given by :

$$T \propto \phi_1 \phi_2 \sin \alpha \qquad \text{[See Art. 21.5]}$$

$$\text{Since} \quad \phi_1 \propto V, \quad \phi_2 \propto I \quad \text{and} \quad \alpha = 90 - \theta$$

$$\therefore \quad T \propto VI \sin (90 - \theta)$$

$$\propto VI \cos \theta$$

$$\propto \text{power in the circuit}$$

It is clear that the direction of driving torque on the disc depends upon the direction of power flow in the circuit to which the relay is associated. When the power in the circuit flows in the normal direction, the driving torque and the restraining torque (due to spring) help each other to turn away the moving contact from the fixed contacts. Consequently, the relay remains inoperative. However, the reversal of current in the circuit reverses the direction of driving torque on the disc. When the reversed driving torque is large enough, the disc rotates in the reverse direction and the moving contact closes the trip circuit. This causes the operation of the circuit breaker which disconnects the faulty section.

21.13 Induction Type Directional Overcurrent Relay

The directional power relay discussed above is unsuitable for use as a directional protective relay under short-circuit conditions. When a short-circuit occurs, the system voltage falls to a low value and there may be *insufficient torque developed in the relay to cause its operation. This difficulty is overcome in the directional overcurrent relay which is designed to be almost independent of system voltage and power factor.

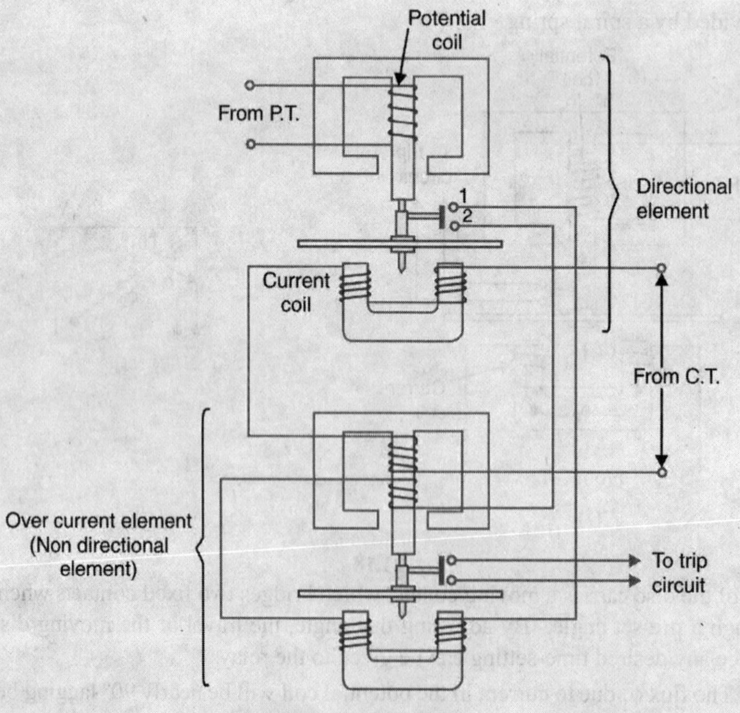

Fig. 21.19

* Directional power relay being of wattmeter type, the driving torque on the disc also depends upon the system voltage.

Constructional details. Fig. 21.19 shows the constructional details of a typical induction type directional ovecurrent relay. It consists of two relay elements mounted on a common case *viz.* (*i*) directional element and (*ii*) non-directional element.

(*i*) *Directional element.* It is essentially a directional power relay which operates when power flows in a specific direction. The potential coil of this element is connected through a potential transformer (P.T.) to the system voltage. The current coil of the element is energised through a C.T. by the circuit current. This winding is carried over the upper magnet of the non-directional element. The trip contacts (1 and 2) of the directional element are connected in series with the secondary circuit of the overcurrent element. Therefore, the latter element cannot start to operate until its secondary circuit is completed. In other words, the directional element must operate first (*i.e.* contacts 1 and 2 should close) in order to operate the overcurrent element.

(*ii*) *Non-directional element.* It is an overcurrent element similar in all respects to a non-directional overcurrent relay described in Art. 21.11. The spindle of the disc of this element carries a moving contact which closes the fixed contacts (trip circuit contacts) after the operation of directional element.

It may be noted that plug-setting bridge is also provided in the relay for current setting but has been omitted in the figure for clarity and simplicity. The tappings are provided on the upper magnet of overcurrent element and are connected to the bridge.

Operation. Under normal operating conditions, power flows in the normal direction in the circuit protected by the relay. Therefore, directional power relay (upper element) does not operate, thereby keeping the overcurrent element (lower element) unenergised. However, when a short-circuit occurs, there is a tendency for the current or power to flow in the reverse direction. Should this happen, the disc of the *upper element rotates to bridge the fixed contacts 1 and 2. This completes the circuit for overcurrent element. The disc of this element rotates and the moving contact attached to it closes the trip circuit. This operates the circuit breaker which isolates the faulty section. The two relay elements are so arranged that final tripping of the current controlled by them is not made till the following conditions are satisfied :

(*i*) current flows in a direction such as to operate the directional element.

(*ii*) current in the reverse direction exceeds the pre-set value.

(*iii*) excessive current persists for a period corresponding to the time setting of overcurrent element.

21.14 Distance or Impedance Relays

The operation of the relays discusssed so far depended upon the magnitude of current or power in the protected circuit. However, there is another group of relays in which the operation is governed by the ratio of applied voltage to current in the protected circuit. Such relays are called distance or **impedance relays. In an impedance relay, the torque produced by a current element is opposed by the torque produced by a voltage element. The relay will operate when the ratio *V/I* is less than a predetermined value.

Fig. 21.20 illustrates the basic principle of operation of an impedance relay. The voltage element of the relay is excited through a potential transformer (P.T.) from the line to be protected. The current element of the relay is excited from a current transformer (C.T.) in series with the line. The portion *AB* of the line is the protected zone. Under normal operating conditions, the impedance of the protected zone is Z_L. The relay is so designed that it closes its contacts whenever impedance of the protected section falls below the pre-determined value *i.e.* Z_L in this case.

* The directional element is made very sensitive so that with the lowest value of voltage which may be anticipated under severe fault conditions, sufficient torque is produced by the current winding to complete the operation of the element and allow its contacts to close.

** Impedance is an electrical measure of distance along a transmission line.

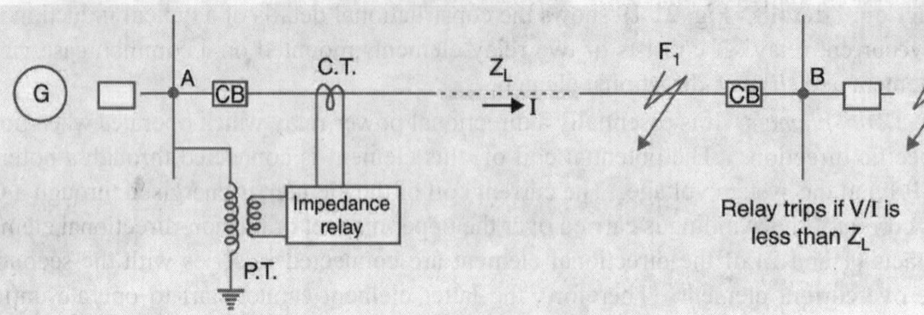

Fig. 21.20

Now suppose a fault occurs at point F_1 in the protected zone. The impedance Z (= *V/I) between the point where the relay is installed and the point of fault will be less than Z_L and hence the relay operates. Should the fault occur beyond the protected zone (say point F_2), the impedance Z will be greater than Z_L and the relay does not operate.

Types. A distance or impedance relay is essentially an ohmmeter and operates whenever the impedance of the protected zone falls below a pre-determined value. There are two types of distance relays in use for the protection of power supply, namely ;

(*i*) *Definite-distance relay* which operates instantaneously for fault upto a pre-determined distance from the relay.

(*ii*) *Time-distance relay* in which the time of operation is proportional to the distance of fault from the relay point. A fault nearer to the relay will operate it earlier than a fault farther away from the relay.

It may be added here that the distance relays are produced by modifying either of two types of basic relays; the balance beam or the induction disc.

21.15 Definite – Distance Type Impedance Relay

Fig. 21.21 shows the schematic arrangement of a definite-distance type impedance relay. It consists of a pivoted beam F and two electromagnets energised respectively by a current and voltage transformer in the protected circuit. The armatures of the two electromagnets are mechanically coupled to the beam on the opposite sides of the fulcrum. The beam is provided with a bridging piece for the trip contacts. The relay is so designed that the torques produced by the two electromagnets are in the opposite direction.

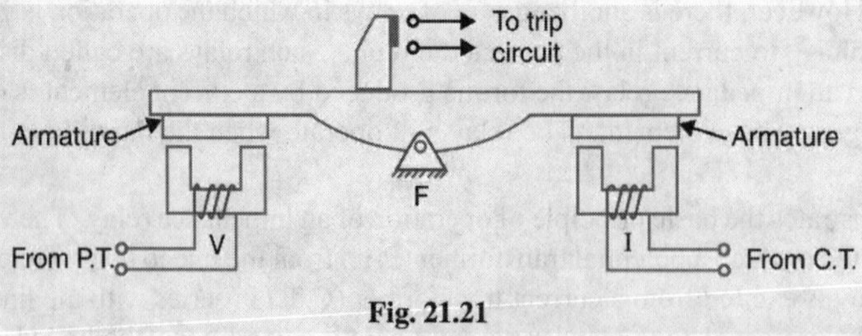

Fig. 21.21

Operation. Under normal operating conditions, the pull due to the voltage element is **greater than that of the current element. Therefore, the relay contacts remain open. However, when a fault occurs in the protected zone, the applied voltage to the relay decreases whereas the current increases. The ratio of voltage to current (*i.e.* impedance) falls below the pre-determined value. Therefore, the

* where I is the fault current and V is the voltage at the sub-station.

** Because the current I corresponds to normal load :

pull of the current element will exceed that due to the voltage element and this causes the beam to tilt in a direction to close the trip contacts.

The pull of the current element is proportional to I^2 and that of voltage element to V^2. Consequently, the relay will operate when

$$k_1 V^2 < k_2 I^2$$

or $$\frac{V^2}{I^2} < \frac{k_2}{k_1}$$

or $$\frac{V}{I} < \sqrt{\frac{k_2}{k_1}}$$

or $$Z < \sqrt{\frac{k_2}{k_1}}$$

The value of the constants k_1 and k_2 depends upon the ampere-turns of the two electromagnets. By providing tappings on the coils, the setting value of the relay can be changed.

21.16 Time-Distance Impedance Relay

A time-distance impedance relay is one which automatically adjusts its operating time according to the distance of the relay from the fault point *i.e.*

$$\text{Operating time, } T \propto V/I$$
$$\propto Z$$
$$\propto \text{distance}$$

Construction. Fig. 21.22 shows the schematic arrangement of a typical induction type time-distance impedance relay. It consists of a current driven induction element similar to the double-winding type induction overcurrent relay (refer back to Fig. 21.8). The spindle carrying the disc of this element is connected by means of a spiral spring coupling to a second spindle which carries the bridging piece of the relay trip contacts. The bridge is normally held in the open position by an armature held against the pole face of an electromagnet excited by the voltage of the circuit to be protected.

Operation. Under normal load conditions, the pull of the armature is more than that of the induction element and hence the trip circuit contacts remain open. However, on the occurence of a short-circuit, the disc of the induction current element starts to rotate at a speed depending upon the operating current. As the rotation of the disc proceeds, the spiral spring coupling is wound up till the tension of the spring is sufficient to pull the armature away from the pole face of the *voltage-excited magnet. Immediately this occurs, the spindle carrying the armature and bridging piece moves rapidly in response to the tension of the spring and trip contacts are closed. This opens the circuit breaker to isolate the faulty section.

The speed of rotation of the disc is approximately proportional to the operating current, neglecting the effect of control spring. Also the time of operation of the relay is directly proportional to the pull of the voltage-excited magnet and hence to the line voltage V at the point where the relay is connected. Therefore, the time of operation of relay would vary as V/I *i.e.* as Z or distance.

---

* The angle through which the disc travels before relay operates depends upon the pull of the voltage-excited magnet. The greater the pull, the greater will be the travel of the disc. The pull of this magnet depends upon the line voltage. The greater the line voltage, the greater the pull and hence larger will be the travel of the disc *i.e.* operating time $\propto V$.

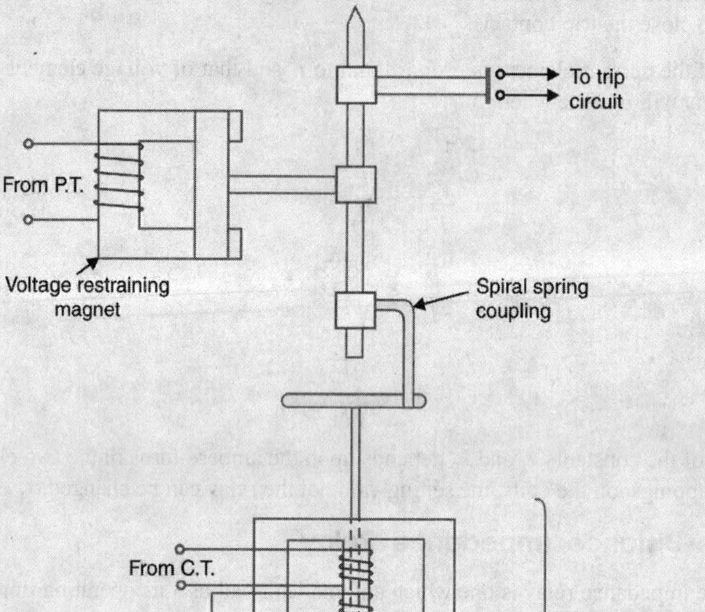

Fig. 21.22

21.17 Differential Relays

Most of the relays discussed so far relied on excess of current for their operation. Such relays are less sensitive because they cannot make correct distinction between heavy load conditions and minor fault conditions. In order to overcome this difficulty, differential relays are used.

***A differential relay** is one that operates when the phasor difference of two or more similar electrical quantities exceeds a pre-determined value.*

Thus a current differential relay is one that compares the current entering a section of the system with the current leaving the section. Under normal operating conditions, the two currents are equal but as soon as a fault occurs, this condition no longer applies. The difference between the incoming and outgoing currents is arranged to flow through the operating coil of the relay. If this differential current is equal to or greater than the pickup value, the relay will operate and open the circuit breaker to isolate the faulty section.

It may be noted that almost any type of relay when connected in a particular way can be made to operate as a differential relay. In other words, it is not so much the relay construction as the way the relay is connected in a circuit that makes it a differential relay. There are two fundamental systems of differential or balanced protection *viz.*

(*i*) Current balance protection (*ii*) Voltage balance protection

21.18 Current Differential Relay

Fig. 21.23 shows an arrangement of an overcurrent relay connected to operate as a differential relay. A pair of identical current transformers are fitted on either end of the section to be protected (alternator winding in this case). The secondaries of CT's are connected in series in such a way that they carry the induced currents in the same direction. The operating coil of the overcurrent relay is connected across the CT secondary circuit. This differential relay compares the current at the two ends of the alternator winding.

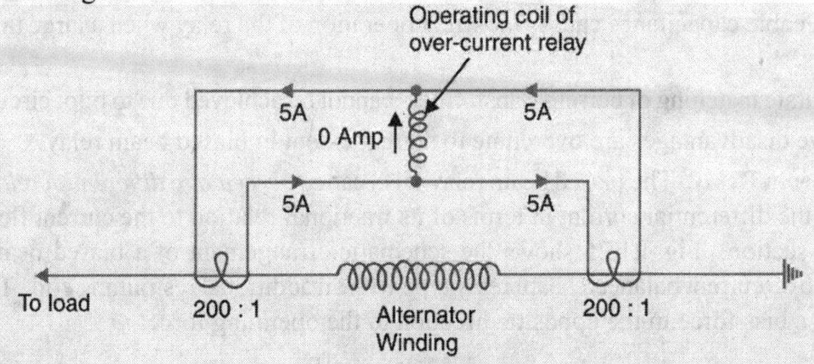

Fig. 21.23

Under normal operating conditions, suppose the alternator winding carries a normal current of 1000 A. Then the currents in the two secondaries of CT's are equal [See Fig. 21.23]. These currents will merely circulate between the two CT's and no current will flow through the differential relay. Therefore, the relay remains inoperative. If a ground fault occurs on the alternator winding as shown in Fig. 21.24 (*i*), the two secondary currents will not be equal and the current flows through the operating coil of the relay, causing the relay to operate. The amount of current flow through the relay will depend upon the way the fault is being fed.

(*i*) If some current (500 A in this case) flows out of one side while a larger current (2000 A) enters the other side as shown in Fig. 21.24 (*i*), then the difference of the CT secondary currents *i.e.* 10 − 2·5 = 7·5 A will flow through the relay.

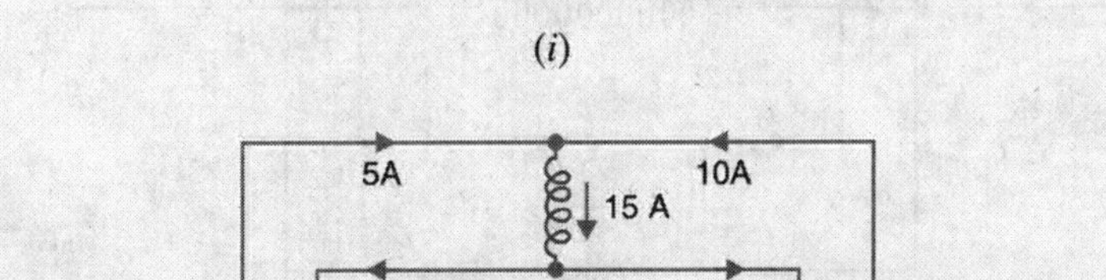

(*i*)

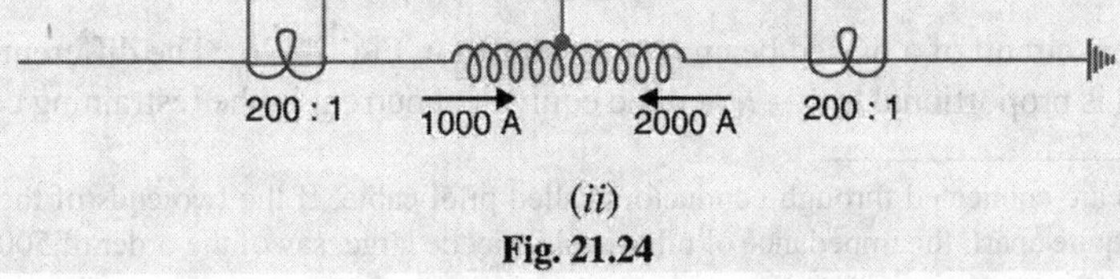

(*ii*)

Fig. 21.24

(*ii*) If current flows to the fault from both sides as shown in Fig. 21.24 (*ii*), then sum of CT secondary currents *i.e.* 10 + 5 = 15 A will flow through the relay.

Disadvantages

(*i*) The impedance of the *pilot cables generally causes a slight difference between the currents at the two ends of the section to be protected. If the relay is very sensitive, then the small differential current flowing through the relay may cause it to operate even under no fault conditions.

(*ii*) Pilot cable capacitance causes incorrect operation of the relay when a large through-current flows.

(*iii*) Accurate matching of current transformers cannot be achieved due to pilot circuit impedance.

The above disadvantages are overcome to a great extent in biased beam relay.

Biased Beam Relay. The biased beam relay (also called *percentage differential relay*) is designed to respond to the differential current in terms of its fractional relation to the current flowing through the protected section. Fig. 21.25 shows the schematic arrangement of a biased beam relay. It is essentially an overcurrent balanced beam relay type with an additional restraining coil. The restraining coil produces a bias force in the opposite direction to the operating force.

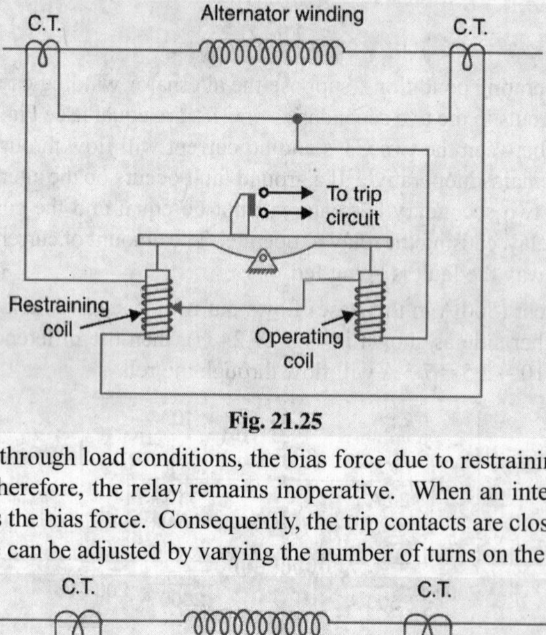

Fig. 21.25

Under normal and through load conditions, the bias force due to restraining coil is greater than the operating force. Therefore, the relay remains inoperative. When an internal fault occurs, the operating force exceeds the bias force. Consequently, the trip contacts are closed to open the circuit breaker. The bias force can be adjusted by varying the number of turns on the restraining coil.

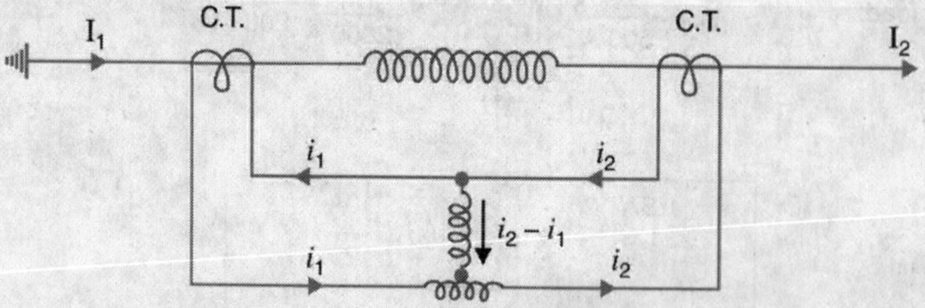

Fig. 21.26

The equivalent circuit of a biased beam relay is shown in Fig. 21.26. The differential current in the operating coil is proportional to $i_2 - i_1$ and the equivalent current in the restraining coil is propor-

* The two CT's are connected through conductors called pilot cable. If the two ends of the section to be protected are quite apart, the impedance of pilot-cable may be large, say of the order of 500 Ω.

tional to *$(i_1 + i_2)/2$ since the operating coil is connected to the mid-point of the restraining coil. It is clear that greater the current flowing through the restraining coil, the higher the value of current required in the operating winding to trip the relay. Thus under a heavy load, a greater differential current through the relay operating coil is required for operation than under light load conditions. This relay is called *percentage relay* because the operating current requried to trip can be expressed as a percentage of load current.

21.19 Voltage Balance Differential Relay

Fig. 21.27 shows the arrangement of voltage balance protection. In this scheme of protection, two similar current transformers are connected at either end of the element to be protected (*e.g.* an alternator winding) by means of pilot wires. The secondaries of current transformers are connected in series with a relay in such a way that under normal conditions, their induced e.m.f.s' are in opposition.

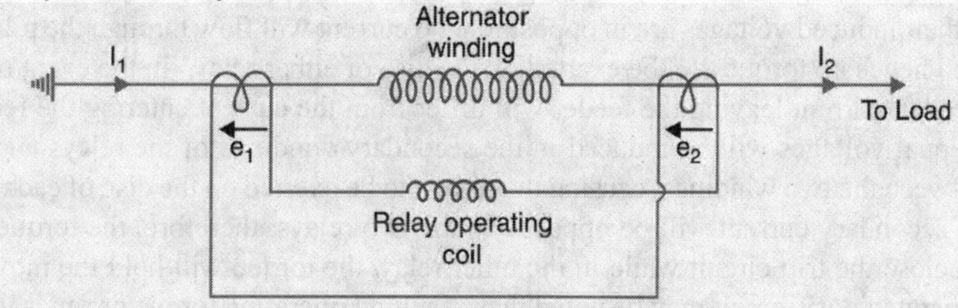

Fig. 21.27

Under healthy conditions, equal currents ($I_1 = I_2$) flow in both primary windings. Therefore, the secondary voltages of the two transformers are balanced against each other and no current will flow through the relay operating coil. When a fault occurs in the protected zone, the currents in the two primaries will differ from one another (*i.e.* $I_1 \neq I_2$) and their secondary voltages will no longer be in balance. This voltage †difference will cause a current to flow through the operating coil of the relay which closes the trip circuit.

Disadvantages

The voltage balance system suffers from the following drawbacks :

(*i*) A multi-gap transformer construction is required to achieve the accurate balance between current transformer pairs.

(*ii*) The system is suitable for protection of cables of relatively short lengths due to the capacitance of pilot wires. On long cables, the charging current may be sufficient to operate the relay even if a perfect balance of current transformers is attained.

The above disadvantages have been overcome in Translay (modified) balanced voltage system.

21.20 Translay System

This system is the modified form of voltage-balance system. Although the principle of balanced (opposed) voltages is retained, it differs from the above voltage-balance system in that the balance or opposition is between voltages induced in the secondary coils wound on the relay magnets and *not* between the secondary voltages of the line current transformers. Since the current transformers used

* Total ampere-turns on restraining coil of N turns

$$\propto i_1 \frac{N}{2} + i_2 \frac{N}{2}$$

$$\propto (i_1 + i_2)\frac{N}{2}$$

This is the same as if current $(i_1 + i_2)/2$ were to flow through the whole restraining coil.

† This is the case when fault is fed in one direction. Should the fault be fed from both ends, the secondary voltages become additive and increase the out-of-balance current operating the relay.

with Translay scheme have only to supply to a relay coil, they can be made of normal design without any air gaps. This permits the scheme to be used for feeders of any voltage.

Constructional details. Fig. 21.28 shows the simplified diagram illustrating the principle of *Translay scheme. It consists of two identical double winding induction type relays fitted at either end of the feeder to be protected. The primary circuits (11, 11*a*) of these relays are supplied through a pair of current transformers. The secondary windings (12, 13 and 12*a*, 13*a*) of the two relays are connected in series by pilot wires in such a way that voltages induced in the former opposes the other. The compensating devices (18, 18*a*) neutralise the effects of pilot-wire capacitance currents and of inherent lack of balance between the two current transformers.

Operation. Under healthy conditions, current at the two ends of the protected feeder is the same and the primary windings (11, 11*a*) of the relays carry the same current. The windings 11 and 11*a* induce equal e.m.f.s in the secondary windings 12, 12*a* and 13, 13*a*. As these windings are so connected that their induced voltages are in opposition, no current will flow through the pilots or operating coils and hence no torque will be exerted on the disc of either relay. In the event of fault on the protected feeder, current leaving the feeder will differ from the current entering the feeder. Consequently, unequal voltages will be induced in the secondary windings of the relays and current will circulate between the two windings, causing the torque to be exerted on the disc of each relay. As the direction of secondary current will be opposite in the two relays, therefore, the torque in one relay will tend to close the trip circuit while in the other relay, the torque will hold the movement in the normal unoperated position. It may be noted that resulting operating torque depends upon the position and nature of the fault in the protected zone and atleast one element of either relay will operate under any fault condition.

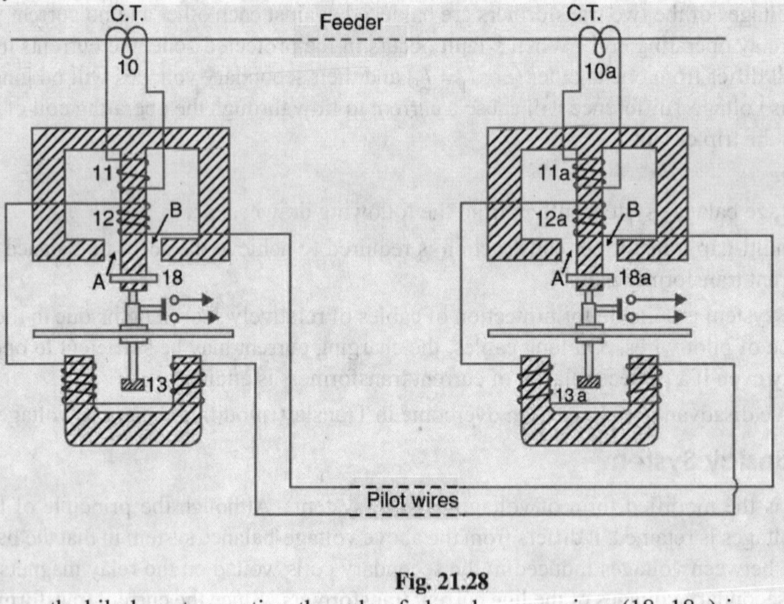

Fig. 21.28

It is worthwhile here to mention the role of closed copper rings (18, 18*a*) in neutralising the effects of pilot capacitive currents. Capacitive currents lead the voltage impressed across the pilots by 90° and when they flow in the operating winding 13 and 13*a* (which are of low inductance), they produce fluxes that also lead the pilot voltage by 90°. Since pilot voltage is that induced in the secondary windings 12 and 12*a*, it lags by a substantial angle behind the fluxes in the field magnet air gaps *A* and *B*. The closed copper rings (18, 18*a*) are so adjusted that this angle is approximately 90°. In this way fluxes acting on the disc are in phase and hence no torque is exerted on the relay disc.

* The relays used embrace the function of transformer as well as relay. Hence the name Translay.

21.21 Types of Protection

When a fault occurs on any part of electric power system, it must be cleared quickly in order to avoid damage and/or interference with the rest of the system. It is a usual practice to divide the protection scheme into two classes *viz.* primary protection and back-up protection.

(*i*) **Primary Protection.** It is the protection scheme which is designed to protect the component parts of the power system. Thus referring to Fig. 21.29, each line has an overcurrent relay that protects the line. If a fault occurs on any line, it will be cleared by its relay and circuit breaker. This forms the primary or main protection and serves as the first line of defence. The service record of primary relaying is very high with well over ninety percent of all operations being correct. However, sometimes faults are not cleared by primary relay system because of trouble within the relay, wiring system or breaker. Under such conditions, back-up protection does the required job.

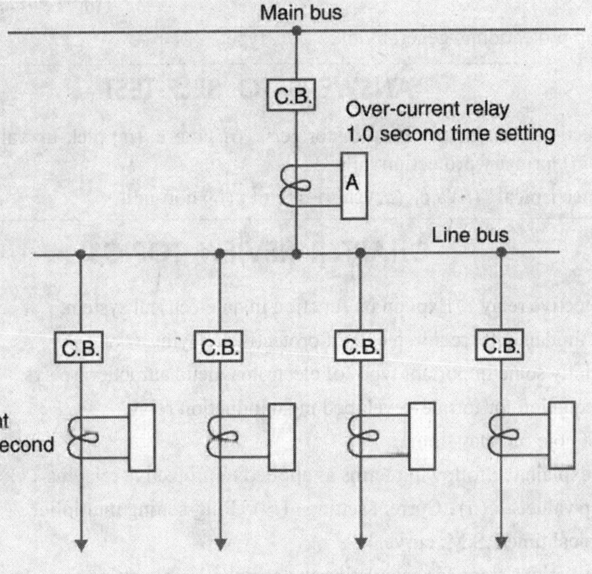

Fig. 21.29

(*ii*) **Back-up protection.** It is the second line of defence in case of failure of the primary protection. It is designed to operate with sufficient time delay so that primary relaying will be given enough time to function if it is able to. Thus referring to Fig. 21.29, relay *A* provides back-up protection for each of the four lines. If a line fault is not cleared by its relay and breaker, the relay *A* on the group breaker will operate after a definite time delay and clear the entire group of lines. It is evident that when back-up relaying functions, a larger part is disconnected than when primary relaying functions correctly. Therefore, greater emphasis should be placed on the better maintenance of primary relaying.

SELF - TEST

1. **Fill in the blanks by inserting appropriate words/figures.**

(*i*) A relay performs the function of ....

(*ii*) The relay operating coil is supplied through ....

(*iii*) A 1 VA relay is .... sensitive than a 3 VA relay.

(*iv*) The minimum relay coil current at which the relay operates is called..............

(*v*) Induction relays .... be used with d.c. quantities.

(*vi*) An overcurrent relay having a current setting of 125% is connected to a supply circuit through a current transformer of ratio 400/5. The pick-up value will be ....

(*vii*) The pick-up value of a relay is 7·5 A and fault current in relay coil is 30A. Its plug-setting mutiplier is ....

(*viii*) Back-up protection functions when ....

2. Pick up the correct words/figures from brackets and fill in the blanks.

(*i*) Most of the relays on service on electric power system are .... relays. *(electronic, electro-mechanical)*

(*ii*) Induction relays are used with .... quantities. *(a.c., d.c., both a.c. and d.c.)*

(*iii*) Back-up protection is generally employed for protection against .... faults. *(short-circuit, other than short-circuit)*

(*iv*) Back-up protection is generally of .... type. *(until, non-unit)*

ANSWERS TO SELF-TEST

1. (*i*) fault detection (*ii*) instrument transformers (*iii*) more (*iv*) pick-up value (*v*) cannot (*vi*) 6·25 A (*vii*) 4 (*viii*) primary protection fails
2. (*i*) electro-mechanical (*ii*) a.c. (*iii*) short-circuit (*iv*) non-unit.

CHAPTER REVIEW TOPICS

1. What is protective relay ? Explain its function in an electrical system.
2. Discuss the fundamental requirements of protective relaying.
3. Describe briefly some important types of electromagnetic attraction relays.
4. Derive the equation for torque developed in an induction relay.
5. Write a brief note on relay timing.
6. Define and explain the following terms as applied to protective relaying :
   (*i*) Pick-up value (*ii*) Current setting (*iii*) Plug-setting multiplier (*iv*) Time-setting multiplier
7. Sketch a typical time/P.S.M. curve.
8. Describe the various steps for calculating the actual relay operating time.
9. Explain with the help of neat diagram the construction and working of :
   (*i*) Non-directional induction type overcurrent relay
   (*ii*) Induction type directional power relay
10. Describe the construction and principle of operation of an induction type directional overcurrent relay.
11. Explain the working principle of distance relays.
12. Write a detailed note on differential relays.
13. Describe the Translay scheme of protection.
14. Write short notes on the following :
    (*i*) Percentage differential relays (*ii*) Definite distance relays (*iii*) Time-distance relays

DISCUSSION QUESTIONS

1. What is the difference between a fuse and a relay ?
2. What is the difference between an overcurrent relay and current differential relay ?
3. Back-up protection is not a good substitute for better maintenance. Discuss.
4. Why are differential relays more sensitive than overcurrent relays ?
5. What is the difference between a balanced voltage relay and a Translay relay ?

CHAPTER

22

Protection of Alternators and Transformers

Introduction

The modern electric power system consists of several elements *e.g.* alternators, transformers, station bus-bars, transmission lines and other equipment. It is desirable and necessary to protect each element from a variety of fault conditions which may occur sooner or later. The protective relays discussed in the previous chapter can be profitably employed to detect the improper behaviour of any circuit element and initiate corrective measures. As a matter of convenience, this chapter deals with the protection of alternators and transformers only.

The most serious faults on alternators which require immediate attention are the stator winding faults. The major faults on transformers occur due to short-circuits in the transformers or their connections. The basic system used for protection against these faults is the differential relay scheme because the differential nature of measurements makes this system much more sensitive than other protective systems.

22.1 Protection of Alternators

The generating units, especially the larger ones, are relatively few in number and higher in individual cost than most other equipments. Therefore, it is desirable and necessary to provide protection to cover the wide range of faults which may occur in the modern generating plant.

Some of the important faults which may occur on an alternator are :

(*i*) failure of prime-mover (*ii*) failure of field
(*iii*) overcurrent (*iv*) overspeed
(*v*) overvoltage (*vi*) unbalanced loading
(*vii*) stator winding faults

(*i*) **Failure of prime-mover.** When input to the prime-mover fails, the alternator runs as a synchronous motor and draws some current from the supply system. This motoring conditions is known as "inverted running".

(*a*) In case of turbo-alternator sets, failure of steam supply may cause inverted running. If the steam supply is gradually restored, the alternator will pick up load without disturbing the system. If the steam failure is likely to be prolonged, the machine can be safely isolated by the control room attendant since this condition is relatively harmless. Therefore, automatic protection is not required.

(*b*) In case of hydro-generator sets, protection against inverted running is achieved by providing mechanical devices on the water-wheel. When the water flow drops to an insufficient rate to maintain the electrical output, the alternator is disconnected from the system. Therefore, in this case also electrical protection is not necessary.

(*c*) Diesel engine driven alternators, when running inverted, draw a considerable amount of power from the supply system and it is a usual practice to provide protection against motoring in order to avoid damage due to possible mechanical seizure. This is achieved by applying reverse power relays to the alternators which *isolate the latter during their motoring action. It is essential that the reverse power relays have time-delay in operation in order to prevent inadvertent tripping during system disturbances caused by faulty synchronising and phase swinging.

(*ii*) **Failure of field.** The chances of field failure of alternators are undoubtedly very rare. Even if it does occur, no immediate damage will be caused by permitting the alternator to run without a field for a short-period. It is sufficient to rely on the control room attendant to disconnect the faulty alternator manually from the system bus-bars. Therefore, it is a universal practice not to provide †automatic protection against this contingency.

(*iii*) **Overcurrent.** It occurs mainly due to partial breakdown of winding insulation or due to overload on the supply system. Overcurrent protection for alternators is considered unnecessary because of the following reasons :

(*a*) The modern tendency is to design alternators with very high values of internal impedance so that they will stand a complete short-circuit at their terminals for sufficient time without serious overheating. On the occurrence of an overload, the alternators can be disconnected manually.

(*b*) The disadvantage of using overload protection for alternators is that such a protection might disconnect the alternators from the power plant bus on account of some momentary troubles outside the plant and, therefore, interfere with the continuity of electric service.

---

* During inverted running (or motoring), there is a reversal of power flow in the stator windings. This causes the operation of reverse power relay.

† This is the case with attendant stations. However, in unattended stations, the use of a field-failure relay may be justified.

(iv) **Overspeed.** The chief cause of overspeed is the sudden loss of all or the major part of load on the alternator. Modern alternators are usually provided with mechanical centrifugal devices mounted on their driving shafts to trip the main valve of the prime-mover when a dangerous overspeed occurs.

(v) **Over-voltage.** The field excitation system of modern alternators is so designed that over-voltage conditions at normal running speeds cannot occur. However, overvoltage in an alternator occurs when speed of the prime-mover increases due to sudden loss of the alternator load.

In case of steam-turbine driven alternators, the control governors are very sensitive to speed variations. They exercise a continuous check on overspeed and thus prevent the occurrence of over-voltage on the generating unit. Therefore, over-voltage protection is not provided on turbo-alternator sets.

In case of hydro-generator, the control governors are much less sensitive and an appreciable time may elapse before the rise in speed due to loss of load is checked. The over-voltage during this time may reach a value which would over-stress the stator windings and insulation breakdown may occur. It is, therefore, a usual practice to provide over-voltage protection on hydro-generator units. The over-voltage relays are operated from a voltage supply derived from the generator terminals. The relays are so arranged that when the generated voltage rises 20% above the normal value, they operate to

(a) trip the main circuit breaker to disconnect the faulty alternator from the system

(b) disconnect the alternator field circuit

(vi) **Unbalanced loading.** Unbalanced loading means that there are different phase currents in the alternator. Unbalanced loading arises from faults to earth or faults between phases on the circuit external to the alternator. The unbalanced currents, if allowed to persist, may either severely burn the mechanical fixings of the rotor core or damage the field winding.

Fig. 22.1 shows the schematic arrangement for the protection of alternator against unbalanced loading. The scheme comprises three line current transformers, one mounted in each phase, having their secondaries connected in parallel. A relay is connected in parallel across the transformer secondaries. Under normal operating conditions, equal currents flow through the different phases of the alternator and their algebraic sum is zero. Therefore, the sum of the currents flowing in the secondaries is also zero and no current flows through the operating coil of the relay. However, if unbalancing occurs, the currents induced in the secondaries will be different and the resultant of these currents will flow through the relay. The operation of the relay will trip the circuit breaker to disconnect the alternator from the system.

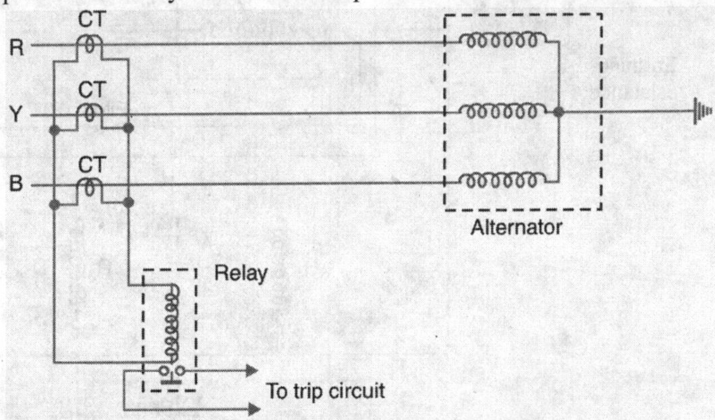

Fig. 22.1

(vii) **Stator winding faults.** These faults occur mainly due to the insulation failure of the stator windings. The main types of stator winding faults, in order of importance are :

(a) fault between phase and ground

(*b*) fault between phases

(*c*) inter-turn fault involving turns of the same phase winding

The stator winding faults are the most dangerous and are likely to cause considerable damage to the expensive machinery. Therefore, automatic protection is absolutely necessary to clear such faults in the quickest possible time in order to minimise the *extent of damage. For protection of alternators against such faults, differential method of protection (also knows as Merz-Price system) is most commonly employed due to its greater sensitivity and reliability. This system of protection is discussed in the following section.

22.2 Differential Protection of Alternators

The most common system used for the protection of stator winding faults employs circulating-current principle (Refer back to Art. 21.18). In this scheme of protection, currents at the two ends of the protected section are compared. Under normal operating conditions, these currents are equal but may become unequal on the occurrence of a fault in the protected section. The difference of the currents under fault conditions is arranged to pass through the operating coil of the relay. The relay then closes its contacts to isolate protected section from the system. This form of protection is also known as *Merz-Price circulating current scheme*.

Schematic arrangement. Fig. 22.2 shows the schematic arrangement of current differential protection for a 3-phase alternator. Identical current transformer pairs CT_1 and CT_2 are placed on either side of each phase of the stator windings. The secondaries of each set of current transformers are connected in star ; the two neutral points and the corresponding terminals of the two star groups being connected together by means of a four-core pilot cable. Thus there is an independent path for the currents circulating in each pair of current transformers and the corresponding pilot *P*.

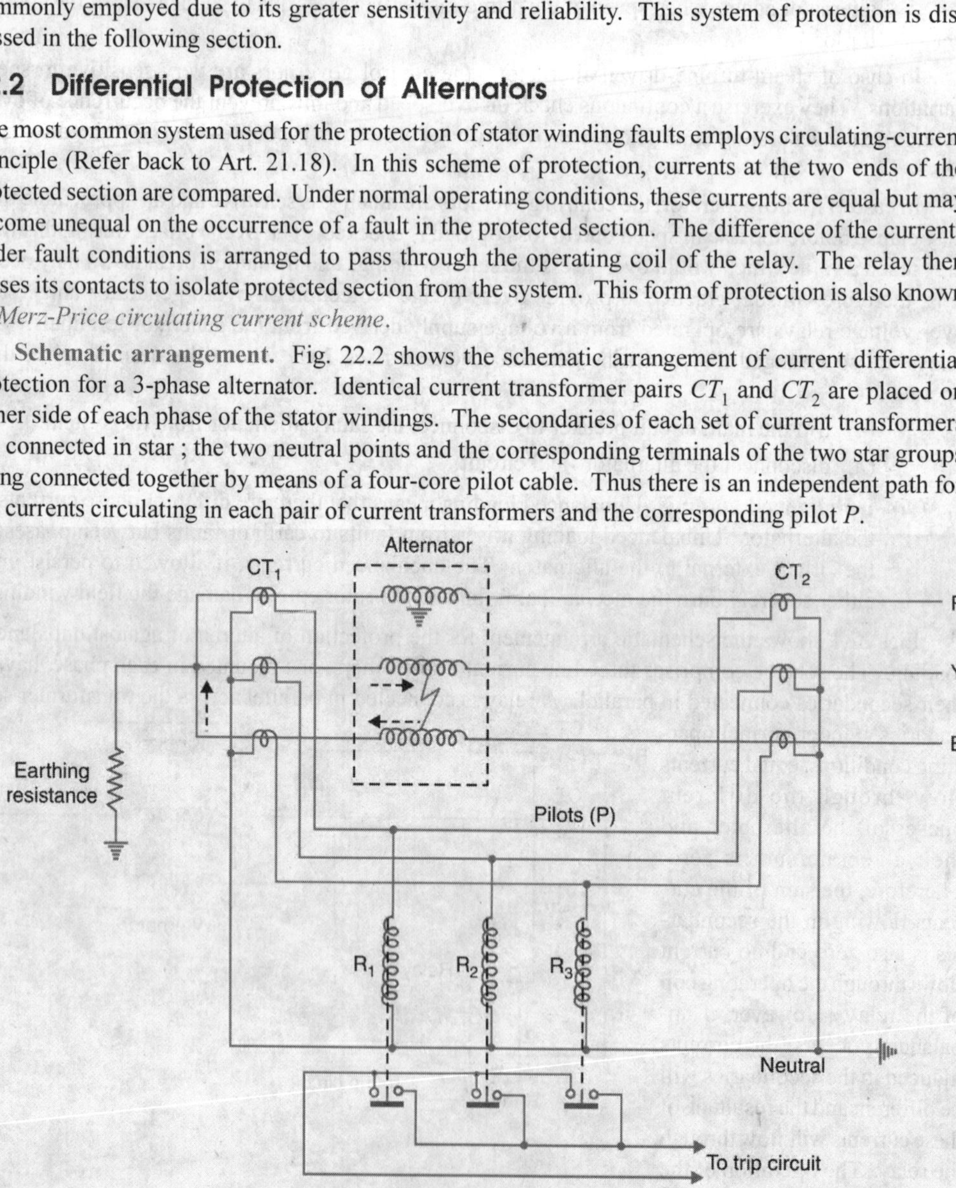

Fig. 22.2

* If the stator winding fault is not cleared quickly, it may lead to

(*i*) burning of stator coils

(*ii*) burning and welding-up of stator laminations

The relay coils are connected in star, the neutral point being connected to the current-transformer common neutral and the outer ends one to each of the other three pilots. In order that burden on each current transformer is the same, the relays are connected across equipotential points of the three pilot wires and these equipotential points would naturally be located at the middle of the pilot wires. The relays are generally of electromagnetic type and are arranged for instantaneous action since fault should be cleared as quickly as possible.

Operation. Referring to Fig. 22.2, it is clear that the relays are connected in shunt across each circulating path. Therefore, the circuit of Fig. 22.2 can be shown in a simpler form in Fig. 22.3. Under normal operating conditions, the current at both ends of each winding will be equal and hence the currents in the secondaries of two CTs connected in any phase will also be equal. Therefore, there is balanced circulating current in the pilot wires and no current flows through the operating coils (R_1, R_2 and R_3) of the relays. When an earth-fault or phase-to-phase fault occurs, this condition no longer holds good and the differential current flowing through the relay circuit operates the relay to trip the circuit breaker.

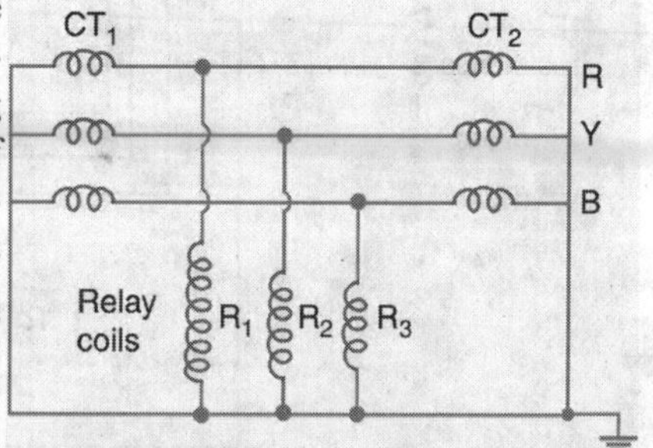

Fig. 22.3

(*i*) Suppose an earth fault occurs on phase *R* due to breakdown of its insulation to earth as shown in Fig. 22.2. The current in the affected phase winding will flow through the core and frame of the machine to earth, the circuit being completed through the neutral earthing resistance. The currents in the secondaries of the two CTs in phase *R* will become unequal and the difference of the two currents will flow through the corresponding relay coil (*i.e.* R_1), returning via the neutral pilot. Consequently, the relay operates to trip the circuit breaker.

(*ii*) Imagine that now a short-circuit fault occurs between the phases *Y* and *B* as shown in Fig. 22.2. The short-circuit current circulates *via* the neutral end connection through the two windings and through the fault as shown by the dotted arrows. The currents in the secondaries of two CTs in each affected phase will become unequal and the differential current will flow through the operating coils of the relays (*i.e.* R_2 and R_3) connected in these phases. The relay then closes its contacts to trip the circuit breaker.

It may be noted that the relay circuit is so arranged that its energising causes (*i*) opening of the breaker connecting the alternator to the bus-bars and (*ii*) opening of the *field circuit of the alternator.

It is a prevailing practice to mount current transformers CT_1 in the neutral connections (usually in the alternator pit) and current transformers CT_2 in the switch-gear equipment. In some cases, the alternator is located at a considerable distance from the switchgear. As the relays are located close to the circuit breaker, therefore, it is not convenient to connect the relay coils to the actual physical mid-points of the pilots. Under these circumstances, balancing resistances are inserted in the shorter lengths of the pilots so that the relay tapping points divide the whole secondary impedance of two sets of CTs into equal portions. This arrangement is shown in Fig. 22.4. These resistances are usually adjustable in order to obtain the exact balance.

Limitations. The two circuits for alternator protection shown above have their own limitations. It is a general practice to use neutral earthing resistance in order to limit the destructive effects of earth-fault currents. In such a situation, it is impossible to protect whole of the stator windings of a star-connected alternator during earth-faults. When an earth-fault occurs near the neutral point, there

* Although disconnection of faulty alternator prevents other alternators on the system feeding into the fault, it is necessary to suppress the field of faulty alternator to stop the machine itself feeding into the fault.

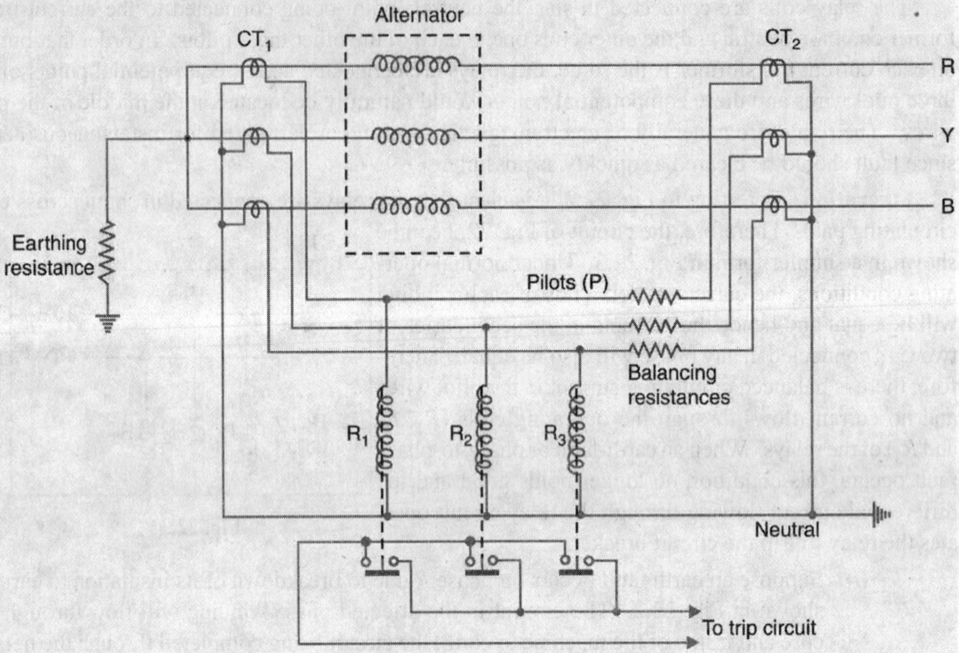

Fig. 22.4

may be insufficient voltage across the short-circuited portion to drive the necessary current round the fault circuit to operate the relay. The magnitude of unprotected zone depends upon the value of earthing resistance and relay setting.

Makers of protective gear speak of "protecting 80% of the winding" which means that faults in the 20% of the winding near the neutral point cannot cause tripping *i.e.* this portion is unprotected. It is a usual practice to protect only 85% of the winding because the chances of an earth fault occurring near the neutral point are very rare due to the uniform insulation of the winding throughout.

22.3 Modified Differential Protection for Alternators

If the neutral point of a star-connected alternator is earthed through a high resistance, protection schemes shown in Fig. 22.2 or 22.4 will not provide sufficient sensitivity for earth-faults. It is because the high earthing resistance will limit the earth-fault currents to a low value, necessitating relays with low current settings if adequate portion of the generator winding is to be protected. However, too low a relay setting is undesirable for reliable stability on heavy through phase-faults. In order to overcome this difficulty, a modified form of differential protection is used in which the setting of earth faults is reduced without impairing stability.

The modified arrangement is shown in Fig. 22.5. The modifications affect only the relay connections and consist in connecting two relays for phase-fault protection and the third for earth-fault protection only. The two phase elements (PC and PA) and balancing resistance (BR) are connected in star and the earth relay (ER) is connected between this star point and the fourth wire of circulating current pilot-circuit.

Operation. Under normal operating conditions, currents at the two ends of each stator winding will be equal. Therefore, there is a balanced circulating current in the phase pilot wires and no current flows through the operating coils of the relays. Consequently, the relays remain inoperative.

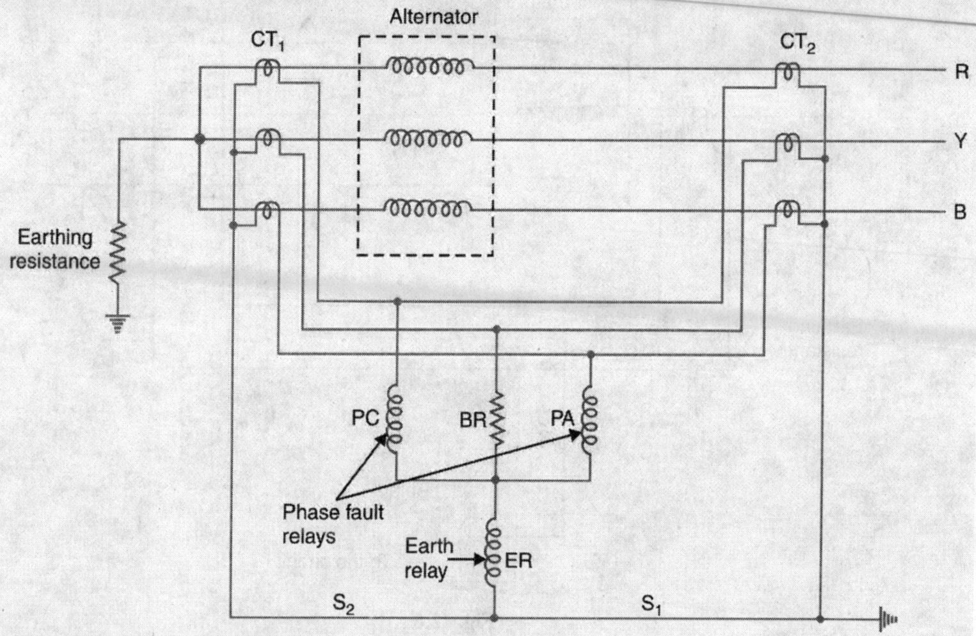

Fig. 22.5

If an earth-fault occurs on any one phase, the out-of-balance secondary current in CTs in that phase will flow through the earth relay *ER* and *via* pilot S_1 or S_2 to the neutral of the current transformers. This will cause the operation of earth relay only. If a fault occurs between two phases, the out-of-balance current will circulate round the two transformer secondaries *via* any two of the coils *PA*, *BR*, *PC* (the pair being decided by the two phases that are faulty) without passing through the earth relay *ER*. Therefore, only the phase-fault relays will operate.

22.4 Balanced Earth-fault Protection

In small-size alternators, the neutral ends of the three-phase windings are often connected internally to a single terminal. Therefore, it is not possible to use Merz-Price circulating current principle described above because there are no facilities for accommodating the necessary current transformers in the neutral connection of each phase winding. Under these circumstances, it is considered sufficient to provide protection against earth-faults only by the use of balanced earth-fault protection scheme. This scheme provides no protection against phase-to-phase faults, unless and until they develop into earth-faults, as most of them will.

Schematic arrangement. Fig. 22.6 shows the schematic arrangement of a balanced earth-fault protection for a 3-phase alternator. It consists of three line current transformers, one mounted in each phase, having their secondaries connected in parallel with that of a single current transformer in the conductor joining the star point of the alternator to earth. A relay is connected across the transformers secondaries. The protection against earth faults is limited to the region between the neutral and the line current transformers.

Operation. Under normal operating conditions, the currents flowing in the alternator leads and hence the currents flowing in secondaries of the line current transformers add to zero and no current flows through the relay. Also under these conditions, the current in the neutral wire is zero and the secondary of neutral current transformer supplies no current to the relay.

If an earth-fault develops at F_2 external to the protected zone, the sum of the currents at the terminals of the alternator is exactly equal to the current in the neutral connection and hence no

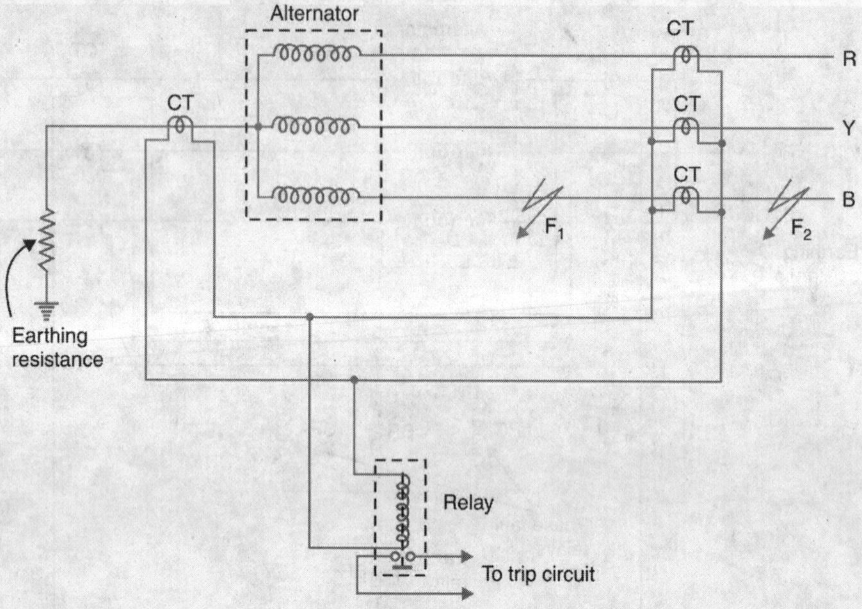

Fig. 22.6

current flows through the relay. When an earth-fault occurs at F_1 or within the protected zone, these currents are no longer equal and the differential current flows through the operating coil of the relay. The relay then closes its contacts to disconnect the alternator from the system.

22.5 Stator Inter-turn Protection

Merz-price circulating-current system protects against phase-to-ground and phase-to-phase faults. It does not protect against turn-to-turn fault on the same phase winding of the stator. It is because the current that this type of fault produces flows in a local circuit between the turns involved and does not create a difference between the currents entering and leaving the winding at its two ends where current transformers are applied. However, it is usually considered unnecessary to provide protection for inter-turn faults because they invariably develop into earth-faults.

In single turn generator (*e.g.* large steam-turbine generators), there is no necessity of protection against inter-turn faults. However, inter-turn protection is provided for multi-turn generators such as hydro-electric generators. These generators have double-winding armatures (*i.e.* each phase wind-

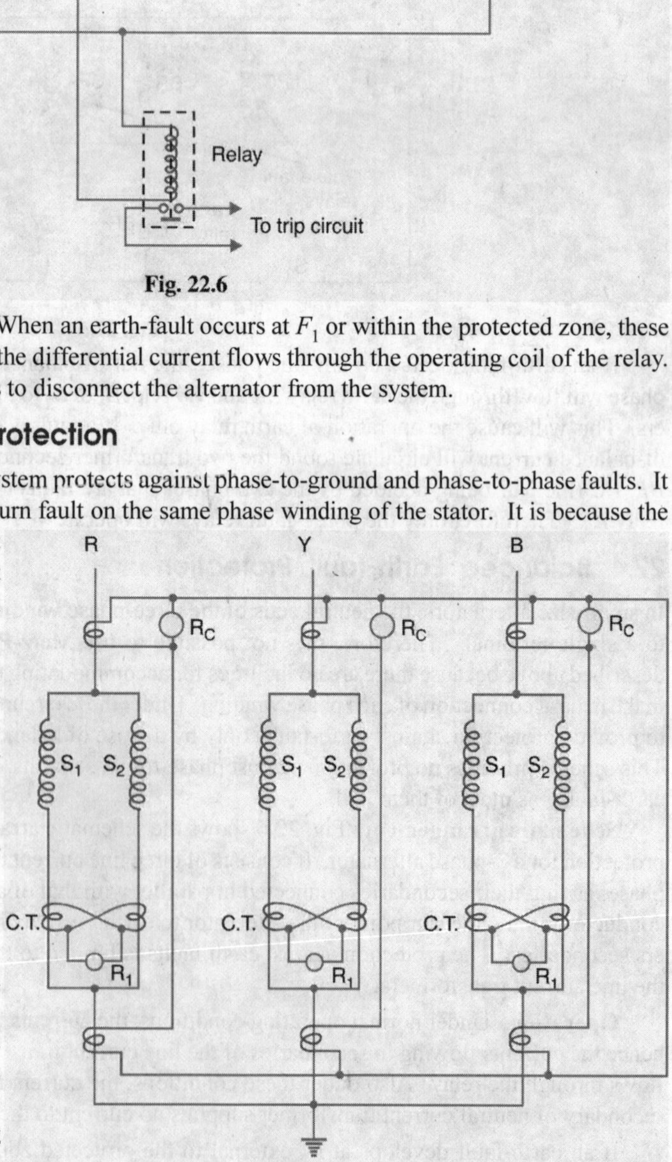

Fig. 22.7

ing is divided into two halves) owing to the very heavy currents which they have to carry. Advantage may be taken of this necessity to protect inter-turn faults on the same winding. Fig. 22.7 shows the schematic arrangement of circulating-current and inter-turn protection of a 3-phase double wound generator. The relays R_C provide protection against phase-to-ground and phase-to-phase faults whereas relays R_1 provide protection against inter-turn faults.

Fig. 22.8 shows the duplicate stator windings S_1 and S_2 of one phase only with a provision against inter-turn faults. Two current transformers are connected on the circulating-current principle. Under normal conditions, the currents in the stator windings S_1 and S_2 are equal and so will be the currents in the secondaries of the two CTs. The secondary current round the loop then is the same at all points and no current flows through the relay R_1. If a short-circuit develops between adjacent turns, say on S_1, the currents in the stator windings S_1 and S_2 will no longer be equal. Therefore, unequal currents will be induced in the secondaries of CTs and the difference of these two currents flows through the relay R_1. The relay then closes its contacts to clear the generator from the system.

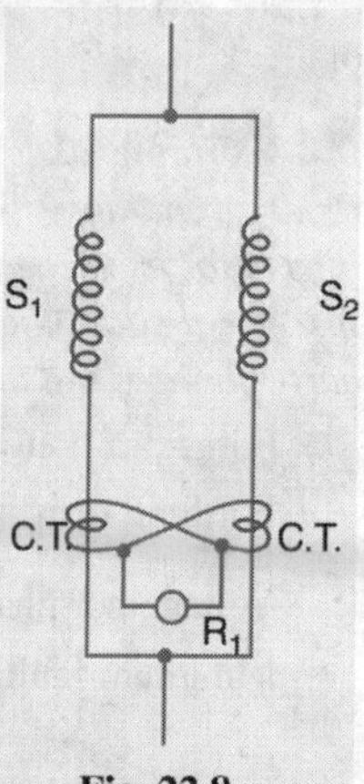

Fig. 22.8

Example 22.1. *A star-connected, 3-phase, 10-MVA, 6·6 kV alternator has a per phase reactance of 10%. It is protected by Merz-Price circulating-current principle which is set to operate for fault currents not less than 175 A. Calculate the value of earthing resistance to be provided in order to ensure that only 10% of the alternator winding remains unprotected.*

Solution. Let r ohms be the earthing resistance required to leave 10% of the winding unprotected (portion NA). The whole arrangement is shown in the simplified diagram of Fig. 22.9.

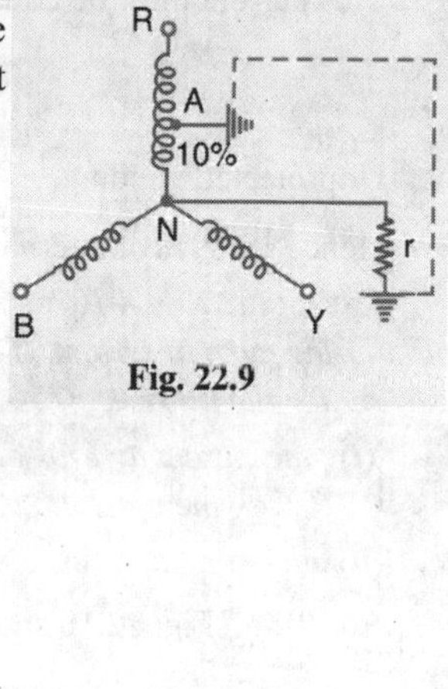

Fig. 22.9

$$\text{Voltage per phase, } V_{ph} = \frac{6\cdot 6\times 10^3}{\sqrt{3}} = 3810 \text{ V}$$

$$\text{Full-load current, } I = \frac{10\times 10^6}{\sqrt{3}\times 6\cdot 6\times 10^3} = 875 \text{ A}$$

Let the reactance per phase be x ohms.

$$\therefore \quad 10 = \frac{\sqrt{3}\times x\times 875}{6600}\times 100$$

$$\text{or} \quad x = 0\cdot 436\ \Omega$$

$$\text{Reactance of 10\% winding} = 0\cdot 436\times 0\cdot 1 = 0\cdot 0436\ \Omega$$

$$\text{E.M.F. induced in 10\% winding} = V_{ph}\times 0\cdot 1 = 3810\times 0\cdot 1 = 381 \text{ V}$$

Impedance offered to fault by 10% winding is

$$Z_f = \sqrt{(0\cdot 0436)^2 + r^2}$$

Earth-fault current due to 10% winding

$$= \frac{381}{Z_f} = \frac{381}{\sqrt{(0\cdot 0436)^2 + r^2}}$$

When this fault current becomes 175 A, the relay will trip.

$$\therefore \quad 175 = \frac{381}{\sqrt{(0\cdot 0436)^2 + r^2}}$$

$$\text{or} \quad (0\cdot 0436)^2 + r^2 = \left(\frac{381}{175}\right)^2$$

or $(0{\cdot}0436)^2 + r^2 = 4{\cdot}715$

or $r = \mathbf{2{\cdot}171\ \Omega}$

Example 22.2. *A star-connected, 3-phase, 10 MVA, 6·6 kV alternator is protected by Merz-Price circulating-current principle using 1000/5 amperes current transformers. The star point of the alternator is earthed through a resistance of 7·5 Ω. If the minimum operating current for the relay is 0·5 A, calculate the percentage of each phase of the stator winding which is unprotected against earth-faults when the machine is operating at normal voltage.*

Solution. Let x % of the winding be unprotected.

Earthing resistance, $r = 7{\cdot}5\ \Omega$

Voltage per phase, $V_{ph} = 6{\cdot}6 \times 10^3/\sqrt{3} = 3810$ V

Minimum fault current which will operate the relay

$$= \frac{1000}{5} \times 0{\cdot}5 = 100 \text{ A}$$

E.M.F. induced in x% winding $= V_{ph} \times (x/100) = 3810 \times (x/100) = 38{\cdot}1\,x$ volts

Earth fault current which x% winding will cause

$$= \frac{38{\cdot}1\,x}{r} = \frac{38{\cdot}1\,x}{7{\cdot}5} \text{ amperes}$$

This current must be equal to 100 A.

$$\therefore \quad 100 = \frac{38{\cdot}1\,x}{7{\cdot}5}$$

or Unprotected winding, $x = \dfrac{100 \times 7{\cdot}5}{38{\cdot}1} = \mathbf{19{\cdot}69\%}$

Hence 19·69% of alternator winding is left unprotected.

Example 22.3. *A 10 MVA, 6·6 kV, 3-phase star-connected alternator is protected by Merz-Price circulating current system. If the ratio of the current transformers is 1000/5, the minimum operating current for the relay is 0·75 A and the neutral point earthing resistance is 6 Ω, calculate :*

(*i*) *the percentage of each of the stator windings which is unprotected against earth faults when the machine is operating at normal voltage.*

(*ii*) *the minimum resistance to provide protection for 90% of the stator winding.*

Solution. Fig. 22.10 shows the circuit diagram.

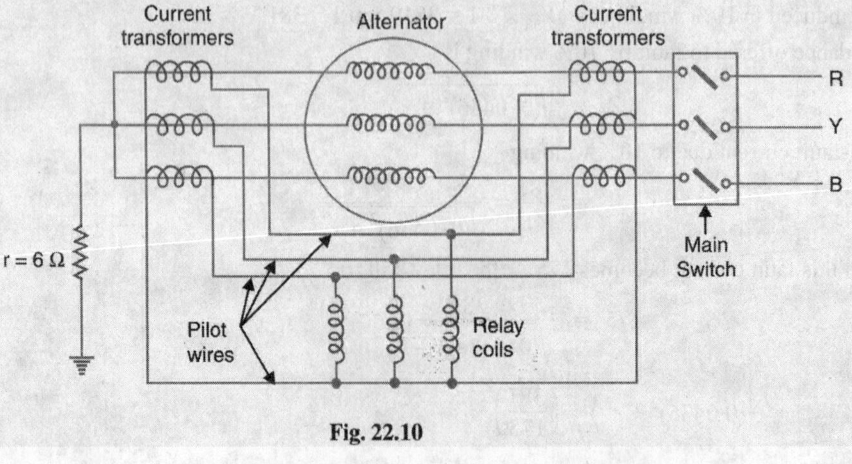

Fig. 22.10

(*i*) Let x% of the winding be unprotected.

Earthing resistance, $r = 6\ \Omega$

Voltage per phase, $V_{ph} = 6 \cdot 6 \times 10^3/\sqrt{3} = 3810$ volts

Minimum fault current which will operate the relay

$$= \frac{1000}{5} \times 0 \cdot 75 = 150 \text{ A}$$

E.M.F. induced in x% of stator winding

$$= V_{ph} \times (x/100) = 3810 \times (x/100) = 38 \cdot 1\, x \text{ volts}$$

Earth fault current which x% winding will cause

$$= \frac{38 \cdot 1\, x}{r} = \frac{38 \cdot 1\, x}{6} \text{ amperes}$$

This must be equal to 150 A.

$$\therefore \quad 150 = \frac{38 \cdot 1\, x}{6}$$

or $\quad x = $ **23·6%**

(*ii*) Let r ohms be the minimum earthing resistance required to provide protection for 90% of stator winding. Then 10% winding would be unprotected *i.e.* $x = 10\%$.

$$\therefore \quad 150 = \frac{38 \cdot 1\, x}{r}$$

or $\quad r = \frac{38 \cdot 1\, x}{150} = \frac{38 \cdot 1 \times 10}{150} = \mathbf{2 \cdot 54\ \Omega}$

Example 22.4. *A star-connected, 3-phase, 10 MVA, 6·6 kV alternator is protected by circulating current protection, the star point being earthed via a resistance r. Estimate the value of earthing resistor if 85% of the stator winding is protected against earth faults. Assume an earth fault setting of 20%. Neglect the impedance of the alternator winding.*

Solution. Since 85% winding is to be protected, 15% would be unprotected. Let r ohms be the earthing resistance required to leave 15% of the winding unprotected.

$$\text{Full-load current} = \frac{10 \times 10^6}{\sqrt{3} \times 6 \cdot 6 \times 10^3} = 876 \text{ A}$$

Minimum fault current which will operate the relay

$$= 20\% \text{ of full-load current}$$

$$= \frac{20}{100} \times 876 = 175 \text{ A}$$

Voltage induced in 15% of winding

$$= \frac{15}{100} \times \frac{6 \cdot 6 \times 10^3}{\sqrt{3}} = 330\sqrt{3} \text{ volts}$$

Earth fault current which 15% winding will cause

$$= \frac{330\sqrt{3}}{r}$$

This current must be equal to 175 A.

$$\therefore \quad 175 = \frac{330\sqrt{3}}{r}$$

or $\quad r = \frac{330\sqrt{3}}{175} = \mathbf{3 \cdot 27\ \Omega}$

TUTORIAL PROBLEMS

1. A 10 MVA, 11 kV, 3-phase star-connected alternator is protected by the Merz-Price balance-current system, which operates when the out-of-balance current exceeds 20% of full-load current. Determine what portion of the alternating winding is unprotected if the star point is earthed through a resistance of 9 Ω. The reactance of the alternator is 2 Ω. **[14·88%]**
2. The neutral point of 25 MVA, 11 kV alternator is grounded through a resistance of 5 Ω, the relay is set to operate when there is an out of balance current of 2A. The CTs used have a ratio of 1000/5. Calculate (neglect reactance of alternator) :
   (*i*) the percentage of stator winding protected against an earth fault
   (*ii*) the minimum value of earthing resistance to protect 95% of the winding **[(*i*) 68·5% (*ii*) 0·8 Ω]**
3. A 3-phase, 20 MVA, 11kV star connected alternator is protected by Merz-Price circulating current system. The star point is earthed through a resistance of 5 ohms. If the CTs have a ratio of 1000/5 and the relay is set to operate when there is an out of balance current of 1·5 A, calculate :
   (*i*) the percentage of each phase of the stator winding which is unprotected
   (*ii*) the minimum value of earthing resistance to protect 90% of the winding **[(*i*) 23·6% (*ii*) 2·12 Ω]**

22.6 Protection of Transformers

Transformers are static devices, totally enclosed and generally oil immersed. Therefore, chances of faults occurring on them are very rare. However, the consequences of even a rare fault may be very serious unless the transformer is quickly disconnected from the system. This necessitates to provide adequate automatic protection for transformers against possible faults.

Small distribution transformers are usually connected to the supply system through series fuses instead of circuit breakers. Consequently, no automatic protective relay equipment is required. However, the probability of faults on power transformers is undoubtedly more and hence automatic protection is absolutely necessary.

Common transformer faults. As compared with generators, in which many abnormal conditions may arise, power transformers may suffer only from :

(*i*) open circuits
(*ii*) overheating
(*iii*) winding short-circuits *e.g.* earth-faults, phase-to-phase faults and inter-turn faults.

An open circuit in one phase of a 3-phase transformer may cause undesirable heating. In practice, relay protection is not provided against open circuits because this condition is relatively harmless. On the occurrence of such a fault, the transformer can be disconnected manually from the system.

Overheating of the transformer is usually caused by sustained overloads or short-circuits and very occasionally by the failure of the cooling system. The relay protection is also not provided against this contingency and thermal accessories are generally used to sound an alarm or control the banks of fans.

Winding short-circuits (also called *internal faults*) on the transformer arise from deterioration of winding insulation due to overheating or mechanical injury. When an internal fault occurs, the transformer must be disconnected quickly from the system because a prolonged arc in the transformer may cause oil fire. Therefore, relay protection is absolutely necessary for internal faults.

22.7 Protection Systems for Transformers

For protection of generators, Merz-Price circulating-current system is unquestionably the most satisfactory. Though this is largely true of transformer protection, there are cases where circulating current system offers no particular advantage over other systems or impracticable on account of the troublesome conditions imposed by the wide variety of voltages, currents and earthing conditions invariably associated with power transformers. Under such circumstances, alternative protective systems are used which in many cases are as effective as the circulating-current system. The principal relays and systems used for transformer protection are :

(*i*) *Buchholz devices* providing protection against all kinds of incipient faults *i.e.* slow-developing faults such as insulation failure of windings, core heating, fall of oil level due to leaky joints etc.

(*ii*) *Earth-fault relays* providing protection against earth-faults only.

(*iii*) *Overcurrent relays* providing protection mainly against phase-to-phase faults and overloading.

(*iv*) *Differential system* (or circulating-current system) providing protection against both earth and phase faults.

The complete protection of transformer usually requires the combination of these systems. Choice of a particular combination of systems may depend upon several factors such as (*a*) size of the transformer (*b*) type of cooling (*c*) location of transformer in the network (*d*) nature of load supplied and (*e*) importance of service for which transformer is required. In the following sections, above systems of protection will be discussed in detail.

22.8 Buchholz Relay

Buchholz relay is a gas-actuated relay installed in oil immersed transformers for protection against all kinds of faults. Named after its inventor, Buchholz, it is used to give an alarm in case of incipient (*i.e.* slow-developing) faults in the transformer and to disconnect the transformer from the supply in the event of severe internal faults. It is usually installed in the pipe connecting the conservator to the main tank as shown in Fig. 22.11. It is a universal practice to use Buchholz relays on all such oil immersed transformers having ratings in *excess of 750 kVA.

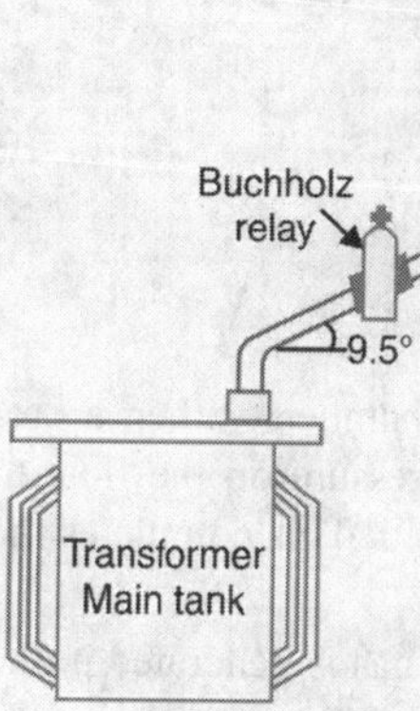

Fig. 22.11

Buchholz Relay

Construction. Fig. 22.12 shows the constructional details of a Buchholz relay. It takes the form of a domed vessel placed in the connecting pipe between the main tank and the conservator. The device has two elements. The upper element consists of a mercury type switch attached to a float. The lower element contains a mercury switch mounted on a hinged type flap located in the direct path of the flow of oil from the transformer to the conservator. The upper element closes an alarm circuit during incipient faults whereas the lower element is arranged to trip the circuit breaker in case of severe internal faults.

Operation. The operation of Buchholz relay is as follows :

(*i*) In case of incipient faults within the transformer, the heat due to fault causes the decomposition of some transformer oil in the main tank. The products of decomposition contain more than 70% of hydrogen gas. The hydrogen gas being light tries to go into the conserva-

* Its use for oil immersed transformers of rating less than 750 kVA is generally uneconomical.

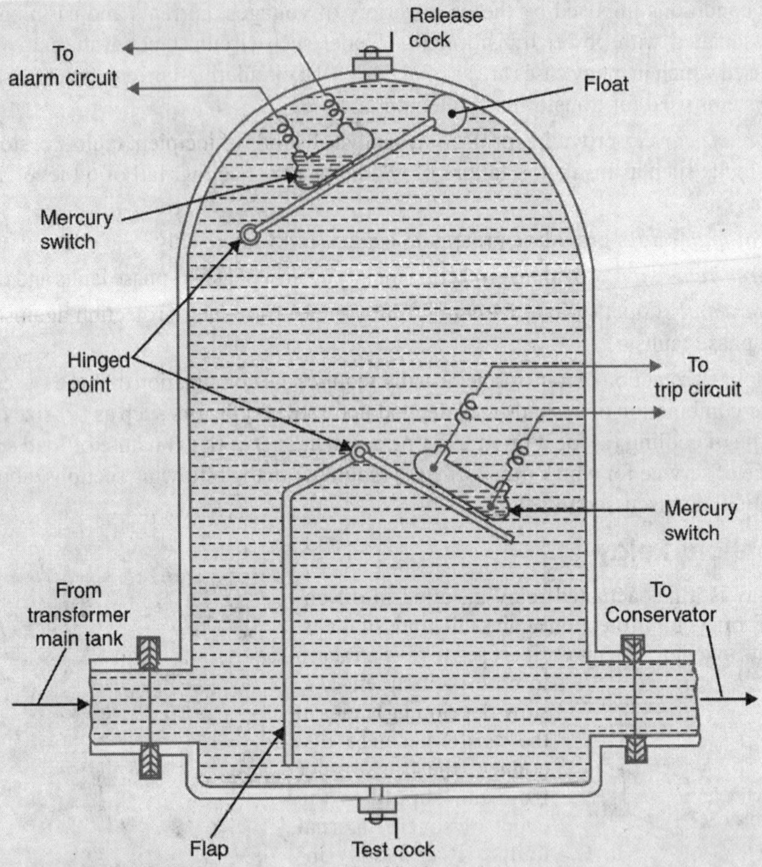

Fig. 22.12

tor and in the process gets entrapped in the upper part of relay chamber. When a pre-determined amount of gas gets accumulated, it exerts sufficient pressure on the float to cause it to tilt and close the contacts of mercury switch attached to it. This completes the alarm circuit to sound an *alarm.

(*ii*) If a serious fault occurs in the transformer, an enormous amount of gas is generated in the main tank. The oil in the main tank rushes towards the conservator *via* the Buchholz relay and in doing so tilts the flap to close the contacts of mercury switch. This completes the trip circuit to open the circuit breaker controlling the transformer.

Advantages

(*i*) It is the simplest form of transformer protection.

(*ii*) It detects the incipient faults at a stage much earlier than is possible with other forms of protection.

Disadvantages

(*i*) It can only be used with oil immersed transformers equipped with conservator tanks.

(*ii*) The device can detect only faults below oil level in the transformer. Therefore, separate protection is needed for connecting cables.

---

* The conditions described do not call for the immediate removal of the faulty transformer. It is because sometimes the air bubbles in the oil circulation system of a healthy transformer may operate the float. For this reason, float is arranged to sound an alarm upon which steps can be taken to verify the gas and its composition.

22.9 Earth-Fault or Leakage Protection

An earth-fault usually involves a partial breakdown of winding insulation to earth. The resulting leakage current is considerably less than the short-circuit current. The earth-fault may continue for a long time and cause considerable damage before it ultimately develops into a short-circuit and removed from the system. Under these circumstances, it is profitable to employ earth-fault relays in order to ensure the disconnection of earth-fault or leak in the early stage. An earth-fault relay is essentially an overcurrent relay of low setting and operates as soon as an earth-fault or leak develops. One method of protection against earth-faults in a transformer is the **core-balance leakage protection* shown in Fig. 22.13.

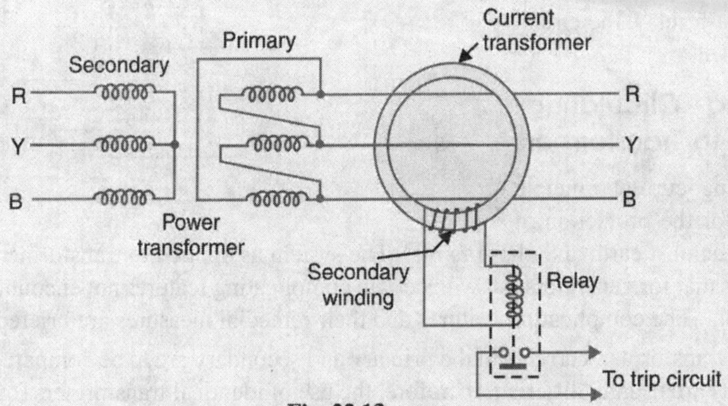

Fig. 22.13

The three leads of the primary winding of power transformer are taken through the core of a current transformer which carries a single secondary winding. The operating coil of a relay is connected to this secondary. Under normal conditions (*i.e.* no fault to earth), the vector sum of the three phase currents is zero and there is no resultant flux in the core of current transformer no matter how much the load is out of balance. Consequently, no current flows through the relay and it remains inoperative. However, on the occurrence of an earth-fault, the vector sum of three phase currents is no longer zero. The resultant current sets up flux in the core of the C.T. which induces e.m.f. in the secondary winding. This energises the relay to trip the circuit breaker and disconnect the faulty transformer from the system.

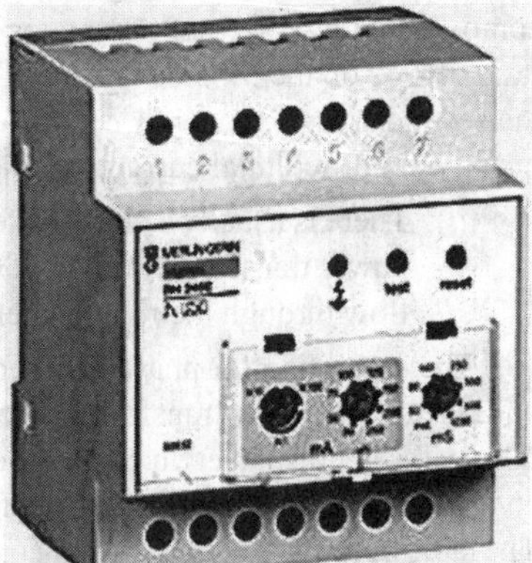

Earth Leakage Relay

22.10 Combined Leakage and Overload Protection

The core-balance protection described above suffers from the drawback that it cannot provide protection against overloads. If a fault or leakage occurs between phases, the core-balance relay will not operate. It is a usual practice to provide combined leakage and overload protection for transformers. The earth relay has low current setting and operates under earth or leakage faults only. The overload relays have high current setting and are arranged to operate against faults between the phases.

* An earth-fault relay is also described as a core-balance relay. Strictly the term 'core-balance' is reserved for the case in which the relay is energised by a 3-phase current transformer and the balance is between the fluxes in the core of the current transformer.

Fig. 22.14 shows the schematic arrangement of combined leakage and overload protection. In this system of protection, two overload relays and one leakage or earth relay are connected as shown. The two overload relays are sufficient to protect against phase-to-phase faults. The trip contacts of overload relays and earth-fault relay are connected in parallel. Therefore, with the energising of either overload relay or earth relay, the circuit breaker will be tripped.

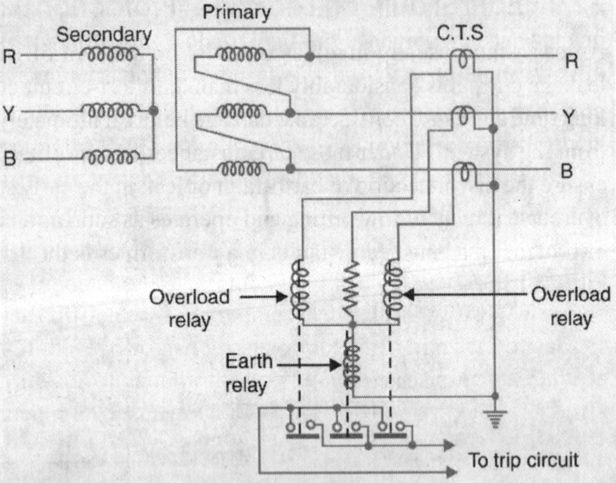

Fig. 22.14

22.11 Applying Circulating-current System to Transformers

Merz-Price circulating -current principle is commonly used for the protection of power transformers against earth and phase faults. The system as applied to transformers is fundamentally the same as that for generators but with certain complicating features not encountered in the generator application. The complicating features and their remedial measures are briefed below :

(*i*) In a power transformer, currents in the primary and secondary are to be compared. As these two currents are usually different, therefore, the use of identical transformers (of same turn ratio) will give differential current and operate the relay even under no load conditions.

The difference in the magnitude of currents in the primary and secondary of power transformer is compensated by different turn ratios of CTs. If T is the turn-ratio of power transformer, then turn-ratio of CTs on the *l.v.* side is made T times that of the CTs on the *h.v.* side. Fulfilled this condition, the secondaries of the two CTs will carry identical currents under normal load conditions. Consequently, no differential current will flow through the relay and it remains inoperative.

(*ii*) There is usually a phase difference between the primary and secondary currents of a 3-phase power transformer. Even if CTs of the proper turn-ratio are used, a differential current may flow through the relay under normal conditions and cause relay operation.

The correction for phase difference is effected by appropriate connections of CTs. The CTs on one side of the power transformer are connected in such a way that the resultant currents fed into the pilot wires are displaced in phase from the individual phase currents in the same direction as, and by an angle equal to, the phase shift between the power-transformers primary and secondary currents. The table below shows the type of connections to be employed for CTs in order to compensate for the phase difference in the primary and secondary currents of power transformer.

S. No.	Power transformer connections		Current transformer connections	
	Primary	Secondary	Primary	Secondary
1	Star with neutral earthed	Delta	Delta	Star
2	Delta	Delta	Star	Star
3	Star	Star with neutral earthed	Delta	Delta
4	Delta	Star with neutral earthed	Star	Delta

Thus referring to the above table, for a delta/star power transformer, the CTs on the delta side must be connected in star and those on the star side in delta.

(*iii*) Most transformers have means for tap changing which makes this problem even more difficult. Tap changing will cause differential current to flow through the relay even under normal operating conditions.

The above difficulty is overcome by adjusting the turn-ratio of CTs on the side of the power transformer provided with taps.

(*iv*) Another complicating factor in transformer protection is the magnetising in-rush current. Under normal load conditions, the magnetising current is very small. However, when a transformer is energised after it has been taken out of service, the magnetising or in-rush current can be extremely high for a short period. Since magnetising current represents a current going into the transformer without a corresponding current leaving, it appears as a fault current to differential relay and may cause relay operation.

In order to overcome above difficulty, differential relays are set to operate at a relatively high degree of unbalance. This method decreases the sensitivity of the relays. In practice, advantage is taken of the fact that the initial in-rush currents contain prominent second-harmonic component. Hence, it is possible to design a scheme employing second-harmonic bias features, which, being tuned to second-harmonic frequency only, exercise restrain during energising to prevent maloperation.

While applying circulating current principle for protection of transformers, above precautions are necessary in order to avoid inadvertent relay operation.

22.12 Circulating-Current Scheme for Transformer Protection

Fig. 22.15 shows Merz-Price circulating-current scheme for the protection of a 3-phase delta/delta power transformer against phase-to-ground and phase-to-phase faults. Note that *CT*s on the two sides of the transformer are connected in star. This compensates for the phase difference between the power transformer primary and secondary. The *CT*s on the two sides are connected by pilot wires and one relay is used for each pair of *CT*s.

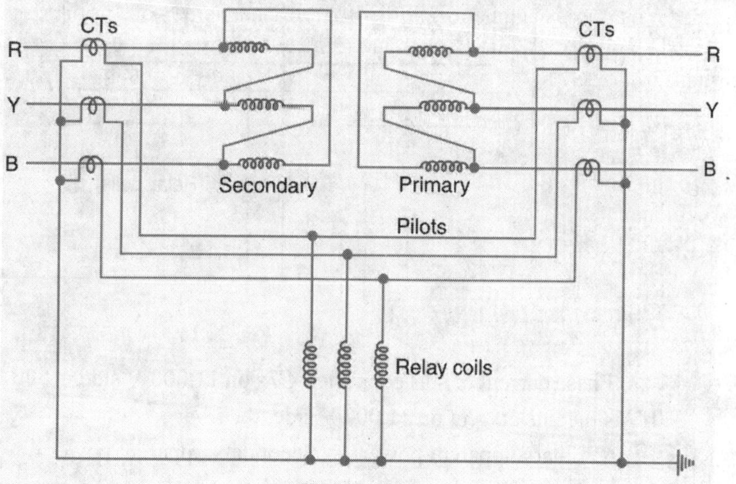

Fig. 22.15

During normal operating conditions, the secondaries of *CT*s carry identical currents. Therefore, the currents entering and leaving the pilot wires at both ends are the same and no current flows through the relays. If a ground or phase-to-phase fault occurs, the currents in the secondaries of *CT*s will no longer be the same and the differential current flowing through the relay circuit will clear the breaker on both sides of the transformer. The-protected zone is limited to the region between *CT*s on the high-voltage side and the *CT*s on the low-voltage side of the power transformer.

It is worthwhile to note that this scheme also provides protection for short-circuits between turns on the same phase winding. When a short-circuit occurs between the turns, the turn-ratio of the power transformer is altered and causes unbalance between current transformer pairs. If turn-ratio of power transformer is altered sufficiently, enough differential current may flow through the relay to cause its operation. However, such short-circuits are better taken care of by Buchholz relays.

Example 22.5. *A 3-phase transformer of 220/11,000 line volts is connected in star/delta. The protective transformers on 220 V side have a current ratio of 600/5. What should be the CT ratio on 11,000 V side ?*

Solution. For star/delta power transformers, *CT*s will be connected in delta on 220 V side (*i.e.* star side of power transformer) and in star on 11,000 V side (*i.e.* delta side of power transformer) as shown in Fig. 22.16.

Suppose that line current on 220 V side is 600 A.

∴ Phase current of delta connected *CT*s on 220V side

$$= 5 \text{ A}$$

Line current of delta connected *CT*s on 220 V side

$$= 5 \times \sqrt{3} = 5\sqrt{3} \text{ A}$$

This current (*i.e.* $5\sqrt{3}$) will flow through the pilot wires. Obviously, this will be the current which flows through the secondary of *CT*s on the 11,000 V side.

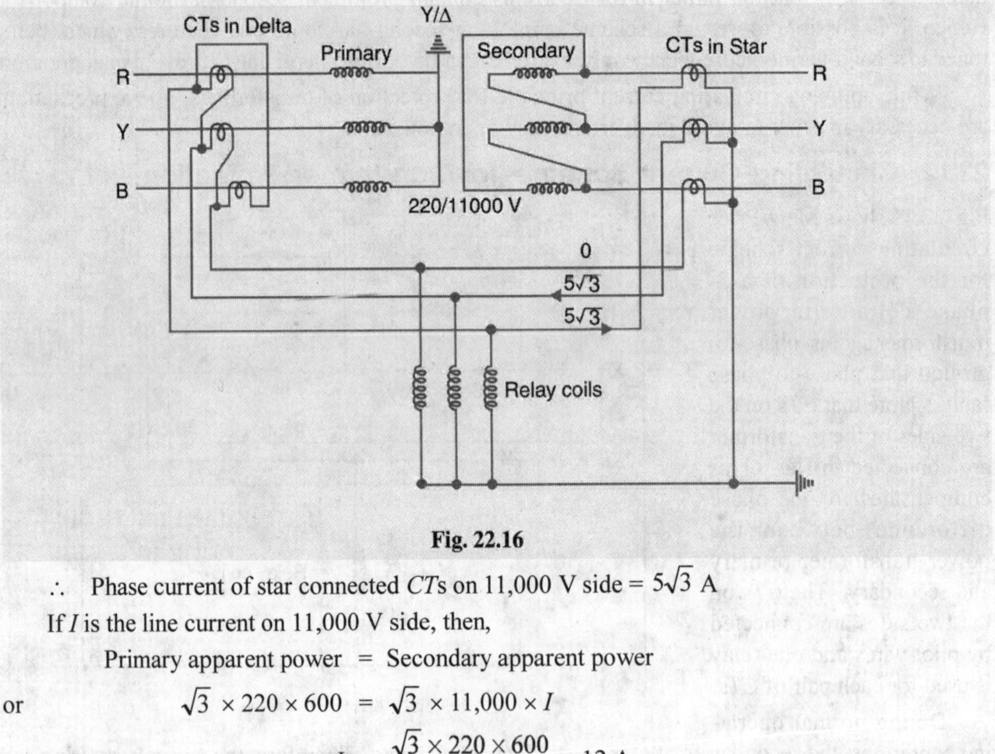

Fig. 22.16

∴ Phase current of star connected *CT*s on 11,000 V side = $5\sqrt{3}$ A

If I is the line current on 11,000 V side, then,

Primary apparent power = Secondary apparent power

or $$\sqrt{3} \times 220 \times 600 = \sqrt{3} \times 11{,}000 \times I$$

or $$I = \frac{\sqrt{3} \times 220 \times 600}{\sqrt{3} \times 11000} = 12 \text{ A}$$

∴ Turn-ratio of *CT*s on 11000 V side

$$= 12 : 5\sqrt{3} = 1{\cdot}385 : 1$$

Example 22.6. *A 3-phase transformer having line-voltage ratio of 0·4 kV/11kV is connected in star-delta and protective transformers on the 400 V side have a current ratio of 500/5. What must be the ratio of the protective transformers on the 11 kV side ?*

Solution. Fig. 22.17 shows the circuit connections. For star/delta transformers, *CTs* will be connected in delta on 400 V side (*i.e.* star side of power transformer) and in star on 11,000 V side (*i.e.* delta side of power transformer).

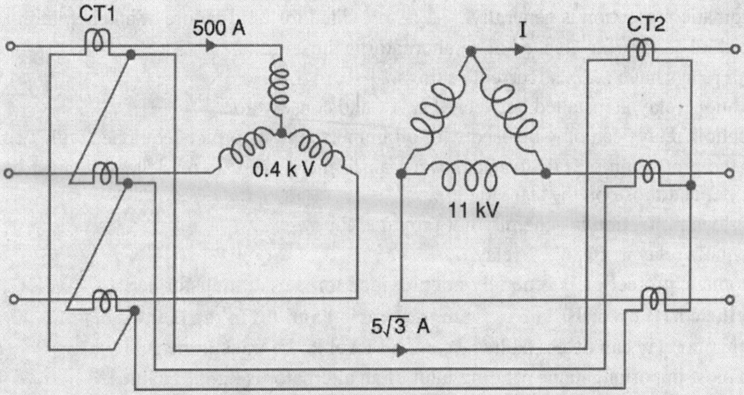

Fig. 22.17

Suppose the line current on 400 V side is 500 A.

∴ Phase current of delta connected *CTs* on 400 V side

$$= 5 \text{ A}$$

Line current of delta connected *CTs* on 400 V side

$$= 5 \times \sqrt{3} = 5\sqrt{3} \text{ A}$$

This current (*i.e.* $5\sqrt{3}$ A) will flow through the pilot wires. Obviously, this will be the current which flows through the secondary of the *CTs* on 11000 V side.

∴ Phase current of star-connected *CTs* on 11000 V side

$$= 5\sqrt{3} \text{ A}$$

If I is the line current on 11000 V side, then,

Primary apparent power = Secondary apparent power

or $$\sqrt{3} \times 400 \times 500 = \sqrt{3} \times 11000 \times I$$

or $$I = \frac{\sqrt{3} \times 400 \times 500}{\sqrt{3} \times 11000} = \frac{200}{11} \text{ A}$$

∴ C.T. ratio of *CTs* on 11000 V side

$$= \frac{200}{11} : 5\sqrt{3} = \frac{200}{11 \times 5\sqrt{3}} = \frac{10 \cdot 5}{5} = \mathbf{10{\cdot}5 : 5}$$

TUTORIAL PROBLEMS

1. A 3-phase, 33/6·6 kV, star/delta connected transformer is protected by Merz-Price circulating current system. If the CTs on the low-voltage side have a ratio of 300/5, determine the ratio of CTs on the high voltage side. **[60 : $5\sqrt{3}$]**
2. A 3-phase, 200 kVA, 11/0·4 kV transformer is connected as delta/star. The protective transformers on the 0·4 kV side have turn ratio of 500/5. What will be the C.T. ratios on the high voltage side ? **[18·18 : 8·66]**

SELF - TEST

1. **Fill in the blanks by inserting appropriate words/figures.**
   (*i*) The most commonly used system for the protection of generator is ....
   (*ii*) Automatic protection is generally .... provided for field failure of an alternator.
   (*iii*) The chief cause of overspeed in an alternator is the ....
   (*iv*) Earth relays have .... current settings.
   (*v*) Buchholz relay is installed between .... and conservator.
   (*vi*) Buchholz relays can only be used with oil immersed transformers equipped with ....
   (*vii*) For the protection of a delta/star power transformers, the CTs on delta side must be connected in .... and those on the star side in ....
   (*viii*) Overload protection is generally not provided for ....
   (*ix*) Buchholz relay is a .... relay.
   (*x*) Automatic protection is generally not provided for .... transformer.
2. **Pick up the correct words/figures from the bracket and fill in the blanks.**
   (*i*) Buchholz relay can detect faults .... oil level in the transformer. (*below, above*)
   (*ii*) The most important stator winding fault of an alternator is .... fault. (*earth, phase-to-phase, inter-turn*)
   (*iii*) Balanced earth-fault protection is generally provided for ....generators. (*small-size, large-size*)
   (*iv*) An earth-fault current is generally .... than short-circuit current. (*less, greater*)
   (*v*) Merz-Price circulating current principle is more suitable for .... than .... (*generators, transformers*)

ANSWERS TO SELF-TEST

1. (*i*) circulating-current system (*ii*) not (*iii*) sudden loss of load (*iv*) lower (*v*) main tank (*vi*) conservator (*vii*) star, delta (*viii*) alternators (*ix*) gas actuated (*x*) small distribution
2. (*i*) below (*ii*) earth (*iii*) small-size (*iv*) less (*v*) gencratuors, transformers

CHAPTER REVIEW TOPICS

1. Discuss the important faults on an alternator.
2. Explain with a neat diagram the application of Merz-Price circulating current principle for the protection of alternator.
3. Describe with a neat diagram the balanced earth protection for small-size generators.
4. How will you protect an alternator from turn-to-turn fault on the same phase winding ?
5. What factors cause difficulty in applying circulating current principle to a power transformer ?
6. Describe the construction and working of a Buchholz relay.
7. Describe the Merz-Price circulating current system for the protection of transformers.
8. Write short notes on the following :
   (*i*) Earth-fault protection for alternator
   (*ii*) Combined leakage and overload protection for transformers
   (*iii*) Earth-fault protection for transformers

DISCUSSION QUESTIONS

1. What is the difference between an earth relay and overcurrent relay ?
2. How does grounding affect relay application ?
3. Why is overload protection not necessary for alternators ?
4. Can relays be used to protect an alternator against (*i*) one-phase open circuits (*ii*) unbalanced loading (*iii*) motoring (*iv*) loss of synchronism ?
5. How many faults develop in a power transformer ?

CHAPTER

23

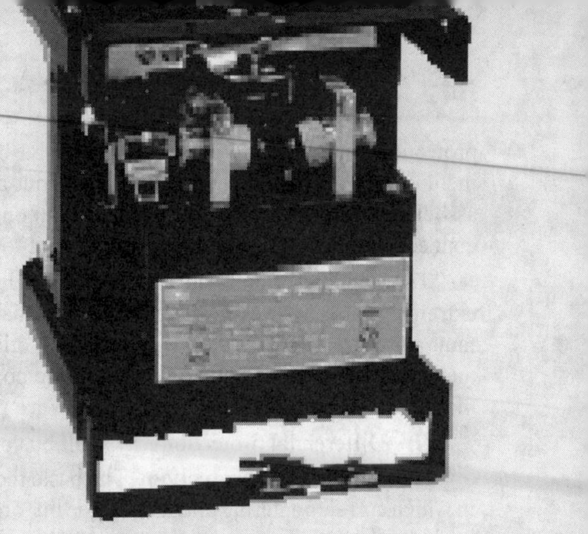

Protection of Busbars and Lines

Introduction

Busbars and lines are important elements of electric power system and require the immediate attention of protection engineers for safeguards against the possible faults occurring on them. The methods used for the protection of generators and transformers can also be employed, with slight modifications, for the busbars and lines. The modifications are necessary to cope with the protection problems arising out of greater length of lines and a large number of circuits connected to a busbar. Although differential protection can be used, it becomes too expensive for longer lines due to the greater length of pilot wires required. Fortunately, less expensive methods are available which are reasonably effective in providing protection for the busbars and lines. In this chapter, we shall focus our attention on the various methods of protection of busbars and lines.

23.1 Busbar Protection

Busbars in the generating stations and sub-stations form important link between the incoming and outgoing circuits. If a fault occurs on a busbar, considerable damage and disruption of supply will occur unless some form of quick-acting automatic protection is provided to isolate the faulty busbar. The busbar zone, for the purpose of protection, includes not only the busbars themselves but also the isolating switches, circuit breakers and the associated connections. In the event of fault on any section of the busbar, all the circuit equipments connected to that section must be tripped out to give complete isolation.

The standard of construction for busbars has been very high, with the result that bus faults are extremely rare. However, the possibility of damage and service interruption from even a rare bus fault is so great that more attention is now given to this form of protection. Improved relaying methods have been developed, reducing the possibility of incorrect operation. The two most commonly used schemes for busbar protection are :

(*i*) Differential protection (*ii*) Fault bus protection

(*i*) **Differential protection.** The basic method for busbar protection is the differential scheme in which currents entering and leaving the bus are totalised. During normal load condition, the sum of these currents is equal to zero. When a fault occurs, the fault current upsets the balance and produces a differential current to operate a relay.

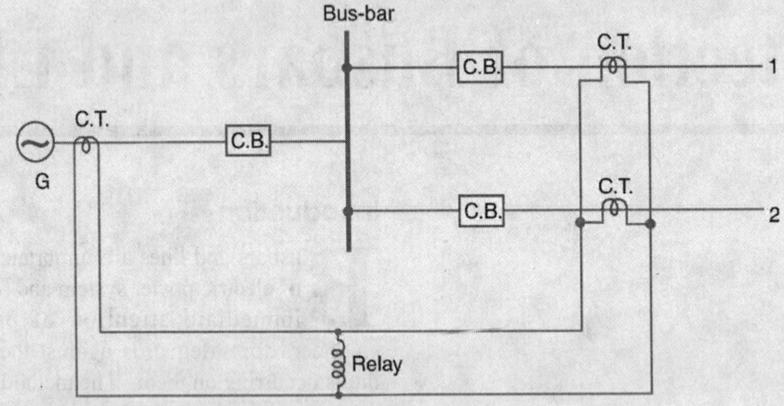

Fig. 23.1

Fig. 23.1 shows the single line diagram of current differential scheme for a station busbar. The busbar is fed by a generator and supplies load to two lines. The secondaries of current transformers in the generator lead, in line 1 and in line 2 are all connected in parallel. The protective relay is connected across this parallel connection. All *CT*s must be of the same ratio in the scheme regardless of the capacities of the various circuits. Under normal load conditions or external fault conditions, the sum of the currents entering the bus is equal to those leaving it and no current flows through the relay. If a fault occurs within the protected zone, the currents entering the bus will no longer be equal to those leaving it. The difference of these currents will flow through the relay and cause the opening of the generator, circuit breaker and each of the line circuit breakers.

(*ii*) **Fault Bus protection.** It is possible to design a station so that the faults that develop are mostly earth-faults. This can be achieved by providing earthed metal barrier (known as *fault bus*) surrounding each conductor throughout its entire length in the bus structure. With this arrangement, every fault that might occur must involve a connection between a conductor and an earthed metal part. By directing the flow of earth-fault current, it is possible to detect the faults and determine their location. This type of protection is known as fault bus protection.

Fig. 23.2 show the schematic arrangement of fault bus protection. The metal supporting structure or fault bus is earthed through a current transformer. A relay is connected across the secondary of this CT. Under normal operating conditions, there is no current flow from fault bus to ground and the relay remains inoperative. A fault involving a connection between a conductor and earthed supporting structure will result in current flow to ground through the fault bus, causing the relay to operate. The operation of relay will trip all breakers connecting equipment to the bus.

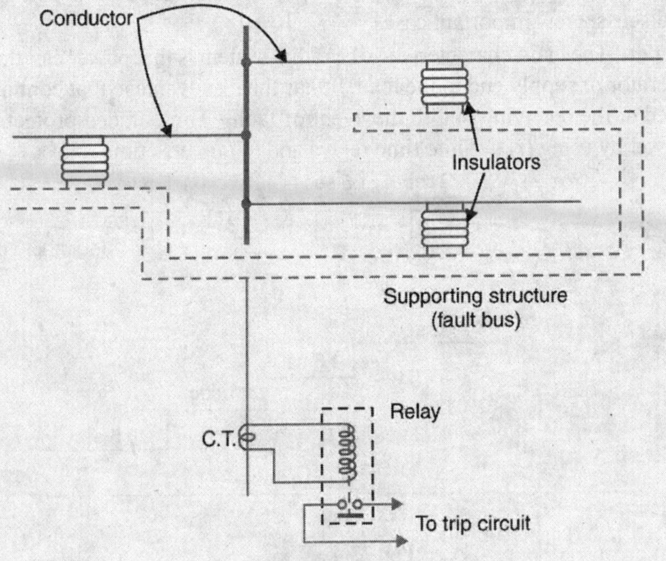

Fig. 23.2

23.2 Protection of Lines

The probability of faults occurring on the lines is much more due to their greater length and exposure to atmospheric conditions. This has called for many protective schemes which have no application to the comparatively simple cases of alternators and transformers. The requirements of line protection are :

(*i*) In the event of a short-circuit, the circuit breaker closest to the fault should open, all other circuit breakers remaining in a closed position.

(*ii*) In case the nearest breaker to the fault fails to open, back-up protection should be provided by the adjacent circuit breakers.

(*iii*) The relay operating time should be just as short as possible in order to preserve system stability, without unnecessary tripping of circuits.

The protection of lines presents a problem quite different from the protection of station apparatus such as generators, transformers and busbars. While differential protection is ideal method for lines, it is much more expensive to use. The two ends of a line may be several kilometres apart and to compare the two currents, a costly pilot-wire circuit is required. This expense may be justified but in general less costly methods are used. The common methods of line protection are :

(*i*) Time-graded overcurrent protection

(*ii*) Differential protection

(*iii*) Distance protection

Fig. 23.3

Fig. 23.3 shows the symbols indicating the various types of relays.

23.3 Time-Graded Overcurrent Protection

In this scheme of overcurrent protection, time discrimination is incorporated. In other words, the time setting of relays is so graded that in the event of fault, the smallest possible part of the system is isolated. We shall discuss a few important cases.

1. Radial feeder. The main characteristic of a radial system is that power can flow only in one direction, from generator or supply end to the load. It has the disadvantage that continuity of supply cannot be maintained at the receiving end in the event of fault. Time-graded protection of a radial feeder can be achieved by using (*i*) definite time relays and (*ii*) inverse time relays.

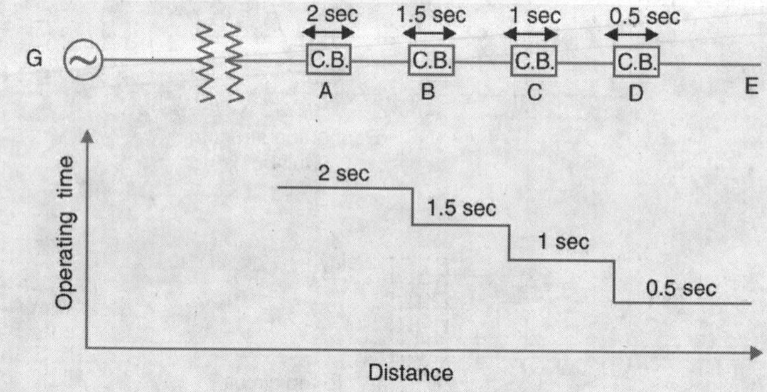

Fig. 23.4

(*i*) *Using definite time relays.* Fig. 23.4 shows the overcurrent protection of a radial feeder by definite time relays. The time of operation of each relay is fixed and is independent of the operating current. Thus relay *D* has an operating time of 0·5 second while for other relays, time delay* is successively increased by 0·5 second. If a fault occurs in the section *DE*, it will be cleared in 0·5 second by the relay and circuit breaker at *D* because all other relays have higher operating time. In this way only section *DE* of the system will be isolated. If the relay at *D* fails to trip, the relay at *C* will operate after a time delay of 0·5 second *i.e.* after 1 second from the occurrence of fault.

The disadvantage of this system is that if there are a number of feeders in series, the tripping time for faults near the supply end becomes high (2 seconds in this case). However, in most cases, it is necessary to limit the maximum tripping time to 2 seconds. This disadvantage can be overcome to a reasonable extent by using inverse-time relays.

(*ii*) *Using inverse time relays.* Fig. 23.5 shows overcurrent protection of a radial feeder using

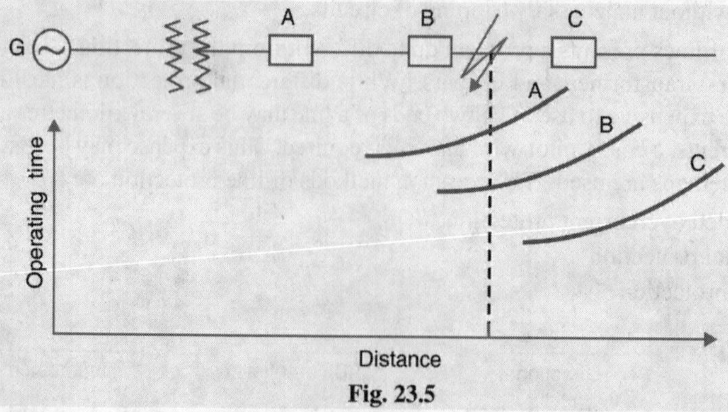

Fig. 23.5

* The amount of time delay depends upon the speed of breaker tripping. Sufficient time delay must be allowed to permit the breaker on the faulted section to clear the fault before the next relay in the sequence trips. The time-delay usually varies from 0·25 second to 0·5 second.

inverse time relays in which operating time is inversely proportional to the operating current. With this arrangement, the farther the circuit breaker from the generating station, the shorter is its relay operating time.

The three relays at *A*, *B* and *C* are assumed to have inverse-time characteristics. A fault in section *BC* will give relay times which will allow breaker at *B* to trip out before the breaker at *A*.

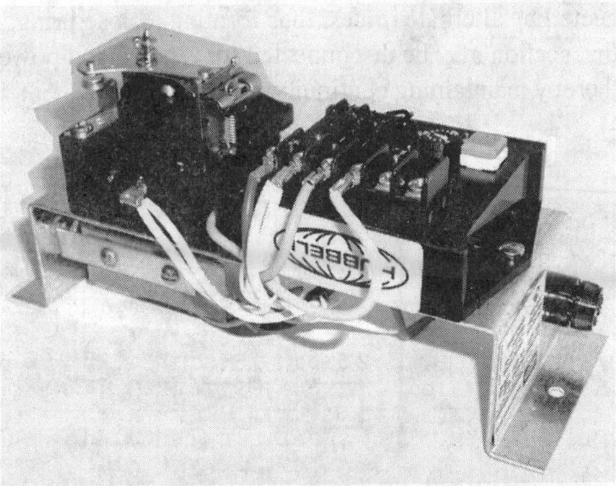

Inverse Time Relay

2. Parallel feeders. Where continuity of supply is particularly necessary, two parallel feeders may be installed. If a fault occurs on one feeder, it can be disconnected from the system and continuity of supply can be maintained from the other feeder. The parallel feeders cannot* be protected by non-directional overcurrent relays only. It is necessary to use directional relays also and to grade the time setting of relays for selective trippings.

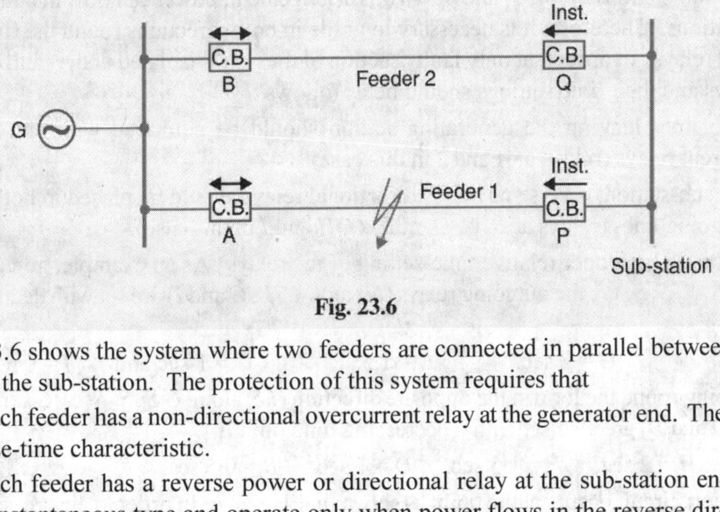

Fig. 23.6

Fig. 23.6 shows the system where two feeders are connected in parallel between the generating station and the sub-station. The protection of this system requires that

(*i*) each feeder has a non-directional overcurrent relay at the generator end. These relays should have inverse-time characteristic.

(*ii*) each feeder has a reverse power or directional relay at the sub-station end. These relays should be instantaneous type and operate only when power flows in the reverse direction *i.e.* in the direction of arrow at *P* and *Q*.

Suppose an earth fault occurs on feeder 1 as shown in Fig. 23.6. It is desired that only circuit breakers at *A* and *P* should open to clear the fault whereas feeder 2 should remain intact to maintain the continuity of supply. In fact, the above arrangement accomplishes this job. The shown fault is fed *via* two routes, *viz*.

(*a*) directly from feeder 1 *via* the relay *A*

(*b*) from feeder 2 *via B*, *Q*, sub-station and *P*

Therefore, power flow in relay *Q* will be in normal direction but is reversed in the relay *P*. This causes the opening of circuit breaker at *P*. Also the relay *A* will operate while relay *B* remains inop-

* Referring to Fig. 23.6, suppose relays at *P* and *Q* are non-directional type and their time settings are lower than relays at *A* and *B*. When a fault occurs at the shown point, the relay at *Q* will operate first and disconnect the feeder 2, and then feeder 1 will be cut off. Thus even the sound feeder (No. 2) is isolated.

erative. It is because these relays have inverse-time characteristics and current flowing in relay *A* is in excess of that flowing in relay *B*. In this way only the faulty feeder is isolated.

3. Ring main system. In this system, various power stations or sub-stations are interconnected by alternate routes, thus forming a closed ring. In case of damage to any section of the ring, that section may be disconnected for repairs, and power will be supplied from both ends of the ring, thereby maintaining continuity of supply.

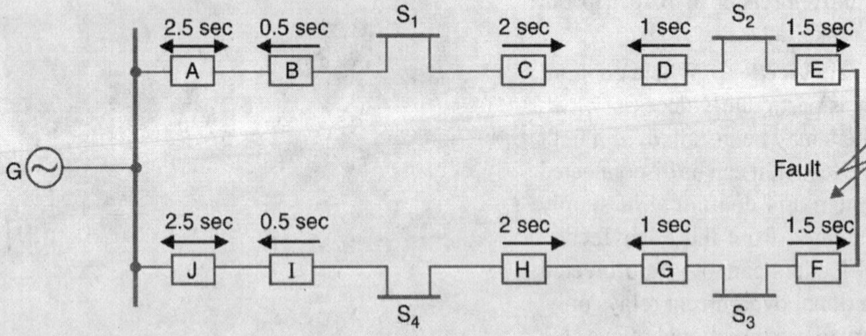

Fig. 23.7

Fig. 23.7 shows the single line diagram of a typical ring main system consisting of one generator *G* supplying four sub-stations S_1, S_2, S_3 and S_4. In this arrangement, power can flow in both directions under fault conditions. Therefore, it is necessary to grade in both directions round the ring and also to use directional relays. In order that only faulty section of the ring is isolated under fault conditions, the types of relays and their time settings should be as follows :

(*i*) The two lines leaving the generating station should be equipped with non-directional overcurrent relays (relays at *A* and *J* in this case).

(*ii*) At each sub-station, reverse power or directional relays should be placed in both incoming and outgoing lines (relays at *B, C, D, E, F, G, H* and *I* in this case).

(*iii*) There should be proper relative time-setting of the relays. As an example, going round the loop $G\ S_1\ S_2\ S_3\ S_4\ G$; the outgoing relays (*viz* at *A, C, E, G* and *I*) are set with decreasing time limits *e.g.*

$A = 2{\cdot}5$ sec, $C = 2$ sec, $E = 1{\cdot}5$ sec $G = 1$ sec and $I = 0{\cdot}5$ sec

Similarly, going round the loop in the opposite direction (*i.e.* along $G\ S_4\ S_3\ S_2\ S_1\ G$), the *outgoing relays* (*J, H, F, D* and *B*) are also set with a decreasing time limit *e.g.*

$J = 2{\cdot}5$ sec, $H = 2$ sec, $F = 1{\cdot}5$ sec, $D = 1$ sec, $B = 0{\cdot}5$ sec.

Suppose a short circuit occurs at the point as shown in Fig. 23.7. In order to ensure selectivity, it is desired that only circuit breakers at *E* and *F* should open to clear the fault whereas other sections of the ring should be intact to maintain continuity of supply. In fact, the above arrangement accomplishes this job. The power will be fed to the fault *via* two routes *viz* (*i*) from *G* around S_1 and S_2 and (*ii*) from *G* around S_4 and S_3. It is clear that relays at *A, B, C* and *D* as well as *J, I, H* and *G* will not trip. Therefore, only relays at *E* and *F* will operate before any other relay operates because of their lower time-setting.

23.4 Differential Pilot-Wire Protection

The differential pilot-wire protection is based on the principle that under normal conditions, the current entering one end of a line is equal to that leaving the other end. As soon as a fault occurs between the two ends, this condition no longer holds and the difference of incoming and outgoing currents is arranged to flow through a relay which operates the circuit breaker to isolate the faulty line. There are several differential protection schemes in use for the lines. However, only the following two schemes will be discussed :

1. Merz-Price voltage balance system
2. Translay scheme

1. **Merz-Price voltage balance system.** Fig. 23.8 shows the single line diagram of Merz-Price voltage balance system for the protection of a 3-phase line. Identical current transformers are placed in each phase at both ends of the line. The pair of *CTs* in each line is connected in series with a relay in such a way that under normal conditions, their secondary voltages are equal and in opposition *i.e.* they balance each other.

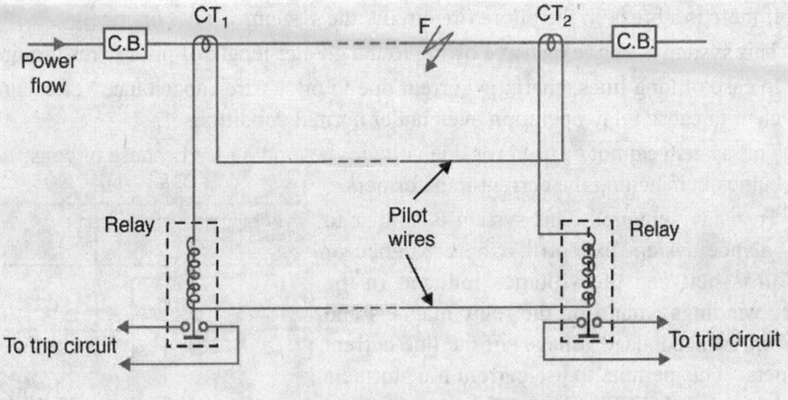

Fig. 23.8

Under healthy conditions, current entering the line at one-end is equal to that leaving it at the other end. Therefore, equal and opposite voltages are induced in the secondaries of the *CTs* at the two ends of the line. The result is that no current flows through the relays. Suppose a fault occurs at point *F* on the line as shown in Fig. 23.8. This will cause a greater current to flow through CT_1 than through CT_2. Consequently, their secondary voltages become unequal and circulating current flows through the pilot wires and relays. The circuit breakers at both ends of the line will trip out and the faulty line will be isolated.

Fig. 23.9 shows the connections of Merz-Price voltage balance scheme for all the three phases of the line.

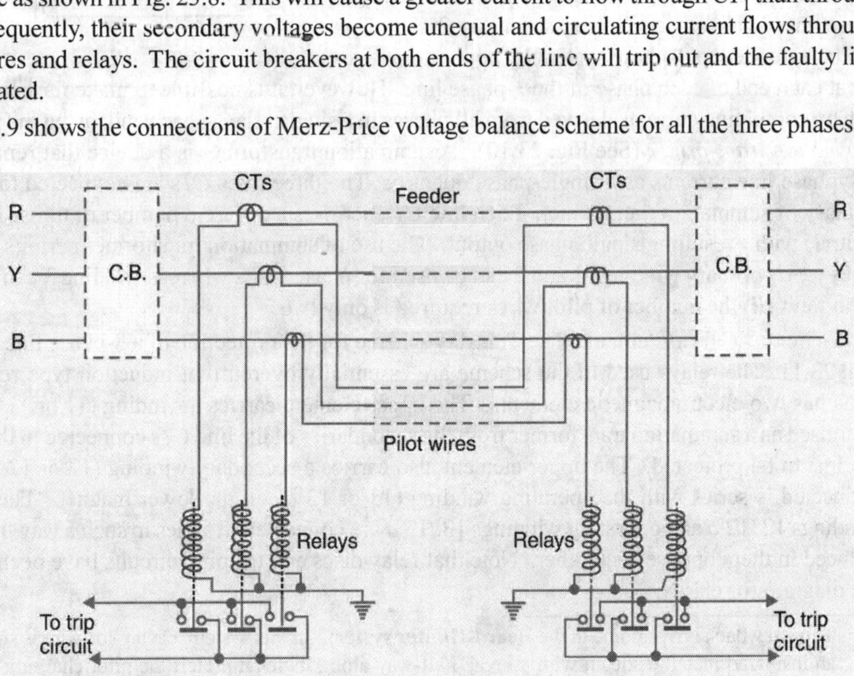

Fig. 23.9

Advantages

(*i*) This system can be used for ring mains as well as parallel feeders.

(*ii*) This system provides instantaneous protection for ground faults. This decreases the possibility of these faults involving other phases.

(*iii*) This system provides instantaneous relaying which reduces the amount of damage to overhead conductors resulting from arcing faults.

Disadvantages

(*i*) Accurate matching of current transformers is very essential.

(*ii*) If there is a break in the pilot-wire circuit, the system will not operate.

(*iii*) This system is very expensive owing to the greater length of pilot wires required.

(*iv*) In case of long lines, charging current due to pilot-wire capacitance* effects may be sufficient to cause relay operation even under normal conditions.

(*v*) This system cannot be used for line voltages beyond 33 kV because of constructional difficulties in matching the current transformers.

2. **Translay scheme.** This system is similar to voltage balance system except that here balance or opposition is between the voltages induced in the secondary windings wound on the relay magnets and *not* between the secondary voltages of the line current transformers. This permits to use current transformers of normal design and eliminates one of the most serious limitations of original voltage balance system, namely ; its limitation to the system operating at voltages not exceeding 33 kV.

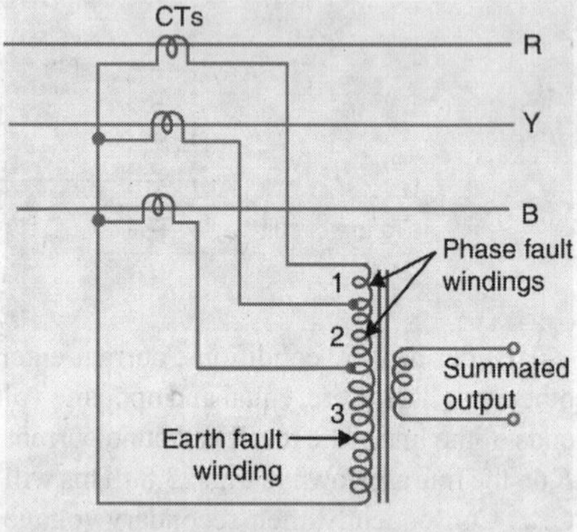

Fig. 23.10

The application of Translay scheme for a single phase line has already been discussed in Art. 21.20. This can be extended to 3-phase system by applying one relay at each end of each phase of the 3-phase line. However, it is possible to make further simplification by combining currents derived from all phases in a single relay at each end, using the principle of *summation transformer* (See Fig. 23.10). A summation transformer is a device that reproduces the polyphase line currents as a single-phase quantity. The three lines *CT*s are connected to the tapped primary of summation transformer. Each line *CT* energises a different number of turns (from line to neutral) with a resulting single phase output. The use of summation transformer permits two advantages *viz* (*i*) primary windings 1 and 2 can be used for phase faults whereas winding 3 can be used for earth fault (*ii*) the number of pilot wires required is only two.

Schematic arrangement. The Translay scheme for the protection of a 3-phase line is shown in Fig. 23.11. The relays used in the scheme are essentially overcurrent induction type relays. Each relay has two electromagnetic elements. The upper element carries a winding (11 or 11 *a*) which is energised as a summation transformer from the secondaries of the line *CT*s connected in the phases of the line to be protected. The upper element also carries a secondary winding (12 or 12 *a*) which is connected is series with the operating winding (13 or 13 *a*) on the lower magnet. The secondary windings 12, 12 *a* and operating windings 13, 13 *a* are connected in series in such a way that voltages induced in them oppose each other. Note that relay discs and tripping circuits have been omitted in the diagram for clarity.

---

* This drawback is overcome in the **Beard-Hunter system.** In this system, each pilot-wire is surrounded by an insulated metallic sheath with a break half-way along its length. Half the pilot charging current thus comes from the sending end and half from the receiving end. Therefore, voltage applied to the relay at the sending end is balanced by an equal voltage at the receiving end.

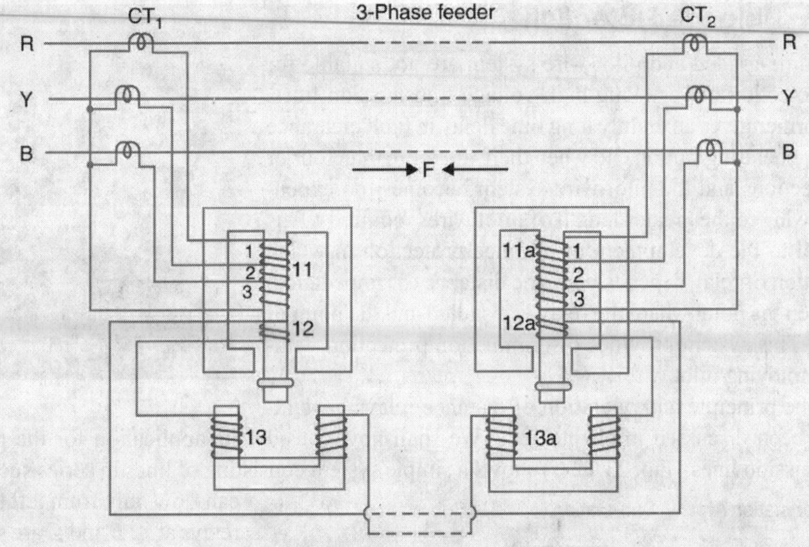

Fig. 23.11

Operation. When the feeder is sound, the currents at its two ends are equal so that the secondary currents in both sets of *CT*s are equal. Consequently, the currents flowing in the relay primary winding 11 and 11 *a* will be equal and they will induce equal voltages in the secondary windings 12 and 12*a*. Since these windings are connected in opposition, no current flows in them or in the operating windings 13 and 13*a*. In the event of a fault on the protected line, the line current at one end must carry a greater current than that at the other end. The result is that voltages induced in the secondary windings 12 and 12 *a* will be different and the current will flow through the operating coils 13, 13*a* and the pilot circuit. Under these conditions, both upper and lower elements of each relay are energised and a forward torque acts on the each relay disc. The operation of the relays will open the circuit breakers at both ends of the line.

(*i*) Suppose a fault *F* occurs between phases *R* and *Y* and is fed from both sides as shown in Fig. 23.11. This will energise only section 1 of primary windings 11 and 11*a* and induce voltages in the secondary windings 12 and 12*a*. As these voltages are now additive*, therefore, current will circulate through operating coils 13, 13*a* and the pilot circuit. This will cause the relay contacts to close and open the circuit breakers at both ends. A fault between phases *Y* and *B* energises section 2 of primary windings 11 and 11*a* whereas that between *R* and *B* will energise the sections 1 and 2.

(*ii*) Now imagine that an earth fault occurs on phase *R*. This will energise sections 1, 2 and 3 of the primary windings 11 and 11*a*. Again if fault is fed from both ends, the voltages induced in the secondary windings 12 and 12*a* are additive and cause a current to flow through the operating coils 13, 13*a*. The relays, therefore, operate to open the circuit breakers at both ends of the line. In the event of earth fault on phase *Y*, sections 2 and 3 of primary winding 11 and 11*a* will be energised and cause the relays to operate. An earth fault on phase B will energise only section 3 of relay primary windings 11 and 11*a*.

Advantages

(*i*) The system is economical as only two pilot wires are required for the protection of a 3-phase line.

(*ii*) Current transformers of normal design can be used.

(*iii*) The pilot wire capacitance currents do not affect the operation of relays.

* Because the fault is being fed from both sides.

23.5 Distance Protection

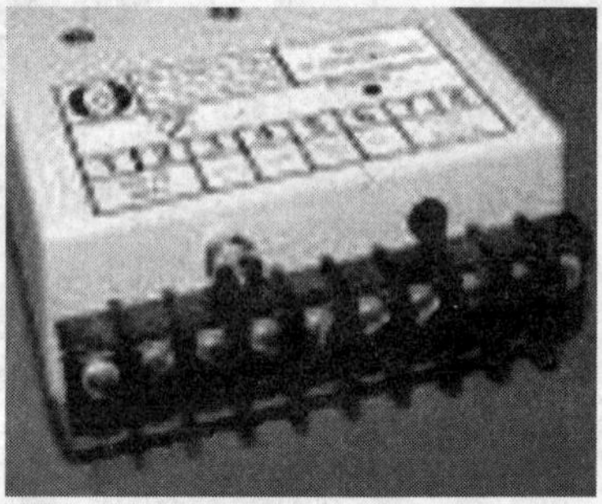

Both time-graded and pilot-wire system are not suitable for the protection of very long high voltage transmission lines. The former gives an unduly long time delay in fault clearance at the generating station end when there are more than four or five sections and the pilot-wire system becomes too expensive owing to the greater length of pilot wires required. This has led to the development of distance protection in which the action of relay depends upon the distance (or impedance) between the point where the relay is installed and the point of fault. This system provides discrimination protection without employing pilot wires.

The principle and operation of distance relays have already been discussed in chapter 21. We shall now consider its application for the protection of transmission lines. Fig. 23.12 (*i*) shows a simple system consisting of lines in series such that power can flow only from left to right. The relays at *A*, *B* and *C* are set to operate for impedance less than Z_1, Z_2 and Z_3 respectively. Suppose a fault occurs between sub-stations *B* and *C*, the fault impedance at power station and sub-station *A* and *B* will be $Z_1 + Z$ and *Z* respectively. It is clear that for the portion shown, only relay at *B* will operate. Similarly, if a fault occurs within section *AB*, then only relay at *A* will operate. In this manner, instantaneous protection can be obtained for all conditions of operation.

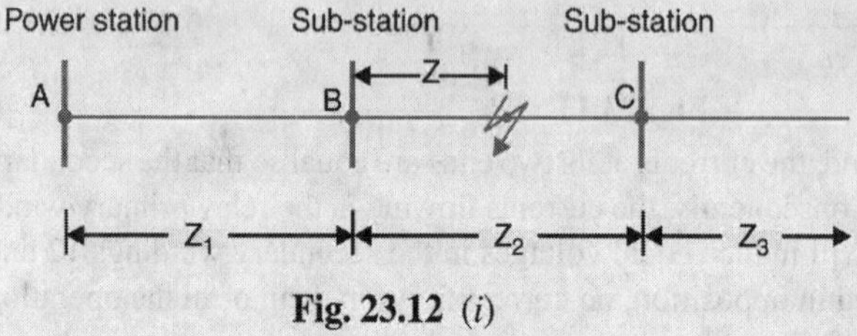

Fig. 23.12 (*i*)

In actual practice, it is not possible to obtain instantaneous protection for complete length of the line due to inaccuracies in the relay elements and instrument transformers. Thus the relay at *A* [See Fig. 23.12 (*i*)] would not be very reliable in distinguishing between a fault at 99% of the distance *AB* and the one at 101% of distance *AB*. This difficulty is overcome by using *'three-zone'* distance protection shown in Fig. 23.12 (*ii*).

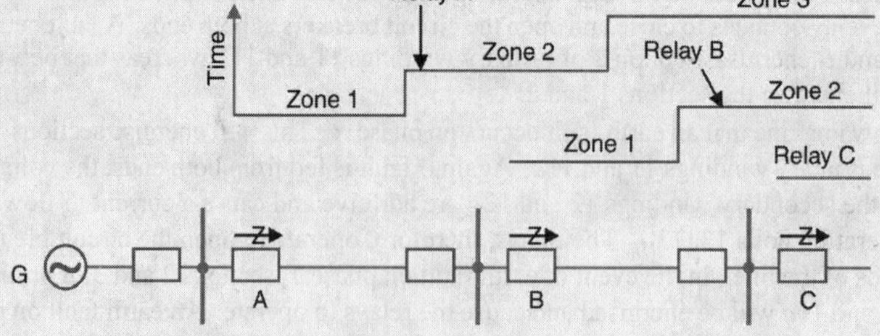

Fig. 23.12 (*ii*)

In this scheme of protection, three distance elements are used at each terminal. The zone 1 element covers first 90% of the line and is arranged to trip instantaneously for faults in this portion. The zone 2 element trips for faults in the remaining 10% of the line and for faults in the next line section, but a time delay is introduced to prevent the line from being tripped if the fault is in the next section. The zone 3 element provides back-up protection in the event a fault in the next section is not cleared by its breaker.

SELF - TEST

1. **Fill in the blanks by inserting appropriate words/figures :**
   (*i*) Differential protection scheme for longer lines is .... costly.
   (*ii*) The bus-bar zone, for the purpose of protection, includes .... , ... and ....
   (*iii*) The two most commonly used schemes for bus-bar protection are ...., and ....
   (*iv*) The probability of faults occurring on the lines is much more due to their .... and ....
   (*v*) In time-graded overcurrent protection, ... discrimination is incorporated.
2. **Pick-up the correct words/figures from the brackets and fill in the blanks :**
   (*i*) The parallel feeders .... be protected by non-directional overcurrent relays alone. (*can, cannot*)
   (*ii*) The Translay scheme is essentially a .... balance system. (*current, voltage*)
   (*iii*) A summation transformer is a device that reproduces the polyphase line currents as a .... phase quantity. (*single, two*)
   (*iv*) The ideal scheme of protection for lines is .... protection. (*differential, distance*)
   (*v*) Accurate matching of current transformers is .... in Merz-Price voltage balance system. (*essential, not essential*)

ANSWERS TO SELF-TEST

1. (*i*) very (*ii*) bus-bars, isolating switches, circuit breakers (*iii*) differential protection, fault bus protection (*iv*) greater length, exposure to atmospheric conditions (*v*) time
2. (*i*) cannot (*ii*) voltage (*iii*) single (*iv*) differential (*v*) essential

CHAPTER REVIEW TOPICS

1. What is the importance of bus-bar protection ?
2. Describe the following systems of bus-bar protection :
   (*i*) Differential protection
   (*ii*) Fault-bus protection
3. What are the requirements of protection of lines ?
4. Discuss the time-graded overcurrent protection for
   (*i*) Radial feeders
   (*ii*) Parallel feeders
   (*iii*) Ring main system
5. Describe the differential pilot wire method of protection of feeders.
6. Explain the Translay protection scheme for feeders.
7. Describe distance protection scheme for the protection of feeders.
8. Write short-notes on the following :
   (*i*) Fault-bus protection
   (*ii*) Merz-Price voltage balance system for protection of feeders
   (*iii*) Translay scheme

DISCUSSION QUESTIONS

1. What methods can be used to prevent saturation of current transformers ?
2. What factors govern choosing pilot-wire installation ?
3. Why must directional relays be used on a ring main system ?
4. How do time-delay overcurrent relays work on a radial system ?
5. Do overhead systems need differential protection schemes than underground systems ?
6. How are pilot-wire relays built for transmission-line protection ?

CHAPTER

24

Protection Against Overvoltages

Introduction

There are several instances when the elements of a power system (*e.g.* generators, transformers, transmission lines, insulators etc.) are subjected to overvoltages *i.e.* voltages greater than the normal value. These overvoltages on the power system may be caused due to many reasons such as lightning, the opening of a circuit breaker, the grounding of a conductor *etc.* Most of the overvoltages are not of large magnitude but may still be important because of their effect on the performance of circuit interrupting equipment and protective devices. An appreciable number of these overvoltages are of sufficient magnitude to cause insulation breakdown of the equipment in the power system. Therefore, power system engineers always device ways and means to limit the magnitude of the overvoltages produced and to control their effects on the operating equipment. In this chapter, we shall confine our attention to the various causes of overvoltages on the power system with special emphasis on the protective devices used for the purpose.

24.1 Voltage Surge

A sudden rise in voltage for a very short dura-

tion on the power system is known as a **voltage surge** *or* **transient voltage.**

Transients or surges are of temporary nature and exist for a very short duration (a few hundred μs) but they cause overvoltages on the power system. They originate from switching and from other causes but by far the most important transients are those caused by lightning striking a transmission line. When lightning strikes a line, the surge rushes along the line, just as a flood of water rushes along a narrow valley when the retaining wall of a reservoir at its head suddenly gives way. In most of the cases, such surges may cause the line insulators (near the point where lightning has struck) to flash over and may also damage the nearby transformers, generators or other equipment connected to the line if the equipment is not suitably protected.

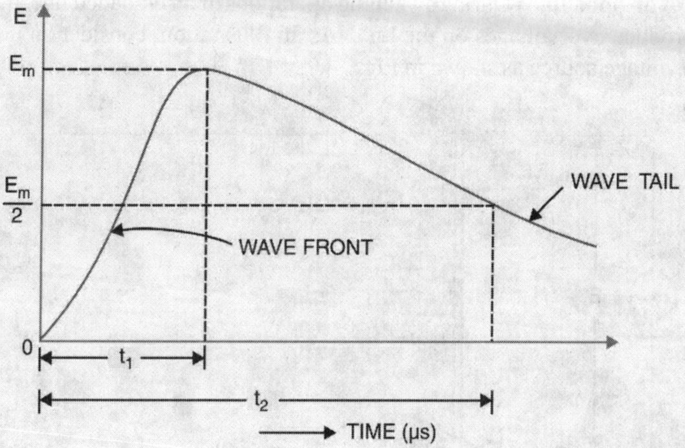

Fig. 24.1

Fig. 24.1 shows the wave-form of a typical lightning surge. The voltage build-up is taken along *y*-axis and the time along *x*-axis. It may be seen that lightning introduces a steep-fronted wave. The steeper the wave front, the more rapid is the build-up of voltage at any point in the network. In most of the cases, this build-up is comparatively rapid, being of the order of 1–5 μs. Voltage surges are generally specified in terms of *rise time t_1 and the time t_2 to decay to half of the peak value. For example, a 1/50 μs surge is one which reaches its maximum value in 1 μs and decays to half of its peak value is 50 μs.

24.2 Causes of Overvoltages

The overvoltages on a power system may be broadly divided into two main categories *viz.*

1. **Internal causes**

(*i*) Switching surges (*ii*) Insulation failure

(*iii*) Arcing ground (*iv*) Resonance

2. **External causes** *i.e.* lightning

Internal causes do not produce surges of large magnitude. Experience shows that surges due to internal causes hardly increase the system voltage to twice the normal value. Generally, surges due to internal causes are taken care of by providing proper insulation to the equipment in the power system. However, surges due to lightning are very severe and may increase the system voltage to several times the normal value. If the equipment in the power system is not protected against lightning surges, these surges may cause considerable damage. In fact, in a power system, the protective devices provided against overvoltages mainly take care of lightning surges.

* It is the time from the beginning of the surge to the peak value.

24.3 Internal Causes of Overvoltages

Internal causes of overvoltages on the power system are primarily due to oscillations set up by the sudden changes in the circuit conditions. This circuit change may be a normal switching operation such as opening of a circuit breaker, or it may be the fault condition such as grounding of a line conductor. In practice, the normal system insulation is suitably designed to withstand such surges. We shall briefly discuss the internal causes of overvoltages.

1. **Switching Surges.** The overvoltages produced on the power system due to switching operations are known as switching surges. A few cases will be discussed by way of illustration :

(*i*) Case of an open line. During switching operations of an unloaded line, travelling waves are set up which produce overvoltages on the line. As an illustration, consider an unloaded line being connected to a voltage source as shown in Fig. 24.2.

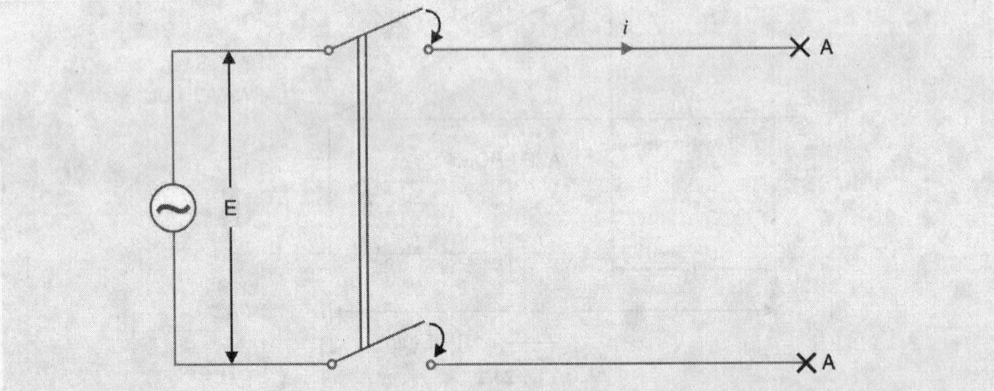

Fig. 24.2

When the unloaded line is connected to the voltage source, a voltage wave is set up which travels along the line. On reaching the terminal point *A*, it is reflected back to the supply end without change of sign. This causes voltage doubling *i.e.* voltage on the line becomes twice the normal value. If $E_{r.m.s.}$ is the supply voltage, then instantaneous voltage which the line will have to withstand will be $2\sqrt{2}\,E$. This overvoltage is of temporary nature. It is because the line losses attenuate the wave and in a very short time, the line settles down to its normal supply voltage *E*. Similarly, if an unloaded line is switched off, the line will attain a voltage of $2\sqrt{2}\,E$ for a moment before settling down to the normal value.

(*ii*) Case of a loaded line. Overvoltages will also be produced during the switching operations of a loaded line. Suppose a loaded line is suddenly interrupted. This will set up a voltage of $2\,Z_n\,i$ across the break (*i.e.* switch) where *i* is the instantaneous value of current at the time of opening of line and *Z_n is the natural impedance of the line. For example, suppose the line having $Z_n = 1000\ \Omega$ carries a current of 100 A (r.m.s.) and the break occurs at the moment when current is maximum. The voltage across the breaker (*i.e.* switch) $= 2\sqrt{2} \times 100 \times 1000/1000 = 282{\cdot}8$ kV. If V_m is the peak value of voltage in kV, the maximum voltage to which the line may be subjected is $= (V_m + 282{\cdot}8)$ kV.

(*iii*) Current chopping. Current chopping results in the production of high voltage transients across the contacts of the air blast circuit breaker as detailed in chapter 19. It is briefly discussed here. Unlike oil circuit breakers, which are independent for the effectiveness on the magnitude of the

---

* It can be shown that natural impedance of the line is given by :

$$Z_n = \sqrt{L/C}$$

where *L* and *C* are the constants of the line. The term 'natural' is used because this impedance has nothing to do with any load, but depends only upon line constants.

current being interrupted, air-blast circuit breakers retain the same extinguishing power irrespective of the magnitude of this current. When breaking low currents (*e.g.* transformer magnetising current) with air-blast breaker, the powerful de-ionising effect of air-blast causes the current to fall abruptly to zero well before the natural current zero is reached. This phenomenon is called current chopping and produces high transient voltage across the breaker contacts. Overvoltages due to current chopping are prevented by resistance switching (See Chapter 19).

2. Insulation failure. The most common case of insulation failure in a power system is the grounding of conductor (*i.e.* insulation failure between line and earth) which may cause overvoltages in the system. This is illustrated in Fig. 24.3.

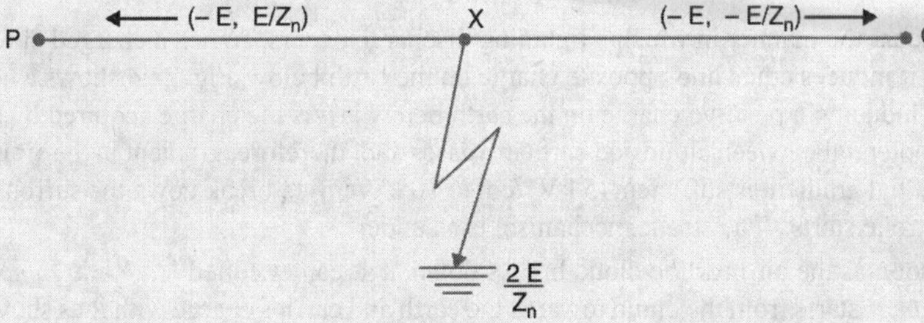

Fig. 24.3

Suppose a line at potential E is earthed at point X. The earthing of the line causes two equal voltages of $-E$ to travel along XQ and XP containing currents $-E/Z_n$ and $+E/Z_n$ respectively. Both these currents pass through X to earth so that current to earth is $2\ E/Z_n$.

3. Arcing ground. In the early days of transmission, the neutral of three phase lines was not earthed to gain two advantages. Firstly, in case of line-to-ground fault, the line is not put out of action. Secondly, the zero sequence currents are eliminated, resulting in the decrease of interference with communication lines. Insulated neutrals give no problem with short lines and comparatively low voltages. However, when the lines are long and operate at high voltages, serious problem called *arcing ground* is often witnessed. The arcing ground produces severe oscillations of three to four times the normal voltage.

The phenomenon of intermittent arc taking place in line-to-ground fault of a 3ϕ system with consequent production of transients is known as **arcing ground.**

The transients produced due to arcing ground are cumulative and may cause serious damage to the equipment in the power system by causing breakdown of insulation. Arcing ground can be prevented by earthing the neutral.

4. Resonance. Resonance in an electrical system occurs when inductive reactance of the circuit becomes equal to capacitive reactance. Under resonance, the impedance of the circuit is equal to resistance of the circuit and the p.f. is unity. Resonance causes high voltages in the electrical system. In the usual transmission lines, the capacitance is very small so that resonance rarely occurs at the fundamental supply frequency. However, if generator *e.m.f.* wave is distorted, the trouble of resonance may occur due to 5th or higher harmonics and in case of underground cables too.

24.4 Lightning

An electric discharge between cloud and earth, between clouds or between the charge centres of the same cloud is known as **lightning.**

Lightning is a huge spark and takes place when clouds are charged to such a high potential (+ve or –ve) with respect to earth or a neighbouring cloud that the dielectric strength of neighbouring medium (air) is destroyed. There are several theories which exist to explain how the clouds acquire charge. The most accepted one is that during the uprush of warm moist air from earth, the friction between the air and the tiny particles of water causes the building up of charges. When drops of water are formed, the larger drops become positively charged and the smaller drops become negatively charged. When the drops of water accumulate, they form clouds, and hence cloud may possess either a positive or a negative charge, depending upon the charge of drops of water they contain. The charge on a cloud may become so great that it may discharge to another cloud or to earth and we call this discharge as lightning. The thunder which accompanies lightning is due to the fact that lightning suddenly heats up the air, thereby causing it to expand. The surrounding air pushes the expanded air back and forth causing the wave motion of air which we recognise as thunder.

24.5 Mechanism of Lightning Discharge

Let us now discuss the manner in which a lightning discharge occurs. When a charged cloud passes over the earth, it induces equal and opposite charge on the earth below. Fig. 24.4 shows a negatively charged cloud inducing a positive charge on the earth below it. As the charge acquired by the cloud increases, the potential between cloud and earth increases and, therefore, gradient in the air increases. When the potential gradient is sufficient (5 kV*/cm to 10 kV/cm) to break down the surrounding air, the lightning stroke starts. The stroke mechanism is as under :

(*i*) As soon as the air near the cloud breaks down, a streamer called *leader streamer* or *pilot streamer* starts from the cloud towards the earth and carries charge with it as shown in Fig. 24.4 (*i*). The leader streamer will continue its journey towards earth as long as the cloud, from which it originates feeds enough charge to it to maintain gradient at the tip of leader streamer above the strength of air. If this gradient is not maintained, the leader streamer stops and the charge is dissipated without the formation of a complete stroke. In other words, the leader streamer will not reach the earth. Fig. 24.4 (*i*) shows the leader streamer being unable to reach the earth as gradient at its end cloud not be maintained above the strength of air. It may be noted that current in the leader streamer is low (<100 A) and its velocity of propagation is about 0·05% that of velocity of light. Moreover, the luminosity of leader is also very low.

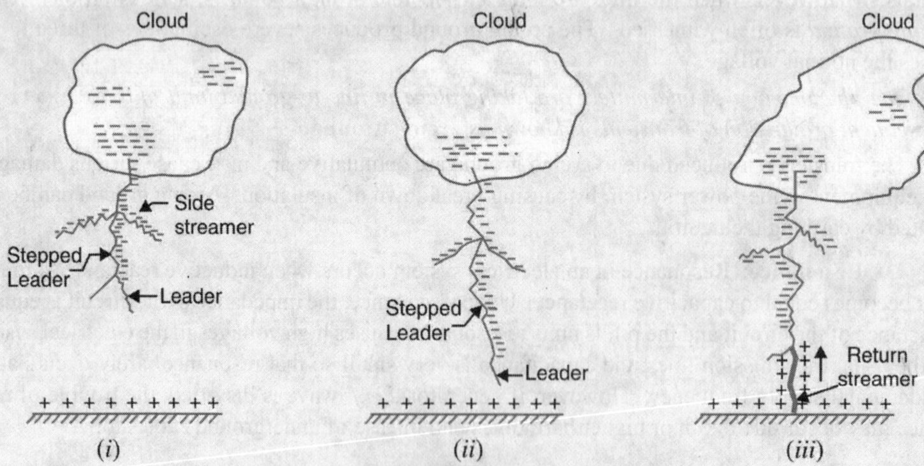

Fig. 24.4

(*ii*) In many cases, the leader streamer continues its journey towards earth [See Fig. 24.4 (*ii*)] until it makes contact with earth or some object on the earth. As the leader streamer moves towards earth, it is accompanied by points of luminescence which travel in jumps giving

* It has been found that in a region occupied by droplets of the size expected in the clouds, the breakdown voltage is 5 kV/cm to 10 kV/cm compared to 30 kV/cm in air without droplets. One reason for this is that the low pressure at high altitudes decreases the breakdown gradient.

rise to stepped leaders. The velocity of stepped leader exceeds one-sixth of that of light and distance travelled in one step is about 50 m. It may be noted that stepped leaders have sufficient luminosity and give rise to first visual phenomenon of discharge.

(*iii*) The path of leader streamer is a path of ionisation and, therefore, of complete breakdown of insulation. As the leader streamer reaches near the earth, a *return streamer* shoots up from the earth [See Fig. 24.4 (*iii*)] to the cloud, following the same path as the main channel of the downward leader. The action can be compared with the closing of a switch between the positive and negative terminals; the downward leader having negative charge and return streamer the positive charge. This phenomenon causes a sudden spark which we call lightning. With the resulting neutralisation of much of the negative charge on the cloud, any further discharge from the cloud may have to originate from some other portion of it.

The following points may be noted about lightning discharge :

(*a*) A lightning discharge which usually appears to the eye as a single flash is in reality made up of a number of separate strokes that travel down the same path. The interval between them varies from 0·0005 to 0·5 second. Each separate stroke starts as a downward leader from the cloud.

(*b*) It has been found that 87% of all lightning strokes result from negatively charged clouds and only 13% originate from positively charged clouds.

(*c*) It has been estimated that throughout the world, there occur about 100 lightning strokes per second.

(*d*) Lightning discharge may have currents in the range of 10 kA to 90 kA.

24.6 Types of Lightning Strokes

There are two main ways in which a lightning may strike the power system (*e.g.* overhead lines, towers, sub-stations etc.), namely;

1. Direct stroke 2. Indirect stroke

1. Direct stroke. In the direct stroke, the lightning discharge (*i.e.* current path) is directly from the cloud to the subject equipment *e.g.* an overhead line. From the line, the current path may be over the insulators down the pole to the ground. The overvoltages set up due to the stroke may be large enough to flashover this path directly to the ground. The direct strokes can be of two types *viz.* (*i*) Stroke *A* and (*ii*) stroke *B*.

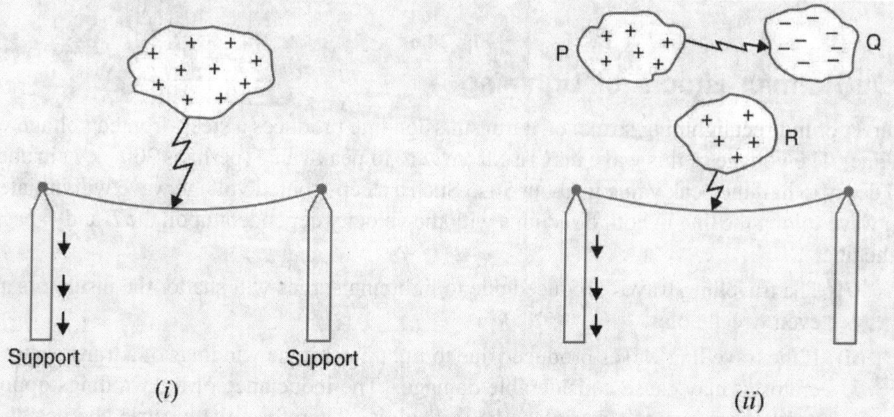

Fig. 24.5

(*i*) In stroke *A*, the lightning discharge is from the cloud to the subject equipment *i.e.* an overhead line in this case as shown in Fig. 24.5 (*i*). The cloud will induce a charge of opposite sign on the tall object (*e.g.* an overhead line in this case). When the potential between the cloud and line exceeds the breakdown value of air, the lightning discharge occurs between the cloud and the line.

(*ii*) In stroke *B*, the lightning discharge occurs on the overhead line as a result of stroke *A* between the clouds as shown in Fig. 24.5 (*ii*). There are three clouds *P*, *Q* and *R* having positive, negative and positive charges respectively. The charge on the cloud *Q* is bound by the cloud *R*. If the cloud *P* shifts too near the cloud *Q*, then lightning discharge will occur between them and charges on both these clouds disappear quickly. The result is that charge on cloud *R* suddenly becomes free and it then discharges rapidly to earth, ignoring tall objects.

Two points are worth noting about direct strokes. Firstly, direct strokes on the power system are very rare. Secondly, stroke *A* will always occur on tall objects and hence protection can be provided against it. However, stroke *B* completely ignores the height of the object and can even strike the ground. Therefore, it is not possible to provide protection against stroke *B*.

2. **Indirect stroke.** Indirect strokes result from the electrostatically induced charges on the conductors due to the presence of charged clouds. This is illustrated in Fig. 24.6. A positively charged cloud is above the line and induces a negative charge on the line by electrostatic induction. This negative charge, however, will be only on that portion of the line right under the cloud and the portions of the line away from it will be positively charged as shown in Fig. 24.6. The induced positive charge leaks slowly to earth *via* the insulators. When the cloud discharges to earth or to another cloud, the negative charge on the wire is isolated as it cannot flow quickly to earth over the insulators. The result is that negative charge rushes along the line is both directions in the form of travelling waves. It may be worthwhile to mention here that majority of the surges in a transmission line are caused by indirect lightning strokes.

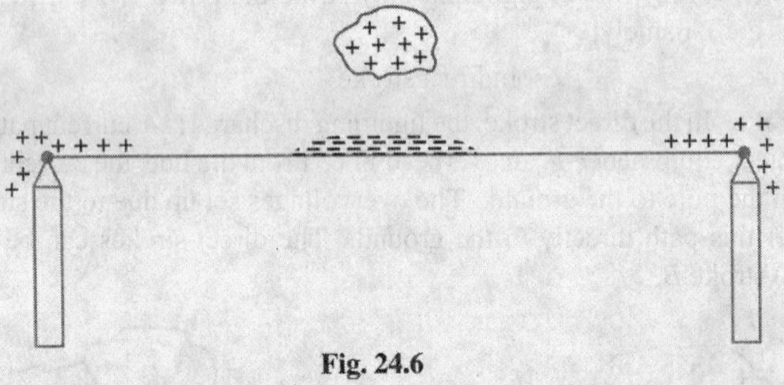

Fig. 24.6

24.7 Harmful Effects of Lightning

A direct or indirect lightning stroke on a transmission line produces a steep-fronted voltage wave on the line. The voltage of this wave may rise from zero to peak value (perhaps 2000 kV) in about 1 μs and decay to half the peak value in about 5μs. Such a steep-fronted voltage wave will initiate travelling waves along the line in both directions with the velocity dependent upon the *L* and *C* parameters of the line.

(*i*) The travelling waves produced due to lightning surges will shatter the insulators and may even wreck poles.

(*ii*) If the travelling waves produced due to lightning hit the windings of a transformer or generator, it may cause considerable damage. The inductance of the windings opposes any sudden passage of electric charge through it. Therefore, the electric charges "piles up" against the transformer (or generator). This induces such an excessive pressure between the windings that insulation may breakdown, resulting in the production of arc. While the normal voltage between the turns is never enough to *start* an arc, once the insulation has broken down and an arc has been started by a momentary overvoltage, the line voltage is usually sufficient to *maintain* the arc long enough to severely damage the machine.

(*iii*) If the arc is initiated in any part of the power system by the lightning stroke, this arc will set up very disturbing oscillations in the line. This may damage other equipment connected to the line.

24.8 Protection Against Lightning

Transients or surges on the power system may originate from switching and from other causes but the most important and dangerous surges are those caused by lightning. The lightning surges may cause serious damage to the expensive equipment in the power system (*e.g.* generators, transformers etc.) either by direct strokes on the equipment or by strokes on the transmission lines that reach the equipment as travelling waves. It is necessary to provide protection against both kinds of surges. The most commonly used devices for protection against lightning surges are :

(*i*) Earthing screen

(*ii*) Overhead ground wires

(*iii*) Lightning arresters or surge diverters

Earthing screen provides protection to power stations and sub-stations against direct strokes whereas overhead ground wires protect the transmission lines against direct lightning strokes. However, lightning arresters or surge diverters protect the station apparatus against both direct strokes and the strokes that come into the apparatus as travelling waves. We shall briefly discuss these methods of protection.

24.9 The Earthing Screen

The power stations and sub-stations generally house expensive equipment. These stations can be protected against direct lightning strokes by providing earthing screen. It consists of a network of copper conductors (generally called shield or screen) mounted all over the electrical equipment in the sub-station or power station. The shield is properly connected to earth on atleast two points through a low impedance. On the occurrence of direct stroke on the station, screen provides a low resistance path by which lightning surges are conducted to ground. In this way, station equipment is protected against damage. The limitation of this method is that it does not provide protection against the travelling waves which may reach the equipment in the station.

24.10 Overhead Ground Wires

The most effective method of providing protection to transmission lines against direct lightning strokes is by the use of overhead ground wires as shown in Fig. 24.7. For simplicity, one ground wire and one line conductor are shown. The ground wires are placed *above* the line conductors at such positions that practically all lightning strokes are intercepted by them (*i.e.* ground wires). The ground wires are grounded at each tower or pole through as low resistance as possible. Due to their proper location, the *ground wires will take up all the lightning strokes instead of allowing them to line conductors.

When the direct lightning stroke occurs on the transmission line, it will be taken up by the ground wires. The heavy lightning current (10 kA to 50 kA) from the ground wire flows to the ground, thus protecting the line from the harmful effects of lightning. It may be mentioned here that the degree of protection provided by the ground wires depends upon the footing resistance of the tower. Suppose, for example, tower-footing resistance is R_1 ohms and that the lightning current from tower to ground

* The degree of protection by ground wires depends upon the shielding angle (*i.e.* the angle subtended by the outermost line conductors at the ground wire). The lower this angle, the greater the protection.

is I_1 amperes. Then the tower *rises to a potential V_t given by ;

$$V_t = I_1 R_1$$

Since V_t (= I_1R_1) is the approximate voltage between tower and line conductor, this is also the voltage that will appear across the string of insulators. If the value of V_t is less than that required to cause insulator flashover, no trouble results. On the other hand, if V_t is excessive, the insulator flashover may occur. Since the value of V_t depends upon tower-footing resistance R_1, the value of this resistance must be kept as low as possible to avoid insulator flashover.

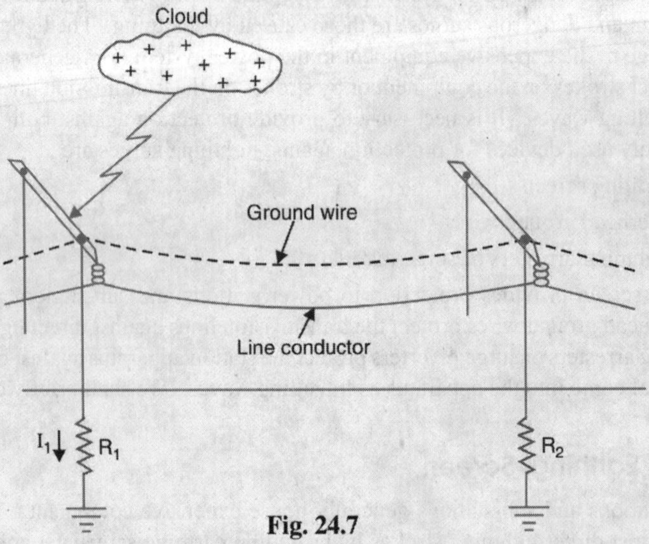

Fig. 24.7

Advantages

(*i*) It provides considerable protection against direct lightning strokes on transmission lines.

(*ii*) A grounding wire provides damping effect on any disturbance travelling along the line as it acts as a short-circuited secondary.

(*iii*) It provides a certain amount of electrostatic shielding against external fields. Thus it reduces the voltages induced in the line conductors due to the discharge of a neighbouring cloud.

Disadvantages

(*i*) It requires additional cost.

(*ii*) There is a possibility of its breaking and falling across the line conductors, thereby causing a short-circuit fault. This objection has been greatly eliminated by using galvanised stranded steel conductors as ground wires. This provides sufficient strength to the ground wires.

24.11 Lightning Arresters

The earthing screen and ground wires can well protect the electrical system against direct lightning strokes but they fail to provide protection against travelling waves which may reach the terminal apparatus. The lightning arresters or surge diverters provide protection against such surges.

***A lightning arrester** or a **surge diverter** is a protective device which conducts the high voltage surges on the power system to the ground.*

* As a numerical illustration, if I_1 = 50 kA and R_1 = 50 Ω, then $V_t = 50 \times 10^3 \times 50$ = 2500 kV. However, if R_1 = 10 Ω, then $V_t = 50 \times 10^3 \times 10$ = 500 kV. Clearly, lesser the tower-footing resistance, smaller the potential to which the tower rises.

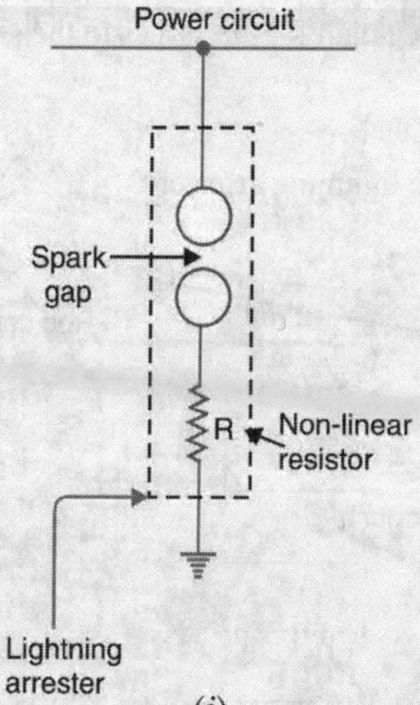

(i)

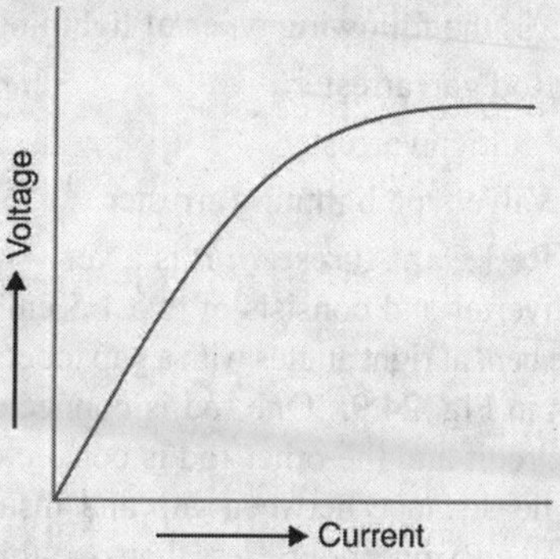

(ii)

Fig. 24.8

Fig. 24.8 (*i*) shows the basic form of a surge diverter. It consists of a spark gap in series with a non-linear resistor. One end of the diverter is connected to the terminal of the equipment to be protected and the other end is effectively grounded. The length of the gap is so set that normal line voltage is not enough to cause an arc across the gap but a dangerously high voltage will break down the air insulation and form an arc. The property of the non-linear resistance is that its resistance decreases as the voltage (or current) increases and vice-versa. This is clear from the *volt/amp characteristic of the resistor shown in Fig. 24.8 (*ii*).

Action. The action of the lightning arrester or surge diverter is as under :

(*i*) Under normal operation, the lightning arrester is off the line *i.e.* it conducts **no current to earth or the gap is non-conducting.

(*ii*) On the occurrence of overvoltage, the air insulation across the gap breaks down and an arc is formed, providing a low resistance path for the surge to the ground. In this way, the excess charge on the line due to the surge is harmlessly conducted through the arrester to the ground instead of being sent back over the line.

(*iii*) It is worthwhile to mention the function of non-linear resistor in the operation of arrester. As the gap sparks over due to overvoltage, the arc would be a short-circuit on the power system and may cause power-follow current in the arrester. Since the characteristic of the resistor is to offer high resistance to high voltage (or current), it prevents the effect of a short-circuit. After the surge is over, the resistor offers high resistance to make the gap non-conducting.

Two things must be taken care of in the design of a lightning arrester. Firstly, when the surge is over, the arc in gap should cease. If the arc does not go out, the current would continue to flow through the resistor and both resistor and gap may be destroyed. Secondly, IR drop (where I is the surge current) across the arrester when carrying surge current should not exceed the breakdown strength of the insulation of the equipment to be protected.

24.12 Types of Lightning Arresters

There are several types of lightning arresters in general use. They differ only in constructional details

---

* The characteristic is drawn between the voltage across the resistance and current through it.

** In actual practice, it may conduct current to ground even at normal supply due to capacitive effects. As the resistance R offers high resistance to normal votlage, this current is extremely small.

but operate on the same principle *viz*. providing low resistance path for the surges to the ground. We shall discuss the following types of lightning arresters :

1. Rod gap arrester
2. Horn gap arrester
3. Multigap arrester
4. Expulsion type lightning arrester
5. Valve type lightning arrester

1. Rod Gap Arrester. It is a very simple type of diverter and consists of two 1·5 cm rods which are bent at right angles with a gap inbetween as shown in Fig. 24.9. One rod is connected to the line circuit and the other rod is connected to earth. The distance between gap and insulator (*i.e.* distance *P*) must not be less than one-third of the gap length so that the arc may not reach the insulator and damage it. Generally, the gap length is so adjusted that breakdown should occur at 80% of spark- over voltage in order to avoid cascading of very steep wave fronts across the insulators. The string of insulators for an overhead line on the bushing of transformer has frequently a rod gap across it. Fig. 24.9 shows the rod gap across the bushing of a transformer.

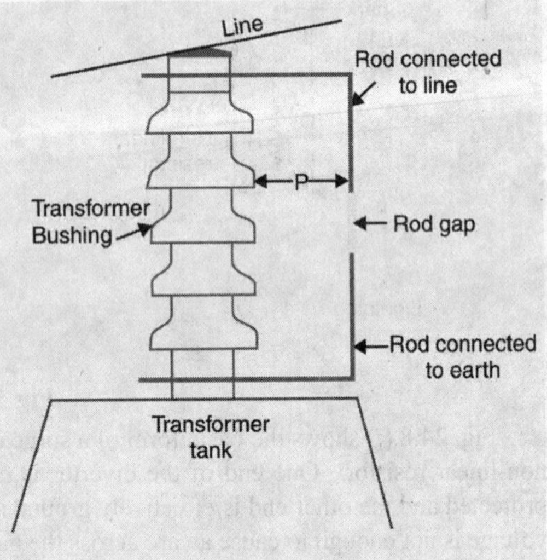

Fig. 24.9

Under normal operating conditions, the gap remains non-conducting. On the occurrence of a high voltage surge on the line, the gap sparks over and the surge current is conducted to earth. In this way, excess charge on the line due to the surge is harmlessly conducted to earth.

Limitations

(*i*) After the surge is over, the arc in the gap is maintained by the † normal supply voltage, leading to a short-circuit on the system.

(*ii*) The rods may melt or get damaged due to excessive heat produced by the arc.

(*iii*) The climatic conditions (*e.g.* rain, humidity, temperature etc.) affect the performance of rod gap arrester.

(*iv*) The polarity of the surge also affects the performance of this arrester.

Due to the above limitations, the rod gap arrester is only used as a 'back-up' protection in case of main arresters.

2. Horn Gap Arrester. Fig. 24.10 shows the horn gap arrester. It consists of two horn shaped metal rods *A* and *B* separated by a small air gap. The horns are so constructed that distance between them gradually increases towards the top as shown. The horns are mounted on porcelain insulators. One end of horn is connected to the line through a resistance *R* and choke coil *L* while the other end is effectively grounded. The resistance *R* helps in limiting the follow current to a small value. The choke coil is so designed that it offers small reactance at normal power frequency but a very high reactance at transient frequency. Thus the choke does not allow the transients to enter the apparatus to be protected. The gap between the horns is so adjusted that normal supply voltage is not enough to cause an arc across the gap.

Under normal conditions, the gap is non-conducting *i.e.* normal supply voltage is insufficient to initiate the arc between the gap. On the occurrence of an overvoltage, spark-over takes place across

† The normal supply voltage may not be able to initiate the arc across the gap. But once the arc is started by the surge, the normal voltage is enough to maintain it.

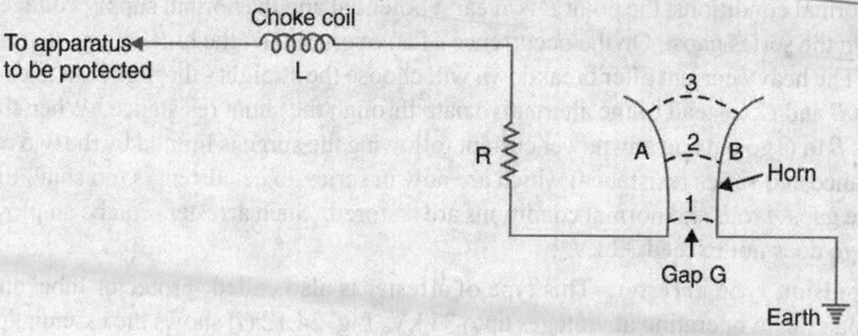

Fig. 24.10

the *small gap G. The heated air around the arc and the magnetic effect of the arc cause the arc to travel up the gap. The arc moves progressively into positions 1, 2 and 3. At some position of the arc (perhaps position 3), the distance may be too great for the voltage to maintain the arc. Consequently, the arc is extinguished. The excess charge on the line is thus conducted through the arrester to the ground.

Advantages

(*i*) The arc is self-clearing. Therefore, this type of arrester does not cause short-circuiting of the system after the surge is over as in the case of rod gap.

(*ii*) Series resistance helps in limiting the follow current to a small value.

Limitations

(*i*) The bridging of gap by some external agency (*e.g.* birds) can render the device useless.

(*ii*) The setting of horn gap is likely to change due to corrosion or pitting. This adversely affects the performance of the arrester.

(*iii*) The time of operation is comparatively long, say about 3 seconds. In view of the very short operating time of modern protective gear for feeders, this time is far long.

Due to the above limitations, this type of arrester is not reliable and can only be used as a second line of defence like the rod gap arrester.

3. **Multigap arrester.** Fig. 24.11 shows the multigap arrester. It consists of a series of metallic (generally alloy of zinc) cylinders insulated from one another and separated by small intervals of air gaps. The first cylinder (*i.e.* A) in the series is connected to the line and the other to the ground through a series resistance. The series resistance limits the power arc. By the inclusion of series resistance, the degree of protection against travelling waves is reduced. In order to overcome this difficulty, some of the gaps (B to C in Fig. 24.11) are shunted by a resistance.

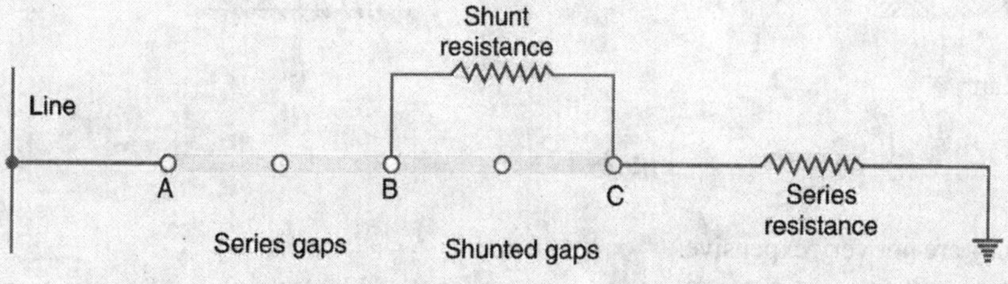

Fig. 24.11

* The arc will occur at the smallest space (*i.e.* gap G) between the horns.

Under normal conditions, the point *B* is at earth potential and the normal supply voltage is unable to break down the series gaps. On the occurrence of an overvoltage, the breakdown of series gaps *A* to *B* occurs. The heavy current after breakdown will choose the straight - through path to earth *via* the shunted gaps *B* and *C*, instead of the alternative path through the shunt resistance. When the surge is over, the arcs *B* to *C* go out and any power current following the surge is limited by the two resistances (shunt resistance and series resistance) which are now in series. The current is too small to maintain the arcs in the gaps *A* to *B* and normal conditions are restored. Such arresters can be employed where system voltage does not exceed 33 kV.

4. Expulsion type arrester. This type of arrester is also called 'protector tube' and is commonly used on system operating at voltages upto 33 kV. Fig. 24.12 (*i*) shows the essential parts of an expulsion type lightning arrester. It essentially consists of a rod gap *AA′* in series with a second gap enclosed within the fibre tube. The gap in the fibre tube is formed by two electrodes. The upper electrode is connected to rod gap and the lower electrode to the earth. One expulsion arrester is placed under each line conductor. Fig. 24.12 (*ii*) shows the installation of expulsion arrester on an overhead line.

On the occurrence of an overvoltage on the line, the series gap *AA′* is spanned and an arc is struck between the electrodes in the tube. The heat of the arc vaporises some of the fibre of tube walls, resulting in the production of a neutral gas*. In an extremely short time, the gas builds up high pressure and is expelled through the lower electrode which is hollow. As the gas leaves the tube violently, it carries away ionised air around the arc. This de-ionising effect is generally so strong that arc goes out at a current zero and will not be re-established.

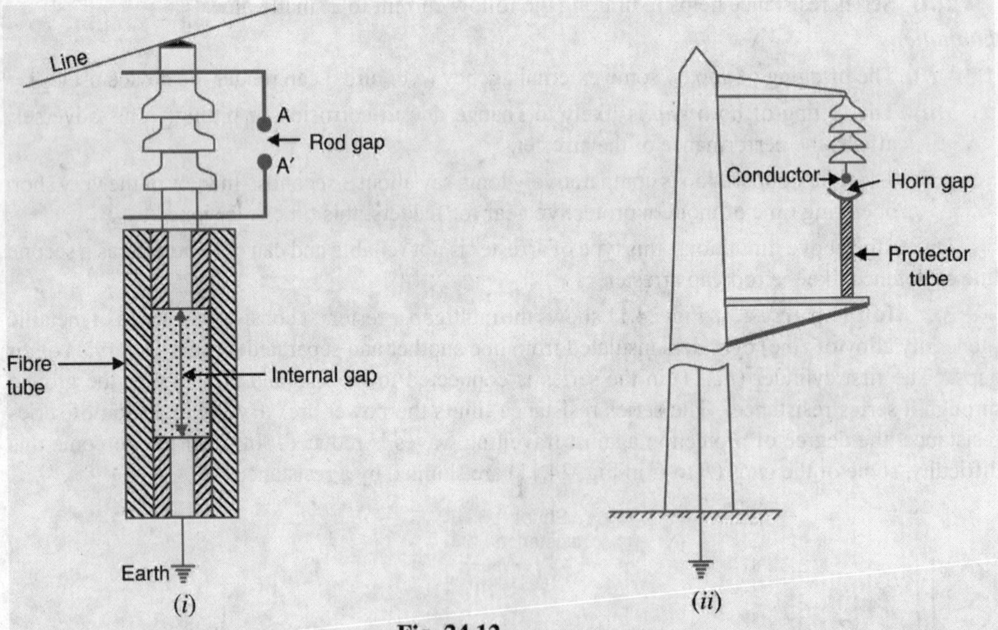

Fig. 24.12

Advantages

(*i*) They are not very expensive.

(*ii*) They are improved form of rod gap arresters as they block the flow of power frequency follow currents.

(*iii*) They can be easily installed.

* The gas evolved is an un-ionised mixture of water vapour and decomposition products of the fibre.

Limitations

(*i*) An expulsion type arrester can perform only limited number of operations as during each operation some of the fibre material is used up.

(*ii*) This type of arrester cannot be mounted in an enclosed equipment due to the discharge of gases during operation.

(*iii*) Due to the poor volt/amp characteristic of the arrester, it is not suitable for the protection of expensive equipment.

5. **Valve type arrester.** Valve type arresters incorporate non-linear resistors and are extensively used on systems operating at high voltages. Fig. 24.13 (*i*) shows the various parts of a valve type arrester. It consists of two assemblies (*i*) series spark gaps and (*ii*) non-linear resistor discs (made of material such as thyrite or metrosil) in series. The non-linear elements are connected in series with the spark gaps. Both the assemblies are accommodated in tight porcelain container.

(*i*) The spark gap is a multiple assembly consisting of a number of identical spark gaps in series. Each gap consists of two electrodes with a fixed gap spacing. The voltage distribution across the gaps is linearised by means of additional resistance elements (called grading resistors) across the gaps. The spacing of the series gaps is such that it will withstand the normal circuit voltage. However, an overvoltage will cause the gap to breakdown, causing the surge current to ground via the non-linear resistors.

(*ii*) The non-linear resistor discs are made of an inorganic compound such as Thyrite or Metrosil. These discs are connected in series. The non-linear resistors have the property of offering a high resistance to current flow when normal system voltage is applied, but a low resistance to the flow of high-surge currents. In other words, the resistance of these non-linear elements decreases with the increase in current through them and *vice-versa*.

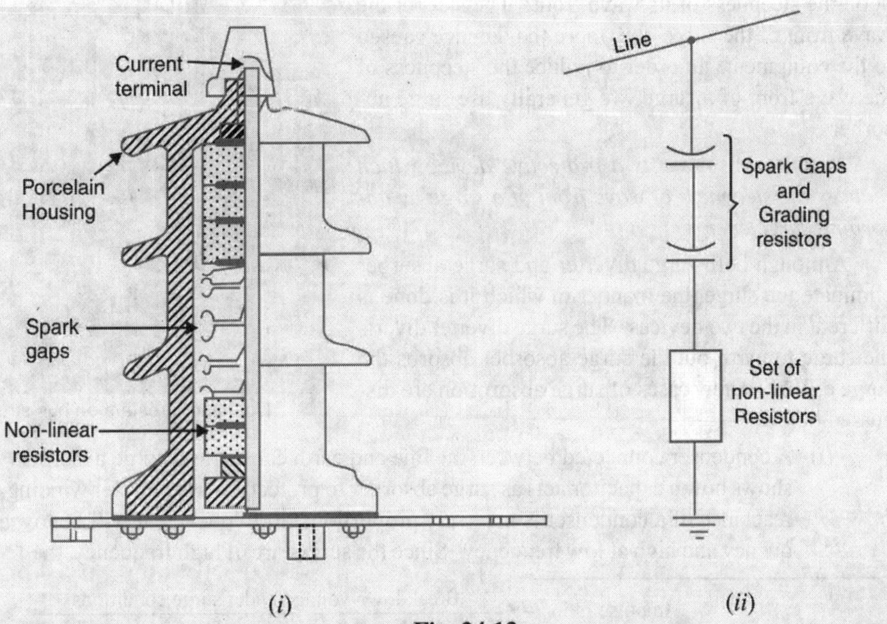

Fig. 24.13

Working. Under normal conditions, the normal system voltage is insufficient to cause the breakdown of air gap assembly. On the occurrence of an overvoltage, the breakdown of the series spark gap takes place and the surge current is conducted to earth *via* the non-linear resistors. Since the magnitude of surge current is very large, the non-linear elements will offer a very low resistance to the passage of surge. The result is that the surge will rapidly go to earth instead of being sent back over the line. When the surge is over, the non-linear resistors assume high resistance to stop the flow of current.

Advantages

(*i*) They provide very effective protection (especially for transformers and cables) against surges.

(*ii*) They operate very rapidly taking less than a second.

(*iii*) The *impulse ratio is practically unity.

Limitations

(*i*) They **may fail to check the surges of very steep wave front from reaching the terminal apparatus. This calls for additional steps to check steep-fronted waves.

(*ii*) Their performance is adversely affected by the entry of moisture into the enclosure. This necessitates effective sealing of the enclosure at all times.

Applications. According to their application, the valve type arresters are classified as (*i*) station type and (*ii*) line type. The station type arresters are generally used for the protection of important equipment in power stations operating on voltages upto 220 kV or higher. The line type arresters are also used for stations handling voltages upto 66 kV.

24.13 Surge Absorber

The travelling waves set up on the transmission lines by the surges may reach the terminals apparatus and cause damage to it. The amount of damage caused not only depends upon the amplitude of the surge but also upon the steepness of its wave front. The steeper the wave front of the surge, the more the damage caused to the equipment. In order to reduce the steepness of the wave front of a surge, we generally use surge absorber.

A **surge absorber** *is a protective device which reduces the steepness of wave front of a surge by absorbing surge energy.*

Although both surge diverter and surge absorber eliminate the surge, the manner in which it is done is different in the two devices. The surge diverter diverts the surge to earth but the surge absorber absorbs the surge energy. A few cases of surge absorption are discussed below :

Lightning arresters on bus structures

(*i*) A condenser connected between the line and earth can act as a surge absorber. Fig. 24.14 shows how a capacitor acts as surge absorber to protect the transformer winding. Since the reactance of a condenser is inversely proportional to frequency, it will be low at high frequency and high at low frequency. Since the surges are of high frequency, the ***capacitor

---

* $\text{Impulse ratio} = \dfrac{\text{Breakdown voltage under surge conditions}}{\text{Breakdown voltage under low frequency conditions}}$

** The normal strokes on transmission lines after travelling along the line are considerably attenuated so that they are well within the reach of protection afforded by such arresters.

*** A pure capacitor, however, cannot dissipate the energy in the wave front of a travelling wave or in a high frequency discharge. It merely reflects the wave energy away from the equipment to be protected and the energy is dissipated in the line resistance and earth resistance.

acts as a short circuit and passes them directly to earth. However, for power frequency, the reactance of the capacitor is very high and practically no current flows to the ground.

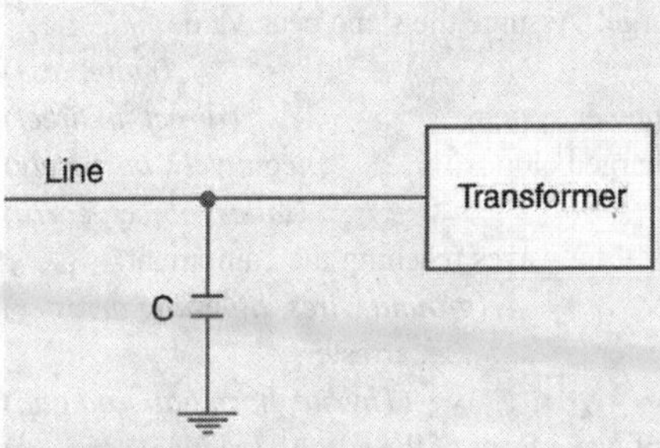

Fig. 24.14

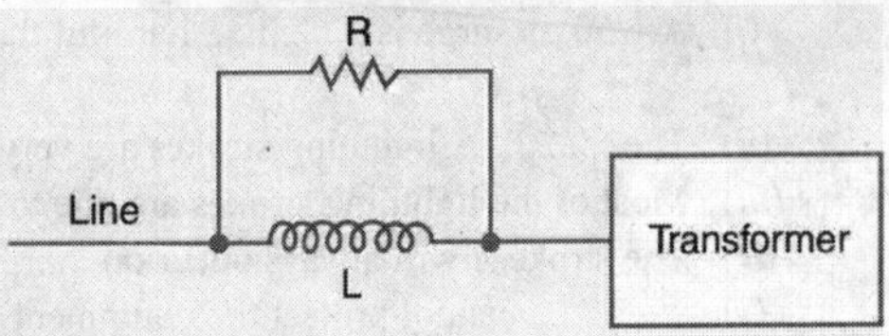

Fig. 24.15

(*ii*) Another type of surge absorber consists of a parallel combination of choke and resistance connected in series with the line as shown in Fig. 24.15. The choke offers high reactance to surge frequencies $(X_L = 2\pi f L)$. The surges are, therefore, forced to flow through the resistance R where they are dissipated.

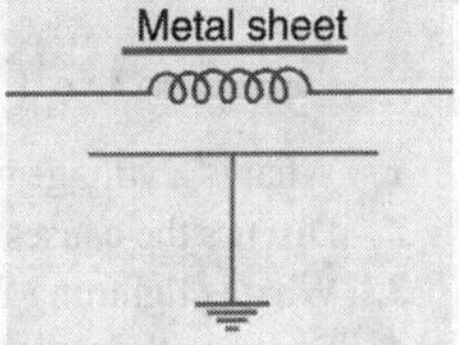

Fig. 24.16

(*iii*) Fig. 24.16 shows the another type of surge absorber. It is called Ferranti surge absorber. It consists of an air cored inductor connected in series with the line. The inductor is surrounded by but insulated from an earthed metallic sheet called dissipator. This arrangement is equivalent to a transformer with short-circuited secondary. The inductor forms the primary whereas the dissipator forms the short-circuited secondary. The energy of the surge is used up in the form of heat generated in the dissipator due to transformer action. This type of surge absorber is mainly used for the protection of transformers.

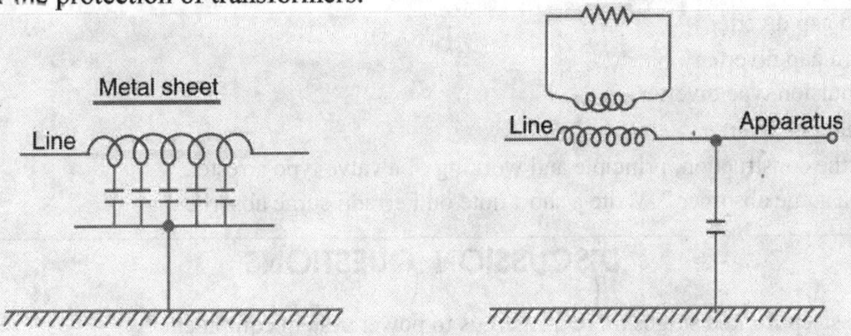

Fig. 24.17

Fig. 24.17 (*i*) shows the schematic diagram of 66 kV Ferranti surge absorber while Fig. 24.17 (*ii*) shows its equivalent circuit.

SELF - TEST

1. **Fill in the blanks by inserting appropriate words/figures :**
   (*i*) The most severe surges on the line are produced by ....
   (*ii*) Lightning produces a .... fronted wave.
   (*iii*) Transients on the power system due to current chopping are taken care of by ....
   (*iv*) Arcing ground can be prevented by ....
   (*v*) The lightning currents range from 10 kA to about .... kA.
   (*vi*) For successful working of ground wire, the footing resistance of tower should be ....

(*vii*) A surge diverter should be located .... to the apparatus to be protected.

2. **Fill in the blanks by picking correct words/figures from brackets :**

(*i*) A 1/50 μs surge is .... harmful than 3/50 μs surge. Assume the same peak value. (*more, less*)

(*ii*) The .... lightning strokes are very rare on the power system. (*direct, indirect*)

(*iii*) Most of the lightning strokes are due to .... charged clouds. (*negatively, positively*)

(*iv*) The stroke *A* will always occur on .... (*tallest object, earth*)

(*v*) .... cannot protect the equipment from the travelling waves reaching the equipment. (*ground wires, lightning arrester*)

(*vi*) In sub-stations, the most commonly used type of arrester is .... arrester. (*Thyrite, horn gap, rod gap*)

(*vii*) Surge absorbers are used to .... the steepness of wave front of the surge. (*reduce, increase*)

ANSWERS TO SELF-TEST

1. (*i*) lightning (*ii*) steep (*iii*) resistance switching (*iv*) earthing the neutral (*v*) 90 (*vi*) low (*vii*) close
2. (*i*) more (*ii*) direct (*iii*) negatively (*iv*) tallest object (*v*) ground wires (*vi*) thyrite (*vii*) reduce

CHAPTER REVIEW TOPICS

1. What is a voltage surge ? Draw a typical lightning voltage surge.
2. Discuss the causes of overvoltages.
3. What is lightning ? Describe the mechanism of lightning discharge.
4. Describe the various types of lightning stroke.
5. What are the harmful effects of lightning ?
6. How do earthing screen and ground wires provide protection against direct lightning strokes ?
7. What is a surge diverter ? What is the basic principle of operation of a surge diverter ?
8. Write short notes on the following surge diverters :
   (*i*) Rod gap diverter
   (*ii*) Horn gap diverter
   (*iii*) Expulsion type diverter
   (*iv*) Multigap diverter
9. Discuss the construction, principle and working of a valve type arrester.
10. What is a surge absorber ? Write a short note on Ferranti surge absorber.

DISCUSSION QUESTIONS

1. Why are steep fronted surges more dangerous to power system equipment ?
2. Why is lightning accompanied by a thunder ?
3. Is the name lightning arrester appropriate ?
4. Why are surge diverters located very close to the equipment to be protected ?
5. Where will you use a surge absorber ?

CHAPTER

25

Sub-Stations

Introduction

The present-day electrical power system is a.c. *i.e.* electric power is generated, trans mitted and distributed in the form of alternating current. The electric power is produced at the power stations which are located at favourable places, generally quite away from the consumers. It is delivered to the consumers through a large network of transmission and distribution. At many places in the line of the power system, it may be desirable and necessary to change some characteristic (*e.g.* voltage, a.c. to d.c., frequency, p.f. etc.) of electric supply. This is accomplished by suitable apparatus called sub-station. For example, generation voltage (11 kV or 6·6 kV) at the power station is stepped up to high voltage (say 220 kV or 132 kV) for transmission of electric power. The assembly of apparatus (*e.g.* transformer etc.) used for this purpose is the sub-station. Similarly, near the consumers localities, the voltage may have to be stepped down to utilisation level. This job is again accomplished by a suitable apparatus called sub-station. Yet at some places in the line of the power system, it may be desirable to convert large quantities of a.c. power to d.c. power *e.g.* for traction, electroplating, *d.c. motors etc. This job is again performed by suitable apparatus (*e.g.* ignitron) called sub-station. It is clear that type of equipment needed in a sub-station will depend upon the service requirement. Although there can be several types of sub-stations, we shall mainly confine our attention to only those sub-stations where the incoming and outgoing supplies are a.c. *i.e.* sub-stations which change the voltage level of the electric supply.

25.1 Sub-Station

The assembly of apparatus used to change some characteristic (e.g. voltage, a.c. to d.c., frequency, p.f. etc.) of electric supply is called a **sub-station.**

Sub-stations are important part of power system. The continuity of supply depends to a considerable extent upon the successful operation of sub-stations. It is, therefore, essential to exercise utmost care while designing and building a sub-station. The following are the important points which must be kept in view while laying out a sub-station :

(*i*) It should be located at a proper site. As far as possible, it should be located at the centre of gravity of load.

(*ii*) It should provide safe and reliable arrangement. For safety, consideration must be given to the maintenance of regulation clearances, facilities for carrying out repairs and maintenance, abnormal occurrences such as possibility of explosion or fire etc. For reliability, consideration must be given for good design and construction, the provision of suitable protective gear *etc.*

(*iii*) It should be easily operated and maintained.

(*iv*) It should involve minimum capital cost.

25.2 Classification of Sub-Stations

There are several ways of classifying sub-stations. However, the two most important ways of classifying them are according to (1) service requirement and (2) constructional features.

1. According to service requirement. A sub-station may be called upon to change voltage level or improve power factor or convert a.c. power into d.c. power etc. According to the service requirement, sub-stations may be classified into :

(*i*) **Transformer sub-stations.** Those sub-stations which change the voltage level of electric supply are called transformer sub-stations. These sub-stations receive power at some voltage and deliver it at some other voltage. Obviously, transformer will be the main component in such sub-stations. Most of the sub-stations in the power system are of this type.

(*ii*) **Switching sub-stations.** These sub-stations do not change the voltage level *i.e.* incoming and outgoing lines have the same voltage. However, they simply perform the switching operations of power lines.

(*iii*) **Power factor correction sub-stations.** Those sub-stations which improve the power factor of the system are called power factor correction sub-stations. Such sub-stations are generally located at the receiving end of transmission lines. These sub-stations generally use synchronous condensers as the power factor improvement equipment.

(*iv*) **Frequency changer sub-stations.** Those sub-stations which change the supply frequency are known as frequency changer sub-stations. Such a frequency change may be required for industrial utilisation.

(*v*) **Converting sub-stations.** Those sub-stations which change a.c. power into d.c. power are called converting sub-stations. These sub-stations receive a.c. power and convert it into d.c. power

* Although most of the motors in operation in the world today are a.c. motors, yet one can find d.c. motors where fine speed control is required or where these were installed before the development of a.c. machinery.

with suitable apparatus (*e.g.* ignitron) to supply for such purposes as traction, electroplating, electric welding etc.

(*vi*) **Industrial sub-stations.** Those sub-stations which supply power to individual industrial concerns are known as industrial sub-stations.

2. **According to constructional features.** A sub-station has many components (*e.g.* circuit breakers, switches, fuses, instruments etc.) which must be housed properly to ensure continuous and reliable service. According to constructional features, the sub-stations are classified as :

(*i*) Indoor sub-station (*ii*) Outdoor sub-station
(*iii*) Underground sub-station (*iv*) Pole-mounted sub-station

(*i*) **Indoor sub-stations.** For voltages upto 11 kV, the equipment of the sub-station is installed indoor because of economic considerations. However, when the atmosphere is contaminated with impurities, these sub-stations can be erected for voltages upto 66 kV.

(*ii*) **Outdoor sub-stations.** For voltages beyond 66 kV, equipment is invariably installed outdoor. It is because for such voltages, the clearances between conductors and the space required for switches, circuit breakers and other equipment becomes so great that it is not economical to install the equipment indoor.

(*iii*) **Underground sub-stations.** In thickly populated areas, the space available for equipment and building is limited and the cost of land is high. Under such situations, the sub-station is created underground. The reader may find further discussion on underground sub-stations in Art. 25.6.

(*iv*) **Pole-mounted sub-stations.** This is an outdoor sub-station with equipment installed overhead on *H*-pole or 4-pole structure. It is the cheapest form of sub-station for voltages not exceeding 11kV (or 33 kV in some cases). Electric power is almost distributed in localities through such sub-stations. For complete discussion on pole-mounted sub-station, the reader may refer to Art. 25.5.

25.3 Comparison between Outdoor and Indoor Sub-Stations

The comparison between outdoor and indoor sub-stations is given below in the tabular form :

S.No.	Particular	Outdoor Sub-station	Indoor Sub-station
1	Space required	More	Less
2	Time required for erection	Less	More
3	Future extension	Easy	Difficult
4	Fault location	Easier beacuse the equipment is in full view	Difficult because the equipment is enclosed
5	Capital cost	Low	High
6	Operation	Difficult	Easier
7	Possibility of fault escalation	Less because greater clearances can be provided	More

From the above comparison, it is clear that each type has its own advantages and disadvantages. However, comparative economics (*i.e.* annual cost of operation) is the most powerful factor influencing the choice between indoor and outdoor sub-stations. The greater cost of indoor sub-station prohibits its use. But sometimes non-economic factors (*e.g.* public safety) exert considerable influence in choosing indoor sub-station. In general, most of the sub-stations are of outdoor type and the indoor sub-stations are erected only where outdoor construction is impracticable or prohibited by the local laws.

25.4 Transformer Sub-Stations

The majority of the sub-stations in the power system are concerned with the changing of voltage level of electric supply. These are known as transformer sub-stations because transformer is the main component employed to change the voltage level. Depending upon the purpose served, transformer sub-stations may be classified into :

(*i*) Step-up sub-station (*ii*) Primary grid sub-station

(*iii*) Secondary sub-station (*iv*) Distribution sub-station

Fig. 25.1 shows the block diagram of a typical electric supply system indicating the position of above types of sub-stations. It may be noted that it is not necessary that all electric supply schemes include all the stages shown in the figure. For example, in a certain supply scheme there may not be secondary sub-stations and in another case, the scheme may be so small that there are only distribution sub-stations.

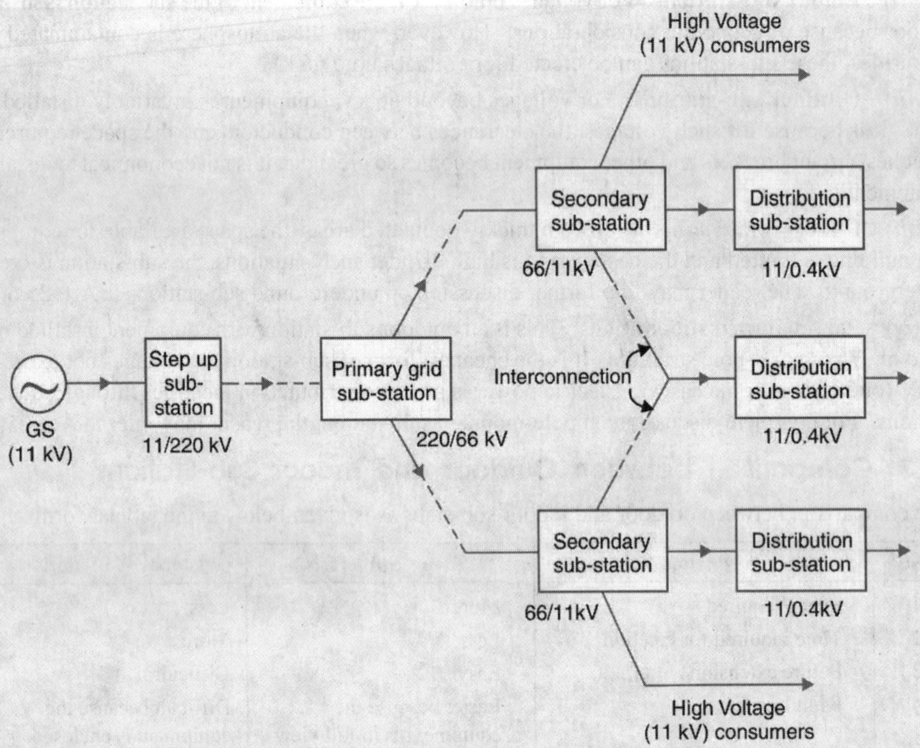

Fig. 25.1

(*i*) **Step-up sub-station.** The generation voltage (11 kV in this case) is stepped up to high voltage (220 kV) to affect economy in transmission of electric power. The sub-stations which accomplish this job are called step-up sub-stations. These are generally located in the power houses and are of outdoor type.

(*ii*) **Primary grid sub-station.** From the step-up sub-station, electric power at 220 kV is transmitted by 3-phase, 3-wire overhead system to the outskirts of the city. Here, electric power is received by the primary grid sub-station which reduces the voltage level to 66 kV for secondary transmission. The primary grid sub-station is generally of outdoor type.

(*iii*) **Secondary sub-station.** From the primary grid sub-station, electric power is transmitted at 66 kV by 3-phase, 3-wire system to various secondary sub-stations located at the strategic points in the city. At a secondary sub-station, the voltage is further stepped down to 11 kV. The 11 kV lines run along the important road sides of the city. It may be noted that big consumers (having demand more than 50 kW) are generally supplied power at 11 kV for further handling with their own sub-stations. The secondary sub-stations are also generally of outdoor type.

(*iv*) **Distribution sub-station.** The electric power from 11 kV lines is delivered to distribution sub-stations. These sub-stations are located near the consumers localities and step down the voltage to 400 V, 3-phase, 4-wire for supplying to the consumers. The voltage between any two phases is 400V and between any phase and neutral it is 230 V. The single phase residential lighting load is connected between any one phase and neutral whereas 3-phase, 400V motor load is connected across 3-phase lines directly. It may be worthwhile to mention here that majority of the distribution sub-stations are of pole-mounted type.

25.5 Pole-Mounted Sub-Station

It is a distribution sub-station placed overhead on a pole. It is the cheapest form of sub-station as it does not involve any building work. Fig 25.2 (*i*) shows the layout of pole-mounted sub-station whereas Fig. 25.2 (*ii*) shows the schematic connections. The transformer and other equipment are mounted on H-type pole (or 4-pole structure).

Sub-Station

The 11 kV line is connected to the transformer (11kV / 400 V) through gang isolator and fuses. The lightning arresters are installed on the H.T. side to protect the sub-station from lightning strokes. The transformer steps down the voltage to 400V, 3-phase, 4-wire supply. The voltage between any two lines is 400V whereas the voltage between any line and neutral is 230 V. The oil circuit breaker (O.C.B.) installed on the L.T. side automatically isolates the transformer from the consumers in the event of any fault. The pole-mounted

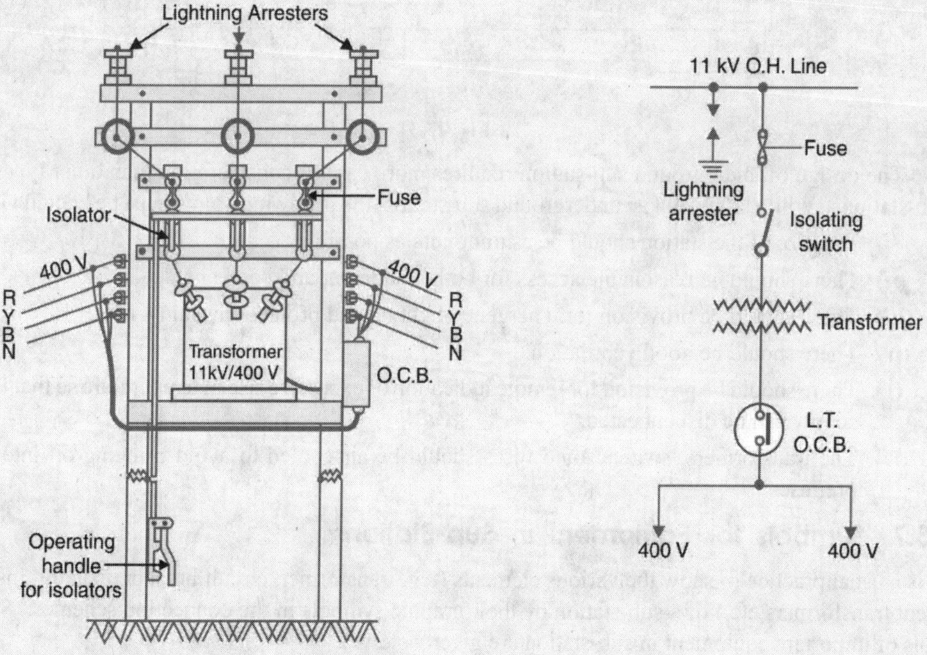

Fig. 25.2

sub-stations are generally used for transformer capacity upto *200 kVA. The following points may be noted about pole-mounted sub-stations :

(*i*) There should be periodical check-up of the dielectric strength of oil in the transformer and O.C.B.

(*ii*) In case of repair of transformer or O.C.B., both gang isolator and O.C.B. should be shut off.

25.6 Underground Sub-Station

In thickly populated cities, there is scarcity of land as well as the prices of land are very high. This has led to the development of underground sub-station. In such sub-stations, the equipment is placed underground. Fig. 25.3 shows a typical underground sub-station.

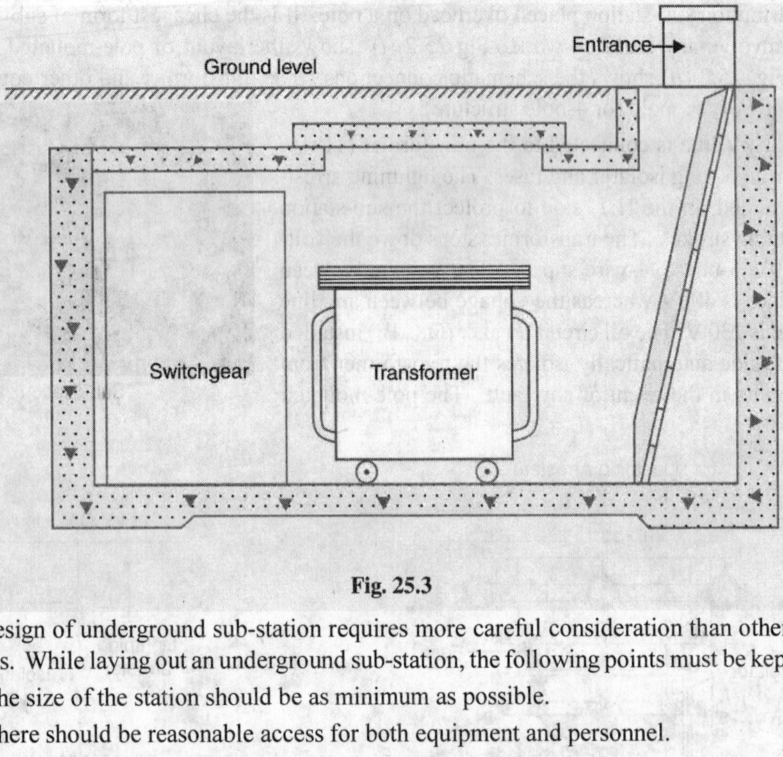

Fig. 25.3

The design of underground sub-station requires more careful consideration than other types of sub-stations. While laying out an underground sub-station, the following points must be kept in view:

(*i*) The size of the station should be as minimum as possible.

(*ii*) There should be reasonable access for both equipment and personnel.

(*iii*) There should be provision for emergency lighting and protection against fire.

(*iv*) There should be good ventilation.

(*v*) There should be provision for remote indication of excessive rise in temperature so that H.V. supply can be disconnected.

(*vi*) The transformers, switches and fuses should be air cooled to avoid bringing oil into the premises.

25.7 Symbols for Equipment in Sub-Stations

It is a usual practice to show the various elements (*e.g.* transformer, circuit breaker, isolator, instrument transformers etc.) of a sub-station by their graphic symbols in the connection schemes. Symbols of important equipment in sub-station are given below :

* For capacity greater than 200kVA, transformer is installed on steel platform adjoining to the *H*-pole construction.

S.No.	Circuit element	Symbol
1	Bus-bar	
2	Single-break isolating switch	
3	Double-break isolating switch	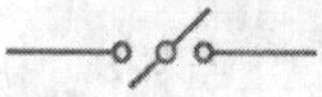
4	On load isolating switch	
5	Isolating switch with earth Blade	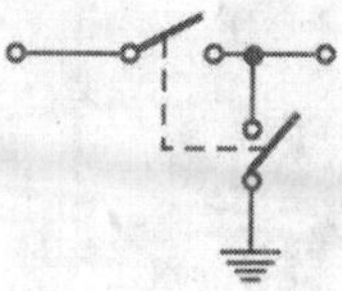
6	Current transformer	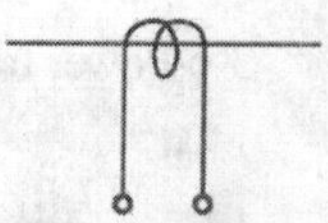
7	Potential transformer	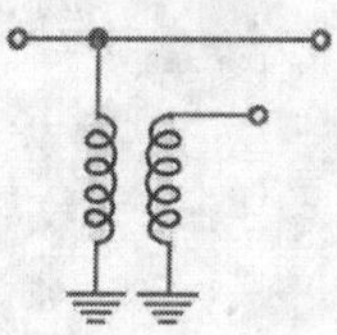
8	Capacitive voltage transformer	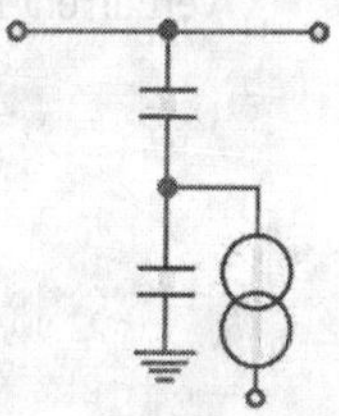
9	Oil circuit breaker	
10	Air circuit breaker with overcurrent tripping device	
11	Air blast circuit breaker	
12	Lightning arrester (active gap)	
13	Lightning arrester (valve type)	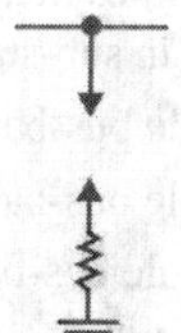

S.No.	Circuit element	Symbol
14	Arcing horn	
15	3-ϕ Power transformer	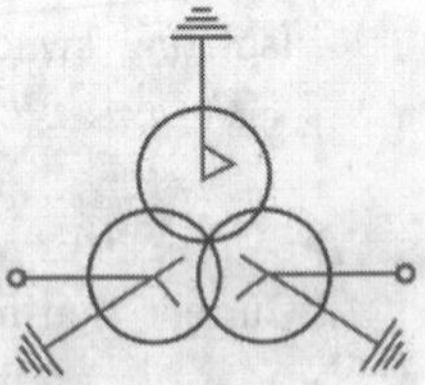
16	Overcurrent relay	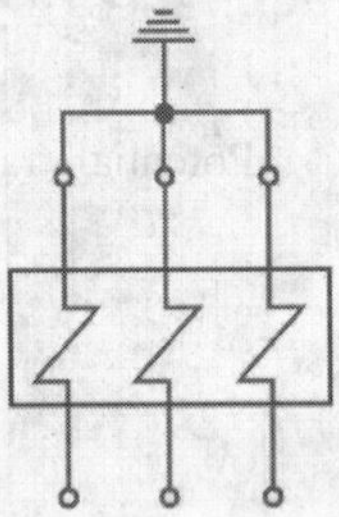
17	Earth fault relay	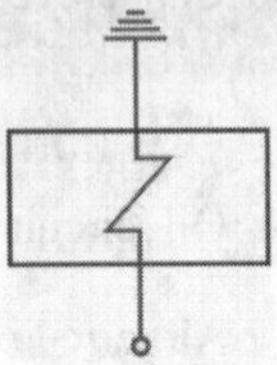

25.8 Equipment in a Transformer Sub-Station

The equipment required for a transformer sub-station depends upon the type of sub-station, service requirement and the degree of protection desired. However, in general, a transformer sub-station has the following main equipment :

1. Bus-bars. When a number of lines operating at the same voltage have to be directly connected electrically, bus-bars are used as the common electrical component. Bus-bars are copper or aluminium bars (generally of rectangular *x*-section) and operate at constant voltage. The incoming and outgoing lines in a sub-station are connected to the bus-bars. The most commonly used bus-bar arrangements in sub-stations are :

(*i*) Single bus-bar arrangement

(*ii*) Single bus-bar system with sectionalisation

(*iii*) Double bus-bar arrangement

A detailed discussion on these bus-bar arrangements has already been made in Art. 16.3. However, their practical applications in sub-stations are discussed in Art. 25.9.

2. Insulators. The insulators serve two purposes. They support the conductors (or bus-bars) and confine the current to the conductors. The most commonly used material for the manufacture of insulators is porcelain. There are several types of insulators (*e.g.* pin type, suspension type, post insulator etc.) and their use in the sub-station will depend upon the service requirement. For example, post insulator is used for bus-bars. A post insulator consists of a porcelain body, cast iron cap and flanged cast iron base. The hole in the cap is threaded so that bus-bars can be directly bolted to the cap.

3. Isolating switches. In sub-stations, it is often desired to disconnect a part of the system for general maintenance and repairs. This is accomplished by an isolating switch or isolator. An isolator is essentially a knife switch and is designed to open a circuit under *no load.* In other words, isolator switches are operated only when the lines in which they are connected carry *no current.

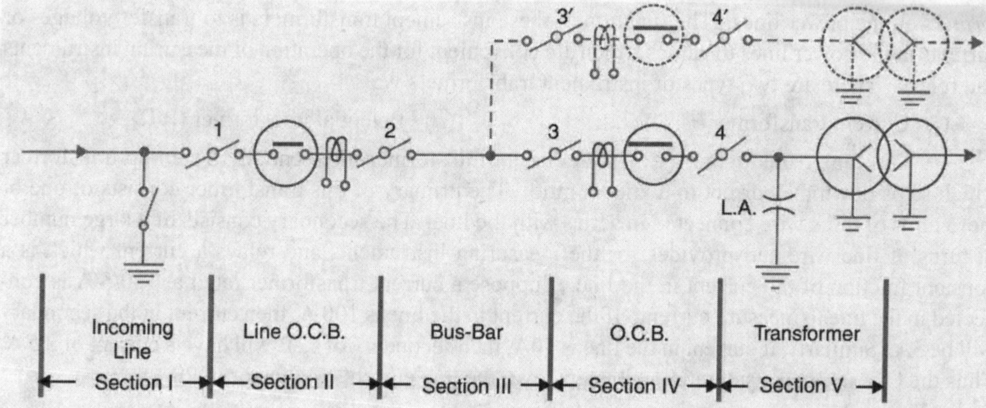

Fig. 25.4

Fig. 25.4 shows the use of isolators in a typical sub-station. The entire sub-station has been divided into V sections. Each section can be disconnected with the help of isolators for repair and maintenance. For instance, if it is desired to repair section No. II, the procedure of disconnecting this section will be as follows. First of all, open the circuit breaker in this section and then open the isolators 1 and 2. This procedure will disconnect section II for repairs. After the repair has been done, close the isolators 1 and 2 first and then the circuit breaker.

4. Circuit breaker. A circuit breaker is an equipment which can open or close a circuit under **normal as well as fault conditions. It is so designed that it can be operated manually (or by remote control) under normal conditions and automatically under fault conditions. For the latter operation, a relay circuit is used with a circuit breaker. Generally, bulk oil circuit breakers are used for voltages upto 66kV while for high (>66 kV) voltages, low oil circuit breakers are used. For still higher voltages, air-blast, vacuum or SF_6 circuit breakers are used. For detailed discussion of these breakers, the reader may refer to chapter 19.

5. Power Transformers. A power transformer is used in a sub-station to step-up or step-down the voltage. Except at the ***power station, all the subsequent sub-stations use step-down transformers to gradually reduce the voltage of electric supply and finally deliver it at utilisation voltage. The modern practice is to use 3-phase transformers in sub-stations ; although 3 single phase bank of

* For example, consider that the isolators are connected on both sides of a circuit breaker. If the isolators are to be opened, the C.B. must be opened first.

** An isolator cannot be used to open a circuit under normal conditions. It is because it has no provision to quench the arc that is produced during opening operation. Hence the use of circuit breaker is essential.

*** where a step-up transformer is used to step-up generation voltage to a high value (say 132 kV or 220 kV or more) for transmission of electric power.

transformers can also be used. The use of 3-phase transformer (instead of 3 single phase bank of transformers) permits two advantages. Firstly, only one 3-phase load-tap changing mechanism can be used. Secondly, its installation is much simpler than the three single phase transformers.

The power transformer is generally installed upon lengths of rails fixed on concrete slabs having foundations 1 to 1·5 m deep. For ratings upto 10 MVA, naturally cooled, oil immersed transformers are used. For higher ratings, the transformers are generally air blast cooled.

6. **Instrument transformers.** The lines in sub-stations operate at high voltages and carry current of thousands of amperes. The measuring instruments and protective devices are designed for low voltages (generally 110 V) and currents (about 5 A). Therefore, they will not work satisfactorily if mounted directly on the power lines. This difficulty is overcome by installing *instrument transformers* on the power lines. The function of these instrument transformers is to transfer voltages or currents in the power lines to values which are convenient for the operation of measuring instruments and relays. There are two types of instrument transformers *viz.*

(*i*) Current transformer (C.T.) (*ii*) Potential transformer (P.T.)

(*i*) **Current transformer (C.T.).** A current transformer in essentially a step-up transformer which steps down the current to a known ratio. The primary of this transformer consists of one or more turns of thick wire connected in series with the line. The secondary consists of a large number of turns of fine wire and provides for the measuring instruments and relays a current which is a constant fraction of the current in the line. Suppose a current transformer rated at 100/5 A is connected in the line to measure current. If the current in the line is 100 A, then current in the secondary will be 5A. Similarly, if current in the line is 50A, then secondary of C.T. will have a current of 2·5 A. Thus the C.T. under consideration will step down the line current by a factor of 20.

(*ii*) **Voltage transformer.** It is essentially a step down transformer and steps down the voltage to a known ratio. The primary of this transformer consists of a large number of turns of fine wire connected across the line. The secondary winding consists of a few turns and provides for measuring instruments and relays a voltage which is a known fraction of the line voltage. Suppose a potential transformer rated at 66kV/110V is connected to a power line. If line voltage is 66kV, then voltage across the secondary will be 110 V.

7. **Metering and Indicating Instruments.** There are several metering and indicating instruments (*e.g.* ammeters, voltmeters, energy meters etc.) installed in a sub-station to maintain watch over the circuit quantities. The instrument transformers are invariably used with them for satisfactory operation.

8. **Miscellaneous equipment.** In addition to above, there may be following equipment in a sub-station :

(*i*) fuses
(*ii*) carrier-current equipment
(*iii*) sub-station auxiliary supplies

25.9 Bus-Bar Arrangements in Sub-Stations

Bus-bars arc the important components in a sub-station. There are several bus-bar arrangements that can be used in a sub-station. The choice of a particular arrangement depends upon various factors such as system voltage, position of sub-station, degree of reliability, cost etc. The following are the important bus-bar arrangements used in sub-stations :

(*i*) **Single bus-bar system.** As the name suggests, it consists of a single bus-bar and all the incoming and outgoing lines are connected to it. The chief advantages of this type of arrangement are low initial cost, less maintenance and simple operation. However, the principal disadvantage of single bus-bar system is that if repair is to be done on the bus-bar or a fault occurs on the bus, there is

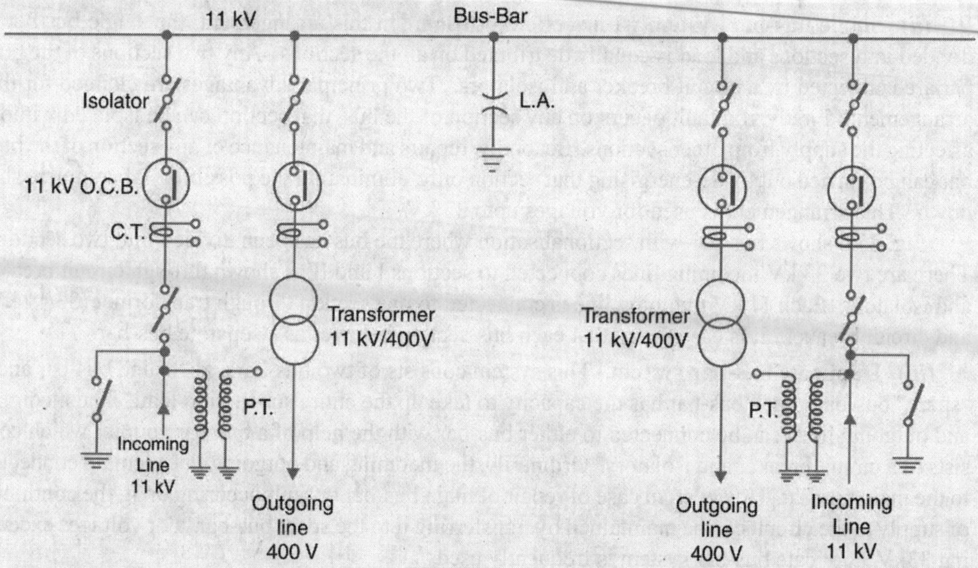

Fig. 25.5

a complete interruption of the supply. This arrangement is not used for voltages exceeding 33kV. The indoor 11kV sub-stations often use single bus-bar arrangement.

Fig. 25.5 shows single bus-bar arrangement in a sub-station. There are two 11 kV incoming lines connected to the bus-bar through circuit breakers and isolators. The two 400V outgoing lines are connected to the bus bars through transformers (11kV/400 V) and circuit breakers.

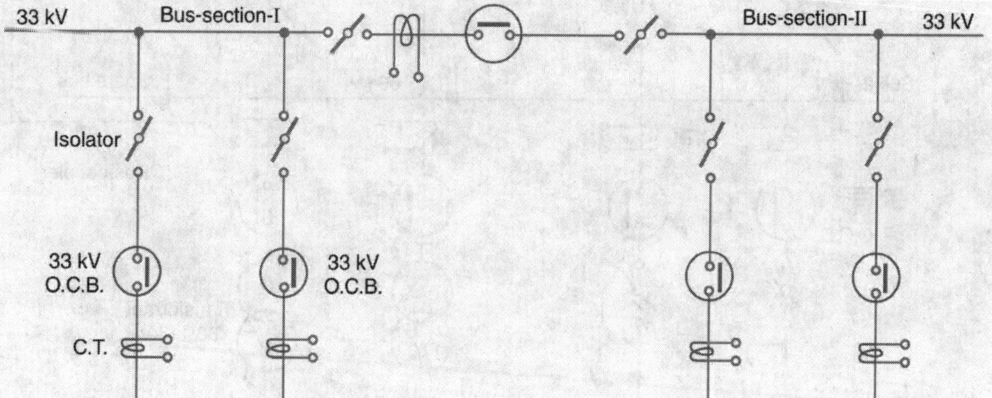

Fig. 25.6

(*ii*) **Single bus-bar system with sectionalisation.** In this arrangement, the single bus-bar is divided into sections and load is equally distributed on all the sections. Any two sections of the bus-bar are connected by a circuit breaker and isolators. Two principal advantages are claimed for this arrangement. Firstly, if a fault occurs on any section of the bus, that section can be isolated without affecting the supply from other sections. Secondly, repairs and maintenance of any section of the bus-bar can be carried out by de-energising that section only, eliminating the possibility of complete shut down. This arrangement is used for voltages upto 33 kV.

Fig. 25.6 shows bus-bar with sectionalisation where the bus has been divided into two sections. There are two 33 kV incoming lines connected to sections I and II as shown through circuit breaker and isolators. Each 11 kV outgoing line is connected to one section through transformer (33/11 kV) and circuit breaker. It is easy to see that each bus-section behaves as a separate bus-bar.

(*iii*) **Duplicate bus-bar system.** This system consists of two bus-bars, a "main" bus-bar and a "spare" bus-bar. Each bus-bar has the capacity to take up the entire sub-station load. The incoming and outgoing lines can be connected to either bus-bar with the help of a bus-bar coupler which consists of a circuit breaker and isolators. Ordinarily, the incoming and outgoing lines remain connected to the main bus-bar. However, in case of repair of main bus-bar or fault occuring on it, the continuity of supply to the circuit can be maintained by transferring it to the spare bus-bar. For voltages exceeding 33kV, duplicate bus-bar system is frequently used.

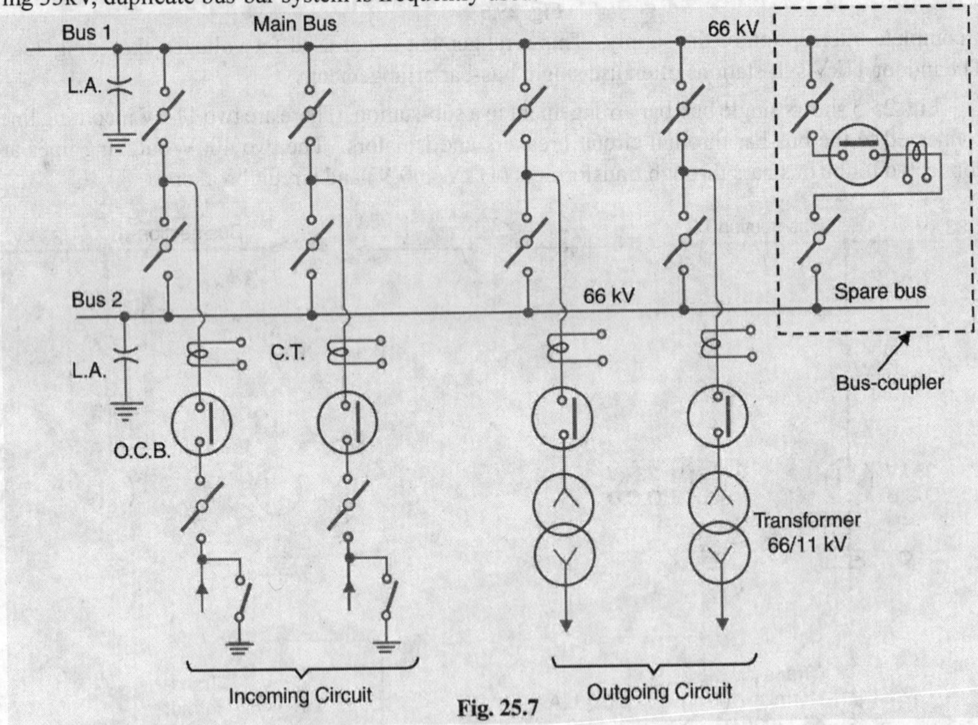

Fig. 25.7

Fig. 25.7 shows the arrrangement of duplicate bus-bar system in a typical sub-station. The two 66kV incoming lines can be connected to either bus-bar by a bus-bar coupler. The two 11 kV outgoing lines are connected to the bus-bars through transformers (66/11 kV) and circuit breakers.

25.10 Terminal and Through Sub-Stations

All the transformer sub-stations in the line of power system handle incoming and outgoing lines. Depending upon the manner of incoming lines, the sub-stations are classified as :

(*i*) Terminal sub-station (*ii*) Through sub-station

(*i*) **Terminal sub-station.** A terminal sub-station is one in which the line supplying to the sub-station terminates or ends. It may be located at the end of the main line or it may be situated at a point away from main line route. In the latter case, a tapping is taken from the main line to supply to the sub-station. Fig. 25.8 shows the schematic connections of a terminal sub-station. It is clear that incoming 11 kV main line terminates at the sub-station. Most of the distribution sub-stations are of this type.

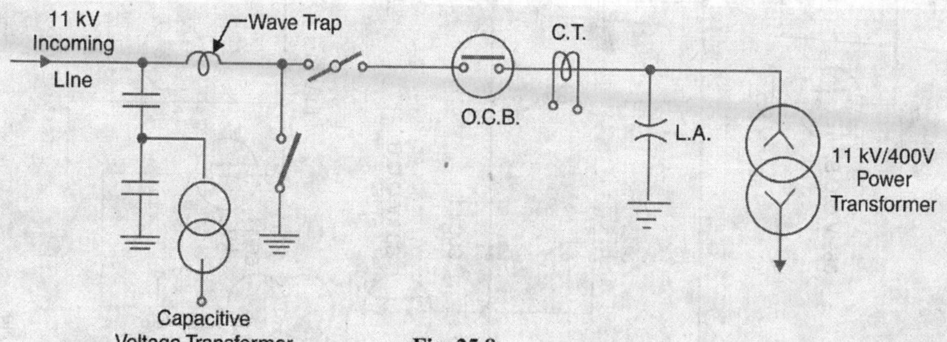

Fig. 25.8

(*ii*) **Through sub-station.** A through sub-station is one in which the incoming line passes 'through' at the same voltage. A tapping is generally taken from the line to feed to the transformer to reduce the voltage to the desired level. Fig. 25.9 shows the schematic connections of a through sub-station. The incoming 66 kV line passes through the sub-station as 66kV outgoing line. At the same time, the incoming line is tapped in the sub-station to reduce the voltage to 11 kV for secondary distribution.

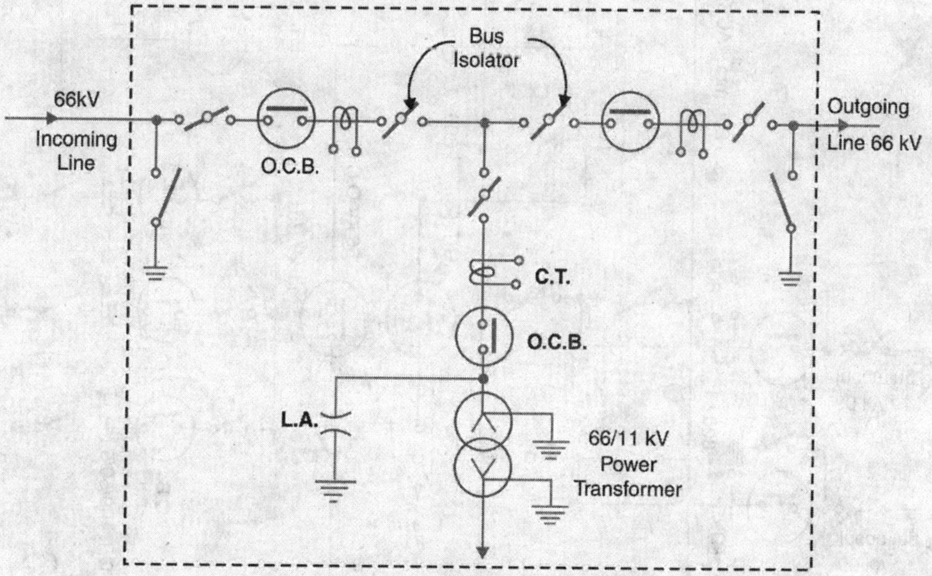

Fig. 25.9

25.11 Key Diagram of 66/11 kV Sub-Station

Fig. 25.10 shows the key diagram of a typical 66/11 kV sub-station. The key diagram of this sub-station can be explained as under :

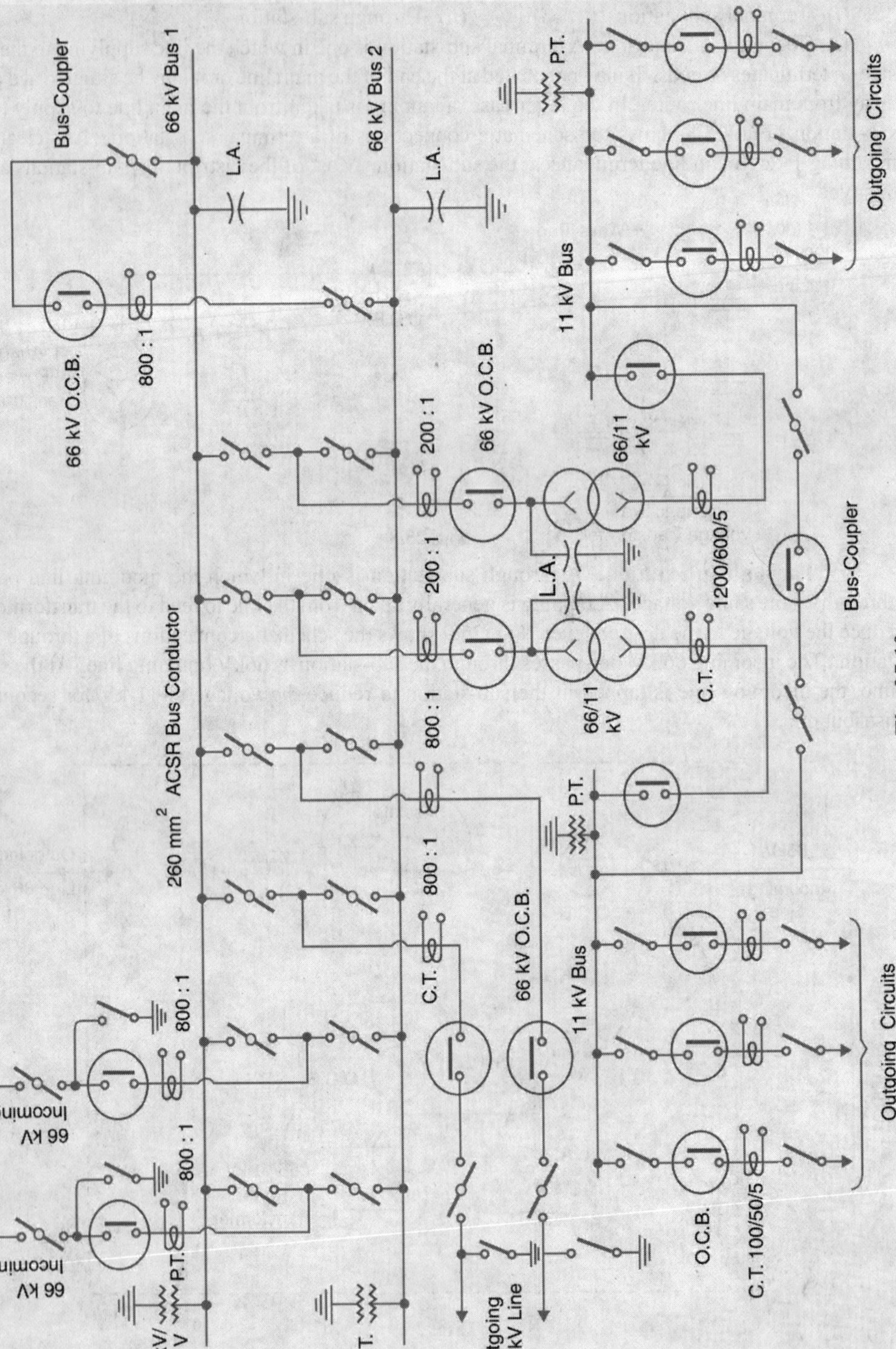
Bus-Coupler
66 kV Bus 1
66 kV Bus 2
L.A.
L.A.
66 kV O.C.B.
800 : 1
200 : 1
200 : 1
800 : 1
800 : 1
66 kV O.C.B.
260 mm² ACSR Bus Conductor
66 kV Incoming 1
66 kV Incoming 2
800 : 1
800 : 1
P.T.
66 kV/ 110 V
P.T.
C.T.
66 kV O.C.B.
Outgoing 66 kV Line
66/11 kV
66/11 kV
L.A.
C.T.
1200/600/5
P.T.
P.T.
11 kV Bus
11 kV Bus
Bus-Coupler
Outgoing Circuits
Outgoing Circuits
O.C.B.
C.T. 100/50/5

Fig. 25.10

(*i*) There are two 66 kV incoming lines marked 'incoming 1' and 'incoming 2' connected to the bus-bars. Such an arrangement of two incoming lines is called a double circuit. Each incoming line is capable of supplying the rated sub-station load. Both these lines can be loaded simultaneously to share the sub-station load or any one line can be called upon to meet the entire load. The double circuit arrangement increases the reliability of the system. In case there is a breakdown of one incoming line, the continuity of supply can be maintained by the other line.

(*ii*) The sub-station has duplicate bus-bar system; one 'main bus-bar' and the other spare bus-bar. The incoming lines can be connected to either bus-bar with the help of a bus-coupler which consists of a circuit breaker and isolators. The advantage of double bus-bar system is that if repair is to be carried on one bus-bar, the supply need not be interrupted as the entire load can be transferred to the other bus.

(*iii*) There is an arrangement in the sub-station by which the same 66 kV double circuit supply is going out *i.e.* 66 kV double circuit supply is passing through the sub-station. The outgoing 66 kV double circuit line can be made to act as incoming line.

(*iv*) There is also an arrangement to step down the incoming 66 kV supply to 11 kV by two units of 3-phase transformers; each transformer supplying to a separate bus-bar. Generally, one transformer supplies the entire sub-station load while the other transformer acts as a standby unit. If need arises, both the transformers can be called upon to share the sub-station load. The 11 kV outgoing lines feed to the distribution sub-stations located near consumers localities.

(*v*) Both incoming and outgoing lines are connected through circuit breakers having isolators on their either end. Whenever repair is to be carried over the line towers, the line is first switched off and then earthed.

(*vi*) The potential transformers (P.T.) and current transformers (C.T.) and suitably located for supply to metering and indicating instruments and relay circuits (not shown in the figure). The P.T. is connected right on the point where the line is terminated. The CTs are connected at the terminals of each circuit breaker.

(*vii*) The lightning arresters are connected near the transformer terminals (on H.T. side) to protect them from lightning strokes.

(*viii*) There are other auxiliary components in the sub-station such as capacitor bank for power factor improvement, earth connections, local supply connections, d.c. supply connections etc. However, these have been omitted in the key diagram for the sake of simplicity.

25.12 Key Diagram of 11 kV/400 V Indoor Sub-Station

Fig. 25.11 shows the key diagram of a typical 11 kV/400 V indoor sub-station. The key diagram of this sub-station can be explained as under :

(*i*) The 3-phase, 3-wire 11 kV line is tapped and brought to the gang operating switch installed near the sub-station. The G.O. switch consists of isolators connected in each phase of the 3-phase line.

(*ii*) From the G.O. switch, the 11 kV line is brought to the indoor sub-station as underground cable. It is fed to the H.T. side of the transformer (11 kV/400 V) *via* the 11 kV O.C.B. The transformer steps down the voltage to 400 V, 3-phase, 4-wire.

(*iii*) The secondary of transformer supplies to the bus-bars *via* the main O.C.B. From the bus-bars, 400 V, 3-phase, 4-wire supply is given to the various consumers *via* 400 V O.C.B. The voltage between any two phases is 400 V and between any phase and neutral it is 230 V. The single phase residential load is connected between any one phase and neutral whereas 3-phase, 400 V motor load is connected across 3-phase lines directly.

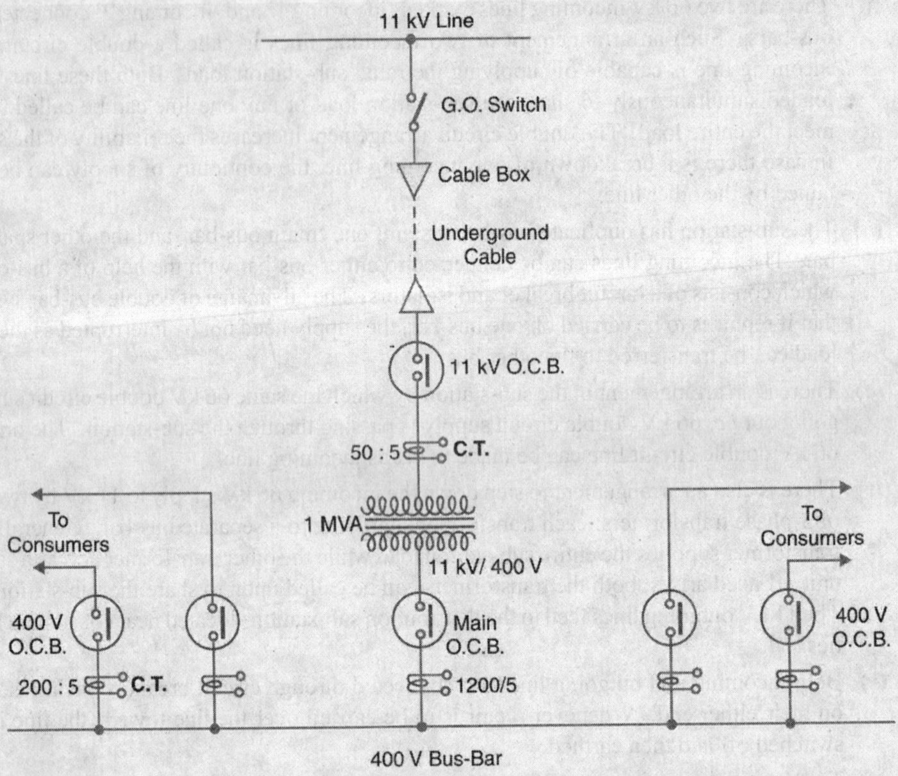

Fig. 25.11

(*iv*) The CTs are located at suitable places in the sub-station circuit and supply for the metering and indicating instruments and relay circuits.

SELF - TEST

1. Fill in the blanks by appropriate words/figures :

(*i*) A sub-station .... some characteristic of electric supply.

(*ii*) Most of the sub-stations in the power system change................. of electric supply.

(*iii*) An ideal location for the sub-station would be at the .... of load.

(*iv*) Pole-mounted sub-stations are used for .... distribution.

(*v*) The voltage rating of the transformer in a pole-mounted sub-station is............. .

(*vi*) Single bus-bar arrangement in sub-stations is used for voltages less then ....

(*vii*) For voltages greater than 33kV, ... bus-bar arrangement is employed.

(*viii*) The kVA rating of transformer in a pole-mounted sub-station does not exceed.................

(*ix*) An indoor sub-station is .... expensive than outdoor sub-station.

(*x*) Fault location is .... in an outdoor sub-station than in indoor sub-station.

2. Pick up the correct words/figures from brackets and fill in the blanks :

(*i*) Outdoor sub-station requires .... space. (*more, less*)

(*ii*) The possibility of fault escalation is .... in outdoor sub-station than that of indoor sub-station. (*more, less*)

(*iii*) Majority of distribution sub-stations are of .... type. (*pole-mounted, indoor, outdoor*) (*iv*) Power factor correction sub-stations are generally located at the .... end of a transmission line. (*sending, receiving*)

(*v*) Underground sub-stations are generally located in................. (*thickly populated areas, villages*)

ANSWERS TO SELF-TEST

1. (*i*) changes (*ii*) voltage level (*iii*) centre of gravity (*iv*) secondary (*v*) 11 kV/400 V (*vi*) 33 kV (*vii*) duplicate (*viii*) 200 (*ix*) more (*x*) easier
2. (*i*) more (*ii*) less (*iii*) pole-mounted (*iv*) receiving (*v*) thickly populated areas

CHAPTER REVIEW TOPICS

1. What is a sub-station ? Name the factors that should be taken care of while designing and erecting a sub-station.
2. Discuss the different ways of classifying the sub-stations.
3. Give the comparison of outdoor and indoor sub-stations.
4. What is a transformer sub-station ? What are the different types of transformer sub-stations ? Illustrate your answer with a suitable block diagram.
5. Draw the layout and schematic connection of a pole-mounted sub-station.
6. Draw the layout of a typical underground sub-station.
7. Write a short note on the sub-station equipment.
8. What are the different types of bus-bar arrangements used in sub-stations ? Illustrate your answer with suitable diagrams.
9. What are terminal and through sub-stations ? What is their purpose in the power system ?
10. Draw the key diagram of a typical 66/11 kV sub-station.
11. Draw the key diagram of a typical 11 kV/400 V indoor sub-station.

DISCUSSION QUESTIONS

1. What is the need of a sub-station in the power system ?
2. Why are pole-mounted sub-stations very popular ?
3. Where we erect a terminal sub-station ?
4. Why do we use isolators on both sides of circuit breaker ?
5. What is the utility of instrument transformers in sub-stations ?

CHAPTER

26

Neutral Grounding

Introduction

In power system, **grounding* or *earthing* means connecting frame of electrical equipment (non-current carrying part) or some electrical part of the system (*e.g.* neutral point in a star-connected system, one conductor of the secondary of a transformer etc.) to earth *i.e.* soil. This connection to earth may be through a conductor or some other circuit element (e.g. a resistor, a circuit breaker etc.) depending upon the situation. Regardless of the method of connection to earth, grounding or earthing offers two principal advantages. First, it provides protection to the power system. For example, if the neutral point of a star-connected system is grounded through a circuit breaker and phase to earth fault occurs on any one line, a large fault current will flow through the circuit breaker. The

* In electrical and electronic circuits, all the points which are electrically the same (called common points) are connected to the metal chassis. This method of grounding is called *chassis ground* or *circuit ground.* Circuit ground may or may not be connected to earth (*i.e.* soil). For example, in most cars, the negative terminal of the battery and one side of all electrical circuits are connected to the metal chassis.

circuit breaker will open to isolate the faulty line. This protects the power system from the harmful effects of the fault. Secondly, earthing of electrical equipment (*e.g.* domestic appliances, hand-held tools, industrial motors etc.) ensures the safety of the persons handling the equipment. For example, if insulation fails, there will be a direct contact of the live conductor with the metallic part (*i.e.* frame) of the equipment. Any person in contact with the metallic part of this equipment will be subjected to a dangerous electrical shock which can be fatal. In this chapter, we shall discuss the importance of grounding or earthing in the line of power system with special emphasis on neutral grounding.

26.1 Grounding or Earthing

The process of connecting the metallic frame (i.e. non-current carrying part) of electrical equipment or some electrical part of the system (e.g. neutral point in a star-connected system, one conductor of the secondary of a transformer etc.) to earth (i.e. soil) is called **grounding or earthing.**

It is strange but true that grounding of electrical systems is less understood aspect of power system. Nevertheless, it is a very important subject. If grounding is done systematically in the line of the power system, we can effectively prevent accidents and damage to the equipment of the power system and at the same time continuity of supply can be maintained. Grounding or earthing may be classified as : (*i*) Equipment grounding (*ii*) System grounding.

Equipment grounding deals with earthing the non-current-carrying metal parts of the electrical equipment. On the other hand, system grounding means earthing some part of the electrical system *e.g.* earthing of neutral point of star-connected system in generating stations and sub-stations.

26.2 Equipment Grounding

The process of connecting non-current-carrying metal parts (i.e. metallic enclosure) of the electrical equipment to earth (i.e. soil) in such a way that in case of insulation failure, the enclosure effectively remains at earth potential is called **equipment grounding.**

We are frequently in touch with electrical equipment of all kinds, ranging from domestic appliances and hand-held tools to industrial motors. We shall illustrate the need of effective equipment grounding by considering a single-phase circuit composed of a 230 V source connected to a motor M as shown in Fig. 26.1. Note that neutral is solidly grounded at the service entrance. In the interest of easy understanding, we shall divide the discussion into three heads *viz.* (*i*) Ungrounded enclosure (*ii*) enclosure connected to neutral wire (*iii*) ground wire connected to enclosure.

(*i*) **Ungrounded enclosure.** Fig. 26.1 shows the case of ungrounded metal enclosure. If a person touches the metal enclosure, nothing will happen if the equipment is functioning correctly. But if the winding insulation becomes faulty, the resistance R_e between the motor and enclosure drops to a low value (a few hundred ohms or less). A person having a body resistance R_b would complete the current path as shown in Fig. 26.1.

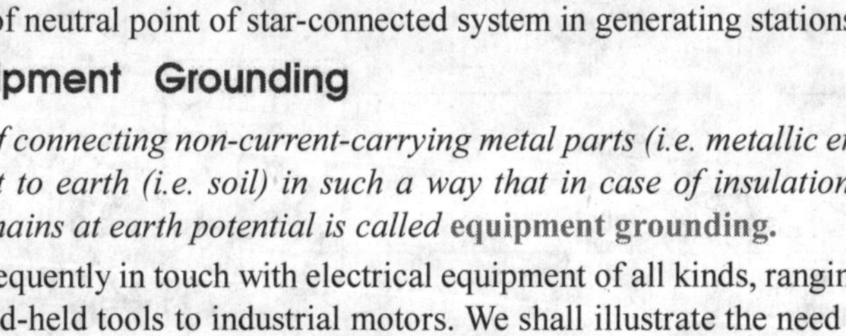

Fig. 26.1

If R_e is small (as is usually the case when insulation failure of winding occurs), the leakage current I_L through the person's body could be dangerously high. As a result, the person would get severe *electric shock which may be fatal. Therefore, this system is unsafe.

(ii) **Enclosure connected to neutral wire.** It may appear that the above problem can be solved by connecting the enclosure to the grounded neutral wire as shown in Fig. 26.2. Now the leakage current I_L flows from the motor, through the enclosure and straight back to the neutral wire (See Fig. 26.2). Therefore, the enclosure remains at earth potential. Consequently, the operator would not experience any electric shock.

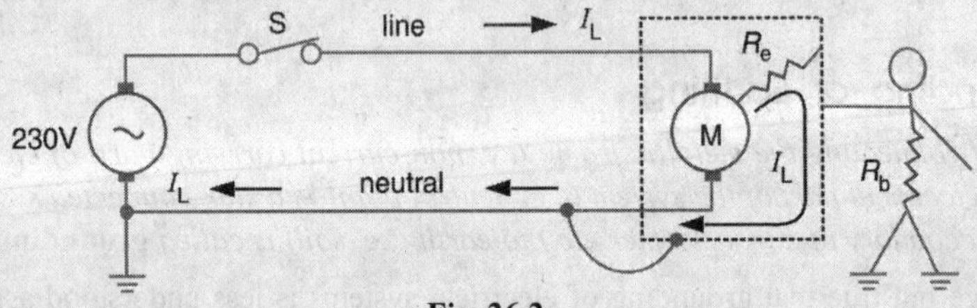

Fig. 26.2

The trouble with this method is that the neutral wire may become open either accidentally or due to a faulty installation. For example, if the switch is inadvertently in series with the neutral rather than the live wire (See Fig. 26.3), the motor can still be turned on and off. However, if someone touched the enclosure while the motor is *off,* he would receive a severe electric shock (See Fig. 26.3). It is because when the motor is off, the potential of the enclosure rises to that of the live conductor.

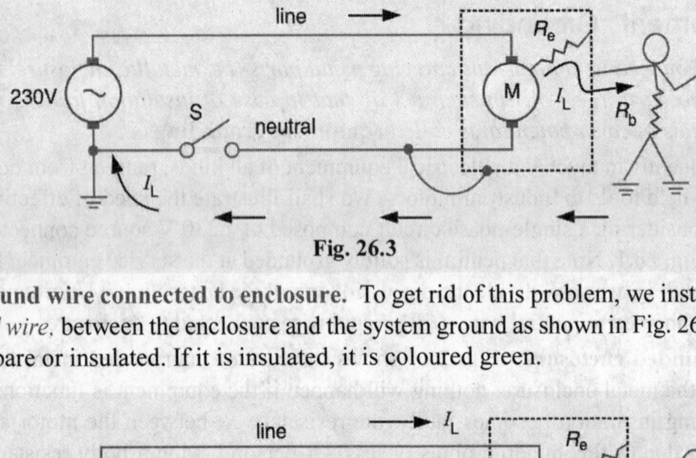

Fig. 26.3

(iii) **Ground wire connected to enclosure.** To get rid of this problem, we install a third wire, called *ground wire,* between the enclosure and the system ground as shown in Fig. 26.4. The ground wire may be bare or insulated. If it is insulated, it is coloured green.

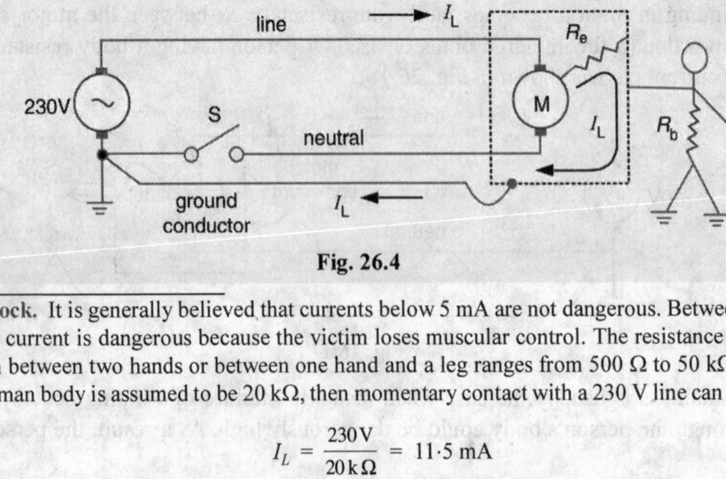

Fig. 26.4

* **Electric shock.** It is generally believed that currents below 5 mA are not dangerous. Between 10 mA and 20 mA, the current is dangerous because the victim loses muscular control. The resistance of the human body, taken between two hands or between one hand and a leg ranges from 500 Ω to 50 kΩ. If the resistance of human body is assumed to be 20 kΩ, then momentary contact with a 230 V line can be potentially fatal.

$$I_L = \frac{230\,\text{V}}{20\,\text{k}\Omega} = 11{\cdot}5\text{ mA}$$

Electrical outlets have three contacts — one for live wire, one for neutral wire and one for ground wire.

26.3 System Grounding

The process of connecting some electrical part of the power system (e.g. neutral point of a star-connected system, one conductor of the secondary of a transformer etc.) to earth (i.e. soil) is called **system grounding.**

The system grounding has assumed considerable importance in the fast expanding power system. By adopting proper schemes of system grounding, we can achieve many advantages including protection, reliability and safety to the power system network. But before discussing the various aspects of *neutral grounding*, it is desirable to give two examples to appreciate the need of system grounding.

(*i*) Fig. 26.5 (*i*) shows the primary winding of a distribution transformer connected between the line and neutral of a 11 kV line. If the secondary conductors are *ungrounded,* it would appear that a person could touch either secondary conductor without harm because there is no ground return. However, this is not true. Referring to Fig. 26.5, there is capacitance C_1 between primary and secondary and capacitance C_2 between secondary and ground. This capacitance coupling can produce a high voltage between the secondary lines and the ground. Depending upon the relative magnitudes of C_1 and C_2, it may be as high as 20% to 40% of the primary voltage. If a person touches either one of the secondary wires, the resulting capacitive current I_C flowing through the body could be dangerous even in case of small transformers [See Fig. 26.5(*ii*)]. For example, if I_C is only 20 mA, the person may get a fatal electric shock.

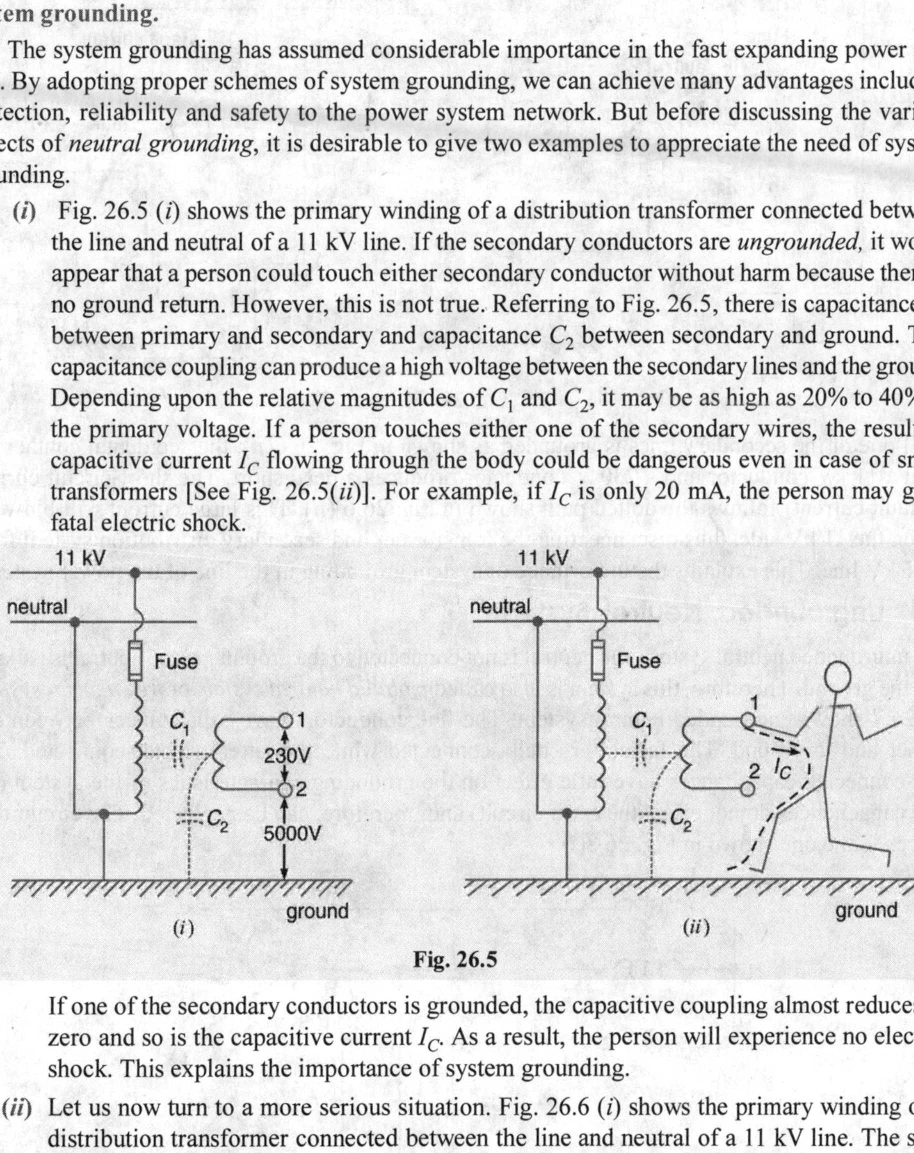

Fig. 26.5

If one of the secondary conductors is grounded, the capacitive coupling almost reduces to zero and so is the capacitive current I_C. As a result, the person will experience no electric shock. This explains the importance of system grounding.

(*ii*) Let us now turn to a more serious situation. Fig. 26.6 (*i*) shows the primary winding of a distribution transformer connected between the line and neutral of a 11 kV line. The secondary conductors are ungrounded. Suppose that the high voltage line (11 kV in this case) touches the 230 V conductor as shown in Fig. 26.6 (*i*). This could be caused by an internal fault in the transformer or by a branch or tree falling across the 11 kV and 230 V lines. Under these circumstances, a very high voltage is imposed between the secondary conductors and ground. This would immediately puncture the 230 V insulation, causing a massive flashover. This flashover could occur anywhere on the secondary network, possibly inside a home or factory. Therefore, ungrounded secondary in this case is a potential fire hazard and may produce grave accidents under abnormal conditions.

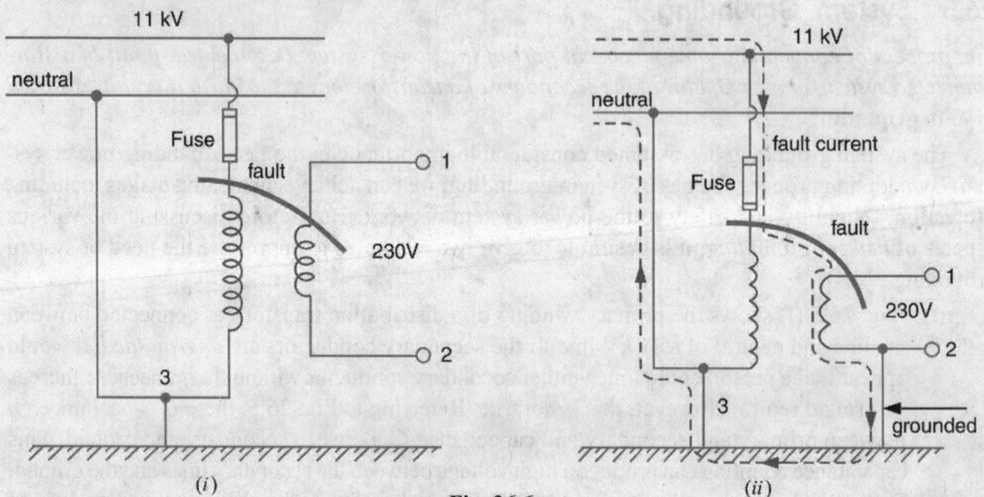

Fig. 26.6

If one of the secondary lines is grounded as shown in Fig. 26.6(*ii*), the accidental contact between a 11 kV conductor and a 230 V conductor produces a dead short. The short-circuit current (*i.e.* fault current) follows the dotted path shown in Fig. 26.6 (*ii*). This large current will blow the fuse on the 11 kV side, thus disconnecting the transformer and secondary distribution system from the 11 kV line. This explains the importance of system grounding in the line of the power system.

26.4 Ungrounded Neutral System

In an ungrounded neutral system, the neutral is not connected to the ground *i.e.* the neutral is isolated from the ground. Therefore, this system is also called *isolated neutral system* or *free neutral system.* Fig. 26.7 shows ungrounded neutral system. The line conductors have capacitances between one another and to ground. The former are delta-connected while the latter are star-connected. The delta-connected capacitances have little effect on the grounding characteristics of the system (*i.e.* these capacitances do not effect the earth circuit) and, therefore, can be neglected. The circuit then reduces to the one shown in Fig. 26.8(*i*).

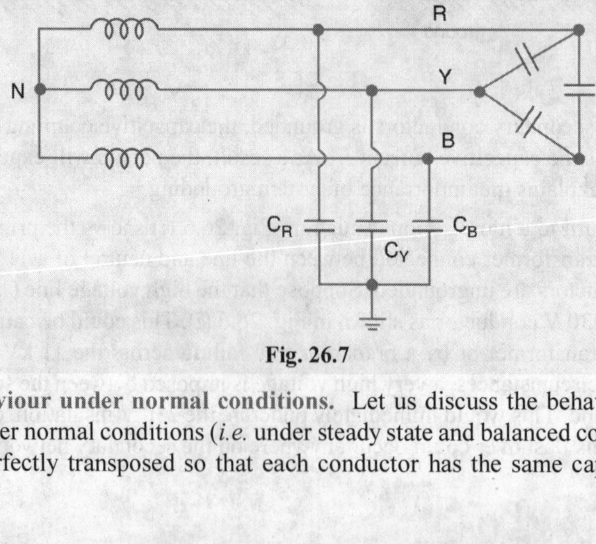

Fig. 26.7

Circuit behaviour under normal conditions. Let us discuss the behaviour of ungrounded neutral system under normal conditions (*i.e.* under steady state and balanced conditions). The line is assumed to be perfectly transposed so that each conductor has the same capacitance to ground.

Therefore, $C_R = C_Y = C_B = C$ (say). Since the phase voltages V_{RN}, V_{YN} and V_{BN} have the same magnitude (of course, displaced 120° from one another), the capacitive currents I_R, I_Y and I_B will have the same value *i.e.*

$$I_R = I_Y = I_B = \frac{V_{ph}}{X_C} \quad \text{.... in magnitude}$$

where V_{ph} = Phase voltage (*i.e.* line-to-neutral voltage)

X_C = Capacitive reactance of the line to ground.

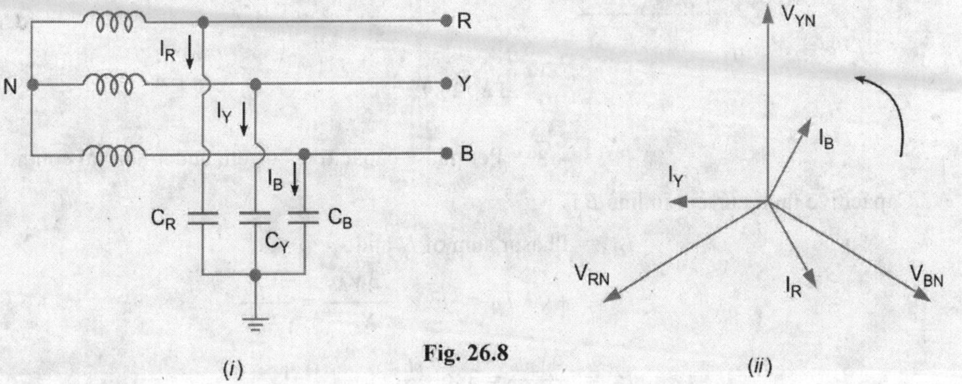

Fig. 26.8

The capacitive currents I_R, I_Y and I_B lead their respective phase voltages V_{RN}, V_{YN} and V_{BN} by 90° as shown in the phasor diagram in Fig. 26.8(*ii*). The three capacitive currents are equal in magnitude and are displaced 120° from each other. Therefore, their phasor sum is zero. As a result, no current flows to ground and the *potential of neutral is the same as the ground potential.* Therefore, ungrounded neutral system poses no problems under normal conditions. However, as we shall see, currents and voltages are greatly influenced during fault conditions.

Circuit behaviour under single line to ground-fault. Let us discuss the behaviour of ungrounded neutral system when single line to ground fault occurs. Suppose line to ground fault occurs in line B at some point F. The *circuit then becomes as shown in Fig. 26.9(*i*). The capacitive currents I_R and I_Y flow through the lines R and Y respectively. The voltages driving I_R and I_Y are V_{BR} and V_{BY} respectively. Note that V_{BR} and V_{BY} are the line voltages [See Fig. 26.9 (*ii*)]. The paths of I_R and I_Y are essentially capacitive. Therefore, I_R leads V_{BR} by 90° and I_Y leads V_{BY} by 90° as shown in Fig. 26.9 (*ii*). The capacitive fault current I_C in line B is the phasor sum of I_R and I_Y.

Fault current in line B, $I_C = I_R + I_Y$ .... Phasor sum

Now, $$I_R = \frac{V_{BR}}{X_C} = \frac{\sqrt{3}\,V_{ph}}{X_C}$$

and $$I_Y = \frac{V_{BY}}{X_C} = \frac{\sqrt{3}\,V_{ph}}{X_C}$$

$$\therefore \quad I_R = I_Y = \frac{\sqrt{3}\,V_{ph}}{X_C}$$

* Due to line-to-ground fault in line B, the potential of phase B becomes equal to the ground potential. This short circuits the capacitance of this line (*i.e.* capacitance C_B). Hence no capacitive current flows through C_B.

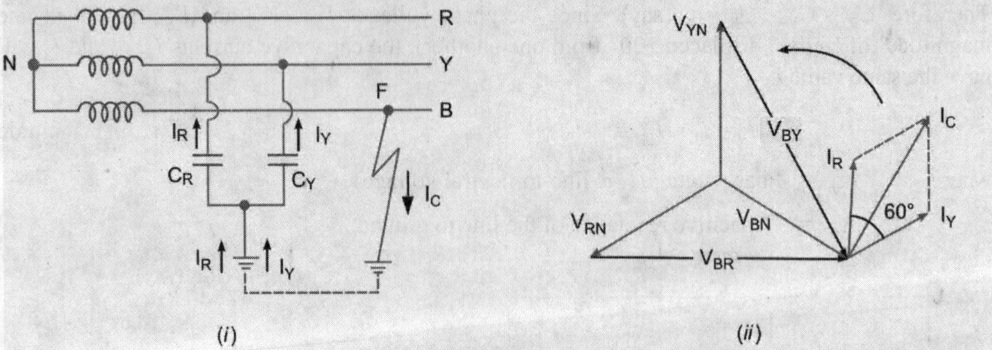

Fig. 26.9

$$= \sqrt{3} \times \text{Per phase capacitive current under normal conditions}$$

Capacitive fault current in line B is

$$I_C = \text{Phasor sum of } I_R \text{ and } I_Y$$

$$= \dagger\sqrt{3}\ I_R = \sqrt{3} \times \frac{\sqrt{3}\,V_{ph}}{X_C} = \frac{3V_{ph}}{X_C}$$

$$\therefore \quad I_C = \frac{3V_{ph}}{X_C} = 3 \times \frac{V_{ph}}{X_C}$$

$$= 3 \times \text{Per phase capacitive current under normal conditions}$$

Therefore, when single line to ground fault occurs on an ungrounded neutral system, the following effects are produced in the system:

(*i*) The potential of the faulty phase becomes equal to ground potential. However, the voltages of the two remaining healthy phases rise from their normal phase voltages to full line value. This may result in insulation breakdown.

(*ii*) The capacitive current in the two healthy phases increase to $\sqrt{3}$ times the normal value.

(*iii*) The capacitive fault current (I_C) becomes 3 times the normal per phase capacitive current.

(*iv*) This system cannot provide adequate protection against earth faults. It is because the capacitive fault current is small in magnitude and cannot operate protective devices.

(*v*) The capacitive fault current I_C flows into earth. Experience shows that I_C in excess of 4A is sufficient to maintain an arc in the ionized path of the fault. If this current is once maintained, it may exist even after the earth fault is cleared. This phenomenon of *persistent arc is called *arcing ground.* Due to arcing ground, the system capacity is charged and discharged in a cyclic order. This sets up high-frequency oscillations on the whole system and the phase voltage of healthy conductors may rise to 5 to 6 times its normal value. The overvoltages in healthy conductors may damage the insulation in the line.

---

† Referring to Fig. 26.9(*ii*), the magnitudes of I_R and I_Y are equal and the angle between them is 60°. Therefore, the resultant capacitive fault current I_C is given by:

$$I_C = 2I_R \cos 60°/2 = 2I_R \cos 30° = 2\,I_R \times \sqrt{3}/2 = \sqrt{3}\,I_R.$$

* When the arc is formed, the voltage across it becomes zero and the arc is extinguished. As a result, the potential of the faulty conductor is restored and the formation of second arc takes place. This phenomenon of intermittent arcing is called arcing ground.

Due to above disadvantages, ungrounded neutral system is not used these days. The modern high-voltage 3-phase systems employ grounded neutral owing to a number of advantages.

26.5 Neutral Grounding

The process of connecting neutral point of 3-phase system to earth (i.e. soil) either directly or through some circuit element (e.g. resistance, reactance etc.) is called **neutral grounding.**

Neutral grounding provides protection to personal and equipment. It is because during earth fault, the current path is completed through the earthed neutral and the protective devices (*e.g.* a fuse etc.) operate to isolate the faulty conductor from the rest of the system. This point is illustrated in Fig. 26.10.

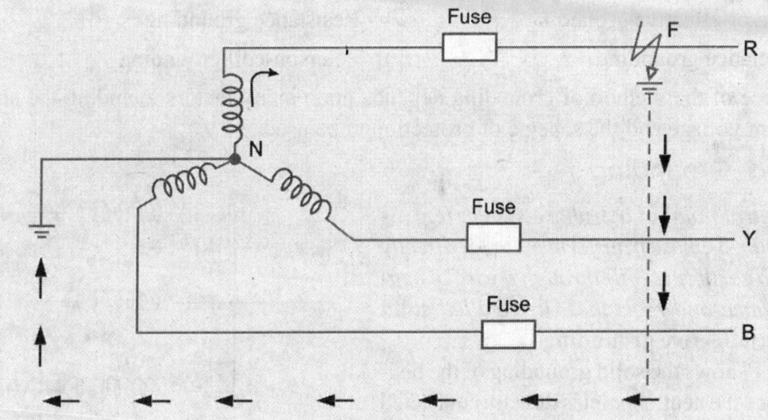

Fig. 26.10

Fig. 26.10 shows a 3-phase, star-connected system with neutral earthed (*i.e.* neutral point is connected to soil). Suppose a single line to ground fault occurs in line *R* at point *F*. This will cause the current to flow through ground path as shown in Fig. 26.10. Note that current flows from *R*-phase to earth, then to neutral point *N* and back to *R*-phase. Since the impedance of the current path is low, a large current flows through this path. This large current will blow the fuse in *R*-phase and isolate the faulty line *R*. This will protect the system from the harmful effects (*e.g.* damage to equipment, electric shock to personnel etc.) of the fault. One important feature of grounded neutral is that the potential difference between the live conductor and ground will not exceed the phase voltage of the system *i.e.* it will remain nearly constant.

26.6 Advantages of Neutral Grounding

The following are the advantages of neutral grounding :

(*i*) Voltages of the healthy phases do not exceed line to ground voltages *i.e.* they remain nearly constant.

(*ii*) The high voltages due to arcing grounds are eliminated.

(*iii*) The protective relays can be used to provide protection against earth faults. In case earth fault occurs on any line, the protective relay will operate to isolate the faulty line.

(*iv*) The overvoltages due to lightning are discharged to earth.

(*v*) It provides greater safety to personnel and equipment.

(*vi*) It provides improved service reliability.

(*vii*) Operating and maintenance expenditures are reduced.

Note : It is interesting to mention here that ungrounded neutral has the following advantages :

(*i*) In case of earth fault on one line, the two healthy phases will continue to supply load for a short period.

(*ii*) Interference with communication lines is reduced because of the absence of zero sequence currents.

The advantages of ungrounded neutral system are of negligible importance as compared to the advantages of the grounded neutral system. Therefore, modern 3-phase systems operate with grounded neutral points.

26.7 Methods of Neutral Grounding

The methods commonly used for grounding the neutral point of a 3-phase system are :

(*i*) Solid or effective grounding (*ii*) Resistance grounding

(*iii*) Reactance grounding (*iv*) Peterson-coil grounding

The choice of the method of grounding depends upon many factors including the size of the system, system voltage and the scheme of protection to be used.

26.8 Solid Grounding

*When the neutral point of a 3-phase system (e.g. 3-phase generator, 3-phase transformer etc.) is directly *connected to earth (i.e. soil) through a wire of negligible resistance and reactance, it is called* **solid grounding** *or* **effective grounding**.

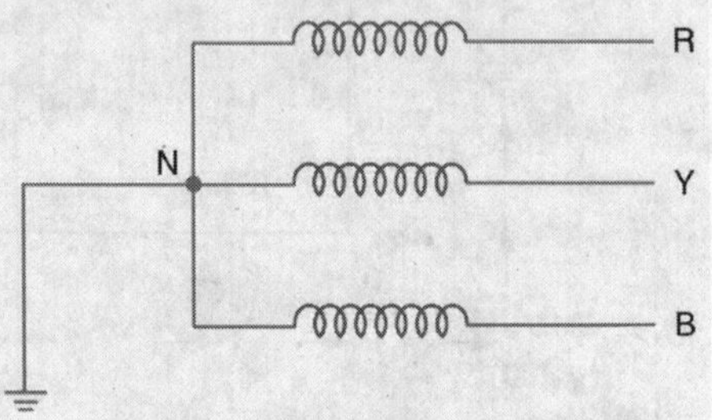

Fig. 26.11

Fig. 26.11 shows the solid grounding of the neutral point. Since the neutral point is directly connected to earth through a wire, the neutral point is held at earth potential under all conditions. Therefore, under fault conditions, the voltage of any conductor to earth will not exceed the normal phase voltage of the system.

Advantages. The solid grounding of neutral point has the following advantages :

(*i*) The neutral is effectively held at earth potential.

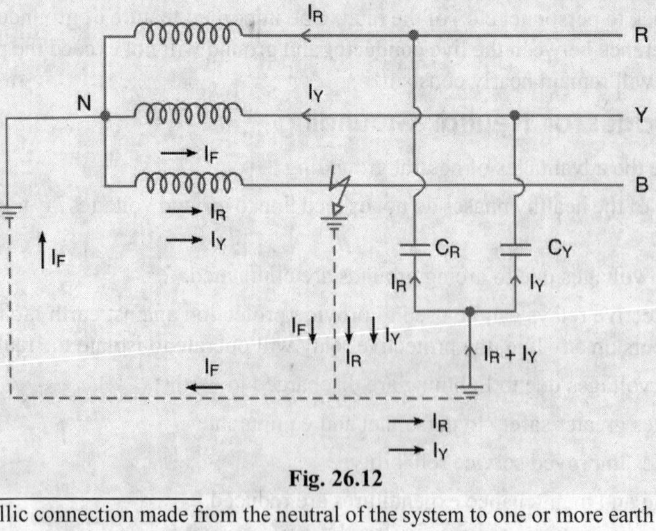

Fig. 26.12

* This is a metallic connection made from the neutral of the system to one or more earth electrodes consisting of plates, rods or pipes buried in the ground.

(*ii*) When earth fault occurs on any phase, the resultant capacitive current I_C is in phase opposition to the fault current I_F. The two currents completely cancel each other. Therefore, no arcing ground or over-voltage conditions can occur. Consider a line to ground fault in line *B* as shown in Fig. 26.12. The capacitive currents flowing in the healthy phases *R* and *Y* are I_R and I_Y respectively. The resultant capacitive current I_C is the phasor sum of I_R and I_Y. In addition to these capacitive currents, the power source also supplies the fault current I_F. This fault current will go from fault point to earth, then to neutral point *N* and back to the fault point through the faulty phase. The path of I_C is capacitive and that of I_F is *inductive. The two currents are in phase opposition and completely cancel each other. Therefore, no arcing ground phenomenon or over-voltage conditions can occur.

(*iii*) When there is an earth fault on any phase of the system, the phase to earth voltage of the faulty phase becomes zero. However, the phase to earth voltages of the remaining two healthy phases remain at normal phase voltage because the potential of the neutral is fixed at earth potential. This permits to insulate the equipment for phase voltage. Therefore, there is a saving in the cost of equipment.

(*iv*) It becomes easier to protect the system from earth faults which frequently occur on the system. When there is an earth fault on any phase of the system, a large fault current flows between the fault point and the grounded neutral. This permits the easy operation of earth-fault relay.

Disadvantages. The following are the disadvantages of solid grounding :

(*i*) Since most of the faults on an overhead system are phase to earth faults, the system has to bear a large number of severe shocks. This causes the system to become unstable.

(*ii*) The solid grounding results in heavy earth fault currents. Since the fault has to be cleared by the circuit breakers, the heavy earth fault currents may cause the burning of circuit breaker contacts.

(*iii*) The increased earth fault current results in greater interference in the neighbouring communication lines.

Applications. Solid grounding is usually employed where the circuit impedance is sufficiently high so as to keep the earth fault current within safe limits. This system of grounding is used for voltages upto 33 kV with total power capacity not exceeding 5000 kVA.

26.9 Resistance Grounding

In order to limit the magnitude of earth fault current, it is a common practice to connect the neutral point of a 3-phase system to earth through a resistor. This is called *resistance grounding*.

When the neutral point of a 3-phase system (e.g. 3-phase generator, 3-phase transformer etc.) is connected to earth (i.e. soil) through a resistor, it is called **resistance grounding.**

Fig. 26.13 shows the grounding of neutral point through a **resistor *R*. The value of *R* should neither be very low nor very high. If the value of earthing resistance *R* is very low, the earth fault current will be large and the system becomes similar to the solid grounding system. On the other hand, if the earthing resistance *R* is very high, the system conditions become similar to ungrounded

---

* By symmetrical components, the fault current I_F is given by :

$$I_F = \frac{3V_{ph}}{Z_1 + Z_2 + Z_0}$$

Since $Z_1 + Z_2 + Z_0$ is predominantly inductive, I_F lags behind the phase to neutral voltage of the faulted phase by nearly 90°.

** It may be a metallic resistor or liquid resistor. Metallic resistors do not change with time and practically require no maintenance. However, a metallic resistor is slightly inductive and this poses a problem with overhead lines exposed to lightning, Liquid resistors are free from this disadvantage.

neutral system. The value of R is so chosen such that the earth fault current is limited to safe value but still sufficient to permit the operation of earth fault protection system. In practice, that value of R is selected that limits the earth fault current to 2 times the normal full load current of the earthed generator or transformer.

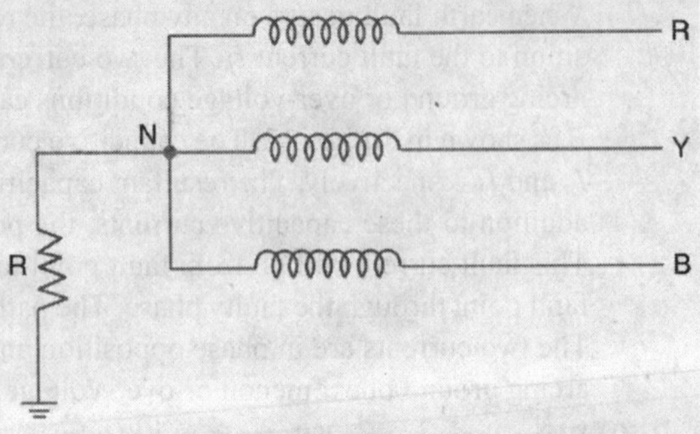

Fig. 26.13

Advantages. The following are the advantages of resistance earthing:

(*i*) By adjusting the value of R, the arcing grounds can be minimised. Suppose earth fault occurs in phase B as shown in Fig. 26.14. The capacitive currents I_R and I_Y flow in the healthy phases R and Y respectively. The fault current I_F lags behind the phase voltage of the faulted phase by a certain angle depending upon the earthing resistance R and the reactance of the system upto the point of fault. The fault current I_F can be resolved into two components *viz.*

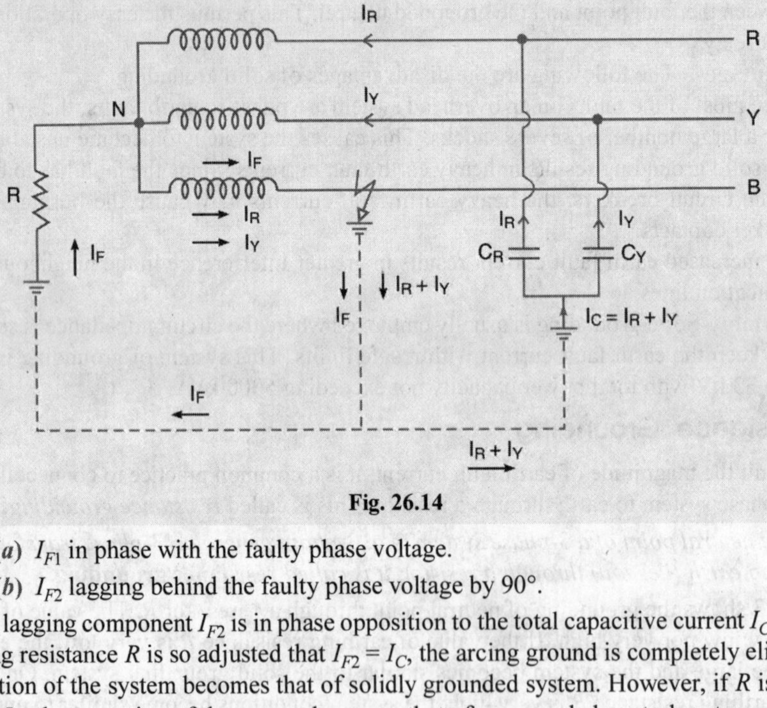

Fig. 26.14

(*a*) I_{F1} in phase with the faulty phase voltage.

(*b*) I_{F2} lagging behind the faulty phase voltage by 90°.

The lagging component I_{F2} is in phase opposition to the total capacitive current I_C. If the value of earthing resistance R is so adjusted that $I_{F2} = I_C$, the arcing ground is completely eliminated and the operation of the system becomes that of solidly grounded system. However, if R is so adjusted that $I_{F2} < I_C$, the operation of the system becomes that of ungrounded neutral system.

(*ii*) The earth fault current is small due to the presence of earthing resistance. Therefore, interference with communication circuits is reduced.

(*iii*) It improves the stability of the system.

Disadvantages. The following are the disadvantages of resistance grounding :

(*i*) Since the system neutral is displaced during earth faults, the equipment has to be insulated for higher voltages.

(*ii*) This system is costlier than the solidly grounded system.

(*iii*) A large amount of energy is produced in the earthing resistance during earth faults. Sometimes it becomes difficult to dissipate this energy to atmosphere.

Applications. It is used on a system operating at voltages between 2.2 kV and 33 kV with power source capacity more than 5000 kVA.

26.10 Reactance Grounding

In this system, a reactance is inserted between the neutral and ground as shown in Fig. 26.15. The purpose of reactance is to limit the earth fault current. By changing the earthing reactance, the earth fault current can to changed to obtain the conditions similar to that of solid grounding. This method is not used these days because of the following disadvantages :

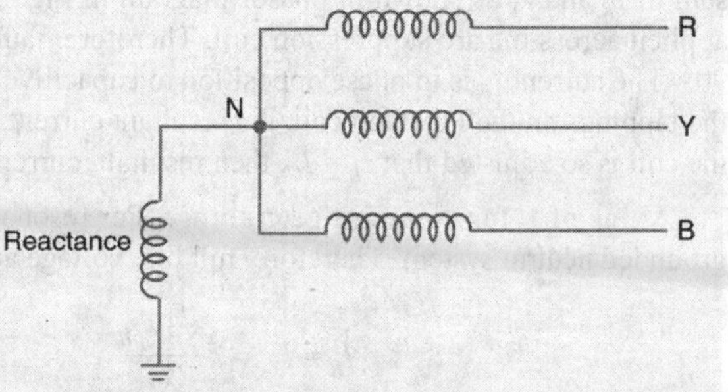

Fig. 26.15

(*i*) In this system, the fault current required to operate the protective device is higher than that of resistance grounding for the same fault conditions.

(*ii*) High transient voltages appear under fault conditions.

26.11 *Arc Suppression Coil Grounding (or Resonant Grounding)

We have seen that capacitive currents are responsible for producing arcing grounds. These capacitive currents flow because capacitance exists between each line and earth. If inductance L of appropriate value is connected in parallel with the capacitance of the system, the fault current I_F flowing through L will be in phase opposition to the capacitive current I_C of the system. If L is so adjusted that $I_L = I_C$, then resultant current in the fault will be zero. This condition is known as *resonant grounding*.

When the value of L of arc suppression coil is such that the fault current I_F exactly balances the capacitive current I_C, it is called **resonant grounding**.

Circuit details. An arc suppression coil (also called Peterson coil) is an iron-cored coil connected between the neutral and earth as shown in Fig. 26.16(*i*). The reactor is provided with tappings to change the inductance of the coil. By adjusting the tappings on the coil, the coil can be tuned with the capacitance of the system i.e. resonant grounding can be achieved.

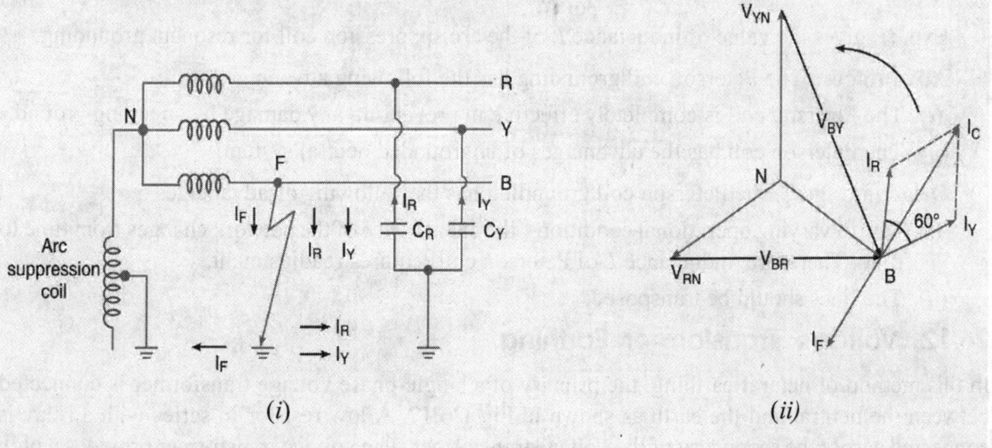

Fig. 26.16

* Also called **Peterson coil grounding.**

Operation. Fig. 26.16(*i*) shows the 3-phase system employing Peterson coil grounding. Suppose line to ground fault occurs in the line *B* at point *F*. The fault current I_F and capacitive currents I_R and I_Y will flow as shown in Fig. 26.16(*i*). Note that I_F flows through the Peterson coil (or Arc suppression coil) to neutral and back through the fault. The total capacitive current I_C is the phasor sum of I_R and I_Y as shown in phasor diagram in Fig. 26.16(*ii*). The voltage of the faulty phase is applied across the arc suppression coil. Therefore, fault current I_F lags the faulty phase voltage by 90°. The current I_F is in phase opposition to capacitive current I_C [See Fig. 26.16(*ii*)]. By adjusting the tappings on the Peterson coil, the resultant current in the fault can be reduced. If inductance of the coil is so adjusted that $I_L = I_C$, then resultant current in the fault will be zero.

Value of L for resonant grounding. For resonant grounding, the system behaves as an ungrounded neutral system. Therefore, full line voltage appears across capacitors C_R and C_Y.

$$\therefore \quad I_R = I_Y = \frac{\sqrt{3}V_{ph}}{X_C}$$

$$\therefore \quad I_C = \sqrt{3}\, I_R = \sqrt{3} \times \frac{\sqrt{3}V_{ph}}{X_C} = \frac{3V_{ph}}{X_C}$$

Here, X_C is the line to ground capacitive reactance.

Fault current, $$I_F = \frac{V_{ph}}{X_L}$$

Here, X_L is the inductive reactance of the arc suppression coil.

For resonant grounding, $I_L = I_C$.

or $$\frac{V_{ph}}{X_L} = \frac{3V_{ph}}{X_C}$$

or $$X_L = \frac{X_C}{3}$$

or $$\omega L = \frac{1}{3\omega C}$$

$$\therefore \quad L = \frac{1}{3\omega^2 C} \qquad \text{...}(i)$$

Exp. (*i*) gives the value of inductance *L* of the arc suppression coil for resonant grounding.

Advantages. The Peterson coil grounding has the following advantages:

(*i*) The Peterson coil is completely effective in preventing any damage by an arcing ground.

(*ii*) The Peterson coil has the advantages of ungrounded neutral system.

Disadvantages. The Peterson coil grounding has the following disadvantages :

(*i*) Due to varying operational conditions, the capacitance of the network changes from time to time. Therefore, inductance *L* of Peterson coil requires readjustment.

(*ii*) The lines should be transposed.

26.12 Voltage Transformer Earthing

In this method of neutral earthing, the primary of a single-phase voltage transformer is connected between the neutral and the earth as shown in Fig. 26.17. A low resistor in series with a relay is connected across the secondary of the voltage transformer. The voltage transformer provides a high reactance in the neutral earthing circuit and operates virtually as an ungrounded neutral system. An earth fault on any phase produces a voltage across the relay. This causes the operation of the protective device.

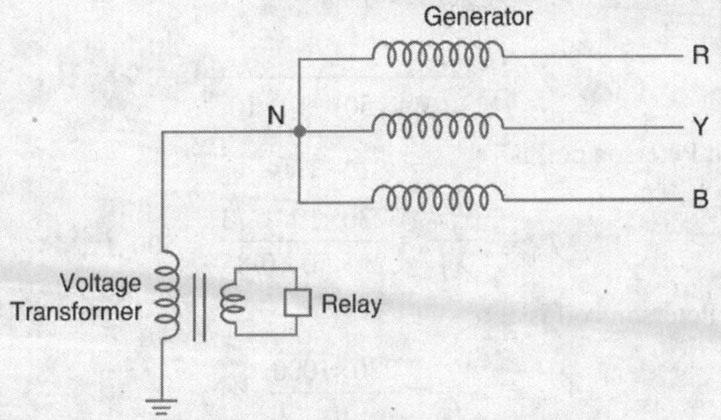

Fig. 26.17

Advantages. The following are the advantages of voltage transformer earthing :

(*i*) The transient overvoltages on the system due to switching and arcing grounds are reduced. It is because voltage transformer provides high reactance to the earth path.

(*ii*) This type of earthing has all the advantages of ungrounded neutral system.

(*iii*) Arcing grounds are eliminated.

Disadvantages. The following are the disadvantages of voltage transformer earthing :

(*i*) When earth fault occurs on any phase, the line voltage appears across line to earth capacitances. The system insulation will be overstressed.

(*ii*) The earthed neutral acts as a reflection point for the travelling waves through the machine winding. This may result in high voltage build up.

Applications. The use of this system of neutral earthing is normally confined to generator equipments which are directly connected to step-up power transformers.

Example 26.1. *Calculate the reactance of Peterson coil suitable for a 33 kV, 3-phase transmission line having a capacitance to earth of each conductor as 4.5 μF. Assume supply frequency to be 50 Hz.*

Solution. Supply frequency, $f = 50$ Hz

Line to earth capacitance, $C = 4{\cdot}5\ \mu\text{F} = 4{\cdot}5 \times 10^{-6}$ F

For Peterson coil grounding, reactance X_L of the Peterson coil should be equal to $X_C/3$ where X_C is line to earth capacitive reactance.

$$\therefore \text{ Reactance of Peterson coil, } X_L = \frac{X_C}{3} = \frac{1}{3\,\omega C} = \frac{1}{3 \times 2\pi f \times C}$$

$$= \frac{1}{3 \times 2\pi \times 50 \times 4{\cdot}5 \times 10^{-6}} = 235{\cdot}8\,\Omega$$

Example 26.2. *A 230 kV, 3-phase, 50 Hz, 200 km transmission line has a capacitance to earth of 0·02 μF/km per phase. Calculate the inductance and kVA rating of the Peterson coil used for earthing the above system.*

Solution. Supply frequency, $f = 50$ Hz

Capacitance of each line to earth, $C = 200 \times 0{\cdot}02 = 4 \times 10^{-6}$ F

Required inductance of Peterson coil is

$$L = \frac{1}{3\omega^2 C}$$

$$= \frac{1}{3\times(2\pi\times 50)^2\times 4\times 10^{-6}} = \mathbf{0{\cdot}85\ H}$$

Current through Peterson coil is

$$I_F = \frac{V_{ph}}{X_L} = \frac{230\times 10^3/\sqrt{3}}{2\pi\times 50\times 0{\cdot}85} = 500\ A$$

Voltage across Peterson coil is

$$V_{ph} = \frac{V_L}{\sqrt{3}} = \frac{230\times 1000}{\sqrt{3}}\ V$$

$\therefore$ Rating of Peterson coil $= V_{ph}\times I_F = \dfrac{230\times 1000}{\sqrt{3}}\times 500\times\dfrac{1}{1000}$ **kVA = 66397 kVA**

Example 26.3. *A 50 Hz overhead line has line to earth capacitance of 1.2 μF. It is desired to use *earth fault neutralizer. Determine the reactance to neutralize the capacitance of (i) 100% of the length of the line (ii) 90% of the length of the line and (iii) 80% of the length of the line.*

Solution.

(*i*) Inductive reactance of the coil to neutralize capacitance of 100% of the length of the line is

$$X_L = \frac{1}{3\omega C} = \frac{1}{3\times 2\pi\times 50\times 1{\cdot}2\times 10^{-6}} = \mathbf{884{\cdot}19\Omega}$$

(*ii*) Inductive reactance of the coil to neutralize capacitance of 90% of the length of the line is

$$X_L = \frac{1}{3\omega\times 0{\cdot}9C} = \frac{1}{3\times 2\pi\times 50\times 0{\cdot}9\times 1{\cdot}2\times 10^{-6}} = \mathbf{982{\cdot}43\Omega}$$

(*iii*) Inductive reactance of the coil to neutralize capacitance of 80% of the length of the line is

$$X_L = \frac{1}{3\omega\times 0{\cdot}8C} = \frac{1}{3\times 2\pi\times 50\times 0{\cdot}8\times 1{\cdot}2\times 10^{-6}} = \mathbf{1105{\cdot}24\Omega}$$

Example 26.4. *A 132 kV, 3-phase, 50 Hz transmission line 200 km long consists of three conductors of effective diameter 20 mm arranged in a vertical plane with 4 m spacing and regularly transposed. Find the inductance and kVA rating of the arc suppression coil in the system.*

Solution. Radius of conductor, $r = 20/2 = 10$ mm $= 0{\cdot}01$ m

Conductor spacing, $d = 4m$

$\therefore$ Capacitance between phase and neutral or earth

$$= \frac{2\pi\varepsilon_0}{\log_e\frac{d}{r}}\ F/m = \frac{2\pi\times 8{\cdot}885\times 10^{-12}}{\log_e\frac{4}{0{\cdot}01}} = 9{\cdot}285\times 10^{-12}\ \text{F/m}$$

$$= 9{\cdot}285\times 10^{-12}\times 10^{3}\ \text{F/km} = 9{\cdot}285\times 10^{-9}\ \text{F/km}$$

$\therefore$ Capacitance C between phase and earth for 200 km line is

* Note that Peterson coil is also known as earth fault neutralizer.

$$C = 200 \times 9{\cdot}285 \times 10^{-9} = 18{\cdot}57 \times 10^{-7} F$$

The required inductance L of the arc suppression coil is

$$L = \frac{1}{3\omega^2 C} = \frac{1}{3\times(2\pi\times50)^2\times18{\cdot}57\times10^{-7}} = \mathbf{1{\cdot}82H}$$

Current through the coil, $I_F = \dfrac{V_{ph}}{X_L} = \dfrac{132\times10^3/\sqrt{3}}{2\pi\times50\times1{\cdot}82} = 132A$

$\therefore$ Rating of the coil $= V_{ph} \times I_F = \dfrac{132}{\sqrt{3}} \times 132 = \mathbf{10060\ kVA}$

TUTORIAL PROBLEMS

1. A 132 kV, 3-phase, 50 Hz transmission line 192 km long consists of three conductors of effective diameter 20 mm, arranged in a vertical plane with 4 m spacing and regularly transposed. Find the inductance and MVA rating of the arc suppression coil in the system. **[1·97H; 9·389 MVA]**
2. A 33 kV, 50 Hz network has a capacitance to neutral of 1·0 μF per phase. Calculate the reactance of an arc suppression coil suitable for the system to avoid adverse effect of arching ground. **[1061Ω]**
3. A transmission line has a capacitance of 0·1 μF per phase. Determine the inductance of Peterson coil to neutralize the effect of capacitance of (*i*) complete length of the line, (*ii*) 97% of the line, (*iii*) 90% length of the line. The supply frequency is 50 Hz. **[(*i*) 33·80H (*ii*) 34·84H (*iii*) 37·55H]**

26.13 Grounding Transformer

We sometimes have to create a neutral point on a 3-phase, 3-wire system (e.g. delta connection etc.) to change it into 3-phase, 4-wire system. This can be done by means of a *grounding transformer.* It is a core type transformer having three limbs built in the same fashion as that of the power transformer. Each limb of the transformer has two identical windings wound differentially (i.e. directions of current in the two windings on each limb arc opposite to each other) as shown in Fig. 26.18. Under normal operating conditions, the total flux in each limb is negligibly small. Therefore, the transformer draws very small magnetising current.

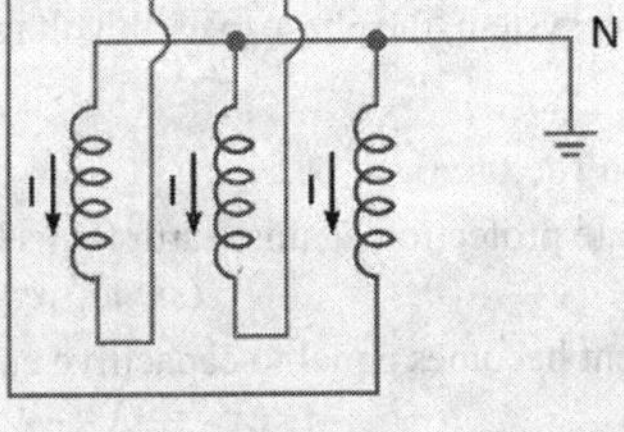

Fig. 26.18

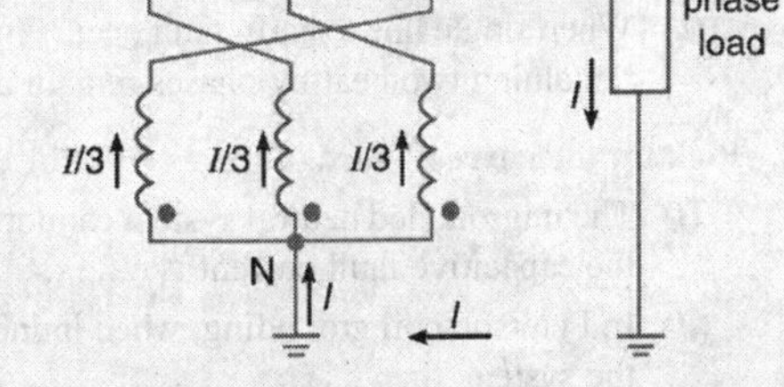

Fig. 26.19

Fig. 26.19 shows the use of grounding transformer to create neutral point N. If we connect a single-phase load between one line and neutral, the load current I divides into three equal currents in each winding. Because the currents are equal, the neutral point stays fixed and the line to neutral voltages remain balanced as they would be on a regular 4-wire system. In practice, the single-phase loads are distributed as evenly as possible between the three phases and neutral so that unbalanced load current I is relatively small.

The impedance of grounding transformer is quite low. Therefore, when line to earth fault occurs, the fault current will be quite high. The magnitude of fault current is limited by inserting a resistance (not shown in the figure) in the neutral circuit. Under normal conditions, only iron losses will be continuously occurring in the grounding transformer. However, in case of fault, the high fault current will also produce copper losses in the transformer. Since the duration of the fault current is generally between 30-60 seconds, the copper losses will occur only for a short interval.

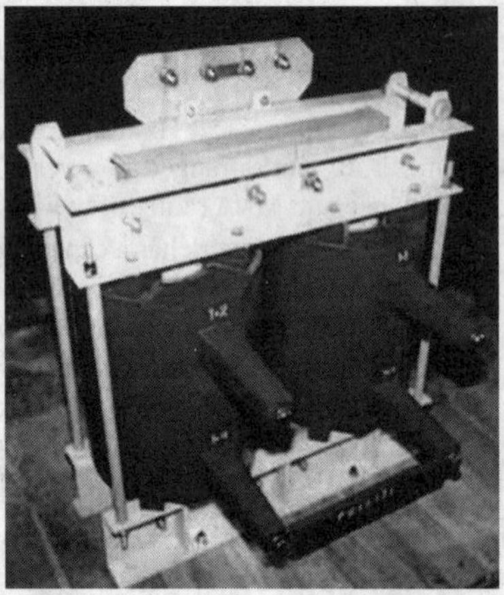

Grounded Transformer

SELF - TEST

1. Fill in the blanks by inserting appropriate words/figures :

(*i*) When single line to earth fault occurs on an ungrounded neutral system, the voltages of the healthy phases (other than the faulty phase) rise from their normal phase voltages to ....

(*ii*) When single line to earth fault occurs on an ungrounded neutral system, the capacitive current in the two healthy phases rises to .... times the normal value.

(*iii*) When single line to earth fault occurs on an ungrounded neutral system, the capacitive fault current becomes .... times the normal per phase capacitive current.

(*iv*) In Peterson coil grounding, inductance L of the coil is related to line to earth capacitance C as ....

(*v*) When single line to earth fault occurs in solid grounding system, the phase to earth voltage of the remaining two healthy phases remain at ....

2. Pick up the correct words/figures from brackets and fill in the blanks.

(*i*) The ungrounded neutral system cannot provide adequate protection against earth faults because the capacitive fault current is .... *(small, very large)*

(*ii*) In Peterson coil grounding, when inductive fault current becomes equal to capacitive current of the system, then .... ($X_C = 3X_L$; $X_L = 3X_C$)

(*iii*) In voltage transformer grounding .... of single phase transformer is connected between neutral and earth. *(secondary, primary)*

(*iv*) In equipment grounding, the enclosure is connected to .... wire. *(ground, neutral)*

(*v*) The ground wire is coloured .... *(black, green)* (*vi*) The neutral wire is coloured .... (*black, green*)

(*vii*) In Peterson coil grounding, the inductance of the coil is .... (*fixed, variable*)

(*viii*) In case of earth fault, the ungrounded neutral system .... lead to arcing ground. (*does, does not*)

(*ix*) Grounding transformer is used where neutral .... available. (*is, is not*)

(*x*) Most of the faults on an overhead system are ....faults.(*phase to earth, phase to phase*)

ANSWERS TO SELF-TEST

1. (*i*) line value (*ii*) $\sqrt{3}$ (*iii*) 3 (*iv*) $L = \dfrac{1}{3\omega^2 C}$ (*v*) normal phase voltage
2. (*i*) small (*ii*) $X_C = 3X_L$ (*iii*) primary (*iv*) ground (*v*) green (*vi*) black (*vii*) variable (*viii*) does (*ix*) is not (*x*) phase to earth

CHAPTER REVIEW TOPICS

1. What do you mean by grounding or earthing? Explain it with an example.
2. Describe ungrounded or isolated neutral system. What are its disadvantages?
3. What do you mean by equipment grounding?
4. Illustrate the need of equipment grounding.
5. What is neutral grounding?
6. What are the advantages of neutral grounding?
7. What is solid grounding? What are its advantages?
8. What are the disadvantages of solid grounding?
9. What is resistance grounding? What are its advantages and disadvantages?
10. Describe Arc suppression coil grounding.
11. What is resonant grounding?

DISCUSSION QUESTIONS

1. Why is ground wire used in equipment grounding?
2. There is 11 kV/230V single phase transformer. One can notice that one of the secondary conductors is grounded. Why?
3. The H.V. line of a single phase transformer accidently falls on L.V. line. There may be massive flash-over in a home or factory. Why?
4. In an overhead system, most of the faults are single line to ground. Why?
5. What are the factors causing arching grounds?
6. What is the importance of arc suppression coil grounding?
7. Where do we use grounding transformer?

INDEX

E

F

G

H

I

K

L